Physics
A Student Companion

Physics
A Student Companion

Lowry A Kirkby
Magdalen College, University of Oxford, Oxford, UK
Currently at Department of Biophysics,
University of California, Berkeley, USA

With a foreword by Frank Close
Professor of Theoretical Physics,
University of Oxford, Oxford, UK

ISBN 978 1904842 68 2

First published in 2011

A CIP catalogue record for this book is available from the British Library.

Scion Publishing Limited
The Old Hayloft, Vantage Business Park, Bloxham Road, Banbury, OX16 9UX, UK
www.scionpublishing.com

Important Note from the Publisher
The information contained within this book was obtained by Scion Publishing Limited from sources believed by us to be reliable. However, while every effort has been made to ensure its accuracy, no responsibility for loss or injury whatsoever occasioned to any person acting or refraining from action as a result of information contained herein can be accepted by the authors or publishers.

Although every effort has been made to ensure that all owners of copyright material have been acknowledged in this publication, we would be pleased to acknowledge in subsequent reprints or editions any omissions brought to our attention.

Typeset by Medlar Publishing Solutions, India
Printed by Charlesworth Press, Wakefield, UK

Contents

Foreword

There are many textbooks of physics, most of which have been written by professors with the purpose of teaching a subject in all its detail. As students we study these, attend the lectures, and face the upcoming examinations. In preparation for the latter, we study what we have already (hopefully) learned. How do we go about this? Reading all the textbooks again is time consuming and, for review, inefficient: if you haven't already done the preparation, trying to play catch up is usually too late. Assiduous students make their own notes, summarizing what they have learned in a convenient aide-memoire of essential formulae and facts. The very act of doing so is invaluable as part of the learning process. Which is what makes this book unusual, and potentially unique.

Lowry Kirkby was until recently a physics undergraduate at Oxford University. She created her own set of study notes and achieved one of the highest scores in her year. Her notes are what have formed the kernel of this book.

For students preparing for examinations, who want access to a handy, and successful, set of study and review notes, look no further. Furthermore, they were originally written by a student, and so they have a "feel" for what real students actually need. I have only two regrets. First, that notes like these were not available when I was a student. Secondly, that I didn't think of doing this myself. This novel approach to exam preparation promises to provide an invaluable aid to all students of physics.

Frank Close
Professor of Theoretical Physics, University of Oxford
Tutor in Physics, Exeter College Oxford

Preface

Shortly after starting my degree in physics at the University of Oxford, it became clear to me that if I wanted to stay on top of my subject, and with a particular eye on examinations, then I would need to find a way to organize the enormous amount of material we were covering into something manageable. Consequently, throughout the four years, I spent a great deal of time going through lecture and class notes, problem sets and textbooks, to produce a comprehensive and consolidated set of study notes. I resumed and organized key concepts, equations and their derivations in such a way to give me total confidence that when it came to exam crunch time, I could safely stow the pile of textbooks and mountains of lecture notes at the back of the closet and focus on my notes alone. In fact, over those four years I found myself using my notes not only for exams but also on a regular basis to remind myself of an equation or derivation. So the initial effort paid off over and over again.

By the end of my degree, I had a complete overview of my four undergraduate years summarized in just a few hundred sheets of paper. Since the notes had proved successful for my exam preparation, I thought other students could benefit from such a resource: something in between lecture notes and a textbook, an accessible reference to key ideas and equations, and a reliable examination companion. And so the idea of writing this book, which had been quietly forming itself for a while, took concrete shape.

The book covers the core concepts for undergraduate physics in an accessible and memory-friendly manner. It is intended as a complement to the standard textbooks and to your own hard work. I hope you will find it both useful and rewarding!

Lowry A. Kirkby
August 2011

About the author

Lowry Kirkby graduated with a First-class Honours Master of Physics degree (MPhys) from the University of Oxford, Magdalen College, in 2007. She is currently pursuing a PhD in Biophysics at the University of California, Berkeley.

Acknowledgements

Many people have helped make it possible for me to write this book. I would like to give warm thanks to Arzhang Ardavan, Giles Barr, John Gregg and Geoff Smith, my physics tutors at Magdalen College, for making my years at Oxford an invaluable and most enjoyable learning experience; to Frank Close, physics tutor at Exeter College, Oxford, for hearing me out and advocating for me when I first came to him with the idea of writing this book; to Matt George for his helpful comments and feedback on an early version of the electromagnetism section; to Tenzing Joshi for his enthusiasm and motivation that encouraged me through the final writing stages; and to Kathryn Nicklin for being my alter-brain during my undergraduate years.

I would also like to thank the anonymous reviewers for their constructive and positive comments, as well as my copy editor Caren Brown, proof-readers John Jarvis and Clare Boomer, and publisher at Scion Publishing Jonathan Ray.

And of course, a special thanks to my family for their constant encouragement and support. In particular, to my sister Eva whose belief in my project helped sustain my stamina throughout; to my mother Rita, a consistent source of tea and moral support; and to my father Jasper, for talking to me about the big bang almost before I was old enough to speak and for showing me the pleasure of physics.

How to use this book

The book combines a consistent narrative together with equations, derivations and worked examples that are set apart from the main text in boxes or panels (see below). Thus the narrative text maintains an uninterrupted flow for ease of reading and comprehension, while the examples in the panels reinforce the main text.

Each of the five Parts consists of self-contained Chapters, which develop a set of topics that build on each other. The reader can go directly to the section of interest, without needing to read earlier chapters. Topics discussed in more than one section are cross-referenced. Figures are simple line drawings and contain the amount of detail that a student could be expected to sketch in an exam.

Equations and derivations: when making my study notes, I followed a consistent layout: topic heading, overview, boxed equation and derivation. The derivation would be a 2-column panel with the left-hand column containing the mathematics and the right-hand column a short explanation of the key mathematical steps. This book has been written with an almost identical framework: equations and derivations are set apart from the main text, and they reference one another, making them stand separately from the main text. Where derivations require some explanation there is an associated commentary to guide the reader. The commentary explains the steps as one student would to another, and are at an appropriate level for annotation in examination answers.

Worked examples and problems: in addition to the derivation panels there are also panels for worked examples, showing how to apply a given concept or equation to a specific problem. The format is similar to that of the derivation panel, with the mathematics and an explanation of the steps. Problem questions, for practical reasons, are not included here. For a set of problems to accompany this book, I recommend:

- W. G. Rees, 1994, *Physics by Example: 200 Problems and Solutions*, Cambridge University Press. The problems in Rees are creative and challenging and cover the same topics as those covered in this book, at the appropriate level. The two books complement each other, and I recommend using them together.

One of the best ways to prepare for an exam is to solve problem after problem of past examination questions. Each examination has its own particular style and problems from a book may not necessarily match that style. Get hold of several years of past papers and work conscientiously through them. Having done so myself I can confirm that my study notes – and hence this book – provide the right level of background and understanding for applying to specific questions.

What this book covers

The book covers the following core topics:

Part I: Newtonian Mechanics and Special Relativity
Part II: Electromagnetism
Part III: Waves and Optics
Part IV: Quantum Physics
Part V: Thermal Physics

These topics are usually covered in the first and second years of a 4-year undergraduate physics degree. They make up the majority of the "Core of Physics" syllabus issued by the Institute of Physics (IoP; the accrediting body for physics degrees in the UK) and the majority of the syllabus for the physics Graduate Record Examination (GRE), which students applying for a PhD program in the USA are required to pass.

At the back of this book there is a special index for the IoP and GRE syllabuses, including the level to which each topic is covered. Please note that these syllabuses also require topics not included in this book due to their more specialized (non-core) nature: condensed matter physics; atomic, nuclear and particle physics; astrophysics and cosmology.

Part I
Newtonian Mechanics and Special Relativity

Introduction

Classical mechanics is the study of the motion of bodies as a result of forces that act on them. At the heart of classical mechanics are Newton's Laws of Motion that describe the effect of a force on the linear motion of a body (Chapter 1, with applications in Chapter 2). A force applied about a given axis exerts a torque on a body, resulting in a change in its *angular* motion; Newton's Laws of Motion thus have an angular equivalent, which describes the angular motion of rotating bodies (Chapter 3). Practical applications of linear and angular motion include gravitational orbits (Chapter 4) and two-body dynamics (Chapter 5). Although Newton's Laws are successful in describing the motion of bodies at low velocities, it was found that they break down for bodies moving at velocities on the order of the speed of light. In these instances, the motion of bodies is described by Einstein's Theory of Special Relativity, which approximates to Newton's Laws in the low-velocity limit (Chapter 6).

Practical applications

The applications of classical mechanics are extensive. We are familiar with applying Newton's Laws to projectiles, collisions, oscillations, and gravitation, for example, but the laws of classical mechanics describe the mechanical properties and motion of all large-scale bodies, such as fluids, stars, galaxies, tides, gyroscopes, living organisms, etc. Perhaps it's more informative to say when we *cannot* use classical mechanics—for the very small (quantum mechanics) and for the very fast (relativistic mechanics). However, conservation of energy, momentum, and angular momentum—concepts that are fundamental to classical mechanics—appear to be universally valid.

My two cents

Classical mechanics is a subject that is taught largely through examples and applications. This may appeal to many students, but others may find it difficult since it gives the impression that one application is independent from another, when in fact they are not. In Newtonian mechanics, there are a few fundamental concepts—most of which are covered in Chapter 1 of this book—that we apply again and again to solve a multitude of different problems, in a variety of different ways. It's worth taking the time to revisit the concepts outlined in Chapter 1 and to appreciate that a lot of classical mechanics problem solving boils down to just a few basic principles, such as energy and momentum conservation and Newton's Second Law.

Recommended books

- Kibble, T. and Berkshire, F., 2004, *Classical Mechanics*, 5th edition. Imperial College Press.
- McCall, M., 2001, *Classical Mechanics: a Modern Introduction*. Wiley.
- Taylor, J., 2005, *Classical Mechanics*. University Science Books.

The book by McCall is a great book for a clear, concise, and manageable introduction to Newtonian mechanics and special relativity. For more detail, both the book by Kibble and Berkshire and the book by Taylor provide clear explanations with numerous applications.

Chapter 1
Linear motion

Linear motion is the motion of a body along a line. We contrast linear motion with *angular* motion (Chapter 3), which is the rotational motion of a body around an axis. Bodies move in response to forces acting on them. The action of a force on a body is encapsulated in Newton's three Laws of Motion. From these laws, we can predict subsequent motion of a body, given known initial conditions.

1.1 Newton's Laws of Motion

Classical motion is based on three laws, known collectively as *Newton's Laws of Motion.* Although you are no doubt familiar with these laws, it's useful to revisit them and to appreciate that many fundamental concepts that we "take for granted" are in fact explained by Newton's Laws, as we shall see throughout this chapter. For example, they underlie the concept of force, momentum, work, and energy, as well as the conservation of momentum and energy.

1.1.1 Newton's First and Second Laws

Force and momentum

Consider a small body of mass m [kg]. In linear mechanics, we treat this body as though all its mass were concentrated at a single point: at its *center of mass.* This is the point that represents the mean position of all the matter in the body. For example, our center of mass is near our stomachs. The *displacement* of a body, $\vec{x}$ or $\vec{r}$ [m], is a vector that represents the position of its center of mass from an arbitrary origin. For a body moving in time, t [s], the *velocity*, $\vec{v}$ [m s^{-1}], describes the rate of change of displacement, and the *acceleration*, $\vec{a}$ [m s^{-2}], describes the rate of change of velocity (Eqs. 1.1–1.3). A common shorthand of the first and second time derivatives of a variable is a single or double dot over the variable, respectively. For example:

$$\frac{d\vec{x}}{dt} \equiv \dot{\vec{x}} \text{ and } \frac{d^2\vec{x}}{dt^2} \equiv \ddot{\vec{x}}$$

The linear motion of a body is characterized by its *momentum*, $\vec{p}$ [kg m s^{-1}], which is defined as a body's mass multiplied by its velocity (Eq. 1.4).

Velocity and acceleration:

Velocity:

$$\vec{v} = \frac{d\vec{x}}{dt} \equiv \dot{\vec{x}} \tag{1.1}$$

Acceleration:

$$\vec{a} = \frac{d\vec{v}}{dt} \equiv \dot{\vec{v}} \tag{1.2}$$

$$= \frac{d^2\vec{x}}{dt^2} \equiv \ddot{\vec{x}} \tag{1.3}$$

Momentum:

$$\vec{p} = m\vec{v} \tag{1.4}$$

Bodies move in response to a net *force*,[1] $\vec{F}$ [N], acting on them. *How* a body's momentum changes as a result of a force acting on it—i.e. the motion of a body—is the question that forms the basis of classical mechanics. The relationship between force and momentum is expressed formally in Newton's First and Second Laws. When no net force acts on a body, its momentum remains constant (Newton's First Law, Eq. 1.5 in Summary box 1.1). The action of a net force, $\vec{F}$, is to change its momentum, where the rate of change of momentum is equal to $\vec{F}$:

$$\vec{F} = \frac{d\vec{p}}{dt}$$

(Newton's Second Law, Eq. 1.6 in Summary box 1.1). Therefore, Newton's First and Second Laws define the concept of a force; the First Law describes it qualitatively, whereas the Second Law describes it quantitatively. Newton's Second Law is at the heart of classical motion. From it, we can predict the time evolution of a system given known initial conditions (see Chapter 2 for examples).

Dynamic equilibrium

Newton's First Law might be thought of as being redundant since it is incorporated into the Second Law in the case where $\vec{F}=0$. However, it is useful to state the First Law explicitly, since it defines the concept of *dynamic equilibrium*. A body in dynamic equilibrium has no net force acting on it, and is therefore in a state of constant motion; it either remains stationary or has constant momentum. A body of constant mass m in dynamic equilibrium therefore travels at a constant velocity, since $\vec{p}=m\vec{v}=$constant.

Inertial and gravitational mass

Another concept that follows from Newton's First Law is the concept of *inertia*. Inertia is a property of a body that describes its tendency to either remain stationary or to continue moving at a constant velocity; it represents resistance to a change in motion. Inertia is quantified by a body's mass. *Inertial mass*, m [kg], is defined as the force required to accelerate a body by 1 m s^{-2}:

$$m = \left|\frac{\vec{F}}{\vec{a}}\right|$$

This definition of mass follows from Newton's Second Law (Eq. 1.6). More massive bodies have a greater resistance to a change in motion (greater force is required to accelerate them by 1 m s^{-2}), which means they have greater inertia.

Conventionally, we define the *gravitational mass*, m_g [kg], as the gravitational force, F_g, required to accelerate a free-falling body by 1 m s^{-2}:

$$m_g = \left|\frac{\vec{F}_g}{\vec{g}}\right|$$

where the acceleration due to gravity on Earth is $|\vec{g}| \simeq 9.81$ m s^{-2}. In the 1600s, Galileo conducted a series of experiments that involved dropping objects of different mass from the same height and allowing them to free-fall under gravity, from which he found that inertial mass and gravitational mass are equivalent. This statement is commonly known as the *Galilean equivalence principle*. Therefore, bodies possess only one "type" of mass, which is equal in magnitude to inertial mass.

1.1.2 Newton's Third Law

Newton's Third Law makes reference to the mutual interaction between two bodies. It states that the force exerted on body 1 by body 2 is equal in magnitude and opposite in direction to that exerted on body 2 by body 1:

$$\vec{F}_{12} = -\vec{F}_{21}$$

(Eq. 1.7 in Summary box 1.1). We use this law to determine the internal dynamics of many-body systems that are not influenced by their external surroundings (see Chapter 5 for examples).

1.2 Force and momentum

1.2.1 Impulse

If a constant force $\vec{F}$ acts on a body over a time interval Δt, then the body experiences an *impulse*, $\vec{I}$ [kg m s^{-1}], which is defined as the product of force and time:

$$\vec{I} = \vec{F}\Delta t$$

Impulse is a vector quantity, whose direction is parallel to the applied force. For a time-dependent force, $\vec{F}(t)$, that acts between times t_1 and t_2, the impulse is the integral of the force with respect to time (Eq. 1.8).

[1] The net force on a body is the vector sum of the individual forces acting on it.

Summary box 1.1: Newton's Laws of Motion

1. Newton's First Law: principle of inertia

A body on which no net external force acts does not change momentum. Therefore, a body of constant mass m remains stationary or moves at constant velocity. A body with no net external force acting on it is in a state of *dynamic equilibrium*.

$$\vec{F} = 0 \rightarrow \vec{p} = \text{constant} \quad \text{or } \vec{v} = \text{constant for constant } m \tag{1.5}$$

2. Newton's Second Law: principle of action

A net external force acting on a body results in a change in its momentum. The rate of change of momentum is equal to the external force.

$$\vec{F} = \frac{d\vec{p}}{dt} = m\vec{a} \quad \text{for constant } m, \text{ where } \vec{a} = d\vec{v}/dt \tag{1.6}$$

3. Newton's Third Law: principle of action and reaction

Two isolated bodies exert equal and opposite forces on one another.

$$\vec{F}_{12} = -\vec{F}_{21} \tag{1.7}$$

Impulse–momentum theorem

Using Newton's Second Law in Eq. 1.8, we see that the impulse of a body equals the change in momentum it experiences from the action of the force:

$$\vec{I} = \int_{t_1}^{t_2} \vec{F}dt = \int_{t_1}^{t_2} \frac{d\vec{p}}{dt}dt = \int_{\vec{p}_1}^{\vec{p}_2} d\vec{p} = \Delta\vec{p}$$

The statement that the change in momentum of a body, $\Delta\vec{p}$, equals its impulse is sometimes known as the *impulse–momentum theorem* (Eq. 1.9).

Impulse:

$$\vec{I} = \int_{t_1}^{t_2} \vec{F}(t)dt \tag{1.8}$$

$$\equiv \Delta\vec{p} \tag{1.9}$$

1.2.2 Conservation of momentum

A powerful corollary of Newton's Laws of Motion is the *conservation of momentum*. Consider an isolated system of N bodies that are not influenced by external forces, but experience two-body mutual interactions, $\vec{F}_{ij}$, where the indices i and j represent two of the bodies. Using Newton's Third Law, we find that the total momentum of all the bodies, $\vec{P}$ (the sum of individual momenta of each body in the system) is a constant quantity (Eq. 1.10 and Derivation 1.1).

Conservation of momentum (Derivation 1.1):

$$\vec{P} = \sum_{i=1}^{N} \vec{p}_i = \text{constant} \tag{1.10}$$

This statement implies that momentum is a conserved quantity in isolated systems (i.e. ones that do not interact with their surroundings). Since the Universe constitutes an isolated system, then the total momentum in the Universe is constant: momentum cannot be created or destroyed. However, momentum *can* be transferred from one body to another. The conservation of momentum is a fundamental result that applies to *all* areas of physics. We shall look at its application to collisions in Section 2.1.

1.3 Force and energy

Since many of us were introduced to the concepts of force and energy at a young age, we never stop to question what energy really is and where it originates from. In its broadest sense, the energy, E [J], possessed by a body characterizes its physical state. In turn, the *state* of a body is a general term that encompasses any physical characteristics used to describe it, such as its position, its velocity, its temperature, its deformation, etc. Energy is classified into different forms, depending on the body's state. For example, bodies in a high velocity state have high *kinetic* energy, and bodies in a high temperature state have high *thermal* energy.

1.3.1 Work done by a force

So how does a body gain energy in the first place? A body gains energy if work, W [J], is done on it. Work is done on a body by a force acting on and displacing it. The element of work done by a force is equal to the dot product of force and distance moved by the body:

$$\mathrm{d}W = \vec{F} \cdot \mathrm{d}\vec{r}$$

The dot product picks out the component of $\vec{F}$ in the same direction as the displacement, $\vec{r}$ (Fig. 1.1). Integrating dW, the total work done on a body by a force

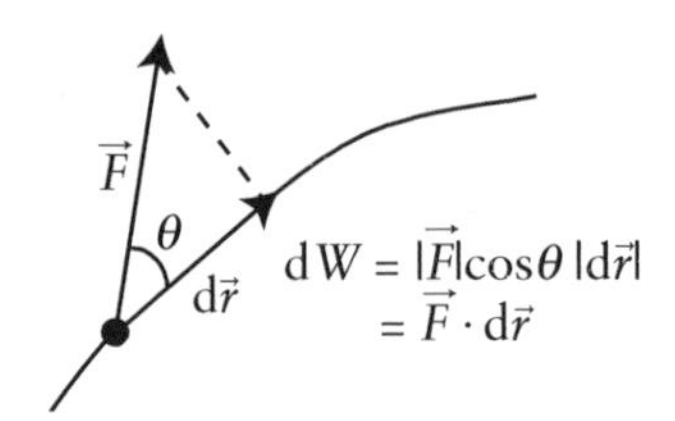

Figure 1.1: Work is done on a body by a force if the force displaces the body. The work done equals the component of force in the direction of displacement, multiplied by the displacement.

displacing it from position a to position b is given by Eqs. 1.11 and 1.12 for motion in one and three dimensions, respectively.

Work done by a force:

$$\text{In one dimension:}\quad W = \int_a^b F(x)\mathrm{d}x \tag{1.11}$$

$$\text{In three dimensions:}\quad W = \int_{\vec{a}}^{\vec{b}} \vec{F}(\vec{r}) \cdot \mathrm{d}\vec{r} = \int_a^b F(r)\cos\theta \mathrm{d}r \tag{1.12}$$

For a one-dimensional plot of force, $F(x)$, against distance x, the work done equals the area under the curve (Fig. 1.2).

Derivation 1.1: Derivation of momentum conservation (Eq. 1.10)

$$\begin{aligned}
\vec{P} &= \sum_{i=1}^{N} \vec{p}_i \\
&= \vec{p}_1 + \vec{p}_2 + \vec{p}_3 + \cdots + \vec{p}_N \\
&= \vec{p}_1 + \vec{p}_2 \quad \text{«««««««««} \\
\therefore \frac{\mathrm{d}\vec{P}}{\mathrm{d}t} &= \frac{\mathrm{d}(\vec{p}_1 + \vec{p}_2)}{\mathrm{d}t} \\
&= \frac{\mathrm{d}\vec{p}_1}{\mathrm{d}t} + \frac{\mathrm{d}\vec{p}_2}{\mathrm{d}t} \\
&= \vec{F}_{12} + \vec{F}_{21} \\
&= 0 \quad \text{from Newton's Third Law} \\
\therefore \mathrm{d}\vec{P} &= 0 \\
\Rightarrow \vec{P} &= \text{constant}
\end{aligned}$$

- Starting point: the total momentum of a system of bodies, $\vec{P}$, is the sum of individual momenta of each body, $\vec{p}_i$. For simplicity, we will consider a system of two bodies.
- From Newton's Second Law, the rate of change of momentum of a body equals the net force acting on it. If we consider an isolated system, on which no external forces act, then the net force on body i is its two-body mutual interaction with body j, $\vec{F}_{ij}$.
- From Newton's Third Law, the mutual forces between two interacting bodies are equal and opposite: $\therefore\ \vec{F}_{12} = -\vec{F}_{21}$.
- Integrating, we find that the total momentum of an isolated system is constant.

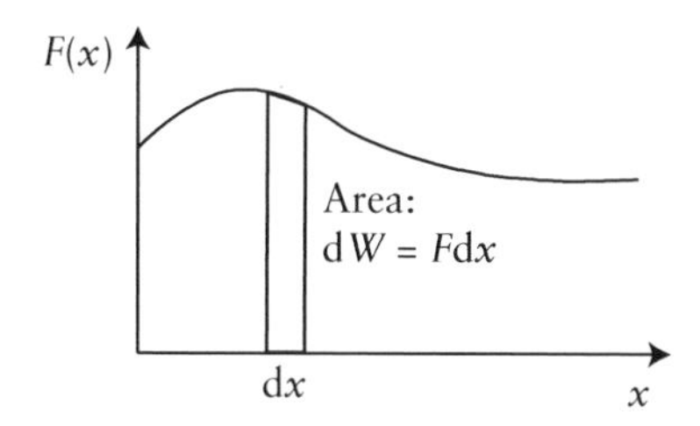

Figure 1.2: Work done is the area under a curve of force against distance.

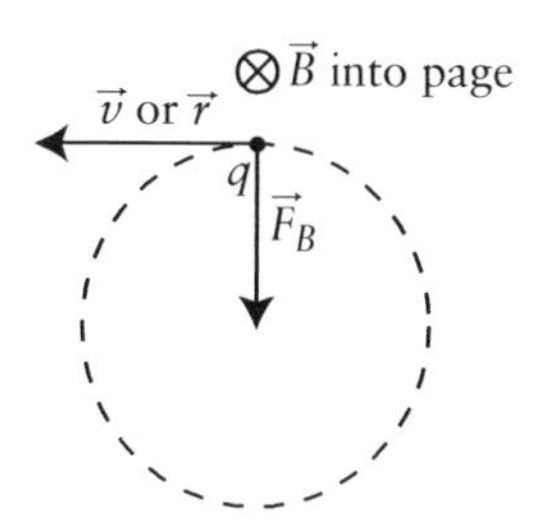

Figure 1.3: Magnetostatic forces act in the direction of $\vec{v}\times\vec{B}$, where $\vec{v}$ is the velocity of a charged particle. Therefore, since the displacement, $\vec{r}$, of the particle is perpendicular to the force ($\vec{r}$ is parallel to $\vec{v}$), then magnetic forces do no work on charged particles.

Work–energy theorem

The *work–energy theorem* states that when an external force acts on a body, the energy of the body changes by an amount equal to the work done on it (Eq. 1.13). Therefore, although work is a scalar quantity measured in Joules, like energy, it is not a form of energy *per se*: it is a form of energy *transfer* to or from a body.

Work–energy theorem:

$$\Delta E = W = \int_{\vec{a}}^{\vec{b}} \vec{F}(\vec{r})\cdot d\vec{r} \quad (1.13)$$

A force does no work on a body if (i) the applied force does not displace the body or if (ii) the direction of displacement is perpendicular to the direction of the force. For example, the magnetostatic force, $\vec{F}_B$, acts perpendicular to the velocity of a charged object:

$$\vec{F}_B = q\vec{v}\times\vec{B}$$

where q is the charge [C] of the object and $\vec{B}$ is the magnetic field [T] experienced by the charge. Since $\vec{F}_B$ is perpendicular to $\vec{v}$, there is no displacement of the charged object along the direction of $\vec{F}_B$, which means that magnetostatic forces do no work on charges—they do not change the energy of moving charges (Fig. 1.3).

Work done *on* versus work done *by* a body

Work is done *on* a body if $W>0$. This results in an *increase* of the body's total energy. Therefore, the total energy of a body is a measure of the amount of work that was done on it by an external force. Conversely, if a body moves in a direction that has a net component in the *opposite* direction to an external force, $\vec{F}$, then the dot product $\vec{F}\cdot d\vec{r}$ is negative, and hence $W<0$. In this case, we say work is done *by* the body, and this results in a *decrease* of the body's total energy. For example work is done by a ball rolling on a flat surface that is slowing under the force of friction with the ground. Therefore, energy is often defined as "the ability of a body to do work": the higher a body's energy, the greater its ability to do work.

Analogy between work and impulse

We saw in Eq. 1.8 that impulse is defined as a force integrated over time. The effect of an impulse is to change the momentum of a body. Analogously, work is defined as a force integrated over *distance*. The effect of doing work is to change the *energy* of a body.

1.3.2 Power

Power, P [W], is the rate of doing work (Eq. 1.14). Using $dW = \vec{F}\cdot d\vec{r}$, and assuming that $\vec{F}$ is constant in time, we can express the rate at which work is done on a body as the external force acting on it multiplied by the velocity of the body (Eq. 1.15):

$$P = \frac{dW}{dt} = \vec{F}\cdot\frac{d\vec{r}}{dt} = \vec{F}\cdot\vec{v}$$

The dot product picks out the component of force acting in the same direction as the body's velocity.

Power:

In terms of work:

$$P = \frac{dW}{dt} \quad (1.14)$$

In terms of force:

$$P = \vec{F}\cdot\vec{v} = |\vec{F}||\vec{v}|\cos\theta \quad (1.15)$$

1.4 Mechanical energy

Mechanical energy is the energy associated with the motion and position of a body, and comprises both kinetic and potential energy, which are described further below.

1.4.1 Kinetic energy (KE)

Kinetic energy, T [J], is the energy associated with the motion of a body. The kinetic energy of a body equals the work done on it when it is accelerated from being stationary to a velocity $\vec{v}$. Using $dW = \vec{F} \cdot d\vec{r}$, together with Newton's Second Law, we find that the kinetic energy of a body of mass m is proportional to the square of its velocity or momentum (Eq. 1.16 and Derivation 1.2). Therefore, the kinetic energy of a body changes if its velocity or momentum changes.

Kinetic energy (Derivation 1.2):

$$T = \frac{1}{2}mv^2 = \frac{p^2}{2m} \quad (1.16)$$

1.4.2 Potential energy (PE)

Potential energy, V [J], is the energy associated with a body's position or configuration. Potential energy is stored in a system, and it "has the potential" to be converted to another form of energy, such as kinetic energy or thermal energy, via work done by the system. For example, a body stores *gravitational* potential energy as a result of its height, and elastic bodies store *elastic* potential energy as a result of being stretched or compressed. Bodies thus store potential energy if they are in a strained or non-equilibrium state. The force that acts to "pull" the system back to its relaxed or equilibrium state is called the *restoring force*, $\vec{F}_R$. The potential energy of a body increases when work is done against the restoring force. This means that the force that must be applied to a body to maintain constant potential energy is equal in magnitude, but opposite in direction to $\vec{F}_R$: $\vec{F}_{app} = -\vec{F}_R$. Therefore, in terms of work done, potential energy is defined as:

$$V = W = \int \vec{F}_{app} \cdot d\vec{r} = -\int \vec{F}_R \cdot d\vec{r}$$

Derivation 1.2: Derivation of kinetic energy (Eq. 1.16)

$$T = W = \int_0^{\vec{r}} \vec{F} \cdot d\vec{r}' = \int_0^t \vec{F} \cdot \vec{v}' dt' = m\int_0^t \frac{d\vec{v}'}{dt'} \cdot \vec{v}' dt' = m\int_0^{\vec{v}} d\vec{v}' \cdot \vec{v}' = m\int_0^v v' dv' = \frac{1}{2}mv^2$$

The primed variables are to distinguish them from the integration limits.

- Starting point: the kinetic energy of a body results from the work done on it during acceleration:

$$T = W = \int \vec{F} \cdot d\vec{r} = \int \vec{F} \cdot \vec{v} dt$$

where $\vec{F}$ is the external force required to accelerate a body, and its velocity is $\vec{v} = d\vec{r}/dt$.
- From Newton's Second Law, the external force required to accelerate a body is equal to the rate of change of its momentum:

$$\vec{F} = \frac{d\vec{p}}{dt} = m\frac{d\vec{v}}{dt} \quad \text{for constant mass, } m$$

- The dot product of a vector with itself is $\vec{v} \cdot \vec{v} = |\vec{v}|^2 \equiv v^2$. Differentiating both sides, we have:

$$d(\vec{v} \cdot \vec{v}) = d\vec{v} \cdot \vec{v} + \vec{v} \cdot d\vec{v} = 2\vec{v} \cdot d\vec{v}$$
$$\text{and } d\left(v^2\right) = 2v dv$$
$$\therefore \vec{v} \cdot d\vec{v} = v dv$$

The potential energy gained by a body moving from $\vec{a}$ to $\vec{b}$ against a restoring force $\vec{F}(\vec{r})$ is given by Eq. 1.17. The nature of the restoring force depends on the system in question. Expressions for elastic and gravitational potential energies are calculated in Worked example 1.1.

Potential energy:

$$V = -\int_{\vec{a}}^{\vec{b}} \vec{F}(\vec{r}) \cdot d\vec{r} \tag{1.17}$$

1.4.3 Conservation of energy

A corollary of Newton's Second Law, $\vec{F} = m\vec{a}$, is that energy is a conserved quantity in isolated systems (i.e. ones that do not interact with their surroundings). If we consider an ideal mechanical system (one in which friction is negligible), then the total energy, E_0, is the sum of kinetic plus potential energy (Eq. 1.20 and Derivation 1.3). Energy can be transferred from one body to another or transformed from one form to another, via work done. For example, if we consider a ball dropped from a height h, its potential energy decreases as it approaches the ground, but its kinetic energy increases, such that $T + V = \text{constant}$.

Conservation of mechanical energy (Derivation 1.3):

$$\begin{aligned} E_0 &= T + V \\ &= \text{constant} \end{aligned} \tag{1.20}$$

Conservation of energy is a fundamental law that applies to all areas of physics. Since the Universe is an isolated system, the total energy in the Universe is constant—it can neither be created nor destroyed. This statement has philosophical implications, regarding the "ultimate fate of the Universe".

1.4.4 Dissipated energy

In many mechanical systems, there is usually some loss of mechanical energy as heat, Q [J]. This energy is irrecoverably lost from the system to its surroundings. Heat is not a form of energy *per se*, but (like work) it is a form of energy *transfer* from one body to another, or from a system to its surroundings. Heat loss from a mechanical system results in an increase in thermal energy of the surroundings.

Forces that cause mechanical energy to be lost from a system are called *dissipative* forces. Common dissipative forces are friction, and air or fluid resistance

WORKED EXAMPLE 1.1: Elastic and gravitational potential energies (Eq. 1.17)

Elastic potential energy:

The restoring force on a spring stretched a small distance, $\vec{r}$, from equilibrium is:

$$\begin{aligned} \vec{F}_{\text{spring}} &= -k\vec{r} \quad \text{in three dimensions} \\ \text{or } F_{\text{spring}} &= -kx \quad \text{in one dimension} \end{aligned}$$

where k is the *spring constant* [N m^{-1}]. Using Eq. 1.17, the potential energy stored in a spring that is stretched a distance x from its equilibrium position is:

$$\begin{aligned} V_{\text{spring}} &= -\int_0^x F_{\text{spring}}\, dx' \\ &= \int_0^x kx'\, dx' = \frac{1}{2}kx^2 \end{aligned} \tag{1.18}$$

The primed variables are to distinguish them from the integration limits.

Gravitational potential energy:

The restoring force on a body lifted vertically in a gravitational field is:

$$F_g = -mg$$

where g is the acceleration due to gravity [9.81 m s^{-2}]. Using Eq. 1.17, the gravitational potential energy, GPE, of a mass m lifted a vertical height h from ground along the z axis is:

$$\begin{aligned} \text{GPE} &= -\int_0^h F_g\, dz \\ &= \int_0^h mg\, dz \\ &= mgh \end{aligned} \tag{1.19}$$

Derivation 1.3: Derivation of energy conservation (Eq. 1.20)

$$\vec{F} = m\ddot{\vec{r}}$$
$$\therefore \vec{F}\cdot\dot{\vec{r}} = m\ddot{\vec{r}}\cdot\dot{\vec{r}}$$
$$\therefore \int \vec{F}\cdot\frac{d\vec{r}}{dt}dt = m\int\frac{d\dot{\vec{r}}}{dt}\cdot\dot{\vec{r}}dt$$
$$= m\int d\dot{\vec{r}}\cdot\dot{\vec{r}}$$
$$= m\int \dot{r}d\dot{r}$$
$$= \frac{1}{2}m\dot{r}^2 + \underbrace{\text{constant}}_{=-E_0}$$
$$\therefore \int \vec{F}\cdot d\vec{r} = \frac{1}{2}m\dot{r}^2 - E_0$$

Rearranging:

$$\therefore E_0 = \underbrace{\frac{1}{2}m\dot{r}^2}_{=T} \underbrace{-\int \vec{F}\cdot d\vec{r}}_{=V}$$
$$= T + V$$
$$= \text{constant}$$

- Starting point: the rate of doing work (power) is:

$$P = \frac{dW}{dt} = \vec{F}\cdot\dot{\vec{r}}$$

where $dW = \vec{F}\cdot d\vec{r}$ (refer to Section 1.3.2). Therefore, the rate at which an external force, $\vec{F} = m\ddot{\vec{r}}$, does work is found by taking the dot product of Newton's Second Law with $\dot{\vec{r}}$:

$$\vec{F}\cdot\dot{\vec{r}} = m\ddot{\vec{r}}\cdot\dot{\vec{r}}$$

(recall that $\dot{\vec{r}} = \vec{v}$ and $\ddot{\vec{r}} = \dot{\vec{v}} = \vec{a}$).

- Integrating this equation with respect to time gives the total work done (the total energy).
- Use the relation:

$$d(\dot{\vec{r}}\cdot\dot{\vec{r}}) = d\left|\dot{\vec{r}}\right|^2 = d(\dot{r}^2)$$
$$\therefore 2\dot{\vec{r}}\cdot d\dot{\vec{r}} = 2\dot{r}d\dot{r}$$

- Let the constant of integration be equal to $-E_0$, where E_0 represents the total energy of the system. Rearranging, we find that the total energy is the sum of kinetic and potential energies.

(drag). For low-velocity bodies, drag forces can often be modeled as being proportional to their velocity. The constant of proportionality is called the *drag constant* or *damping coefficient*, Γ [$\mathrm{Nsm^{-1}}$]. Drag forces act in a direction opposite to the object's velocity, therefore:

$$\vec{F}_\Gamma = -\Gamma\vec{v}$$

In Section 2.2, we will look at the effects of drag forces on projectile motion and oscillating springs.

1.5 Conservative forces

The action of a dissipative force on a system results in an irreversible loss of mechanical energy from the system via heat loss. In contrast, the action of a *conservative* force is to conserve mechanical energy. Below, we shall look at some properties of conservative forces.

Potential energy functions

Differentiating the energy conservation equation, $E_0 = T + V$ (Eq. 1.20), we have:

$$\frac{d}{dt}E_0 = \frac{d}{dt}\left(\frac{1}{2}m\dot{x}^2 + V(x)\right)$$
$$\therefore 0 = m\dot{x}\ddot{x} + \frac{dV(x)}{dt}$$
$$= m\dot{x}\ddot{x} + \frac{dV(x)}{dx}\frac{dx}{dt}$$
$$= m\dot{x}\ddot{x} + \dot{x}\frac{dV(x)}{dx}$$
$$\therefore m\ddot{x} = -\frac{dV(x)}{dx}$$

Therefore, using $F = ma = m\ddot{x}$, we see that a conservative force is related to the gradient of a potential energy function:

$$\therefore F(x) = -\frac{dV(x)}{dx}$$

(Eq. 1.21). This result is not very surprising, since we saw earlier that potential energy is defined as the work done against a restoring force:

$$V(x) = -\int F(x)dx \quad \text{in one dimension}$$
$$\text{or} \quad V(\vec{r}) = -\int \vec{F}(\vec{r})\cdot d\vec{r} \quad \text{in three dimensions}$$

(refer to Eq. 1.17). Generalizing to three dimensions, the space derivative in Eq. 1.21 is replaced by the gradient operator (Eq. 1.22).

Conservative forces:

In one dimension: $F(x) = -\dfrac{dV(x)}{dx}$ (1.21)

In three dimensions: $\vec{F}(\vec{r}) = -\nabla V(\vec{r})$ (1.22)

A conservative force has zero curl:

$\nabla \times \vec{F} = 0$ (1.23)

So, what do Eqs. 1.21 and 1.22 mean? A good conceptual example of a potential energy function is a landscape with hills and valleys. In Worked example 1.1, we calculated the gravitational potential energy as a function of vertical height, z:

$$V(z) = mgz$$

Dividing by mass, the *gravitational potential*, ϕ, is a scalar function of z, defined as:

$$\phi(z) = \frac{V(z)}{m} = gz$$

Therefore, in a landscape of hills and valleys, hill-tops represent regions of high gravitational potential (high z), valleys those of low gravitational potential (low z), and contours are lines of constant potential (equipotential lines, constant z).

The steepness of a hill represents the gradient of the gravitational potential function—it is a measure of the change in potential over distance. We know from experience that on steep hills, we experience the force of gravity more strongly than on horizontal planes. Therefore, the force of gravity is related to (proportional to) the gradient of the gravitational potential function. Expressing this statement mathematically, we have:

$$F \propto \frac{d\phi}{dz}$$

Therefore, from this intuitive example, we can get a "feel" for what Eqs. 1.21 and 1.22 represent.

The curl of conservative forces

Taking the curl of Eq. 1.22, we see that the curl of a conservative force is zero (since mathematically, the curl of a gradient is always zero):

$$\begin{aligned}\nabla \times \vec{F} &= -\nabla \times \nabla V(\vec{r}) \\ &= 0\end{aligned}$$

(Eq. 1.23). In fact, this statement is more commonly the "starting point," and a conservative force is *defined* as a force whose curl equals zero. It is then *because* of this property that we are able to define a scalar potential field, $V(\vec{r})$, that equals the path integral of $\vec{F}(\vec{r})$. We'll see why this is below.

From Stokes's theorem, the closed line integral around a curl-free vector field is equal to zero:

$$\begin{aligned}\int_S \nabla \times \vec{F} \cdot d\vec{s} &= \oint_L \vec{F} \cdot d\vec{l} \\ &= 0 \ \text{ if } \nabla \times \vec{F} = 0\end{aligned}$$

Therefore, the integral between any two points in such a vector field is path independent, and hence has a unique value (refer to Fig. 1.4):

$$\begin{aligned}\therefore \oint_L \vec{F} \cdot d\vec{l} &= \int_a^b \vec{F} \cdot d\vec{l}_{L_1} + \int_b^a \vec{F} \cdot d\vec{l}_{L_2} \\ &= \int_a^b \vec{F} \cdot d\vec{l}_{L_1} - \int_a^b \vec{F} \cdot d\vec{l}_{L_2} \\ &= 0\end{aligned}$$

$$\therefore \int_a^b \vec{F} \cdot d\vec{l}_{L_1} = \int_a^b \vec{F} \cdot d\vec{l}_{L_2}$$

Therefore, we can define a scalar function, $V(\vec{r})$, whose value at a point $\vec{r}$ is equal to the line integral of the corresponding vector field from a fixed reference point, O, to $\vec{r}$:

$$V(\vec{r}) \stackrel{\text{def}}{=} -\int_O^{\vec{r}} \vec{F}(\vec{r}\,') \cdot d\vec{r}\,'$$

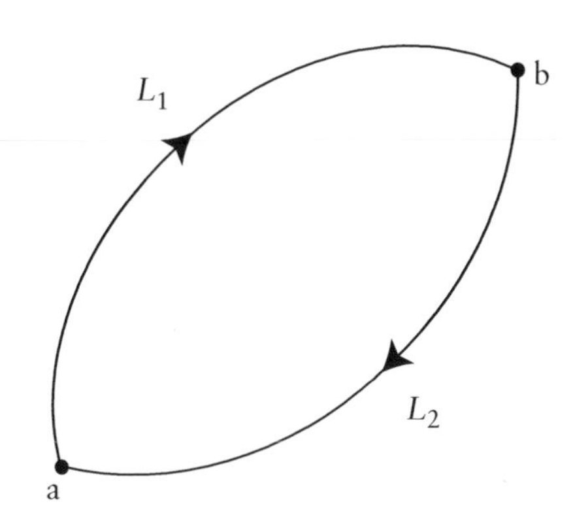

Figure 1.4: Path integral of a conservative force around a closed loop.

The negative sign is for convention. Dissipative forces such as friction have non-zero curl, which means that their action is not path independent. Therefore, for dissipative forces, we cannot define a scalar potential energy field or express the force as the gradient of a potential.

Equilibria and stability

Consider an arbitrary one-dimensional potential energy function that has two turning points, as illustrated in Fig. 1.5. Imagine that this function represents a potential energy landscape $V(x) = mgh(x)$, where the peak represents a hill and the trough a valley, and consider placing a ball in the landscape. We know from experience that a ball placed at the top of the hill is in unstable equilibrium: a small displacement from the peak will cause the ball to roll downhill. In contrast, a ball placed in the valley is in stable equilibrium: a small displacement from the trough results in the ball rolling back towards its starting position in the valley floor.

Eq. 1.21 illustrates stable and unstable equilibrium mathematically. If the ball is displaced by a small amount either side of the local maximum, then from $F = -\mathrm{d}V/\mathrm{d}x$, the external force acts in the same direction as the displacement, and the ball moves further from the maximum. (For example, for a small displacement $\mathrm{d}x$ in the $+x$-direction, $\mathrm{d}V/\mathrm{d}x$ is negative. This means that F is positive, and thus acts in the $+x$-direction, pushing the ball further from the local maximum.) If on the other hand a ball at x_{min} is displaced either side of the local minimum, then the force acts in the *opposite* direction to the displacement, thus pulling it back towards x_{min}. Forces that act in this way are called *restoring* forces, since they act to "restore" the system to its equilibrium configuration. Therefore, systems have a tendency to minimize their potential energies, since such states correspond to states of stable equilibrium.

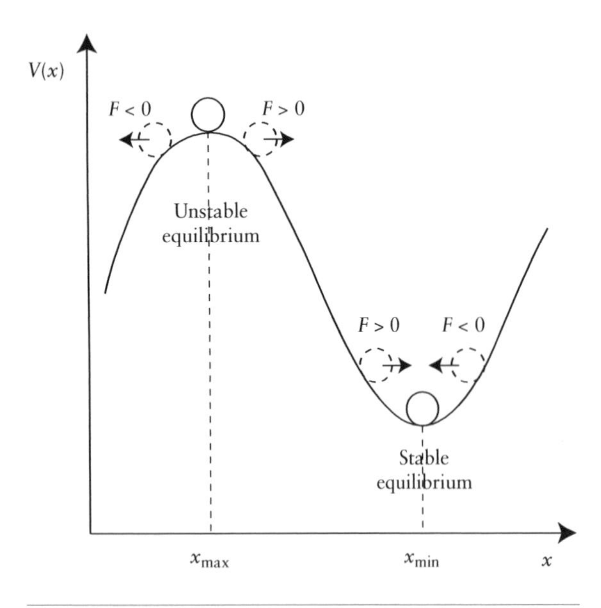

Figure 1.5: Stable and unstable equilibria, and potential energy functions. The negative of the gradient of the potential energy function equals the external force that acts on a body: $F = -\mathrm{d}V/\mathrm{d}x$. Local potential energy maxima correspond to regions of unstable equilibria, whereas local minima correspond to regions of stable equilibria.

Restoring forces and spring constants

What is the strength of the restoring force on a ball displaced a small distance from its equilibrium position? From Eq. 1.21, force is related to potential energy by:

$$F(x) = -\frac{\mathrm{d}V(x)}{\mathrm{d}x}$$

Therefore, a knowledge of the x-dependence of $V(x)$ about the local minimum will provide us with an expression for the restoring force. To approximate the functional dependence of $V(x)$ on x about the minimum, we can perform a Taylor expansion of $V(x)$ about x_{min} (Derivation 1.4). Doing this, we find that the restoring force is proportional to the displacement from equilibrium, and acts in the opposite direction to the displacement. Therefore, for a small displacement Δx from x_{min}, the restoring force is:

$$F(\Delta x) = -k\Delta x$$

where the proportionality constant is:

$$k = \left(\frac{\mathrm{d}^2 V}{\mathrm{d}x^2}\right)_{x_{\text{min}}}$$

(Eqs. 1.24 and 1.25, Derivation 1.4). This linear dependence of restoring force on displacement should look familiar: it is *Hooke's Law*. The restoring force of a mass on a spring, a swinging pendulum, and vibrating atoms in a solid all obey—or can be approximated to—Hooke's Law. The constant of proportionality, k [$\mathrm{N\,m^{-1}}$], is called the *spring constant* or force constant, and is a property of the system in question. Motion about a point of stable equilibrium under Hooke's Law is oscillatory and is known as *simple harmonic motion* (see Section 2.3).

Derivation 1.4: Derivation of Hooke's Law approximation (Eqs. 1.24 and 1.25)

Taylor expansion of $V(x)$ about a local minimum, x_0:

$$\begin{aligned} V(x) &= V(x_0 + \Delta x) \\ &= V(x_0) + \underbrace{\Delta x \left(\frac{\mathrm{d}V}{\mathrm{d}x}\right)_{x_0}}_{=0} + \frac{\Delta x^2}{2}\left(\frac{\mathrm{d}^2V}{\mathrm{d}x^2}\right)_{x_0} + \cdots \\ &\simeq V(x_0) + \frac{\Delta x^2}{2}\left(\frac{\mathrm{d}^2V}{\mathrm{d}x^2}\right)_{x_0} \end{aligned}$$

The force is the gradient of the potential:

$$\begin{aligned} \therefore F(\Delta x) &= -\frac{\mathrm{d}V(x)}{\mathrm{d}\Delta x} \\ &= -\frac{\mathrm{d}}{\mathrm{d}\Delta x}\left(\underbrace{V(x_0)}_{=\text{constant}} + \frac{\Delta x^2}{2}\underbrace{\left(\frac{\mathrm{d}^2V}{\mathrm{d}x^2}\right)_{x_0}}_{=\text{constant}}\right) \\ &= -\Delta x \underbrace{\left(\frac{\mathrm{d}^2V}{\mathrm{d}x^2}\right)_{x_0}}_{\equiv k} \\ &\equiv -k\Delta x \end{aligned}$$

$$\text{where } k = \left(\frac{\mathrm{d}^2V}{\mathrm{d}x^2}\right)_{x_0}$$

- Starting point: consider a point of stable equilibrium, x_0 (local minimum). Define a small displacement from equilibrium as:

$$\Delta x = x - x_0$$

- Perform a Taylor expansion on the potential energy function about the equilibrium position x_0. The first derivative of $V(x)$ at x_0 equals zero since x_0 is a turning point.
- The force as a function of the displacement from equilibrium, $F(\Delta x)$, is equal to the negative gradient of the potential. $V(x_0)$ and the second derivative evaluated at x_0 are both constants.
- If we define the second derivative at x_0 as the force constant, k, then we see that the restoring force about equilibrium is proportional to the displacement from equilibrium, and therefore obeys Hooke's Law ($F(x) = -kx$).

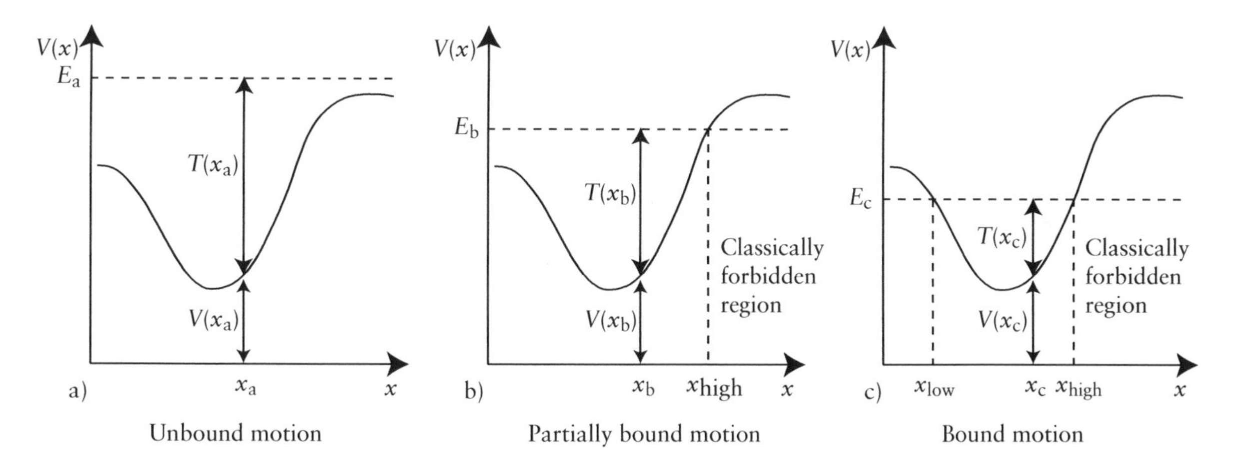

Figure 1.6: a) Unbound, b) partially bound, and c) bound motion. The regions where the total energy of the system falls below the potential energy is classically forbidden.

Hooke's Law approximation (Derivation 1.4):

Restoring force for a small displacement from equilibrium, Δx:

$$F(\Delta x) = -k\Delta x \tag{1.24}$$

where the spring constant is:

$$k = \left(\frac{d^2V(x)}{dx^2}\right)_{x_0} \tag{1.25}$$

($\Delta x = x - x_0$, where x_0 is the equilibrium position).

Bound and unbound motion

The motion of an object is classified as *bound motion* if it is restricted to a particular region in space, and *unbound* if it is not. For example, consider the potential energy function shown in Fig. 1.6. Let the total mechanical energy of the object be E_a, E_b, or E_c for Figs. 1.6a, b, and c, respectively. This is divided into potential energy $V(x)$ plus kinetic energy $T(x)$ such that the total energy is conserved:

$$E_{a,b,c} = V(x) + T(x)$$

The distribution of energy between these two forms changes as a function of x. If the potential energy function does not exceed the total energy of the object, as illustrated in Fig. 1.6a, then there are no restrictions on the body's displacement (it can take any value of x), thus the motion is *unbound*. If however the total energy falls *below* $V(x)$ for a given range of x, as illustrated in Figs. 1.6b and c, then the object cannot have displacements that correspond to these regions. Regions with $E < V(x)$ are called *classically forbidden* regions.[2] In these cases, the body is confined to regions where $E > V(x)$, thus its motion is *bound*. A body at x_{low} or x_{high} (refer to Figs. 1.6b and c) has zero kinetic energy. Therefore, at these positions, the body is stationary and all its energy is stored as potential energy.

[2] In quantum mechanics, it is possible for a particle to enter into the classically forbidden region by a process known as *quantum tunneling* (see Section 17.2.2).

Chapter 2
Practical applications of linear motion

In the previous chapter, we visited Newton's Laws and saw that they underlie the concepts of force, momentum, work, and energy, as well as the conservation of energy and momentum. In this chapter, we will look at some practical applications of Newton's Laws. We'll start by using momentum and energy conservation to solve two-body collision problems, and then use Newton's Second Law to determine the behavior of systems under the action of external forces. All examples, except where specified, involve motion in one dimension; therefore, we will not use vector notation in these cases.

2.1 Two-body collisions

Collision problems provide a good illustration of using energy and momentum conservation to determine the motion of bodies. As a first example, let's consider a system of two spherical bodies, with masses m_1 and m_2. Before the collision, the masses have velocities u_1 and u_2, respectively. Upon collision, kinetic energy and momentum are transferred from one body to the other, which results in final velocities v_1 and v_2 (see figure in Worked example 2.1).

2.1.1 Elastic and inelastic collisions

Collisions are defined as either *elastic* or *inelastic*, depending on whether kinetic energy is lost to the surroundings on impact. For example, energy is lost if the colliding objects deform.

- Elastic collisions: the total kinetic energy of the system is conserved. If we assume that the colliding bodies are hard objects, such as billiard balls, then they do not deform on impact and their internal potential energy remains constant. Therefore, from the conservation of energy, the total kinetic energy before the collision equals the total kinetic energy after. Worked example 2.1 provides an example of solving the final velocities of two bodies after a one-dimensional elastic collision.
- Inelastic collisions: during an inelastic collision, some kinetic energy of the colliding objects is lost on impact, for example, as heat or due to deformation. Therefore, the total kinetic energy before the collision is greater than the total kinetic energy after. A collision is said to be *totally inelastic* if the colliding bodies stick together on impact. This means that both bodies have the same velocity as one another after the collision. This form of collision corresponds to the maximum possible kinetic energy loss during a two-body collision.

Momentum is conserved for both elastic and inelastic collisions alike.

The coefficient of restitution

The *coefficient of restitution*, ε, is a dimensionless parameter that characterizes the extent to which a collision is inelastic. It is defined as the ratio of the velocity of separation of the masses after the collision, $v_2 - v_1$, to their velocity of approach before the collision, $u_1 - u_2$:

$$\varepsilon = \frac{v_2 - v_1}{u_1 - u_2}$$

(Eq. 2.3; refer to diagram in Worked example 2.1 for velocity definitions.). For perfectly elastic collisions, the velocity of approach is equal and opposite to the velocity of separation, $\therefore \varepsilon = 1$. In the opposite extreme, for totally inelastic collisions, the two bodies merge on impact to form a single body traveling at a common

WORKED EXAMPLE 2.1: Two-body elastic collision in one dimension

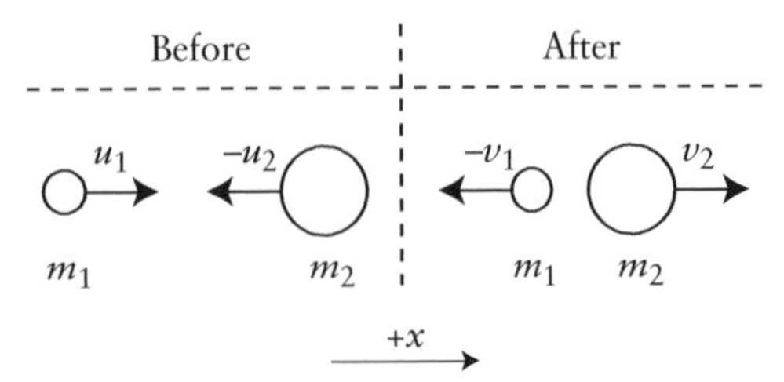

1. Conserve momentum. Positive velocity is defined in the direction of the positive x-axis.
2. Conserve kinetic energy. Use Eq. 2.1 to simplify the expression.
3. Solve the simultaneous Eqs. 2.1 and 2.2 for final velocities v_1 and v_2, given known initial velocities u_1 and u_2.

1. Conservation of momentum:

$$\underbrace{m_1u_1 + m_2u_2}_{\text{momentum before}} = \underbrace{m_1v_1 + m_2v_2}_{\text{momentum after}}$$

$$\therefore m_1(v_1 - u_1) = m_2(u_2 - v_2) \qquad (2.1)$$

2. Conservation of kinetic energy:

$$\underbrace{\frac{1}{2}m_1u_1^2 + \frac{1}{2}m_2u_2^2}_{\text{kinetic energy before}} = \underbrace{\frac{1}{2}m_1v_1^2 + \frac{1}{2}m_2v_2^2}_{\text{kinetic energy after}}$$

$$\therefore m_1\left(v_1^2 - u_1^2\right) = m_2\left(u_2^2 - v_2^2\right)$$

$$m_1(v_1 - u_1)(v_1 + u_1) = m_2(u_2 - v_2)(u_2 + v_2)$$

$$\therefore (v_1 + u_1) = (u_2 + v_2) \qquad (2.2)$$

3. Solve simultaneous equations for v_1 and v_2:

$$\therefore v_1 = \left(\frac{m_1 - m_2}{m_1 + m_2}\right)u_1 + \left(\frac{2m_2}{m_1 + m_2}\right)u_2$$

$$\text{and } v_2 = \left(\frac{2m_1}{m_1 + m_2}\right)u_1 - \left(\frac{m_1 - m_2}{m_1 + m_2}\right)u_2$$

Special cases: consider the special case where m_2 is initially stationary, $\therefore u_2 = 0$. In this case, the final velocities are:

$$v_1 = \left(\frac{m_1 - m_2}{m_1 + m_2}\right)u_1 \text{ and } v_2 = \left(\frac{2m_1}{m_1 + m_2}\right)u_1$$

- Stationary mass is less than incident mass: both bodies travel in the positive x-direction. The velocity of the incident body decreases after collision since some of its kinetic energy is transferred to the stationary body. $m_1 > m_2 \rightarrow v_1 > 0$, $v_2 > 0$
- Both masses are equal: complete transfer of momentum from the moving body to the stationary body; therefore, the incident body is stationary after the collision. $m_1 = m_2 \rightarrow v_1 = 0$, $v_2 = u_1$
- Stationary mass is greater than incident mass: incident body rebounds and travels in the negative x-direction, while the initially stationary mass travels in the positive x-direction. $m_1 < m_2 \rightarrow v_1 < 0$, $v_2 > 0$
- Stationary mass much greater than incident mass: the incident body reflects off the stationary body with approximately equal and opposite velocity to its incoming velocity, while the massive body remains approximately stationary. (Note that this does not violate momentum conservation—the mass m_2 recoils with finite momentum, but with negligible velocity due to its large mass). $m_1 \ll m_2 \rightarrow v_1 \simeq -u_1$, $v_2 \simeq 0$

velocity; therefore, $v_2 - v_1 = 0$ and $\varepsilon = 0$. Inelastic collisions, therefore, have a coefficient of restitution in between 0 and 1.

Coefficient of restitution:

$$\varepsilon = \frac{v_2 - v_1}{u_1 - u_2} \tag{2.3}$$

$$\text{Elastic collisions} : \varepsilon = 1$$
$$\text{Inelastic collisions} : 0 < \varepsilon < 1$$
$$\text{Totally inelastic collisions} : \varepsilon = 0$$

Using Eq. 2.3 together with the conservation of momentum ($m_1u_1 + m_2u_2 = m_1v_1 + m_2v_2$), provides us with a set of simultaneous equations from which we can express the final velocities in terms of the initial velocities and ε. Doing this, we find:

$$v_1 = \left(\frac{m_1 - \varepsilon m_2}{m_1 + m_2}\right)u_1 + \left(\frac{(1+\varepsilon)m_2}{m_1 + m_2}\right)u_2$$
$$v_2 = \left(\frac{(1+\varepsilon)m_1}{m_1 + m_2}\right)u_1 + \left(\frac{m_2 - \varepsilon m_1}{m_1 + m_2}\right)u_2$$

When $\varepsilon = 1$, we recover the expressions for v_1 and v_2 for elastic collisions from Worked example 2.1, as expected.

2.1.2 Elastic collisions in two dimensions

We can solve collision problems in two or three dimensions in a similar manner to one-dimensional problems: by conserving momentum and energy of the colliding bodies before and after the collision. Since momentum is a vector, momentum is conserved along each orthogonal axis, x, y, and z, independently. For now, let's consider an elastic collision in two dimensions between two bodies of equal mass m, where body 2 is initially stationary, and body 1 is incident at a velocity $\vec{u}$ (see figure in Derivation 2.1). The final velocities of the bodies have magnitudes v_1 and v_2, and they travel at angles θ_1 and θ_2 relative to the direction of $\vec{u}$, as shown in the figure in Derivation 2.1. Conserving horizontal and vertical components of momentum together with kinetic energy, the final velocities of the bodies are found to satisfy the condition:

$$v_1v_2\cos(\theta_1 + \theta_2) = 0$$

(Eq. 2.4 and Derivation 2.1).

Elastic collisions in two dimensions (Derivation 2.1):

$$v_1v_2\cos(\theta_1 + \theta_2) = 0 \tag{2.4}$$

$$\therefore \text{ either } v_1 = 0 \text{ for head-on collisions}$$
$$(\text{since } \theta_1 + \theta_2 = 0)$$
$$\text{or } \theta_1 + \theta_2 = \frac{\pi}{2} \text{ for oblique collisions}$$

If the collision is exactly head on, then body 2 travels in the same direction as $\vec{u}$ after the impact ($\theta_1 + \theta_2 = 0$), and the problem reduces to a one-dimensional problem. In this case, for two bodies of equal mass, body 1 transfers all its kinetic energy to body 2; therefore, the final velocity of body 1 is $v_1 = 0$, and the final velocity of body 2 is $v_2 = u$. This calculation agrees with that from Worked example 2.1 in the special case where $m_1 = m_2 = m$. For oblique collisions, body 1 imparts some of its kinetic energy to body 2, such that neither is stationary after the collision. In this case, body 2 travels in the direction of the line joining the center of the two balls at their point of contact, as shown in Fig. 2.1. From Eq. 2.4, body 1 travels at 90° to this direction. This is the physics behind billiard ball collisions and the trick to being a strong pool player.

2.2 Equations of motion

An *equation of motion* describes how the motion of a body, such at its displacement or velocity, changes as a function of time. Newton's Second Law states that the action of a net force on a body results in a change of its momentum, where the rate of change of momentum equals the force:

$$\vec{F} = \frac{d\vec{p}}{dt}$$

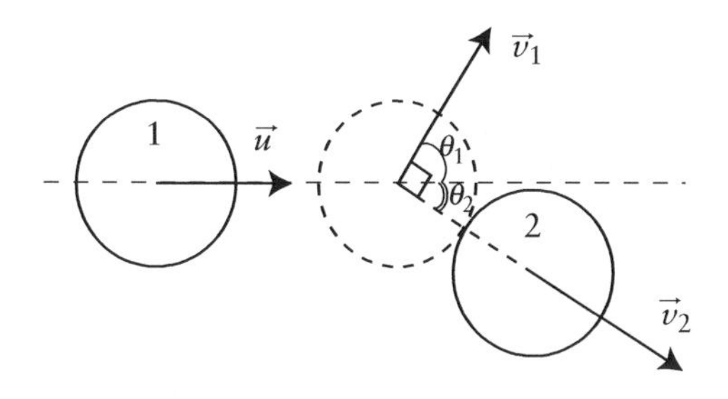

Figure 2.1: Elastic collision between two bodies of equal mass, at oblique incidence. The direction of $\vec{v}_2$ is along the line joining the center of the two bodies at their point of contact, and, from Eq. 2.4, the direction of $\vec{v}_1$ is perpendicular to $\vec{v}_2$.

Derivation 2.1: Derivation of conditions for two-body elastic collisions in two dimensions (Eq. 2.4)

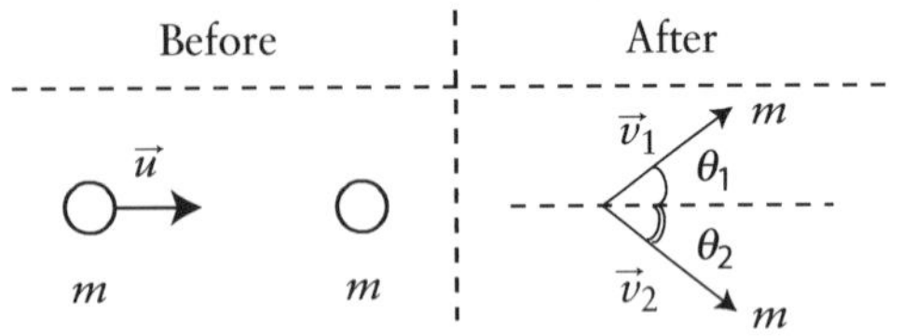

Before | After

m $\vec{u}$ m | $\vec{v}_1$ m θ_1 θ_2 $\vec{v}_2$ m

Solution 1: in terms of momentum components:

$$\begin{pmatrix} u \\ 0 \end{pmatrix} = \begin{pmatrix} v_1 \cos\theta_1 \\ v_1 \sin\theta_1 \end{pmatrix} + \begin{pmatrix} v_2 \cos\theta_2 \\ -v_2 \sin\theta_2 \end{pmatrix}$$

$$\therefore u^2 = v_1^2 \cos^2\theta_1 + v_2^2 \cos^2\theta_2 + 2v_1v_2 \cos\theta_1 \cos\theta_2$$

$$\text{and } 0 = v_1^2 \sin^2\theta_1 + v_2^2 \sin^2\theta_2 - 2v_1v_2 \sin\theta_1 \sin\theta_2$$

adding the two expressions:

$$\begin{aligned} u^2 &= v_1^2\left(\cos^2\theta_1 + \sin^2\theta_1\right) + v_2^2\left(\cos^2\theta_2 + \sin^2\theta_2\right) \\ &\quad + 2v_1v_2\left(\cos\theta_1 \cos\theta_2 - \sin\theta_1 \sin\theta_2\right) \\ &= v_1^2 + v_2^2 + 2v_1v_2 \cos\left(\theta_1 + \theta_2\right) \end{aligned}$$

Solution 2: in terms of velocity vectors:

$$\begin{aligned} u^2 &= \vec{u} \cdot \vec{u} \\ &= \left(\vec{v}_1 + \vec{v}_2\right) \cdot \left(\vec{v}_1 + \vec{v}_2\right) \\ &= \vec{v}_1 \cdot \vec{v}_1 + \vec{v}_2 \cdot \vec{v}_2 + 2\vec{v}_1 \cdot \vec{v}_2 \\ &= v_1^2 + v_2^2 + 2v_1v_2 \cos\phi \\ &= v_1^2 + v_2^2 + 2v_1v_2 \cos\left(\theta_1 + \theta_2\right) \end{aligned}$$

Therefore, using kinetic energy conservation to eliminate u^2:

$$\therefore v_1v_2 \cos\left(\theta_1 + \theta_2\right) = 0$$

- Momentum conservation for equal masses:

$$\begin{aligned} m\vec{u} &= m\vec{v}_1 + m\vec{v}_2 \\ \therefore \vec{u} &= \vec{v}_1 + \vec{v}_2 \end{aligned} \qquad (2.5)$$

Horizontal and vertical components of momentum are conserved separately.

- Use momentum conservation to find an expression for u^2. We can do this in two ways: we can consider conservation of each component separately (solution 1), or we can work directly in terms of vectors and take the dot product of Eq. 2.5 with itself (solution 2). For the latter case, let the angle between vectors $\vec{v}_1$ and $\vec{v}_2$ be $\phi = \theta_1 + \theta_2$.
- Kinetic energy conservation for equal masses:

$$\begin{aligned} \frac{1}{2}mu^2 &= \frac{1}{2}mv_1^2 + \frac{1}{2}mv_2^2 \\ \therefore u^2 &= v_1^2 + v_2^2 \end{aligned}$$

- Equate the expression of u^2 derived from momentum conservation to that from kinetic energy conservation to determine the condition for elastic collisions.

Newton's Second Law is thus an equation of motion since it describes the time evolution of a system's momentum, or using $\vec{p} = m\vec{v}$, its velocity (Eq. 2.6). If mass is constant, we can express Newton's Second Law as either a first-order differential equation of velocity or a second-order differential equation of position, where $\vec{v} = \mathrm{d}\vec{r}/\mathrm{d}t$ (Eqs. 2.7 and 2.8, respectively). If mass is not constant, for example, for a raindrop or a rocket, then we can use the product rule to differentiate $\vec{p} = m\vec{v}$ (Eq. 2.9).

Position or velocity solutions of the differential equations 2.7–2.9 are also equations of motion. Given known boundary conditions, these solutions predict future motion of a body under a given net external

Equation of motion:

$$\vec{F} = \frac{\mathrm{d}\vec{p}}{\mathrm{d}t} = \frac{\mathrm{d}(m\vec{v})}{\mathrm{d}t} \qquad (2.6)$$

Constant mass:

$$\vec{F} = m\frac{\mathrm{d}\vec{v}}{\mathrm{d}t} = m\dot{\vec{v}} \qquad (2.7)$$

$$\text{or } \vec{F} = m\frac{\mathrm{d}^2\vec{r}}{\mathrm{d}t^2} = m\ddot{\vec{r}} \qquad (2.8)$$

Variable mass:

$$\vec{F} = m\frac{\mathrm{d}\vec{v}}{\mathrm{d}t} + \vec{v}\frac{\mathrm{d}m}{\mathrm{d}t} \qquad (2.9)$$

force. In this section, we will look at several examples of solving Newton's Second Law to establish the behavior of a system as a function of time.

2.2.1 Constant force and projectiles

A body moving under a constant force has a constant acceleration, $\vec{a}$. Therefore, the force acting on the body is $\vec{F}=m\vec{a}$. Let's consider the one-dimensional problem: $F=ma$. In one dimension, Eqs. 2.7 and 2.8 are:

$$F = m\frac{dv}{dt}$$

$$\text{and } F = m\frac{d^2x}{dt^2}$$

Equating these expressions to $F=ma$, we can solve differential equations for the velocity and displacement of a body moving under constant force, as functions of time (Eqs. 2.10 and 2.11, Derivation 2.2). We can further combine Eqs. 2.10 and 2.11 to express the equations of motion in terms of different combinations of variables (Eqs. 2.12–2.14, Derivation 2.2).

Motion under constant external force (Derivation 2.2):

$$v(t) = u_0 + at \tag{2.10}$$

$$s(t) = u_0 t + \frac{1}{2}at^2 \tag{2.11}$$

Combining the above equations, we have:

$$s(t) = \frac{1}{2}(u_0 + v(t))t \tag{2.12}$$

$$\text{or } s(t) = v(t)t - \frac{1}{2}at^2 \tag{2.13}$$

$$v^2(t) = u_0^2 + 2as(t) \tag{2.14}$$

where u_0 is the initial velocity.

Many of you will recognize these five equations from having used them at high school. They are usually derived without making reference to force or Newton's Second Law, but instead directly from integrating the definitions of a and v:

$$a = \frac{dv}{dt}$$

$$\text{and } v = \frac{dx}{dt}$$

However, it's useful to appreciate that these equations of motion correspond to solutions of Newton's Second Law in the case where the net external force, and therefore the acceleration of a body, are constant.

Projectile motion

A common example of motion under a constant force is *projectile* motion, where bodies move under gravity. The acceleration due to gravity on Earth is constant at $g \simeq 9.81\,\text{m s}^{-2}$. Worked example 2.2 shows how to solve Newton's Second Law for projectile motion, under the condition that air resistance is negligible. Doing this, we find that the horizontal (x) and vertical (y) displacements of a projectile as a function of time are:

$$x(t) = vt\cos\phi$$

$$y(t) = vt\sin\phi - \frac{1}{2}gt^2$$

where v is the initial velocity and ϕ is its angle with respect to the ground (refer to the figure in Worked example 2.2). Using these expressions for $x(t)$ and $y(t)$, we can deduce properties of the projectile motion such as time of flight and projectile range (refer to Worked example 2.2).

If air resistance is *not* negligible, then the differential equation from Newton's Second Law must also take into account the resistive force. For a resistive force that is proportional to the projectile velocity, $\vec{F}_\Gamma = -\Gamma\vec{v}$, the net external force is a combination of air resistance and gravity: $\vec{F} = -\Gamma\vec{v} + m\vec{g}$. Equating this expression to Newton's Second Law, $\vec{F} = m d\vec{v}/dt$, provides a differential equation in $\vec{v}$, which can be solved to determine the velocity of a body as a function of time. An example of solving Newton's Second Law for this type of air resistance is provided in Worked example 2.3.

2.2.2 Variable mass problems

If the mass of a body is not constant, then Newton's Second Law has a term that includes the rate of change of mass as well as the acceleration term:

$$\vec{F} = \frac{d\vec{p}}{dt} = m\frac{d\vec{v}}{dt} + \vec{v}\frac{dm}{dt}$$

Below, we will apply this form of Newton's Law to a falling raindrop and a rocket.

Raindrop accumulation of mass

A raindrop forms around a nucleation center and accumulates mass as it falls. If we define downwards as positive, and let the raindrop fall at velocity v under gravity, then $F=mg$ and our differential equation of motion in one dimension (downward) is:

$$m\frac{dv}{dt} + v\frac{dm}{dt} = mg$$

Derivation 2.2: Derivation of equations of motion under constant acceleration (Eqs. 2.10–2.14)

Final velocity:

$$\frac{dv}{dt} = a \rightarrow dv = a dt$$

$$\int_{u_0}^{v(t)} dv = a\int_0^t dt$$

$$\therefore v(t) = u_0 + at$$

- Final velocity: equating $F = ma$ to $F = m dv/dt$, where a is constant, we have:

$$a = \frac{dv}{dt}$$

Integrating this expression, we can determine the velocity of a body as a function of time, $v(t)$. Let the initial velocity be u_0.

Displacement:

$$\frac{d^2x}{dt^2} = a \rightarrow \frac{dx}{dt} = u_0 + at$$

$$\therefore \int_0^{s(t)} dx = u_0\int_0^t dt + a\int_0^t t\,dt$$

$$s(t) = u_0 t + \frac{1}{2}at^2$$

$$\equiv v(t)t - \frac{1}{2}at^2$$

$$\equiv \underbrace{\frac{1}{2}\left(u_0 + v(t)\right)}_{\text{average velocity, } \bar{v}} t$$

- Displacement: equating $F = ma$ to $F = m d^2x/dt^2$, where a is constant, we have:

$$a = \frac{d^2x}{dt^2}$$

Integrating twice with respect to time, we can determine the displacement of a body as a function of time, $s(t)$. Let the initial velocity be u_0 and the initial displacement be $s_0 = 0$.
- Use the expression for $v(t)$ to eliminate either u_0 or a from $s(t)$, to obtain two additional equivalent expressions for $s(t)$. The expression $s(t) = (u_0 + v(t))t/2$ defines the average velocity over a time t: $\bar{v} = (u_0 + v(t))/2 = s(t)/t$.

Velocity squared:

$$v^2(t) = u_0^2 + 2u_0 at + (at)^2$$

$$= u_0^2 + 2a(u_0 t + \frac{1}{2}at^2)$$

$$= u_0^2 + 2as(t)$$

- Velocity squared: square the expression for $v(t)$ and substitute in for $s(t)$.

Assume that the raindrop accumulates mass exponentially with distance:

$$\frac{dm}{dz} = \alpha m$$

where α is a constant. Using this, together with the differential equation of motion, the rate of change of velocity is therefore:

$$\frac{dv}{dt} = g - \frac{v}{m}\frac{dm}{dt}$$

$$= g - \frac{v}{m}\frac{dm}{dz}\frac{dz}{dt}$$

$$= g - \frac{v^2}{m}\frac{dm}{dz} = g - \alpha v^2$$

If we wish to solve the velocity as a function of distance, we can convert the above differential equation to one in dv/dz using:

$$\frac{dv}{dt} = \frac{dz}{dt}\frac{dv}{dz} = v\frac{dv}{dz}$$

Therefore, the differential equation of velocity as a function of distance is:

$$\therefore \frac{dv}{dz} = \frac{g}{v} - \alpha v$$

$$\therefore \int_0^z dz = \int_0^{v(z)} \frac{dv}{g/v - \alpha v}$$

WORKED EXAMPLE 2.2: Projectile motion

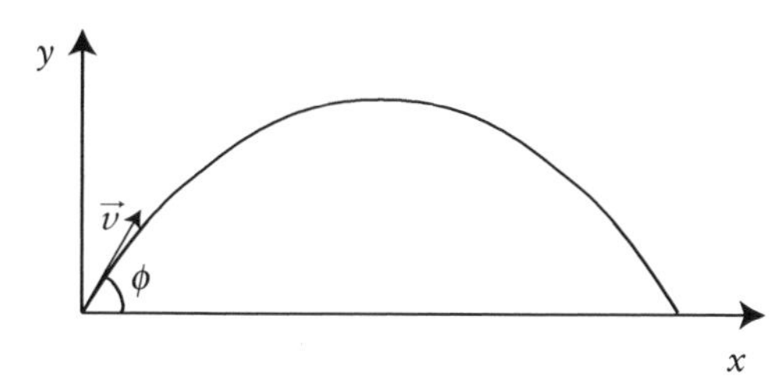

A projectile of mass m is launched from a flat surface, with velocity $\vec{v}$, at an angle ϕ to the horizontal. Assume that air resistance is negligible.

1. **Equation of motion:** gravity is the only force that acts on the projectile, therefore the external force is $\vec{F}=m\vec{g}$. Using this, together with Newton's Second Law, the equation of motion is $\vec{F}=m\ddot{\vec{r}}=m\vec{g}$. We can split the two-dimensional motion into two components by considering x and y motion separately. Let the acceleration due to gravity have magnitude g in the negative y-direction (downwards):

$$m\ddot{\vec{r}} = m\vec{g}$$

$$\therefore m\begin{pmatrix}\ddot{x}\\ \ddot{y}\end{pmatrix} = m\begin{pmatrix}0\\ -g\end{pmatrix}$$

2. **Solutions with arbitrary constants:** integrating the horizontal and vertical components of motion twice, the x and y solutions are:

$$\ddot{x} = \frac{\mathrm{d}^2x}{\mathrm{d}t^2} = 0 \rightarrow x(t) = a + bt \qquad (2.15)$$

$$\ddot{y} = \frac{\mathrm{d}^2y}{\mathrm{d}t^2} = -g \rightarrow y(t) = c + dt - \frac{1}{2}gt^2 \qquad (2.16)$$

where a, b, c, and d are the integration constants. Notice that these equations agree with $s(t)=s_0+u_0t+at^2/2$ (refer to Eq. 2.11), with acceleration $a=-g$ and initial displacement s_0.

3. **Boundary conditions and solutions:** the arbitrary constants are fixed by initial positions and velocities. If we assume the initial position is at the origin, then the boundary conditions are:

$$x(0) = 0 \qquad \therefore a = 0$$
$$y(0) = 0 \qquad \therefore c = 0$$
$$\dot{x}(0) = v\cos\phi \quad \therefore b = v\cos\phi$$
$$\dot{y}(0) = v\sin\phi \quad \therefore d = v\sin\phi$$

Eliminating the integration constants from Eqs. 2.15 and 2.16, the solutions are therefore:

$$x(t) = vt\cos\phi$$
$$y(t) = vt\sin\phi - \frac{1}{2}gt^2$$

- The time of flight is found by solving for t when $y=0$:

$$\therefore t_{\mathrm{tot}} = \frac{2v\sin\phi}{g}$$

- The projectile range is found by substituting t_{tot} into x:

$$x_{\max} = \frac{2v^2\sin\phi\cos\phi}{g}$$

Integrating (test by substitution), the velocity solution is:

$$z = \left[-\frac{1}{2\alpha}\ln\left(g-\alpha v^2\right)\right]_0^{v(z)}$$

$$= -\frac{1}{2\alpha}\ln\left(\frac{g-\alpha v^2(z)}{g}\right)$$

$$\therefore v(z) = \sqrt{\frac{g}{\alpha}\left(1-\mathrm{e}^{-2\alpha z}\right)} \qquad (2.17)$$

$v(z)$ is plotted in Fig. 2.2. The velocity of the raindrop increases with distance until it reaches terminal velocity, $v_{\mathrm{t}}=\sqrt{g/\alpha}$.

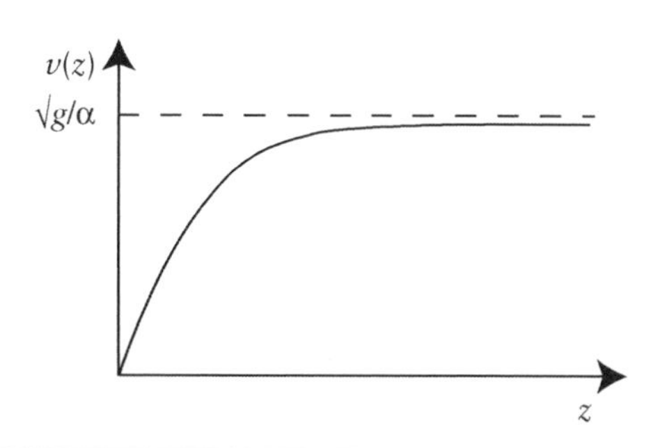

Figure 2.2: Velocity of a falling raindrop (Eq. 2.17). We assume a raindrop accumulates mass exponentially with distance: $\mathrm{d}m/\mathrm{d}z=\alpha m$. Its velocity increases until it reaches terminal velocity, $v_{\mathrm{t}}=\sqrt{g/\alpha}$, where g is the acceleration due to gravity.

WORKED EXAMPLE 2.3: Projectile motion with air resistance

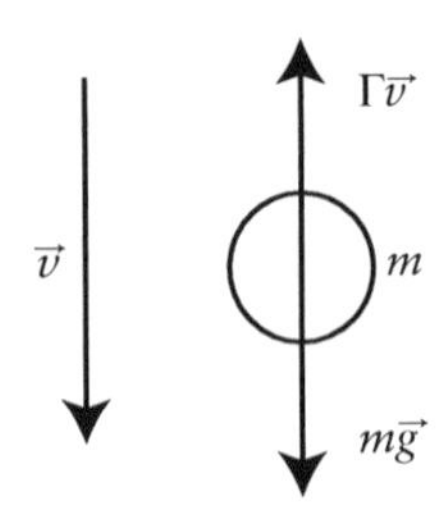

A projectile of mass m falls with velocity $\vec{v}$. The air resistance on the mass is proportional in magnitude to its velocity and opposes the motion: $\vec{F}_\Gamma = -\Gamma\vec{v}$.

1. **Equation of motion:** the net external force on the body is $\vec{F} = m\vec{g} - \Gamma\vec{v}$. Using this, together with Newton's Second Law, the equation of motion is $\vec{F} = m\dot{\vec{v}} = m\vec{g} - \Gamma\vec{v}$. We can split the two-dimensional motion into two components by considering the x and y motion separately. Let the acceleration due to gravity have magnitude g in the negative y-direction (downwards):

$$m\dot{\vec{v}} = -\Gamma\vec{v} + m\vec{g}$$

$$\therefore m\begin{pmatrix}\dot{v}_x \\ \dot{v}_y\end{pmatrix} = -\Gamma\begin{pmatrix}v_x \\ v_y\end{pmatrix} + m\begin{pmatrix}0 \\ -g\end{pmatrix}$$

2. **Solutions with arbitrary constants:** define $\gamma = \Gamma/m$. Rearranging the above equations and integrating once, we find that the velocity components v_x and v_y are exponentially decaying solutions:

$$\dot{v}_x = \frac{dv_x}{dt} = -\gamma v_x \to \ln(v_x(t)) = -\gamma t + a$$

$$\therefore v_x(t) = Ae^{-\gamma t}$$

$$\dot{v}_y = \frac{dv_y}{dt} = -\gamma v_y - g \to \text{let } p = \gamma v_y + g$$

$$\therefore \frac{dp}{dt} = -\gamma p$$

$$\therefore p = Be^{-\gamma t}$$

$$\therefore v_y(t) = \frac{1}{\gamma}\left(Be^{-\gamma t} - g\right)$$

where a, A and B are integration constants.

3. **Boundary conditions:** consider a mass m dropped vertically from a large height. The initial velocity is 0 in both the x- and y-directions. In this case, the boundary conditions and solutions are:

$$v_x(0) = 0 \quad \therefore A = 0 \to v_x(t) = 0$$

$$\text{and} \quad v_y(0) = 0 \quad \therefore B = g \to v_y(t) = -\frac{g}{\gamma}\left(1 - e^{-\gamma t}\right)$$

The body reaches *terminal velocity as* $t \to \infty$:

$$\therefore v_t = -\frac{g}{\gamma}$$

This corresponds to the velocity at which the air resistance equals gravity: $\Gamma v_t = -mg \therefore v_t = -g/\gamma$. The approach to terminal velocity is exponential.

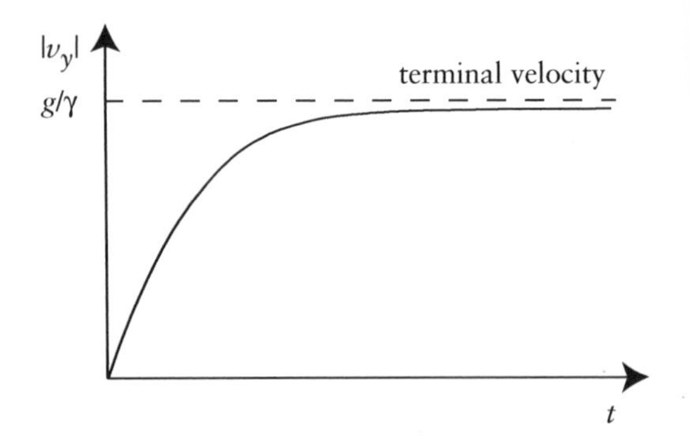

Rocket thrust

Rockets are propelled forwards by ejecting hot gas behind them, as illustrated in Fig. 2.3. From momentum conservation, the backwards momentum of the gas is balanced by an equal forward momentum of the rocket. Movement arising from momentum conservation in this way is called *recoil*, similar to the recoil a person experiences on firing a bullet from a gun. The force that acts on the rocket due to its recoil is called *thrust*.

Let's set up the equation of motion by considering momentum conservation. At a time t, let the velocity of the rocket be $v(t)$ and its mass be $m(t)$. Therefore, its initial momentum, before fuel ejection, is:

$$p_{\text{initial}} = m(t)v(t)$$

The rocket then ejects a small mass of fuel, dm, at a velocity $-v_0$ relative to the rocket, which increases the velocity of the rocket by an amount dv (refer to Fig. 2.3).

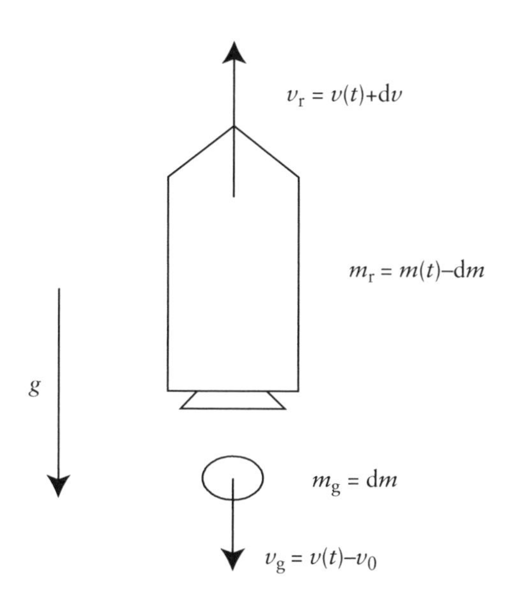

Figure 2.3: Rocket thrust. The mass of ejected gas is $m_g = dm$. The instantaneous velocity of the rocket is $v_r = v(t) + dv$, where dv is its gain in velocity from recoil. The instantaneous velocity of the gas is $v_g = v(t) - v_0$, where v_0 is the velocity at which fuel is ejected from the rocket relative to the rocket. During take off, gravity acts downwards on the system.

The new mass of the rocket is then $m_r = m(t) - dm$, and its velocity is $v_r = v(t) + dv$. The mass of the ejected fuel is $m_g = dm$, and its velocity is $v_g = v(t) - v_0$. Therefore, the final momentum of the rocket and fuel, after fuel ejection, is:

$$\begin{aligned} p_{\text{final}} &= m_r v_r + m_g v_g \\ &= \big(m(t) - dm\big)\big(v(t) + dv\big) + dm\big(v(t) - v_0\big) \\ &\simeq m(t)v(t) + m(t)dv - v_0 dm \end{aligned}$$

where the second-order term $dm\,dv$ has been ignored. Therefore, the change in momentum is $dp = m(t)dv - v_0 dm$, which means that the rate of change of momentum is therefore:

$$\begin{aligned} F &= \frac{dp}{dt} \\ &= m(t)\frac{dv}{dt} - \underbrace{v_0\frac{dm}{dt}}_{\text{thrust}} \end{aligned} \tag{2.18}$$

This is the differential equation of motion for the rocket. It has the form of Newton's Second Law for a variable-mass problem, as expected. The minus sign in front of the second term is because the rocket mass decreases with time, therefore, dm/dt is negative. This force is the *thrust* on the rocket.

We can solve Eq. 2.18 for two different regimes: (i) during take off, where the external force that acts on the rocket is gravity, $\therefore F = -m(t)g$, and (ii) in space, where there is no external force that acts on the rocket, $\therefore F = 0$. The solutions of the rocket velocity and mass as a function of time for both these regimes are provided in Eqs. 2.19–2.22 (Derivation 2.3).

Rocket motion (Derivation 2.3):

During take off (under gravity):

$$v(t) = \ln\left(\frac{m_0}{m(t)}\right)v_0 - gt \tag{2.19}$$

$$m(t) = m_0 e^{-(v(t)+gt)/v_0} \tag{2.20}$$

where $m(t) = m_0 - \dot{m}t$.

In space (no gravity):

$$v_f = v_i + \ln\left(\frac{m_i}{m_f}\right)v_0 \tag{2.21}$$

$$m_f = m_i e^{-(v_f - v_i)/v_0} \tag{2.22}$$

where $m_f = m_i - \dot{m}\Delta t$.

2.3 Oscillating systems

In Section 1.5, we saw that points of stable equilibria correspond to local minima in the system's potential energy function. The system experiences a *restoring force* upon a displacement from equilibrium, which acts in the direction to "pull" the system back to equilibrium. For sufficiently small displacements from equilibrium, the restoring force is proportional to the displacement from equilibrium, x (Hooke's Law):

$$F = -kx$$

The constant of proportionality, k [N m^{-1}], is called the *force constant*,[1] and can be approximated to the second space derivative of the potential energy function calculated at the point of stable equilibrium, x_0:

$$k = \left(\frac{d^2V(x)}{dx^2}\right)_{x_0}$$

(refer to Section 1.5, Eq. 1.25).

In this section, we will use Newton's Second Law, together with Hooke's Law, to construct a differential equation that governs the motion of bodies displaced

1 k is also sometimes called the *spring constant* since a common application of Hooke's Law is a mass on a spring.

Derivation 2.3: Derivation of rocket motion equations (Eqs. 2.19–2.22)

- Starting point: the equation of motion for a rocket is:

$$F = m(t)\frac{dv}{dt} - v_0\frac{dm}{dt}$$
$$= m(t)\frac{dv}{dt} - v_0\dot{m}$$

where v_0 is the velocity at which fuel is ejected from the rocket relative to the rocket (refer to text for derivation). Assume that the rate of mass ejection is constant, $\therefore \dot{m} = \text{constant}$.

- Motion under gravity: during take off, gravity acts on the rocket $\therefore F = -m(t)g$, and the equation of motion is:

$$m(t)\frac{dv}{dt} - v_0\dot{m} = -m(t)g$$

- Let the mass of the rocket after a time t be $m(t) = m_0 - \dot{m}t$, where m_0 is the starting rocket mass.
- Motion in space, without gravity: in space, no external forces act on the rocket $\therefore F = 0$ and the equation of motion is:

$$m(t)\frac{dv}{dt} - v_0\dot{m} = 0$$

- Let the mass of the rocket after a time t be $m(t) = m_i - \dot{m}t$, where m_i is the initial rocket mass in the zero-gravity region.
- Let the final mass of the velocity after a time Δt be $m_f = m_i - \dot{m}\Delta t$. The velocity changes from initial velocity v_i to final velocity v_f during this time.

Motion under gravity:

$$m(t)\frac{dv}{dt} = v_0\dot{m} - m(t)g$$
$$\therefore \frac{dv}{dt} = \frac{\dot{m}}{m(t)}v_0 - g$$
$$= \frac{\dot{m}}{m_0 - \dot{m}t}v_0 - g$$
$$\int_0^{v(t)} dv = \int_0^t \left(\frac{\dot{m}}{m_0 - \dot{m}t'}v_0 - g\right)dt'$$
$$\therefore v(t) = \ln\left(\frac{m_0}{m_0 - \dot{m}t}\right)v_0 - gt$$
$$= \ln\left(\frac{m_0}{m(t)}\right)v_0 - gt$$

Motion in space, without gravity:

$$m(t)\frac{dv}{dt} = v_0\dot{m}$$
$$\therefore \frac{dv}{dt} = \frac{\dot{m}}{m(t)}v_0$$
$$= \frac{\dot{m}}{m_i - \dot{m}t}v_0$$
$$\int_{v_i}^{v_f} dv = \int_0^{\Delta t}\left(\frac{\dot{m}}{m_i - \dot{m}t}v_0\right)dt$$
$$\therefore v_f = v_i + \ln\left(\frac{m_i}{m_i - \dot{m}\Delta t}\right)v_0$$
$$= v_i + \ln\left(\frac{m_i}{m_f}\right)v_0$$
$$\therefore m_f = m_i e^{-(v_f - v_i)/v_0}$$

The primed variables are to distinguish them from the integration limits.

a small distance from equilibrium, and find solutions of the differential equation under various conditions. Doing this, we will see that such systems undergo small oscillations about the point of equilibrium; this motion is called *simple harmonic motion*. Many physical systems undergo simple harmonic motion, for example, a pendulum, a mass on a spring, or the vibrations of atoms bound in a solid.

2.3.1 Simple harmonic motion

From Newton's Second Law (refer to Section 1.1.1), the external force on a body of constant mass, in one dimension, is:

$$F = \frac{dp}{dt}$$
$$= m\frac{dv}{dt}$$
$$= m\frac{d^2x}{dt^2}$$
$$= m\ddot{x}$$

Equating this expression to Hooke's Law ($F = -kx$), the differential equation of motion for a body of mass m

displaced a small distance x from equilibrium is:

$$m\ddot{x} = -kx$$

(Eq. 2.23). Our goal is to find a solution, $x(t)$, that satisfies this differential equation. This solution describes the displacement of the body as a function of time—i.e. its *motion*. The question we ask ourselves is "what function is equal to the negative of its second derivative?" We know from experience that sinusoidal functions satisfy this criteria (if $x = \sin\theta$, $x' = -\cos\theta$ $\therefore x'' = -\sin\theta = -x$). Therefore, let's try a solution:

$$x(t) = a\sin\alpha t$$

where a and α are constants (amplitude and frequency, respectively). Trying this solution in Eq. 2.23, we have:

$$-m\alpha^2 a\sin\alpha t = -ka\sin\alpha t$$
$$\therefore \alpha = \pm\sqrt{\frac{k}{m}}$$

Since α represents a frequency, the negative solution is unphysical. Therefore, the angular frequency[2] is:

$$\therefore \alpha = \omega_0 = \sqrt{\frac{k}{m}}\ \mathrm{s}^{-1}$$

Similarly, $x(t) = a\cos\omega_0 t$ satisfies Eq. 2.23, as do $x(t) = a\sin(\omega_0 t + \phi)$ or $a\cos(\omega_0 t + \phi)$, where ϕ is an arbitrary phase shift (check by substitution) (Eq. 2.24). The constants a and ϕ are the two integration constants, which are set by boundary conditions of the system in question. Using trigonometric identities, we can express the solution as the sum of a sine and cosine:

$$\begin{aligned} x(t) &= a\sin(\omega_0 t + \phi) \\ &= \underbrace{a\cos\phi}_{=A}\sin\omega_0 t + \underbrace{a\sin\phi}_{=B}\cos\omega_0 t \\ &= A\sin\omega_0 t + B\cos\omega_0 t \end{aligned}$$

where A and B are the integration constants set by boundary conditions (Eq. 2.25). These solutions illustrate that a mass displaced a small distance from its equilibrium position will oscillate about the point of equilibrium with a sinusoidal time dependence (Fig. 2.4). This form of oscillatory motion is called *simple harmonic motion*, which is often abbreviated to *SHM*. The angular frequency ω_0 is the *natural frequency* of the oscillation (Eq. 2.27); it is the frequency at which the system oscillates provided there are no frictional or additional driving forces.

[2] From wave motion, the angular frequency, ω, is related to the frequency, f, by $\omega = 2\pi f$. The time period for one oscillation is then $T = 1/f = 2\pi/\omega$.

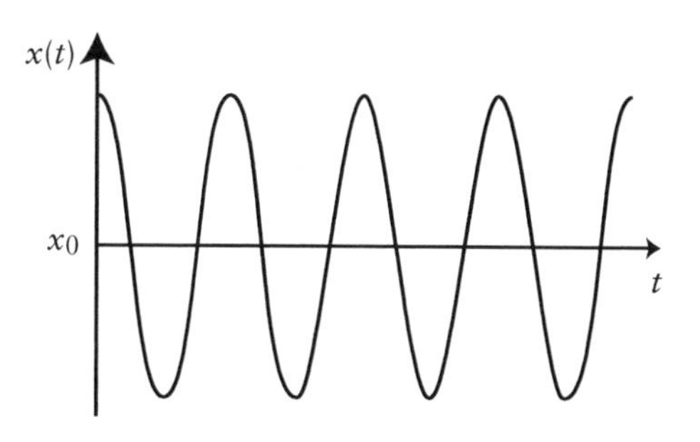

Figure 2.4: Simple harmonic motion (sinusoidal oscillations) about equilibrium position, x_0.

Simple harmonic motion:

Differential equation of motion:

$$m\ddot{x} = -kx \qquad (2.23)$$

Oscillatory solutions:

$$\begin{aligned} x(t) &= a\sin(\omega_0 t + \phi) && (2.24) \\ &\equiv A\sin\omega_0 t + B\cos\omega_0 t && (2.25) \\ &\equiv A'e^{i\omega_0 t} + B'e^{-i\omega_0 t} && (2.26) \end{aligned}$$

where the natural frequency of oscillation is:

$$\omega_0 = \sqrt{\frac{k}{m}} \qquad (2.27)$$

The coefficients are related by:

$$A = a\cos\phi$$
$$B = a\sin\phi$$
and $$A = i(A' - B')$$
$$B = A' + B'$$

These are fixed by boundary conditions.

Complex exponential form

Another function that equals the negative of its second derivative, and therefore satisfies the differential equation of motion from Eq. 2.23, is the complex exponential function. Trying the solution $x(t) = Ce^{i\beta t}$, where C is an arbitrary constant, in Eq. 2.23, we have:

$$\begin{aligned} -m\beta^2 Ce^{i\beta t} &= -kCe^{i\beta t} \\ \therefore \beta^2 &= \frac{k}{m} \\ &\equiv \omega_0^2 \\ \therefore \beta = \pm\omega_0 &= \pm\sqrt{\frac{k}{m}} \end{aligned}$$

Therefore, the general solution is:

$$x(t) = A'e^{i\omega_0 t} + B'e^{-i\omega_0 t}$$

(Eq. 2.26). Using Euler's formula ($e^{i\theta} = \cos\theta + i\sin\theta$) in $x(t)$, we recover the sinusoidal solution:

$$x(t) = A\sin\omega_0 t + B\cos\omega_0 t$$
$$\text{where } A = i(A' - B')$$
$$\text{and } B = A' + B'$$

Therefore, sinusoidal and complex exponential solutions are equivalent to one another. Often, it is mathematically more straightforward to work with complex exponential functions, therefore it's good to be comfortable using these.

Oscillator energy

The total mechanical energy of an oscillating system is the sum of its kinetic and potential energy: $E_0 = T + V$. Although the total energy remains constant (provided there are no frictional forces that act on the system), the relative amounts of kinetic or potential energies vary with time, depending on the system's displacement, $x(t)$. Expressions for the kinetic and potential energies in terms of $x(t)$ and its derivatives are:

$$T(t) = \frac{1}{2}mv^2(t) = \frac{1}{2}m\dot{x}^2(t)$$
$$\text{and } V(t) = -\int F(t)\mathrm{d}x = \int kx(t)\mathrm{d}x = \frac{1}{2}kx^2(t)$$

(Eqs. 2.28 and 2.29 and Fig. 2.5). The energy of the body changes from maximum kinetic energy and zero potential energy near equilibrium to zero kinetic energy and maximum potential energy at the extremities (turning points) of the oscillation.

Oscillator energy:

$$\text{Kinetic}: \quad T(t) = \frac{1}{2}m\dot{x}^2 \qquad (2.28)$$

$$\text{Potential}: \quad V(t) = \frac{1}{2}kx^2 \qquad (2.29)$$

$$\text{Total}: \quad E_0 = T(t) + V(t) \qquad (2.30)$$

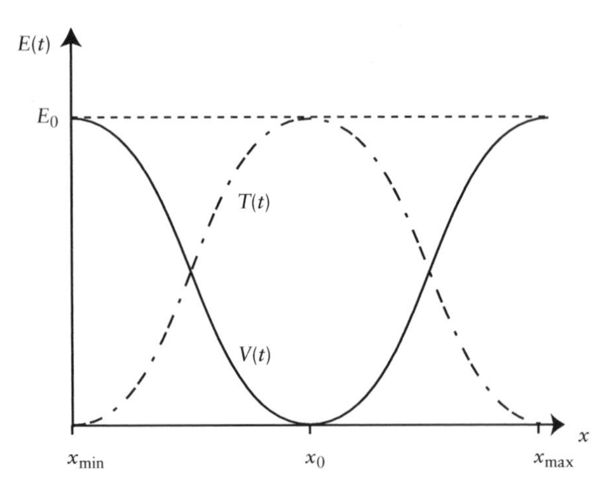

Figure 2.5: Kinetic energy, $T(t)$, and potential energy, $V(t)$, of a mass undergoing simple harmonic motion between x_{min} and x_{max}, with equilibrium position x_0. The sum of kinetic and potential energy is constant at E_0.

2.3.2 Damped simple harmonic motion

Now let's consider a body moving under simple harmonic motion that is also subject to a damping force, F_Γ. For bodies moving at low velocity, F_Γ can be approximated as being proportional to the velocity of the body:

$$F_\Gamma = -\Gamma v = -\Gamma\dot{x}$$

The proportionality constant, Γ [N s m^{-1}], is the *damping coefficient*. The total external force on the body is then the sum of the restoring force and the damping force:

$$F = \underbrace{-kx}_{\text{Restoring force}} - \underbrace{\Gamma\dot{x}}_{\text{Damping force}}$$

Equating this to Newton's Second Law ($F = ma = m\ddot{x}$), the differential equation of motion for damped simple harmonic motion is:

$$\therefore m\ddot{x} = -kx - \Gamma\dot{x}$$

The solution to this differential equation is sinusoidal oscillation (simple harmonic motion), whose amplitude is multiplied by an exponential damping factor:

$$x(t) = \underbrace{ae^{-\gamma t/2}}_{\text{damping}}\underbrace{\sin(\omega t + \phi)}_{\text{SHM}}$$

(Eqs. 2.31–2.34 and Derivation 2.4). The amplitude, a, and phase, ϕ, are constants of integration whose values are set by boundary conditions.

> **Damped simple harmonic motion (Derivation 2.4):**
>
> Differential equation of motion:
>
> $$m\ddot{x} + \Gamma\dot{x} + kx = 0$$
>
> $$\text{or}\quad \ddot{x} + \gamma\dot{x} + \omega_0^2 x = 0 \qquad (2.31)$$
>
> where $\omega_0^2 = k/m$ (natural frequency) and $\gamma = \Gamma/m$.
>
> Solutions (damped oscillations):
>
> $$x(t) = ae^{-\gamma t/2}\sin(\omega t + \phi) \qquad (2.32)$$
>
> $$\equiv e^{-\gamma t/2}(A\sin\omega t + B\cos\omega t) \qquad (2.33)$$
>
> $$\equiv e^{-\gamma t/2}\left(A'e^{i\omega t} + B'e^{-i\omega t}\right) \qquad (2.34)$$
>
> where the oscillation frequency is:
>
> $$\omega = \sqrt{\omega_0^2 - \gamma^2/4} \qquad (2.35)$$
>
> The coefficients are related by:
>
> $$A = a\cos\phi$$
> $$B = a\sin\phi$$
> $$\text{and}\quad A = i(A' - B')$$
> $$B = A' + B'$$
>
> These are fixed by boundary conditions.

Types of damping

The oscillation frequency depends on both the natural frequency, ω_0, and the damping coefficient per unit mass, γ:

$$\omega = \sqrt{\omega_0^2 - \gamma^2/4}$$

(see Derivation 2.4). There are three types of damped motion depending on the relative magnitudes of ω_0 and γ (refer to Fig. 2.6):

1. **Underdamped motion:**

   $$\omega_0^2 > \gamma^2/4 \qquad \therefore \omega \text{ is real}$$

   The system oscillates at frequency ω and the amplitude of the oscillations decreases exponentially, with time constant $\tau = 2/\gamma$:

   $$x(t) = e^{-\gamma t/2}\left(A'e^{i\omega t} + B'e^{-i\omega t}\right)$$

2. **Critically damped motion:**

   $$\omega_0^2 = \gamma^2/4 \qquad \therefore \omega = 0$$

   The system does not oscillate. It returns exponentially to equilibrium with time constant $\tau = 2/\gamma$:

   $$x(t) = e^{-\gamma t/2}(A' + B')$$

3. **Overdamped motion:**

   $$\omega_0^2 < \gamma^2/4 \qquad \therefore \omega \text{ is imaginary}$$

   The system does not oscillate. It returns exponentially to equilibrium, with a time constant slower than the critical damping limit:

   $$x(t) = e^{-\gamma t/2}\left(A'e^{-\omega t} + B'e^{\omega t}\right)$$

Quality factor

The *quality factor*, Q, is a dimensionless parameter that describes the extent to which an oscillating system

x(t)
Underdamped motion
x0
t
SHM (no damping)

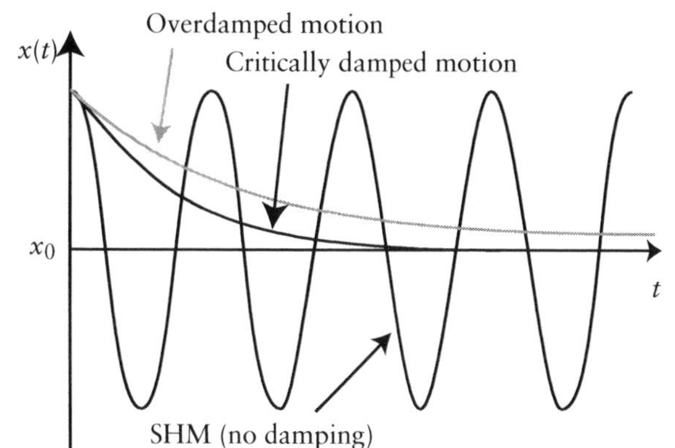

Figure 2.6: Underdamped, critically damped, and overdamped simple harmonic motion (SHM).

Derivation 2.4: Derivation of damped simple harmonic motion (Eqs. 2.32–2.35)

- Starting point: the differential equation of motion for damped simple harmonic oscillations is:

$$\ddot{x} + \gamma\dot{x} + \omega_0^2 x = 0$$

where the natural oscillation frequency is $\omega_0^2 = k/m$ and the damping coefficient per unit mass is $\gamma = \Gamma/m$. (Refer to text for derivation.)
- Try a solution of the form $x = Ce^{i\alpha t}$.
- Solve the quadratic for α and define the oscillation frequency as:

$$\omega^2 = \omega_0^2 - \frac{\gamma^2}{4}$$

- The solutions are therefore sinusoidal oscillations with an exponentially decaying amplitude. When there is no damping ($\gamma = 0$), we recover non-damped simple harmonic motion and the frequency is the natural frequency of oscillation, $\omega = \omega_0$.

$$-\alpha^2 + i\alpha\gamma + \omega_0^2 = 0$$

$$\therefore \alpha = \frac{-i\gamma \pm \sqrt{-\gamma^2 + 4\omega_0^2}}{-2} = \frac{i\gamma}{2} \pm \sqrt{\omega_0^2 - \gamma^2/4} = \frac{i\gamma}{2} \pm \omega$$

Therefore, the solution is the sum of two complex exponential terms:

$$\therefore x = Ce^{i\alpha t} = e^{-\gamma t/2}\left(A'e^{i\omega t} + B'e^{-i\omega t}\right)$$

Using Euler's formula ($e^{i\theta} = \cos\theta + i\sin\theta$):

$$x = e^{-\gamma t/2}\left(A'(\cos\omega t + i\sin\omega t) + B'(\cos\omega t - i\sin\omega t)\right)$$

$$= e^{-\gamma t/2}\left(\underbrace{i(A' - B')}_{=A}\sin\omega t + \underbrace{(A' + B')}_{=B}\cos\omega t\right)$$

$$= e^{-\gamma t/2}\left(A\sin\omega t + B\cos\omega t\right)$$

$$\equiv ae^{-\gamma t/2}\sin(\omega t + \phi)$$

is damped (for underdamped motion only). If damping is small, then Q is large, and vice versa. Q is defined as the ratio of the system's natural frequency ω_0 to the damping coefficient γ:

$$Q \stackrel{\text{def}}{=} \frac{\omega_0}{\gamma}$$

(Eq. 2.36). We can get an intuitive feeling for what Q represents if we consider the two natural time scales associated with damped simple harmonic motion: (i) the *natural period* of the oscillations, $T_\omega = 2\pi/\omega_0$, and (ii) the *decay constant*, $\tau = 2/\gamma$. Therefore, the approximate number of periods in a relaxation time[3] is:

$$N_p \sim \frac{\tau}{T_\omega} \sim \frac{\omega_0}{\pi\gamma} \sim \frac{Q}{\pi}$$

(Eq. 2.37). Therefore, oscillators with a higher Q (lower damping) go through more cycles of oscillation before the oscillation amplitude has significantly decayed. For example, a tuning fork will have a higher Q-value than a mechanical pendulum.

Quality factor for underdamped motion:

$$Q \stackrel{\text{def}}{=} \frac{\omega_0}{\gamma} \qquad (2.36)$$

Number of periods in a relaxation time:

$$N_p \sim \frac{Q}{\pi} \qquad (2.37)$$

2.3.3 Forced oscillations

In order to overcome unwanted damping of oscillations, for example in pendulum clocks, it's necessary to apply an external *driving force* to the system. If we assume the driving force is periodic with amplitude

[3] The relaxation time of a system is the time constant of an exponential return of the system to equilibrium after a disturbance, which in this case corresponds to the decay constant, τ.

F_d' and frequency ω, then the total external force on the system is:

$$F = \underbrace{-kx}_{\text{Restoring force}} - \underbrace{\Gamma\dot{x}}_{\text{Friction}} + \underbrace{F_d' \cos \omega t}_{\text{Driving force}}$$

Therefore, the differential equation of motion in one dimension is:

$$m\ddot{x} = -kx - \Gamma\dot{x} + F_d' \cos \omega t$$

(Eq. 2.38). To solve for x, it is convenient to express the driving force as the real part of a complex exponential term:

$$F_d' \cos \omega t = \mathrm{Re}\left\{F_d' e^{i\omega t}\right\}$$

We can then try a solution of the form $x = \mathrm{Re}\{Ae^{i\omega t}\}$, and solve for the amplitude, A. Doing this, we find that A is a function of the driving frequency, ω, which means that displacement x is also a function of ω (Eq. 2.39 and Derivation 2.5). The system oscillates at the same frequency as the driving frequency, and there is a phase lag, ϕ, between the system's response and the driving force, which is also a function of ω (Eq. 2.40 and Derivation 2.5).

Forced simple harmonic motion (Derivation 2.5):

Differential equation of motion:

$$m\ddot{x} + \Gamma\dot{x} + kx = F_d' \cos \omega t$$

$$\text{or} \quad \ddot{x} + \gamma\dot{x} + \omega_0^2 x = F_d \cos \omega t \equiv \mathrm{Re}\left\{F_d e^{i\omega t}\right\} \qquad (2.38)$$

where $\omega_0^2 = k/m$ (natural frequency), $\gamma = \Gamma/m$, and $F_d = F_d'/m$

Oscillatory solution for driving frequency ω:

$$x(t) = \frac{F_d \cos(\omega t - \phi)}{\sqrt{\left(\omega_0^2 - \omega^2\right)^2 + (\omega\gamma)^2}} \qquad (2.39)$$

where the phase lag between the response and the driving force is:

$$\tan \phi = \frac{\omega\gamma}{\omega_0^2 - \omega^2} \qquad (2.40)$$

Resonance

The amplitude of the oscillations is a function of the driving frequency, ω:

$$x(t) = \underbrace{\frac{F_d}{\sqrt{\left(\omega_0^2 - \omega^2\right)^2 + (\omega\gamma)^2}}}_{\text{Amplitude, } A} \cos(\omega t - \phi)$$

Fig. 2.7a shows A as a function of ω. We see that there is a frequency at which the amplitude of oscillations is a maximum, which corresponds to the point at which the denominator of A is a minimum. Therefore, differentiating the denominator of A and setting it equal to zero, we find that the amplitude is a maximum when:

$$-2\omega\left(\omega_0^2 - \omega^2\right) + \omega\gamma^2 = 0$$

This corresponds to a driving frequency of:

$$\omega^2 = \omega_0^2 - \frac{\gamma^2}{2}$$

This is called the *resonant frequency*, ω_R (Eq. 2.41). It is equal to the the natural frequency when $\gamma = 0$ (no frictional forces).

Resonant frequency:

$$\omega_R^2 = \omega_0^2 - \frac{\gamma^2}{2} \qquad (2.41)$$

The phase lag ϕ between the response of the system and the driving force is a function of ω (Eq. 2.40), and it passes through $\pi/2$ at resonance (Fig. 2.7b).

Quality factor and energy

We saw above that the quality factor, Q, describes the extent to which a system is damped. For resonant systems, it is often defined in terms of the inverse of the fractional energy loss per cycle:

$$Q = 2\pi\left(\frac{\text{energy stored in system}}{\text{energy loss per cycle at resonance}}\right) = 2\pi\frac{E}{\Delta E} \qquad (2.42)$$

This is equivalent to the definition of Q as $Q = \omega_0/\gamma$ for underdamped motion (refer to Eq. 2.36), as we shall see by the following calculation. Consider an underdamped

Derivation 2.5: Derivation of forced oscillations (Eqs. 2.39 and 2.40)

$$-\omega^2 A + i\omega\gamma A + \omega_0^2 A = F_d$$

$$\therefore A = \frac{F_d}{\underbrace{\omega_0^2 - \omega^2 + i\omega\gamma}_{z=\beta e^{i\phi}}}$$

$$\equiv \frac{F_d e^{-i\phi}}{\sqrt{\left(\omega_0^2 - \omega^2\right)^2 + (\omega\gamma)^2}}$$

$$\text{where } \tan\phi = \frac{\omega\gamma}{\omega_0^2 - \omega^2}$$

$$\therefore x(t) = \mathrm{Re}\left\{A e^{i\omega t}\right\}$$

$$= \mathrm{Re}\left\{\frac{F_d e^{i(\omega t - \phi)}}{\sqrt{\left(\omega_0^2 - \omega^2\right)^2 + (\omega\gamma)^2}}\right\}$$

$$= \frac{F_d \cos(\omega t - \phi)}{\sqrt{\left(\omega_0^2 - \omega^2\right)^2 + (\omega\gamma)^2}}$$

- Starting point: the differential equation of motion for damped simple harmonic oscillations with an oscillatory driving force $F_d \cos \omega t$, is:

$$\ddot{x} + \gamma\dot{x} + \omega_0^2 x = F_d \cos \omega t$$
$$\equiv \mathrm{Re}\left\{F_d e^{i\omega t}\right\}$$

The natural oscillation frequency is $\omega_0^2 = k/m$ and the damping coefficient per unit mass is $\gamma = \Gamma/m$. (Refer to text for derivation.)

- Try a solution of the form $x(t) = \mathrm{Re}\{Ce^{i\omega t}\}$. Substitute this into the differential equation of motion and solve for A.
- Express the complex denominator of A as a complex exponential:

$$z = \underbrace{\omega_0^2 - \omega^2}_{\mathrm{Re}\{z\}} + i\underbrace{\omega\gamma}_{\mathrm{Im}\{z\}} \equiv \beta e^{i\phi}$$

$$\text{where } \beta = \sqrt{(\mathrm{Re}\{z\})^2 + (\mathrm{Im}\{z\})^2}$$
$$= \sqrt{\left(\omega_0^2 - \omega^2\right)^2 + (\omega\gamma)^2}$$

$$\text{and } \tan\phi = \frac{\mathrm{Im}\{z\}}{\mathrm{Re}\{z\}}$$
$$= \frac{\omega\gamma}{\omega_0^2 - \omega^2}$$

- Substitute A into $x(t) = \mathrm{Re}\{Ae^{i\omega t}\}$. The amplitude of x is a function of the driving frequency, ω.

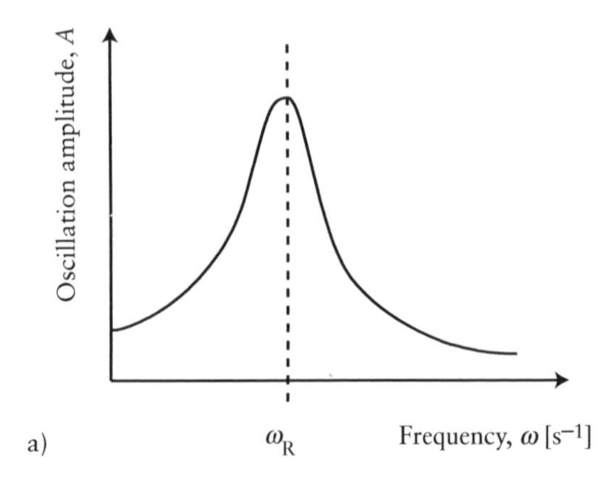

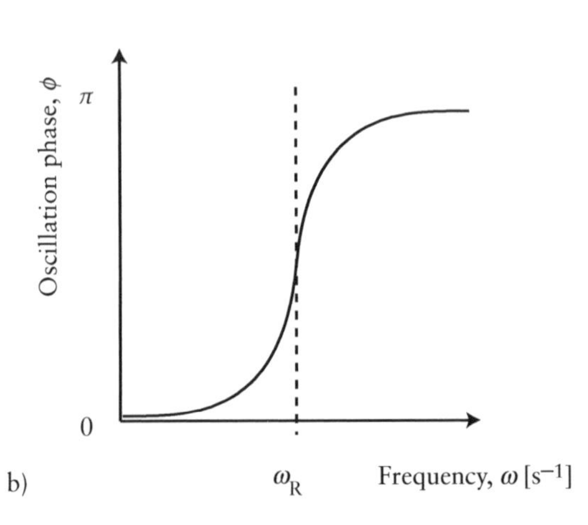

Figure 2.7: Amplitude and phase of oscillations as a function of driving frequency, ω. The resonant frequency, ω_R, corresponds to the driving frequency at which the amplitude of oscillations is a maximum.

system with displacement given by:

$$x(t) = ae^{-\gamma t/2}\sin(\omega t + \phi)$$

Differentiating, we have:

$$\dot{x}(t) = -\frac{\gamma}{2}ae^{-\gamma t/2}\sin(\omega t + \phi) + \omega ae^{-\gamma t/2}\cos(\omega t + \phi)$$

For an underdamped system at resonance with light damping, $\gamma \ll \omega_0$ and $\therefore \omega \simeq \omega_0$. In this case, the total energy of the system is:

$$\begin{aligned}
E &= \frac{1}{2}m\dot{x}^2(t) + \frac{1}{2}kx^2(t) \\
&= \frac{1}{2}m\dot{x}^2(t) + \frac{1}{2}m\omega_0^2x^2(t) \\
&\simeq \frac{1}{2}mA^2\omega_0^2e^{-\gamma t}\left(\underbrace{\frac{\gamma^2}{4\omega_0^2}\sin^2(\omega_0 t + \phi)}_{\to 0 \text{ for } \gamma \ll \omega_0} + \cos^2(\omega_0 t + \phi)\right. \\
&\qquad \left. \underbrace{-\frac{\gamma}{\omega_0}\sin(\omega_0 t + \phi)\cos(\omega_0 t + \phi)}_{\to 0 \text{ for } \gamma \ll \omega_0}\right) \\
&\qquad + \frac{1}{2}mA^2\omega_0^2e^{-\gamma t}\sin^2(\omega_0 t + \phi) \\
&\simeq \frac{1}{2}mA^2\omega_0^2e^{-\gamma t}
\end{aligned}$$

The energy loss per cycle at resonance is:

$$\Delta E = E(t - \pi/\omega_0) - E(t + \pi/\omega_0)$$

Therefore, in the limit $\gamma \ll \omega_0$, the Q-value of the system is:

$$\begin{aligned}
Q &= 2\pi\frac{E}{\Delta E} \\
&\simeq 2\pi\frac{mA^2\omega_0^2e^{-\gamma t}/2}{mA^2\omega_0^2e^{-\gamma t}\left(e^{\gamma\pi/\omega_0} - e^{-\gamma\pi/\omega_0}\right)/2} \\
&\simeq 2\pi\frac{1}{(1 + \gamma\pi/\omega_0) - (1 - \gamma\pi/\omega_0)} \\
&\simeq 2\pi\frac{\omega_0}{2\pi\gamma} \\
&\simeq \frac{\omega_0}{\gamma}
\end{aligned}$$

The Q-value describes the sharpness of the resonance peak (refer to Fig. 2.10a). Systems with a high Q-value have narrow and sharply peaked resonance curves, whereas systems with a low Q-value have flatter and broader curves. The width of the curve at half-maximum amplitude is $\Delta\omega \simeq \gamma$. Therefore, another definition of Q is:

$$Q = \frac{\omega_0}{\Delta\omega}$$

where $\Delta\omega$ is the called the *bandwidth* of the system. This definition is useful in electrical systems, where the bandwidth corresponds to the range of output frequencies of a given alternating current (AC) electrical circuit (refer to Section 9.4.4 for more detail).

2.3.4 Coupled oscillators

As a final example of oscillating systems, we'll look at *coupled oscillators*, which are bodies that are joined together in some way. Therefore, the oscillation of one body influences the motion of the others.

Two coupled oscillators

Consider two masses joined together by springs, as shown in Fig. 2.8. For simplicity, assume that the masses are equal and spring constants of all the springs are equal. Let the displacement of the first mass from equilibrium be x_1, and that of the second mass be x_2. Therefore, the forces acting on the first body are:

$$\begin{aligned}
&F_1 = -kx_1 \\
\text{and } &F_2 = k(x_2 - x_1)
\end{aligned}$$

and the forces acting on the second body are:

$$\begin{aligned}
&F_3 = -k(x_2 - x_1) \\
\text{and } &F_4 = -kx_2
\end{aligned}$$

Therefore, the equations of motion for each body are:

$$\begin{aligned}
m\ddot{x}_1 &= F_1 + F_2 \\
&= kx_2 - 2kx_1 \\
\text{and } m\ddot{x}_2 &= F_3 + F_4 \\
&= kx_1 - 2kx_2
\end{aligned}$$

These equations are symmetric if we exchange the subscripts, as expected. If we try simple harmonic solutions of the form:

$$\begin{aligned}
&x_1 = X_1\cos(\omega t) \\
\text{and } &x_2 = X_2\cos(\omega t)
\end{aligned}$$

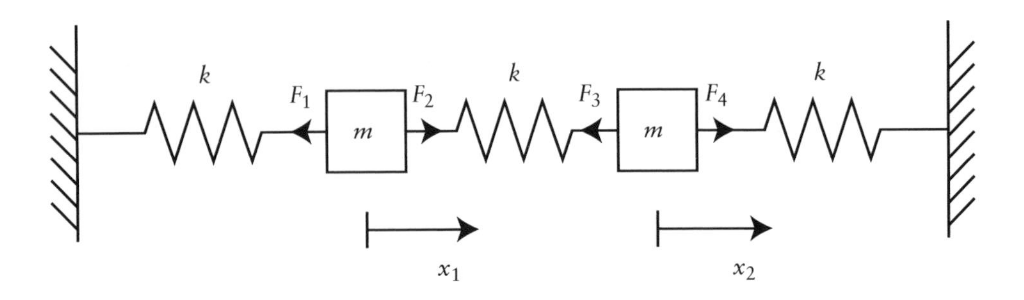

Figure 2.8: System of two coupled masses on springs.

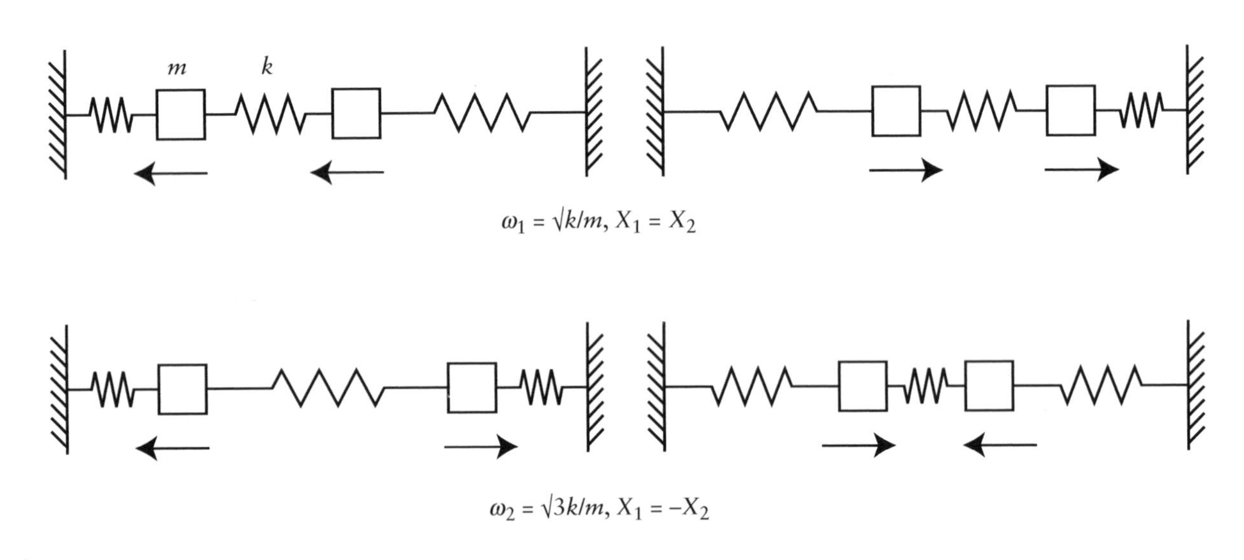

Figure 2.9: The two normal modes of two coupled masses on springs. The masses are equal (m), as are the spring constants of each spring (k). The two normal mode frequencies are $\omega_1 = \sqrt{k/m}$, where the masses oscillate in phase (equal amplitude displacements, $X_1 = X_2$), and $\omega_2 = \sqrt{3k/m}$, where the masses oscillate in anti-phase (equal and opposite amplitude displacements, $X_1 = -X_2$).

then we find two simultaneous equations for X_1 and X_2:

$$(2k - m\omega^2)X_1 = kX_2 \tag{2.43}$$

$$\text{and}\ \ (2k - m\omega^2)X_2 = kX_1 \tag{2.44}$$

Normal mode oscillations

Eliminating X_2 from Eq. 2.43 using Eq. 2.44, we find:

$$\begin{aligned}(2k - m\omega^2)^2 &= k^2 \\ \therefore\ \omega^2 &= 2\frac{k}{m} \pm \frac{k}{m} \\ &= \frac{k}{m}\ \text{or}\ 3\frac{k}{m}\end{aligned}$$

When $\omega_1^2 = k/m$, the amplitude ratio is:

$$\frac{X_1}{X_2} = 1$$

and when $\omega_2^2 = 3k/m$, the amplitude ratio is:

$$\frac{X_1}{X_2} = -1$$

These two solutions are called the *normal modes* of the system. When a system is in one of its normal modes, then all the components undergo simple harmonic motion at the same frequency as one another, with a fixed amplitude ratio. The two normal modes of the system described above are illustrated in Fig. 2.9. In one normal mode, when $\omega_1^2 = k/m$, the two masses oscillate in the same direction (in phase): $X_1 = X_2$; in the other normal mode, when $\omega_2^2 = 3k/m$, the two masses oscillate in opposite directions (in anti-phase): $X_1 = -X_2$. Since the equations of motion are linear, then the general solution is a linear superposition of normal modes:

$$\begin{aligned} x_1 &= A\cos\omega_1 t + B\sin\omega_1 t + C\cos\omega_2 t + D\sin\omega_2 t \\ x_2 &= \underbrace{A\cos\omega_1 t + B\sin\omega_1 t}_{X_1/X_2=1} \underbrace{-C\cos\omega_2 t - D\sin\omega_2 t}_{X_1/X_2=-1} \end{aligned}$$

The constants A, B, C, and D are fixed by initial conditions (two initial positions plus two initial velocities). The method described above can be used to solve various configurations of coupled oscillators, for example, two coupled pendula or two masses on a vibrating string.

N coupled oscillators

Normal modes exist for any linear system of springs and masses, with any number of components. For a system of N coupled oscillators, there are N normal modes; in three dimensions, there are $3N$ normal modes. The normal modes of masses on a vibrating string in the limit $N \to \infty$ describes the propagation of a transverse wave on the string. Similarly, the propagation of waves through matter are equivalent to the vibrations of $N \to \infty$ coupled atoms in the solid.

Chapter 3
Angular motion

So far, we have treated masses as point particles and considered only the linear motion of their center of mass. Now we'll consider the *angular* motion of a point particle about an axis, and of extended mass distributions (rigid bodies), where the shape and mass of the body must also be taken into account. Angular equations of motion are for the most part equivalent to those in linear mechanics. Therefore, where appropriate, we will compare the angular case with its linear counterpart.

3.1 Angular motion

The parameters used to describe linear motion have an angular equivalent.

Angular displacement

Linear displacement, $\vec{x}$, is the distance moved by a body along the line of travel. Analogously, *angular displacement*, $\vec{\theta}$, is the *angle* swept out by a body about a given axis. This is illustrated for a body traveling in a circle in Fig. 3.1. The relationship between linear displacement along the arc (the *arc length*), s, and angular displacement θ, for a circle of radius r, is:

$$s = r\theta$$

Since θ has a direction, it is a vector that points in the direction of $\hat{\theta}$. $\hat{\theta}$ is always perpendicular to $\hat{r}$ (Fig. 3.1).

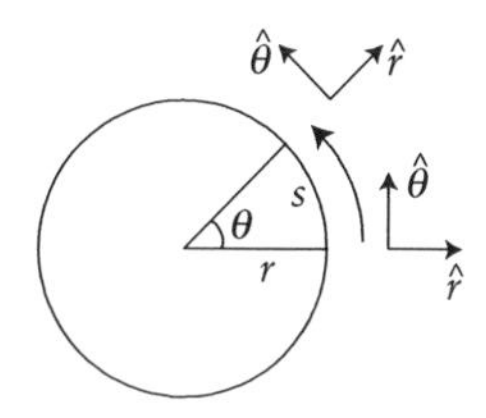

Figure 3.1: Direction and magnitude of angular displacement, $\vec{\theta}$. In polar coordinates, $\hat{r}$ points radially outward, and $\hat{\theta}$ is perpendicular to $\hat{r}$ and tangential to the direction of motion. The direction of the axes changes as $\vec{\theta}$ changes.

Angular velocity and acceleration

The rate of change of angular displacement, $\dot{\vec{\theta}}$, is called the *angular velocity*, $\vec{\omega}$ [s^{-1}], which is the angular equivalent of linear velocity. Similarly, the rate of change of angular velocity, $\dot{\vec{\omega}}$, is called the *angular acceleration*, $\vec{\alpha}$ [s^{-2}]. The magnitudes of $\vec{\theta}$, $\vec{\omega}$, and $\vec{\alpha}$ are related to one another in the same way as $\vec{x}$, $\vec{v}$, and $\vec{a}$ are (Eqs. 3.1–3.3).

Linear → angular motion:

Angular displacement, $\vec{\theta}$:

$$\vec{x} \rightarrow \vec{\theta} \tag{3.1}$$

Angular velocity, $\vec{\omega}$:

$$\vec{v} = \dot{\vec{x}} \rightarrow \vec{\omega} = \dot{\vec{\theta}} \tag{3.2}$$

Angular acceleration, $\vec{\alpha}$:

$$\vec{a} = \dot{\vec{v}} = \ddot{\vec{x}} \rightarrow \vec{\alpha} = \dot{\vec{\omega}} = \ddot{\vec{\theta}} \tag{3.3}$$

3.1.1 Polar and axial vectors

The variables that are vectors in linear motion are also vectors in angular motion, and likewise for scalars. Therefore, since $\vec{x}$, $\vec{v}$, and $\vec{a}$ are vectors, so are $\vec{\theta}$, $\vec{\omega}$, and $\vec{\alpha}$. In linear motion, we are used to dealing with *polar* vectors. Polar vectors are usually defined as such if they are transformed to their negative under inversion of the coordinate axes. For example, a body traveling in the $+x$-direction is transformed to traveling in the $-x$-direction under inversion of Cartesian coordinate axes. There is another type of vector, called *axial* or *pseudovectors*. The direction of an axial vector specifies the *axis of rotation* around which a changing polar vector points (Fig. 3.2). Unlike polar vectors, axial vectors do not transform to their negative under inversion of the

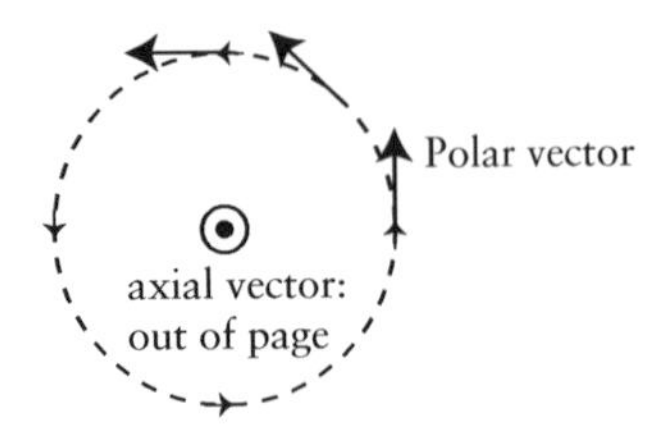

Figure 3.2: Polar and axial vectors for a particle traveling in a circle. The polar vector points in the direction of its displacement, which is tangential to the circle, whereas the axial vector points along the axis of rotation (out of the page in this case).

coordinate axes. Many vector quantities in angular motion are axial vectors, such as angular velocity, angular momentum, and torque, as we shall see throughout this chapter.

Right-hand screw rule

The direction of an axial vector is obtained by applying the right-hand screw rule: let your fingers on your right hand follow the rotation of the object, and the thumb then points along the axis of rotation. This axis corresponds to the direction of the axial vector (refer to Fig. 3.2).

Relation between axial and polar vectors

Axial and polar vectors that are related to one another (such as velocity and angular velocity, as we shall see next) are related by cross products. For a polar vector, $\vec{p}$, and an axial vector, $\vec{a}$, the vector types following a cross product are related by:

$$\vec{p} \times \vec{p} = \vec{a} \tag{3.4}$$
$$\vec{a} \times \vec{a} = \vec{a} \tag{3.5}$$
$$\vec{a} \times \vec{p} = \vec{p} \tag{3.6}$$

This is useful for figuring out whether a given quantity is a polar or axial vector if the nature of the two other vectors in the cross product are known. For example, the magnetic force is described by the cross product $\vec{F}_B = q\vec{v} \times \vec{B}$. We know that $\vec{F}_B$ and $\vec{v}$ are polar vectors, and therefore, from Eq. 3.6, can deduce that $\vec{B}$ is an axial vector.

3.1.2 Angular velocity

The arc length swept out by a body traveling in a circle is $s = r\theta$. Differentiating, the tangential velocity is:

$$\begin{aligned} v_\theta &= \frac{ds}{dt} \\ &= r\frac{d\theta}{dt} \\ &= r\omega \end{aligned}$$

where ω is the *angular velocity*. The parameters $\vec{v}$, $\vec{r}$, and $\vec{\omega}$ are all vectors, so what are their respective directions? Let's consider an arbitrary position vector $\vec{r}$ that is not in the plane of circular motion, as illustrated in Fig. 3.3. For a small change in $\vec{\theta}$, the small change in arc length is:

$$d\vec{s} = |\vec{r}| \sin\phi \, d\vec{\theta}$$

where ϕ is the angle between $\vec{\omega}$ and $\vec{r}$. Differentiating, the instantaneous velocity is therefore:

$$\begin{aligned} \vec{v} &= \frac{d\vec{s}}{dt} \\ &= |\vec{r}| \sin\phi \frac{d\vec{\theta}}{dt} \\ &= |\vec{\omega}||\vec{r}| \sin\phi \hat{n} \\ &\equiv \vec{\omega} \times \vec{r} \end{aligned}$$

where $\hat{n}$ is a unit vector in the direction of $\vec{v}$. (This equation has the general form of the cross product from Eq. 3.6.) Therefore, $\vec{v}$ is perpendicular to both $\vec{\omega}$ and $\vec{r}$ (Eq. 3.7).

Angular velocity, $\vec{\omega}$:

$$\vec{v} = \vec{\omega} \times \vec{r} \tag{3.7}$$

$\vec{\omega}$ is often referred to as the *angular frequency* [rad s^{-1}]. The magnitude of the angular frequency is related to temporal frequency, f [s^{-1}], by:

$$\omega = 2\pi f$$

3.2 Torque and angular momentum

A force that acts on a body with a tendency to rotate it about an axis produces a *torque*, $\vec{\tau}$ [N m]. Torque is an axial vector whose direction is along the axis of rotation and whose magnitude describes the strength of the turning effect. Torque is related to linear force by the

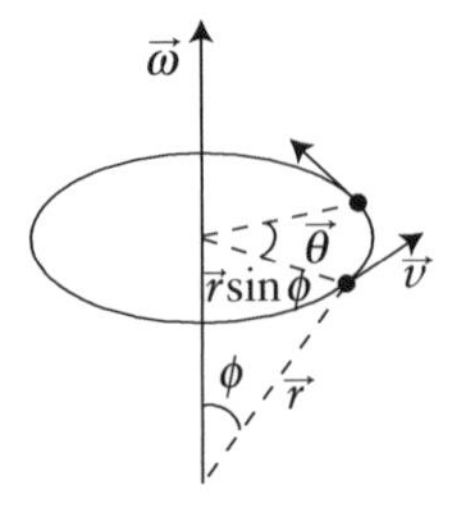

Figure 3.3: Angular and linear velocity.

cross product:

$$\vec{\tau} \overset{\text{def}}{=} \vec{r} \times \vec{F} = |\vec{r}||\vec{F}_\perp|\hat{n}$$

where $\vec{F}_\perp$ is the component of force perpendicular to the position vector, $\vec{r}$, of the body from the pivot point and $\hat{n}$ is a unit vector in the direction of $\vec{\tau}$ (Eq. 3.8 and Fig. 3.4). Therefore, only components of force that do not act along the direction of $\vec{r}$ produce a torque. A common example of a force that produces a torque is pushing on a spanner to loosen a bolt.

Angular momentum

A body rotating about a pivot point has an *angular momentum*, $\vec{L}$ [$\text{kg m}^2\text{s}^{-1}$]. Angular momentum is also an axial vector, and is related to linear momentum in a similar way that torque is related to linear force:

$$\vec{L} \overset{\text{def}}{=} \vec{r} \times \vec{p} = |\vec{r}||\vec{p}_\perp|\hat{n}$$

where $\vec{p}_\perp$ is the component of linear momentum perpendicular to the position vector, $\vec{r}$, of the body from the axis of rotation and $\hat{n}$ is a unit vector in the direction of $\vec{L}$ (Eq. 3.9 and Fig. 3.4). Torque and angular momentum can be thought of as the angular analogs of force and linear momentum, respectively.

> **Torque and angular momentum:**
>
> Torque:
>
> $$\vec{\tau} \overset{\text{def}}{=} \vec{r} \times \vec{F} \quad (3.8)$$
>
> Angular momentum:
>
> $$\vec{L} \overset{\text{def}}{=} \vec{r} \times \vec{p} \quad (3.9)$$

Newton's Second Law

Fortunately for us, many of the linear equations of motion also hold in terms of their angular counterparts.

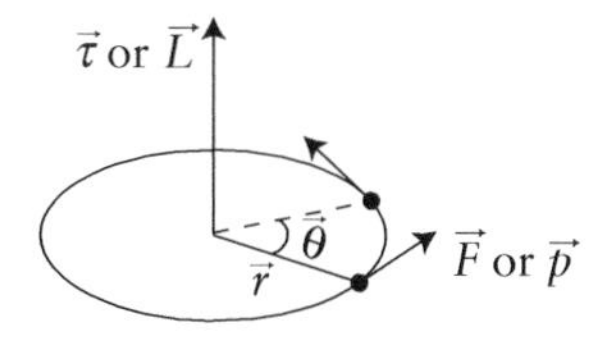

Figure 3.4: Torque and angular momentum.

Let's start with Newton's Second Law. The rate of change of angular momentum is:

$$\begin{aligned}\frac{d\vec{L}}{dt} &= \frac{d}{dt}(\vec{r} \times \vec{p}) \\ &= \vec{r} \times \dot{\vec{p}} + \dot{\vec{r}} \times \vec{p} \\ &= \vec{r} \times \dot{\vec{p}} + m\dot{\vec{r}} \times \dot{\vec{r}} \\ &= \vec{r} \times \dot{\vec{p}} \\ &= \vec{r} \times \vec{F} \\ &= \vec{\tau}\end{aligned}$$

which of course, is analogous to $\vec{F} = d\vec{p}/dt$ (Eq. 3.10). Therefore, the angular momentum of an object remains constant unless acted on by a torque. Rearranging Eq. 3.10 and integrating, we see that the torque integrated over a period of time—the angular impulse—equals the total change in angular momentum of the body. Therefore, the impulse–momentum theorem also holds for angular motion (Eq. 3.11).

> **Newton's Second Law (linear → angular):**
>
> $$\vec{F} = \frac{d\vec{p}}{dt} \rightarrow \vec{\tau} = \frac{d\vec{L}}{dt} \quad (3.10)$$
>
> Impulse:
>
> $$\Delta\vec{p} = \int \vec{F} dt \rightarrow \Delta\vec{L} = \int \vec{\tau} dt \quad (3.11)$$

Conservation of angular momentum

Consider an isolated system of two bodies, interacting under mutual forces $\vec{F}_{12}$ and $\vec{F}_{21}$, where:

$$\vec{F}_{12} = -\vec{F}_{21}$$

(Newton's Third Law). We find that an analogous relation holds for mutual torque (Eq. 3.12 and Derivation 3.1). This, in turn, implies that the total angular momentum of an isolated system is constant and, therefore, that angular momentum is a conserved quantity (Eq. 3.13 and Derivation 3.1). Conservation of angular momentum is what makes a pivoting figure skater spin faster upon bringing their arms closer to their chest: $\vec{r}$ decreases, therefore, $\vec{v}$ must increase in order to keep $\vec{L}$ constant.

Newton's Third Law (linear → angular) (Derivation 3.1):

$$\vec{F}_{12} = -\vec{F}_{21} \rightarrow \vec{\tau}_{12} = -\vec{\tau}_{21} \quad (3.12)$$

Conservation of angular momentum:

$$\vec{L} = \sum \vec{L}_i \\ = \text{constant} \quad (3.13)$$

Work and power (linear → angular) (Derivation 3.2):

$$W = \int \vec{F} \cdot d\vec{r} \rightarrow W = \int \vec{\tau} \cdot d\vec{\theta} \quad (3.14)$$

$$P = \vec{F} \cdot \vec{v} \rightarrow P = \vec{\tau} \cdot \vec{\omega} \quad (3.15)$$

3.2.1 Work and power

The angular work done when rotating an object about an axis is also analogous to its linear counterpart. In linear motion, work is done when a force acts on and displaces an object; in angular motion, work is done when a torque acts on and rotates an object about a given axis. Therefore, an element of work done by a torque $\vec{\tau}$ through an angular displacement $d\vec{\theta}$ is:

$$dW = \vec{\tau} \cdot d\vec{\theta}$$

and the total work is found by integrating over all angles (Eq. 3.14 and Derivation 3.2). Power is the rate of doing work ($P = dW/dt$). Using this together with Eq. 3.14, the power of a rotating object is equal to the torque multiplied by its angular velocity (Eq. 3.15).

3.3 Rigid-body rotation

The previous section described the angular motion of a point mass about an axis. Now, we'll consider the angular motion of an extended solid body, where the shape and mass of the body must also be taken into account. A so-called *rigid body* is an extended solid body that does not deform or change shape when acted on by a force. Rigid bodies are idealizations, since all bodies deform slightly when subject to an external force.

3.3.1 The center of mass

In Chapter 1, we treated bodies of mass m as point particles, as though all their mass were located at their *center of mass*, $\vec{R}$ [m]. This position vector corresponds to the point that represents the mean position of all mass in a body. It's useful at this stage to determine how to calculate the center of mass of a given object.

Derivation 3.1: Derivation of Newton's Third Law for angular motion (Eqs. 3.12 and 3.13)

$$\vec{\tau}_{tot} = \vec{\tau}_{12} + \vec{\tau}_{21} \\ = (\vec{r}_1 - \vec{r}_2) \times \vec{F}_{12} \\ = -\vec{r}_s \times \vec{F}_{12} \\ = 0$$

$$\therefore \vec{\tau}_{12} = -\vec{\tau}_{21}$$

Conservation of angular momentum:

$$\frac{d\vec{L}}{dt} = \vec{\tau}_{tot} = 0$$

$$\therefore \vec{L} = \text{constant}$$

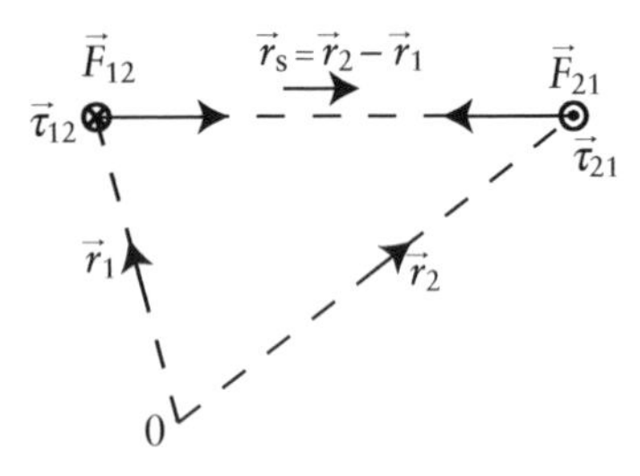

- Starting point: from the diagram, the mutual torques on either body are:

$$\vec{\tau}_{12} = \vec{r}_1 \times \vec{F}_{12}$$
$$\vec{\tau}_{21} = \vec{r}_2 \times \vec{F}_{21} \\ = -\vec{r}_2 \times \vec{F}_{12}$$

- The total torque of the two-body system, $\vec{\tau}_{tot}$, is the sum of mutual torques. The separation vector, $\vec{r}_s = \vec{r}_2 - \vec{r}_1$, is parallel to $\vec{F}12$, as shown in the diagram, therefore their cross product is zero.

Derivation 3.2: Derivation of work done by a torque (Eq. 3.14)

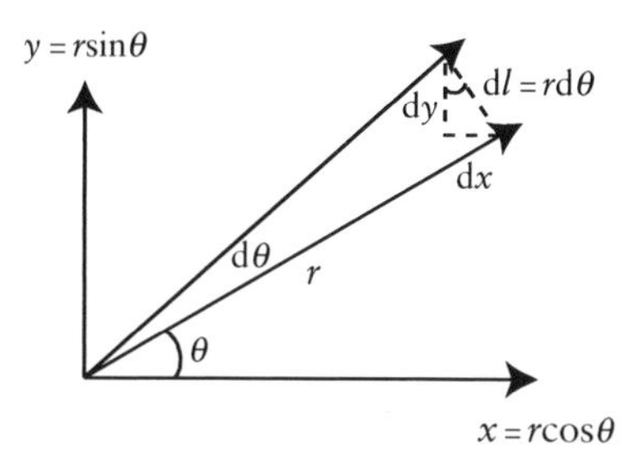

- Starting point: work done by a force is defined as the force multiplied by the component of displacement along the direction of the force:

$$\mathrm{d}W = \vec{F}\cdot\mathrm{d}\vec{l}$$

- Convert from Cartesian to polar coordinates using:

$$\begin{aligned} x &= r\cos\theta \\ \text{and}\quad y &= r\sin\theta \\ \therefore \mathrm{d}x &= -r\sin\theta\mathrm{d}\theta = -y\mathrm{d}\theta \\ \text{and}\quad \mathrm{d}y &= r\cos\theta\mathrm{d}\theta = x\mathrm{d}\theta \end{aligned}$$

for constant r.

- Doing this, we see that work is done by a torque to rotate an object about a pivot point.

$$\begin{aligned} \mathrm{d}W &= \vec{F}\cdot\mathrm{d}\vec{l} \\ &= F_x\mathrm{d}x + F_y\mathrm{d}y \\ &= \underbrace{\left(-F_x y + F_y x\right)}_{=\vec{r}\times\vec{F}=\vec{\tau}}\mathrm{d}\theta \\ &= \vec{\tau}\cdot\mathrm{d}\vec{\theta} \\ \therefore W &= \int\vec{\tau}\cdot\mathrm{d}\vec{\theta} \end{aligned}$$

For a body of total mass M, with center of mass $\vec{R}$, Newton's Second Law can be expressed as:

$$\vec{F} = M\ddot{\vec{R}} \tag{3.16}$$

Now consider breaking up the body into small elements of mass, $\mathrm{d}m$. The momentum of an element of mass is $\mathrm{d}\vec{p} = \dot{\vec{r}}\mathrm{d}m$, where $\vec{r}$ is the position vector and $\dot{\vec{r}}$ is the velocity of element $\mathrm{d}m$. The total momentum of the body is therefore:

$$\vec{p} = \int\dot{\vec{r}}\mathrm{d}m$$

Using this in Newton's Second Law, we have:

$$\vec{F} = \dot{\vec{p}} = \int\ddot{\vec{r}}\mathrm{d}m$$

Therefore, equating this to Eq. 3.16, the center of mass is defined as:

$$\begin{aligned} M\ddot{\vec{R}} &= \int\ddot{\vec{r}}\mathrm{d}m \\ \therefore \ddot{\vec{R}} &= \frac{1}{M}\int\ddot{\vec{r}}\mathrm{d}m \\ \therefore \vec{R} &= \frac{1}{M}\int\vec{r}\mathrm{d}m \end{aligned}$$

(Eq. 3.17). The center of mass is therefore a mass-weighted average of all position vectors, $\vec{r}$. For a symmetric mass distribution, the center of mass is located at the center of the body. For example, the center of mass of human beings is approximately in our stomach. If we were to carry a large weight above our heads, our center of mass would move up towards our head.

The total mass is the integral of all elements of mass:

$$M = \int\mathrm{d}m$$

It is straightforward to express this as a volume integral, over a volume V. If the density of a body as a function of $\vec{r}$ is $\rho(\vec{r})$ [kg m^{-3}], then $\mathrm{d}m = \rho(\vec{r})\mathrm{d}V$ and:

$$\therefore M = \int_V\rho(\vec{r})\mathrm{d}V$$

Center of mass:

$$\vec{R} = \frac{1}{M}\int\vec{r}\mathrm{d}m \tag{3.17}$$

$$\begin{aligned} \text{where}\quad M &= \int\mathrm{d}m \\ &= \int_V\rho(\vec{r})\mathrm{d}V \end{aligned}$$

3.3.2 Moment of inertia

Inertia is a property of a body that describes its tendency to either remain stationary or to continue moving at a constant velocity; it represents resistance to a change in motion. More massive bodies have a greater resistance to a change in motion, which means they have greater inertia. The angular analog of mass and inertia is *moment of inertia*, I [$\mathrm{kg\,m^2}$]. Moment of inertia represents resistance to a change in *angular* motion, such as rotation of a body about a given axis. I is a function of the geometry and mass distribution of a body.

If we consider breaking up a body into small elements of mass dm, we can find an expression for I by considering the angular momentum of dm:

$$d\vec{L} = (\vec{r} \times \vec{v})dm$$

where $\vec{r}$ is the position vector of dm from the rotation axis. For simplicity, let's take $\vec{r}$ to be in the plane of rotation, such that we can drop the cross product.[1] Using $v = \omega r$, the magnitude of $d\vec{L}$ is therefore:

$$\begin{aligned} \left|d\vec{L}\right| &= |\vec{r}||\vec{v}|\, dm \\ &= r^2\, |\vec{\omega}|\, dm \end{aligned}$$

Integrating, the total angular momentum of the body is:

$$\begin{aligned} \therefore \left|\vec{L}\right| &= |\vec{\omega}| \underbrace{\int r^2 dm}_{=I} \\ &= I\,|\vec{\omega}| \end{aligned}$$

This is the angular analog of $\vec{p} = m\vec{v}$, where the moment of inertia is defined by the integral:

$$I \overset{\text{def}}{=} \int r^2 dm$$

Using $dm = \rho(\vec{r})dV$, where ρ is the density, we can express I as a volume integral (Eq. 3.18). The resistance of a body to rotation depends on the axis of rotation, therefore, the moment of inertia of a body must always be specified with respect to a given axis. For example, the moment of inertia of a thin cylindrical rod of radius a and length L depends on whether rotation is about the central axis ($I = Ma^2/2$), about one end of the rod ($I = ML^2/3$), or about the middle of the rod ($I = ML^2/12$), as shown in Worked example 3.1.

[1] If $\vec{r}$ is *not* in the plane of motion, then $\vec{\omega}$ and $\vec{L}$ do not point along the same axis, which gives rise to a precession of $\vec{L}$ about $\vec{\omega}$. This applies to the motion of gyroscopes.

Moment of inertia:

$$\begin{aligned} I &\overset{\text{def}}{=} \int r^2 dm \\ &\equiv \int_V \rho(\vec{r}) r^2\, dV \end{aligned} \tag{3.18}$$

Radius of gyration

Moments of inertia take the general form: $I = Mk^2$, where M is the total mass of the object and k [m] is a quantity called the *radius of gyration*. From Worked example 3.1, we see that the radius of gyration takes a different form for the same object depending on the axis of rotation. Other common moments of inertia are summarized in Eqs. 3.19–3.25 (Derivation 3.3 for moment of inertia of a sphere). The general procedure for calculating moments of inertia are as outlined in Worked example 3.1 (i.e. determine an expression for dV; determine an expression for ρ; plug into $I = \int \rho r^2 dV$, where r is the distance from the axis of rotation, and integrate).

Common moments of inertia:

Thin cylindrical shell: (radius a)

$$\text{About central axis: } I = Ma^2 \tag{3.19}$$

Thin cylindrical rod (Worked example 3.1): (radius a, length L)

$$\text{About central axis: } I = \frac{1}{2}Ma^2 \tag{3.20}$$

$$\text{About one end: } I = \frac{1}{3}ML^2 \tag{3.21}$$

$$\text{About middle: } I = \frac{1}{12}ML^2 \tag{3.22}$$

Sphere (Derivation 3.3): (radius a)

$$\text{Shell: } I = \frac{2}{3}Ma^2 \tag{3.23}$$

$$\text{Solid: } I = \frac{2}{5}Ma^2 \tag{3.24}$$

Thin rectangular plate: (length a, width b)

$$\text{About center: } I = \frac{1}{12}M(a^2 + b^2) \tag{3.25}$$

WORKED EXAMPLE 3.1: Calculating moment of inertia of a thin cylindrical rod

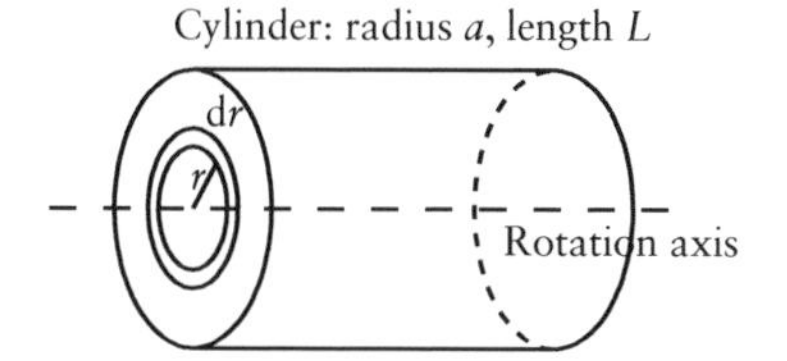

About central axis:

$$dV = 2\pi r L dr$$
$$\rho = \frac{M}{V} = \frac{M}{\pi a^2 L}$$
$$\therefore I = \rho \int r^2 dV$$
$$= \frac{M}{\pi a^2 L} \int_0^a 2\pi L r^3 dr$$
$$= \frac{1}{2} M a^2$$

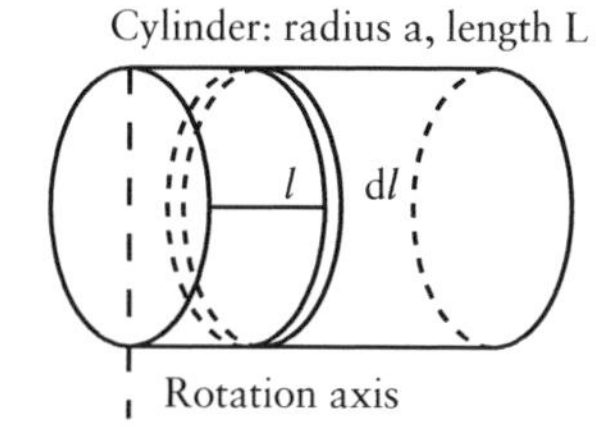

About one end:

$$dV = \pi a^2 dl$$
$$\rho = \frac{M}{V} = \frac{M}{\pi a^2 L}$$
$$\therefore I = \rho \int l^2 dV$$
$$= \frac{M}{\pi a^2 L} \int_0^L \pi a^2 l^2 dl$$
$$= \frac{1}{3} M L^2$$

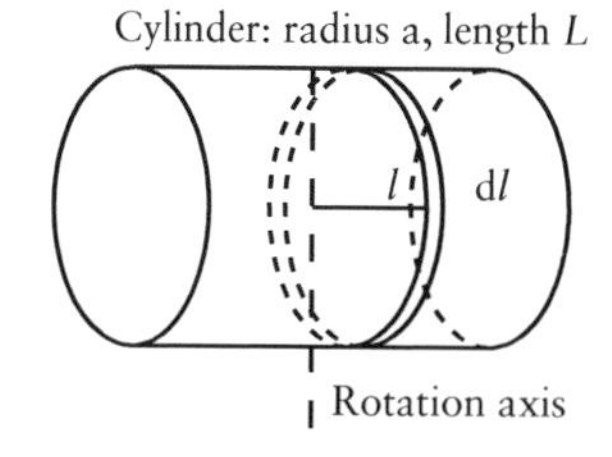

About middle:

$$dV = \pi a^2 dl$$
$$\rho = \frac{M}{V} = \frac{M}{\pi a^2 L}$$
$$\therefore I = \rho \int l^2 dV$$
$$= \frac{M}{\pi a^2 L} \int_{-L/2}^{L/2} \pi a^2 l^2 dl$$
$$= \frac{1}{12} M L^2$$

Parallel and perpendicular axis theorems

There are two theorems that help to calculate the moment of inertia of an object about a given axis. These are called the *parallel* and *perpendicular* axis theorems. The parallel axis theorem relates the moment of inertia about the center-of-mass axis, I_{cm}, to that about an axis parallel to the center-of-mass axis, which is a perpendicular distance h away from it (Eq. 3.26 and Derivation 3.4). The perpendicular axis theorem can only be used for flat or thin objects. If the object lies in the x–y plane, and the moments of inertia about each axis are known, I_x and I_y, then the moment of inertia about the perpendicular z-axis is given by their sum (Eq. 3.27 and Derivation 3.5).

Axis theorems (Derivations 3.4 and 3.5):

Parallel axis theorem:

$$I_0 = I_{cm} + Mh^2 \tag{3.26}$$

Perpendicular axis theorem :

$$I_z = I_x + I_y \tag{3.27}$$

3.3.3 Equation of motion

We saw in Section 3.2 that Newton's Second Law holds for both linear and angular dynamics. For angular dynamics, the torque is therefore equal to the rate of change of angular momentum:

$$\vec{\tau} = \frac{d\vec{L}}{dt}$$

Using $\vec{L} = I\vec{\omega}$ (the angular equivalent of $\vec{p} = m\vec{v}$), we can express Newton's Second Law for constant I in terms of the rate of change of $\vec{\omega}$ or $\vec{\theta}$ (Eqs. 3.28 and 3.29). These equations are angular equations of motion, which can be solved given known initial conditions to predict the subsequent angular motion of a system. Worked example 3.2 shows how to use Eq. 3.29 to determine the oscillation frequency of a compound pendulum.

Equations of motion (linear → angular):

$$\vec{F} = m\frac{d\vec{v}}{dt} \rightarrow \vec{\tau} = I\frac{d\vec{\omega}}{dt} \tag{3.28}$$

$$\equiv m\frac{d^2\vec{x}}{dt^2} \quad \equiv I\frac{d^2\vec{\theta}}{dt^2} \tag{3.29}$$

Derivation 3.3: Derivation of moment of inertia of a sphere (Eqs. 3.23 and 3.24)

Shell:

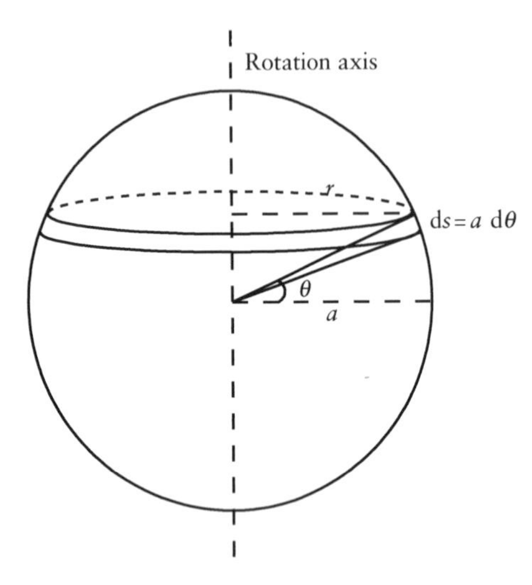

Solid:

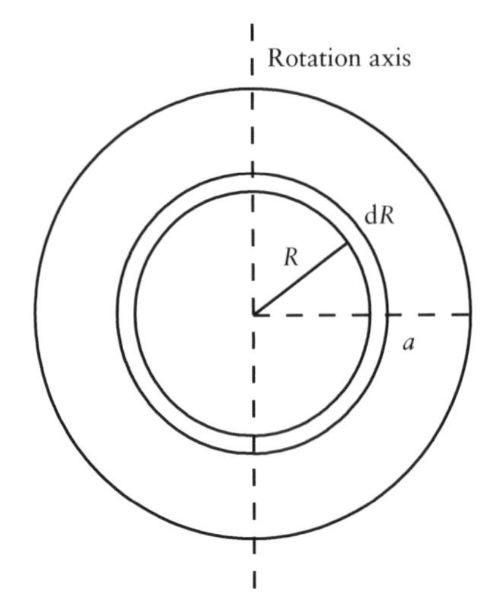

The density (mass per unit area) of a spherical shell of radius a is:

$$\rho_{\text{shell}} = \frac{M}{4\pi a^2}$$

To determine the moment of inertia of the shell, consider circular strips a perpendicular radius r from the rotation axis of thickness ds. The area of the strip is:

$$\begin{aligned} dA &= 2\pi r ds \\ &= 2\pi r a d\theta \end{aligned}$$

Therefore, the moment of inertia of the shell is:

$$\begin{aligned} I_{\text{shell}} &= \rho\int r^2 dA \\ &= 2\pi a\rho\int r^3 d\theta \\ &= 2\pi a\rho\int (a\cos\theta)^3\, d\theta \\ &= 2\pi a^4 \frac{M}{4\pi a^2} \int_{-\pi/2}^{\pi/2} \cos^3\theta d\theta \\ &= \frac{1}{2}Ma^2 \int_{-\pi/2}^{\pi/2} \cos\theta\left(1-\sin^2\theta\right) d\theta \\ &= \frac{1}{2}Ma^2\left[\sin\theta - \frac{1}{3}\sin^3\theta\right]_{-\pi/2}^{\pi/2} \\ &= \frac{2}{3}Ma^2 \end{aligned}$$

The density (mass per unit volume) of a solid sphere of radius a is:

$$\rho_{\text{solid}} = \frac{M}{4\pi a^3/3} = \frac{3M}{4\pi a^3}$$

To determine the moment of inertia of the solid sphere, treat the sphere as a sum of spherical shells of increasing radius. The volume of a shell at radius R from the center of the sphere is:

$$dV = 4\pi R^2 dR$$

Therefore, the moment of inertia of the solid sphere is:

$$\begin{aligned} I_{\text{solid}} &= \int_{R=0}^{R=a} I_{\text{shell}} \\ &= \frac{2}{3}\int R^2 dm \\ &= \frac{2}{3}\rho\int R^2 dV \\ &= \frac{2}{3}\frac{3M}{4\pi a^3}\int 4\pi R^4 dR \\ &= \frac{2M}{a^3}\int_0^a R^4 dR \\ &= \frac{2}{5}Ma^2 \end{aligned}$$

Derivation 3.4: Derivation of the parallel axis theorem (Eq. 3.26)

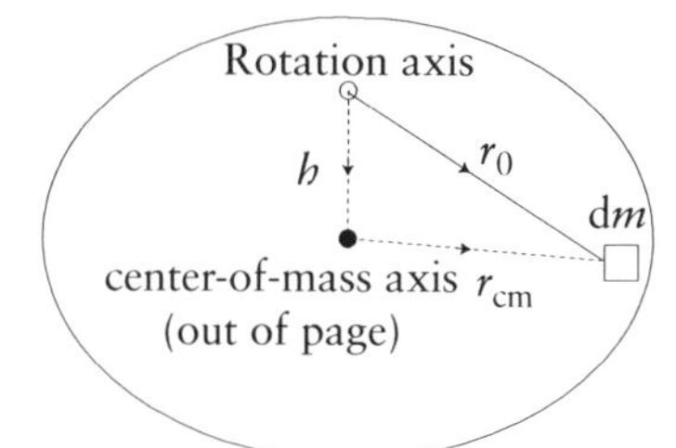

$$
\begin{aligned}
I_0 &= \int r_0^2 \mathrm{d}m \\
&= \int \left(\vec{h} + \vec{r}_{\mathrm{cm}}\right) \cdot \left(\vec{h} + \vec{r}_{\mathrm{cm}}\right) \mathrm{d}m \\
&= \underbrace{\int r_{\mathrm{cm}}^2 \mathrm{d}m}_{=I_{\mathrm{cm}}} + h^2 \underbrace{\int \mathrm{d}m}_{=M} + 2\vec{h} \cdot \underbrace{\int \vec{r}_{\mathrm{cm}} \mathrm{d}m}_{=0} \\
&= I_{\mathrm{cm}} + Mh^2
\end{aligned}
$$

- Starting point: the moment of inertia about a given rotation axis is:

  $$I_0 = \int r_0^2 \mathrm{d}m$$

  where $\vec{r}_0$ is the position vector of an element of mass $\mathrm{d}m$ from the rotation axis.
- The distance between the rotation axis and $\mathrm{d}m$ is:

  $$\vec{r}_0 = \vec{h} + \vec{r}_{\mathrm{cm}}$$

  where $\vec{h}$ is the position vector of the center-of-mass axis from the rotation axis, and $\vec{r}_{\mathrm{cm}}$ the position vector of $\mathrm{d}m$ from the center-of-mass axis.
- From the definition of the center of mass:

  $$\int \vec{r}_{\mathrm{cm}} \mathrm{d}m = 0$$

Derivation 3.5: Derivation of the perpendicular axis theorem (Eq. 3.27)

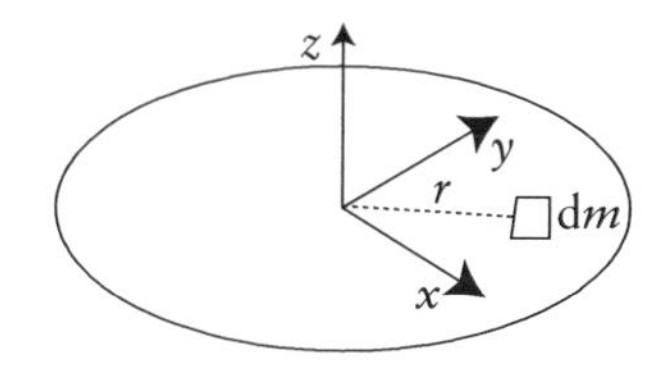

$$
\begin{aligned}
I_z &= \int r^2 \mathrm{d}m \\
&= \int \left(x^2 + y^2\right) \mathrm{d}m \\
&= \underbrace{\int x^2 \mathrm{d}m}_{=I_y} + \underbrace{\int y^2 \mathrm{d}m}_{=I_x} \\
&= I_y + I_x
\end{aligned}
$$

- Starting point: the moment of inertia about the z-axis is:

  $$I_z = \int r^2 \mathrm{d}m$$

  where r is the radial distance of element of mass $\mathrm{d}m$ from the z-axis.
- The magnitude of r is related to x and y by:

  $$r = \sqrt{x^2 + y^2}$$

- Since the object is planar, the moment of inertia about the x-axis is:

  $$I_x = \int y^2 \mathrm{d}m$$

  and we have a similar expression for I_y.

WORKED EXAMPLE 3.2: Oscillation frequency of a compound pendulum (Eq. 3.29)

A thin cylindrical rod of length l and mass m is suspended at one end and allowed to pivot about point P. Determine the frequency of oscillations from the equation of motion of the pendulum.

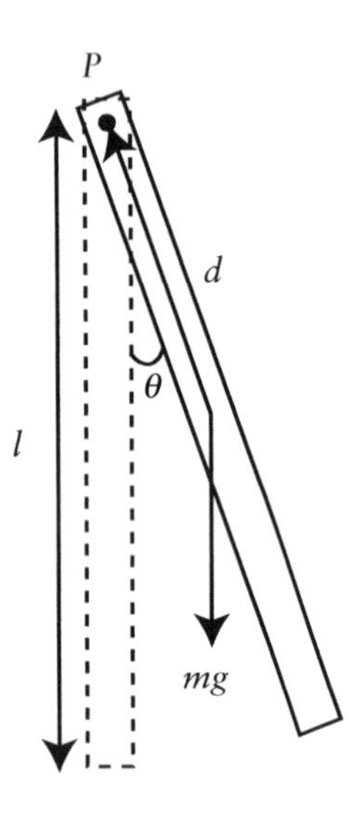

Torque of center of mass about point P:

$$\vec{\tau} = \vec{r} \times \vec{F}$$
$$\therefore \tau = -dmg\sin\theta$$
$$= -\frac{lmg}{2}\sin\theta$$
$$\simeq -\frac{lmg}{2}\theta \text{ in small angle approx.}$$

Moment of inertia (refer to Eq. 3.21):

$$I = \frac{1}{3}ml^2$$

Equation of motion:

$$I\frac{d^2\theta}{dt^2} = \tau$$
$$\therefore \frac{1}{3}ml^2\frac{d^2\theta}{dt^2} = -\frac{lmg}{2}\theta$$
$$\therefore \frac{d^2\theta}{dt^2} = -\underbrace{\frac{3g}{2l}}_{=\omega^2}\theta$$

This corresponds to simple harmonic oscillations with oscillation frequency:

$$\omega = \sqrt{\frac{3g}{2l}}$$

Angular kinetic energy

In calculating the work required to accelerate a body through an angle θ, we find that angular kinetic energy also has an analogous form to its linear counterpart:

$$T = \frac{1}{2}I\omega^2$$

(Eq. 3.30 and Derivation 3.6). Many problems can be broken down into the motion *of* the center of mass (linear motion) and motion *about* the center of mass (angular motion). Worked example 3.3 shows how to calculate the final velocity of a sphere rolling down an inclined plane by considering linear motion of the center of mass plus angular motion about the center of mass.

Kinetic energy (linear → angular) (Derivation 3.6):

$$T = \frac{1}{2}mv^2 \rightarrow T = \frac{1}{2}I\omega^2$$
$$= \frac{p^2}{2m} \qquad = \frac{L^2}{2I} \tag{3.30}$$

Derivation 3.6: Derivation of angular kinetic energy (Eq. 3.30)

$$
\begin{aligned}
T &= W \\
&= \int_0^{\theta} \vec{\tau} \cdot \mathrm{d}\vec{\theta}' \\
&= I \int_0^{\theta} \frac{\mathrm{d}\vec{\omega}'}{\mathrm{d}t} \cdot \mathrm{d}\vec{\theta}' \\
&= I \int_0^{\omega} \frac{\mathrm{d}\vec{\theta}'}{\mathrm{d}t} \cdot \mathrm{d}\vec{\omega}' \\
&= I \int_0^{\omega} \vec{\omega}' \cdot \mathrm{d}\vec{\omega}' \\
&= I \int_0^{\omega} \omega' \mathrm{d}\omega' \\
&= \frac{1}{2} I \omega^2
\end{aligned}
$$

The primed variables are to distinguish them from the integration limits.

- Starting point: the work required to accelerate a body through an angle $\vec{\theta}$ is:

  $$W = \int_0^{\theta} \vec{\tau} \cdot \mathrm{d}\vec{\theta}$$

  From the work–energy theorem, this is equal to the gain in angular kinetic energy, T, of the body.
- Newton's Second Law for angular motion is:

  $$\vec{\tau} = I\vec{\alpha} = I\dot{\vec{\omega}}$$

  (analogous to $\vec{F} = m\vec{a} = m\dot{\vec{v}}$).
- The dot product of a vector with itself is $\vec{\omega} \cdot \vec{\omega} = |\vec{\omega}|^2 = \omega^2$. Differentiating both sides, we have:

  $$\mathrm{d}(\vec{\omega} \cdot \vec{\omega}) = \mathrm{d}\vec{\omega} \cdot \vec{\omega} + \vec{\omega} \cdot \mathrm{d}\vec{\omega} = 2\vec{\omega} \cdot \mathrm{d}\vec{\omega}$$

  and $\mathrm{d}\left(\omega^2\right) = 2\omega \mathrm{d}\omega$

  $$\therefore \vec{\omega} \cdot \mathrm{d}\vec{\omega} = \omega \mathrm{d}\omega$$

- Integrating the expression for T, we see that the expression for angular kinetic energy is analogous to linear kinetic energy, $T = mv^2/2$.

WORKED EXAMPLE 3.3: Velocity of sphere rolling down inclined plane

A spherical shell and a solid sphere roll without slipping down a plane inclined at an angle θ. Show that the ratio of their velocities at the foot of the plane is: $v_{solid}/v_{shell} = 5/\sqrt{21}$.

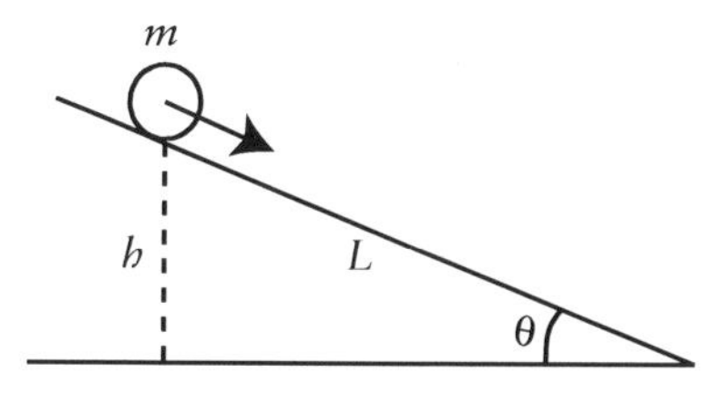

Conservation of energy:

$$V = T_{\text{linear}} + T_{\text{rotational}}$$

$$\therefore mgh = \frac{1}{2}mv^2 + \frac{1}{2}I\omega^2$$

Moment of inertia of objects (refer to Eqs. 3.23 and 3.24):

$$I_{\text{shell}} = \frac{2}{3}mr^2$$

$$I_{\text{solid}} = \frac{2}{5}mr^2$$

Velocity of spherical shell:

$$
\begin{aligned}
mgL\sin\theta &= \frac{1}{2}mv^2 + \frac{1}{2}\left(\frac{2}{3}mr^2\right)\left(\frac{v}{r}\right)^2 \\
&= \frac{1}{2}mv^2 + \frac{1}{3}mv^2 = \frac{5}{6}mv^2
\end{aligned}
$$

$$\therefore v_{\text{shell}} = \sqrt{\frac{6}{5}gL\sin\theta}$$

Velocity of solid sphere:

$$
\begin{aligned}
mgL\sin\theta &= \frac{1}{2}mv^2 + \frac{1}{2}\left(\frac{2}{5}mr^2\right)\left(\frac{v}{r}\right)^2 \\
&= \frac{1}{2}mv^2 + \frac{1}{5}mv^2 = \frac{7}{10}mv^2
\end{aligned}
$$

$$\therefore v_{\text{solid}} = \sqrt{\frac{10}{7}gL\sin\theta}$$

Therefore, the ratio of their velocities at the foot is:

$$\Rightarrow \frac{v_{\text{solid}}}{v_{\text{shell}}} = \frac{5}{\sqrt{21}}$$

Chapter 4
Gravity and orbits

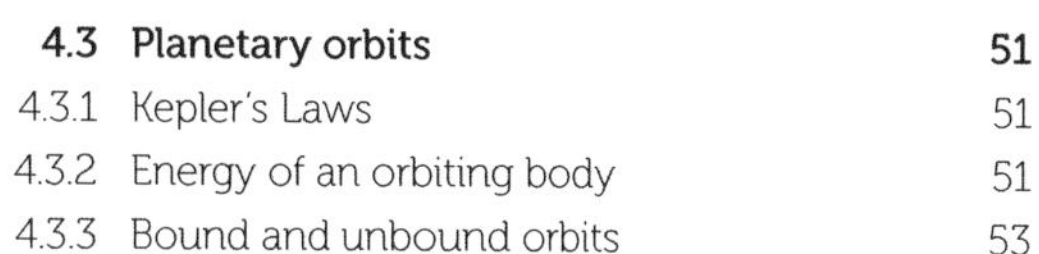

Another application of Newton's Laws is in the gravitational orbit of one celestial body around another. Gravity is the weakest of the four fundamental forces of nature.[1] It is an attractive force, and depends only on the masses of the interacting bodies and the distance between them. Newton discovered that the gravitational force on a body of mass m_1 by another of mass m_2 falls off with the square of the distance between them, r_{12}: $F_G \propto m_1 m_2 / r_{12}^2$. In this chapter, we shall look at the properties of gravity and planetary orbits. There is essentially no "new physics" in this chapter—it's simply an application of the concepts covered in Chapters 1 and 3.

4.1 Equation of motion

Before looking specifically at the properties of gravity and orbits, let's determine the general equation of motion for a body undergoing circular or elliptical motion in a two-dimensional plane. The best way to do this is to work in *polar* coordinates, (r, θ), rather than Cartesian coordinates, (x, y).

4.1.1 Polar coordinates

In polar coordinates, the motion of a body is defined by its position vector $\vec{r}$, and its angle θ relative to the positive Cartesian x-axis (Fig. 4.1). The polar unit vectors are $\hat{r}$, which points radially outward and $\hat{\theta}$, which is perpendicular to $\hat{r}$. They are related to Cartesian unit vectors, $\hat{i}$ and $\hat{j}$ by:

$$\hat{r} = \hat{i}\cos\theta + \hat{j}\sin\theta \tag{4.1}$$

and $$\hat{\theta} = -\hat{i}\sin\theta + \hat{j}\cos\theta \tag{4.2}$$

(refer to Fig. 4.2).

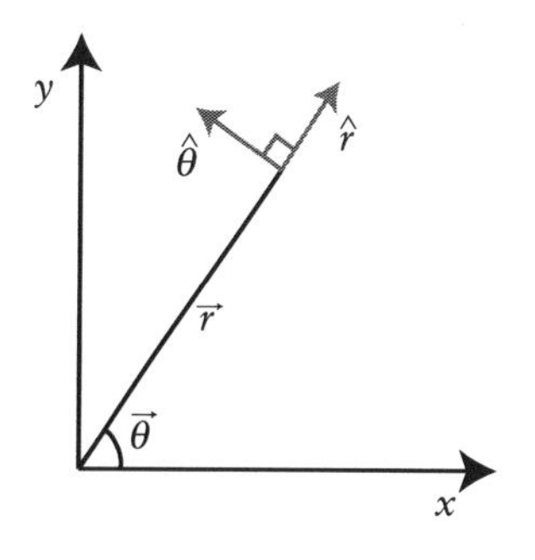

Figure 4.1: Cartesian and polar position vectors.

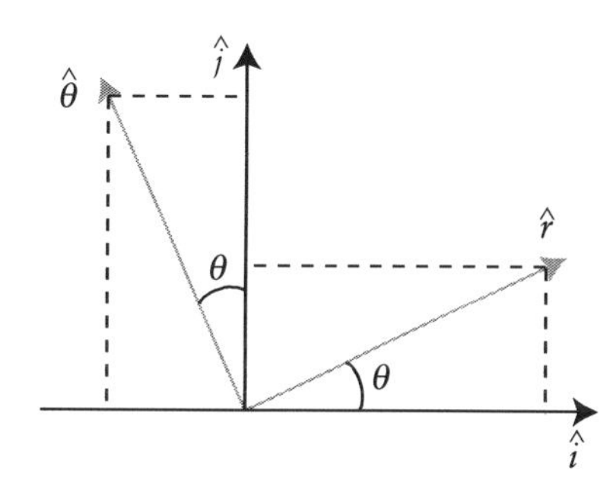

Figure 4.2: Cartesian and polar unit vectors.

The rate of change of the unit vectors are found by differentiating Eqs. 4.1 and 4.2 with respect to time:

$$\begin{aligned}\frac{d\hat{r}}{dt} &= -\hat{i}\sin\theta\dot{\theta} + \hat{j}\cos\theta\dot{\theta} \\ &= \dot{\theta}\left(-\hat{i}\sin\theta + \hat{j}\cos\theta\right) \\ &= \dot{\theta}\hat{\theta}\end{aligned} \tag{4.3}$$

and

$$\begin{aligned}\frac{d\hat{\theta}}{dt} &= -\hat{i}\cos\theta\dot{\theta} - \hat{j}\sin\theta\dot{\theta} \\ &= -\dot{\theta}\left(\hat{i}\cos\theta + \hat{j}\sin\theta\right) \\ &= -\dot{\theta}\hat{r}\end{aligned} \tag{4.4}$$

[1] The other three are the weak force, the electromagnetic force, and the strong force.

The position vector $\vec{r}$ changes its value if either its length changes (a change along $\hat{r}$) or if its direction changes (a change along $\hat{\theta}$). Therefore, velocity, $\dot{\vec{r}}$, and acceleration, $\ddot{\vec{r}}$, have both an $\hat{r}$- and a $\hat{\theta}$-component—i.e. both a radial and an angular component. We can deduce expressions for these components by differentiating $\vec{r}$, as shown in Derivation 4.1. Doing this, we can separate Newton's Second Law, $\vec{F} = m\ddot{\vec{r}}$, into a radial component of force, F_r (Eq. 4.5 and Derivation 4.1) plus an angular component of force, F_θ (Eq. 4.6 and Derivation 4.1). The total force is then given by the vector sum of both components (Eq. 4.7 and Fig. 4.3). We will look at the interpretation of each component below.

Equation of motion in polar coordinates (Derivation 4.1):

Radial component of force:

$$F_r = m\left(\ddot{r} - r\dot{\theta}^2\right) \tag{4.5}$$

Angular component of force:

$$F_\theta = m\left(r\ddot{\theta} + 2\dot{r}\dot{\theta}\right) \tag{4.6}$$

Total force:

$$\vec{F} = F_r\hat{r} + F_\theta\hat{\theta} \tag{4.7}$$

Radial component

The radial component of acceleration in Eq. 4.5 has two terms. The first term is the linear acceleration, $\ddot{r}$, that we are familiar with from linear motion. The second term, $-r\dot{\theta}^2$, points radially inwards (from the − sign), and is only non-zero for particles rotating about an axis with angular velocity $\omega = \dot{\theta}$. Writing this in terms of $v_\theta = \omega r$, where v_θ is the tangential component of velocity, we have:

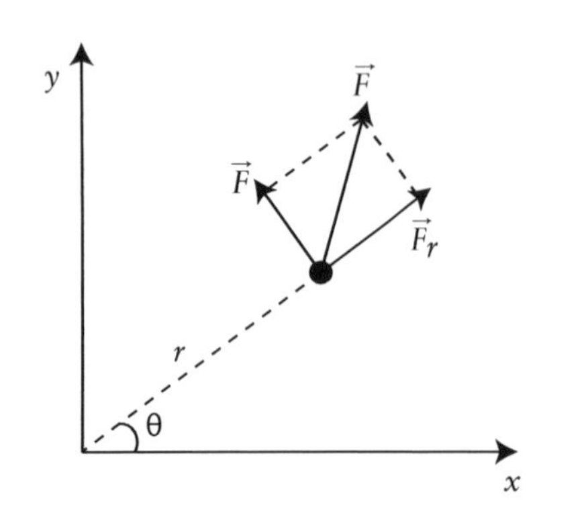

Figure 4.3: Separating a force into its radial and tangential (angular) components, F_r and F_θ.

$$\begin{aligned} r\dot{\theta}^2 &= r\omega^2 \\ &= \frac{v_\theta^2}{r} \end{aligned}$$

Some of you will recognize this as the inward *centripetal acceleration* for bodies moving in a curved path. The external force that gives rise to this acceleration is called the *centripetal force*, and is given by:

$$\vec{F} = -\frac{mv_\theta^2}{r}\hat{r}$$

Bodies that are rotating at a constant distance from the axis of rotation have no linear acceleration, meaning the length r = constant and $\ddot{r} = 0$. Therefore, such bodies move under the centripetal force only. This force could be provided by, for example, string tension for a rotating mass on a string, or friction between car wheels and the road when a car turns a corner.

Angular component

It's a bit less obvious to see intuitively what the angular component of force represents. Newton's Second Law states that the torque on a body equals its change in angular momentum, $\tau = \dot{L}$. For circular motion, the magnitude of angular momentum is:

$$L = mrv_\theta = m\omega r^2 = m\dot{\theta}r^2$$

Substituting this into Newton's Second Law, we have:

$$\begin{aligned} \tau &= \frac{dL}{dt} \\ &= m\frac{d}{dt}\left(r^2\dot{\theta}\right) \\ &= m\left(r^2\ddot{\theta} + 2r\dot{r}\dot{\theta}\right) \\ &= mr\left(r\ddot{\theta} + 2\dot{r}\dot{\theta}\right) \\ &\equiv rF_\theta \end{aligned}$$

Therefore, F_θ (in Eq. 4.6) represents the linear force at r that gives rise to a torque about an axis.

4.1.2 Central forces

A force that depends only on the magnitude of the distance between interacting bodies, r, and acts radially outward or inward is called a *central force*:

$$\vec{F} = f(r)\hat{r} \tag{4.8}$$

Derivation 4.1: Derivation of the equation of motion in polar coordinates (Eqs. 4.5 and 4.6)

- Starting point: in polar coordinates, the radial position vector, $\vec{r}$, is defined as:

$$\vec{r} = r\hat{r}$$

- Differentiating with respect to time, we have:

$$\begin{aligned}\dot{\vec{r}} &= \dot{r}\hat{r} + r\frac{d\hat{r}}{dt} \\ &= \underbrace{\dot{r}}_{\text{Radial velocity}}\hat{r} + \underbrace{r\dot{\theta}}_{\text{Angular velocity}}\hat{\theta}\end{aligned}$$

The radial velocity term is $v_r = \dot{r}$, and the angular velocity term is $v_\theta = r\dot{\theta} = r\omega$, as expected.

- The acceleration vector, $\ddot{\vec{r}}$, is then found by differentiating $\dot{\vec{r}}$ with respect to time (left-hand panel).
- Doing this, we see that acceleration has a radial and an angular component, from which we define a radial and an angular component of force (F_r and F_θ, respectively).

$$\left(\frac{d\hat{r}}{dt} = \dot{\theta}\hat{\theta} \text{ and } \frac{d\hat{\theta}}{dt} = -\dot{\theta}\hat{r}\right)$$

Differentiating $\dot{\vec{r}}$:

$$\begin{aligned}\ddot{\vec{r}} &= \ddot{r}\hat{r} + \dot{r}\frac{d\hat{r}}{dt} + \dot{r}\dot{\theta}\hat{\theta} + r\ddot{\theta}\hat{\theta} + r\dot{\theta}\frac{d\hat{\theta}}{dt} \\ &= \ddot{r}\hat{r} + \dot{r}\dot{\theta}\hat{\theta} + \dot{r}\dot{\theta}\hat{\theta} + r\ddot{\theta}\hat{\theta} - r\dot{\theta}^2\hat{r} \\ &= \underbrace{\left(\ddot{r} - r\dot{\theta}^2\right)}_{\text{Radial component}}\hat{r} + \underbrace{\left(r\ddot{\theta} + 2\dot{r}\dot{\theta}\right)}_{\text{Angular component}}\hat{\theta}\end{aligned}$$

$\therefore$ Radial force:

$$F_r = m\ddot{\vec{r}} \cdot \hat{r} = m\left(\ddot{r} - r\dot{\theta}^2\right)$$

Angular force:

$$F_\theta = m\ddot{\vec{r}} \cdot \hat{\theta} = m\left(r\ddot{\theta} + 2\dot{r}\dot{\theta}\right)$$

where $f(r)$ is an arbitrary function of r. Equating this to Eq. 4.7, the radial and angular components of a central force are:

$$\begin{aligned}F_r &= m(\ddot{r} - r\dot{\theta}^2) \\ &\equiv f(r)\end{aligned}$$

and

$$\begin{aligned}F_\theta &= m(r\ddot{\theta} + 2\dot{r}\dot{\theta}) \\ &= 0\end{aligned}$$

Systems governed by central forces conserve mechanical energy and angular momentum, as shown in Derivation 4.2. Some common central forces are:

- **Gravitational force**: the force of gravity between two masses M and m obeys an inverse square law in distance:

$$\vec{F}_G = -\frac{GMm}{r^2}\hat{r}$$

  Since mass is always positive, the gravitational force is always attractive.

- **Coulomb force**: the electrostatic force between two charges Q and q also obeys an inverse square law in distance:

$$\vec{F}_C = \frac{qQ}{4\pi\varepsilon_0 r^2}\hat{r}$$

  The force is attractive for charges of opposite sign and repulsive for like charges.

- **Restoring force of a spring**: for small extensions, the restoring force of a spring is proportional to the displacement:

$$\vec{F}_R = -kr\hat{r}$$

- **Centripetal force**: bodies undergoing circular motion are acted on by the radially inward centripetal force:

$$\vec{F} = -\frac{mv^2}{r}\hat{r}$$

4.2 Gravitational force and energy

We'll now turn our attention to the *gravitational force*, although the following discussion holds for all central forces.

Newton discovered that the attractive gravitational force on a body of mass m by another of mass M is inversely proportional to the square of the distance between them, r, and acts along the line separating them:

$$\vec{F}_G = -\frac{GMm}{r^2}\hat{r}$$

(Eq. 4.9). The constant of proportionality, G [$6.67 \cdot 10^{-11}\,\text{m}^3\,\text{kg}^{-1}\,\text{s}^{-2}$], is called the *Gravitational constant.* Its small magnitude means that the gravitational force is relatively weak, and the strength of F_G becomes more significant for more massive bodies.

Comparing Eq. 4.9 with Eq. 4.8, we see that the gravitational force is a central force, where the magnitude of the force is:

$$f(r) = -\frac{GMm}{r^2}$$

Gravity as a conservative force

By definition, a conservative force (i.e. one that conserves mechanical energy) has zero curl:

$$\nabla \times \vec{F} = 0$$

(refer to Section 1.5). Since gravity is a radial force, and the curl of a vector is a measure of how much it rotates around an axis, we can guess that the curl of the gravitational force must equal zero. This is shown using Stokes's theorem in Derivation 4.3. Therefore, since gravity is curl-free, it is a *conservative* force, and mechanical energy is conserved.

Gravitational potential energy

Since gravity is a conservative force, we can define a *gravitational potential energy* function (GPE) or V, which is a function of position only. For a body of mass m in the gravitational field of a body of mass M a distance r away from its center, the gravitational potential energy function is:

$$\begin{aligned}\therefore V(\vec{r}) &= -\int f(r')\mathrm{d}r' \\ &= \int_O^r \frac{GMm}{(r')^2}\mathrm{d}r' \\ &= -\frac{GMm}{r}\end{aligned}$$

(Eq. 4.10) where O is a fixed reference point at which $V(\mathrm{O}) = 0$.

Gravitational force and potential energy:

$$\vec{F}_G(\vec{r}) = -\frac{GMm}{r^2}\hat{r} \quad (4.9)$$

$$V(\vec{r}) = -\frac{GMm}{r} \quad (4.10)$$

Derivation 4.2: Derivation of conservation of energy and momentum for central forces

Conservation of energy:
The dot product of a central force ($\vec{F} = f(r)\hat{r} = m\ddot{\vec{r}}$) with $\dot{\vec{r}}$ gives the rate at which the force does work ($\dot{W} = P = \vec{F} \cdot \vec{v}$):

$$\therefore m\ddot{\vec{r}} \cdot \dot{\vec{r}} = f(r)\hat{r} \cdot \dot{\vec{r}}$$

Integrating this with respect to time, and letting the constant of integration equal the total energy, E_0, we recover the conservation of energy equation:

$$\begin{aligned}\frac{1}{2}m\dot{r}^2 &= \int f(r)\mathrm{d}r + \text{constant} \\ \therefore E_0 &= \underbrace{\frac{1}{2}m\dot{r}^2}_{\text{KE}} - \underbrace{\int f(r)\mathrm{d}r}_{\text{PE}} \\ &= T + V\end{aligned}$$

Conservation of angular momentum:
Bodies moving under a central force experience no torque since there is no tangential component of force ($F_\theta = 0$):

$$\begin{aligned}\vec{\tau} &= \vec{r} \times \vec{F} \\ &= f(r)\vec{r} \times \hat{r} \\ &= 0\end{aligned}$$

Therefore, angular momentum is conserved:

$$\begin{aligned}\therefore \vec{L} &= \int \vec{\tau}\mathrm{d}t \\ &= \text{constant}\end{aligned}$$

Derivation 4.3: Derivation of the curl of the gravitational force, $\vec{F}_G$

$$\int_S \left(\nabla \times \vec{F}_G\right)\cdot d\vec{s} = \oint_L \vec{F}_G \cdot d\vec{l}$$
$$= -\oint_L \frac{GMm}{r^2}\hat{r}\cdot d\vec{l}$$
$$= -\oint_L \frac{GMm}{r^2}\hat{r}\cdot\hat{r}dr$$
$$= -\oint_L \frac{GMm}{r^2}dr$$
$$= GMm\left[\frac{1}{r}\right]_a^{b=a}$$
$$= GMm\left(\frac{1}{a}-\frac{1}{a}\right)$$
$$= 0$$
$$\therefore \nabla \times \vec{F}_G = 0$$

- Starting point: Stokes's theorem relates the curl of a vector $\vec{F}$ to its closed line integral around a loop:

$$\int_S \left(\nabla \times \vec{F}\right)\cdot d\vec{s} = \oint_L \vec{F}\cdot d\vec{l}$$

- The gravitational force is a central force (radial component only):

$$\vec{F}_G = -\frac{GMm}{r^2}\hat{r}$$

- A line element in spherical polar coordinates is:

$$d\vec{l} = dr\hat{r} + rd\theta\hat{\theta} + r\sin\theta d\phi\hat{\phi}$$

Since $\hat{r}$ is orthogonal to $\hat{\theta}$ and $\hat{\phi}$, only the $\hat{r}$-component in $d\vec{l}$ remains in the integral.

- For a closed line integral, the integration limits are equal.

4.3 Planetary orbits

4.3.1 Kepler's Laws

Kepler's Laws of Planetary Motion describe the orbits of planets around the sun. These were determined empirically by Johannes Kepler in 1609, and can be easily derived from Newtonian mechanics. His three laws are summarized in Summary box 4.1.

4.3.2 Energy of an orbiting body

The total energy of a body of mass m in a gravitational field of another body of mass M is given by the sum of its kinetic and potential energies (Eq. 4.13 and Derivation 4.4).

Energy of an orbiting body (Derivation 4.4):

$$E_0 = \frac{1}{2}m\dot{r}^2 + V_{eff} \tag{4.13}$$

where the effective potential energy is:

$$V_{eff} = \frac{L^2}{2mr^2} - \frac{GMm}{r} \tag{4.14}$$

V_{eff} is the *effective potential energy* of the body (Eq. 4.14 and Derivation 4.4). It is composed of two terms: the *angular kinetic energy* ($L^2/2mr^2$) and the *gravitational potential* energy ($-GM\,m/r$). The angular kinetic energy term is sometimes called *centrifugal potential energy* because its form corresponds to the negative integral of the centrifugal force. Whereas the centri*petal* force points radially inwards and keeps an object moving in a curved path, the centri*fugal* force points radially outwards (Fig. 4.4). It is equal in magnitude, but opposite in direction, to the centripetal force.[2] Recall that the centripetal force is:

$$\vec{F}_{in} = -\frac{mv_\theta^2}{r}\hat{r}$$

Therefore, the centrifugal force is:

$$\vec{F}_{out} = \frac{mv_\theta^2}{r}\hat{r}$$
$$\equiv mr\omega^2\hat{r}$$
$$\equiv mr\dot{\theta}^2\hat{r}$$
$$\equiv \frac{L^2}{mr^3}\hat{r}$$

[2] Note that a centripetal and centrifugal force do not act simultaneously on the same body (otherwise they would cancel and the body would move in a straight line). They correspond instead to the action–reaction pair of forces for circular motion (recall from Newton's Third Law that for every force of action there is an equal and opposite force of reaction). For example, consider a ball tethered to a rotating axis by a string, which is undergoing circular motion in the horizontal plane. The tension in the string provides the centripetal force on the ball to keep it moving in a curved path. The force exerted *by* the ball *on* the string then corresponds to the centrifugal force, which is equal in magnitude and opposite in direction to the centripetal force.

Summary box 4.1: Kepler's Laws of Planetary Motion

1. **Kepler's First Law: law of elliptical orbits**

Planets rotate around the sun in elliptical orbits, with the sun at one of the focal points of the ellipse. The equation of an ellipse with eccentricity e in polar coordinates is:

$$\frac{l}{r} = 1 + e\cos\theta \quad \text{where} \quad l = \frac{L^2}{GMm^2} \tag{4.11}$$

l is the *length parameter* [m], which is a constant that depends on the total angular momentum, L, of the system. The Earth's orbit is almost circular $\therefore e \simeq 0$.

2. **Kepler's Second Law: law of equal areas**

A radius vector from the sun to a planet sweeps out equal areas in equal time intervals: $A_1 = A_2$ in a time Δt.

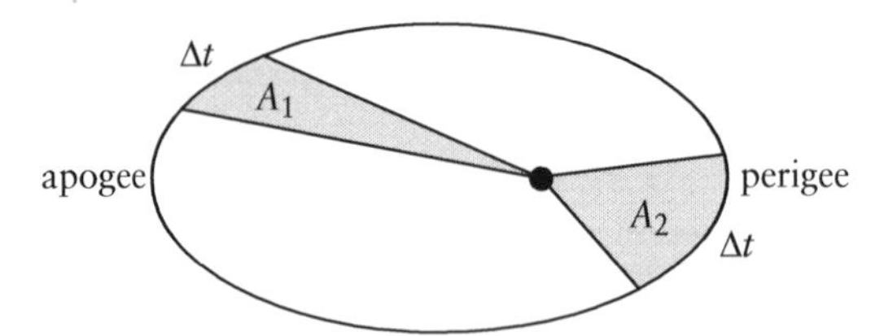

Proof: the element of area swept out in an angle $d\theta$ is approximately the area of a triangle of length r and height $rd\theta$, $\therefore dA \simeq r^2 d\theta/2$. Therefore, the rate at which the area is swept out is:

$$\frac{dA}{dt} = \frac{1}{2}r^2\frac{d\theta}{dt} = \frac{L}{2m} = \text{constant}$$

where the total angular momentum, which is constant, is $L = rp = mrv_\theta = m\omega r^2$, and angular velocity is $\omega = \dot{\theta}$.

3. **Kepler's Third Law: law of time periods**

The square of the period of revolution about the sun, T, is proportional to the cube of the semi-major orbital axis, a:

$$T^2 \propto a^3 \tag{4.12}$$

Proof: For a circular orbit, the force of gravity equals the centripetal force. For motion with angular velocity $\omega = 2\pi/T$, where T is the time period, we have:

$$\frac{GMm}{r^2} = \frac{mv^2}{r} = mr\omega^2 = mr\left(\frac{2\pi}{T}\right)^2$$

$$\therefore T^2 \propto r^3$$

Although this proof is for a circular orbit, the law also holds for elliptical orbits, where r is the semi-major axis.

Derivation 4.4: Derivation of the total energy of a mass m in orbit (Eqs. 4.13 and 4.14)

$$
\begin{aligned}
E &= T + V \\
&= \frac{1}{2}m\dot{\vec{r}}\cdot\dot{\vec{r}} - \frac{GMm}{r} \\
&= \frac{1}{2}m(\dot{r}\hat{r} + r\dot{\theta}\hat{\theta})\cdot(\dot{r}\hat{r} + r\dot{\theta}\hat{\theta}) - \frac{GMm}{r} \\
&= \frac{1}{2}m\dot{r}^2 + \frac{1}{2}mr^2\dot{\theta}^2 - \frac{GMm}{r} \\
&= \underbrace{\frac{1}{2}m\dot{r}^2}_{\text{Linear KE}} + \underbrace{\underbrace{\frac{L^2}{2mr^2}}_{\text{Angular KE}} \underbrace{- \frac{GMm}{r}}_{\text{potential energy}}}_{\text{Effective PE}}
\end{aligned}
$$

- Starting point: the total energy of a body of mass m orbiting a body of mass M is the sum of its kinetic energy, T, and gravitational potential energy, V, where:

  $$V = -\frac{GMm}{r}$$

- The velocity of the body is:

  $$
  \begin{aligned}
  \dot{\vec{r}} &= v_r\hat{r} + v_\theta\hat{\theta} \\
  &= \dot{r}\hat{r} + r\dot{\theta}\hat{\theta}
  \end{aligned}
  $$

  (refer to Derivation 4.1).

- The angular momentum of the body is:

  $$L = mrv_\theta = mr^2\omega = mr^2\dot{\theta}$$

  Use this to eliminate $\dot{\theta}^2$ from the expression for E.

- Doing this, the kinetic energy can be written as the sum of its linear and angular components. Combining the angular kinetic energy term with the gravitational potential energy, we define the *effective potential energy* as:

  $$V_{\text{eff}} = \frac{L^2}{2mr^2} - \frac{GMm}{r}$$

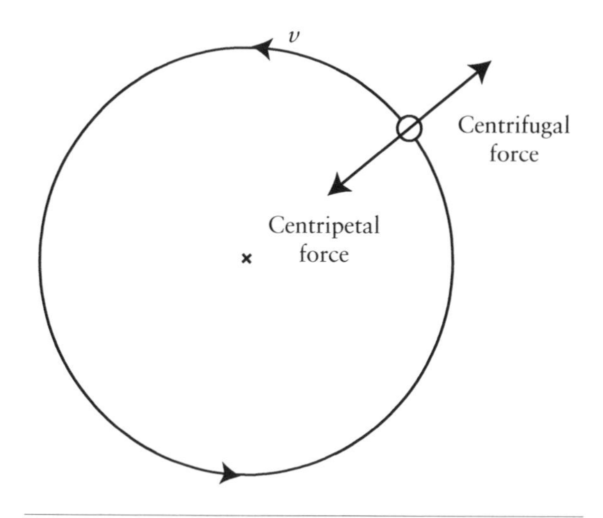

Figure 4.4: Centripetal and centrifugal force.

Therefore, we see that the potential energy function corresponding to this force is:

$$
\begin{aligned}
V_{\text{out}} &= -\int \frac{L^2}{mr^3}\,\mathrm{d}r \\
&= \frac{L^2}{2mr^2}
\end{aligned}
$$

which is the first term of the effective potential energy.

4.3.3 Bound and unbound orbits

A body of mass m will orbit a higher mass body, of mass M, if it enters its gravitational field. The type of orbit is classified as either *bound* or *unbound*, depending on the ratio of the kinetic energy of the orbiting body to its potential energy. If the kinetic energy of the orbiting body is *less* than its potential energy, then the body remains in orbit about the central mass and follows a *bound* orbit. The shape of the orbit is either elliptical or circular, and obeys Kepler's Laws. If the kinetic energy of the orbiting body is *greater* than its potential energy, then it does not remain in a bound orbit around it, but its trajectory is deflected by the large body. This type of orbit is therefore called an *unbound* orbit, and its shape is either a parabola or a hyperbola.

From Kepler's First Law, the equation describing the path of an orbiting body is:

$$\frac{l}{r} = 1 + e\cos\theta$$

For bound orbits, the eccentricity is:

$$e = 0 \quad \text{for circular orbits}$$
$$\text{and} \quad 0 < e < 1 \quad \text{for elliptical orbits}$$

whereas for unbound orbits it is:

$$e = 1 \quad \text{for parabolic orbits}$$
$$\text{and} \quad e > 1 \quad \text{for hyperbolic orbits}$$

Circular orbits

For circular orbits, the radius is constant, therefore, $\dot{r} = 0$. This means that the radial kinetic energy term in Eq. 4.13 (the first term) is zero. Therefore, the total energy is equal to the effective potential energy, which is the sum of angular kinetic energy and gravitational potential energy:

$$E_0 = \frac{L_0^2}{2mr_0^2} - \frac{GMm}{r_0}$$

We can determine an expression for the radius of the orbit, r_0, or the orbital velocity, v_θ, by equating the magnitude of the inward gravitational force with the outward centrifugal force:

$$\frac{GMm}{r_0^2} = \frac{mv_\theta^2}{r_0}$$

(Eqs. 4.15 and 4.16). Using the expression for r_0 from Eq. 4.15, we can find expressions for the magnitude of the total angular momentum, L_0, and the total energy, E_0, in terms of the orbital velocity, v_θ (Eqs. 4.17 and 4.18, respectively).

Circular orbits:

Radius and velocity of orbit:

$$\frac{GMm}{r_0^2} = \frac{mv_\theta^2}{r_0} \rightarrow r_0 = \frac{GM}{v_\theta^2} \quad (4.15)$$

$$\text{and} \quad v_\theta = \sqrt{\frac{GM}{r_0}} \quad (4.16)$$

Angular momentum:

$$L_0 = mr_0v_\theta \rightarrow L_0 = \frac{GMm}{v_\theta} \quad (4.17)$$

Total energy:

$$\begin{aligned} E_0 &= \frac{L_0^2}{2mr_0^2} - \frac{GMm}{r_0} \\ &= \frac{1}{2}mv_\theta^2 - \frac{GMm}{r_0} \\ &= -\frac{1}{2}mv_\theta^2 \end{aligned} \quad (4.18)$$

Elliptical orbits

The distance of closest approach in an elliptical orbit is called the *perigee*, r_p, and the distance of furthest approach is the *apogee*, r_a. At r_p and r_a, the radial component of velocity is zero, $\dot{r} = 0$ and therefore (using Eq. 4.13), the total energy at these points is:

$$E_e = \frac{L_e^2}{2mr_{p,a}^2} - \frac{GMm}{r_{p,a}}$$

E_e is the total energy of the body in the elliptical orbit and L_e is its total angular momentum. Rearranging the above, we find a quadratic in r:

$$r_{p,a}^2 + \frac{GMm}{E_e}r_{p,a} - \frac{L_e^2}{2mE_e} = 0$$

The two solutions for $r_{p,a}$ give us the radius of the perigee and the apogee.

Unbound orbits

To determine the distance of closest approach for unbound orbits, we define the *impact parameter*, b, which is the perpendicular distance between the incoming velocity vector of the orbiting body, $\vec{v}_0$, and the center of the deflecting mass M (Fig. 4.5). The radius of closest approach is the point at which the radial component

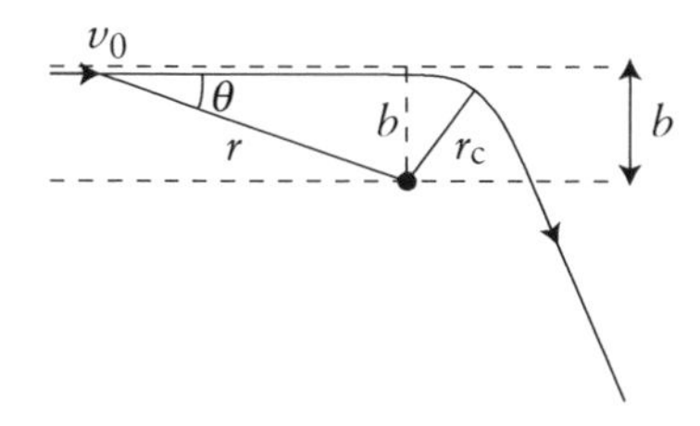

Figure 4.5: Unbound hyperbolic orbit. The impact parameter, b, is the perpendicular distance between the center of the deflecting object and the incoming velocity vector, $\vec{v}_0$, and r_c is the distance of closest approach.

of velocity is zero, $\dot{r} = 0$. Therefore (using Eq. 4.13), the total energy at the distance of closest approach is:

$$E_0 = \frac{L^2}{2mr_c^2} - \frac{GMm}{r_c} \tag{4.19}$$

If we assume that the initial distance of the orbiting body from the mass M is very large, then its initial potential energy is negligible and its total energy equals its kinetic energy:

$$E_0 = \frac{1}{2}mv_0^2$$

From Fig. 4.5, the magnitude of the angular momentum of the body is:

$$\begin{aligned} L &= mv_0 r \sin\theta \\ &= mv_0 b \end{aligned}$$

Therefore, using the above expressions to eliminate E_0 and L from Eq. 4.19, we can solve a quadratic to find the radius of closest approach, r_c:

$$\frac{1}{2}mv_0^2 = \frac{(mv_0 b)^2}{2mr_c^2} - \frac{GMm}{r_c}$$

$$r_c^2 = b^2 - \frac{2GM}{v_0^2}r_c$$

$$\therefore r_c = -\frac{GM}{v_0^2} + \sqrt{\left(\frac{GM}{v_0^2}\right)^2 + b^2}$$

(the positive root is the required solution).

Chapter 5
Two-body dynamics

In this chapter, we will combine the concepts of linear and angular motion to look at the mechanics of two-body systems that are subject to both mutual interactions and to a net external force. We find in all cases that we can break the system up into the linear motion *of* its center of mass plus the angular motion *about* its center of mass.

5.1 Frames of reference

How we (the "observer") perceive motion of a body depends on our point of reference. For example, a person sitting on a train is considered stationary to an observer sitting on the same train, but is considered to be moving at the velocity of the train to an observer sitting at the station. Therefore, we must define a coordinate system with respect to which the motion of other bodies can be measured. This coordinate system is called a *frame of reference*.

Inertial frames of reference

Frames of reference are classified as either *inertial* or *non-inertial*. To understand the distinction between an inertial and a non-inertial frame, we first need to define two frames: the *laboratory* frame, denoted S, which is the frame in which the observer is "in", and the *moving* frame,[1] denoted S*, which is the frame that the observer is "observing". Let's assume that the moving frame is moving at a velocity $\vec{v}_\text{f}$ relative to the laboratory frame. An *inertial* frame is one in which $\vec{v}_\text{f}$ is constant whereas a *non-inertial* frame is one in which $\vec{v}_\text{f}$ is not constant (i.e. a non-inertial frame is an accelerating frame of reference).

Now consider an object of mass m traveling at velocity $\vec{v}^*$ in S*. What velocity does this mass appear to have to an observer in S? Well, it's simply the sum of its velocity in the moving frame, $\vec{v}^*$, plus the velocity of the frame $\vec{v}_\text{f}$:

$$\vec{v} = \vec{v}^* + \vec{v}_\text{f}$$

(refer to Fig. 5.1). Applying Newton's Second Law to this, we find:

$$m\frac{\mathrm{d}\vec{v}}{\mathrm{d}t} = m\frac{\mathrm{d}\vec{v}^*}{\mathrm{d}t} + m\frac{\mathrm{d}\vec{v}_\text{f}}{\mathrm{d}t}$$

For an inertial frame, $\vec{v}_\text{f}$ is constant (by definition):

$$\therefore m\frac{\mathrm{d}\vec{v}}{\mathrm{d}t} = m\frac{\mathrm{d}\vec{v}^*}{\mathrm{d}t}$$
$$\Rightarrow \vec{F} = \vec{F}^*$$

Therefore, the force on a mass m as measured in the laboratory frame is the same as the apparent force on it as measured from another inertial frame of reference—force "looks" the same in all inertial frames of reference.

Invariant quantities

The statement that force "looks" the same in all inertial frames of reference means that force is an *invariant* quantity—it is a quantity that remains unchanged when a particular transformation is applied (in this case, a transformation of reference frame). Mass is also an invariant quantity.

However, since the apparent velocity of a body depends on the frame of reference, then velocity is not an invariant quantity. Similarly, neither is momentum—its value depends on the frame of reference:

$$\vec{v} = \vec{v}^* + \vec{v}_\text{f}$$
$$\therefore m\vec{v} = m\vec{v}^* + m\vec{v}_\text{f}$$
$$\text{or}\quad \vec{p} = \vec{p}^* + \vec{p}_\text{f}$$

In general, any quantity q, as measured in the laboratory frame S, is related to that in a moving frame S* by:

$$q_\text{S} = q_{\text{S}*} + q_\text{f}$$

[1] From the train example above, the frame of reference of the observer at the station corresponds to the laboratory frame, S, and the frame of reference of the person in the train corresponds to the moving frame, S*.

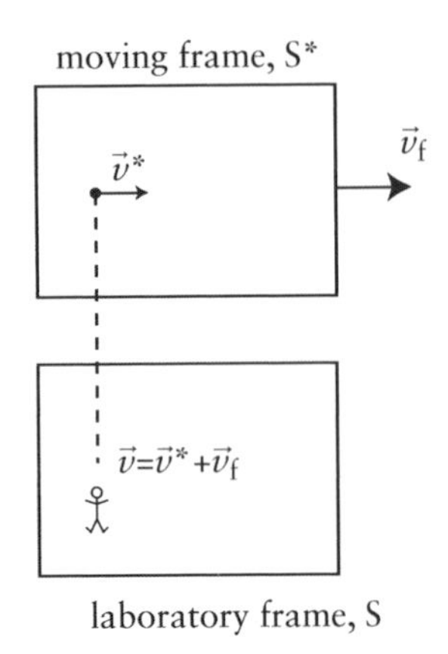

Figure 5.1: Apparent velocity of a body in a moving frame S* to an observer in the laboratory frame, S, where $\vec{v}_f$ is the frame velocity.

where q_f is a property of the frame itself. q is invariant if $q_f = 0$, and not invariant if $q_f \neq 0$ (Eqs. 5.1–5.3).

Non-inertial frames of reference

For *non-inertial* frames, $\vec{v}_f$ is *not* constant, which means that the force on a mass m is not equal in either frame:

$$\therefore \vec{F} = \vec{F}^* + \vec{F}_f$$

$$\text{where } \vec{F}_f = m\frac{d\vec{v}_f}{dt}$$

Therefore, force is not an invariant quantity for non-inertial frames (Eqs. 5.4 and 5.5). The effect of a non-inertial frame is to introduce an additional force on a body, which results from the fact that the frame itself is accelerating. This is sometimes called a *fictitious* force since it is not an external force acting on the body, but it nonetheless influences the apparent motion of the body.

Frames of reference:

For a quantity q:

$$q_S = q_{S^*} + q_f \tag{5.1}$$

Definition of invariant quantity:

$$q \text{ invariant}: \; q_S = q_{S^*} \tag{5.2}$$

$$q \text{ not invariant}: \; q_S \neq q_{S^*} \tag{5.3}$$

Definition of inertial frame:

$$\text{Inertial frame}: \; \vec{F} = \vec{F}^* \tag{5.4}$$

$$\text{Non-inertial frame}: \; \vec{F} \neq \vec{F}^* \tag{5.5}$$

where S corresponds to the laboratory frame and S* to the moving frame.

As an example, consider person A sitting on a moving carousel (a non-inertial moving frame relative to the ground). Person B on the ground (the "laboratory frame") throws a ball in a straight line. To person A, the ball appears to follow a curved trajectory, suggesting a lateral force is acting on it. This apparent lateral force is the so-called fictitious force that is a consequence of the non-inertial nature of the carousel frame relative to the laboratory frame.

5.1.1 The center-of-mass frame

For many-body systems, it is useful to define a frame of reference called the *center-of-mass* frame.

The center of mass

The *center of mass* of a system of point particles, $\vec{r}_{cm}$ [m], represents the point at which the mass of the system appears to be concentrated. It is defined in a similar way to the center of mass of a rigid body, which we saw in Section 3.3.1.

Consider a system of N bodies. The momentum of the ith body is $\vec{p}_i = m_i\dot{\vec{r}}_i$, where $\vec{r}_i$ is its position vector and m_i its mass. The total momentum of all bodies in the system is therefore:

$$\vec{p} = \sum_{i=1}^{N} m_i\dot{\vec{r}}_i$$

Using this in Newton's Second Law (assuming mass is constant), we have:

$$\vec{F} = \dot{\vec{p}} = \sum_{i=1}^{N} m_i\ddot{\vec{r}}_i \tag{5.6}$$

If we assume that we can treat the system as a single body of mass $M = \sum m_i$ located at the center of mass, $\vec{r}_{cm}$, then Newton's Second Law for this body can be expressed as:

$$\vec{F} = M\ddot{\vec{r}}_{cm}$$

Equating this to Eq. 5.6, the center of mass is then defined as:

$$M\ddot{\vec{r}}_{cm} = \sum_i m_i\ddot{\vec{r}}_i$$

$$\therefore \ddot{\vec{r}}_{cm} = \frac{1}{M}\sum_i m_i\ddot{\vec{r}}_i$$

$$\therefore \vec{r}_{cm} = \frac{1}{M}\sum_i m_i\vec{r}_i$$

Therefore, the center of mass is the position vector that represents the mass-weighted average of individual position vectors, $\vec{r}_i$ (Eq. 5.7 and Fig. 5.2).

Center of mass of a multi-body system:

$$\vec{r}_{\text{cm}} = \frac{\sum m_i \vec{r}_i}{M} \quad (5.7)$$

where $M = \sum m_i$

The center-of-mass frame

The *center-of-mass frame*, S^*_{cm}, is defined as the frame of reference in which the center of mass is stationary, and therefore the position vector of the center of mass in the center-of-mass frame is constant:[2]

$$\dot{\vec{r}}^*_{\text{cm}} = 0$$
$$\therefore \vec{r}^*_{\text{cm}} = \text{constant}$$

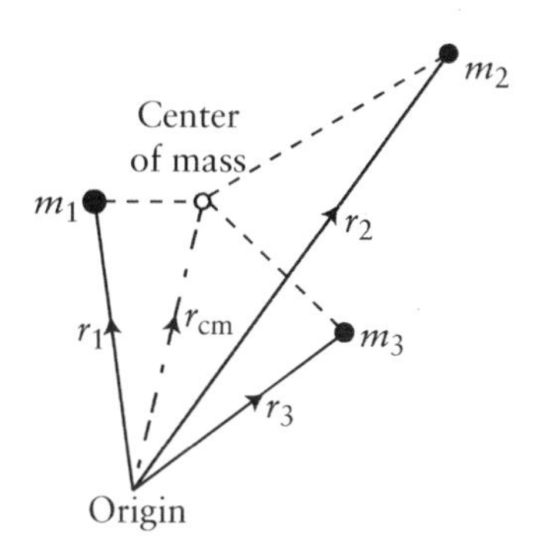

Figure 5.2: The center of mass of a multi-body system corresponds to the mass-weighted mean position of individual bodies.

Therefore, the velocity of the center-of-mass frame relative to the laboratory frame is:

$$\dot{\vec{r}}_{\text{f}} = \dot{\vec{r}}_{\text{cm}} - \dot{\vec{r}}^*_{\text{cm}}$$
$$= \dot{\vec{r}}_{\text{cm}}$$

i.e. the center-of-mass frame moves with the center of mass. The transformation of velocity and position vectors between the laboratory frame and center-of-mass frame therefore involves $\vec{r}_{\text{cm}}$ and its derivative (Eqs. 5.8 and 5.9).

Laboratory frame ↔ center-of-mass frame:

Velocity of mass m_i:

$$\dot{\vec{r}}_i = \dot{\vec{r}}^*_i + \dot{\vec{r}}_{\text{cm}} \quad (5.8)$$

Position vector of mass m_i:

$$\vec{r}_i = \vec{r}^*_i + \vec{r}_{\text{cm}} \quad (5.9)$$

The center-of-mass frame is sometimes also referred to as the *center-of-momentum frame*, since it corresponds to the frame in which the sum of all momenta is zero:

$$\sum \vec{p}^*_i = \sum m_i \dot{\vec{r}}^*_i$$
$$= M\dot{\vec{r}}^*_{\text{cm}}$$
$$= 0$$

5.1.2 Separation vector

Consider two point masses m_1 and m_2 in the laboratory frame, which have position variables $\vec{r}_1$ and $\vec{r}_2$, respectively. The center of mass falls on the line connecting the two masses, at $\vec{r}_{\text{cm}}$ (Fig. 5.3a). We can define the vector that separates the two masses, which we will call

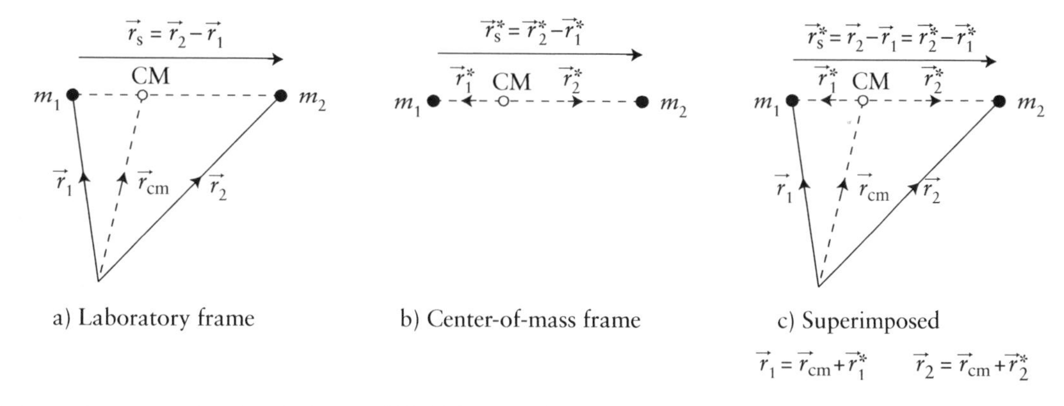

Figure 5.3: Definition of position vectors in a) the laboratory frame, b) the center-of-mass frame (with center of mass at the origin), and c) a superposition of laboratory and center-of-mass frames.

[2] Strictly speaking, the center-of-mass frame is stationary at the *origin*, i.e. $\vec{r}^*_{\text{cm}} = 0$. However, the choice of origin is usually arbitrary, therefore, we can define a coordinate system such that $\vec{r}^*_{\text{cm}} = 0$.

the *separation vector* $\vec{r}_S$, as the difference between the individual position vectors:

$$\vec{r}_s = \vec{r}_2 - \vec{r}_1$$

Using $\vec{r}_S$ in problems has the advantage that the choice of origin is arbitrary, since it cancels out.

If we now imagine sitting on the center of mass, then we are viewing the system from the center-of-mass frame. In this frame, mass m_1 has position vector $\vec{r}_1^*$ and m_2 has position vector $\vec{r}_2^*$ (Fig. 5.3b). The separation vector in the center-of-mass frame is then:

$$\vec{r}_s^* = \vec{r}_2^* - \vec{r}_1^*$$

Using Eq. 5.9, we see that $\vec{r}_S$ is an invariant quantity:

$$\begin{aligned}\vec{r}_s &= \vec{r}_2 - \vec{r}_1 \\ &= (\vec{r}_2^* + \vec{r}_{cm}) - (\vec{r}_1^* + \vec{r}_{cm}) \\ &= \vec{r}_s^*\end{aligned}$$

(Fig. 5.3c).

Center of mass and separation vector (Fig. 5.3):

Center of mass (not invariant):

Laboratory frame: $$\vec{r}_{cm} = \frac{m_1\vec{r}_1 + m_2\vec{r}_2}{M} \quad (5.10)$$

Center-of-mass frame: $$\vec{r}_{cm}^* = 0 \text{ (or constant)} \quad (5.11)$$

Separation vector (invariant):

Laboratory frame: $$\vec{r}_s = \vec{r}_2 - \vec{r}_1 \quad (5.12)$$

Center-of-mass frame: $$\vec{r}_s^* = \vec{r}_2^* - \vec{r}_1^* = \vec{r}_s \quad (5.13)$$

Motion of and about the center of mass

Using expressions for the center of mass together with expressions for the separation vector (Eqs. 5.10–5.13), we have two sets of simultaneous equations, from which we can express the position vectors of m_1 and m_2 in terms of $\vec{r}_{cm}$ and $\vec{r}_s$ (Eqs. 5.14–5.17). Doing this separates a system into motion *of* the center of mass ($\vec{r}_{cm}$) and motion *about* the center of mass ($\vec{r}_s$). We shall see in the next section that it is useful to break the motion down in this way when transforming quantities such as kinetic energy and angular momentum between frames.

Motion of and about the center of mass (from simultaneous Eqs. 5.10–5.13, Fig. 5.3):

Laboratory frame:

$$\vec{r}_1 = \vec{r}_{cm} - \frac{m_2}{M}\vec{r}_s \quad (5.14)$$

$$\vec{r}_2 = \vec{r}_{cm} + \frac{m_1}{M}\vec{r}_s \quad (5.15)$$

Center-of-mass frame:

$$\vec{r}_1^* = -\frac{m_2}{M}\vec{r}_s \quad (5.16)$$

$$\vec{r}_2^* = \frac{m_1}{M}\vec{r}_s \quad (5.17)$$

5.2 Two-body dynamics

Isolated systems

A system is classified as *isolated* if it does not interact with its surroundings. Isolated systems are not acted on by external forces, and neither energy nor matter is exchanged between the system and its surroundings. For example, the Universe constitutes an isolated system. Bodies in an isolated system are therefore only subject to internal forces—i.e. those that originate from within the system itself, such as mutual interactions between bodies. Newton's Third Law states that the force exerted on body 1 by body 2 is equal in magnitude and opposite in direction to that exerted on body 2 by body 1:

$$\vec{F}_{12} = -\vec{F}_{21}$$

In isolated systems, moving bodies can collide with each other, transferring energy and momentum from one body to another in the process. The total energy and momentum in an isolated system is always conserved.

Reduced mass

Consider an isolated system of two bodies of mass m_1 and m_2. In the laboratory frame, the equations of motion for either body are:

$$m_1\ddot{\vec{r}}_1 = \vec{F}_{12}$$
$$m_2\ddot{\vec{r}}_2 = \vec{F}_{21} = -\vec{F}_{12}$$

From this, the acceleration of the separation vector, $\ddot{\vec{r}}_s$ is:

$$\begin{aligned}\ddot{\vec{r}}_s &= \ddot{\vec{r}}_2 - \ddot{\vec{r}}_1 \\ &= -\left(\frac{1}{m_2} + \frac{1}{m_1}\right)\vec{F}_{12} \\ &\equiv -\frac{1}{\mu}\vec{F}_{12}\end{aligned}$$

μ [kg] is the *reduced mass* of the system (Eq. 5.18). Like the total mass, M (Eq. 5.19), reduced mass is an invariant quantity.

Total mass and reduced mass:

Reduced mass (invariant):

$$\frac{1}{\mu} = \frac{1}{m_2} + \frac{1}{m_1} \rightarrow \mu = \frac{m_1 m_2}{M} \quad (5.18)$$

Total mass (invariant):

$$M = m_1 + m_2 \quad (5.19)$$

Therefore, an isolated system under no external forces behaves like a single particle of mass μ, accelerating at $\ddot{\vec{r}}_s$:

$$\vec{F}_{12} = -\mu \ddot{\vec{r}}_s$$

Force is an invariant quantity, therefore $\vec{F} = \mu \ddot{\vec{r}}_s$ in both the laboratory frame and the center-of-mass frame (Eqs. 5.20 and 5.21).

Force on a body (invariant quantity):

Laboratory frame: $\vec{F}_{12} = -\mu \ddot{\vec{r}}_s$, $\vec{F}_{21} = \mu \ddot{\vec{r}}_s$ (5.20)

Center-of-mass frame: $\vec{F}^*_{12} = \vec{F}_{12} = -\mu \ddot{\vec{r}}_s$, $\vec{F}^*_{21} = \vec{F}_{21} = \mu \ddot{\vec{r}}_s$ (5.21)

5.2.1 Transforming between frames

Since velocity is not an invariant quantity, neither is momentum or kinetic energy. Below, we will look at how momentum and kinetic energy transform between the laboratory and the center-of-mass frames.

Momentum transformations

We can calculate the total momentum in the laboratory frame by summing the individual momenta of each body. Doing this, we see that the total momentum is:

$$\vec{P} = M\dot{\vec{r}}_{cm}$$

(Eq. 5.22 and Derivation 5.1). Therefore, the total momentum of the system is equivalent to the momentum of a single particle of mass $M = m_1 + m_2$ moving at the velocity of the center of mass, $\dot{\vec{r}}_{cm}$. In the center-of-mass frame, $\dot{\vec{r}}^*_{cm} = 0$, therefore, $\vec{P}^* = 0$ (Eq. 5.23 and Derivation 5.1). So, the momenta of two bodies in the center-of-mass frame are always equal and opposite.

Total momentum (not invariant) (Derivation 5.1):

Laboratory frame: $\vec{P} = M\dot{\vec{r}}_{cm}$ (5.22)

Center-of-mass frame: $\vec{P}^* = 0$ (5.23)

Momentum transformation:

$$\vec{P} = \vec{P}^* + \vec{P}_{cm}$$

where $\vec{P}_{cm} = M\dot{\vec{r}}_{cm}$

Derivation 5.1: Derivation of the total momentum of a two-body system (Eqs. 5.22 and 5.23)

Laboratory frame:

$$\begin{aligned}\vec{P} &= \vec{p}_1 + \vec{p}_2 \\ &= m_1\dot{\vec{r}}_1 + m_2\dot{\vec{r}}_2 \\ &= (m_1 + m_2)\dot{\vec{r}}_{cm} \\ &= M\dot{\vec{r}}_{cm}\end{aligned}$$

Center-of-mass frame:

$$\begin{aligned}\vec{P}^* &= \vec{p}^*_1 + \vec{p}^*_2 \\ &= m_1\dot{\vec{r}}^*_1 + m_2\dot{\vec{r}}^*_2 \\ &= -\frac{m_1 m_2}{M}\dot{\vec{r}}_s + \frac{m_1 m_2}{M}\dot{\vec{r}}_s \\ &= 0\end{aligned}$$

- Starting point: the total momentum is the sum of individual momenta of m_1 and m_2.
- Change coordinates to $\vec{r}_{cm}$ and $\vec{r}_s$ using the transformations:

$$\vec{r}_1 = \vec{r}_{cm} - \frac{m_2}{M}\vec{r}_s \text{ and } \vec{r}_2 = \vec{r}_{cm} + \frac{m_1}{M}\vec{r}_s$$

$$\text{or } \vec{r}^*_1 = -\frac{m_2}{M}\vec{r}_s \text{ and } \vec{r}^*_2 = \frac{m_1}{M}\vec{r}_s$$

(refer to Section 5.1.2).

- The momenta in the laboratory and center-of-mass frames are related to one another by:

$$\vec{P} = \vec{P}^* + \vec{P}_{cm}$$

where $\vec{P}_{cm} = M\dot{\vec{r}}_{cm}$

Kinetic energy transformations

We can calculate the total kinetic energy in the laboratory frame by summing the individual kinetic energies of each body (Eq. 5.24 and Derivation 5.2). Doing this, we see that the total kinetic energy in the laboratory frame can be expressed as the sum of two components: the kinetic energy *of* the center of mass, $M\dot{r}_{cm}^2/2$, plus the kinetic energy *about* the center of mass, $\mu\dot{r}_S^2/2$. The second term in Eq. 5.24 represents the total kinetic energy in the center-of-mass frame (Eq. 5.25 and Derivation 5.2).

Total kinetic energy (not invariant) (Derivation 5.2):

$$\text{Laboratory frame: } T = \frac{1}{2}M\dot{r}_{cm}^2 + \frac{1}{2}\mu\dot{r}_s^2 \quad (5.24)$$

$$\text{Center-of-mass frame: } T^* = \frac{1}{2}\mu\dot{r}_s^2 \quad (5.25)$$

Kinetic energy transformation:

$$T = T^* + T_{cm}$$

$$\text{where } T_{cm} = \frac{1}{2}M\dot{r}_{cm}^2$$

The kinetic energy of the center-of-mass frame ($\mu\dot{r}_S^2/2$) corresponds to the maximum possible kinetic energy lost in a totally inelastic collision (i.e. one in which the bodies stick together and move at the same velocity after the collision). It is easy to see this intuitively by considering a totally inelastic collision in the center-of-mass frame. The final velocity of the combined bodies in the center-of-mass frame would be zero (from momentum conservation), therefore, $T^*=0$ after a totally inelastic collision. Thus, the maximum kinetic energy loss is indeed $\mu\dot{r}_S^2/2$. In the laboratory frame, the "remainder" of the energy, $M\dot{r}_{cm}^2/2$, is "locked up" in the system, and it cannot be dissipated in an inelastic collision.

Angular momentum transformations

Similarly, we can calculate the total angular momentum of a two-body system by summing the individual angular momenta of each body (Eq. 5.26 and Derivation 5.3). Just like kinetic energy, the total angular momentum can be broken down into the angular momentum *of* the center of mass $\left(M\left(\vec{r}_{cm} \times \dot{\vec{r}}_{cm}\right)\right)$ plus the angular momentum about the center of mass $\left(\mu\left(\vec{r}_S \times \dot{\vec{r}}_S\right)\right)$. In the center-of-mass frame, the total angular momentum equals the angular momentum about the center of mass (Eq. 5.27 and Derivation 5.3).

Derivation 5.2: Derivation of the total kinetic energy of a two-body system (Eqs. 5.24 and 5.25)

Laboratory frame:

$$\begin{aligned} T &= T_1 + T_2 \\ &= \frac{1}{2}m_1\dot{r}_1^2 + \frac{1}{2}m_2\dot{r}_2^2 \\ &= \frac{1}{2}(m_1+m_2)\dot{r}_{cm}^2 + \frac{1}{2}\frac{m_1 m_2}{M}\dot{r}_s^2 \\ &= \frac{1}{2}M\dot{r}_{cm}^2 + \frac{1}{2}\mu\dot{r}_s^2 \end{aligned}$$

Center-of-mass frame:

$$\begin{aligned} T^* &= T_1^* + T_2^* \\ &= \frac{1}{2}m_1\left(\dot{r}_1^*\right)^2 + \frac{1}{2}m_2\left(\dot{r}_2^*\right)^2 \\ &= \frac{1}{2}\frac{m_1 m_2}{M}\dot{r}_s^2 \\ &= \frac{1}{2}\mu\dot{r}_s^2 \end{aligned}$$

- Starting point: the total kinetic energy is the sum of individual kinetic energies of m_1 and m_2.
- Change coordinates to $\vec{r}_{cm}$ and $\vec{r}_S$ using the transformations:

$$\vec{r}_1 = \vec{r}_{cm} - \frac{m_2}{M}\vec{r}_s \text{ and } \vec{r}_2 = \vec{r}_{cm} + \frac{m_1}{M}\vec{r}_s$$

$$\text{or } \vec{r}_1^* = -\frac{m_2}{M}\vec{r}_s \text{ and } \vec{r}_2^* = \frac{m_1}{M}\vec{r}_s$$

(refer to Section 5.1.2).

- The kinetic energy in the laboratory and center-of-mass frames are related to one another by:

$$T = T^* + T_{cm}$$

$$\text{where } T_{cm} = \frac{1}{2}M\dot{r}_{cm}^2$$

Total angular momentum (not invariant) (Derivation 5.3):

Laboratory frame: $\vec{L} = M\left(\vec{r}_{\text{cm}} \times \dot{\vec{r}}_{\text{cm}}\right) + \mu\left(\vec{r}_{\text{s}} \times \dot{\vec{r}}_{\text{s}}\right)$ (5.26)

Center-of-mass frame: $\vec{L}^* = \mu\left(\vec{r}_{\text{s}} \times \dot{\vec{r}}_{\text{s}}\right)$ (5.27)

Angular momentum transformation:

$$\vec{L} = \vec{L}^* + \vec{L}_{\text{cm}}$$

where $\vec{L}_{\text{cm}} = M\left(\vec{r}_{\text{cm}} \times \dot{\vec{r}}_{\text{cm}}\right)$

5.2.2 Action of an external force

The action of an external force, $\vec{F}^{\text{ext}}$, is to change the total momentum of a system (and the angular momentum if $\vec{F}^{\text{ext}}$ produces a torque).

Linear motion

In the presence of an external force, the equations of motion for either body are:

$$\dot{\vec{p}}_1 = m_1\ddot{\vec{r}}_1 = \vec{F}_{12} + \vec{F}_1^{\text{ext}}$$
$$\text{and } \dot{\vec{p}}_2 = m_2\ddot{\vec{r}}_2 = \vec{F}_{21} + \vec{F}_2^{\text{ext}}$$

where $\vec{F}_{ij}$ represents the internal force (interaction between bodies i and j) and $\vec{F}_i^{\text{ext}}$ the external force on body i. Summing these equations and using Newton's Third Law, we find that rate of change of the total momentum of the system, $\dot{\vec{P}}$, equals the external force:

$$\dot{\vec{p}}_1 + \dot{\vec{p}}_2 = \vec{F}_1^{\text{ext}} + \vec{F}_2^{\text{ext}}$$
$$\therefore \dot{\vec{P}} = \vec{F}^{\text{ext}}$$

We saw previously that total momentum in the laboratory frame is $\vec{P} = M\dot{\vec{r}}_{\text{cm}}$ (Eq. 5.22). Therefore, for constant total mass, the external force acts on the system as though it were a single body of mass M located at the center of mass:

$$\vec{F}^{\text{ext}} = M\ddot{\vec{r}}_{\text{cm}}$$

Said another way, the motion of the center of mass responds to external forces (Eq. 5.28).

Momentum and external force:

$$\dot{\vec{P}} = \vec{F}^{\text{ext}}$$
$$= M\ddot{\vec{r}}_{\text{cm}} \quad (5.28)$$

Angular motion

The total angular momentum in the laboratory frame is:

$$\vec{L} = M\left(\vec{r}_{\text{cm}} \times \dot{\vec{r}}_{\text{cm}}\right) + \mu\left(\vec{r}_{\text{s}} \times \dot{\vec{r}}_{\text{s}}\right)$$

Derivation 5.3: Derivation of the total angular momentum of a two-body system (Eqs. 5.26–5.27)

Laboratory frame:

$$\vec{L} = \vec{L}_1 + \vec{L}_2$$
$$= m_1\left(\vec{r}_1 \times \dot{\vec{r}}_1\right) + m_2\left(\vec{r}_2 \times \dot{\vec{r}}_2\right)$$
$$= (m_1 + m_2)\left(\vec{r}_{\text{cm}} \times \dot{\vec{r}}_{\text{cm}}\right) + \frac{m_1 m_2}{M}\left(\vec{r}_{\text{s}} \times \dot{\vec{r}}_{\text{s}}\right)$$
$$= M\left(\vec{r}_{\text{cm}} \times \dot{\vec{r}}_{\text{cm}}\right) + \mu\left(\vec{r}_{\text{s}} \times \dot{\vec{r}}_{\text{s}}\right)$$

Center-of-mass frame:

$$\vec{L}^* = \vec{L}_1^* + \vec{L}_2^*$$
$$= m_1\left(\vec{r}_1^* \times \dot{\vec{r}}_1^*\right) + m_2\left(\vec{r}_2^* \times \dot{\vec{r}}_2^*\right)$$
$$= \frac{m_1 m_2}{M}\left(\vec{r}_{\text{s}} \times \dot{\vec{r}}_{\text{s}}\right)$$
$$= \mu\left(\vec{r}_{\text{s}} \times \dot{\vec{r}}_{\text{s}}\right)$$

- Starting point: the total angular momentum is the sum of individual angular momenta of m_1 and m_2.
- Change coordinates to $\vec{r}_{\text{cm}}$ and $\vec{r}_{\text{s}}$ using the transformations:

$$\vec{r}_1 = \vec{r}_{\text{cm}} - \frac{m_2}{M}\vec{r}_{\text{s}} \quad \text{and} \quad \vec{r}_2 = \vec{r}_{\text{cm}} + \frac{m_1}{M}\vec{r}_{\text{s}}$$
$$\text{or} \quad \vec{r}_1^* = -\frac{m_2}{M}\vec{r}_{\text{s}} \quad \text{and} \quad \vec{r}_2^* = \frac{m_1}{M}\vec{r}_{\text{s}}$$

(refer to Section 5.1.2).

- The angular momentum in the laboratory and center-of-mass frame are related to one another by:

$$\vec{L} = \vec{L}^* + \vec{L}_{\text{cm}}$$

where $\vec{L}_{\text{cm}} = M\left(\vec{r}_{\text{cm}} \times \dot{\vec{r}}_{\text{cm}}\right)$

(Eq. 5.26). Differentiating and using $\vec{F}^{\text{ext}} = M\ddot{\vec{r}}_{\text{cm}}$ and $\vec{F}_{21} = \mu\ddot{\vec{r}}_{\text{S}}$, the rate of change of angular momentum is:

$$\begin{aligned}\dot{\vec{L}} &= M\left(\vec{r}_{\text{cm}} \times \ddot{\vec{r}}_{\text{cm}}\right) + \mu\left(\vec{r}_{\text{s}} \times \ddot{\vec{r}}_{\text{s}}\right) \\ &= \underbrace{\vec{r}_{\text{cm}} \times \vec{F}^{\text{ext}}}_{=\vec{\tau}^{\text{ext}}} + \underbrace{\vec{r}_{\text{s}} \times \vec{F}_{21}}_{=0} \\ &= \vec{\tau}^{\text{ext}}\end{aligned}$$

(The second term equals zero since $\vec{r}_{\text{S}}$ and $\vec{F}_{21}$ are parallel.) Therefore, if $\vec{F}^{\text{ext}}$ produces a torque about the center of mass, then the total angular momentum of the system changes, where its rate of change equals the external torque (Eq. 5.29).

Angular momentum and external torque:

$$\begin{aligned}\dot{\vec{L}} &= \vec{r}_{\text{cm}} \times \vec{F}^{\text{ext}} \\ &= \vec{\tau}^{\text{ext}}\end{aligned} \tag{5.29}$$

Chapter 6
Introduction to special relativity

Newton's Laws are successful in describing the motion of bodies at low velocities. However, it was found that they break down for bodies moving at velocities on the order of the speed of light. In 1905, Einstein proposed the Theory of Special Relativity, which describes the properties and behavior of bodies traveling at velocities approaching the speed of light. This theory approximates to Newton's Laws in the low-velocity limit. Practical applications of special relativity include experimental particle physics, where high-energy particles travel at velocities approaching the speed of light.

6.1 Postulates of special relativity

Special relativity is a theory that applies to inertial frames of reference—i.e. frames in which the velocity of one relative to another is constant (refer to Section 5.1). It is based on Einstein's two fundamental postulates:

1. The laws of physics are the same in all inertial frames.
2. The speed of light in a vacuum is the same in all inertial frames.

The latter is incompatible with classical velocity transformations from Newtonian mechanics. In Section 5.1, we saw that to convert between the velocity v of an object in the laboratory frame S, and its apparent velocity v' in a moving inertial frame S′, we add or subtract the frame velocity, v_f (refer to Fig. 5.1):

$$\text{Laboratory frame}: \; v = v' + v_f$$
$$\text{Moving frame}: \; v' = v - v_f$$

These classical velocity transformations are called *Galilean transformations*. Consider a photon in a moving frame traveling at $v' = c$. Using the Galilean transformations, its velocity in the laboratory frame would be $v = c + v_f > c$, which violates the second postulate of special relativity. Therefore, Galilean transformations only hold true in the low-velocity limit, $v \ll c$. (For objects traveling at velocities approaching the speed of light, we must instead use *Lorentz transformation* to convert between frames, which are derived in Section 6.2.)

The Michelson–Morley experiment

Before Einstein postulated the constancy of the speed of light, c, the *aether hypothesis* was proposed. This hypothesis postulated that the Earth was moving through a medium called the *aether*, and that there was a preferred frame in which c was constant through this medium. This hypothesis, in turn, is suggestive of an absolute frame of reference. The *Michelson–Morley* experiment was designed to measure the speed of light through the aether. This experiment consisted of using an interferometer to measure the interference pattern of coherent light (Fig. 6.1). It was thought that the position of constructive and destructive interference fringes would shift upon rotating the apparatus, as the velocity of light through the aether would change depending on its direction of propagation relative to the aether wind. However, the experiment yielded a null result—no shift in the interference pattern was observed. These observations disproved the aether hypothesis and the notion of an absolute frame of reference.

6.1.1 Time dilation and length contraction

To maintain the constancy of the speed of light, special relativity gives rise to some counterintuitive phenomena. For example, in inertial frames traveling at speeds approaching the speed of light, time appears to tick slower relative to the laboratory frame, and lengths

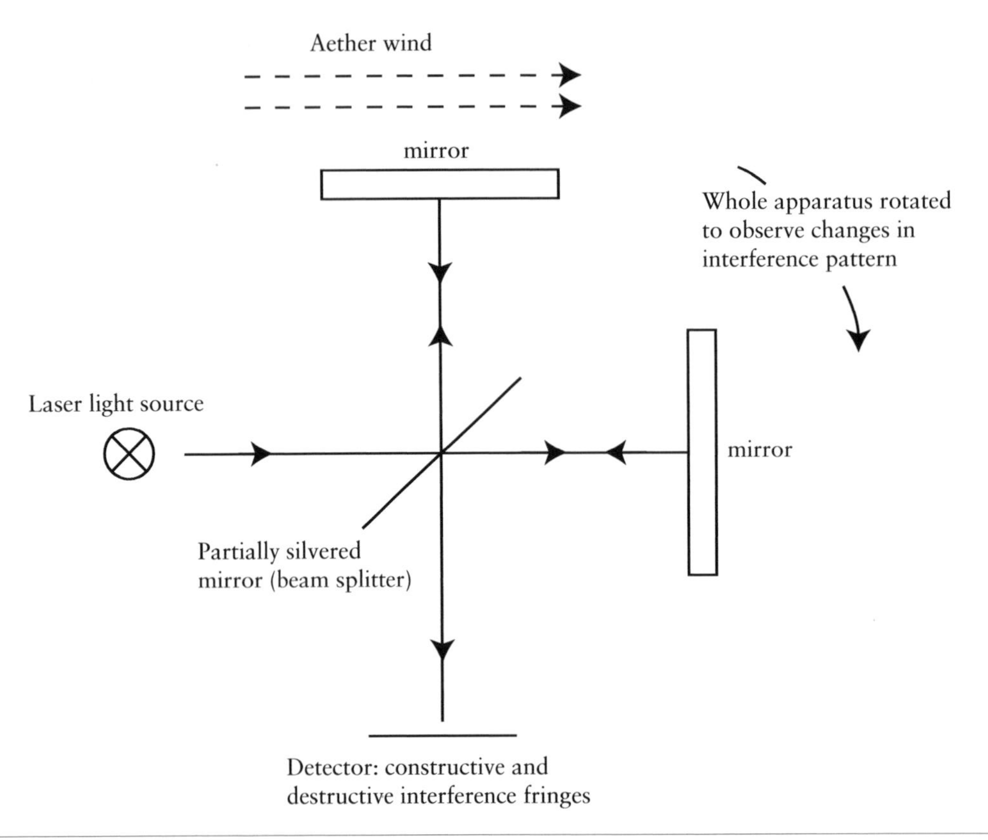

Figure 6.1: Michelson–Morley experimental setup.

appear shorter. These two phenomena are known as *time dilation* and *length contraction*, respectively.
One way to think about these phenomena is to consider a hypothetical device called a *light clock*. In this device, light bounces back and forth between two parallel mirrors a distance d' apart (refer to figure in Derivation 6.1). The time for one "tick" of the clock (for light to travel from one mirror to the other) is:

$$t' = \frac{d'}{c}$$

Imagine that the light clock is traveling at a velocity, v, that is on the order of the speed of light. Using simple geometry, we find that the apparent time for one tick for an observer in the laboratory frame, t, is longer than t' by a factor that depends on the velocity of the light clock:

$$t = \frac{t'}{\sqrt{1-(v/c)^2}} \equiv \gamma t'$$

(Eq. 6.1 and Derivation 6.1). Therefore, time appears *dilated* relative to the moving object by an amount:

$$\gamma = \frac{1}{\sqrt{1-(v/c)^2}}$$

where γ is called the *Lorentz factor* (see below). To an observer in the laboratory frame, the moving light clock ticks n times in a distance l_0, where:

$$l_0 = nv\Delta t$$

However, if we imagine sitting on the light clock such that we are traveling with it at velocity v, then the length over which n ticks of the clock occur appears *shorter* than l_0:

$$\begin{aligned}\Delta l' &= nv\Delta t' \\ &= nv\frac{\Delta t}{\gamma} \\ &= \frac{l_0}{\gamma}\end{aligned}$$

(Eq. 6.2)[1]. Therefore, moving objects observe distances that are *contracted* relative to the "stationary" distance

[1] In Eq. 6.2, we have used $l_0 \to l'$ and $\Delta l' \to l$. The use of primed and non-primed variables when considering time

(in this case, this corresponds to the distance as observed in the laboratory frame) by an amount γ.

> **Time dilation and length contraction (Derivation 6.1):**
>
> $$t = \gamma t' \tag{6.1}$$
>
> $$l = \frac{l'}{\gamma} \tag{6.2}$$
>
> where $\gamma = \dfrac{1}{\sqrt{1-(v/c)^2}}$

The Lorentz factor

The *Lorentz factor* is defined as:

$$\gamma = \frac{1}{\sqrt{1-\beta^2}}$$

where the *beta factor* is:

$$\beta = \frac{v}{c}$$

(Eqs. 6.3 and 6.4). These are dimensionless quantities that define the velocity of moving frames of reference.

> **Lorentz factor:**
>
> $$\gamma = \frac{1}{\sqrt{1-\beta^2}} \tag{6.3}$$
>
> $$\text{where } \beta = \frac{v}{c} \tag{6.4}$$

In classical frames, $v \ll c$. Therefore $\beta \ll 1$ and $\gamma \simeq 1$. As v approaches c, γ increases. The Lorentz factor incorporates the requirement that all objects must travel slower than the speed of light; if $v = c$, then γ becomes infinite, which is unphysical.

dilation and length contraction together can be confusing since the "stationary" frame corresponds to a different frame in either case; the frame that measures a time t' (the "moving" clock frame) measures contracted distance l, whereas the frame that measures a dilated time t (the "stationary" laboratory frame) measures the distance l'. It is more straightforward to think in terms of the *rest frame* of the object, which is further described in Sections 6.3.2 and 6.3.4. (In the example in the text, the rest frame of the clock is not the same as the rest frame of the distance l.)

6.2 Lorentz transformation

We saw above that neither time nor space are invariant quantities[2] for frames moving at velocities on the order of the speed of light. To transform time and space coordinates from one frame to another, we use the *Lorentz transformation*. This is a transformation that converts the coordinate (x', y', z') of a body in the moving frame to the coordinate (x, y, z) as observed from the laboratory frame, and the time t' in the moving frame to t in the laboratory frame:

$$(x', y', z') \rightarrow (x, y, z)$$
$$\text{and } t' \rightarrow t$$

For motion in one dimension, say along x, length contraction occurs only along this axis, and distances *transverse* to the direction of motion are invariant:

$$\therefore y' = y$$
$$z' = z$$

Below, we will derive the Lorentz transformation for motion in one dimension.

Derivation of the Lorentz transformation

Let's assume that the Lorentz transformation is linear. Written in matrix form, the transformation is then:

$$\underbrace{\begin{pmatrix} ct' \\ x' \end{pmatrix}}_{\text{moving frame}} = \underbrace{\begin{pmatrix} A & B \\ C & D \end{pmatrix}}_{\text{transformation matrix}} \underbrace{\begin{pmatrix} ct \\ x \end{pmatrix}}_{\text{laboratory frame}}$$

where A, B, C, and D are constants to be determined. By convention, the time coordinate is expressed as ct or ct' rather than t or t'.
For a moving frame S′ traveling at velocity v relative to the laboratory frame S, the origin in S′ transforms to vt in S: $x' = 0 \rightarrow x = vt$. Therefore, the transformation for this coordinate is:

$$\begin{pmatrix} ct' \\ 0 \end{pmatrix} = \begin{pmatrix} A & B \\ C & D \end{pmatrix} \begin{pmatrix} ct \\ vt \end{pmatrix}$$

[2] Recall that an invariant quantity is a quantity that takes the same value in all inertial frames of reference (refer to Section 5.1).

Derivation 6.1: Derivation of the time dilation and the Lorentz factor (Eq. 6.1)

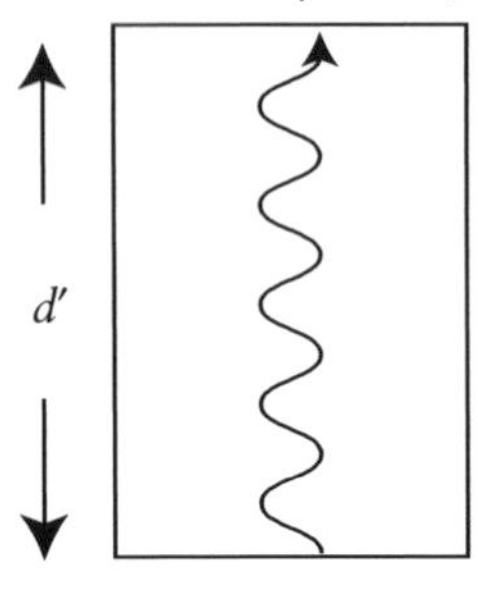

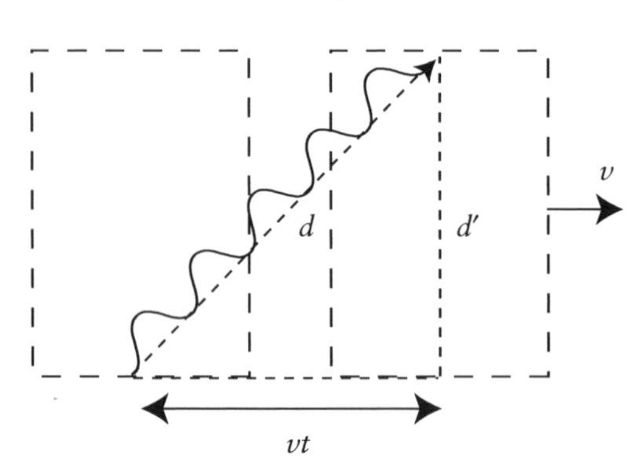

$$t = \frac{d}{c}$$
$$= \frac{\sqrt{(vt)^2 + (d')^2}}{c}$$
$$= \frac{\sqrt{(vt)^2 + (ct')^2}}{c}$$
$$\therefore (ct)^2 = (vt)^2 + (ct')^2$$
$$\therefore t^2 = \left(\frac{1}{1-(v/c)^2}\right)t'^2$$
$$\therefore t = \underbrace{\frac{1}{\sqrt{1-(v/c)^2}}}_{=\gamma}t'$$
$$\equiv \gamma t'$$

- Starting point: consider a photon in its rest frame (moving frame relative to laboratory frame). In the rest frame of the photon, the time taken for it to travel a distance d' is:

$$t' = \frac{d'}{c}$$

- If the rest frame of the photon is moving at a speed v relative to the laboratory frame, then to an observer in a laboratory frame, the time taken for it to travel a distance d is $t = d/c$. Using geometry (refer to figure above), we find that:

$$d^2 = (vt)^2 + (d')^2$$

- Using the above expressions to solve for t, we find that the apparent time in the laboratory frame relative to the moving frame is $t = \gamma t'$, where the proportionality factor is the *Lorentz factor*:

$$\gamma = \frac{1}{\sqrt{1-(v/c)^2}}$$

Expanding, we find that the ratio of C to D is related to the frame's beta factor:

$$0 = Cct + Dvt$$
$$\therefore \frac{C}{D} = -\frac{v}{c} = -\beta \qquad (6.5)$$

Using similar reasoning, the coordinate $-vt'$ in S′ transforms to the origin in $S: x' = -vt' \rightarrow x = 0$. The transformation for this coordinate is then:

$$\begin{pmatrix} ct' \\ -vt' \end{pmatrix} = \begin{pmatrix} A & B \\ C & D \end{pmatrix}\begin{pmatrix} ct \\ 0 \end{pmatrix}$$

Expanding, we find that the ratio of C to A is also related to the frame's beta factor:

$$ct' = Act$$
$$\text{and } -vt' = Cct$$
$$\therefore \frac{C}{A} = -\frac{v}{c} = -\beta \qquad (6.6)$$

Using Eqs. 6.5 and 6.6 and letting $F = B/A$, where F is a constant, the transformation can be written as:

$$\begin{pmatrix} ct' \\ x' \end{pmatrix} = A\begin{pmatrix} 1 & F \\ -\beta & 1 \end{pmatrix}\begin{pmatrix} ct \\ x \end{pmatrix} \qquad (6.7)$$

Now consider a third frame, S″, with beta factor $\beta' = v'/c$ relative to frame S′. The transformation in this case is:

$$\begin{pmatrix} ct'' \\ x'' \end{pmatrix} = A' \begin{pmatrix} 1 & F' \\ -\beta' & 1 \end{pmatrix} \begin{pmatrix} ct' \\ x' \end{pmatrix}$$

Combining the two transformations, we have:

$$\begin{aligned} \begin{pmatrix} ct'' \\ x'' \end{pmatrix} &= AA' \begin{pmatrix} 1 & F' \\ -\beta' & 1 \end{pmatrix} \begin{pmatrix} 1 & F \\ -\beta & 1 \end{pmatrix} \begin{pmatrix} ct \\ x \end{pmatrix} \\ &= AA' \begin{pmatrix} (1-\beta F') & (F+F') \\ (-\beta'-\beta) & (1-\beta' F) \end{pmatrix} \begin{pmatrix} ct \\ x \end{pmatrix} \end{aligned}$$

The matrix transformation above is another Lorentz transformation, and therefore must take the same form as the Lorentz transformation from Eq. 6.7, where the terms in the leading diagonal are equal:

$$\begin{aligned} \therefore 1-\beta F' &= 1-\beta' F \\ \therefore \frac{\beta}{F} &= \frac{\beta'}{F'} \\ &= \text{constant} \end{aligned}$$

The ratio β/F must be constant since the left-hand side is a function of v only and the right-hand side is a function of v' only. If we let the constant equal −1, by defining $F = B/A = -\beta$, then the transformation from Eq. 6.7 can be written as:

$$\begin{pmatrix} ct' \\ x' \end{pmatrix} = A \begin{pmatrix} 1 & -\beta \\ -\beta & 1 \end{pmatrix} \begin{pmatrix} ct \\ x \end{pmatrix} \qquad (6.8)$$

So now all we have left to do is find an expression for A. We can do this by considering the inverse of the above transformation. Doing this, we have:

$$\begin{pmatrix} ct \\ x \end{pmatrix} = \frac{1}{A\left(1-\beta^2\right)} \begin{pmatrix} 1 & \beta \\ \beta & 1 \end{pmatrix} \begin{pmatrix} ct' \\ x' \end{pmatrix} \qquad (6.9)$$

What does the inverse transformation represent physically? It's essentially the same transformation, but the relative velocity is now $-v$ rather than v. Therefore, another way of writing the inverse transformation is to substitute β for $-\beta$ (or $-\beta \rightarrow \beta$) in Eq. 6.8:

$$\therefore \begin{pmatrix} ct \\ x \end{pmatrix} = A \begin{pmatrix} 1 & \beta \\ \beta & 1 \end{pmatrix} \begin{pmatrix} ct' \\ x' \end{pmatrix} \qquad (6.10)$$

Equating Eqs. 6.9 and 6.10, we therefore have:

$$\begin{aligned} A &= \frac{1}{A\left(1-\beta^2\right)} \\ \therefore A &= \frac{1}{\sqrt{1-\beta^2}} \\ &= \gamma \end{aligned}$$

Finally, substituting A back into Eqs. 6.8 and 6.10, the Lorentz transformation and its inverse are:

$$\begin{pmatrix} ct' \\ x' \end{pmatrix} = \gamma \begin{pmatrix} 1 & -\beta \\ -\beta & 1 \end{pmatrix} \begin{pmatrix} ct \\ x \end{pmatrix}$$

$$\text{and} \begin{pmatrix} ct \\ x \end{pmatrix} = \gamma \begin{pmatrix} 1 & \beta \\ \beta & 1 \end{pmatrix} \begin{pmatrix} ct' \\ x' \end{pmatrix}$$

(Eqs. 6.11 and 6.12).

Lorentz transformation in one dimension:

Moving frame coordinates, S′:

$$\begin{pmatrix} ct' \\ x' \end{pmatrix} = \gamma \begin{pmatrix} 1 & -\beta \\ -\beta & 1 \end{pmatrix} \begin{pmatrix} ct \\ x \end{pmatrix} \qquad (6.11)$$

Laboratory frame coordinates, S:
(inverse transformation)

$$\begin{pmatrix} ct \\ x \end{pmatrix} = \gamma \begin{pmatrix} 1 & \beta \\ \beta & 1 \end{pmatrix} \begin{pmatrix} ct' \\ x' \end{pmatrix} \qquad (6.12)$$

Expanding the Lorentz transformation, we find:

$$\begin{aligned} ct' &= \gamma(ct-\beta x) \\ \text{and } x' &= \gamma(x-\beta ct) \end{aligned}$$

Letting $\beta = v/c$, the transformations can be expressed as:

$$\begin{aligned} t' &= \gamma\left(t-\frac{vx}{c^2}\right) \\ \text{and } x' &= \gamma(x-vt) \end{aligned}$$

The inverse transformations are then:

$$\begin{aligned} t &= \gamma\left(t'+\frac{vx'}{c^2}\right) \\ \text{and } x &= \gamma(x'+vt') \end{aligned}$$

This is a common way of writing the Lorentz transformation, which we will use later in this chapter.

Low-velocity limit

In the low-velocity limit, $v \ll c$, we have $\beta \ll 1$ and $\gamma \simeq 1$. Therefore, the Lorentz transformations become:

$$t' \to t - \frac{vx}{c^2} \simeq t$$

and $x' \to x - vt$

Differentiating the x' equation, the velocity transformation is:

$$u' = u - v$$

Therefore, in the low-velocity limit, we recover the Galilean transformations, where time is an invariant quantity, as required.

6.3 Using the Lorentz transformation

6.3.1 The invariant interval

Consider two *events*; the first event occurs at time t_1 and position x_1, and the second occurs at time t_2 and position x_2. For example, the first event could correspond to throwing a ball, and the second could correspond to catching it. The *space–time interval*, $(\Delta s)^2$, is defined as:

$$(\Delta s)^2 = (c\Delta t)^2 - (\Delta x)^2$$

where $\Delta t = t_2 - t_1$ is the time interval and $\Delta x = x_2 - x_1$ is the space interval between the two events. Using the Lorentz transformation, we find that $(\Delta s)^2$ is an invariant quantity—it takes the same value in all inertial frames. Therefore, this interval is referred to as the *invariant interval* (Eq. 6.13 and Derivation 6.2).

Invariant interval (Derivation 6.2):

$$(\Delta s)^2 = (c\Delta t)^2 - (\Delta x)^2 \qquad (6.13)$$

Types of space–time interval

We define three different types of space–time interval depending on the sign of $(\Delta s)^2$.

1. Time-like interval:

$$(\Delta s)^2 > 0 \to \therefore (c\Delta t)^2 > (\Delta x)^2$$

For two events occurring a distance Δx apart with a time Δt between them, enough time passes such that the event at x_1 can cause the event at x_2. Therefore, the events are connected by cause and effect. Said another way, in a time Δt, information travels from x_1 to x_2 at a speed $v = \Delta x / \Delta t < c$. The order of events connected by time-like intervals is the same in all frames (for example, we could never catch a ball before it is thrown, therefore throwing and catching are related by a time-like interval).

2. Light-like interval:

$$(\Delta s)^2 = 0 \to \therefore (c\Delta t)^2 = (\Delta x)^2$$

Now, in a time Δt, information travels from x_1 to x_2 at a speed $v = \Delta x / \Delta t = c$. Therefore, events with light-like intervals are connected by photons traveling at the speed of light.

3. Space-like interval:

$$(\Delta s)^2 < 0 \to \therefore (c\Delta t)^2 < (\Delta x)^2$$

Events separated by space-like intervals do not have a cause-and-effect relationship, since not enough time passes such that the event at x_1 can cause the event at x_2: $v = \Delta x / \Delta t > c$. Therefore, if we change frames, the order of the two events might also change.

6.3.2 Time dilation revisited

In Section 6.1.1, we derived an expression for time dilation by considering the motion of a photon and using geometrical arguments (refer to Derivation 6.1). Here, we will derive the same result using the Lorentz transformation. We find that, for an event that lasts a time $\Delta t'$ in a moving frame S′, an observer in the laboratory frame measures a longer time interval: $\Delta t = \gamma \Delta t'$ (Eq. 6.14 and Derivation 6.3).

Time dilation (Derivation 6.3):

$$\Delta t = \gamma \Delta t' \qquad (6.14)$$

In Derivation 6.3, we calculate the time interval of an event in S′, as measured from the laboratory frame. If, however, the event took place in the laboratory frame, such that $x_2 - x_1 = 0$, then the laboratory frame

Derivation 6.2: Derivation of the invariant interval (Eq. 6.13)

$$
\begin{aligned}
(\Delta s')^2 &= (c\Delta t')^2 - (\Delta x')^2 \\
&= (ct_2' - ct_1')^2 - (x_2' - x_1')^2 \\
&= (\gamma(ct_2 - \beta x_2) - \gamma(ct_1 - \beta x_1))^2 \\
&\quad - (\gamma(x_2 - \beta ct_2) - \gamma(x_1 - \beta ct_1))^2 \\
&= \gamma^2 (c\Delta t - \beta\Delta x)^2 - \gamma^2 (\Delta x - \beta c\Delta t)^2 \\
&= \gamma^2 \Big((c\Delta t)^2 + (\beta\Delta x)^2 - 2\beta c\Delta t\Delta x \\
&\quad - (\Delta x)^2 - (\beta c\Delta t)^2 + 2\beta c\Delta t\Delta x\Big) \\
&= \gamma^2 \left((c\Delta t)^2 - (\Delta x)^2\right)\left(1 - \beta^2\right) \\
&= (c\Delta t)^2 - (\Delta x)^2 \\
&= (\Delta s)^2
\end{aligned}
$$

- Starting point: the space–time interval in a frame S′ is defined as:

  $$(\Delta s')^2 = (c\Delta t')^2 - (\Delta x')^2$$

  where $\Delta t' = t_2' - t_1'$ and $\Delta x' = x_2' - x_1'$.
- Convert to the laboratory frame, S, using the Lorentz transformations:

  $$ct' = \gamma(ct - \beta x)$$
  and $$x' = \gamma(x - \beta ct)$$
  where $$\gamma = \frac{1}{\sqrt{1-\beta^2}}$$
- Rearranging, we find that the space–time interval in the moving frame equals the space–time interval in the laboratory frame, therefore $(\Delta s)^2$ is an invariant quantity:

  $$(\Delta s')^2 = (\Delta s)^2 = \text{invariant}$$

Derivation 6.3: Derivation of time dilation (Eq. 6.14)

$$
\begin{aligned}
\Delta t &= t_2 - t_1 \\
&= \gamma\left(t_2' + \frac{vx_2'}{c^2}\right) - \gamma\left(t_1' + \frac{vx_1'}{c^2}\right) \\
&= \gamma(t_2' - t_1') + \frac{\gamma v}{c^2}\underbrace{(x_2' - x_1')}_{=0} \\
&= \gamma\Delta t'
\end{aligned}
$$

- Starting point: the time interval between two events occurring in the laboratory frame S is given by $\Delta t = t_2 - t_1$.
- Use the inverse Lorentz transformation to convert to a moving frame, S′:

  $$t = \gamma\left(t' + \frac{vx'}{c^2}\right)$$
- Define the frame S′ as the frame in which both events correspond to the same position $\therefore x_2' = x_1' \rightarrow x_2' - x_1' = 0$. This corresponds to the rest frame of the object/event.
- Doing this, we find that the time interval as measured in the laboratory frame is longer than the corresponding interval in the moving frame (rest frame of the event) ⇒ *time dilation.*

corresponds to the rest frame, and an observer in S′ would measure the same temporal dilation factor:

$$
\begin{aligned}
\Delta t' &= t_2' - t_1' \\
&= \gamma\left(t_2 - \frac{vx_2}{c^2}\right) - \gamma\left(t_1 - \frac{vx_1}{c^2}\right) \\
&= \gamma(t_2 - t_1) - \frac{\gamma v}{c^2}\underbrace{(x_2 - x_1)}_{=0} \\
&= \gamma\Delta t
\end{aligned}
$$

Therefore, the apparent time interval in a moving frame is always dilated relative to the rest frame of the event.

Proper time

The frame in which $\Delta x' = 0$ is called the *rest frame* of the event. The time measured in the rest frame is called the *proper time*, τ. Since $\Delta x > 0$ in all other frames, then the proper time is always the shortest time interval. Using the space–time interval from Eq. 6.13, we have:

Derivation 6.4: Derivation of the relativistic Doppler effect (Eqs. 6.15 and 6.16)

- Starting point: the apparent wavelength of an electromagnetic wave emitted from a source moving towards an observer at a relativistic velocity v is:

$$\lambda' = \lambda - vT = \frac{c}{f} - \frac{v}{f} = T(c-v)$$

where T is the wave's time period as measured in the laboratory frame. The apparent frequency of the wave is $f' = c/\lambda'$.
- Due to time dilation, the time period of a wave in the laboratory frame, T, is related to the time period in the frame of the source, T_0, by $T = \gamma T_0$. The frequency of the wave emitted by the source in the frame of the source is then $f_0 = 1/T_0$.
- The beta factor of the source is $\beta = v/c$, and the Lorentz factor is:

$$\gamma = \frac{1}{\sqrt{1-\beta^2}} = \frac{1}{\sqrt{(1-\beta)(1+\beta)}}$$

- The apparent wavelength of an electromagnetic wave emitted from a source moving away from an observer at a relativistic velocity v is:

$$\lambda' = T(c+v)$$

Motion of observer towards source:

$$f' = \frac{c}{\lambda'} = \frac{c}{(c-v)T} = \frac{c}{(c-v)\gamma T_0} = \frac{f_0}{(1-\beta)\gamma} = f_0\frac{\sqrt{(1-\beta)(1+\beta)}}{(1-\beta)} = f_0\sqrt{\frac{(1+\beta)}{(1-\beta)}}$$

$f' > f_0 \Rightarrow$ blue–shift

Motion of observer away from source:

$$f' = \frac{c}{\lambda'} = \frac{c}{(c+v)T} = \frac{f_0}{(1+\beta)\gamma} = f_0\sqrt{\frac{(1-\beta)}{(1+\beta)}}$$

$f' < f_0 \Rightarrow$ red–shift

$$(c\tau)^2 = (c\Delta t)^2 - (\Delta x)^2$$
$$(c\tau)^2 < (c\Delta t)^2$$
$$\therefore \tau < \Delta t$$
$$\text{or } \Delta t = \gamma\tau$$

For example, consider a muon[3] traveling through the atmosphere at a relativistic velocity. If the lifetime of a stationary muon before it decays is τ (i.e. in the rest frame of the muon), then an observer on Earth measures a lifetime of $t = \gamma\tau$. This means that a moving muon lives longer before it decays than a stationary muon.

[3] A muon is an elementary particle similar to the electron, but has a mass ~200 times greater than the electron. Muons are produced in the atmosphere when cosmic rays (high-energy particles from outer space) hit the atmosphere, and they travel through it at relativistic velocities.

6.3.3 Relativistic Doppler effect

The *Doppler effect* is the apparent change in frequency of a wave due to the relative motion between a source and an observer. If the relative velocity approaches the speed of light, we use the *relativistic* Doppler effect to account for time dilation of the wave's time period. For an observer moving towards the source, the frequency of the wave appears to be higher than the frequency of the emitted wave, f_0:

$$f_B = f_0\sqrt{\frac{(1+\beta)}{(1-\beta)}}$$

(Eq. 6.15 and Derivation 6.4). For visible light, this corresponds to a shift in frequency towards the blue end of the spectrum, therefore, the wave is said to be *blue-shifted*. For an observer moving away from the source, the frequency of the wave appears to be lower than the frequency of the emitted wave:

$$f_R = f_0\sqrt{\frac{(1-\beta)}{(1+\beta)}}$$

(Eq. 6.16 and Derivation 6.4). In this case, the wave is said to be *red-shifted*, since the effect corresponds to a shift towards the red end of the spectrum.

Relativistic Doppler effect (Derivation 6.4):

Towards source (blue-shift):

$$f_B = f_0\sqrt{\frac{(1+\beta)}{(1-\beta)}} \qquad (6.15)$$

Away from source (red-shift):

$$f_R = f_0\sqrt{\frac{(1-\beta)}{(1+\beta)}} \qquad (6.16)$$

$\beta = v/c$ where v is the velocity of the source.

6.3.4 Length contraction revisited

We can perform a similar calculation to the time-dilation calculation to see that, for an object of length l' in a moving frame S′, an observer in the laboratory frame measures a shorter length, $l<l'$. This phenomenon is known as *length contraction*, and the contraction factor is γ (Eq. 6.17 and Derivation 6.5).

Length contraction (Derivation 6.5):

$$l = \frac{l'}{\gamma} \qquad (6.17)$$

In a similar way, if S is the rest frame of the object, its length, as measured by an observer in S′, is *also* contracted. This can be seen by performing the *inverse* Lorentz transformation on l:

$$\begin{aligned} l &= x_2 - x_1 \\ &= \gamma(x_2' + vt_2') - \gamma(x_1' + vt_1') \\ &= \gamma(x_2' - x_1') + \gamma v\underbrace{(t_2' - t_1')}_{=0} \\ &= \gamma l' \\ \therefore l' &= \frac{l}{\gamma} \end{aligned}$$

Therefore, the apparent length in a moving frame is always contracted relative to the rest frame of the object.

Proper length

The *proper length*, l_0, of an object is its length as measured in the rest frame of the object. (If we imag-

Derivation 6.5: Derivation of length contraction (Eq. 6.17)

$$\begin{aligned} l' &= x_2' - x_1' \\ &= \gamma(x_2 - vt) - \gamma(x_1 - vt) \\ &= \gamma(x_2 - x_1) - \gamma v\underbrace{(t_2 - t_1)}_{=0} \\ &= \gamma l \end{aligned}$$

- Starting point: the length of an object in its rest frame S′ is the difference between two arbitrary end points, $l' = x_2' - x_1'$.
- Use the Lorentz transformation to convert to the laboratory frame:

$$x' = \gamma(x - vt)$$

- The corresponding length of l' in the laboratory frame is determined by the coordinates of the corresponding end points of the object at a given time, t, which means that $t_2 - t_1 = \Delta t = 0$.
- Therefore, the length of an object as measured in the laboratory frame appears shorter than its length as measured in its own rest frame, by a factor of $\gamma \Rightarrow$ *length contraction*:

$$l = \frac{l'}{\gamma}$$

ine sitting on the object in question, then we are in the rest frame of that object.) The proper length always corresponds to the longest length of the object; when measured from any frame that's moving relative to the object's rest frame, its length appears shorter than l_0:

$$l = \frac{l_0}{\gamma}$$

For example, if we consider a muon traveling through the atmosphere, the proper length of the atmosphere is that as measured in the laboratory frame, L_{atm} (the atmosphere is not moving relative to Earth). However, in the rest frame of the muon, the atmosphere is traveling towards it at a relativistic velocity, and therefore it appears to be shorter than L_{atm} due to length contraction. Therefore, muons are able to reach the Earth before they decay due to a combination of time dilation of their lifetime and length contraction of the distance that they travel.

6.3.5 Transformation of velocities

The relativistic velocity transformations are found by differentiating the Lorentz transformations (Eqs. 6.18 and 6.19 and Derivation 6.6).

Velocity transformations (Derivation 6.6):

Velocity in moving frame:

$$u' = \frac{u - v}{1 - uv / c^2} \tag{6.18}$$

Velocity in laboratory frame:

$$u = \frac{u' + v}{1 + u'v / c^2} \tag{6.19}$$

where v is the relative frame velocity.

If $u = c$, then from Eq. 6.18:

$$u' = \frac{c - v}{1 - v / c} = c\frac{(c - v)}{(c - v)} = c$$

whereas if $v = c$:

$$u' = \frac{u - c}{1 - u / c} = c\frac{(u - c)}{(c - u)} = -c$$

Derivation 6.6: Derivation of velocity transformations (Eqs. 6.18 and 6.19)

Velocity in moving frame:

$$u' = \frac{dx'}{dt'} = \frac{dx - vdt}{dt - vdx/c^2} = \frac{u - v}{1 - uv/c^2}$$

Velocity in laboratory frame:

$$u = \frac{dx}{dt} = \frac{dx' + vdt'}{dt' + vdx'/c^2} = \frac{u' + v}{1 + u'v/c^2}$$

- Starting point: the velocity of an object in a moving frame S′ is:

$$u' = \frac{dx'}{dt'}$$

- Differentiate the Lorentz transformation:

$$x' = \gamma(x - vt)$$
$$\therefore dx' = \gamma(dx - vdt)$$
$$\text{and } t' = \gamma\left(t - \frac{vx}{c^2}\right)$$
$$\therefore dt' = \gamma\left(dt - \frac{vdx}{c^2}\right)$$

and substitute into u'. The corresponding velocity of the object in the laboratory frame is $u = dx/dt$.
- Use the inverse Lorentz transformations to determine the velocity of an object in the laboratory frame, u, in terms of its velocity in the moving frame, u', and the frame velocity, v.

Similarly from Eq. 6.19, $u = c$ if either $u' = c$ or if $v = c$. Therefore, no object can travel faster than the speed of light.

Low-velocity limit

When either the object or the frame velocity are much less than c, or their product is such that $uv \ll c$ or $u'v \ll c$, we recover the Galilean transformations:

$$u' = u - v$$
$$\text{and } u = u' + v$$

Therefore, relativistic theory approximates to classical theory in the low-velocity limit.

6.4 Relativistic energy and momentum

In Newtonian mechanics, the kinetic energy and momentum of a particle are:

$$T = \frac{1}{2}mv^2$$
$$\text{and } p = mv$$

From Newton's Second Law, the momentum of a body increases if an external force is applied to it. Therefore, classically, we would expect there to be no upper limit to the momentum of a body since there is no upper limit to the theoretical applied force. However, since the velocity of a body cannot exceed the speed of light, then there *is* an upper limit to the momentum of a body. Therefore, at relativistic speeds, the expressions for kinetic energy and momentum are modified from their classical values to incorporate the requirement that a body's velocity must be $v < c$.

Invariant interval

We saw in Section 6.3.2 that the space–time interval for a proper time separation $d\tau$ in the rest frame of an event is:

$$(c\,d\tau)^2 = (c\,dt)^2 - (dx)^2 \tag{6.20}$$

where dt is the time separation and dx is the spatial separation in the laboratory frame. Rearranging, we have the expression for time dilation:

$$d\tau^2 = dt^2\left(1 - \frac{(dx/dt)^2}{c^2}\right)$$
$$= dt^2\underbrace{\left(1 - v^2/c^2\right)}_{=1/\gamma^2}$$
$$\therefore d\tau = \frac{dt}{\gamma}$$

Using this in Eq. 6.20, we have:

$$c^2 = \left(c\frac{dt}{d\tau}\right)^2 - \left(\frac{dx}{d\tau}\right)^2$$
$$= (\gamma c)^2 - \left(\gamma\frac{dx}{dt}\right)^2$$
$$= (\gamma c)^2 - (\gamma v)^2$$

Multiplying this expression through by m^2c^2, where m is the rest mass of a body (i.e. the mass of a body in its rest frame), we can express the invariant interval in terms of the rest mass, energy, and momentum of a body:

$$\underbrace{\left(mc^2\right)^2}_{\text{rest-mass term}} = \underbrace{\left(\gamma mc^2\right)^2}_{\text{energy term}} - \underbrace{(\gamma mvc)^2}_{\text{momentum term}}$$

Below, we will look at each term in turn.

6.4.1 Mass–energy equivalence

Mass–energy equivalence is the concept that the mass of a body is equivalent to its energy content. This concept is described by Einstein's famous equation:

$$E = mc^2$$

When expressed in natural units, whereby $c = 1$, the expression becomes $E = m$, which plainly illustrates mass–energy equivalence.

Rest mass and relativistic mass

In the rest frame of a particle, m is its *rest mass* and $E_0 = mc^2$ is its *rest-mass energy* (Eq. 6.21). In a frame where the particle is *not* stationary relative to an observer, its mass appears greater by a factor γ. The *relativistic mass* of a moving particle is defined as:

$$m_\gamma = \gamma m$$

The total energy of a relativistic particle in this frame is then:

$$E = \gamma mc^2$$

(Eq. 6.22).

Relativistic momentum

The momentum of a relativistic particle is:

$$p = \gamma mv$$

Or, letting $\beta = v/c$, we have $p = \gamma\beta mc$ (Eq. 6.23). In the low-velocity limit, the first term in v corresponds to the expression for classical momentum:

$$\begin{aligned} p &= \left(\frac{1}{\sqrt{1-(v/c)^2}}\right) mv \\ &= \left(\left(1-(v/c)^2\right)^{-1/2}\right) mv \\ &\simeq \left(1+\frac{(v/c)^2}{2}+\cdots\right) mv \\ &\simeq mv \quad \text{for } v \ll c \end{aligned}$$

Relativistic kinetic energy

The kinetic energy of a relativistic particle is the difference between its total energy and its rest-mass energy:

$$\begin{aligned} T &= E - E_0 \\ &= \gamma mc^2 - mc^2 \\ &= (\gamma - 1)mc^2 \end{aligned}$$

We arrive at the same expression if we calculate the work required to accelerate a particle to a relativistic velocity, as shown in Derivation 6.7 (Eq. 6.24 and Derivation 6.7). In the low-velocity limit, we recover the expression for classical kinetic energy:

$$\begin{aligned} T &= \left(\frac{1}{\sqrt{1-(v/c)^2}} - 1\right) mc^2 \\ &= \left(\left(1-(v/c)^2\right)^{-1/2} - 1\right) mc^2 \\ &\simeq \left(1+\frac{(v/c)^2}{2}+\cdots-1\right) mc^2 \\ &\simeq \frac{1}{2}mv^2 \quad \text{for } v \ll c \end{aligned}$$

Relativistic energy and momentum:

Rest-mass energy:

$$E_0 = mc^2 \tag{6.21}$$

Total energy:

$$E = \gamma mc^2 \tag{6.22}$$

Momentum:

$$\begin{aligned} p &= \gamma mv \\ &= \gamma\beta mc \end{aligned} \tag{6.23}$$

Kinetic energy (Derivation 6.7):

$$\begin{aligned} T &= E - E_0 \\ &= (\gamma - 1)mc^2 \end{aligned} \tag{6.24}$$

where m is the rest mass of the body.

Velocity of a relativistic particle

Using Eqs. 6.22 and 6.23, some useful identities for finding the velocity (γ and β) of a particle are:

$$\gamma = \frac{E}{mc^2} \tag{6.25}$$

$$\gamma\beta = \frac{pc}{mc^2} \tag{6.26}$$

$$\therefore \beta = \frac{pc}{E} \tag{6.27}$$

6.5 Lorentz transformations for *E* and *p* in one dimension

The total energy and momentum of a particle depend on its Lorentz factor (i.e. on its velocity). Therefore, energy and momentum are not invariant quantities—their value depends on the frame of reference from which we measure them. We transform from one frame to another using the same transformation matrix as for x and t transformations: the Lorentz transformation. To transform energy and momentum from the laboratory frame S to a moving frame S′, we therefore have:

$$\begin{pmatrix} cp' \\ E' \end{pmatrix} = \gamma_0 \begin{pmatrix} 1 & -\beta_0 \\ -\beta_0 & 1 \end{pmatrix} \begin{pmatrix} cp \\ E \end{pmatrix}$$

(Eq. 6.28). We write the momentum as the product cp instead of just p, since this has units of energy, and then the transformation matrix is symmetric. Similarly,

Derivation 6.7: Derivation of the relativistic kinetic energy (Eq. 6.24)

$$
\begin{aligned}
T &= \int_0^x F \mathrm{d}x' \\
&= \int_0^t \frac{\mathrm{d}(\gamma' m v')}{\mathrm{d}t'} v' \mathrm{d}t' \\
&= \int_0^t \left(\gamma' m \frac{\mathrm{d}v'}{\mathrm{d}t'} + m v' \frac{\mathrm{d}\gamma'}{\mathrm{d}t'} \right) v' \mathrm{d}t' \\
&= \int_{\substack{v'=0 \\ \gamma'=1}}^{\substack{v \\ \gamma}} \left(\gamma' m v' \mathrm{d}v' + m v'^2 \mathrm{d}\gamma' \right) \\
&= \int_1^{\gamma} \left(\gamma' m c^2 \frac{\mathrm{d}\gamma'}{\gamma'^3} + m c^2 \left(1 - \frac{1}{\gamma'^2} \right) \mathrm{d}\gamma' \right) \\
&= \int_1^{\gamma} m c^2 \mathrm{d}\gamma' \\
&= (\gamma - 1) m c^2
\end{aligned}
$$

The primed variables are to distinguish them from the integration limits.

- Starting point: the kinetic energy of a particle results from the work done on it during acceleration:

$$
\begin{aligned}
T = W &= \int F \mathrm{d}x \\
&= \int F v \mathrm{d}t
\end{aligned}
$$

  where $v = \mathrm{d}x/\mathrm{d}t$.
- From Newton's Second Law, the external force required to accelerate a body is equal to the rate of change of its momentum:

$$F = \frac{\mathrm{d}p}{\mathrm{d}t}$$

  where $p = \gamma m v$ for relativistic velocities. The rest mass m is constant.
- The Lorentz factor is given by:

$$\gamma = \frac{1}{\sqrt{1 - (v/c)^2}}$$

  Rearranging γ and differentiating, we find:

$$
\begin{aligned}
&\left(\frac{v}{c} \right)^2 = 1 - \frac{1}{\gamma^2} \\
&\therefore \frac{2v\mathrm{d}v}{c^2} = \frac{2\mathrm{d}\gamma}{\gamma^3}
\end{aligned}
$$

  Use these expressions to eliminate v^2 and $v\mathrm{d}v$ from the expression for T.
- Integrating over γ, we see that:

$$T = (\gamma - 1) m c^2 = E - E_0$$

to transform from a moving frame to the laboratory frame, we use the inverse transformation:

$$\begin{pmatrix} cp \\ E \end{pmatrix} = \gamma_0 \begin{pmatrix} 1 & \beta_0 \\ \beta_0 & 1 \end{pmatrix} \begin{pmatrix} cp' \\ E' \end{pmatrix}$$

(Eq. 6.29). The Lorentz factor and beta factor used in the transformation, γ_0 and β_0 respectively, correspond to those of the moving *frame*. The energy and momentum of a particle in the laboratory frame are then $E = \gamma m c^2$ and $p = \gamma m v$, and in the moving frame are $E' = \gamma' m c^2$ and $p' = \gamma' m v'$, where γ, γ', and γ_0 correspond to three different Lorentz factors.

Lorentz transformation for *E* and *p*:

Moving frame, S′:

$$\begin{pmatrix} cp' \\ E' \end{pmatrix} = \gamma_0 \begin{pmatrix} 1 & -\beta_0 \\ -\beta_0 & 1 \end{pmatrix} \begin{pmatrix} cp \\ E \end{pmatrix} \tag{6.28}$$

Laboratory frame, S:
(inverse transformation)

$$\begin{pmatrix} cp \\ E \end{pmatrix} = \gamma_0 \begin{pmatrix} 1 & \beta_0 \\ \beta_0 & 1 \end{pmatrix} \begin{pmatrix} cp' \\ E' \end{pmatrix} \tag{6.29}$$

The transformations in Eqs. 6.28 and 6.29 are for motion in one dimension. Momenta transverse to the direction of motion are invariant. For example, for motion along the x-axis, the transverse momenta are:

$$p_{y'} = p_y$$
$$\text{and } p_{z'} = p_z$$

6.5.1 Rest-mass-energy invariant

In space–time coordinates, we found that the quantity:

$$(\Delta s)^2 = (c\Delta t)^2 - (\Delta x)^2$$

is an invariant quantity (it takes the same value in all inertial frames). Therefore, through symmetry between the E–p and the x–t transformation, we can infer that the quantity $(c\Delta p)^2 - (\Delta E)^2$ is also invariant (refer to Derivation 6.2).

Using $E = \gamma mc^2$ and $p = \gamma mv$, we find that the quantity $E^2 - (pc)^2$ is equal to the square of the rest-mass energy of a particle:

$$\begin{aligned} E^2 - (pc)^2 &= \left(\gamma mc^2\right)^2 - (\gamma mvc)^2 \\ &= (\gamma m)^2 c^2 \left(c^2 - v^2\right) \\ &= \left(mc^2\right)^2 \gamma^2 \left(1 - \beta^2\right) \\ &= \left(mc^2\right)^2 \end{aligned}$$

Therefore, the rest-mass energy of a particle is an invariant quantity (Eq. 6.30).

Rest-mass-energy invariant:

$$\begin{aligned} (mc^2)^2 &= E^2 - (pc)^2 \\ &= \text{invariant} \end{aligned} \qquad (6.30)$$

6.5.2 The center-of-mass frame

The center-of-mass frame (or center-of-momentum frame)[4] is defined as the frame in which the total momentum of all particles is zero. Below we will determine the Lorentz factor and beta factor of the center-of-mass frame for both a single particle and for a system of N particles.

[4] The center-of-mass frame is a special case of a center-of-momentum frame, in which the center of mass is stationary at the origin. However, the choice of origin is usually arbitrary, therefore, we can often define a coordinate system such that this is true.

For a single particle

For a single particle, the center-of-mass frame corresponds to the *rest frame* of the particle, i.e. the frame in which the particle is stationary. (If we imagine sitting on a particle, then we are in the rest frame of that particle.) In the rest frame of a particle of mass m, the particle momentum is $p' = 0$ and its rest-mass energy is $E' = mc^2$:

$$\therefore \begin{pmatrix} cp' \\ E' \end{pmatrix} = \begin{pmatrix} 0 \\ mc^2 \end{pmatrix}$$

The Lorentz transformation from the laboratory frame to the particle's rest frame is therefore:

$$\begin{pmatrix} 0 \\ mc^2 \end{pmatrix} = \gamma_0 \begin{pmatrix} 1 & -\beta_0 \\ -\beta_0 & 1 \end{pmatrix} \begin{pmatrix} cp \\ E \end{pmatrix}$$

where γ_0 and β_0 are the Lorentz factor and beta factor of the rest frame relative to the laboratory frame. Expanding the transformation, we have:

$$0 = \gamma_0 cp - \beta_0 \gamma_0 E \qquad (6.31)$$
$$\text{and } mc^2 = -\beta_0 \gamma_0 cp + \gamma_0 E \qquad (6.32)$$

Rearranging Eq. 6.31, we find an expression for the frame's beta factor:

$$\beta_0 = \frac{cp}{E}$$

(Eq. 6.33). Using this together with Eq. 6.32, we find an expression for the frame's Lorentz factor:

$$\begin{aligned} mc^2 &= -\beta_0^2 \gamma_0 E + \gamma_0 E \\ &= \gamma_0 E \left(1 - \beta_0^2\right) \\ &= \frac{E}{\gamma_0} \\ \therefore \gamma_0 &= \frac{E}{mc^2} \end{aligned}$$

(Eq. 6.34). Comparing these expressions to Eqs. 6.25 and 6.27, we see that the center-of-mass frame γ and β factors (γ_0 and β_0) correspond to those measured in the laboratory frame (γ and β), as expected.

Single-particle center-of-mass frame velocity:

$$\beta_0 = \frac{cp}{E} \tag{6.33}$$

$$\gamma_0 = \frac{E}{mc^2} \tag{6.34}$$

where m is the rest mass of the body.

For a system of particles

For a system of N particles, the total momentum and total energy of the system are the sum of momenta and energies of individual particles:

$$p_{\text{tot}} = \sum_{i=1}^{N} p_i = \sum_{i=1}^{N} \gamma_i m_i v_i$$

$$E_{\text{tot}} = \sum_{i=1}^{N} E_i = \sum_{i=1}^{N} \gamma_i m_i c^2$$

The Lorentz transformation of the total energy and momentum is then:

$$\begin{pmatrix} cp'_{\text{tot}} \\ E'_{\text{tot}} \end{pmatrix} = \gamma_0 \begin{pmatrix} 1 & -\beta_0 \\ -\beta_0 & 1 \end{pmatrix} \begin{pmatrix} cp_{\text{tot}} \\ cp_{\text{tot}} \end{pmatrix}$$

or

$$\begin{pmatrix} \sum cp'_i \\ \sum E'_i \end{pmatrix} = \gamma_0 \begin{pmatrix} 1 & -\beta_0 \\ -\beta_0 & 1 \end{pmatrix} \begin{pmatrix} \sum cp_i \\ \sum E_i \end{pmatrix}$$

In the center-of-mass frame, the total momentum equals zero and the total energy equals the center-of-mass energy, E_{cm} (the sum of individual particle energies in the center-of-mass frame):

$$\therefore \begin{pmatrix} \sum cp'_i \\ \sum E'_i \end{pmatrix} = \begin{pmatrix} 0 \\ E_{\text{cm}} \end{pmatrix}$$

The Lorentz transformation from the laboratory frame to the center-of-mass frame is therefore:

$$\begin{pmatrix} 0 \\ E_{\text{cm}} \end{pmatrix} = \gamma_{\text{cm}} \begin{pmatrix} 1 & -\beta_{\text{cm}} \\ -\beta_{\text{cm}} & 1 \end{pmatrix} \begin{pmatrix} \sum cp_i \\ \sum E_i \end{pmatrix}$$

Expanding the transformation, we have:

$$0 = \gamma_{\text{cm}} \sum cp_i - \beta_{\text{cm}} \gamma_{\text{cm}} \sum E_i$$
$$\text{and } E_{\text{cm}} = -\beta_{\text{cm}} \gamma_{\text{cm}} \sum cp_i + \gamma_{\text{cm}} \sum E_i$$

Rearranging these equations, we find expressions for the center-of-mass frame's velocity (β_{cm} and γ_{cm}):

$$\beta_{\text{cm}} = \frac{\sum cp_i}{\sum E_i}$$
$$\text{and } \gamma_{\text{cm}} = \frac{\sum E_i}{E_{\text{cm}}}$$

(Eqs. 6.35 and 6.36).

Several particle center-of-mass frame velocity:

$$\beta_{\text{cm}} = \frac{\sum cp_i}{\sum E_i} \tag{6.35}$$

$$\gamma_{\text{cm}} = \frac{\sum E_i}{E_{\text{cm}}} \tag{6.36}$$

where $p_i = \gamma_i m_i v_i$ and $E_i = \gamma_i m_i c^2$.

Center-of-mass energy and invariance

The invariant quantity for a system of several particles is:

$$\text{invariant} = \left(\sum E_i\right)^2 - \left(\sum p_i c\right)^2$$

In the center-of-mass frame, $\sum E'_i = E_{\text{cm}}$ and $\sum p'_i c = 0$. Therefore, the invariant quantity is:

$$\text{invariant} = E_{\text{cm}}^2$$

Said another way, the invariant quantity corresponds to the center-of-mass energy:

$$\therefore E_{\text{cm}}^2 = \left(\sum E_i\right)^2 - \left(\sum p_i c\right)^2$$

(Eq. 6.37). For a two-particle system, expanding this equation gives:

$$
\begin{aligned}
E_{\text{cm}}^2 &= \left(\sum E_i\right)^2 - \left(\sum cp_i\right)^2 \\
&= (E_1 + E_2)^2 - (cp_1 + cp_2)^2 \\
&= \underbrace{E_1^2 - (cp_1)^2}_{=(m_1c^2)^2} + \underbrace{E_2^2 - (cp_2)^2}_{=(m_2c^2)^2} \\
&\quad + 2E_1E_2 - 2cp_1cp_2 \\
&= (m_1c^2)^2 + (m_2c^2)^2 + 2E_1E_2(1 - \beta_1\beta_2)
\end{aligned}
$$

where the beta factor of the ith particle is:

$$\beta_i = \frac{cp_i}{E_i} = \frac{v_i}{c}$$

Center-of-mass energy and effective mass

If all particles are *stationary* in the center-of-mass frame, then E_{cm} equals the sum of their rest-mass energies:

$$
\begin{aligned}
E_{\text{cm}} &= m_1c^2 + m_2c^2 + \cdots m_Nc^2 \\
&= \sum_{i=1}^{N} m_ic^2 \\
&= M_{\text{tot}}c^2
\end{aligned}
$$

In this case, the center-of-mass frame is equivalent to a single body of mass $M_{\text{tot}} = \Sigma m_i$ at rest at the center of mass. If the particles are not stationary in the center-of-mass frame, but are traveling in directions such that their momenta cancel, then E_{cm} also includes the kinetic energy of the individual particles. In this case, E_{cm} is greater than the sum of individual rest-mass energies. E_{cm} is sometimes referred to as the *effective mass*, M_{eff}, of the system since it corresponds to a single particle with effective rest-mass energy of:

$$E_{\text{cm}} = M_{\text{eff}}c^2$$

(Eq. 6.38). In terms of the invariant quantity, we therefore have:

$$(M_{\text{eff}}c^2)^2 = (E_{\text{tot}})^2 - (p_{\text{tot}}c)^2$$

where E_{tot} and p_{tot} are the total energy and momenta of individual particles in the laboratory frame. Worked example 6.2 shows how to use E_{cm} to determine the threshold mass of a particle produced in a high-energy electron–positron collision.

Center-of-mass energy:

$$E_{\text{cm}}^2 = \left(\sum E_i\right)^2 - \left(\sum p_ic\right)^2 \quad (6.37)$$

and $E_{\text{cm}} = M_{\text{eff}}c^2$ (6.38)

where M_{eff} is the effective mass of the system.

6.5.3 Photons and Compton scattering

Photons have zero mass and therefore have zero rest-mass energy:

$$(mc^2)^2 = E^2 - (pc)^2 = 0$$

The energy and momentum of a photon are therefore related by:

$$E = pc$$

Compton scattering is the scattering of an X-ray or a γ-ray off an electron. During the collision, some of the incident photon's energy is imparted to the electron, resulting in a decrease in the scattered photon's energy and therefore an increase in its wavelength. By considering energy and momentum conservation before and after the collision, we find that the photon energies are related by:

$$\frac{1}{E'} - \frac{1}{E} = \frac{(1 - \cos\theta)}{m_ec^2}$$

where E' is the scattered photon energy, E is the incident photon energy, θ is the scattering angle, and m_e is the electron mass [$9.11 \cdot 10^{-31}$ kg]. (Eq. 6.39 and Derivation 6.8). Using $E = hf = hc/\lambda$, where h is *Planck's constant* [$6.63 \cdot 10^{-34}$ Js], we find that this decrease in photon energy corresponds to an increase in photon wavelength of:

$$\Delta\lambda = \lambda' - \lambda = \frac{h(1 - \cos\theta)}{m_ec}$$

(Eq. 6.40).

WORKED EXAMPLE 6.1: Electron–positron collision (Eqs. 6.37 and 6.38)

An electron and a positron, each of mass m_e, collide to produce a stationary particle. Determine the threshold mass of the new particle (its minimum mass) in terms of the energies of the colliding particles. Assume $E_e \gg m_e c^2$.

Before collision:

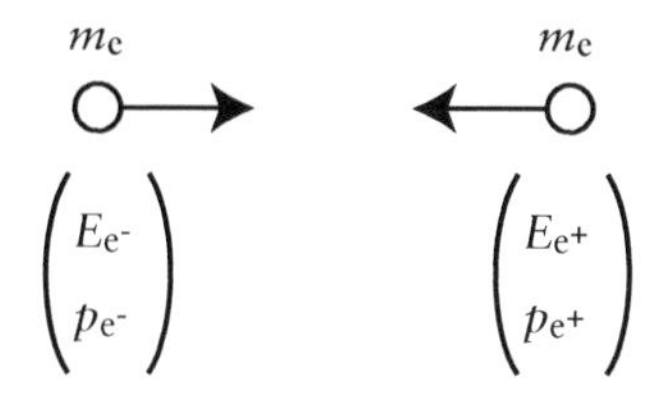

After collision:

m_0 $\begin{pmatrix} m_0 c^2 \\ 0 \end{pmatrix}$

The threshold mass of a stationary particle produced in a collision is:

$$E_{cm} = m_0 c^2$$

The invariant quantity is $(E_{cm})^2 = (\Sigma E_i)^2 - (\Sigma p_i c)^2$. Assuming the energy and momentum of the electron and positron are much greater than their rest-mass energy, we have:

$$\begin{aligned}
(E_{cm})^2 &= \left(\sum E_i\right)^2 - \left(\sum p_i c\right)^2 \\
&= \left(E_{e^-} + E_{e^+}\right)^2 - \left(cp_{e^-} - cp_{e^+}\right)^2 \\
&= E_{e^-}^2 + E_{e^+}^2 + 2E_{e^-}E_{e^+} - \left(c^2 p_{e^-}^2 + c^2 p_{e^+}^2 - 2c^2 p_{e^-} p_{e^+}\right) \\
&= \underbrace{E_{e^-}^2 - (cp_{e^-})^2}_{=(m_e c^2)^2} + \underbrace{E_{e^+}^2 - (cp_{e^+})^2}_{=(m_e c^2)^2} + 2E_{e^-}E_{e^+} + 2c^2 p_{e^-} p_{e^+} \\
&\simeq 2E_{e^-}E_{e^+} + 2c^2 p_{e^-} p_{e^+} \\
&\simeq 4E_{e^-}E_{e^+}
\end{aligned}$$

where the last step assumes $E_e \simeq p_e c$ for $E_e \gg m_e c^2$. Therefore, the threshold mass of a stationary particle produced in the collision is:

$$\begin{aligned}
4E_{e^-}E_{e^+} &\simeq \left(m_0 c^2\right)^2 \\
\therefore m_0 c^2 &= 2\sqrt{E_{e^-}E_{e^+}}
\end{aligned}$$

Compton scattering (Derivation 6.8):

$$\frac{1}{E'} - \frac{1}{E} = \frac{(1-\cos\theta)}{m_e c^2} \quad (6.39)$$

$$\therefore \lambda' - \lambda = \frac{h(1-\cos\theta)}{m_e c} \quad (6.40)$$

The energy lost by the photon is transferred to the scattering electron, giving it a recoil kinetic energy of:

$$\begin{aligned}
T_e &= E - E' \\
&= E - \frac{m_e c^2 E}{(1-\cos\theta)E + m_e c^2} \\
&= \frac{(1-\cos\theta)E^2}{(1-\cos\theta)E + m_e c^2}
\end{aligned}$$

This is a maximum when the photon is scattered backwards, i.e. when $\theta = \pi$:

$$\therefore T_e^{max} = \frac{2E^2}{2E + m_e c^2}$$

Derivation 6.8: Derivation of Compton scattering photon energies (Eq. 6.39)

A photon (γ-ray) of energy E_γ and momentum $\vec{p}_\gamma$ is incident on a stationary electron, of rest-mass energy $m_e c^2$. The photon scatters off the electron at a scattering angle θ. The energy and momentum of the scattered photon are E'_γ and $\vec{p}'_\gamma$, respectively, and those of the recoil electron are E'_e and $\vec{p}'_e$.

Conservation of momentum:

$$\vec{p}_\gamma = \vec{p}'_\gamma + \vec{p}'_e$$

$$\begin{aligned}(p'_e)^2 &= \vec{p}'_e \cdot \vec{p}'_e \\ &= (\vec{p}_\gamma - \vec{p}'_\gamma)\cdot(\vec{p}_\gamma - \vec{p}'_\gamma) \\ &= p_\gamma^2 + p'^2_\gamma - 2p_\gamma p'_\gamma \cos\theta \\ &= \frac{1}{c^2}\left(E_\gamma^2 + E'^2_\gamma - 2E_\gamma E'_\gamma \cos\theta\right)\end{aligned}$$

Before scattering

$$\begin{pmatrix} E_\gamma \\ \vec{p}_\gamma \end{pmatrix} \quad \begin{pmatrix} m_e c^2 \\ 0 \end{pmatrix}$$

m_e

Conservation of energy:

$$E_\gamma + m_e c^2 = E'_\gamma + E'_e$$

$$\begin{aligned}\therefore (E'_e)^2 &= \left(m_e c^2 + E_\gamma - E'_\gamma\right)^2 \\ &= \left(m_e c^2\right)^2 + \left(E_\gamma - E'_\gamma\right)^2 \\ &\quad + 2m_e c^2\left(E_\gamma - E'_\gamma\right) \\ &= \left(m_e c^2\right)^2 + \left(E_\gamma^2 + E'^2_\gamma - 2E_\gamma E'_\gamma\right) \\ &\quad + 2m_e c^2\left(E_\gamma - E'_\gamma\right)\end{aligned}$$

After scattering

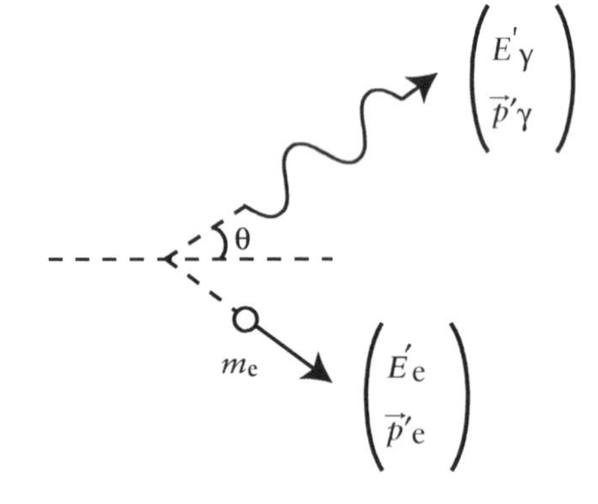

Rest-mass energy of electron:

$$\begin{aligned}\left(m_e c^2\right)^2 &= (E'_e)^2 - (cp'_e)^2 \\ &= \left(m_e c^2\right)^2 + \left(E_\gamma^2 + E'^2_\gamma - 2E_\gamma E'_\gamma\right) + 2m_e c^2\left(E_\gamma - E'_\gamma\right) - \left(E_\gamma^2 + E'^2_\gamma - 2E_\gamma E'_\gamma \cos\theta\right) \\ &= \left(m_e c^2\right)^2 - 2E_\gamma E'_\gamma(1-\cos\theta) + 2m_e c^2\left(E_\gamma - E'_\gamma\right)\end{aligned}$$

Compton scattering formula:

$$\therefore m_e c^2\left(E_\gamma - E'_\gamma\right) = E_\gamma E'_\gamma(1-\cos\theta)$$

$$\therefore \frac{1}{E'_\gamma} - \frac{1}{E_\gamma} = \frac{(1-\cos\theta)}{m_e c^2}$$

Part II
Electromagnetism

Introduction

Electromagnetism is the study of the behavior and properties of electric and magnetic fields and electromagnetic waves, and their interactions with matter. Stationary electric charge produces a constant electric field, the study of which is called *electrostatics* (Chapter 7); steady current produces a constant magnetic field, the study of which is called *magnetostatics* (Chapter 8). We can apply electrostatic and magnetostatic concepts to determine the properties of electrical circuits (Chapter 9). Things get more interesting when electric and magnetic fields are not constant in time; a changing magnetic field produces an electric field, and a changing electric field produces a magnetic field. These phenomena constitute the subject of *electromagnetic induction* (Chapter 10). All properties of electric and magnetic fields are encapsulated elegantly in the four fundamental equations of electromagnetism: Maxwell's equations. From these four equations, we can understand and predict the properties of electric and magnetic fields in matter (Chapter 11), as well as properties of electromagnetic waves in both a vacuum and in matter (Chapter 12).

Practical applications

The electromagnetic force is one of the four fundamental forces of nature.[1] It holds atoms and materials together, and it governs many of the phenomena we encounter in our day-to-day lives. The electromagnetic spectrum alone provides countless applications—TV, radio, microwave, visible light, X-rays, etc. Electromagnetism provides us with electricity to power homes and industries. And of course natural phenomena, such as lightning and the magnetic field of the Earth, are electromagnetic.

My two cents

The difficulty of electromagnetism for a new student is precisely that it is so *well* understood. In particular, there are so many different ways to solve most problems or to prove a given relation that it seems like there is an endless amount of physics to know. It *is* a vast subject—there are countless applications of electromagnetic theory—but it's not as exhaustive as we're led to believe. For example, electric dipoles do not constitute *new* laws of physics—they are simply applying what we know about two point charges in close proximity.

In fact, electromagnetism is often thought of as one of the most elegant and concise disciplines in physics, since the spatial and temporal properties of electromagnetic fields and waves are all contained within Maxwell's equations. However, to understand what these equations represent, we first need to understand the static behavior of electric and magnetic fields, and how charged particles and currents produce them. Therefore, in an introductory electromagnetism course, we first cover electrostatics and magnetostatics, and Maxwell's equations usually constitute the end rather than the means. So be patient—with more familiarity, things will eventually start falling into place, and you'll see that there aren't a seemingly endless number of facts to know; there are just four!

Recommended books

- Fleisch, D., 2008, *A Student's Guide to Maxwell's Equations*. Cambridge University Press.
- Griffiths, D., 1999, *Introduction to Electrodynamics*, 3rd edition. Prentice Hall.
- Lorrain, P., Corson, D., and Lorrain, F., 1988, *Electromagnetic Fields and Waves*, 3rd edition. Freeman.

There are a number of good electromagnetism books, but the ones listed above are my personal favorites. The book by Griffiths is especially good for an introduction to the subject; it's written in a dynamic and very clear way. The book by Lorrain *et al.* also covers the introductory material, but is good for a more advanced discussion of electromagnetic fields and waves in matter. The book by Fleisch was published after I finished my undergraduate studies, but it provides a short yet in depth description, focusing just on Maxwell's equations. It's very good for a clear and conceptual overview of the subject.

[1] The other three are gravity, the weak force and the strong force.

Chapter 7 Electrostatics

Electrostatics is the study of systems with stationary electric charges. These produce electric fields whose properties remains constant in time. In this chapter, we will look at force and energy in electrostatic systems, and at the properties of static electric fields.

7.1 Electric charge

Electric charge, Q [C], is an inherent property of particles. A stationary distribution of electric charge gives rise to electrostatic phenomena. The *elementary charge*, e [$1.602 \cdot 10^{-19}$ C], is a fundamental constant, where the charge of a proton is $Q_p = +e$ and the charge of an electron is $Q_e = -e$. Neutrons[1] have zero electric charge; therefore, the charge of a neutron is $Q_n = 0$. Electrically neutral atoms have an equal number of protons and electrons, whereas charged atoms (ions) do not. An object becomes charged when it has an imbalance of positive and negative charges, meaning it has an imbalance of protons and electrons.

Charge distributions

A *point charge*, denoted q, is an idealized charged object that has negligible physical size. We refer to any system larger than a single point charge as a *charge distribution*. These are usually classified as either *discrete* or *continuous* distributions. A discrete distribution is modeled as a collection of individual point charges (Fig. 7.1a), whereas a continuous distribution is modeled as charge spread continuously over a region (Fig. 7.1b). The total charge of a discrete distribution is the sum of individual point charges:

$$Q = \sum_i q_i$$

For a continuous distribution, we split the system into elements of charge, dq, and treat these elements as point charges. The total charge is then found by integrating over the elements dq:

$$Q = \int dq$$

Charge densities

It's convenient to express the element dq in terms of a *charge density*. A *line charge*, λ [Cm^{-1}], is a charge per unit length:

$$\lambda = \frac{dq}{dl}$$

(Eq. 7.1); a *surface charge*, σ [Cm^{-2}], is a charge per unit area:

$$\sigma = \frac{dq}{ds}$$

[1] Note that the electric charge of quarks, which are the fundamental particles that make up protons and neutrons, is a fraction of e : quarks have a charge of either $+2e/3$ or $-e/3$. Protons and neutrons are each composed of three quarks, where $Q_p = 2 \cdot (2e/3) - (e/3) = e$ and $Q_n = (2e/3) - 2 \cdot (e/3) = 0$.

a) Discrete distribution:

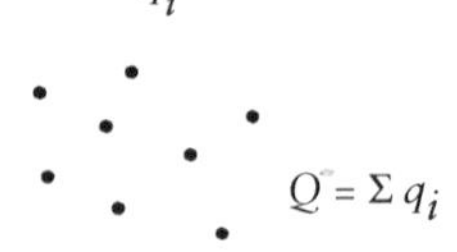

b) Continuous distributions:

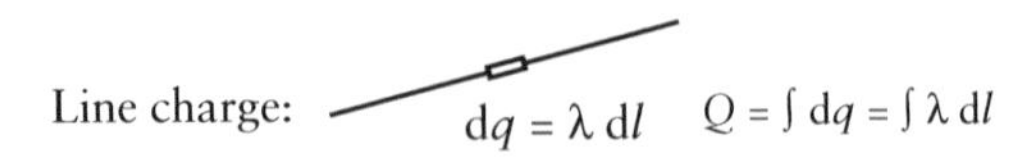

Surface charge: $dq = \sigma\, ds$ $Q = \int dq = \int \sigma\, ds$

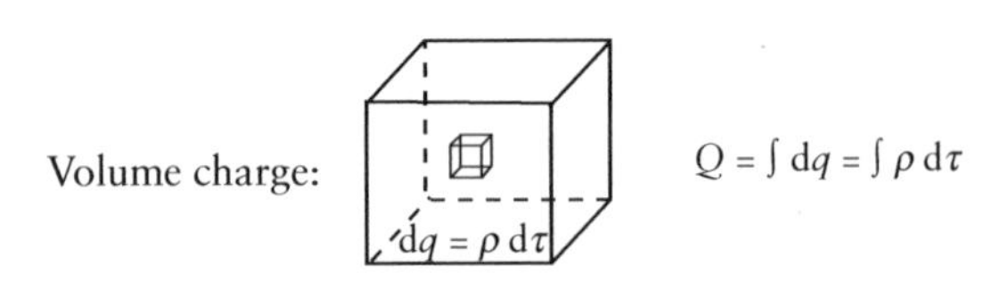

Figure 7.1: a) Discrete charge distribution. The total charge is the sum of individual point charges. b) Continuous charge distributions. The total charge is found by integrating over elements of charge, dq.

Charge densities (refer to Fig. 7.1):

Line charge: $$\lambda = \frac{dq}{dl} \tag{7.1}$$

Surface charge: $$\sigma = \frac{dq}{ds} \tag{7.2}$$

Volume charge: $$\rho = \frac{dq}{d\tau} \tag{7.3}$$

Element of charge and total charge:

$$dq = \begin{cases} \lambda\, dl & \text{in one dimension (1D)} \\ \sigma\, ds & \text{in two dimensions (2D)} \\ \rho\, d\tau & \text{in three dimensions (3D)} \end{cases} \tag{7.4}$$

$$\therefore Q = \int dq = \begin{cases} \int \lambda dl & \text{in 1D} \\ \int \sigma ds & \text{in 2D} \\ \int \rho d\tau & \text{in 3D} \end{cases} \tag{7.5}$$

(Eq. 7.2); and a *volume charge*, ρ [$\mathrm{Cm^{-3}}$], is a charge per unit volume:

$$\rho = \frac{dq}{d\tau}$$

(Eq. 7.3). These definitions allow us to express the total charge as an integral over space, rather than charge, which is particularly useful in calculations, as we shall see in later examples (Eqs. 7.4 and 7.5).

7.1.1 Coulomb's Law

A stationary charge exerts an *electrostatic force* on another charge. Consider two point charges, q_1 and q_2, a distance r_{12} apart, where the vector $\vec{r}_{12}$ points from q_1 to q_2: $\vec{r}_{12} = \vec{r}_2 - \vec{r}_1$ (Fig. 7.2). The force exerted on charge q_2 by q_1 is found to obey an inverse square law:

$$\vec{F}_{12} = k\frac{q_1 q_2}{r_{12}^2}\hat{r}_{12}$$

where k is a constant. Similarly, the force on q_1 by q_2 is:

$$\begin{aligned} \vec{F}_{21} &= k\frac{q_1 q_2}{r_{21}^2}\hat{r}_{21} \\ &= -k\frac{q_1 q_2}{r_{12}^2}\hat{r}_{12} \\ &= -\vec{F}_{12} \end{aligned}$$

This result is, of course, Newton's Third Law. The constant, k, is:

$$k = \frac{1}{4\pi\varepsilon_0}$$

where ε_0 [$8.85 \cdot 10^{-12}\ \mathrm{Fm^{-1}}$] is a fundamental electric parameter called the *permittivity of free space*.

The expression for $\vec{F}_{12}$ is known as *Coulomb's Law* (Eq. 7.6), and it was determined empirically by Charles Coulomb in the 1780s. It shows that (i) the direction of

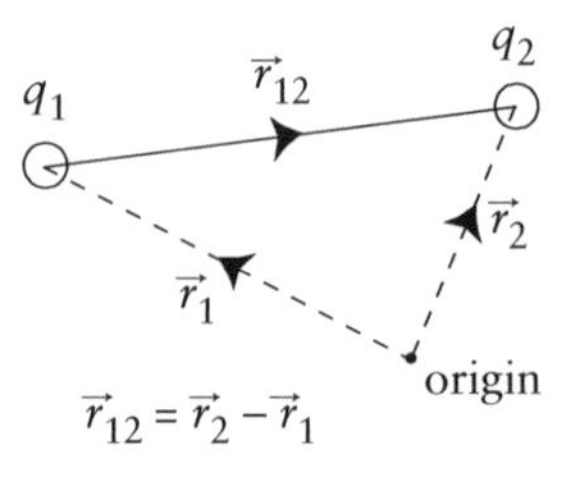

Figure 7.2: Definitions of charges and position vectors for Coulomb's Law.

Coulomb's Law (refer to Fig. 7.2):

$$\vec{F}_{12} = \frac{q_1 q_2}{4\pi\varepsilon_0 r_{12}^2}\hat{r}_{12} \quad (7.6)$$

where $\vec{r}_{12} = \vec{r}_2 - \vec{r}_1$

the electrostatic force on a charge by another is along the line joining the charges ($\vec{F}_{12} \propto \hat{r}_{12}$) and (ii) for any two charges of given magnitude, the strength of the force depends only on the distance between them, and obeys an inverse square law ($|\vec{F}_{12}| \propto 1/r_{12}^2$). Two charges of the same sign repel ($\vec{F}_{12} \propto \hat{r}_{12}$), whereas two charges of opposite sign attract ($\vec{F}_{12} \propto -\hat{r}_{12}$).

Comparison with gravity

Coulomb's Law has a similar form to Newton's Law of gravity, which gives the gravitational force, $\vec{F}_G$, exerted on a mass m_2 by a mass m_1, a distance r_{12} apart:

$$\vec{F}_G = -\frac{Gm_1 m_2}{r_{12}^2}\hat{r}_{12}$$

The strength of the force is characterized by the *gravitational constant*, G [$6.67 \cdot 10^{-11}\ \mathrm{m^3\,kg^{-1}\,s^{-2}}$]. Since mass is always positive, the minus sign in the expression for $\vec{F}_G$ means that the gravitational force is always attractive. If we calculate the ratio of the magnitudes of the gravitational force to the electrostatic force, $\vec{F}_E$, given by Coulomb's Law, for two protons ($m = m_p$ and $q = +e$), we find that gravity is approximately 36 orders of magnitude weaker than the electrostatic force:

$$\left|\frac{\vec{F}_G}{\vec{F}_E}\right| = 4\pi\varepsilon_0 G\left(\frac{m_p}{e}\right)^2 \sim 10^{-36}$$

This means that we can usually safely neglect gravitational effects in many electrostatic calculations.

The principle of superposition

What is the electrostatic force on a charge q_1 that's in the vicinity of *several* other charges, q_2, q_3, q_4? It was found experimentally that the force on q_1 is the vector sum of the individual Coulomb forces on that charge due to each other charge. That is to say, the total force on q_1 is:

$$\vec{F}_1^{\mathrm{tot}} = \vec{F}_{21} + \vec{F}_{31} + \vec{F}_{41} + \ldots$$

This type of linear addition of individual components to determine the total applies to many so-called *linear systems*, and is referred to as the *principle of superposition*.[2] The total force on a charge Q for a discrete charge distribution is therefore given by the *sum* of the individual Coulomb interactions over charges q_i:

$$\begin{aligned}\vec{F}_Q &= \vec{F}_{1Q} + \vec{F}_{2Q} + \vec{F}_{3Q} + \ldots \\ &= \frac{Qq_1}{4\pi\varepsilon_0 r_{1Q}^2}\hat{r}_{1Q} + \frac{Qq_2}{4\pi\varepsilon_0 r_{2Q}^2}\hat{r}_{2Q} + \ldots \\ &= \sum_i \frac{Qq_i}{4\pi\varepsilon_0 r_{iQ}^2}\hat{r}_{iQ}\end{aligned}$$

(Eq. 7.7 and Fig. 7.3a). For a continuous charge distribution, the sum becomes an *integral* over elements of charge dq (Eq. 7.8 and Fig. 7.3b).

Net electrostatic force on a charge Q (Fig. 7.3):

Discrete distribution:

$$\vec{F}_Q = \sum_i \frac{Qq_i}{4\pi\varepsilon_0 r_{iQ}^2}\hat{r}_{iQ} \quad (7.7)$$

where $\vec{r}_{iQ} = \vec{r}_Q - \vec{r}_i$

Continuous distribution:

$$\vec{F}_Q = \int \frac{Q\,dq}{4\pi\varepsilon_0 r_{dqQ}^2}\hat{r}_{dqQ} \quad (7.8)$$

where $\vec{r}_{dqQ} = \vec{r}_Q - \vec{r}_{dq}$

(see Eq. 7.4 for expressions for dq).

7.1.2 The electric field

Electric field, $\vec{E}$ [$\mathrm{NC^{-1}}$ or $\mathrm{Vm^{-1}}$], is a fundamental property of electric charge. It is defined as the electrostatic force on a test charge Q, in the limit that Q tends to zero:

$$\vec{E} \overset{\mathrm{def}}{=} \lim_{Q\to 0}\frac{\vec{F}}{Q}$$

Another perhaps more intuitive way of defining $\vec{E}$ is as the force that a unit test charge $Q = 1\mathrm{C}$ would experience if placed in the vicinity of another charge:

$$\vec{E} = \frac{\vec{F}}{Q}$$

[2] For example, the principle of superposition also holds for the interference of coherent waves (the net amplitude of two interfering waves equals the sum of their individual amplitudes).

a) Discrete distribution:

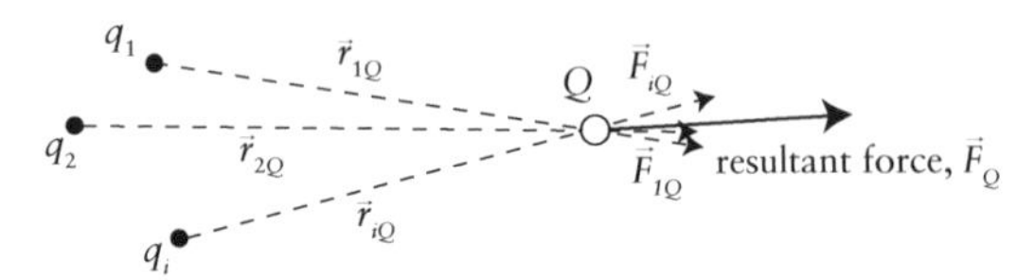

(Charges $q_1, q_2 \ldots q_i$ and Q have position vectors $\vec{r}_1, \vec{r}_2 \ldots \vec{r}_i$ and $\vec{r}_Q$, respectively, therefore $\vec{r}_{iQ} = \vec{r}_Q - \vec{r}_i$)

b) Continuous distribution:

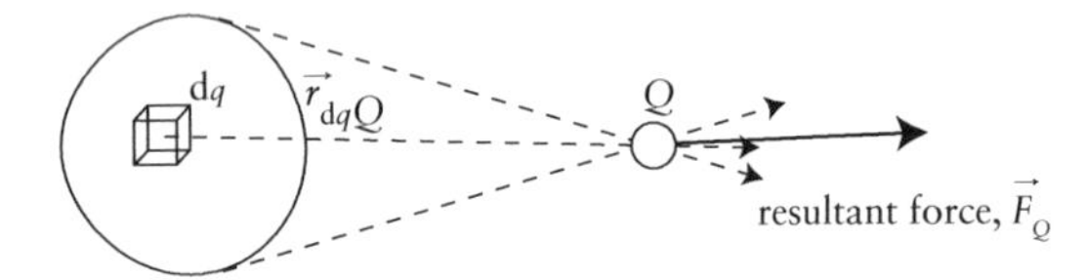

(Elements of charge dq and charge Q have position vectors $\vec{r}_{dq}$ and $\vec{r}_Q$, respectively, therefore $\vec{r}_{dqQ} = \vec{r}_Q - \vec{r}_{dq}$)

Figure 7.3: a) Discrete charge distribution. The total force is the vector sum of individual forces (Eq. 7.7). b) Continuous charge distribution. The total force is found by integrating over elements of force (Eq. 7.8).

However, the former definition shows that $\vec{E}$ is defined even in the absence of a test charge with which to experience $\vec{F}$. Furthermore, the limit $Q \to 0$ ensures that the presence of the test charge Q does not alter the charge distribution, such that the electric field in the presence of Q is the same as that in the absence of Q.

For a charge q placed at the origin, the Coulomb force experienced by a test charge Q a radial distance r from q is:

$$\vec{F}_{qQ} = \frac{qQ}{4\pi\varepsilon_0 r^2}\hat{r}$$

The electric field of q as a function of $\vec{r}$ is then:

$$\vec{E}(\vec{r}) = \left(\frac{1}{Q}\right) \cdot \left(\frac{qQ}{4\pi\varepsilon_0 r^2}\hat{r}\right)$$
$$= \frac{q}{4\pi\varepsilon_0 r^2}\hat{r}$$

Thus, electric field depends only on the magnitude of the charge producing the field, q, and its value varies as a function of distance from q, r.

Aside: choice of terminology for position vectors

Many variables in electromagnetism are functions of *position*. Position is usually denoted by a position vector, $\vec{r}$, where the value of $\vec{r}$ is defined with respect to a given origin, which is usually arbitrary for any given system. As we shall see in the following sections, we must distinguish between three different position vectors: (i) those that correspond to the position vector describing properties of a *field*, (ii) those that correspond to the position vector of a charge *source* producing a field (such as a charge q_i or an element of charge dq), and (iii) those that correspond to the separation vector between the source and the field it produces.

Unfortunately there is little consistency across textbooks, and a good terminology is not well defined. The book by Griffiths has an intuitive notation for distinguishing between field and source coordinates, whereby field coordinates correspond to unprimed symbols, $\vec{r}$, and source coordinates correspond to primed symbols, $\vec{r}'$. The separation between field and source coordinates is then represented by the *separation vector*, $\mathscr{r} = \vec{r} - \vec{r}'$. While this is a good notation, we have chosen not to adopt it for the following reasons: (i) we have reserved primed coordinates to distinguish integration limits from their counterparts within an integral (see for example, Eq. 7.20; the integration limits correspond to unprimed variables while the variables in the integral are primed); (ii) the symbol for the separation vector, $\mathscr{r}$, can be difficult to reproduce by hand, and could easily be mistaken for $\vec{r}$.

With this in mind, we have chosen to use subscripts together with the position vector $\vec{r}$ to distinguish between field, source, and separation coordinates (refer to Fig. 7.4).

- **Field coordinates** are represented simply by $\vec{r}$. This corresponds to the position vector of the point of interest from the origin (which is usually arbitrary).
- **Source coordinates** are represented by a single subscript. The position vector of a charge q is $\vec{r}_q$, that of a charge Q is $\vec{r}_Q$, that of a charge q_i is $\vec{r}_i$, and that of an element of charge, dq, is $\vec{r}_{dq}$.
- **Separation coordinates** are represented by a double subscript, where the first subscript corresponds to the initial position vector and the second to the final position vector (Fig. 7.4). For example, the separation vector that points from a charge q to Q is $\vec{r}_{qQ} = \vec{r}_Q - \vec{r}_q$, one that points from a charge q_1 to q_2 is $\vec{r}_{12} = \vec{r}_2 - \vec{r}_1$ and one that points from an element of charge dq to Q is $\vec{r}_{dqQ} = \vec{r}_Q - \vec{r}_{dq}$ (Fig. 7.4a). For a separation vector that points from a source coordinate to a *field* coordinate, we use a slightly non-standard notation: the separation vector is still represented by a double subscript, but now the second subscript is just "r" to represent position vector $\vec{r}$. For example, the separation vector from charge q to a position $\vec{r}$ is $\vec{r}_{qr} = \vec{r} - \vec{r}_q$, etc. (Fig. 7.4b). Using this terminology, if q is located at the origin, then $\vec{r}_q = 0$ and $\vec{r}_{qr} = \vec{r}$.

a) Charge to charge (source → source)

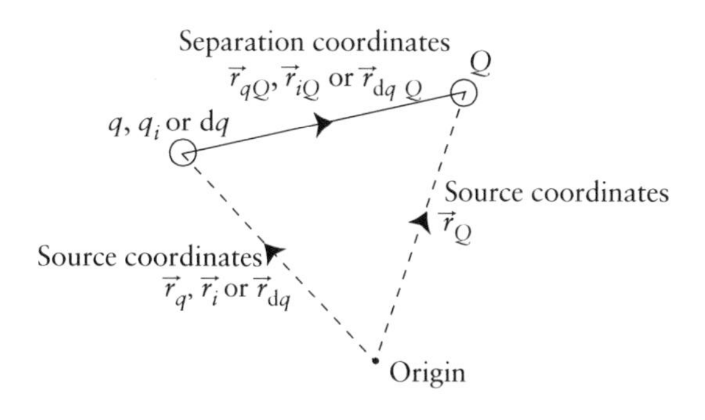

$\vec{r}_{xQ} = \vec{r}_Q - \vec{r}_x$, where $x = q, i$ or dq

b) Charge to point P (source → field)

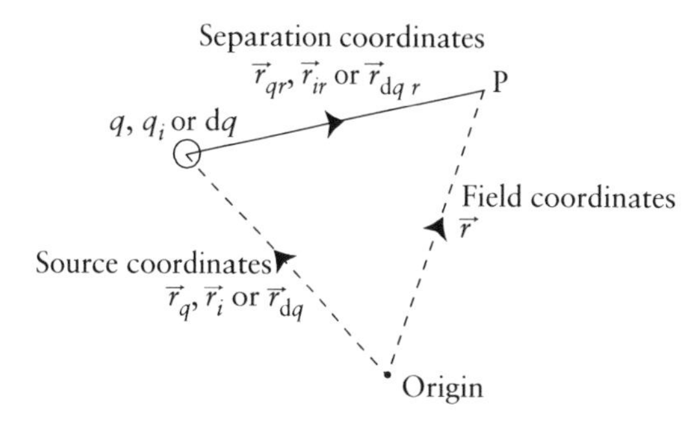

$\vec{r}_{xr} = \vec{r} - \vec{r}_x$, where $x = q, i$ or dq

Figure 7.4: Terminology for position vectors. Field coordinates are denoted by $\vec{r}$ and source coordinates are denoted by a single subscript, where the subscript corresponds to the charge, for example, $\vec{r}_q$ for a charge q or $\vec{r}_i$ for a charge q_i. Separation vectors are denoted by a double subscript, where the first subscript corresponds to the initial position vector and the second to the final position vector. a) Separation vector from a source coordinate to another source coordinate and b) separation vector from a source coordinate to a field coordinate.

Electric field of a point charge

We saw above that the electric field of a point charge q located at the origin is:

$$\vec{E}(\vec{r}) = \frac{q}{4\pi\varepsilon_0 r^2}\hat{r}$$

For a point charge *not* located at the origin, but located instead at $\vec{r}_q$, the electric field as a function of $\vec{r}$ is now:

$$\vec{E}(\vec{r}) = \frac{q}{4\pi\varepsilon_0 r_{qr}^2}\hat{r}_{qr}$$

where $\vec{r}_{qr} = \vec{r} - \vec{r}_q$

(Eq. 7.9).

The principle of superposition

Like the Coulomb force, the electric field also obeys the principle of superposition. Let's first consider a discrete distribution of point charges, q_1, q_2, q_3, etc. located at position vectors $\vec{r}_1$, $\vec{r}_2$, $\vec{r}_3$, etc. The electric field produced by each point charge separately is:

$$\vec{E}_1(\vec{r}) = \frac{q_1}{4\pi\varepsilon_0 r_{1r}^2}\hat{r}_{1r}$$

where $\vec{r}_{1r} = \vec{r} - \vec{r}_1$

$$\vec{E}_2(\vec{r}) = \frac{q_2}{4\pi\varepsilon_0 r_{2r}^2}\hat{r}_{2r}$$

where $\vec{r}_{2r} = \vec{r} - \vec{r}_2$

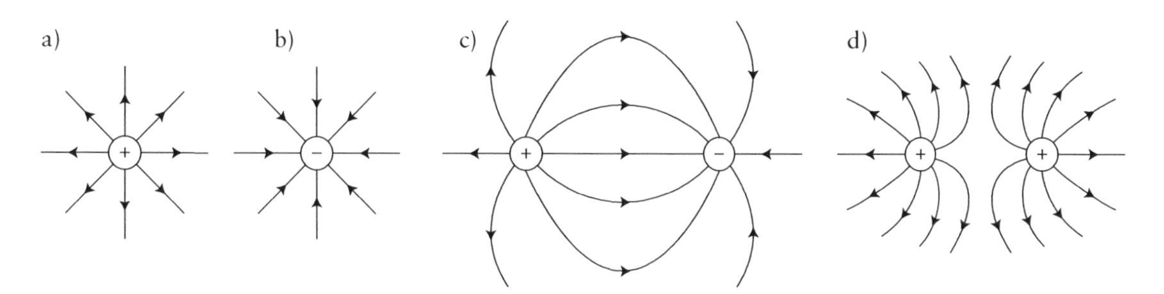

Figure 7.5: Electric field lines for a) a positive point charge, b) a negative point charge, c) a pair of point charges of opposite sign (attractive), and d) a pair of point charges of the same sign (repulsive).

and so on. The net electric field produced by the distribution is then:

$$\vec{E}_{\text{tot}}(\vec{r}) = \vec{E}_1(\vec{r}) + \vec{E}_2(\vec{r}) + \vec{E}_3(\vec{r}) + \ldots$$
$$= \frac{q_1}{4\pi\varepsilon_0 r_{1r}^2}\hat{r}_{1r} + \frac{q_2}{4\pi\varepsilon_0 r_{2r}^2}\hat{r}_{2r} + \ldots$$
$$= \sum_i \frac{q_i}{4\pi\varepsilon_0 r_{ir}^2}\hat{r}_{ir}$$
where $\vec{r}_{ir} = \vec{r} - \vec{r}_i$

(Eq. 7.10). For a continuous distribution, the sum becomes an integral over elements of charge dq, where an element dq is at position vector $\vec{r}_{dq}$ (Eq. 7.11). Worked example 7.1 shows how to use Eq. 7.11 to determine the electric field produced by an infinite planar surface charge.

Electric field at position $\vec{r}$ (refer to diagram below for definitions of position vectors):

Point charge at $\vec{r}_q$:

$$\vec{E}(\vec{r}) = \frac{q}{4\pi\varepsilon_0 r_{qr}^2}\hat{r}_{qr} \quad (7.9)$$

where $\vec{r}_{qr} = \vec{r} - \vec{r}_q$

and $r_{qr} \stackrel{\text{def}}{=} |\vec{r}_{qr}|$

(for a point charge at the origin, $\vec{r}_q = 0$ $\therefore$ $\vec{r}_{qr} = \vec{r}$).

Discrete distribution
(the ith charge has position vector $\vec{r}_i$):

$$\vec{E}(\vec{r}) = \sum_i \frac{q_i}{4\pi\varepsilon_0 r_{ir}^2}\hat{r}_{ir} \quad (7.10)$$

where $\vec{r}_{ir} = \vec{r} - \vec{r}_i$

and $r_{ir} \stackrel{\text{def}}{=} |\vec{r}_{ir}|$

Continuous distribution
(element of charge dq has position vector $\vec{r}_{dq}$):

$$\vec{E}(\vec{r}) = \int \frac{dq}{4\pi\varepsilon_0 r_{dqr}^2}\hat{r}_{dqr} \quad (7.11)$$

where $\vec{r}_{dqr} = \vec{r} - \vec{r}_{dq}$

and $r_{dqr} \stackrel{\text{def}}{=} |\vec{r}_{dqr}|$

(see Eq. 7.4 for expressions for dq).

Definitions of respective position vectors, for electric field at point P (P has position vector $\vec{r}$ with respect to the origin):

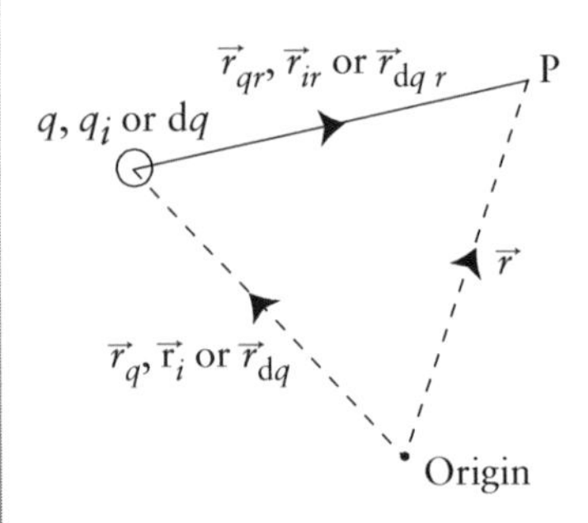

$\vec{r}_{xr} = \vec{r} - \vec{r}_x$, where $x = q$, i or dq

Electric field lines

From the expressions for $\vec{E}$ (Eqs. 7.9–7.11), we see that a vector indicating the magnitude and direction of electric field can be associated with each point in space. This means that electric fields are *vector fields*, and can thus be represented schematically by *field lines*. The direction of field lines represents the direction of the electric field, and the density of field lines represents its strength. From Eq. 7.9, the field lines of a point charge are radially outward for a positive charge or radially inward for a negative charge, and the strength falls off with the inverse square law, $1/r^2$ (Fig. 7.5a and b for a positive and negative point charge, respectively). Field lines begin on positive charges and end on negative ones (they can also begin and end at infinity). From the superposition principle, the combined field lines of two or more nearby point charges are given by the vector sum of the individual field lines of each component, at each point in space. For example, the field lines of two opposite charges sum to result in field lines illustrated in Fig. 7.5c, and those of two like charges cancel to result in field lines illustrated in Fig. 7.5d. These figures illustrate that opposite charges attract (their field lines "join together"), whereas like charges repel. Field lines never cross, since this would imply that an electric field would have two different directions at a single point in space.

Electric flux

Electric flux, Φ_E [N C^{-1} m^2], is a scalar quantity that gives a measure of the electric field across a given area. It can be thought of as being proportional to the number of field lines passing through a surface. It is defined

WORKED EXAMPLE 7.1: Electric field of an infinite surface charge using Eq. 7.11

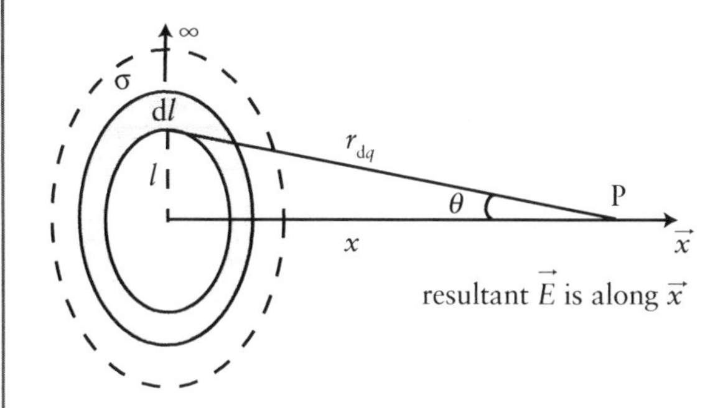

To determine the electric field at a point P, resolve $\vec{E}$ into its components parallel and perpendicular to $\vec{x}$. Doing this, we find that the perpendicular components cancel, and the parallel component is:

$$E_x = \vec{E}\cdot\hat{x} = \left|\vec{E}\right|\cos\theta = \left|\vec{E}\right|\frac{x}{r_{dq}}$$

Using the integral from Eq. 7.11, the x-component of the electric field is then:

$$E_x = \int \frac{dq}{4\pi\varepsilon_0 r_{dq}^2}\cos\theta$$

where, at a perpendicular radius l along the surface (refer to diagram):

$$\begin{aligned} dq &= \sigma ds \\ &= \sigma 2\pi l\, dl \end{aligned}$$

and $r_{dq}^2 = l^2 + x^2$

Use the expressions for dq and r_{dq}^2 in the integral for E_x; to solve the integral, let $m = l^2 + x^2$:

$$\begin{aligned} E_x &= \int \frac{dq}{4\pi\varepsilon_0 r_{dq}^2}\cos\theta \\ &= \int \frac{\sigma ds}{4\pi\varepsilon_0 r_{dq}^2}\cos\theta \\ &= \int \frac{\sigma x ds}{4\pi\varepsilon_0 r_{dq}^3} \\ &= \frac{2\pi\sigma x}{4\pi\varepsilon_0}\int_0^\infty \frac{l dl}{(l^2+x^2)^{3/2}} \\ &= \frac{\sigma x}{2\varepsilon_0}\int_{x^2}^\infty \frac{dm}{2m^{3/2}} \\ &= \frac{\sigma x}{2\varepsilon_0}\left[\frac{-1}{\sqrt{m}}\right]_{x^2}^\infty \\ &= \frac{\sigma}{2\varepsilon_0} \end{aligned}$$

$$\therefore \vec{E} = \frac{\sigma}{2\varepsilon_0}\hat{x}$$

Therefore, the electric field produced by an infinite surface charge is uniform (it does not vary with distance from the surface), and is perpendicular to the surface.

as the perpendicular component of electric field over a surface, multiplied by the surface area. For an electric field $\vec{E}$ passing through an element of area $d\vec{s}$, the element of electric flux is then:

$$d\Phi_E = \vec{E}\cdot d\vec{s}$$

The dot product picks out the perpendicular component of $\vec{E}$ through the element of area, as illustrated in Fig. 7.6. The total electric flux through a surface S is then found by integrating over all elements of flux:

$$\Phi_E = \int d\Phi_E = \int_S \vec{E}\cdot d\vec{s}$$

(Eq. 7.12). Intuitively, flux can be thought of as being analogous to the interaction of wind and the sail of a boat, where the interaction is greatest when the wind is perpendicular to the sail.

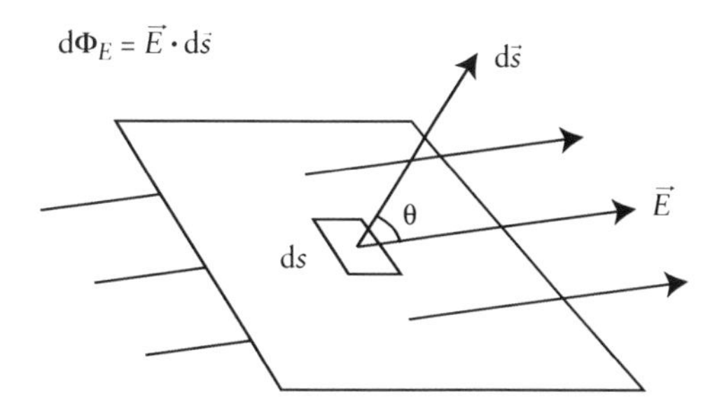

Figure 7.6: Electric flux. $d\vec{s}$ is the vector normal to the surface element ds with magnitude equal to the area of the element, and θ is the angle between $\vec{E}$ and $d\vec{s}$.

We shall see below that the electric flux through a closed surface is proportional to the total charge enclosed within it. This statement is the basis of Gauss's Law, which is covered in the following section.

Electric flux:

$$\Phi_E = \int_S \vec{E} \cdot d\vec{s} \tag{7.12}$$

7.2 Divergence and curl of $\vec{E}$

Electric field is a vector field; this means that a vector that gives the magnitude and direction of the electric field is associated with each point in space. Two quantities that characterize vector fields are their *divergence* and their *curl*. The divergence of a vector field provides a measure of the sources and sinks of the field (i.e. electric charges for an electric field), whereas the curl describes the rotation of the field about a given axis. The divergence and curl of $\vec{E}$ are important quantities in electromagnetism, since they correspond to two out of four of *Maxwell's equations*. (The other two correspond to the divergence and curl of the magnetic field, $\vec{B}$, covered in Section 8.4.) Below, we will determine the divergence and curl of $\vec{E}$.

7.2.1 Gauss's Law

To start with, let's determine the electric flux through a closed surface S:

$$\Phi_E = \oint_S \vec{E} \cdot d\vec{s}$$

Using the expression for $\vec{E}$ from Eq. 7.11 in Φ_E, as shown in Derivation 7.1, we find that Φ_E is proportional to the charge enclosed within the surface:

$$\Phi_E = \oint_S \vec{E} \cdot d\vec{s} = \frac{Q_{enc}}{\varepsilon_0}$$

(Eq. 7.13 and Derivation 7.1). This is an important equation; it is one of Maxwell's four equations, and is known as *Gauss's Law*. Eq. 7.13 holds true for all closed surfaces and for all charge distributions. By invoking the divergence theorem, we can express Gauss's Law as the divergence of $\vec{E}$ (Eq. 7.14 and Derivation 7.2). Thus, Gauss's Law has both an integral and a differential form, as do the other three Maxwell equations. Since the divergence of a vector field represents its sources and sinks, Eq. 7.14 illustrates that electric charge densities provide the sources and sinks of an electric field, as expected.

Gauss's Law (Derivations 7.1 and 7.2):

Integral form:

$$\oint_S \vec{E} \cdot d\vec{s} = \frac{Q_{enc}}{\varepsilon_0} \tag{7.13}$$

Differential form:

$$\nabla \cdot \vec{E} = \frac{\rho}{\varepsilon_0} \tag{7.14}$$

Using Gauss's Law

For systems that possess geometrical symmetry, such as spherical, cylindrical, or planar symmetry, we can use the integral form of Gauss's Law to calculate the electric field of a charge distribution from the charge enclosed in a given surface (Worked example 7.2). In such calculations, the symmetrical surface is called a *Gaussian surface* (Eqs. 7.15–7.18).

Common Gaussian surfaces (refer to Worked example 7.2:)

Spherical symmetry:

Concentric sphere, radius r:

$$\oint_S d\vec{s} = 4\pi r^2 \tag{7.15}$$

Cylindrical symmetry:

Coaxial cylinder, radius r, length l:

$$\oint_S d\vec{s} = 2\pi rl \tag{7.16}$$

Planar symmetry:

Gaussian pillbox,* cross-sectional area A:

$$\begin{aligned} &\text{Thick sheet}: \oint_S d\vec{s} = A \\ &\text{Thin sheet}: \oint_S d\vec{s} = 2A \end{aligned} \tag{7.17}$$

*A Gaussian pillbox has a cylindrical shape, and is oriented such that the end disks of area A are perpendicular to electric field lines (refer to Worked example 7.2 for diagram).

Derivation 7.1: Derivation of the integral form of Gauss's Law (Eq. 7.13)

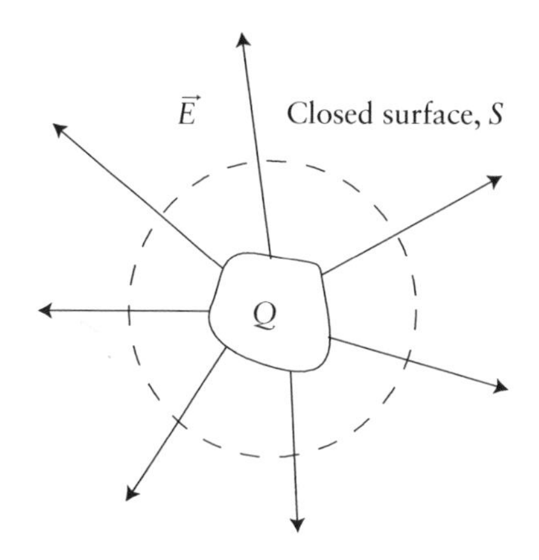

- Starting point: consider a charge distribution centered at the origin, enclosed within a surface S. The electric flux through the closed surface is:

$$\Phi_E = \oint_S \vec{E} \cdot d\vec{s}$$

- The electric field is given by the integral:

$$\vec{E} = \int_Q \frac{dq}{4\pi\varepsilon_0 r^2} \hat{r}$$

(Eq. 7.11), and an element of surface area in spherical polar coordinates is:

$$d\vec{s} = r^2 \sin\theta d\theta d\phi \hat{r} + r \sin\theta dr d\phi \hat{\theta} + r dr d\theta \hat{\phi}$$

- Substitute $\vec{E}$ and $d\vec{s}$ into the expression for Φ_E. Since $\hat{r}$ is orthogonal to $\hat{\theta}$ and $\hat{\phi}$, only the $\hat{r}$ term in $d\vec{s}$ remains in the integral.
- Integrate over all solid angles ($d\Omega = \sin\theta d\theta d\phi$) and charge, where the total charge enclosed within the surface is Q_{enc}.

$$\begin{aligned}
\oint_S \vec{E} \cdot d\vec{s} &= \oint_S \left(\int_Q \frac{dq}{4\pi\varepsilon_0 r^2} \right) \hat{r} \cdot d\vec{s} \\
&= \oint_S \left(\int_Q \frac{dq}{4\pi\varepsilon_0 r^2} \hat{r} \cdot \hat{r} \left(r^2 \sin\theta d\theta d\phi \right) \right) \\
&= \frac{1}{4\pi\varepsilon_0} \int_{q=0}^{Q_{enc}} \int_{\phi=0}^{2\pi} \int_{\theta=0}^{\pi} \sin\theta d\theta d\phi dq \\
&= \frac{2\pi Q_{enc}}{4\pi\varepsilon_0} \left[-\cos\theta \right]_0^{\pi} \\
&= \frac{4\pi Q_{enc}}{4\pi\varepsilon_0} \\
&= \frac{Q_{enc}}{\varepsilon_0}
\end{aligned}$$

The differential form of Gauss's Law is used when we have the inverse problem to solve, namely, we know the electric field and wish to determine the charge density:

$$\rho = \varepsilon_0 \nabla \cdot \vec{E}$$

7.2.2 The curl of $\vec{E}$

We can use Stokes's theorem to find the curl of $\vec{E}$. Stokes's theorem states that the curl of a vector field is related to its closed line integral by:

$$\int_S \left(\nabla \times \vec{E} \right) \cdot d\vec{s} = \oint_L \vec{E} \cdot d\vec{l}$$

where S is the surface enclosed by L. Using the expression for $\vec{E}$ from Eq. 7.11 in Stokes's theorem, as shown in Derivation 7.3, we find that curl of a static electric field is zero (Eq. 7.19 and Derivation 7.3). This property makes intuitive sense, since we know that the field lines of a point charge point radially outward, and thus do not rotate about any axis.

The curl of $\vec{E}$ (Derivation 7.3):

$$\nabla \times \vec{E} = 0 \qquad (7.19)$$

7.3 The electric potential

From Stokes's theorem, the closed line integral around a curl-free vector field is equal to zero:

$$\begin{aligned}
\int_S \left(\nabla \times \vec{E} \right) \cdot d\vec{s} &= \oint_L \vec{E} \cdot d\vec{l} \\
&= 0 \text{ if } \nabla \times \vec{E} = 0
\end{aligned}$$

Derivation 7.2: Derivation of the differential form of Gauss's Law (divergence of E) (Eq. 7.14)

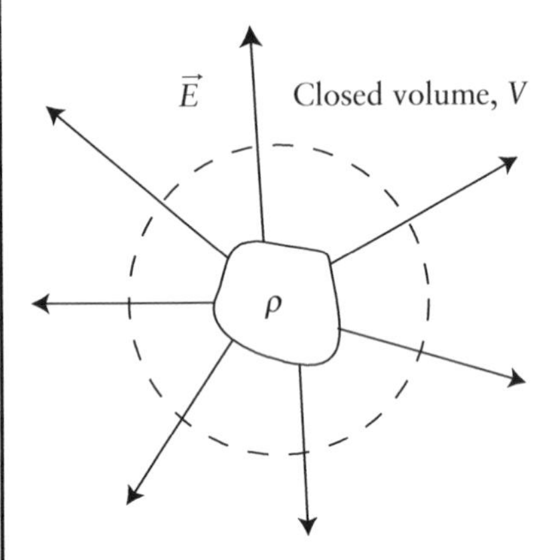

- Starting point: the divergence theorem states that the flux of a vector field through a closed surface is equal to the volume integral of the divergence of the field within that surface:

$$\oint_S \vec{E}\cdot d\vec{s} = \int_V \nabla\cdot\vec{E}\,d\tau$$

- Use Gauss's Law in integral form (Eq. 7.13) to express the surface integral in terms of Q_{enc}, where Q_{enc} is the charge enclosed within a closed surface, S:

$$\oint_S \vec{E}\cdot d\vec{s} = \frac{Q_{enc}}{\varepsilon_0}$$

- Charge and volume charge density are related by:

$$Q = \int_V \rho\,d\tau$$

$$\int_V \nabla\cdot\vec{E}\,d\tau = \oint_S \vec{E}\cdot d\vec{s} = \frac{Q_{enc}}{\varepsilon_0} = \int_V \frac{\rho\,d\tau}{\varepsilon_0}$$

This holds for any volume, so the integrands must be equal:

$$\therefore \nabla\cdot\vec{E} = \frac{\rho}{\varepsilon_0}$$

Therefore, since static electric fields have zero curl:

$$\oint_L \vec{E}\cdot d\vec{l} = 0$$

This means that the integral between any two points in an electric field is path independent, and hence has a unique value (refer to Fig. 7.7):

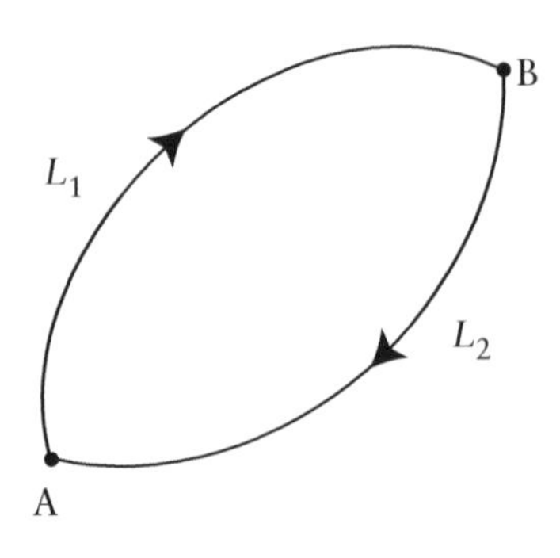

Figure 7.7: Line integral of a closed loop

$$\oint_L \vec{E}\cdot d\vec{l} = \int_A^B \vec{E}\cdot d\vec{l}_{L_1} + \int_B^A \vec{E}\cdot d\vec{l}_{L_2} = 0$$

$$\therefore \int_A^B \vec{E}\cdot d\vec{l}_{L_1} = -\int_B^A \vec{E}\cdot d\vec{l}_{L_2} = \int_A^B \vec{E}\cdot d\vec{l}_{L_2}$$

Therefore, the line integral of $\vec{E}$ from A to B is the same whether we integrate over path L_1 or L_2, and hence is independent of the integration path taken. This property means that we can define a *scalar* field, say $V(\vec{r})$, whose value at a point $\vec{r}$ is equal to the line integral of electric field from a fixed reference point, O, to $\vec{r}$. Expressed mathematically:

$$\int_O^{\vec{r}} \vec{E}\cdot d\vec{l} \overset{\text{def}}{=} V(\vec{r}) - V(O)$$

WORKED EXAMPLE 7.2: Determining $\vec{E}$-fields using Gauss's Law (Eq. 7.13)

The integral form of Gauss's Law is:

$$\oint_S \vec{E}\cdot d\vec{s} = \frac{Q_{enc}}{\varepsilon_0}$$

For systems that possess geometrical symmetry, this form of Gauss's Law allows us to calculate the electric field from just the charge enclosed in a given surface. In the examples below (systems with spherical, cylindrical, and planar symmetry, respectively), we find that (i) $\vec{E}$ and $d\vec{s}$ are either perpendicular or parallel, meaning that we can drop the dot product, and that (ii) the magnitude of $\vec{E}$ is constant over the surface, so it can be taken out of the integral:

$$\Rightarrow |\vec{E}| = \frac{Q_{enc}}{\varepsilon_0 \oint d\vec{s}}$$

The surface over which the integral $\oint d\vec{s}$ is performed is referred to as a *Gaussian surface* (Eqs. 7.15–7.18).

Spherical Gaussian surface
(radius r)

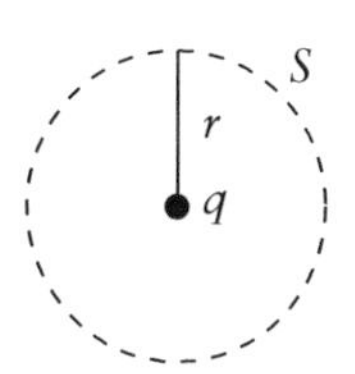

a) Point charge
(spherical Gaussian surface):

$$\oint d\vec{s} = 4\pi r^2$$

$$Q_{enc} = q$$

$$\therefore |\vec{E}| = \frac{Q_{enc}}{\varepsilon_0 \oint d\vec{s}} = \frac{q}{4\pi\varepsilon_0 r^2}$$

Cylindrical Gaussian surface
(radius r, length l)

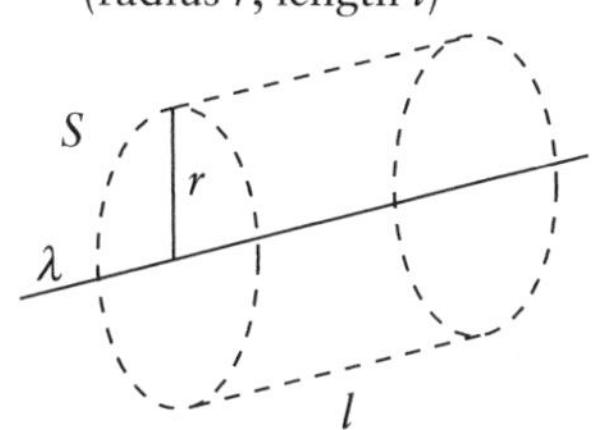

b) Infinite line charge
(cylindrical Gaussian surface):

$$\oint d\vec{s} = 2\pi r l$$

$$Q_{enc} = \lambda l$$

$$\therefore |\vec{E}| = \frac{Q_{enc}}{\varepsilon_0 \oint d\vec{s}} = \frac{\lambda}{2\pi\varepsilon_0 r}$$

Gaussian pillbox
(area A)

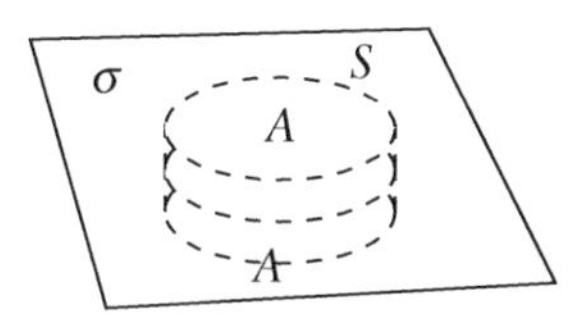

c) Surface charge
(Gaussian pillbox):

$$\oint d\vec{s} = A \text{ or } 2A \text{ (thick or thin)}$$

$$Q_{enc} = \sigma A$$

$$\therefore |\vec{E}| = \frac{Q_{enc}}{\varepsilon_0 \oint d\vec{s}} = \frac{\sigma}{\varepsilon_0} \text{ or } \frac{\sigma}{2\varepsilon_0}$$

Note that the electric field of a surface charge (part c) is the same as that calculated in Worked example 7.1. The direction of the field in a) and b) is radically outward while the direction of the field in c) is perpendicular to the surface.

This scalar field corresponds to the *electric potential*, V [V]. For convenience, the reference point is often taken to be the point at which $V(\text{O})=0$:[3]

$$\therefore \int_{\text{O}}^{\vec{r}} \vec{E}\cdot d\vec{l} \stackrel{\text{def}}{=} -V(\vec{r})$$

(Eq. 7.20). (The minus sign is convention—see *Electric potential and work done* below.) Examples of using Eq. 7.20 to determine the electric potential of a point charge, an infinite line charge, and a surface charge are provided in Worked example 7.3.

Eq. 7.20 expresses V as the integral of $\vec{E}$; can we therefore express $\vec{E}$ as the derivative of V? In Cartesian coordinates, the exact differential (also called the total differential) of V is:

$$dV = \frac{\partial V}{\partial x}dx + \frac{\partial V}{\partial y}dy + \frac{\partial V}{\partial z}dz$$

[3] For most charge distributions, this occurs when $\text{O} \equiv \infty$, but other reference points are also possible (see Worked example 7.3 for examples).

Derivation 7.3: Derivation of the curl of $\vec{E}$ (Eq. 7.19)

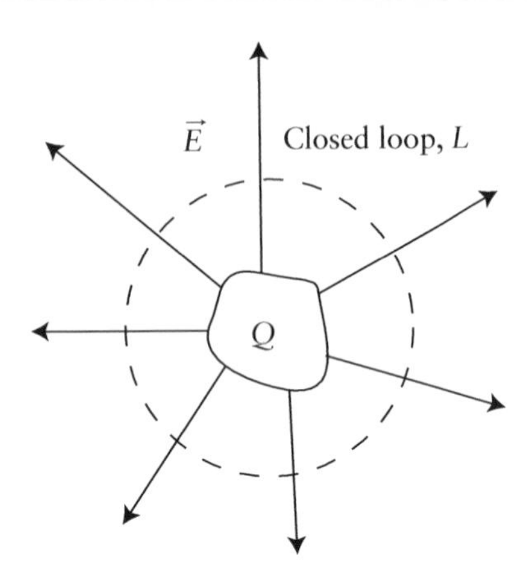

- Starting point: Stokes's theorem states that the curl of a vector field is related to its closed line integral by:

$$\int_S (\nabla \times \vec{E}) \cdot \mathrm{d}\vec{s} = \oint_L \vec{E} \cdot \mathrm{d}\vec{l}$$

where S is the surface enclosed by L.

- The electric field is given by the integral:

$$\vec{E} = \int_Q \frac{\mathrm{d}q}{4\pi\varepsilon_0 r^2}\hat{r}$$

(Eq. 7.11), and a line element in spherical polar coordinates is:

$$\mathrm{d}\vec{l} = \mathrm{d}r\hat{r} + r\mathrm{d}\theta\hat{\theta} + r\sin\theta\mathrm{d}\phi\hat{\phi}$$

- Substitute $\vec{E}$ and $\mathrm{d}\vec{l}$ into Stokes's theorem. Since $\hat{r}$ is orthogonal to $\hat{\theta}$ and $\hat{\phi}$, only the $\hat{r}$ term in $\mathrm{d}\vec{l}$ remains in the integral.
- For a closed line integral, the integration limits are equal.

$$\begin{aligned}
\int_S (\nabla \times \vec{E}) \cdot \mathrm{d}\vec{s} &= \oint_L \vec{E} \cdot \mathrm{d}\vec{l} \\
&= \oint_L \left(\int_Q \frac{\mathrm{d}q}{4\pi\varepsilon_0 r^2} \right) \hat{r} \cdot \mathrm{d}\vec{l} \\
&= \oint_L \int_{q=0}^{Q} \frac{\mathrm{d}q}{4\pi\varepsilon_0 r^2} \hat{r} \cdot \hat{r}\mathrm{d}r \\
&= \oint_L \frac{Q}{4\pi\varepsilon_0 r^2} \mathrm{d}r \\
&= \frac{Q}{4\pi\varepsilon_0} \left[-\frac{1}{r} \right]_a^{b=a} \\
&= \frac{Q}{4\pi\varepsilon_0} \left(-\frac{1}{a} + \frac{1}{a} \right) \\
&= 0
\end{aligned}$$

$$\therefore \nabla \times \vec{E} = 0$$

$$= \underbrace{\begin{pmatrix} \partial/\partial x \\ \partial/\partial y \\ \partial/\partial z \end{pmatrix}}_{=\nabla} V \cdot \underbrace{\begin{pmatrix} \mathrm{d}x \\ \mathrm{d}y \\ \mathrm{d}z \end{pmatrix}}_{=\mathrm{d}\vec{l}}$$

$$= \nabla V \cdot \mathrm{d}\vec{l}$$

(this expression is known as the *gradient theorem*). Additionally, from Eq. 7.20, we have $\mathrm{d}V = -\vec{E}\cdot\mathrm{d}\vec{l}$. Therefore, equating these expressions for $\mathrm{d}V$, we see that $\vec{E}$ indeed can be expressed as the gradient of V:

$$\vec{E} \cdot \mathrm{d}\vec{l} = -\nabla V \cdot \mathrm{d}\vec{l}$$

$$\therefore \vec{E} = -\nabla V$$

$$\text{or } \vec{E}(\vec{r}) = -\frac{\mathrm{d}V(\vec{r})}{\mathrm{d}r}$$

(Eqs. 7.21 and 7.22). The curl of a gradient is equal to zero, therefore the curl of $\vec{E}$ is:

$$\begin{aligned}
\nabla \times \vec{E} &= \nabla \times (-\nabla V) \\
&= 0
\end{aligned}$$

as expected.

WORKED EXAMPLE 7.3: Calculating electric potential from electric field (Eq. 7.20)

a) Point charge:

$$\vec{E}(\vec{r}') = \frac{q}{4\pi\varepsilon_0 (r')^2}\hat{r}'$$

$$V(\vec{r}) = -\int_{\infty}^{r} \vec{E}(\vec{r}')\cdot d\vec{r}' = \frac{q}{4\pi\varepsilon_0 r}$$

b) Infinite line charge:

$$\vec{E}(\vec{r}') = \frac{\lambda}{2\pi\varepsilon_0 r'}\hat{r}'$$

$$V(\vec{r}) = -\int_{O}^{r} \vec{E}(\vec{r}')\cdot d\vec{r}' = V_O - \frac{\lambda \ln r}{2\pi\varepsilon_0}$$

c) Surface charge:

$$\vec{E}(\vec{r}') = \frac{\sigma}{2\varepsilon_0}\hat{n}'$$

$$V(\vec{r}) = -\int_{O}^{l} \vec{E}(\vec{r}')\cdot d\vec{r}' = V_O - \frac{\sigma l}{2\varepsilon_0}$$

Refer to Worked example 7.2 for determining the electric field from Gauss's Law, and for diagram of charge configurations. $\hat{r}'$ in a) and b) corresponds to radial unit vector; $\hat{n}'$ in c) corresponds to unit vector perpendicular to the surface. For a), the variable r corresponds to the radial distance from the point charge q, and the reference point of the integral is at ∞, since this corresponds to the point at which V = 0. For b), the variable r corresponds to the radial distance from a line charge density λ, and the reference point of the integral is O, which corresponds to an arbitrary reference point at which V_O is constant. For c), the variable l corresponds to the perpendicular distance from a surface charge density σ, and the reference point of the integral is O, which corresponds to an arbitrary reference point at which V_O is constant. The primed variables are to distinguish them from the integration limits.

Electric field and potential:

Since $\nabla \times \vec{E}=0$, we can define a scalar potential field, V:

V as the integral of $\vec{E}$:

$$V(\vec{r}) = -\int_{O}^{\vec{r}} \vec{E}(\vec{r}')\cdot d\vec{l}' \quad (7.20)$$

The primed variables are to distinguish them from the integration limits.

$\vec{E}$ as the gradient of V:

$$\vec{E}(\vec{r}) = -\nabla V(\vec{r}) \quad (7.21)$$

$$\text{or } \vec{E}(\vec{r}) = -\frac{dV(\vec{r})}{dr} \quad (7.22)$$

O represents the integration reference point, usually defined as the point at which $V(O) = 0$ (often $O \equiv \infty$).

Electric potential and work done

By considering that electric force and electric field are related by $\vec{F} = q\vec{E}$, we can understand what electric potential represents on a more conceptual basis. Using $\vec{F} = q\vec{E}$ in Eq. 7.20, we have:

$$V(\vec{r}) = -\int_{O}^{\vec{r}} \vec{E}'\cdot d\vec{l}' = -\int_{O}^{\vec{r}} \frac{\vec{F}'}{q}\cdot d\vec{l}' = \frac{W}{q}$$

where W is the work done *against* the Coulomb force. Therefore, the electric potential is a measure of the work done per unit charge in displacing a charge from position O to $\vec{r}$ in an electric field, against electrostatic forces. Using the expression of $\vec{F}$ from Coulomb's Law, we can express $V(\vec{r})$ in terms of q and r for a point charge, a discrete distribution, and a continuous distribution (Eqs. 7.23–7.25 and Derivation 7.4). Worked example 7.4 shows how to use Eq. 7.25 to determine the electric potential of an infinite planar surface charge. Doing this, we find that the electric potential at a perpendicular distance x from the plane is:

$$V = V_O - \frac{\sigma x}{2\varepsilon_0}$$

where V_O corresponds to the potential at an arbitrary reference point. The minus sign indicates that the electric potential decreases the further away from the charged surface. Notice that we get the same expression if we integrate the expression for electric field of an infinite planar surface charge from Worked example 7.1:

WORKED EXAMPLE 7.4: Electric potential of an infinite surface charge using Eq. 7.25

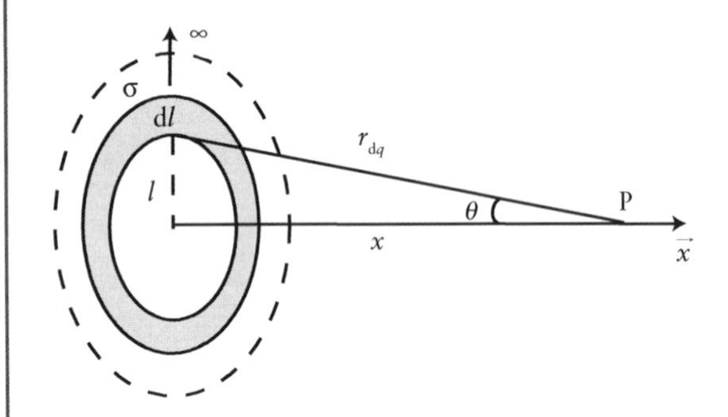

The electric potential at a point P is:

$$V = \int \frac{dq}{4\pi\varepsilon_0 r_{dq}}$$

where, at a perpendicular radius l along the surface (refer to diagram):

$$dq = \sigma ds$$
$$= \sigma 2\pi l dl$$

and $r_{dq} = \sqrt{l^2 + x^2}$

Use the expressions for dq and r_{dq} in the integral for V; to solve the integral, let $m = l^2 + x^2$:

$$V = \int \frac{dq}{4\pi\varepsilon_0 r_{dq}}$$
$$= \int \frac{\sigma ds}{4\pi\varepsilon_0 r_{dq}}$$
$$= \frac{2\pi\sigma}{4\pi\varepsilon_0}\int_0^\infty \frac{l dl}{\sqrt{l^2 + x^2}}$$
$$= \frac{\sigma}{2\varepsilon_0}\int_{x^2}^\infty \frac{dm}{2\sqrt{m}}$$
$$= \frac{\sigma}{2\varepsilon_0}\left[\sqrt{m}\right]_{x^2}^\infty$$
$$= V_O - \frac{\sigma x}{2\varepsilon_0}$$

Therefore, the electric potential produced by an infinite surface charge decreases the further away from the charge (V decreases with increasing x). The potential V_O corresponds to the potential at an arbitrary reference point.

$$V = -\int_O^x \vec{E}' \cdot d\vec{x}'$$
$$= -\int_O^x \frac{\sigma}{2\varepsilon_0} dx'$$
$$= V_O - \frac{\sigma x}{2\varepsilon_0}$$

Electric potential at position $\vec{r}$ (Derivation 7.4):

Point charge at $\vec{r}_q$:

$$V(\vec{r}) = \frac{q}{4\pi\varepsilon_0 r_{qr}} \quad (7.23)$$

where $\vec{r}_{qr} = \vec{r} - \vec{r}_q$

and $r_{qr} \overset{\text{def}}{=} |\vec{r}_{qr}|$

(for a point charge at the origin, $\vec{r}_q = 0 \therefore \vec{r}_{qr} = \vec{r}$).

Discrete distribution

(the ith charge has position vector $\vec{r}_i$):

$$V(\vec{r}) = \sum_i \frac{q_i}{4\pi\varepsilon_0 r_{ir}} \quad (7.24)$$

where $\vec{r}_{ir} = \vec{r} - \vec{r}_i$

and $r_{ir} \overset{\text{def}}{=} |\vec{r}_{ir}|$

Continuous distribution

(element of charge dq has position vector $\vec{r}_{dq}$):

$$V(\vec{r}) = \int \frac{dq}{4\pi\varepsilon_0 r_{dqr}} \quad (7.25)$$

where $\vec{r}_{dqr} = \vec{r} - \vec{r}_{dq}$

and $r_{dqr} \overset{\text{def}}{=} |\vec{r}_{dqr}|$

(see Eq. 7.4 for expressions for dq).

Definitions of respective position vectors for electric potential at point P (P has position vector $\vec{r}$ with respect to the origin):

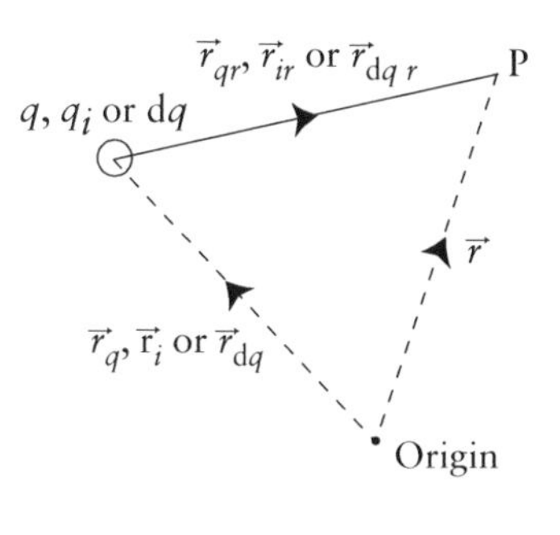

$\vec{r}_{xr} = \vec{r} - \vec{r}_x$, where $x = q$, i or dq

Potential difference

The electric potential at a point $\vec{r}$ is defined as the integral of the electric field from a fixed reference point to $\vec{r}$. Similarly, the *potential difference* between two points $\vec{a}$ and $\vec{b}$ is the line integral of the electric field between these two points (Eq. 7.26 and Derivation 7.5). The work done in moving a charge q from $\vec{a}$ to $\vec{b}$ is then found by multiplying the potential difference by q (Eq. 7.27). Remember however that while electric potential is a scalar *field* (i.e. a function of $\vec{r}$), work is a scalar *quantity* (not a function of $\vec{r}$). The work done on or by q is equal to its change in potential energy. We will see in Section 7.4 that a charge distribution stores electric potential energy as a result of the work required to assemble the system of charges.

Potential difference (Derivations 7.5):

Potential difference between $\vec{a}$ and $\vec{b}$:

$$\Delta V_{ab} = -\int_{\vec{a}}^{\vec{b}} \vec{E} \cdot \mathrm{d}\vec{l} \tag{7.26}$$

Work done on q between $\vec{a}$ and $\vec{b}$:

$$W_{ab} = q\,\Delta V_{ab} \tag{7.27}$$

Equipotential lines

We can represent an electric potential field schematically using *equipotential contour lines*, which represent surfaces or lines of constant potential. Since $\vec{E} = -\nabla V$, equipotential contours lie perpendicular to electric field lines. This is illustrated for a positive point charge in Fig. 7.8.

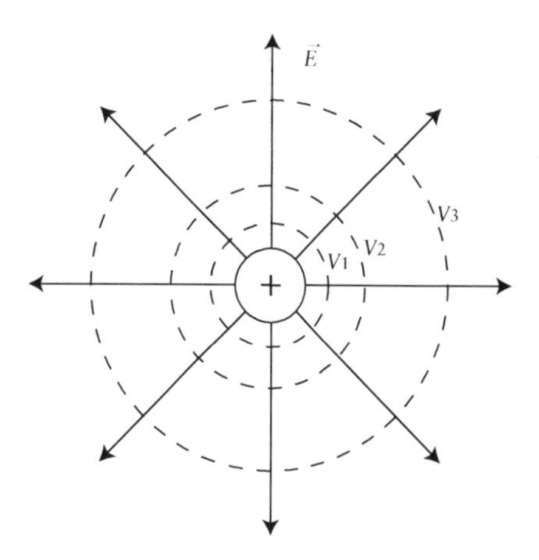

Figure 7.8: Electric field lines (solid) and equipotential contours (dashed) for a point charge.

These lines are conceptually equivalent to contour lines that mark constant elevation on geographical maps.

Laplace's and Poisson's equations

Using $\vec{E} = -\nabla V$ to eliminate $\vec{E}$ from Gauss's Law, we can express Gauss's Law in terms of electric potential:

$$\begin{aligned}\frac{\rho}{\varepsilon_0} &= \nabla \cdot \vec{E} \\ &= \nabla \cdot (-\nabla V) \\ &= -\nabla^2 V\end{aligned}$$

(Eq. 7.28). This equation is called *Poisson's equation*. If there is no free charge in a given region (such that $\rho = 0$), the potential in that region satisfies *Laplace's equation*, $\nabla^2 V = 0$ (Eq. 7.29). It is sometimes useful to express ρ in terms of V instead of $\vec{E}$ since scalars are usually mathematically easier to work with than vectors. We will come across examples of solving Laplace's equation in Sections 7.7 and 11.3. The solution to Poisson's equation, however, we have already come across. It's simply:

$$\begin{aligned}V(\vec{r}) &= \int \frac{\mathrm{d}q}{4\pi\varepsilon_0 r_{\mathrm{d}qr}} \quad \text{from Eq. 7.25} \\ &\equiv \int \frac{\rho\,\mathrm{d}\tau}{4\pi\varepsilon_0 r_{\mathrm{d}\tau r}}\end{aligned}$$

$$\begin{aligned}\text{where } \vec{r}_{\mathrm{d}qr} &= \vec{r} - \vec{r}_{\mathrm{d}q} \\ \vec{r}_{\mathrm{d}\tau r} &= \vec{r} - \vec{r}_{\mathrm{d}\tau} \equiv \vec{r}_{\mathrm{d}qr} \\ \text{and } r_{\mathrm{d}\tau r} &\stackrel{\text{def}}{=} \left|\vec{r}_{\mathrm{d}\tau r}\right|\end{aligned}$$

provided ρ goes to zero at infinity. This result will be useful to keep in mind when we come across the magnetic analog of the electric potential (the *magnetic vector potential*, $\vec{A}$) in Section 8.4.

Derivation 7.4: Derivation of the electric potential from work done (Eqs. 7.23–7.25)

Potential at distance r_{qr} from a point charge:

$$V(\vec{r}) = \frac{W}{Q}$$
$$= \frac{1}{Q}\int_{\infty}^{r_{qr}} \vec{F}' \cdot d\vec{r}'_{qQ}$$
$$= -\frac{1}{Q}\int_{\infty}^{r_{qr}} \frac{qQ dr'_{qQ}}{4\pi\varepsilon_0 (r')^2_{qQ}} \underbrace{\hat{r}'_{qQ} \cdot \hat{r}'_{qQ}}_{=1}$$
$$= -\frac{q}{4\pi\varepsilon_0}\int_{\infty}^{r_{qr}} \frac{dr'_{qQ}}{(r')^2_{qQ}}$$
$$= \frac{q}{4\pi\varepsilon_0 r_{qr}}$$

where $\vec{r}_{qr} = \vec{r} - \vec{r}_q$

and $r_{qr} \stackrel{\text{def}}{=} |\vec{r}_{qr}|$

Discrete distribution:

$$V(\vec{r}) = \sum_i V_i(\vec{r})$$
$$= \sum_i \frac{q_i}{4\pi\varepsilon_0 r_{ir}}$$

where $\vec{r}_{ir} = \vec{r} - \vec{r}_i$

and $r_{ir} \stackrel{\text{def}}{=} |\vec{r}_{ir}|$

Continuous distribution:

$$V(\vec{r}) = \int dV_{dq}(\vec{r})$$
$$= \int \frac{dq}{4\pi\varepsilon_0 r_{dqr}}$$

where $\vec{r}_{dqr} = \vec{r} - \vec{r}_{dq}$

and $r_{dqr} \stackrel{\text{def}}{=} |\vec{r}_{dqr}|$

- Starting point: the electric potential is defined as the work required to bring a charge Q from ∞ to a radial distance r from another point charge q, per unit charge:

$$V = \frac{W}{Q}$$
$$= \frac{1}{Q}\int_{\infty}^{r} \vec{F}' \cdot d\vec{r}'_{qQ}$$

where $\vec{r}'_{qQ}$ is the separation vector that points from q to Q $(\vec{r}'_{qQ} = \vec{r}'_Q - \vec{r}'_q)$.

- By definition of the unit vector, $d\vec{r}'_{qQ} = dr'_{qQ}\,\hat{r}'_{qQ}$.
- The force required to move a charge through an electric field is equal in magnitude and opposite in direction to the electrostatic Coulomb force, $\vec{F}_C$. Therefore, for a system of two point charges, q and Q, the force doing the work is:

$$\vec{F}' = -\vec{F}'_C$$
$$= -\frac{qQ}{4\pi\varepsilon_0 (r')^2_{qQ}} \hat{r}'_{qQ}$$

- Since the Coulomb force obeys the principle of superposition, so does electric potential. For example, the potential at position vector $\vec{r}$ in the vicinity of a discrete distribution is:

$$V(\vec{r}) = V_1(\vec{r}) + V_2(\vec{r}) + V_3(\vec{r}) + \ldots$$

where $V_i(\vec{r}) = \frac{q_i}{4\pi\varepsilon_0 r_{ir}}$

and $\vec{r}_{ir} = \vec{r} - \vec{r}_i$ $(r_{ir} \stackrel{\text{def}}{=} |\vec{r}_{ir}|)$

for $i = 1, 2, 3\ldots$

For a continuous distribution, the sum turns into an integral over elements of charge dq.

Poisson's and Laplace's equations:

Poisson's equation:

$$\nabla^2 V = -\frac{\rho}{\varepsilon_0} \quad (7.28)$$

Laplace's equation ($\rho = 0$):

$$\nabla^2 V = 0 \quad (7.29)$$

7.4 Electric energy of a system

Electric energy, U_E [J], is a form of potential energy that is stored in charge distributions. This potential energy is equal to the work required to assemble the charge distribution (against electrostatic interactions), and would be recovered if the system were to be dismantled.

First, let's consider the simple example of the work required to assemble a system of two point charges, q_1 and q_2, a distance r_{12} apart. The work required to

Derivation 7.5: Derivation of the potential difference between two points (Eq. 7.26)

$$\begin{aligned}\Delta V_{ab} &= V(\vec{b}) - V(\vec{a}) \\ &= -\int_{O}^{\vec{b}} \vec{E}\cdot d\vec{l} + \int_{O}^{\vec{a}} \vec{E}\cdot d\vec{l} \\ &= -\int_{O}^{\vec{b}} \vec{E}\cdot d\vec{l} - \int_{\vec{a}}^{O} \vec{E}\cdot d\vec{l} \\ &= -\int_{\vec{a}}^{\vec{b}} \vec{E}\cdot d\vec{l}\end{aligned}$$

Work required to move q from $\vec{a}$ to $\vec{b}$:

$$\begin{aligned}W_{ab} &= -\int_{\vec{a}}^{\vec{b}} \vec{F}\cdot d\vec{l} \\ &= -q\int_{\vec{a}}^{\vec{b}} \vec{E}\cdot d\vec{l} \\ &= q\,\Delta V_{ab}\end{aligned}$$

- Starting point: the potential difference between points $\vec{a}$ and $\vec{b}$ is:

  $$\Delta V_{ab} = V(\vec{b}) - V(\vec{a})$$

- Using $V = -\int \vec{E}\cdot d\vec{l}$ (Eq. 7.20), we can express $V(\vec{a})$ as:

  $$V(\vec{a}) = -\int_{O}^{\vec{a}} \vec{E}\cdot d\vec{l}$$

  and similarly for $V(\vec{b})$. The integration limit O corresponds to a standard reference point.

- Definite integrals satisfy the property:

  $$\int_a^b f(x)dx = -\int_b^a f(x)dx$$

- Since the definite integrals share a common integration limit, we can combine the two integrals to form a single integral.

- The work required to move a charge q from $\vec{a}$ to $\vec{b}$ in an electric field is done against a force $\vec{F} = q\vec{E}$.

assemble the first charge is $W_1 = 0$, since there is no initial electric field. The work required to then assemble q_2 a distance r_{12} away from q_1 is:

$$\begin{aligned}W_2 &= q_2 V_1(r_{12}) \\ &= \frac{q_1 q_2}{4\pi\varepsilon_0 r_{12}}\end{aligned}$$

Therefore, the electric potential energy of the system is:

$$\begin{aligned}U_E &= \frac{q_1 q_2}{4\pi\varepsilon_0 r_{12}} \\ &= q_2 V_1(r_{12}) \\ &\equiv q_1 V_2(r_{12})\end{aligned}$$

Adding a 3rd charge to the system, q_3, we have to add the work required to bring q_3 a distance r_{13} from q_1 and a distance r_{23} from q_2:

$$\therefore U_E = \frac{q_1 q_2}{4\pi\varepsilon_0 r_{12}} + \frac{q_1 q_3}{4\pi\varepsilon_0 r_{13}} + \frac{q_2 q_3}{4\pi\varepsilon_0 r_{23}}$$

Generalizing to a system of n charges, we sum over charge pairs:

$$U_E = \sum_{i=1}^{n} \sum_{\substack{j=1 \\ j>i}}^{n} \frac{q_i q_j}{4\pi\varepsilon_0 r_{ij}} \quad (7.30)$$

(also Eq. 7.33). The limits of the second sum ($j > i$) ensure that we do not double count charge pairs. Alternatively, we can perform the sum over all i and all j, and divide by 2 to eliminate double counting:

$$\begin{aligned}U_E &= \sum_{i=1}^{n} \sum_{\substack{j=1 \\ j>i}}^{n} \frac{q_i q_j}{4\pi\varepsilon_0 r_{ij}} \\ &\equiv \frac{1}{2}\sum_{i=1}^{n} \sum_{\substack{j=1 \\ j\neq i}}^{n} \frac{q_i q_j}{4\pi\varepsilon_0 r_{ij}}\end{aligned} \quad (7.31)$$

7.4.1 Energy and electric potential

Now, instead of assembling a distribution from scratch, let's consider the electric energy of a fully-assembled charge distribution. We yield the same results as Eq. 7.31, but the approach is slightly different, and it can be easily generalized to continuous charge distributions.

In terms of electric potential, Eq. 7.31 can be written as:

$$U_E = \frac{1}{2}\sum_{i=1}^{n} \sum_{\substack{j=1 \\ j\neq i}}^{n} \frac{q_i q_j}{4\pi\varepsilon_0 r_{ij}}$$

$$= \frac{1}{2} \sum_{i=1}^{n} \sum_{\substack{j=1 \\ j \neq i}}^{n} q_i V_j(r_{ij}) \qquad (7.32)$$

From the principle of superposition, the electric potential of the fully-assembled charge distribution at position $\vec{r}_i$ (i.e. at the position of the ith charge) is:

$$V(\vec{r}_i) = \sum_{\substack{j=1 \\ j \neq i}}^{n} V_j(r_{ij})$$

Therefore, using this in Eq. 7.32, the electric energy of the fully-assembled charge distribution is:

$$\therefore U_E = \frac{1}{2} \sum_{i=1}^{n} q_i V(\vec{r}_i)$$

(Eq. 7.34). For a continuous distribution, the sum becomes an integral over elements of charge dq:

$$U_E = \frac{1}{2} \int V(\vec{r}) dq$$

(Eq. 7.35, see Derivation 7.6 for an alternate treatment).[4]

Electric energy (Derivation 7.6):

Discrete distribution:

$$U_E = \sum_{i=1}^{n} \sum_{\substack{j=1 \\ j > i}}^{n} \frac{q_i q_j}{4\pi\varepsilon_0 r_{ij}} \qquad (7.33)$$

$$\equiv \frac{1}{2} \sum_{i=1}^{n} q_i V(\vec{r}_i) \qquad (7.34)$$

Continuous distribution:

$$U_E = \frac{1}{2} \int V(\vec{r}) dq \qquad (7.35)$$

(see Eq. 7.4 for expressions for dq).

$V(\vec{r}_i)$ and $V(\vec{r})$ are the electric potentials of the fully-assembled distributions, at positions $\vec{r}_i$ or $\vec{r}$.

[4] Note that, according to the terminology used earlier in this chapter, the position vector $\vec{r}$ in Eq. 7.35 was expressed as $\vec{r}_{dq}$ (i.e. the position vector corresponding to the position of element of charge dq). However, since the integral for U_E is defined as being over dq, and $dq = 0$ for all coordinates outside of the charge distribution itself, then we can write $\vec{r}_{dq} \equiv \vec{r}$.

7.4.2 Energy and electric field

For continuous charge distributions in which a charge density function, ρ, exists, we can use $dq = \rho d\tau$ in Eq. 7.35 to determine an alternate expression for U_E in terms of electric field:

$$U_E = \frac{1}{2} \varepsilon_0 \int_{\substack{\text{all} \\ \text{space}}} \left|\vec{E}\right|^2 d\tau$$

(Eq. 7.36 and Derivation 7.7). This expression illustrates that electric potential energy is stored in the electric field of a charge distribution. Dividing U_E by volume, we get the *electric energy density*, u_E [J m^{-3}] (Eq. 7.37).

Electric energy stored in an electric field (Derivation 7.7):

Total energy:

$$U_E = \frac{1}{2} \varepsilon_0 \int_{\substack{\text{all} \\ \text{space}}} \left|\vec{E}\right|^2 d\tau \qquad (7.36)$$

Energy density (energy per unit volume):

$$u_E = \frac{1}{2} \varepsilon_0 \left|\vec{E}\right|^2 \qquad (7.37)$$

Caution: note that since U_E is quadratic in $\vec{E}$, it does not obey the principle of superposition; the cross terms in the electric fields must be taken into account. For example, the energy stored in two combined electric fields $\vec{E}_1$ and $\vec{E}_2$ is:

$$\begin{aligned} U_{E_{\text{tot}}} &= \frac{1}{2} \varepsilon_0 \int \left(\vec{E}_1 + \vec{E}_2\right)^2 d\tau \\ &= \frac{1}{2} \varepsilon_0 \int \left(\left|\vec{E}_1\right|^2 + \left|\vec{E}_2\right|^2 + 2\vec{E}_1 \cdot \vec{E}_2\right) d\tau \\ &= U_{E_1} + U_{E_2} + \underbrace{\varepsilon_0 \int \vec{E}_1 \cdot \vec{E}_2 d\tau}_{\text{cross term}} \end{aligned}$$

7.5 Electric dipoles

An *electric dipole* consists of two point charges of equal and opposite magnitude, $-q$ and $+q$, separated by a distance $\vec{l}$, as shown in Fig. 7.9. Electric dipoles are defined by their *electric dipole moment*, $\vec{p}$ [C m], which is a vector that points from $-q$ to $+q$, where:

$$\vec{p} = q\vec{l}$$

Derivation 7.6: Derivation of the energy stored in a system of charges (Eqs. 7.34 and 7.35)

$$
\begin{aligned}
\mathrm{d}W_i &= V_i^{\alpha}\mathrm{d}q \\
&= \alpha V(\vec{r}_i)q_i\mathrm{d}\alpha \\
\therefore \mathrm{d}W &= \sum_{i=0}^{n}\mathrm{d}W_i \\
&= \sum_{i=0}^{n} V(\vec{r}_i)q_i\alpha\mathrm{d}\alpha \\
\therefore W &= \sum_{i=0}^{n} V(\vec{r}_i)q_i\int_0^1\alpha\mathrm{d}\alpha \\
&= \frac{1}{2}\sum_{i=0}^{n} V(\vec{r}_i)q_i \\
\text{or}\quad W &= \frac{1}{2}\int V(\vec{r})\mathrm{d}q
\end{aligned}
$$

- Starting point: consider a fully-assembled system of n point charges, where the ith charge is at position vector $\vec{r}_i$. Let $V(\vec{r}_i)$ be the potential at the position of the ith charge, q_i.
- Now consider assembling each charge from zero to q_i, in infinitesimal increments, $\mathrm{d}q$. When each charge is a fraction α of their total charge, where $0 < \alpha < 1$, then the potential at $\vec{r}_i$ is $V_i^{\alpha} = \alpha V(\vec{r}_i)$. If we increase the charge by an infinitesimal amount $\mathrm{d}q = q_i\mathrm{d}\alpha$, then the element of work required is $\mathrm{d}W_i = V_i^{\alpha}\mathrm{d}q$.
- The element of work required to increase all n charges by an amount $q_i\mathrm{d}\alpha$ is found by summing over all charges:

$$\mathrm{d}W = \sum_{i=0}^{n}\mathrm{d}W_i$$

- The total work required is then found by integrating over α.
- For a continuous distribution, the sum becomes an integral over elements of charge, $\mathrm{d}q$, where $\mathrm{d}q = \lambda\,\mathrm{d}l$, $\sigma\,\mathrm{d}s$, or $\rho\,\mathrm{d}\tau$, depending on the dimension of the charge distribution (refer to Eq. 7.4).

Derivation 7.7: Derivation of the energy stored in an electric field (Eq. 7.36)

$$
\begin{aligned}
U_E &= \frac{1}{2}\int_Q V\mathrm{d}q \\
&= \frac{1}{2}\int_V \rho V\mathrm{d}\tau \\
&= \frac{\varepsilon_0}{2}\int_V(\nabla\cdot\vec{E})V\,\mathrm{d}\tau \\
&= \frac{\varepsilon_0}{2}\left(\oint_S V\vec{E}\cdot\mathrm{d}\vec{s} - \int_V \vec{E}\cdot(\nabla V)\mathrm{d}\tau\right) \\
&= \frac{\varepsilon_0}{2}\left(\oint_S V\vec{E}\cdot\mathrm{d}\vec{s} + \int_V \left|\vec{E}\right|^2\mathrm{d}\tau\right) \\
&= \frac{\varepsilon_0}{2}\int_{\substack{\text{all}\\\text{space}}}\left|\vec{E}\right|^2\mathrm{d}\tau
\end{aligned}
$$

- Starting point: using $\mathrm{d}q = \rho\mathrm{d}\tau$ in the expression for U_E (Eq. 7.35), we can express U_E as a volume integral.
- From Gauss's Law, $\rho = \varepsilon_0\nabla\cdot\vec{E}$.
- Integrate by parts using the vector identity:

$$\int_V(\nabla\cdot\vec{F})\phi\mathrm{d}\tau = \oint_S\phi\vec{F}\cdot\mathrm{d}\vec{s} - \int_V\vec{F}\cdot(\nabla\phi)\mathrm{d}\tau$$

where $\vec{F}$ is an arbitrary vector function and ϕ is an arbitrary scalar function.

- Use $\nabla V = -\vec{E}$ to eliminate ∇V, along with $\vec{E}\cdot\vec{E} = |\vec{E}|^2$.
- When integrating over all space, the volume integral dominates over the surface integral.

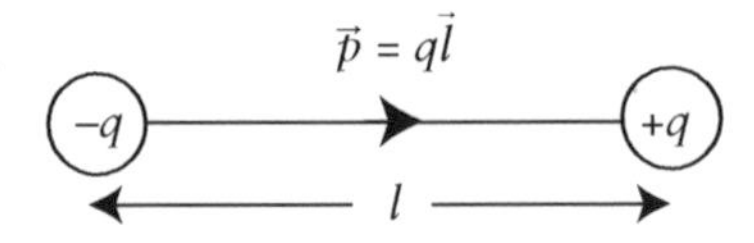

Figure 7.9: Electric dipole.

(Eq. 7.38). Some common physical examples of electric dipoles are polar molecules, antennae, or electrified clouds, so it's useful to understand their properties and behavior. Here, we will treat dipoles as *ideal dipoles*, meaning that the distance between the charges, $|\vec{l}|$, is negligible compared with the position vector $|\vec{r}|$.

Electric dipole moment:

$$\vec{p} = q\vec{l} \tag{7.38}$$

7.5.1 Electric field and potential

Although electric dipoles are electrically neutral overall, they produce an electric field and therefore interact with other charged particles. We can calculate the electric potential of a dipole using the superposition principle applied to two point charges in close proximity (Eqs. 7.39 and 7.40 and Derivation 7.8). The electric potential of a dipole obeys an inverse square law ($V \propto 1/r^2$), whereas we saw earlier that the electric potential of a point charge varies as $V \propto 1/r$. Furthermore, the electric potential of a dipole is not spherically symmetric, as it is for a point charge, but depends on the angle, θ, relative to the dipole.
By differentiating Eq. 7.40, we find that the electric field of a dipole, in spherical coordinates, has a radial and a tangential component (Eq. 7.41 and Derivation 7.9). The shape of the field is illustrated in Fig. 7.10a. Fig. 7.10b shows the electric field of a *real* dipole, in which $|\vec{l}|$ is not necessarily $\ll|\vec{r}|$. Real and ideal dipoles have the same features at large distances from the dipole, but their shapes deviate near the center. Therefore, the electric potential in Eq. 7.39 and electric field in Eq. 7.41 correspond to approximations for dipoles in which the assumption that $|\vec{l}| \ll |\vec{r}|$ is valid.

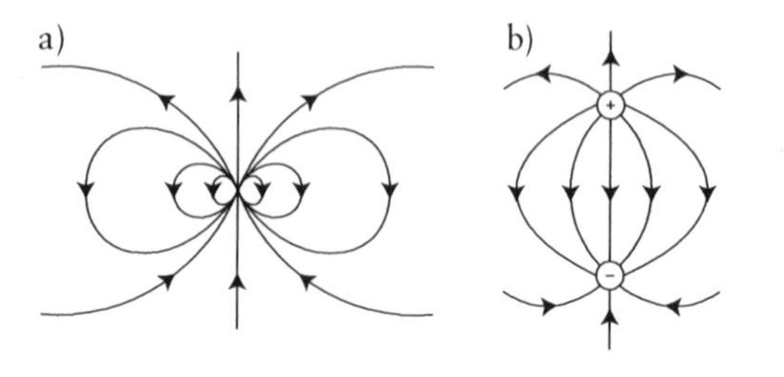

Figure 7.10: Electric field of a) an ideal and b) a real dipole.

Electric potential of an ideal dipole (Derivation 7.8):

(dipole placed at the origin):

$$V(r,\theta) = \frac{\vec{p}\cdot\hat{r}}{4\pi\varepsilon_0 r^2} \tag{7.39}$$

$$= \frac{p\cos\theta}{4\pi\varepsilon_0 r^2} \tag{7.40}$$

Electric field of an ideal dipole (Derivation 7.9):

$$\vec{E}(r,\theta) = \frac{p}{4\pi\varepsilon_0 r^3}(2\cos\theta\hat{r} + \sin\theta\hat{\theta})$$

$$\equiv \frac{p}{4\pi\varepsilon_0 r^3}\begin{pmatrix}2\cos\theta\\ \sin\theta\end{pmatrix} \tag{7.41}$$

7.5.2 Dipole in an external electric field

A dipole in an external electric field experiences a *torque*, $\vec{\tau}_p$ [N m], since opposing forces act on each side of the dipole. The torque is given by the cross product of $\vec{p}$ and $\vec{E}$:

$$\vec{\tau}_p = \vec{p}\times\vec{E}$$

(Eq. 7.42 and Derivation 7.10). The torque tends to align the dipole with the external field, resulting in a dipole potential energy of:

$$U_p = -\vec{p}\cdot\vec{E}$$

(Eq. 7.43 and Derivation 7.11). The net *linear* force on the dipole is then:

$$\vec{F}_p = -\nabla U_p$$
$$= -\nabla(\vec{p}\cdot\vec{E})$$
$$= -(\vec{p}\nabla)\cdot\vec{E}$$

(Eq. 7.44).[5] If the field is uniform, the net linear force is zero since the forces on the positive and negative charges are equal and opposite:

$$\vec{F}_p = q\vec{E} - q\vec{E} = 0$$

[5] The Del operator does not act on the coordinates of the electric dipole, $\vec{p}$, only on the coordinates of the electric field, $\vec{E}$. Therefore, $\nabla(\vec{p}\cdot\vec{E}) = (\vec{p}\nabla)\cdot\vec{E}$.

Derivation 7.8: Derivation of the electric potential of an ideal dipole (Eqs. 7.39 and 7.40)

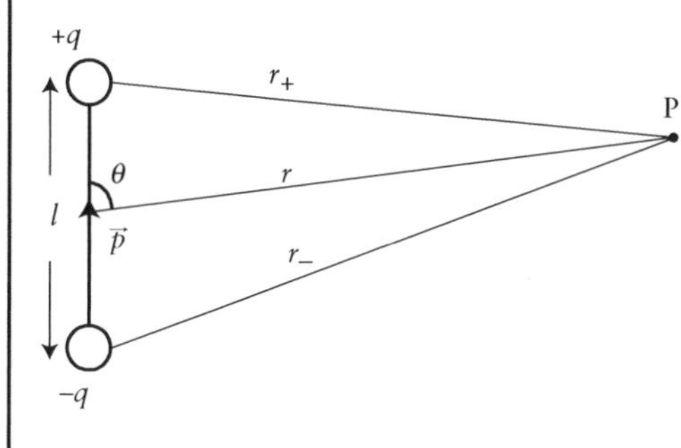

- Starting point: the electric potential at point P is the superposition of the electric potentials of two point charges, $+q$ and $-q$.
- From the cosine rule, we can find expressions for r_+ and r_-:

$$
\begin{aligned}
r_\pm^2 &= r^2 + \left(\frac{l}{2}\right)^2 \mp rl\cos\theta \\
&= r^2\left(1 + \frac{l^2}{4r^2} \mp \frac{l}{r}\cos\theta\right) \\
&\simeq r^2\left(1 \mp \frac{l}{r}\cos\theta\right) \quad \text{for } l << r
\end{aligned}
$$

$$
\begin{aligned}
V(\vec{r}) &= V_+(\vec{r}_+) + V_-(\vec{r}_-) \\
&= \frac{q}{4\pi\varepsilon_0 r_+} - \frac{q}{4\pi\varepsilon_0 r_-} \\
&= \frac{q}{4\pi\varepsilon_0}\left(\frac{1}{r_+} - \frac{1}{r_-}\right) \\
&= \frac{q}{4\pi\varepsilon_0}\left\{\left(\frac{1}{r} + \frac{l}{2r^2}\cos\theta\right) - \left(\frac{1}{r} - \frac{l}{2r^2}\cos\theta\right)\right\} \\
&= \frac{q}{4\pi\varepsilon_0}\left(\frac{l}{r^2}\cos\theta\right) \\
&= \frac{ql\cos\theta}{4\pi\varepsilon_0 r^2} \\
&= \frac{p\cos\theta}{4\pi\varepsilon_0 r^2} \\
&= \frac{\vec{p}\cdot\hat{r}}{4\pi\varepsilon_0 r^2}
\end{aligned}
$$

- For $l << r$ (ideal dipole approximation), we can use a binomial expansion to determine expressions for the reciprocals, $1/r_+$ and $1/r_-$:

$$
\begin{aligned}
\therefore \frac{1}{r_\pm} &= \frac{1}{r}\left(1 \mp \frac{l}{r}\cos\theta\right)^{-1/2} \\
&\simeq \frac{1}{r}\left(1 \pm \frac{l}{2r}\cos\theta\right)
\end{aligned}
$$

- Substitute these expressions into $V(\vec{r})$.
- The *electric dipole moment* is defined as $\vec{p} = q\vec{l}$, therefore the magnitude of $\vec{p}$ is $p = |\vec{p}| = q|\vec{l}| = ql$.

Derivation 7.9: Derivation of the electric field of a dipole (Eq. 7.41)

$$
\begin{aligned}
\vec{E} &= -\nabla V \\
\therefore E_r &= -\frac{\partial V}{\partial r} \\
&= \frac{2p\cos\theta}{4\pi\varepsilon_0 r^3} \\
E_\theta &= -\frac{1}{r}\frac{\partial V}{\partial \theta} \\
&= \frac{p\sin\theta}{4\pi\varepsilon_0 r^3} \\
E_\phi &= -\frac{1}{r\sin\theta}\frac{\partial V}{\partial \phi} \\
&= 0
\end{aligned}
$$

- Starting point: the electric field is the gradient of the electric potential:

$$\vec{E} = -\nabla V$$

- Differentiate $V = p\cos\theta/4\pi\varepsilon_0 r^2$ (Eq. 7.40) using the Del operator in spherical polar coordinates:

$$\nabla V = \frac{\partial V}{\partial r}\hat{r} + \frac{1}{r}\frac{\partial V}{\partial \theta}\hat{\theta} + \frac{1}{r\sin\theta}\frac{\partial V}{\partial \phi}\hat{\phi}$$

Dipole in an external electric field (Derivations 7.10–7.11):

$$\text{Torque}: \quad \vec{\tau}_p = \vec{p} \times \vec{E} \qquad (7.42)$$

$$\text{Energy}: \quad U_p = -\vec{p} \cdot \vec{E} \qquad (7.43)$$

$$\text{Force}: \quad \vec{F}_p = (\vec{p}\nabla) \cdot \vec{E} \qquad (7.44)$$

7.6 Capacitors

A *capacitor* is composed of two conductors separated by an electric insulator. Since charge cannot flow through the insulator, it remains stored on the conductors. A capacitor is assumed to have zero net charge, which means that the conductors have equal and opposite charges, $\pm Q$. The conductors produce an electric field, whose field lines pass through the insulator and point from the positive to the negative conductor. The two conductors therefore have a potential difference, V, between them. Examples of common capacitors are parallel-plate capacitors, coaxial cables, and spherical capacitors (refer to figure in Worked example 7.5).

7.6.1 Capacitance

The *capacitance*, C [F ≡ C V^{-1}], is a measure of the ability of a capacitor to store charge. For a capacitor whose conductors have a charge $\pm Q$ and a potential difference V between them, the capacitance is defined by:

$$C = \frac{Q}{V}$$

(Eq. 7.45). The higher the capacitance, the more charge can be held at a given voltage. The capacitance depends on the sizes, shape, and separation of the conductors (i.e. on the geometry of the capacitor). Example calculations of the capacitance of some common capacitors are provided in Worked example 7.5. For capacitors whose capacitance is voltage dependent, the *differential* or *small signal* capacitance, C_d, is defined by the derivative:

$$C_d = \frac{dQ}{dV}$$

(Eq. 7.46). $C = C_d$ provided the capacitance (the capacitor geometry) is voltage-independent.

Capacitance:

$$C \overset{\text{def}}{=} \frac{Q}{V} \qquad (7.45)$$

Differential capacitance:

$$C_d \overset{\text{def}}{=} \frac{dQ}{dV} \qquad (7.46)$$

7.6.2 Energy of a capacitor

Capacitors store electric potential energy in the electric field that is produced by the charged conductors. The potential energy equals the work required to charge the conductors, and is recovered if the capacitor is

Derivation 7.10: Derivation of the torque on a dipole in an external $\vec{E}$- field (Eq. 7.42)

$$\begin{aligned}
\vec{\tau}_p &= \vec{\tau}_+ + \vec{\tau}_- \\
&= \left(\vec{r}_+ \times \vec{F}_+\right) + \left(\vec{r}_- \times \vec{F}_-\right) \\
&= \frac{\vec{l}}{2} \times q\vec{E} + \left(-\frac{\vec{l}}{2}\right) \times \left(-q\vec{E}\right) \\
&= \vec{l} \times q\vec{E} \\
&= q\vec{l} \times \vec{E} \\
&= \vec{p} \times \vec{E} \\
\therefore \left|\vec{\tau}_p\right| &= \left|\vec{p}\right|\left|\vec{E}\right| \sin\theta
\end{aligned}$$

- Starting point: the net torque on a dipole of length $\vec{l}$ and dipole moment $\vec{p} = q\vec{l}$ about its center is the sum of torques on either charge: $\vec{\tau}_p = \vec{\tau}_+ + \vec{\tau}_-$.
- The force on q is $\vec{F} = q\vec{E}$.
- The torque acts in a sense to decrease θ (the angle between $\vec{p}$ and $\vec{E}$), and is zero when the dipole is parallel to the field.

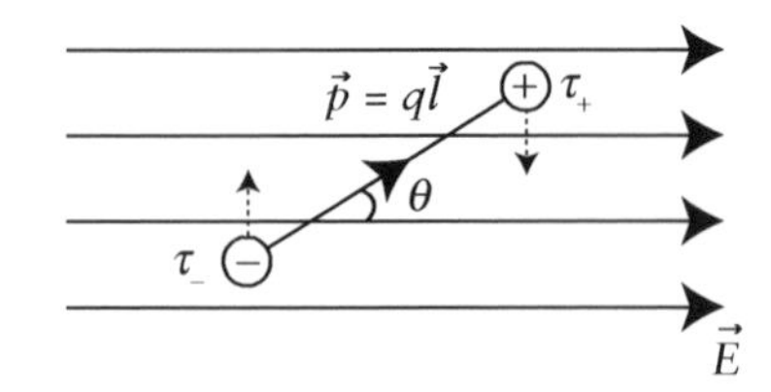

Derivation 7.11: Derivation of the potential energy of a dipole in an external $\vec{E}$-field (Eq. 7.43)

$$U_p = \int_{\pi/2}^{\theta} \left|\vec{\tau}_p\right| d\theta'$$
$$= \int_{\pi/2}^{\theta} \left|\vec{p}\right|\left|\vec{E}\right| \sin\theta' d\theta'$$
$$= \left[-\left|\vec{p}\right|\left|\vec{E}\right|\cos\theta'\right]_{\pi/2}^{\theta}$$
$$= -\left|\vec{p}\right|\left|\vec{E}\right|\cos\theta$$
$$= -\vec{p}\cdot\vec{E}$$

- Starting point: the energy of a dipole in an external electric field equals the work necessary to rotate the dipole from an initial equipotential field line (perpendicular to $\vec{E} \rightarrow \therefore\ \theta' = \pi/2$) to an angle θ.
- The magnitude of torque on a dipole in an external field is:

$$\left|\vec{\tau}_p\right| = \left|\vec{p}\times\vec{E}\right| = \left|\vec{p}\right|\left|\vec{E}\right|\sin\theta'$$

- The potential energy of the dipole is minimum when it is parallel to the external field.

WORKED EXAMPLE 7.5: Determining the capacitance of some common configurations from their electric field

a) Spherical capacitor

outer radius, b
inner radius, a

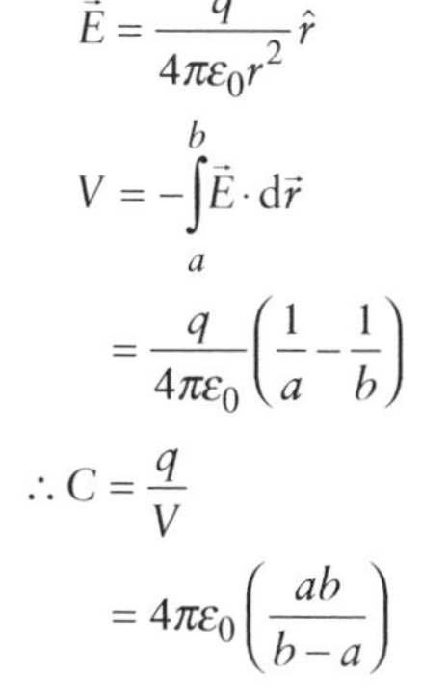

$$\vec{E} = \frac{q}{4\pi\varepsilon_0 r^2}\hat{r}$$
$$V = -\int_a^b \vec{E}\cdot d\vec{r}$$
$$= \frac{q}{4\pi\varepsilon_0}\left(\frac{1}{a}-\frac{1}{b}\right)$$
$$\therefore C = \frac{q}{V}$$
$$= 4\pi\varepsilon_0\left(\frac{ab}{b-a}\right)$$

b) Cylindrical or coaxial capacitor

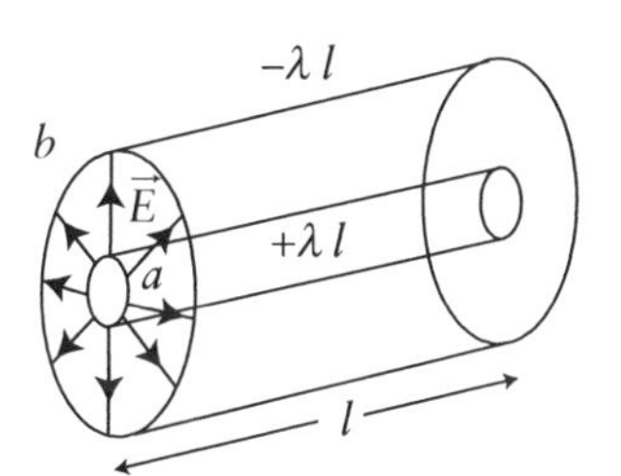

outer radius, b
inner radius, a
length, l

$$\vec{E} = \frac{\lambda}{2\pi\varepsilon_0 r}\hat{r}$$
$$V = -\int_a^b \vec{E}\cdot d\vec{r}$$
$$= \frac{\lambda\ln(b/a)}{2\pi\varepsilon_0}$$
$$\therefore C = \frac{q}{V} = \frac{\lambda l}{V}$$
$$= \frac{2\pi\varepsilon_0 l}{\ln(b/a)}$$

c) Parallel-plate capacitor

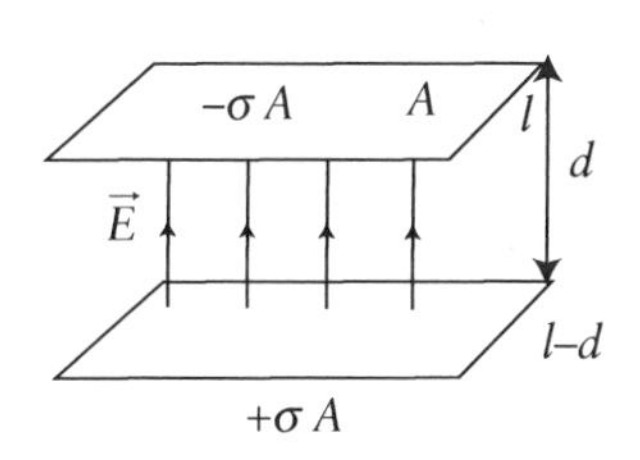

upper plate, l
lower plate, l-d
area, A

$$\vec{E} = \frac{\sigma}{\varepsilon_0}\hat{n}$$
$$V = -\int_l^{l-d} \vec{E}\cdot d\vec{l}$$
$$= \frac{\sigma d}{\varepsilon_0}$$
$$\therefore C = \frac{q}{V} = \frac{\sigma A}{V}$$
$$= \frac{\varepsilon_0 A}{d}$$

Refer to Worked example 7.2 for determining the electric field of each configuration from Gauss's Law.

discharged. The electric energy stored in a capacitor, U_C, is given by:

$$U_C = \frac{1}{2}QV$$

(Eq. 7.47 and Derivation 7.12). The force exerted on the conductor plates is then:

$$\begin{aligned}\vec{F} &= -\nabla U_C \\ &= -\frac{1}{2}Q\nabla V \\ &= \frac{1}{2}Q\vec{E}\end{aligned}$$

Potential energy of a capacitor (Derivation 7.12):

$$U_C = \frac{1}{2}QV,\ \frac{1}{2}CV^2,\ \text{or}\ \frac{Q^2}{2C} \qquad (7.47)$$

7.7 Conductors

Conductors are materials that have mobile charge. For example, metals such as aluminum and copper are conductors; they consist of a fixed lattice of positive ions (the atomic nuclei) and mobile valence electrons.

Electric properties of conductors

Consider a conductor that has an excess or deficit of electrons, such that it has a net charge. In the steady state, the mobile charges experience no net force: $\vec{F} = q\,\vec{E} = 0$. Therefore, the net electric field inside a conductor in the steady state is equal to zero: $\vec{E} = 0$.

Since $\vec{E} = 0$ inside a conductor, it follows that conductors are equipotentials; the potential difference between two points in a conductor is:

$$\begin{aligned}V(\vec{b}) - V(\vec{a}) &= -\int_{\vec{a}}^{\vec{b}} \vec{E}\cdot d\vec{l} = 0 \\ \therefore V(\vec{b}) &= V(\vec{a})\end{aligned}$$

Furthermore, it follows from Gauss's Law that the charge density inside a conductor is equal to zero if $\vec{E} = 0$: $\rho \propto \nabla \cdot \vec{E} = 0$. Therefore, if a conductor is charged, net charge resides on its surface, resulting in a surface charge density of σ. The electric field produced by σ is perpendicular to the conductor's surface since there are no tangential components of $\vec{E}$ at the conductor surface ($\vec{E} = 0$). The surface charge distributes itself such that tangential components cancel, leaving a net radial field. These properties of charged conductors are summarized below.

Electric properties of charged conductors

1. $\vec{E} = 0$ inside a conductor.
2. $\rho = 0$ inside a conductor.
3. Conductors are at a constant potential.
4. Net charge resides on the surface of a conductor.
5. The electric field at the conductor surface is perpendicular to the surface.

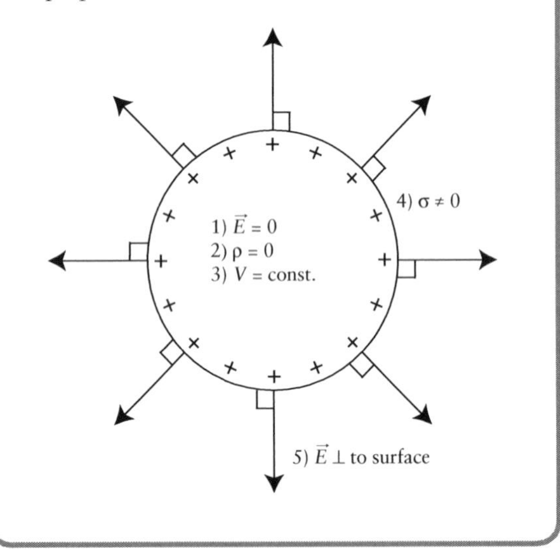

Conductor in an external electric field

If a conductor is placed in an external electric field, the mobile charge drifts into a configuration such that $\vec{E} = 0$ inside the conductor (no net force on a charge). For example, if a conducting rod is placed in an external electric field $\vec{E}_0$, charge redistributes to generate an electric field of $-\vec{E}_0$ such that the net electric field inside the conductor is zero (Fig. 7.11). Redistribution of charge by an external electric field is often referred to as *induced charge separation*.[6]

Laplace's equation and conductors

We saw in Section 7.3 that electric potential and charge density are related to one another by Poisson's equation:

6 Induced charge separation also occurs in non-conducting materials; however, the effect is most pronounced in conductors.

Derivation 7.12: Derivation of the electric energy stored in a capacitor (Eq. 7.47)

$$
\begin{aligned}
U_C &= W \\
&= \int_0^Q V(q)\mathrm{d}q \\
&= \int_0^Q \frac{q}{C}\mathrm{d}q \\
&= \frac{1}{2}\frac{Q^2}{C} \\
&\equiv \frac{1}{2}QV \\
&\equiv \frac{1}{2}CV^2
\end{aligned}
$$

- Starting point: imagine starting with two uncharged conductors, and transferring an element of charge $\mathrm{d}q$ from one to the other, until a total charge Q has been transferred (leaving one conductor at $+Q$ and the other at $-Q$). For a constant capacitance, C, the potential difference between the conductors is a function of the instantaneous charge, q: $V(q) = q/C$.
- The element of work required to move an element of charge $\mathrm{d}q$ through a potential difference $V(q)$ is:

$$
\begin{aligned}
\mathrm{d}W &= V(q)\mathrm{d}q \\
&= \frac{q}{C}\mathrm{d}q
\end{aligned}
$$

- Integrate to determine the total work done, which equals the potential energy stored in a capacitor, $U_C = W$.
- Use the expression for capacitance ($C = Q/V$) to obtain different expressions for U_C.

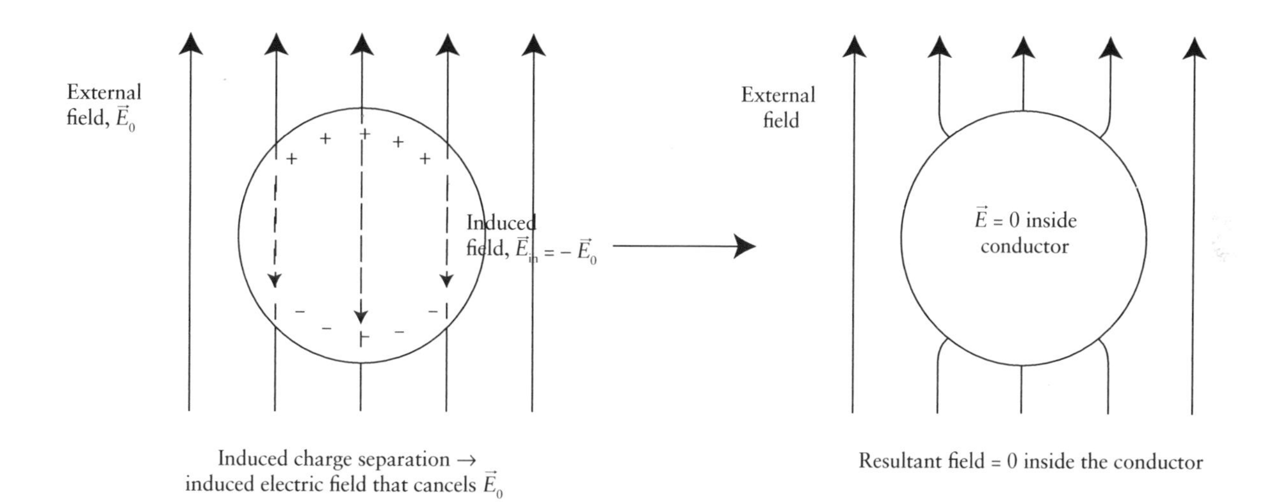

Figure 7.11: Placing a conductor in an external electric field induces charge separation, which, in turn, induces an internal electric field that cancels the external field inside the conductor, such that the resultant field inside the conductor equals zero.

$$\nabla^2 V = -\frac{\rho}{\varepsilon_0}$$

Since $\rho = 0$ inside a conductor, the electric potential of a conductor satisfies Laplace's equation (Poisson's equation with $\rho = 0$):

$$\nabla^2 V = 0$$

An example of solving Laplace's equation to determine the electric potential of a cylindrical conductor in an external electric field is provided in Worked example 7.6.

The uniqueness theorem

The *uniqueness theorem* states that:

if a solution satisfies Poisson's or Laplace's equations and assumes the correct boundary values, then it corresponds to the only solution; it is a unique solution.

It's straightforward to prove this theorem through the following argument. Assume that there are two electric-potential solutions to Laplace's equation:

$$\nabla^2 V_1 = 0 \text{ and } \nabla^2 V_2 = 0$$

WORKED EXAMPLE 7.6: Using Laplace's equation to determine the electric potential of a conductor in an external electric field

Consider an uncharged, conducting rod of radius a, and infinite length, placed in a uniform electric field $\vec{E}_0$ such that the rod axis is perpendicular to the field (refer to figure). Since $\rho = 0$ inside a conductor, the electric potential satisfies Laplace's equation:

$$\nabla^2 V(r,\theta) = 0$$

In cylindrical polar coordinates, the two-dimensional Laplace equation is:

$$r\frac{\partial}{\partial r}\left(r\frac{\partial V}{\partial r}\right) + \frac{\partial^2 V}{\partial \theta^2} = 0$$

It's straightforward to show by substitution that:

$$V(r,\theta) = Ar\cos\theta + \frac{B\cos\theta}{r}$$

is a possible solution.

Boundary conditions

Applying the following boundary conditions, we find expressions for A and B:

1. At radial distances far from the rod ($r >> a$), the potential tends to the potential of a uniform electric field:

$$V(r >> a,\theta) = -E_0 r\cos\theta$$
$$\therefore A = -E_0$$

2. The rod is initially uncharged, so the potential of the conductor at the boundary is equal to zero:

$$V(a,\theta) = 0$$
$$\therefore B = \frac{E_0}{a^2}$$

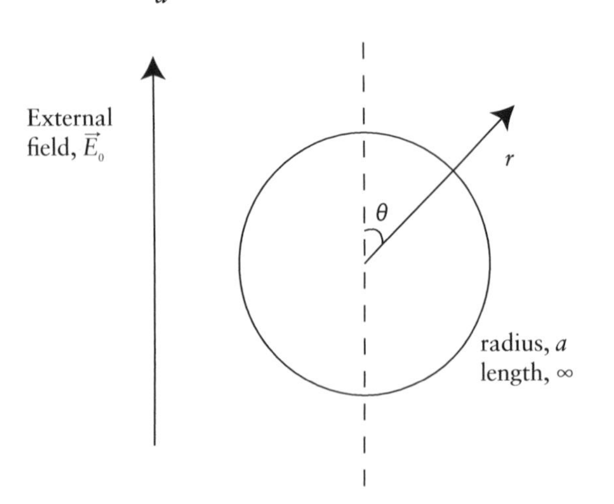

Electric potential of the rod

The electric potential of the charge distribution is:

$$V(r,\theta) = E_0\left(\frac{a^2}{r} - r\right)\cos\theta$$

Surface charge distribution

Using Gauss's Law in integral form, together with the property that $\vec{E}$ is parallel to $d\vec{s}$ at a conductor surface, we find that:

$$\oint_S \vec{E}\cdot d\vec{s} = \frac{q}{\varepsilon_0}$$
$$\therefore |\vec{E}|A = \frac{q}{\varepsilon_0} \Rightarrow |\vec{E}| = \frac{\sigma}{\varepsilon_0}$$
$$\therefore \vec{E} = \frac{\sigma}{\varepsilon_0}\hat{r}$$

where $\hat{r}$ is the radial vector. Electric field is related to the radial part of the electric potential by:

$$\vec{E} = -\frac{\partial V}{\partial r}\hat{r} \quad (\vec{E} = -\nabla V)$$
$$\therefore \frac{\partial V}{\partial r} = -\frac{\sigma}{\varepsilon_0}$$

Rearranging and differentiating, we can find an expression for the surface charge distribution that produces the electric potential of the conductor:

$$\sigma = -\varepsilon_0\left(\frac{\partial V}{\partial r}\right)_{r=a}$$
$$= -\varepsilon_0\frac{\partial}{\partial r}\left(E_0\left(\frac{a^2}{r} - r\right)\cos\theta\right)_{r=a}$$
$$= 2\varepsilon_0 E_0\cos\theta$$

Due to boundary conditions, both V_1 and V_2 must assume the same value on the conductor surface, i.e $V_1 = V_2$ at the boundaries. Now consider a potential that equals the difference between V_1 and V_2: $\Phi = V_2 - V_1$. Since the Laplace operator is linear, then Φ also satisfies Laplace's equation:

$$\begin{aligned}\nabla^2\Phi &= \nabla^2(V_2 - V_1)\\ &= \nabla^2 V_2 - \nabla^2 V_1\\ &= 0\end{aligned}$$

On the boundaries, $V_1 = V_2$, therefore, $\Phi = 0$ in these regions. However, Laplace's equation does not allow any local maxima or minima $\therefore$ $\Phi_{max} = 0$ and $\Phi_{min} = 0$. This implies that $\Phi = 0$ everywhere, which, in turn, implies that $V_1 = V_2$ everywhere. Said another way, there is only one electric-potential solution that satisfies Laplace's equation under given boundary conditions. The uniqueness theorem underlies an electrostatics problem-solving tool known as the *method of image charges*.

The method of image charges

We saw above that an applied (external) electric field induces a redistribution of charge in a conductor, such that the resultant electric field is zero inside the conductor and is perpendicular to the conductor surface. In the same way, the electric field produced by a charge in the vicinity of a conductor induces charge separation in the conductor to produce a field that satisfies these boundary conditions. To determine the resultant electric field of such a system would be difficult since it would require a knowledge of the spatial distribution of induced charge on the conductor surface.

One technique for solving the electric potential of a charge near a conductor is called the *method of image charges*. This technique involves "replacing" the conductor with one or more "image charges," which, together with the original charge, produce an electric potential that satisfies the boundary conditions of the original system. Then, by the uniqueness theorem, this potential is the only solution. This method is best understood through example.

Consider a positive point charge a perpendicular distance d from an infinite conducting plate (Fig. 7.12a). For a plane conductor, the induced charge is equal in magnitude and opposite in sign to the original charge. The image charge of this system is a negative point charge located at the "mirror image" position of the original charge, where the conductor surface acts as the mirror (Fig. 7.12b). The

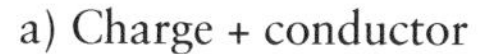

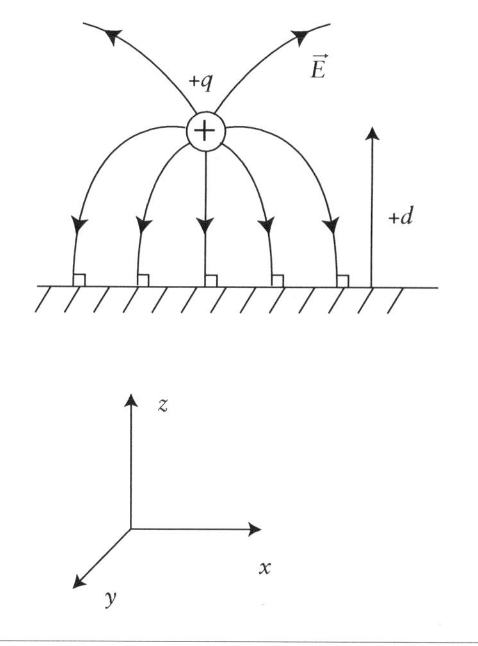

z = 0

b) Charge + image charge
(equivalent charge configuration)

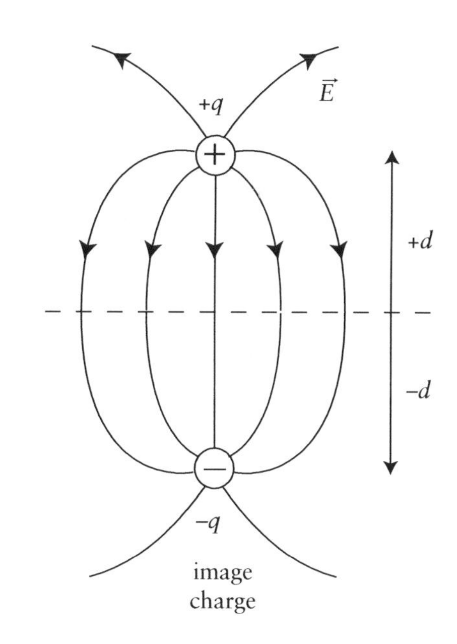

Figure 7.12 Method of image charges example: the image charge of a positive point charge near an infinite conducting plane is another point charge of equal magnitude and opposite sign located in the "mirror image position" of the original charge.

electric potential of this system then corresponds to the superposition of the fields of two point charges:

$$V(x,y,z) = \frac{q}{4\pi\varepsilon_0}\frac{1}{\sqrt{x^2+y^2+(z-d)^2}} - \frac{q}{4\pi\varepsilon_0}\frac{1}{\sqrt{x^2+y^2+(z+d)^2}} \quad (7.48)$$

The boundary conditions of the charge plus conductor system are:

1. The conductor is grounded: $V\ (z=0)=0$.
2. The potential tends to zero at large distance from the charge: V $(x,\ y,\ z \to \infty)=0$.

Since these are both satisfied by Eq. 7.48, then by the uniqueness theorem, Eq. 7.48 corresponds to the electric potential of the original charge plus conductor system. The electric field at the conductor surface (which is perpendicular to the surface) can then be found from the gradient of the potential:

$$E_\perp = -\left(\frac{\partial V}{\partial z}\right)_{z=0} = -\frac{qd}{2\pi\varepsilon_0(x^2+y^2+d^2)^{3/2}}$$

From Gauss's Law, the electric field is related to the surface charge density by:

$$E_\perp = \frac{\sigma}{\varepsilon_0}$$

Therefore, the surface charge density of the conductor is:

$$\sigma = -\frac{qd}{2\pi(x^2+y^2+d^2)^{3/2}}$$

As a cross-check, we can calculate the magnitude of the induced charge by integrating σ over the entire plate. Letting $r^2 = x^2 + y^2$ and integrating over rings of area $2\pi r dr$, we find that the magnitude of the induced charge is:

$$Q_{\text{ind}} = \int_0^\infty \sigma 2\pi r dr = -qd\int_0^\infty \frac{r}{(r^2+d^2)^{3/2}}dr = -qd\left[-\frac{1}{(r^2+d^2)^{1/2}}\right]_0^\infty = -q$$

as expected.

Chapter 8
Magnetostatics

Electric charge moving at a constant speed produces a steady (constant) electric current. Magnetostatics is the study of systems with steady electric current. These produce magnetic fields whose properties remain constant in time. In this chapter, we will look at force and energy in magnetostatic systems, and at the properties of static magnetic fields.

8.1 Electric currents

Current, $I\,[\mathrm{A} \equiv \mathrm{Cs}^{-1}]$, is the rate of flow of electric charge passing a given point (Eq. 8.1). A *steady* current is charge that flows at a constant rate, i.e. the charge per unit time passing a given point is constant in time. Steady currents give rise to magnetostatic phenomena.

Electric current:

$$\text{Scalar form}: \; I = \frac{dq}{dt} \tag{8.1}$$

$$\text{Vector form}: \; \vec{I} = \lambda\vec{v} \tag{8.2}$$

Recall from Section 7.1 that an element of charge dq can be expressed in terms of a charge density, where a line charge, λ, surface charge, σ, and volume charge, ρ, are defined as:

$$\text{Line charge}: \lambda = \frac{dq}{dl}$$

$$\text{Surface charge}: \sigma = \frac{dq}{ds}$$

$$\text{Volume charge}: \rho = \frac{dq}{d\tau}$$

Expressing dq as a line charge density, $dq = \lambda dl$, we have:

$$\begin{aligned} I &= \frac{dq}{dt} \\ &= \lambda\frac{dl}{dt} \\ &= \lambda v \end{aligned}$$

Therefore, current can be thought of as a line charge traveling at a velocity v. Since velocity is a vector, so is current—it flows in a specific direction:

$$\therefore \vec{I} = \lambda\vec{v}$$

(Eq. 8.2). By convention, the direction of current refers to the flow of positive charge (even though electric current in a wire is due to the flow of mobile electrons, which are negatively charged). Thus, so-called "conventional current" (the flow of positive charge) is in the opposite direction to electron current (the flow of negative electrons).

Current densities

Since there are only two directions in which current can flow in a thin wire, we usually ignore the vector nature of $\vec{I}$. The vector nature becomes important to specify when we are dealing with current flowing across a surface or through a volume. Charge flowing across a surface produces a *surface current density*, $\vec{K}\,[\mathrm{A\,m}^{-1}]$, which is defined as the current per unit width-perpendicular-to-the-flow:

$$\vec{K} = \frac{d\vec{I}}{dl_\perp}$$

Using $\vec{I} = \lambda\vec{v}$ and $\sigma = d\lambda/\,dl_\perp$, where σ is the surface charge density, in the above definition, we find $\vec{K} = \sigma\vec{v}$ (Eq. 8.3 and Fig. 8.1a). Similarly, charge flowing through a volume produces *a volume current density*,[1]

[1] Throughout this text, we will often refer to $\vec{J}$ as the "current density" rather than the "volume current density," but will always specify $\vec{K}$ as the surface current density.

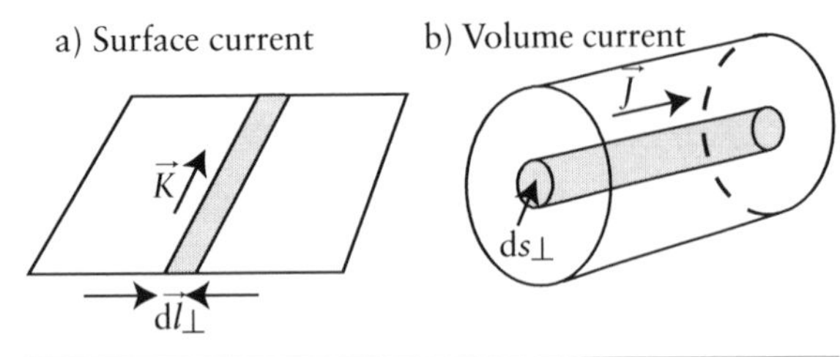

Figure 8.1: a) Surface current density (Eq.8.3) and b) volume current density (Eq. 8.4).

$\vec{J}$ [A m^{-2}], which is the current per unit area-perpendicular-to-the-flow:

$$\vec{J} = \frac{d\vec{I}}{ds_\perp}$$

Letting $\vec{I} = \lambda\vec{v}$ and $\rho = d\lambda/ds_\perp$, we have $\vec{J} = \rho\vec{v}$ (Eq. 8.4 and Fig. 8.1b). The magnitude of total current is found by integrating over elements of current $d\vec{I}$ (Eqs. 8.5 and 8.6).

Current densities:

$$\text{Surface current}: \ \vec{K} = \frac{d\vec{I}}{dl_\perp} = \sigma\vec{v} \quad (8.3)$$

$$\text{Volume current}: \ \vec{J} = \frac{d\vec{I}}{ds_\perp} = \rho\vec{v} \quad (8.4)$$

Magnitude of total current:

$$I = \int_L \vec{K} \cdot d\vec{l} \quad (8.5)$$

$$\text{or } I = \int_S \vec{J} \cdot d\vec{s} \quad (8.6)$$

Magnetic monopoles

An important asymmetry between electricity and magnetism is that magnetic monopoles—the magnetic equivalent of a point charge—do not seem to exist. Although some *Grand Unified Theories* predict the existence of magnetic monopoles, none have been discovered so far. This apparent absence leads to an asymmetry in Maxwell's equations (see Section 10.4).

8.2 Magnetic fields

The Biot–Savart Law

A flowing charge—a current—produces a *magnetic field*, $\vec{B}$ [T]. $\vec{B}$ is the magnetic equivalent of the electric field, $\vec{E}$. Consider a current I flowing through a wire. The element of magnetic field, $d\vec{B}$, produced a distance $\vec{r}_{dlr}$ away from a segment of current-carrying wire, $d\vec{l}$, is given by the *Biot–Savart Law*:

$$d\vec{B} = \frac{\mu_0}{4\pi}\frac{I d\vec{l} \times \hat{r}_{dlr}}{r_{dlr}^2}$$

$$\text{where } \vec{r}_{dlr} = \vec{r} - \vec{r}_{dl}$$

(Eq. 8.7).[2] The constant in the Biot–Savart Law is the *permeability of free space*, μ_0 [$4\pi \cdot 10^{-7}$ N A^{-2}], which is the magnetic equivalent of the electric permittivity of free space, ε_0. The total magnetic field is then found by integrating over all elements of field:

$$\vec{B}(\vec{r}) = \frac{\mu_0}{4\pi} I \int_L \frac{d\vec{l} \times \hat{r}_{dlr}}{r_{dlr}^2}$$

(Eq. 8.8). For a steady current, I is not a function of position and therefore can be taken out of the integral. Expressing I as a vector, such that $Id\vec{l} = \vec{I}dl$ since $\vec{I}$ and $d\vec{l}$ are in the same direction, we can write $\vec{B}$ as:

$$\vec{B}(\vec{r}) = \frac{\mu_0}{4\pi}\int_L \frac{\vec{I} \times \hat{r}_{dlr}}{r_{dlr}^2} dl$$

We can easily generalize this expression to determine the magnetic field produced by a surface or volume current density by replacing $\vec{I}dl$ for $\vec{K}ds$ or $\vec{J}d\tau$, respectively (Eqs. 8.9 and 8.10). Worked example 8.1 shows how to use Eq. 8.8 to determine the magnetic field of an infinite straight current-carrying wire.

The Biot–Savart Law:

$$d\vec{B} = \frac{\mu_0}{4\pi}\frac{I d\vec{l} \times \hat{r}_{dlr}}{r_{dlr}^2} \quad (8.7)$$

$$\text{where } \vec{r}_{dlr} = \vec{r} - \vec{r}_{dl}$$

Magnetic field:

$$\vec{B}(\vec{r}) = \frac{\mu_0}{4\pi} I \int_L \frac{d\vec{l} \times \hat{r}_{dlr}}{r_{dlr}^2} \equiv \frac{\mu_0}{4\pi}\int_L \frac{\vec{I} \times \hat{r}_{dlr}}{r_{dlr}^2} dl \quad (8.8)$$

$$\text{where } \vec{r}_{dlr} = \vec{r} - \vec{r}_{dl}$$

For a surface current, $\vec{K}$:

$$\vec{B}(\vec{r}) = \frac{\mu_0}{4\pi}\int_S \frac{\vec{K} \times \hat{r}_{dsr}}{r_{dsr}^2} ds \quad (8.9)$$

$$\text{where } \vec{r}_{dsr} = \vec{r} - \vec{r}_{ds}$$

2 Refer to Section 7.1.2 for a description on choice of terminology for position vectors.

For a volume current, $\vec{J}$:

$$\vec{B}(\vec{r}) = \frac{\mu_0}{4\pi}\int_V \frac{\vec{J}\times\hat{r}_{\mathrm{d}\tau r}}{r^2_{\mathrm{d}\tau r}}\mathrm{d}\tau \quad (8.10)$$

where $\vec{r}_{\mathrm{d}\tau r} = \vec{r} - \vec{r}_{\mathrm{d}\tau}$

The separation vectors $\vec{r}_{\mathrm{d}l\,r}$, $\vec{r}_{\mathrm{d}s\,r}$, and $\vec{r}_{\mathrm{d}\tau\,r}$ point from the element of current at dl, ds, or dτ, respectively, to the point P at position vector $\vec{r}$ in the $\vec{B}$-field.

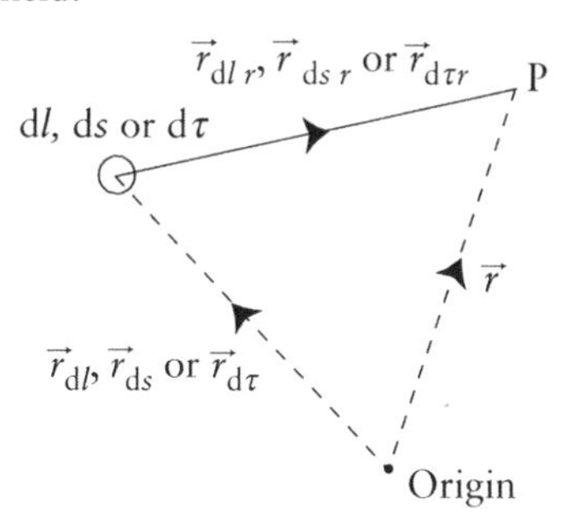

$\vec{r}_{xr} = \vec{r} - \vec{r}_x$, where x = dl, ds or dτ

Magnetic field lines

From the expressions for $\vec{B}$ (Eqs. 8.8–8.10), we see that a vector indicating the magnitude and direction of the magnetic field can be associated with each point in space. This means that magnetic fields are *vector fields* and can thus be represented schematically by *field lines*. The direction of field lines represents the direction of the magnetic field and the density of field lines represents its strength.

The cross product in the Biot–Savart Law means that the direction of d$\vec{B}$ at a point P is perpendicular to the plane containing both the element of current producing d$\vec{B}$ (d$\vec{l}$), and the vector joining the element of current to point P ($\vec{r}_{\mathrm{d}lr}$, where the direction of $\vec{r}_{\mathrm{d}lr}$ is from d$\vec{l}$ to P—refer to figure in Worked example 8.1). The net direction of $\vec{B}$ at P is then given by the vector sum of the contributing elements, d$\vec{B}$. For example, the field lines of an infinite, straight wire are concentric circles (Fig. 8.2). Therefore, whereas electric field lines begin and end on charges, magnetic field lines "wrap around" currents, forming closed loops. The *right-hand screw rule* is useful for determining the direction of magnetic field produced by a current (or vice versa): point the thumb of your right hand in the direction of the current vector, and your fingers then curl in the direction of the magnetic field (Fig. 8.3).

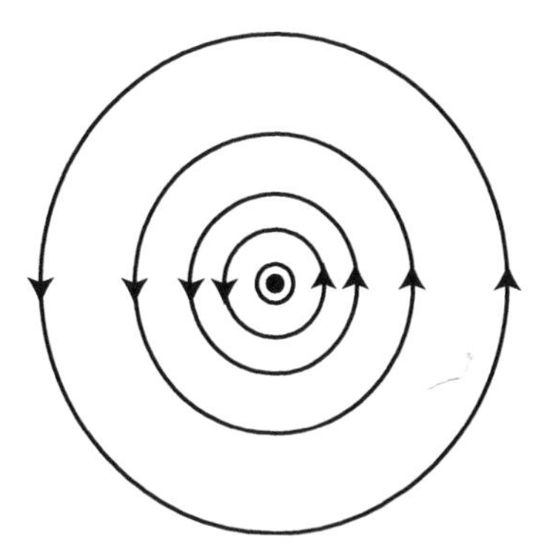

Figure 8.2: Magnetic field lines of a straight wire with the current pointing out of the page. The field strength falls off as l/r (Worked example 8.1).

Magnetic flux

Magnetic flux, Φ_B [Wb], is the magnetic equivalent of electric flux, Φ_E. It is a scalar quantity that gives a measure of the magnetic field across a given area, and can be thought of as being proportional to the number of field lines passing through a surface. The magnetic flux through a surface is defined as the perpendicular component of magnetic field passing through the surface, multiplied by the surface area. Therefore, for a magnetic field $\vec{B}$ passing through an element of area d$\vec{s}$, the element of magnetic flux is:

$$\mathrm{d}\Phi_B = \vec{B}\cdot\mathrm{d}\vec{s}$$

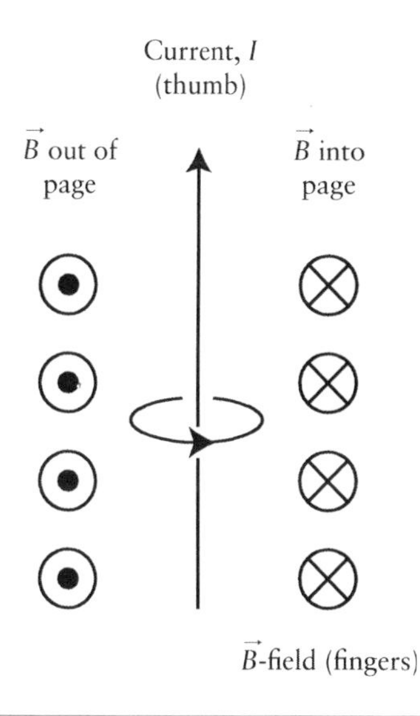

Figure 8.3: Direction of current flow and $\vec{B}$-field using the right-hand screw rule.

The dot product picks out the perpendicular component of $\vec{B}$ through the element of area, as illustrated in

WORKED EXAMPLE 8.1: Magnetic field of an infinite straight wire carrying a steady current using Eq. 8.8

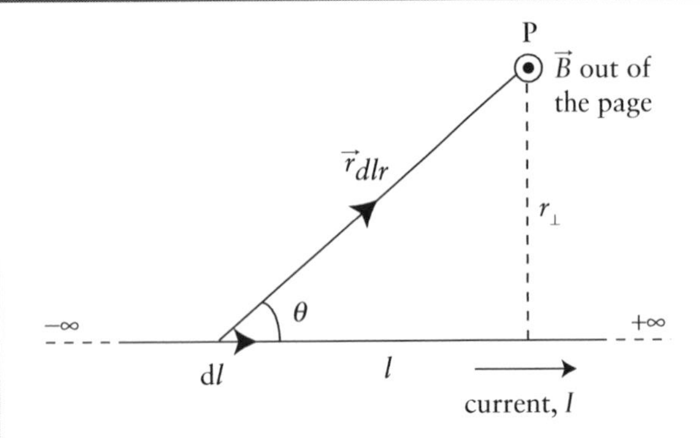

The magnetic field produced at point P (at position vector $\vec{r}$ from the origin) from the Biot–Savart Law is:

$$\vec{B}(\vec{r}) = \frac{\mu_0}{4\pi} I \int_L \frac{d\vec{l} \times \hat{r}_{dlr}}{r_{dlr}^2}$$

where, at a perpendicular distance $r_\perp$ from the wire (refer to diagram):

$$d\vec{l} \times \hat{r}_{dlr} = \sin\theta dl \quad \text{out of the page}$$

$$\text{and } r_{dlr}^2 = r_\perp^2 + l^2$$

Use the expressions for $d\vec{l} \times \hat{r}_{dlr}$ and r_{dlr}^2 in the Biot–Savart Law. To solve the integral, let $l = r_\perp \cot\theta \therefore dl = -r_\perp \csc^2\theta d\theta$ and use the trigonometric identity:

$$\sin^2\theta + \cos^2\theta = 1$$

$$\therefore 1 + \cot^2\theta = \csc^2\theta$$

$$\therefore \vec{B}(\vec{r}) = \frac{\mu_0}{4\pi} I \int_L \frac{d\vec{l} \times \hat{r}_{dlr}}{r_{dlr}^2}$$

$$= \frac{\mu_0 I}{4\pi} \int_{-\infty}^{\infty} \frac{\sin\theta dl}{r_\perp^2 + l^2}$$

$$= -\frac{\mu_0 I r_\perp}{4\pi r_\perp^2} \int_{\pi}^{0} \frac{\sin\theta \csc^2\theta d\theta}{(1+\cot^2\theta)}$$

$$= -\frac{\mu_0 I}{4\pi r_\perp} \int_{\pi}^{0} \sin\theta d\theta$$

$$= \frac{\mu_0 I}{4\pi r_\perp} [\cos\theta]_{\pi}^{0}$$

$$= \frac{\mu_0 I}{2\pi r_\perp}$$

Direction of $\vec{B}$: looking down the wire against the direction of current flow, $\vec{B}$ forms counter-clockwise concentric rings about the wire (refer to Figs. 8.2 and 8.3).

Fig. 8.4. The total magnetic flux through a surface S is then found by integrating over all elements of flux:

$$\Phi_B = \int d\Phi_B = \int_S \vec{B} \cdot d\vec{s}$$

(Eq. 8.11). We will come across applications of using Eq. 8.11 in electromagnetic induction (Chapter 10). Intuitively, flux can be thought of as being analogous to the interaction of wind and the sail of a boat, where the interaction is greatest when the wind is perpendicular to the sail.

Magnetic flux:

$$\Phi_B = \int_S \vec{B} \cdot d\vec{s} \tag{8.11}$$

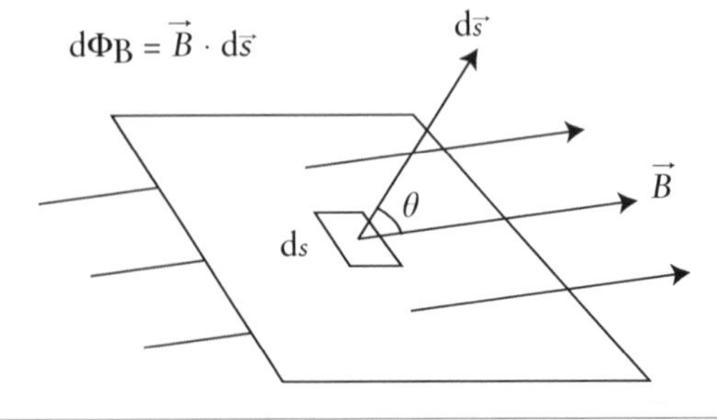

Figure 8.4: Magnetic flux. $d\vec{s}$ is the vector normal to the surface element ds, with magnitude equal to the area of the element, and θ is the angle between $\vec{B}$ and $d\vec{s}$.

8.3 Magnetic forces

Charges moving in magnetic fields experience a magnetic force, $\vec{F}_B$ [N]. The force on a charged particle q

moving at velocity $\vec{v}$ in a magnetic field $\vec{B}$ is perpendicular to both $\vec{v}$ and $\vec{B}$:

$$\vec{F}_B = q(\vec{v} \times \vec{B})$$

(Eq. 8.12). The force $\vec{F}_B$ acts to change the direction of $\vec{v}$, which then means that the direction of $\vec{F}_B$ changes. Therefore, a charged particle follows a circular trajectory in a constant magnetic field, and $\vec{F}_B$ points radially inwards (Fig. 8.5). (This is the fundamental design principle of circular particle accelerators.) By considering the element of force on an element of charge dq, we can manipulate the above expression to determine the force on a segment $d\vec{l}$ of current-carrying wire:

$$\begin{aligned} d\vec{F}_B &= dq(\vec{v} \times \vec{B}) \\ &= dq\left(\frac{d\vec{l}}{dt} \times \vec{B}\right) \\ &= \frac{dq}{dt}\left(d\vec{l} \times \vec{B}\right) \\ &= I\left(d\vec{l} \times \vec{B}\right) \end{aligned}$$

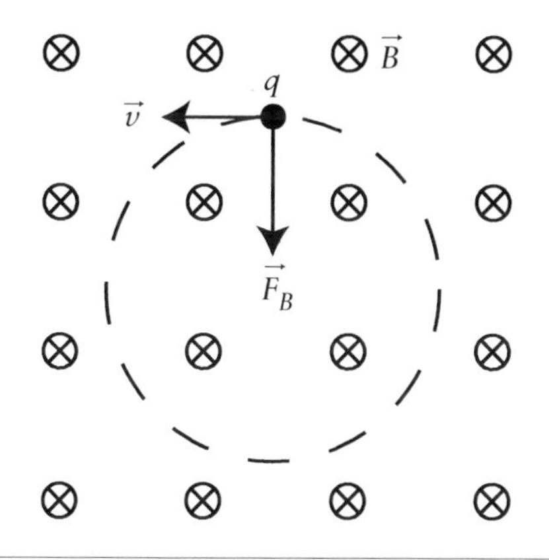

Figure 8.5 : Magnetic force on charged particle moving in a uniform $\vec{B}$-field. The particle has a circular trajectory.

Integrating, the net force on the wire is:

$$\begin{aligned} \vec{F}_B &= I\int_L \left(d\vec{l} \times \vec{B}\right) \\ &\equiv \int_L \left(\vec{I} \times \vec{B}\right) dl \end{aligned}$$

This expression is easily generalized to either a surface or volume current density, by integrating over a two-dimensional surface or three-dimensional volume, respectively (Eq. 8.13). The force on a straight segment of current-carrying wire of length L, where I is constant over L, is therefore:

$$\vec{F}_B = I\left(\vec{L} \times \vec{B}\right)$$

So, if we have two currents each flowing down two parallel wires, the wires will either attract if the currents are flowing in the same direction, or repel if the currents are flowing in the opposite direction (Fig. 8.6). It's straightforward to see this by drawing the combined magnetic field lines of the parallel wires—the wires are "pulled together" when the density of lines is low (weak B-field), and are "pushed apart" when the density is high (strong B-field).

> **Magnetic force:**
> Force on a charged particle:
>
> $$\vec{F}_B = q(\vec{v} \times \vec{B}) \quad (8.12)$$
>
> Force on a current:
>
> $$\vec{F}_B = \begin{cases} \int_L \left(\vec{I} \times \vec{B}\right) dl & \text{in 1D} \\ \int_S \left(\vec{K} \times \vec{B}\right) ds & \text{in 2D} \\ \int_V \left(\vec{J} \times \vec{B}\right) d\tau & \text{in 3D} \end{cases} \quad (8.13)$$

8.3.1 The Lorentz Force Law

The force on a charged particle moving in a magnetic field *and* an electric field is found by summing the magnetic force, $\vec{F}_B = q\vec{v} \times \vec{B}$, and the electric force, $\vec{F}_E = q\vec{E}$ (Eq. 8.14).

> **The Lorentz Force Law:**
>
> $$\vec{F}_L = q(\vec{E} + \vec{v} \times \vec{B}) \quad (8.14)$$

This is called the *Lorentz Force Law*,[3] and can be used along with Newton's Second Law of Motion to determine the trajectory of a charged particle in an electromagnetic field:

$$\begin{aligned} \text{Newton}: \ \vec{F} &= m\vec{a} = m\ddot{\vec{r}} \\ \text{Lorentz}: \ \vec{F} &= q(\vec{E} + \vec{v} \times \vec{B}) \\ &= q\vec{E} + q\dot{\vec{r}} \times \vec{B} \\ \therefore m\ddot{\vec{r}} &= q\vec{E} + q\dot{\vec{r}} \times \vec{B} \end{aligned}$$

Writing this in terms of its Cartesian components, we have:

$$m\begin{pmatrix} \ddot{x} \\ \ddot{y} \\ \ddot{z} \end{pmatrix} = q\begin{pmatrix} E_x \\ E_y \\ E_z \end{pmatrix} + q\begin{pmatrix} \dot{y}B_z - \dot{z}B_y \\ \dot{z}B_x - \dot{x}B_z \\ \dot{x}B_y - \dot{y}B_x \end{pmatrix}$$

[3] Some texts use this term to refer only to the magnetic component, $q\vec{v} \times \vec{B}$.

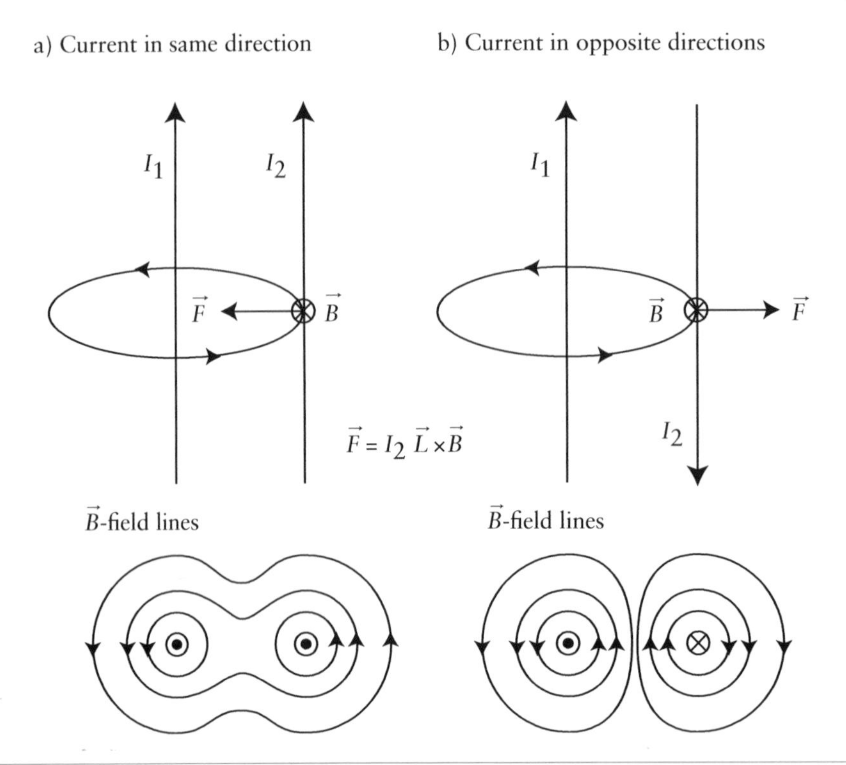

Figure 8.6: Magnetic force on and magnetic field lines of parallel wires with current flowing in a) the same or b) opposite directions.

This provides us with three simultaneous differential equations to solve. Worked example 8.2 solves these equations for a charged particle in a uniform electric and magnetic field.

8.3.2 Work done by magnetic forces

Due to the cross product in the expression for $\vec{F}_B$, the work done on a charge in a magnetic field is zero:

$$\begin{aligned} W_B &= \int \vec{F}_B \cdot d\vec{l} \\ &= \int q(\vec{v} \times \vec{B}) \cdot \vec{v} dt \\ &= 0 \end{aligned}$$

since $\vec{v} \times \vec{B}$ is perpendicular to $\vec{v}$. The direction of a charged particle is altered in a magnetic field, but its energy remains constant—the particle neither speeds up nor slows down.

8.4 Divergence and curl of $\vec{B}$

Now let's calculate the divergence and curl of $\vec{B}$, as we did for $\vec{E}$ in Section 7.2.

8.4.1 The divergence of $\vec{B}$

In contrast with the electrostatic case, a static magnetic field has zero divergence (Eq. 8.16 and Derivation 8.1). This is a statement of the absence of magnetic monopoles: magnetic fields have no sources or sinks. Eq. 8.16 is another of Maxwell's equations, which is sometimes known as *Gauss's Law for magnetism*, but is often referred to as *the divergence of B*.

The divergence of $\vec{B}$ (Derivation 8.1):

$$\nabla \cdot \vec{B} = 0 \qquad (8.16)$$

8.4.2 Ampère's Law

We can use Stokes's theorem to find the curl of $\vec{B}$. Stokes's theorem states that the curl of a vector field is related to its closed line integral by:

$$\int_S \left(\nabla \times \vec{B}\right) \cdot d\vec{s} = \oint_L \vec{B} \cdot d\vec{l}$$

Derivation 8.1: Derivation of the divergence of $\vec{B}$ (Eq. 8.16)

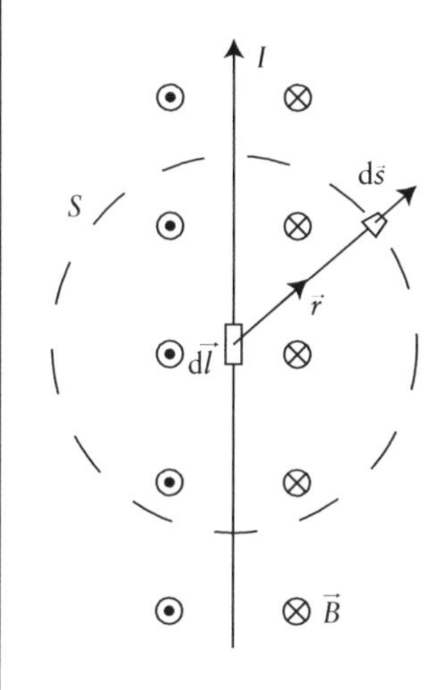

$$\Phi_B = \oint_S \vec{B} \cdot d\vec{s} = \frac{\mu_0}{4\pi} I \oint_S \int \left(\frac{d\vec{l} \times \hat{r}}{r^2} \right) \cdot d\vec{s}$$
$$= 0$$

From the divergence theorem:

$$\int_V (\nabla \cdot \vec{B}) d\tau = \oint_S \vec{B} \cdot d\vec{s} = 0$$
$$\therefore \nabla \cdot \vec{B} = 0$$

- Starting point: consider part of a current-carrying wire enclosed within a closed surface S (refer to diagram). The magnetic flux through S is:

$$\Phi_B = \oint_S \vec{B} \cdot d\vec{s}$$

- The magnetic field is given by the integral:

$$\vec{B} = \frac{\mu_0 I}{4\pi} \int \left(\frac{d\vec{l} \times \hat{r}}{r^2} \right) \cdot d\vec{s}$$

where, in this case, we have defined $\vec{r}$ as the vector pointing from $d\vec{l}$ to a point in the B-field produced by I (this variable corresponds to $\vec{r}_{dl\,r}$ in Eq. 8.8) .
- For each element $+d\vec{B}$ entering the surface, there is an equal and opposite element $-d\vec{B}$ emerging from it. Therefore, the net entering and emerging flux cancels, hence the magnetic flux through the entire closed surface is zero: $\Phi_B = 0$.
- From the divergence theorem, zero flux through a closed surface implies that the divergence of $\vec{B}$ is zero.

where S is the surface enclosed by L. To start with, let's determine the closed line integral of $\vec{B}$ for current flowing down an infinite straight wire, as shown in Derivation 8.2.[4] Doing this, we find that the integral is proportional to the current enclosed within the loop:

$$\oint_L \vec{B} \cdot d\vec{l} = \mu_0 I_{enc}$$

(Eq. 8.17 and Derivation 8.2). This is another of Maxwell's equations, and is known as *Ampère's Law*. It describes the magnetic field solely in terms of the current producing the field. By invoking Stokes's theorem, we can then express Ampère's Law as the curl of $\vec{B}$:

$$\nabla \times \vec{B} = \mu_0 \vec{J}$$

[4] Derivation 8.2 proves a general result from the specific and idealized case of an infinite straight wire, since this treatment is more straightforward and intuitive. However, it holds true for all current configurations.

(Eq. 8.18 and Derivation 8.3). Maxwell's four equations for electrostatic and magnetostatics are summarized in Eqs. 8.19–8.22.

Ampère's Law (Derivations 8.2 and 8.3):

Integral form:

$$\oint_L \vec{B} \cdot d\vec{l} = \mu_0 I_{enc} \tag{8.17}$$

Differential form:

$$\nabla \times \vec{B} = \mu_0 \vec{J} \tag{8.18}$$

Using Ampère's Law

For systems that possess geometrical symmetry, such as radial or planar symmetry, we can use the integral form of Ampère's Law to calculate the magnetic field of a system from the current enclosed in a given

WORKED EXAMPLE 8.2: Motion of a charged particle in a uniform electric and magnetic field (Eq. 8.14)

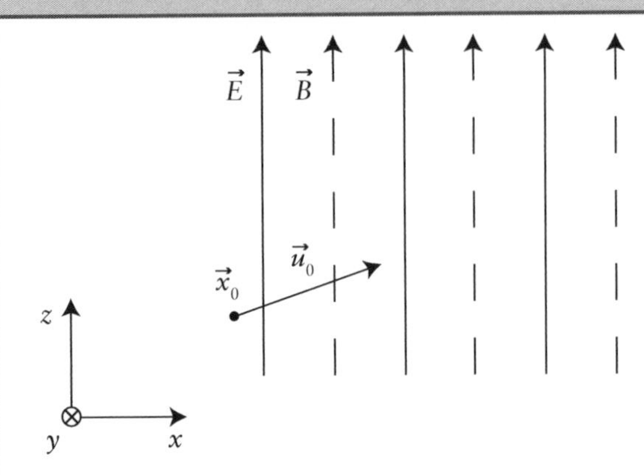

Consider a charged particle q of mass m injected at time $t=0$ into a region in which there is a uniform electric and magnetic field, both along the z-direction:

$$\vec{B} = \begin{pmatrix} 0 \\ 0 \\ B \end{pmatrix} \quad \text{and} \quad \vec{E} = \begin{pmatrix} 0 \\ 0 \\ E \end{pmatrix}$$

Define the axes such that the initial position, $\vec{x}_0$, corresponds to the origin and the initial particle velocity along the y-axis is $u_0^y=0$. Therefore, the initial velocity is u_0^x along the x-axis and u_0^z along the z-axis:

$$\vec{x}_0 = \begin{pmatrix} 0 \\ 0 \\ 0 \end{pmatrix} \quad \text{and} \quad \vec{u}_0 = \begin{pmatrix} u_0^x \\ 0 \\ u_0^z \end{pmatrix}$$

The force on the particle is then:

$$\vec{F} = q(\vec{E} + \vec{v} \times \vec{B}) = q\begin{pmatrix} v_y B \\ -v_x B \\ E \end{pmatrix}$$

Using this, together with Newton's Second Law, $\vec{F}=m\vec{a}$, the equation of motion of the particle is:

$$m\begin{pmatrix} \dot{v}_x \\ \dot{v}_y \\ \dot{v}_z \end{pmatrix} = q\begin{pmatrix} v_y B \\ -v_x B \\ E \end{pmatrix} \tag{8.15}$$

z-solution:

Integrating the z-equation and applying initial conditions, we see that the particle has uniform acceleration in the z-direction:

$$\frac{dv_z}{dt} = \frac{qE}{m}$$

$$\therefore v_z = \frac{qE}{m}t + u_0^z$$

$$\text{and} \quad z = \frac{qE}{2m}t^2 + u_0^z t$$

x- and y-solutions:

The x- and y-equations are a pair of coupled differential equations. We can decouple them by differentiating the equations of motion from Eq. 8.15 and eliminating $\dot{v}_y$ from the x-equation and $\dot{v}_x$ from the y-equation, using expressions for $\dot{v}_y$ and $\dot{v}_x$ from Eq. 8.15:

$$\begin{pmatrix} \ddot{v}_x \\ \ddot{v}_y \end{pmatrix} = \frac{qB}{m}\begin{pmatrix} \dot{v}_y \\ -\dot{v}_x \end{pmatrix} = -\left(\frac{qB}{m}\right)^2\begin{pmatrix} v_x \\ v_y \end{pmatrix} = -\omega^2\begin{pmatrix} v_x \\ v_y \end{pmatrix}$$

where the *gyro-frequency* is defined as:

$$\omega = \frac{qB}{m}$$

WORKED EXAMPLE 8.2: Motion of a charged particle in a uniform electric and magnetic field (Eq. 8.14) (cont.)

We recognize these as equations of simple harmonic motion. The general solutions are therefore:

$$\begin{pmatrix} v_x \\ v_y \end{pmatrix} = \begin{pmatrix} a \cos \omega t + b \sin \omega t \\ c \cos \omega t + d \sin \omega t \end{pmatrix}$$

(refer to Section 2.3.1 for simple harmonic motion solutions.) Applying the initial velocity conditions, we find $a = u_0^x, b = 0$, $c = 0$, and $d = -u_0^x$ (solve for b and d by differentiating the velocity solutions and using Eq. 8.15). Therefore, the x- and y-velocity solutions are:

$$\begin{pmatrix} v_x \\ v_y \end{pmatrix} = u_0^x \begin{pmatrix} \cos \omega t \\ -\sin \omega t \end{pmatrix}$$

Integrating, we find that the x- and y-displacement solutions are:

$$\begin{pmatrix} x \\ y \end{pmatrix} = \frac{u_0^x}{\omega} \begin{pmatrix} \sin \omega t \\ \cos \omega t \end{pmatrix}$$

These solutions correspond to circular motion at a radius of u_0^x/ω in the x–y plane:

$$x^2 + y^2 = \left(\frac{u_0^x}{\omega}\right)^2 \left(\sin^2 \omega t + \cos^2 \omega t\right)$$

$$= \left(\frac{u_0^x}{\omega}\right)^2$$

Combining this with the uniform acceleration in the z-direction, we find that the overall motion of the charged particle is helical and has a time-dependent pitch.

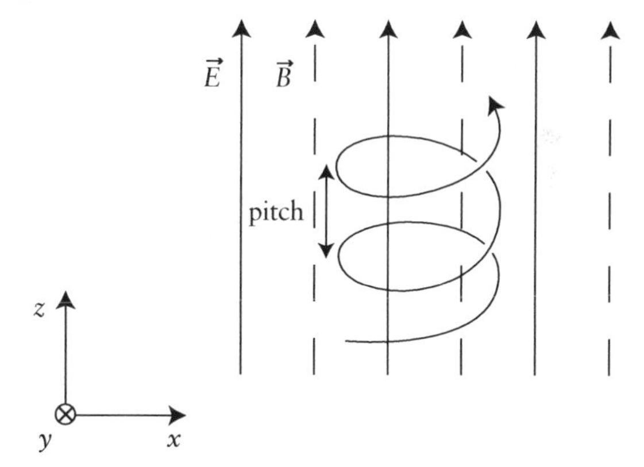

Derivation 8.2: Derivation of the integral form of Ampère's Law from an infinite straight wire (Eq. 8.17)

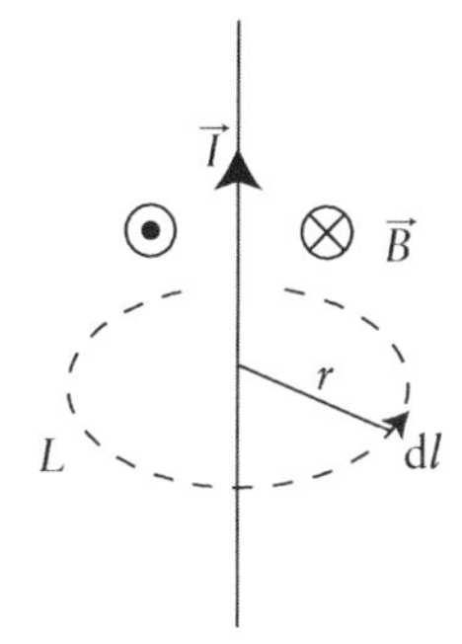

$$\oint_L \vec{B} \cdot d\vec{l} = \oint_L \frac{\mu_0 I}{2\pi r} \hat{\phi} \cdot d\vec{l}$$

$$= \frac{\mu_0 I}{2\pi r} \int_{\phi=0}^{2\pi} \underbrace{\hat{\phi} \cdot \hat{\phi}}_{=1} r \, d\phi$$

$$= \frac{\mu_0 I}{2\pi r} 2\pi r$$

$$= \mu_0 I_{\text{enc}}$$

- Starting point: consider an infinite current-carrying wire. Its magnetic field wraps around the wire in concentric circles, such that at a radius r from the wire, the magnetic field is:

  $$\vec{B} = \frac{\mu_0 I}{2\pi r} \hat{\phi}$$

  (refer to Worked example 8.1).
- A line element in cylindrical polar coordinates is:

  $$d\vec{l} = dr\hat{r} + r d\phi\hat{\phi} + dz\hat{z}$$

- Substitute $\vec{B}$ and $d\vec{l}$ into the closed line integral for $\vec{B}$. Since $\hat{\phi}$ is orthogonal to $\hat{r}$ and $\hat{z}$, only the $\hat{\phi}$ term in $d\vec{l}$ remains in the integral.
- Integrating around a closed loop, we see that the closed line integral of $\vec{B}$ is proportional to the current enclosed within in the loop.

Derivation 8.3: Derivation of the differential form of Ampère's Law (Eq. 8.18)

$$\int_S (\nabla\times\vec{B})\cdot d\vec{s} = \oint_L \vec{B}\cdot d\vec{l} = \mu_0 I_{enc} = \mu_0 \int_S \vec{J}\cdot d\vec{s}$$

This holds for any surface, therefore the integrands must be equal:

$$\therefore \nabla\times\vec{B} = \mu_0\vec{J}$$

- Starting point: Stokes's theorem states that the curl of a vector field is related to its closed line integral by:

  $$\int_S (\nabla\times\vec{B})\cdot d\vec{s} = \oint_L \vec{B}\cdot d\vec{l}$$

  where S is the surface enclosed by L.
- From Ampère's Law in integral form (Eq. 8.17), we have:

  $$\oint_L \vec{B}\cdot d\vec{l} = \mu_0 I_{enc}$$

- Current can be expressed as a volume current density by:

  $$I = \int_S \vec{J}\cdot d\vec{s}$$

Maxwell's equations for electrostatics and magnetostatics:

$$\text{Gauss's Law}: \quad \nabla\cdot\vec{E} = \frac{\rho}{\varepsilon_0} \qquad \oint_S \vec{E}\cdot d\vec{s} = \frac{Q_{enc}}{\varepsilon_0} \tag{8.19}$$

$$\text{Divergence of } \vec{B}: \quad \nabla\cdot\vec{B} = 0 \qquad \oint_S \vec{B}\cdot d\vec{s} = 0 \tag{8.20}$$

$$\text{Curl of } \vec{E}: \quad \nabla\times\vec{E} = 0 \qquad \oint_L \vec{E}\cdot d\vec{l} = 0 \tag{8.21}$$

$$\text{Ampère's Law}: \quad \nabla\times\vec{B} = \mu_0\vec{J} \qquad \oint_L \vec{B}\cdot d\vec{l} = \mu_0 I_{enc} \tag{8.22}$$

loop (Worked example 8.3). In such calculations, the symmetrical loop is called an *Amperian loop* (Eqs. 8.23–8.25). To determine the direction of $\vec{B}$ from the direction of current flow, we can use the right-hand screw rule described in Section 8.2.

Common Amperian loops (refer to Worked example 8.4):

Radial symmetry:
Circular loop, radius r:

$$\oint_L d\vec{l} = 2\pi r \tag{8.23}$$

Planar symmetry:
Rectangular loop, length l:

$$\text{Short width}: \oint_L d\vec{l} = 2l \tag{8.24}$$

$$\text{Infinite width}: \oint_L d\vec{l} = l \tag{8.25}$$

The differential form of Ampère's Law is used when we have the inverse problem to solve, where we know the magnetic field and wish to determine the current density:

$$\vec{J} = \frac{1}{\mu_0}\nabla\times\vec{B}$$

8.4.3 The magnetic vector potential

In electrostatics, the curl-free nature of $\vec{E}$ enabled the definition of the *scalar* electric potential field, V, where:

$$\vec{E} \overset{\text{def}}{=} -\nabla V$$

Similarly, the divergence-free nature of $\vec{B}$ enables the definition of a *vector* potential field: the *magnetic vector potential*, $\vec{A}$[Tm]. Magnetic vector potential is related to magnetic field by:

$$\vec{B} \overset{\text{def}}{=} \nabla\times\vec{A}$$

WORKED EXAMPLE 8.3: Determining $\vec{B}$-fields using Ampère's Law (Eq. 8.17)

The integral form of Ampère's Law is:

$$\oint_L \vec{B} \cdot \mathrm{d}\vec{l} = \mu_0 I_{\text{enc}}$$

For systems that possess geometrical symmetry, this form of Ampère's Law allows us to calculate the magnetic field from just the current enclosed in a given loop. In the examples below (systems with radial and planar symmetry), we find that (i) $\vec{B}$ and $\mathrm{d}\vec{l}$ are either perpendicular or parallel, meaning that we can drop the dot product, and that (ii) the magnitude of $\vec{B}$ is constant around the loop, so it can be taken out of the integral:

$$\Rightarrow |\vec{B}| = \frac{\mu_0 I_{\text{enc}}}{\oint \mathrm{d}\vec{l}}$$

The loop around which the integral $\oint \mathrm{d}\vec{l}$ is performed is referred to as an *Amperian loop* (Eqs. 8.23–8.25).

Circular Amperian loop (radius r)

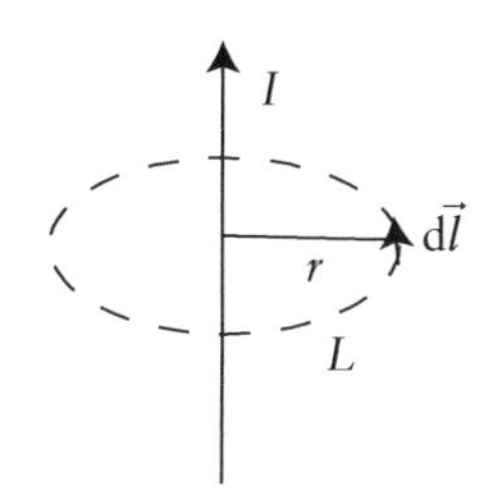

Rectangular Amperian loop (length l)

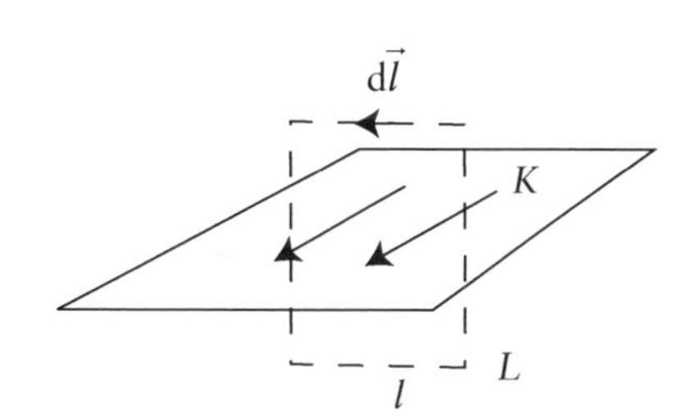

Rectangular Amperian loop (length l)

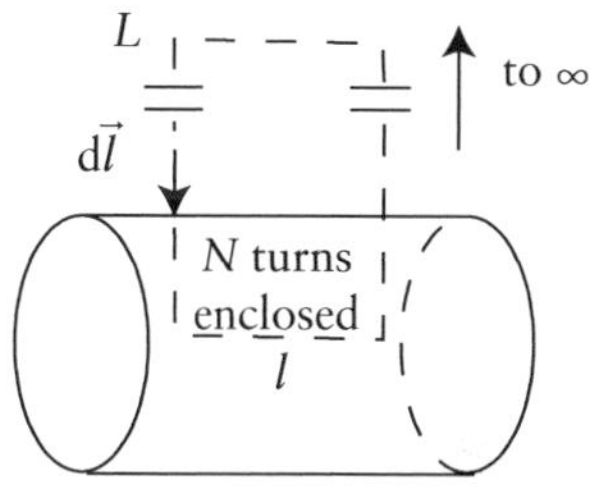

current I per turn,
n turns per unit length

a) Infinite wire (circular Amperian loop):

$$\oint \mathrm{d}\vec{l} = 2\pi r$$

$$I_{\text{enc}} = I$$

$$\therefore |\vec{B}| = \frac{\mu_0 I_{\text{enc}}}{\oint \mathrm{d}\vec{l}} = \frac{\mu_0 I}{2\pi r}$$

b) Infinite surface current (rectangular Amperian loop):

$$\oint \mathrm{d}\vec{l} = 2l \text{ (finite loop)}$$

$$I_{\text{enc}} = Kl$$

$$\therefore |\vec{B}| = \frac{\mu_0 I_{\text{enc}}}{\oint \mathrm{d}\vec{l}} = \frac{\mu_0 K}{2}$$

c) Infinite solenoid (rectangular Amperian loop):

$$\oint \mathrm{d}\vec{l} = l \text{ (infinite loop)}$$

$$I_{\text{enc}} = NI$$

$$\therefore |\vec{B}| = \frac{\mu_0 I_{\text{enc}}}{\oint \mathrm{d}\vec{l}} = \frac{\mu_0 I N}{l} = \mu_0 I n$$

Note that the magnetic field of an infinite wire (part a) is the same as that calculated in Worked example 8.1

The divergence of $\vec{B}$ is then:

$$\nabla\cdot\vec{B} = \nabla\cdot\nabla\times\vec{A} \\ = 0$$

since the divergence of a curl equals zero. While $\vec{A}$ does not have an intuitive physical interpretation like V does (recall that V represents the work done per unit charge in an electric field, $V = W/q$), it has many elegant mathematical parallels with V. Since $\vec{A}$ is a vector, then for every one equation for V, there are three analogous equations for $\vec{A}$ (one for each spatial component). Let's look at an example.

Poisson's equation for $\vec{A}$

Poisson's equation ($\nabla^2 V = -\rho/\varepsilon_0$) comes from substituting $\vec{E} = -\nabla V$ into Gauss's Law. Now, let's substitute $\vec{B} = \nabla\times\vec{A}$ into Ampère's Law, using the vector identity for the curl of a curl:

$$\mu_0\vec{J} = \nabla\times\vec{B} \\ = \nabla\times\nabla\times\vec{A} \\ = \nabla(\nabla\cdot\vec{A}) - \nabla^2\vec{A}$$

It turns out that a solution for $\vec{A}$ always exists such that $\nabla\cdot\vec{A} = 0$ where $\vec{A}$ still satisfies $\vec{B} = \nabla\times\vec{A}$. For a proof, the reader is referred to Griffith's book, *Introduction to Electrodynamics*. Therefore, we have:

$$\therefore \nabla^2\vec{A} = -\mu_0\vec{J} \qquad (8.26)$$

which is Poisson's equation for $\vec{A}$. Or, written in terms of its Cartesian components:

$$\begin{pmatrix} \nabla^2 A_x \\ \nabla^2 A_y \\ \nabla^2 A_z \end{pmatrix} = -\mu_0 \begin{pmatrix} J_x \\ J_y \\ J_z \end{pmatrix}$$

We saw in Section 7.3 that the solution for Poisson's equation (provided $\rho\to 0$ at $\vec{r} = \infty$) is:

$$V(\vec{r}) = \frac{1}{4\pi\varepsilon_0}\int \frac{\rho}{r_{d\tau r}}\,d\tau$$

where $\vec{r}_{d\tau r} = \vec{r} - \vec{r}_{d\tau}$ is the separation vector from the volume element $d\tau$ to the position vector $\vec{r}$. Therefore by analogy, the solution for Eq. 8.26 (provided $\vec{J}\to 0$ at $\vec{r} = \infty$) is:

$$\vec{A}(\vec{r}) = \frac{\mu_0}{4\pi}\int \frac{\vec{J}}{r_{d\tau r}}\,d\tau$$

where $\vec{r}_{d\tau r} = \vec{r} - \vec{r}_{d\tau}$

Similarly, $\vec{A}$ is related to $\vec{I}$ or $\vec{K}$ by:

$$\vec{A}(\vec{r}) = \frac{\mu_0}{4\pi}\int \frac{\vec{I}}{r_{dlr}}\,dl$$

$$\text{or } \vec{A}(\vec{r}) = \frac{\mu_0}{4\pi}\int \frac{\vec{K}}{r_{dsr}}\,ds$$

where $\vec{r}_{dlr} = \vec{r} - \vec{r}_{dl}$

and $\vec{r}_{dsr} = \vec{r} - \vec{r}_{ds}$

The magnetic vector potential has theoretical applications, which are beyond the scope of this book. However, its symmetry with V provides us with an elegant mathematical tool for solving for $\vec{B}$ using $\vec{B} = \nabla\times\vec{A}$, without using the Biot–Savart equation.

8.5 Magnetic dipoles

A *magnetic dipole* is the magnetic equivalent of the electric dipole (refer to Section 7.5). A magnetic dipole can be represented as a single-turn current loop of area $\vec{s}$ (Fig. 8.7). For current I flowing around the loop, the *magnetic dipole moment*, $\vec{m}$[A m^2], is defined as:

$$\vec{m} = I\vec{s}$$

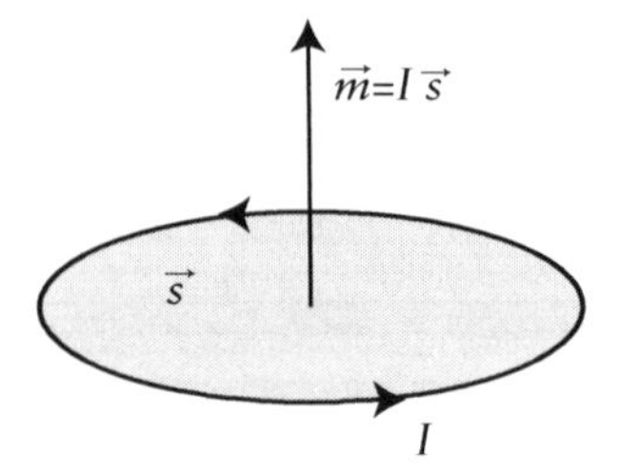

Figure 8.7 : Magnetic dipole.

(Eq. 8.27). The direction of $\vec{m}$ can be found using the right-hand screw rule: curl the fingers of your right hand in the direction of I, and your thumb then points along $\vec{m}$.

Magnetic dipole moment:

$$\vec{m} = I\vec{s} \qquad (8.27)$$

where $\vec{s}$ is the area of the current loop.

8.5.1 Magnetic field and potential

Let's exploit the parallel between V and $\vec{A}$ to write the magnetic vector potential of a dipole. For an ideal electric dipole, we have:

$$\begin{aligned} V(\vec{r}) &= \frac{\vec{p}\cdot\hat{r}}{4\pi\varepsilon_0 r^2} \\ &= \frac{p\cos\theta}{4\pi\varepsilon_0 r^2} \end{aligned}$$

(refer to Section 7.5). It turns out that the magnetic vector potential of an ideal magnetic dipole indeed has an analogous form, as given by Eq. 8.28.

Magnetic vector potential of an ideal dipole:

$$\begin{aligned} \vec{A}(\vec{r}) &= \frac{\mu_0 \vec{m}\times\hat{r}}{4\pi r^2} \\ &= \frac{\mu_0 m \sin\theta}{4\pi r^2}\hat{\phi} \end{aligned} \tag{8.28}$$

The magnetic field of a magnetic dipole is then found by taking the curl of $\vec{A}$ (Eq. 8.29, Derivation 8.4, and Fig. 8.8). From this, we see that the magnetic field produced by a magnetic dipole has an identical form to that of the *electric* field produced by an electric dipole (compare Eq. 8.29 with Eq. 7.41, and Fig. 8.8 with Fig. 7.10). Furthermore, the magnetic field of a dipole has the same form as the magnetic field of a bar magnet.

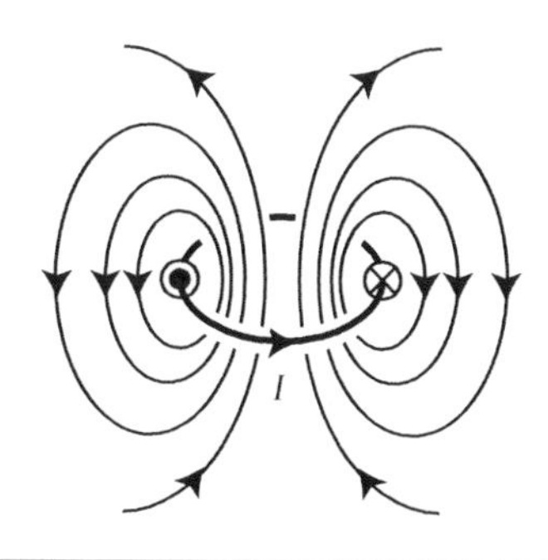

Figure 8.8 : Magnetic field of a magnetic dipole.

Magnetic field of an ideal dipole (Derivation 8.4):

$$\begin{aligned} \vec{B}(\vec{r}) &= \frac{\mu_0 m}{4\pi r^3}(2\cos\theta\hat{r} + \sin\theta\hat{\theta}) \\ &\equiv \frac{\mu_0 m}{4\pi r^3}\begin{pmatrix} 2\cos\theta \\ \sin\theta \end{pmatrix} \end{aligned} \tag{8.29}$$

8.5.2 Dipole in an external magnetic field

A dipole in an external magnetic field experiences a *torque*, $\vec{\tau}_m$[N m], since opposing forces act on each direction of current in the current loop. The torque is given by the cross product of $\vec{m}$ and $\vec{B}$:

$$\vec{\tau}_m = \vec{m}\times\vec{B}$$

Derivation 8.4: Derivation of the magnetic field of a dipole (Eq. 8.29)

$$\begin{aligned} \vec{B} &= \nabla\times\vec{A} \\ \therefore B_r &= \frac{1}{r\sin\theta}\left(\frac{\partial}{\partial\theta}(A_\phi\sin\theta) - \frac{\partial A_\theta}{\partial\phi}\right)\hat{r} \\ &= \frac{2\mu_0 m\cos\theta}{4\pi r^3} \\ B_\theta &= \frac{1}{r}\left(\frac{1}{\sin\theta}\frac{\partial A_r}{\partial\phi} - \frac{\partial}{\partial r}(rA_\phi)\right)\hat{\theta} \\ &= \frac{\mu_0 m\sin\theta}{4\pi r^3} \\ B_\phi &= \frac{1}{r}\left(\frac{\partial}{\partial r}(rA_\theta) - \frac{\partial A_r}{\partial\phi}\right)\hat{\phi} \\ &= 0 \end{aligned}$$

- Starting point: the magnetic field is the curl of the magnetic vector potential:

$$\vec{B} = \nabla\times\vec{A}$$

- Take the curl of $\vec{A} = \mu_0 m\sin\theta/4\pi r^2\hat{\phi}$ (Eq. 8.28) using the Del operator in spherical polar coordinates:

$$\nabla\times A = \begin{pmatrix} \frac{1}{r\sin\theta}\left(\frac{\partial}{\partial\theta}(A_\theta\sin\theta) - \frac{\partial A_\theta}{\partial\phi}\right) \\ \frac{1}{r}\left(\frac{1}{\sin\theta}\frac{\partial A_r}{\partial\phi} - \frac{\partial}{\partial r}(rA_\phi)\right) \\ \frac{1}{r}\left(\frac{\partial}{\partial r}(rA_\theta) - \frac{\partial A_r}{\partial\theta}\right) \end{pmatrix}$$

(Eq. 8.30 and Derivation 8.5). The torque tends to align the dipole with the external field. This is the underlying principle of how a compass works. The resulting potential energy stored in the dipole is then:

$$U_m = -\vec{m}\cdot\vec{B}$$

(Eq. 8.31 and Derivation 8.6). The net *linear* force on the dipole is then:

$$\begin{aligned}\vec{F}_m &= -\nabla U_m \\ &= \nabla(\vec{m}\cdot\vec{B})\end{aligned}$$

(Eq. 8.32). If the field is uniform, the net linear force is zero since the forces on currents moving in either $+d\vec{l}$ or $-d\vec{l}$ are equal and opposite:

$$\vec{F}_m = Id\vec{l}\times\vec{B} - Id\vec{l}\times\vec{B} = 0$$

Eqs. 8.30–8.32 are analogous to their electric counterparts (Eqs. 7.42–7.44).

Dipole in an external magnetic field (Derivations 8.5 and 8.6):

$$\text{Torque: } \vec{\tau}_m = \vec{m}\times\vec{B} \quad (8.30)$$

$$\text{Energy: } U_m = -\vec{m}\cdot\vec{B} \quad (8.31)$$

$$\text{Force: } \vec{F}_m = \nabla(\vec{m}\cdot\vec{B}) \quad (8.32)$$

Derivation 8.5: Derivation of the torque on a dipole in an external $\vec{B}$-field (Eq. 8.30)

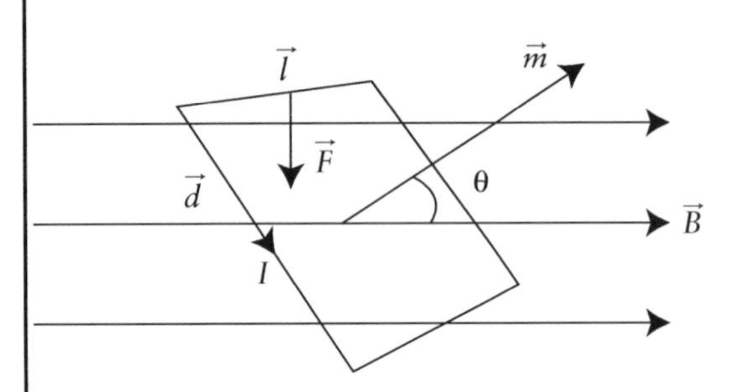

$$\begin{aligned}\vec{\tau}_m &= \vec{r}\times\vec{F} \\ &= \vec{d}\times I\vec{l}\times\vec{B} \\ &= I\vec{s}\times\vec{B} \\ &= \vec{m}\times\vec{B}\end{aligned}$$

$$\therefore |\vec{\tau}_m| = |\vec{m}||\vec{B}|\sin\theta$$

- Starting point: consider a rectangular magnetic dipole, where the dipole moment is $\vec{m} = I\vec{s}$, with sides of length $\vec{d}$ and $\vec{l}$, such that the area of the dipole is $\vec{s} = \vec{d}\times\vec{l}$.
- Taking one of the sides of length $\vec{l}$ as a pivot axis, the force on the adjacent side of length $\vec{l}$ is $\vec{F} = I\vec{l}\times\vec{B}$.
- The torque acts in a sense to decrease θ (the angle between $\vec{m}$ and $\vec{B}$, and is zero when the dipole is parallel to the field.

Derivation 8.6: Derivation of the energy of a dipole in an external $\vec{B}$-field (Eq. 8.31)

$$\begin{aligned}U_m &= \int_{\pi/2}^{\theta} |\vec{\tau}_m|\,d\theta' \\ &= \int_{\pi/2}^{\theta} |\vec{m}||\vec{B}|\sin\theta' d\theta' \\ &= \Big[-|\vec{m}||\vec{B}|\cos\theta'\Big]_{\pi/2}^{\theta} \\ &= -|\vec{m}||\vec{B}|\cos\theta \\ &= -\vec{m}\cdot\vec{B}\end{aligned}$$

- Starting point: the energy of a dipole in an external magnetic field equals the work necessary to rotate the dipole from an initial equipotential field line (perpendicular to $\vec{B} \rightarrow \therefore\ \theta' = \pi/2$) to an angle θ .
- The magnitude of torque on a dipole in an external field is:

$$|\vec{\tau}_m| = |\vec{m}\times\vec{B}| = |\vec{m}||\vec{B}|\sin\theta'$$

- The potential energy of the dipole is minimum when it is parallel to the external field.

Chapter 9
Introduction to circuit theory

Moving electric charge produces an electric current. Circuit theory is the study of the properties of electric current as it flows through a closed loop comprising electrical components, such as resistors and capacitors, producing work in the process. In this chapter, we will look at the laws that apply to electrical circuits, and use them to determine how components in a circuit respond to a direct current (DC response), a transient current (transient response), and an alternating current (AC response).

9.1 Electrical laws

9.1.1 The continuity equation

Current, I [$\text{A} \equiv \text{C s}^{-1}$], is the rate of flow of electric charge passing a given point:

$$I = \frac{\mathrm{d}Q}{\mathrm{d}t}$$

(Eq. 9.3). Charge flowing through a volume produces a *current density*, $\vec{J}$ [A m^{-2}], which is the current per unit area-perpendicular-to-the-flow:

$$\vec{J} = \frac{\mathrm{d}I}{\mathrm{d}s_{\perp}}$$

(refer to Section 8.1). Invoking the divergence theorem, the net current flowing out of a volume V is then:

$$\begin{aligned} I &= \oint_S \vec{J} \cdot \mathrm{d}\vec{s} \\ &= \int_V \nabla \cdot \vec{J}\,\mathrm{d}\tau \end{aligned} \tag{9.1}$$

Charge is related to volume charge density, ρ, by:

$$Q = \int_V \rho\,\mathrm{d}\tau$$

Therefore, the rate of change of electric charge flowing *out* of a volume V is:

$$\frac{\mathrm{d}Q}{\mathrm{d}t} = -\int_V \frac{\partial \rho}{\partial t}\mathrm{d}\tau \tag{9.2}$$

where the partial derivative is because ρ is a function of space as well as time. Equating Eqs. 9.1 and 9.2, we have:

$$\begin{aligned} I &= \frac{\mathrm{d}Q}{\mathrm{d}t} \\ \therefore \int_V \nabla \cdot \vec{J}\,\mathrm{d}\tau &= -\int_V \frac{\partial \rho}{\partial t}\mathrm{d}\tau \end{aligned}$$

This holds for all volumes:

$$\therefore \nabla \cdot \vec{J} = -\frac{\partial \rho}{\partial t}$$

(Eq. 9.4). This is called the *continuity equation*, and it is a statement of the conservation of charge: charge cannot be created or destroyed.

The continuity equation:

$$\text{In one dimension:} \quad I = \frac{\mathrm{d}Q}{\mathrm{d}t} \tag{9.3}$$

$$\text{In three dimensions:} \quad \nabla \cdot \vec{J} = -\frac{\partial \rho}{\partial t} \tag{9.4}$$

9.1.2 Resistance and Ohm's Law

When current flows through a medium, collisions between the moving charges and other atoms lead to a *resistance*, $R\,[\Omega]$, against the flow. The *resistivity*, ρ [Ω m], of a material is defined as the resistance per unit length for a uniform cross section, and is a constant for a given material:[1]

$$\rho = \frac{RA}{L}$$

where L is the length and A is the cross-sectional area of the material over which its resistance is R.

(Eq. 9.5). The inverse of resistivity is the *conductivity*, σ [S m^{-1}], which is a measure of the material's ability to conduct electricity (Eq. 9.6). (Be careful not to confuse these parameters with the volume charge density, ρ, and surface charge density, σ.)

Resistivity and conductivity:

$$\text{Resistivity}: \ \rho = \frac{RA}{L} \tag{9.5}$$

$$\text{Conductivity}: \ \sigma = \frac{1}{\rho} \tag{9.6}$$

Ohm's Law

Current flow through many materials obeys *Ohm's Law*. This law states that, for an electrical component of constant resistance R, the voltage, V, across the component is related to the current, I, through it by:

$$V = IR$$

(Eq. 9.7). Consider current flow through a cylindrical volume, of length L, cross-sectional area A, and resistance $R = L/\sigma A$, as shown in Fig. 9.1. Assuming a constant potential difference V across the length, the electric field produced is:

$$\vec{E} = \frac{V}{L}$$

Using $V = \vec{E}L$ and $R = L/\sigma A$ in Ohm's Law, we derive a more general, three-dimensional form of Ohm's Law:

$$V = IR$$
$$\therefore \vec{E}L = I\frac{L}{\sigma A} = \frac{\vec{J}L}{\sigma}$$
$$\therefore \vec{J} = \sigma\vec{E}$$

[1] A mnemonic for the definition of ρ is to remember rowing: "row = right arm over leg" ($\rho = RA/L$).

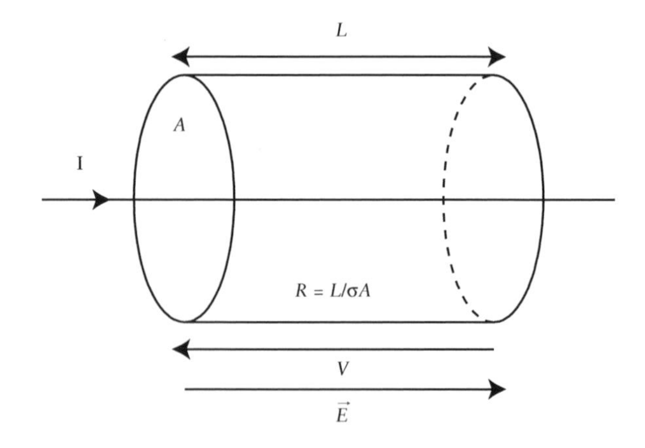

Figure 9.1: Current flow through a cylindrical volume with a constant potential difference across its length.

(Eq. 9.8). For isotropic materials (materials that have the same properties in all directions), the conductivity, σ, is a scalar, and for anisotropic materials, it is a tensor.

Ohm's Law:

$$\text{In one dimension}: \ V = IR \tag{9.7}$$

$$\text{In three dimensions}: \ \vec{J} = \sigma\vec{E} \tag{9.8}$$

where σ is the conductivity.

Electric power

The work done in moving a charge Q through a potential difference V is $W = QV$. The electric power (rate at which electrical energy is transferred by a circuit) for a constant potential is then:

$$P = \frac{\mathrm{d}W}{\mathrm{d}t} = V\frac{\mathrm{d}Q}{\mathrm{d}t} = IV$$

(Eq. 9.9). This expression is known as *Joule's Law*. Electrical energy carried by a circuit can be converted to other forms of energy, for example, to light a bulb, heat an oven, or produce sound from speakers. Using Joule's Law together with Ohm's Law, the power dissipated in a resistive component is:

$$P_R = I^2R = \frac{V^2}{R}$$

(Eq. 9.10).

Electric power and Joule's Law:

$$P = IV \quad (9.9)$$

$$\therefore P_R = I^2 R = \frac{V^2}{R} \quad (9.10)$$

9.1.3 Kirchhoff's Laws

Kirchhoff's Circuit Laws describe the conservation of charge (Kirchhoff's Current Law) and conservation of energy (Kirchhoff's Voltage Law) in electrical circuits (Summary box 9.1). Kirchhoff's Current Law states that current (charge) is conserved at a junction in a circuit, such that the net flow of current into the junction equals the net flow out of it:

$$\therefore \sum_{i=1}^{N} I_i = 0 \text{ at a circuit node}$$

(Eq. 9.11 in Summary box 9.1). Kirchhoff's Voltage Law states that the net change in potential around a closed loop of a circuit is zero:

$$\sum_{i=1}^{\text{closed loop}} V_i = 0$$

(Eq. 9.12 in Summary box 9.1).

Resistors in series and in parallel

Using Kirchhoff's Laws together with Ohm's Law, we find that resistors in series produce a net resistance that is equal to their sum:

$$V = V_1 + V_2 = I\underbrace{(R_1 + R_2)}_{=R}$$

$$\therefore R = R_1 + R_2$$

whereas resistors in parallel produce a net resistance whose reciprocal is equal to the sum of the reciprocal of individual resistors:

$$I = I_1 + I_2 = V\underbrace{\left(\frac{1}{R_1} + \frac{1}{R_2}\right)}_{=1/R}$$

$$\therefore \frac{1}{R} = \frac{1}{R_1} + \frac{1}{R_2}$$

(Eqs. 9.13 and 9.14 and Fig. 9.2).

Summary box 9.1: Kirchhoff's Circuit Laws

1. **Kirchhoff's Current Law: conservation of charge.** At any junction (node) in a circuit, the net flow of current into the junction equals the net flow out of the junction:

$$\sum_{i=1}^{N} I_i = 0 \quad (9.11)$$

2. **Kirchhoff's Voltage Law: conservation of energy.** Around any closed loop of a circuit, the net change in potential is zero:

$$\sum_{i=1}^{\text{closed loop}} V_i = 0 \quad (9.12)$$

a) Current law

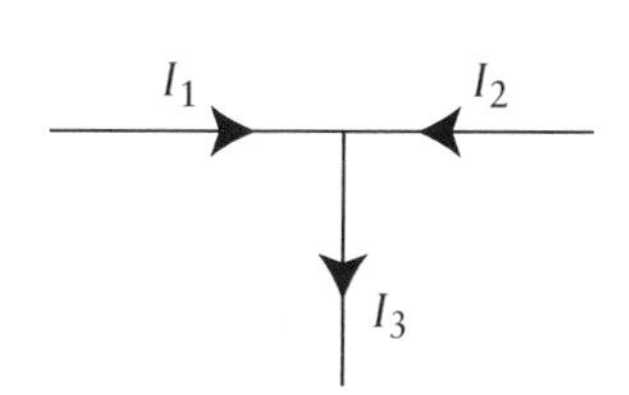

$I_1 + I_2 - I_3 = 0$

b) Voltage law

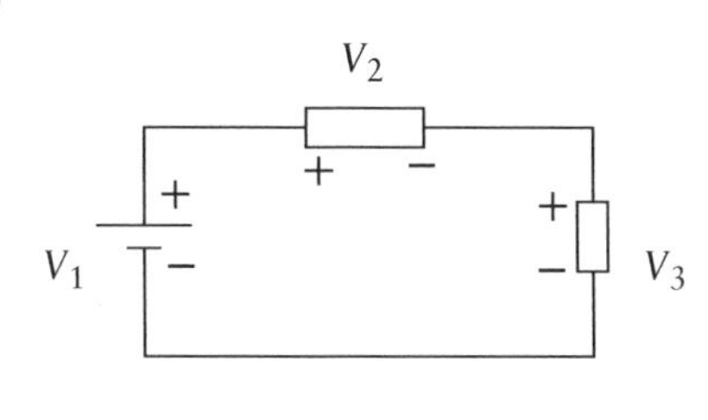

$-V_1 + V_2 + V_3 = 0$

a) Resistors in series

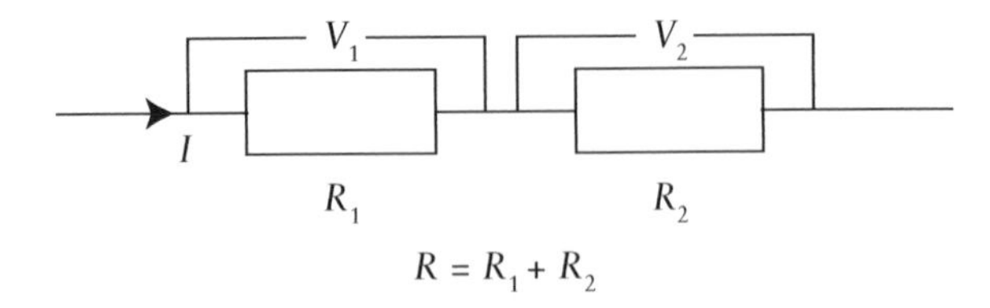

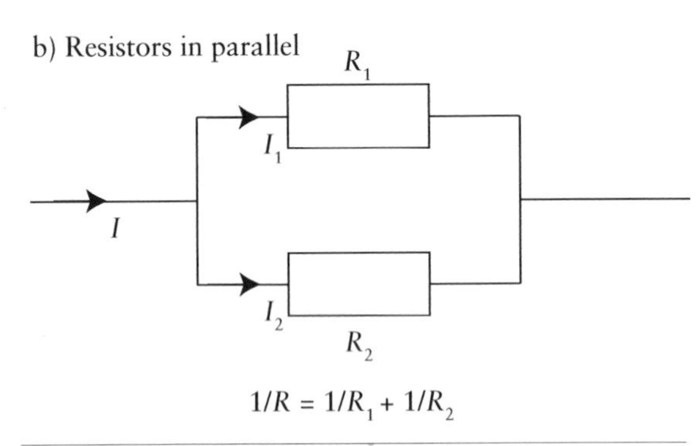

Figure 9.2: Sum of resistors in series and parallel.

Resistors in circuits:

In series:

$$R = \sum_i R_i \tag{9.13}$$

In parallel:

$$\frac{1}{R} = \sum_i \frac{1}{R_i} \tag{9.14}$$

9.2 Capacitance and inductance

Along with resistors, two other common electrical components are *capacitors* and *inductors*. A capacitor consists of two conductors separated by an electrical insulator (Fig. 9.3a). Charge does not flow through the insulator; therefore, capacitors store charge on the conductors. The charge stored in a capacitor is:

$$Q = CV$$

where V [V] is the potential difference between the two conductors, and C [F] is the *capacitance*, which is a measure of the ability of the capacitor to store charge. Capacitance depends on the geometry of the capacitor (refer to Section 7.6 for more detail). If the voltage across a capacitor changes with time, the current that flows through the capacitor is:

$$I_C = \frac{dQ}{dt} = C\frac{dV}{dt}$$

(Eq. 9.15). An inductor consists of a coil of wire (Fig. 9.3 b). A potential difference is induced across the coil as a result of a change in current flowing through the wire (refer to Section 10.2 for more detail). The voltage across an inductor in a circuit is found to be proportional to the change in current flow through it:

$$V_L = L\frac{dI}{dt}$$

(Eq. 9.16), where L [H] is the *inductance*. This is a measure of the "resistance" of an inductor to a change in current through it. Inductance depends on the geometry of the inductor (refer to Section 10.2.1 for examples).

a) Capacitor:

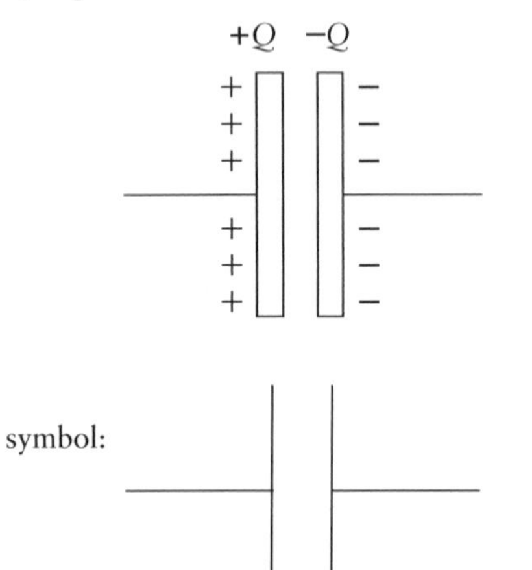

b) Inductor:

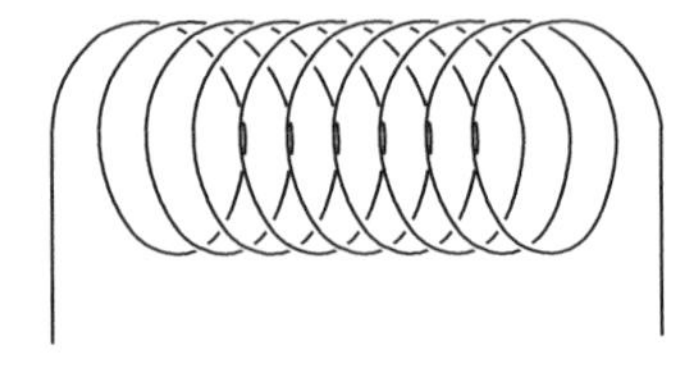

Figure 9.3: a) Capacitor and b) inductor.

Capacitance and inductance:

$$I_C = C\frac{dV}{dt} \quad (9.15)$$

$$V_L = L\frac{dI}{dt} \quad (9.16)$$

Energy stored in electrical components

Joule's Law states that the electric power carried by a circuit is:

$$P = IV$$

We can use this to determine the potential energy stored in a capacitor and an inductor (Eqs. 9.17 and 9.18 and Derivation 9.1).

Energy stored in a capacitor and inductor: (Derivation 9.1):

$$U_C = \frac{1}{2}CV^2 \quad (9.17)$$

$$U_L = \frac{1}{2}LI^2 \quad (9.18)$$

This stored energy is recovered if a capacitor is discharged, or if the current through an inductor is slowly reduced to zero.

Addition of components

To determine the total capacitance or inductance of components in series or in parallel, we use Kirchhoff's Laws together with $Q = CV$ for capacitors or $V = LdI/dt$ for inductors. Consider first two capacitors in series. The total potential difference across the two components is:

$$V = V_1 + V_2$$
$$\therefore \frac{Q}{C} = \frac{Q_1}{C_1} + \frac{Q_2}{C_2}$$

The negative charge on C_1 must balance the positive charge on C_2, such that there is zero potential difference in the wire connecting the two plates:

$$-Q_1 + Q_2 = 0$$
$$\therefore Q_1 = Q_2 = Q$$
$$\therefore \frac{Q}{C} = \frac{Q}{C_1} + \frac{Q}{C_2}$$
$$\Rightarrow \frac{1}{C} = \frac{1}{C_1} + \frac{1}{C_2}$$

Therefore, capacitors in series add by their reciprocals (Eq. 9.19). For capacitors in parallel, the total charge equals the sum of charge on either capacitor:

$$Q = Q_1 + Q_2$$
$$\therefore CV = C_1V + C_2V$$
$$\Rightarrow C = C_1 + C_2$$

Therefore, capacitors in parallel add linearly (Eq. 9.21). Inductors in *series* also add linearly:

$$V = V_1 + V_2$$
$$\therefore L\frac{dI}{dt} = L_1\frac{dI}{dt} + L_2\frac{dI}{dt}$$
$$\Rightarrow L = L_1 + L_2$$

(Eq. 9.20), whereas inductors in parallel add through their reciprocal:

Derivation 9.1: Derivation of the energy stored in a capacitor and an inductor (Eqs. 9.17 and 9.18)

Capacitor energy:

$$U_C = \int P dt = \int IV dt = \int CV\frac{dV}{dt}dt = \frac{1}{2}CV^2$$

Inductor energy:

$$U_L = \int P dt = \int IV dt = \int LI\frac{dI}{dt}dt = \frac{1}{2}LI^2$$

$$I = I_1 + I_2$$
$$\therefore \frac{dI}{dt} = \frac{dI_1}{dt} + \frac{dI_2}{dt}$$
$$\therefore \frac{V}{L} = \frac{V}{L_1} + \frac{V}{L_2}$$
$$\Rightarrow \frac{1}{L} = \frac{1}{L_1} + \frac{1}{L_2}$$

(Eq. 9.22).

Components in circuits:

In series:

$$\frac{1}{C} = \sum_i \frac{1}{C_i} \quad (9.19)$$

$$L = \sum_i L_i \quad (9.20)$$

In parallel:

$$C = \sum_i C_i \quad (9.21)$$

$$\frac{1}{L} = \sum_i \frac{1}{L_i} \quad (9.22)$$

9.3 Transient response

A current that changes over a short period of time before steady state is achieved is called a *transient* current. For example, a transient current flows immediately after closing a switch that completes a circuit. This results in a time-dependent change in voltage across components in a circuit, as well as a time-dependent current flow through them. The nature of a transient response depends on the electrical components that are present in the circuit. Below, we will look at the transient response of three arrangements: (i) a resistor and a capacitor in series (series RC circuit), (ii) a resistor and an inductor in series (series RL circuit), and (iii) a resistor, an inductor, and a capacitor in series (series RLC circuit).

9.3.1 Series RC circuit

Let's first consider an RC circuit involving a resistor and a capacitor in series (refer to figure in Derivation 9.2). At a time $t = 0$, the switch is moved from B to A, and the battery provides a potential drop V_0 across the components in the circuit. By considering the change in potential across either component with time, we can establish a differential equation for current flow through the circuit. Doing this, as shown in Derivation 9.2, we find that the current evolves according to the differential equation:

$$\frac{dI}{dt} = -\frac{I}{RC}$$

The solution to this differential equation is a decaying exponential with initial amplitude $I_0 = V_0/R$ and a time constant $\tau = RC$:

$$I = I_0 e^{-t/\tau}$$

(Eq. 9.23 and Derivation 9.2). The voltage across the resistor follows the current, and therefore also decays exponentially:

$$V_R = IR = V_0 e^{-t/\tau}$$

(Eq. 9.24), while the voltage across the capacitor increases following an inverse exponential decay:

$$V_C = V_0 - V_R = V_0(1 - e^{-t/\tau})$$

(Eq. 9.25). Therefore, current flow exponentially decays to zero as the capacitor becomes charged, arriving at a final capacitor voltage of $V_C = V_0$ (Fig. 9.4). The capacitor charges more slowly the bigger the time constant, $\tau = RC$. A capacitor resists a sudden change in voltage since, from $I_C = C dV/dt$, such a change would imply $I_C \to \infty$ which is not physically possible.

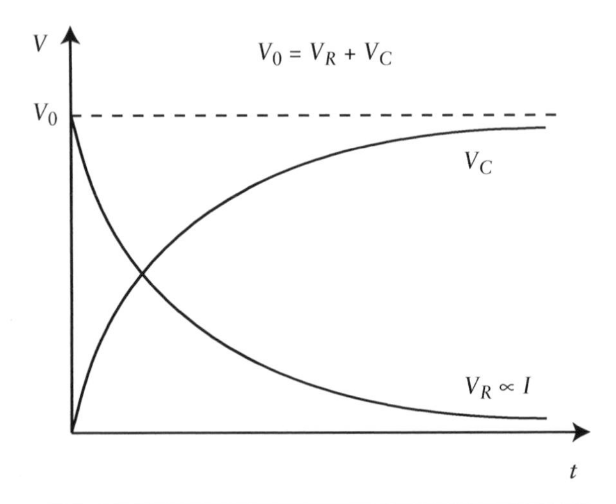

Figure 9.4: Voltage changes across components in a series RC circuit.

Derivation 9.2: Derivation of the current transient through a series RC circuit (Eq. 9.23)

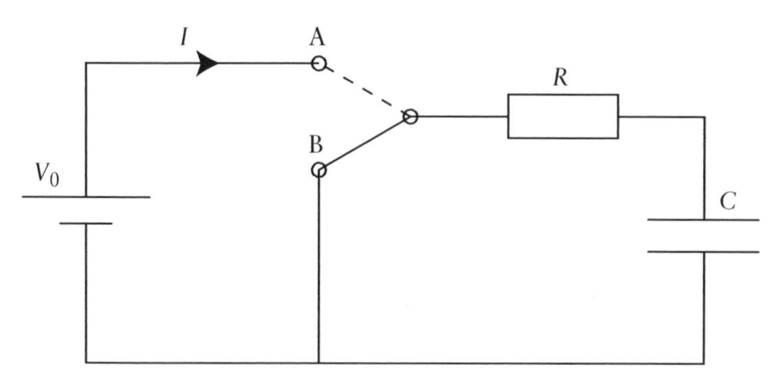

$$V_0 = V_R + V_C = IR + \frac{Q}{C}$$

$$\frac{dV_0}{dt} = \frac{d}{dt}\left(IR + \frac{Q}{C}\right) = R\frac{dI}{dt} + \frac{1}{C}\frac{dQ}{dt} = R\frac{dI}{dt} + \frac{I}{C}$$

$$\text{and } \frac{dV_0}{dt} = 0$$

$$\therefore \frac{dI}{dt} = -\frac{I}{RC}$$

$$\therefore I = I_0 e^{-t/RC} \equiv I_0 e^{-t/\tau}$$

where $\tau = RC$

- Starting point: at a time $t = 0$, the switch is moved from B to A, and the potential provided by the battery is:

$$V_0 = V_R + V_C$$

$$\text{where } V_R = IR$$

$$\text{and } V_C = \frac{Q}{C}$$

- The battery potential is constant $\therefore dV_0/dt = 0$.
- Differentiating $V_0 = V_R + V_C$ and setting it equal to zero, we see that the current through the circuit obeys the differential equation:

$$\frac{dI}{dt} = -\frac{I}{RC}$$

- The solution to this equation is a decaying exponential, with time constant $\tau = RC$ and maximum amplitude I_0.
- The initial conditions are $V_C = 0$ when $t = 0$. Therefore, the maximum amplitude is:

$$I_0 = \frac{V_0}{R}$$

Series RC circuit (Derivation 9.2):

Current through circuit:

$$I = I_0 e^{-t/\tau} \qquad (9.23)$$

where $\tau = RC$

Voltage across resistor and capacitor:

$$V_R = V_0 e^{-t/\tau} \qquad (9.24)$$

$$V_C = V_0(1 - e^{-t/\tau}) \qquad (9.25)$$

9.3.2 Series RL circuit

Now let's consider an RL circuit involving a resistor and an inductor in series (refer to figure in Derivation 9.3). At a time $t = 0$, the switch is moved from B to A, and the battery provides a potential drop V_0 across the components in the circuit. As before, by considering the change in potential across either component with time, as shown in Derivation 9.3, we can establish a differential equation for current flow through the circuit. In this case, we find that the current evolves according to the differential equation:

$$\frac{dI}{dt} = \frac{V_0 - IR}{L}$$

The solution to this differential equation is:

$$I = I_0\left(1 - e^{-t/\tau}\right)$$

$$\text{where } \tau = \frac{L}{R}$$

(Eq. 9.26 and Derivation 9.3). Therefore, the current flow through the circuit increases with time following an inverse exponential decay, with a final current of $I_0 = V_0/R$ and a time constant of $\tau = L/R$. The voltage across the resistor follows the current:

$$\therefore V_R = IR = V_0(1 - e^{-t/\tau})$$

(Eq. 9.27), whereas voltage across the inductor decays exponentially:

$$V_L = V_0 - V_R = V_0 e^{-t/\tau}$$

(Eq. 9.28). Therefore, the potential of the battery, V_0, initially drops solely across the inductor, and this voltage decays to zero as the current through the circuit approaches steady state (Fig. 9.5). An inductor resists a sudden change in current since, from $V_L = L\mathrm{d}I/\mathrm{d}t$, such a change would imply $V_L \to \infty$, which is not physically possible.

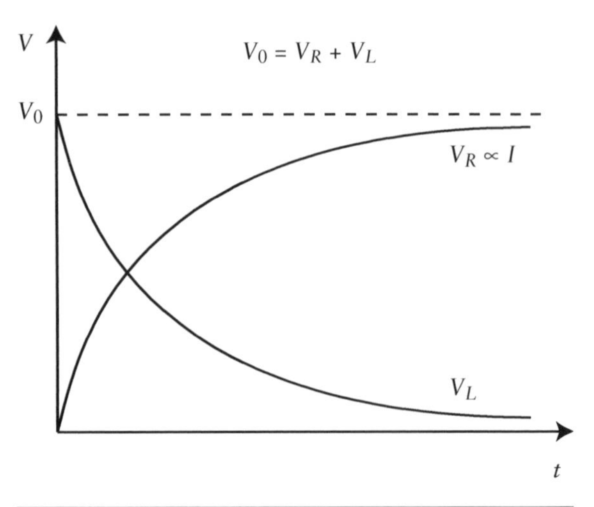

Figure 9.5: Voltage changes across components in a series RL circuit.

Derivation 9.3: Derivation of the current transient through a series RL circuit (Eq. 9.26)

$$
\begin{aligned}
V_0 &= V_R + V_L \\
&= IR + L\frac{\mathrm{d}I}{\mathrm{d}t} \\
\therefore \int_0^t \mathrm{d}t' &= \int_0^I \frac{L}{V_0 - I'R}\mathrm{d}I' \\
t &= -\frac{L}{R}\left[\ln\left(V_0 - I'R\right)\right]_0^I \\
&= -\frac{L}{R}\ln\left(\frac{V_0 - IR}{V_0}\right) \\
\therefore V_0 - IR &= V_0 e^{-tR/L} \\
\therefore I &= \frac{V_0}{R}\left(1 - e^{-tR/L}\right) \\
&= I_0\left(1 - e^{-tR/L}\right) \\
&\equiv I_0\left(1 - e^{-t/\tau}\right)
\end{aligned}
$$

where $\tau = \dfrac{L}{R}$

The primed variables are to distinguish them from the integration limits.

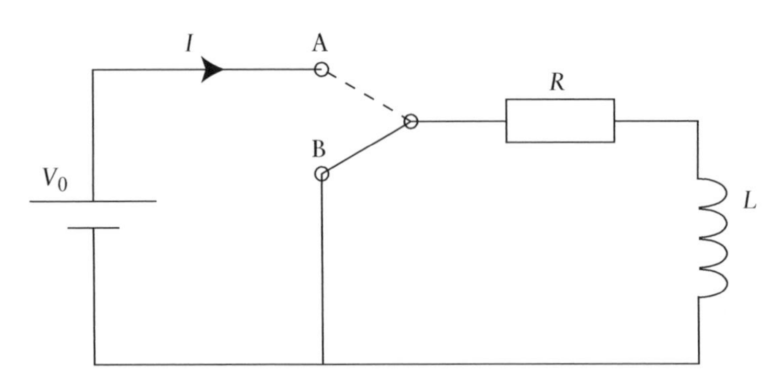

- Starting point: at a time $t = 0$, the switch is moved from B to A and the potential provided by the battery is:

$$V_0 = V_R + V_L$$

where $V_R = IR$

and $V_L = L\dfrac{\mathrm{d}I}{\mathrm{d}t}$

- Rearranging and integrating, we find that the current through the circuit increases following an inverse exponential decay, to a maximum current $I_0 = V_0/R$, with a time constant $\tau = L/R$:

$$I = I_0\left(1 - e^{-t/\tau}\right)$$

Series RL circuit (Derivation 9.3):

Current through circuit:

$$I = I_0\left(1 - e^{-t/\tau}\right) \quad (9.26)$$

where $\tau = \dfrac{L}{R}$

Voltage across resistor and inductor:

$$V_R = V_0(1 - e^{-t/\tau}) \quad (9.27)$$

$$V_L = V_0 e^{-t/\tau} \quad (9.28)$$

9.3.3 Series RLC circuit

Finally, let's consider an RLC circuit involving a resistor, an inductor, and a capacitor in series (refer to figure in Derivation 9.4). At a time $t = 0$, the switch is moved from B to A, and the battery provides a potential drop V_0 across the components in the circuit. The differential equation that governs the time evolution of current in this case is a second-order differential equation:

$$\frac{d^2I}{dt^2} + \frac{R}{L}\frac{dI}{dt} + \frac{I}{LC} = 0$$

$$\text{or} \quad \frac{d^2I}{dt^2} + \gamma\frac{dI}{dt} + \omega_0^2 I = 0$$

where $\gamma = R/L$ and $\omega_0^2 = 1/LC$ (Derivation 9.4). We recognize this standard differential equation as having damped, oscillatory solutions, where γ is the damping factor and ω_0 is the natural frequency of oscillations (solutions to this type of differential equation are covered in detail in Section 2.3.1, and in particular in Section 2.3.2). The general solution is found to be a sum of two complex exponential terms, multiplied by an exponential damping factor (test by substitution):

$$I = \underbrace{e^{-\gamma t/2}}_{\text{damping}}\underbrace{\left(Ae^{i\omega t} + Be^{-i\omega t}\right)}_{\text{simple harmonic oscillation}}$$

(Eq. 9.29 and Derivation 9.4). The frequency of oscillations depends on both the natural frequency of oscillations, ω_0, and the damping factor, γ:

$$\omega = \sqrt{\omega_0^2 - \gamma^2/4}$$

Therefore, there are three different types of solution, depending on whether ω is real, zero, or imaginary:

1. **Underdamped response:**

$$\omega_0^2 > \gamma^2/4 \therefore \omega \text{ is real}$$

$$\Rightarrow I = e^{-\gamma t/2}\left(Ae^{i\omega t} + Be^{-i\omega t}\right)$$

The current through the circuit oscillates, flowing in one direction during capacitor charging, and then in the other direction during capacitor discharging. When the current through the circuit passes through zero (upon a change in direction), the voltage across the inductor is a maximum ($V_L = L\mathrm{d}I/\mathrm{d}t =$ max since $\mathrm{d}I/\mathrm{d}t$ is a maximum); therefore, current continues to be driven through the circuit in a cyclic way. Electrical energy is thus converted between stored energy in the capacitor into producing a field across the inductor. The magnitude of the oscillatory current decays with time as power is lost via Joule heating in the resistor (as governed by the damping term, γ).

Using Euler's formula ($e^{i\theta} = \cos\theta + i\sin\theta$), we have:

$$I = e^{-\gamma t/2}\left(Ae^{i\omega t} + Be^{-i\omega t}\right)$$

$$= e^{-\gamma t/2}\left(A'\sin\omega t + B'\cos\omega t\right)$$

$$\equiv I_0 e^{-\gamma t/2}\sin(\omega t + \phi)$$

(Eq. 9.30). The initial conditions at $t = 0$ are (i) $I = 0$ and (ii) $V_R = 0$, $V_C = 0$, and $V_L = V_0$. From condition (i), we have $\phi = 0$, and from condition (ii), we have:

$$V_0 = L\left(\frac{dI}{dt}\right)_{t=0} = LI_0\omega$$

$$\therefore I_0 = \frac{V_0}{\omega L}$$

2. **Critically damped response:**

$$\omega_0^2 = \gamma^2/4 \therefore \omega = 0$$

$$\Rightarrow I = e^{-\gamma t/2}(A + B)$$

(Eq. 9.31). In this case, current through the circuit does not oscillate, but falls exponentially to zero with a time constant $\tau = 2/\gamma$. A and B are constants to be set by initial conditions.

3. **Overdamped response:**

$$\omega_0^2 < \gamma^2/4 \quad \therefore \omega \text{ is imaginary}$$

$$\Rightarrow I = e^{-\gamma t/2}\left(Ae^{-\omega t} + Be^{\omega t}\right)$$

In this case, current through the circuit does not oscillate either. Starting at zero, it reaches a maximum and

Derivation 9.4: Derivation of the current transient through a series RLC circuit (Eq. 9.29)

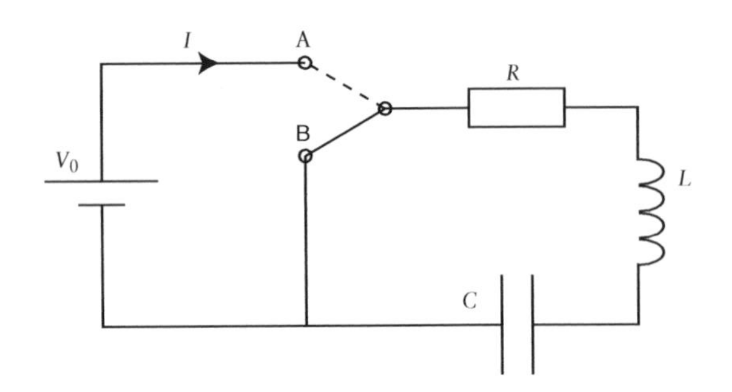

$$
\begin{aligned}
V_0 &= V_R + V_L + V_C \\
&= IR + L\frac{dI}{dt} + \frac{Q}{C} \\
\frac{dV_0}{dt} &= \frac{d}{dt}\left(IR + L\frac{dI}{dt} + \frac{Q}{C}\right) \\
&= R\frac{dI}{dt} + L\frac{d^2I}{dt^2} + \frac{1}{C}\frac{dQ}{dt} \\
&= R\frac{dI}{dt} + L\frac{d^2I}{dt^2} + \frac{I}{C}
\end{aligned}
$$

and $\frac{dV_0}{dt} = 0$

$$\therefore \frac{d^2I}{dt^2} + \frac{R}{L}\frac{dI}{dt} + \frac{I}{LC} = 0$$

or $\frac{d^2I}{dt^2} + \gamma\frac{dI}{dt} + \omega_0^2 I = 0$

where $\gamma = \frac{R}{L}$ (damping term)

and $\omega_0 = \frac{1}{\sqrt{LC}}$ (natural frequency)

Try a solution $I = ae^{i\alpha t}$:

$$-\alpha^2 + i\alpha\gamma + \omega_0^2 = 0$$

$$
\begin{aligned}
\therefore \alpha &= \frac{-i\gamma \pm \sqrt{-\gamma^2 + 4\omega_0^2}}{-2} \\
&= \frac{i\gamma}{2} \pm \sqrt{\omega_0^2 - \gamma^2/4} \\
&= \frac{i\gamma}{2} \pm \omega
\end{aligned}
$$

Therefore, the solution is the sum of two complex exponential terms multiplied by a decaying exponential term:

$$
\begin{aligned}
\therefore I &= ae^{i\alpha t} \\
&= e^{-\gamma t/2}\left(Ae^{i\omega t} + Be^{-i\omega t}\right)
\end{aligned}
$$

- Starting point: at a time $t = 0$, the switch is moved from B to A, and the potential provided by the battery is:

$$V_0 = V_R + V_L + V_C$$

where $V_R = IR$

$$V_L = L\frac{dI}{dt}$$

and $V_C = \frac{Q}{C}$

- The battery potential is constant $\therefore dV_0/dt = 0$.
- Differentiating $V_0 = V_R + V_L + V_C$ and setting it equal to zero, we see that the current through the circuit obeys a standard second-order differential equation, with decay constant $\gamma = R/L$ and natural frequency $\omega_0^2 = 1/LC$:

$$\frac{d^2I}{dt^2} + \gamma\frac{dI}{dt} + \omega_0^2 I = 0$$

- Try a solution of the form $I = ae^{i\alpha t}$.
- Solve the quadratic for α and define the oscillation frequency as:

$$\omega^2 = \omega_0^2 - \frac{\gamma^2}{4}$$

- The solutions correspond to sinusoidal oscillations (complex exponential solutions) multiplied by an exponentially decaying amplitude:

$$I = e^{-\gamma t/2}\left(Ae^{i\omega t} + Be^{-i\omega t}\right)$$

There are three types of solution, depending on whether ω is real, zero, or imaginary:

$$
\begin{aligned}
\omega \text{ is real}: \quad I &= e^{-\gamma t/2}\left(Ae^{i\omega t} + Be^{-i\omega t}\right) \\
\omega \text{ is zero}: \quad I &= e^{-\gamma t/2}(A + B) \\
\omega \text{ is imaginary}: \quad I &= e^{-\gamma t/2}\left(Ae^{-\omega t} + Be^{\omega t}\right)
\end{aligned}
$$

The interpretation of these solutions is covered in the main text.

falls to zero with a time constant slower than the critical damping limit.

Using the initial conditions (i) $I = 0$ and (ii) $V_R = 0$, $V_C = 0$, and $V_L = V_0$, at $t = 0$, then from condition (i), we have , $A = -B$ and from condition (ii), we have:

$$V_0 = L\left(\frac{\mathrm{d}I}{\mathrm{d}t}\right)_{t=0} = -2AL\omega$$
$$\therefore A = -\frac{V_0}{2\omega L} = -\frac{I_0}{2}$$

(Eq. 9.32).

RLC circuit (Derivation 9.4):

Overall solution:

$$I = \mathrm{e}^{-\gamma t/2}\left(A\mathrm{e}^{\mathrm{i}\omega t} + B\mathrm{e}^{-\mathrm{i}\omega t}\right) \quad (9.29)$$

where $\omega = \sqrt{\omega_0^2 - \gamma^2/4}$

$$\omega_0 = \frac{1}{\sqrt{LC}} \text{ (natural frequency)}$$

$$\gamma = \frac{R}{L} \text{ (decay constant)}$$

Underdamped (ω is real):

$$\omega_0^2 > \gamma^2/4$$
$$\therefore I = I_0\mathrm{e}^{-\gamma t/2}\sin(\omega t) \quad (9.30)$$

Critically damped ($\omega = 0$):

$$\omega_0^2 = \gamma^2/4$$
$$\therefore I = \mathrm{e}^{-\gamma t/2}(A + B) \quad (9.31)$$

Overdamped (ω is imaginary):

$$\omega_0^2 < \gamma^2/4$$
$$\therefore I = \frac{I_0}{2}\mathrm{e}^{-\gamma t/2}\left(\mathrm{e}^{\omega t} - \mathrm{e}^{-\omega t}\right) \quad (9.32)$$

The current I_0, set by initial conditions, is found to be:

$$I_0 = \frac{V_0}{\omega L}$$

A and B are constants of integration to be set by initial conditions.

9.4 AC circuits

Currents through circuits can either flow in one direction (direct current, DC), or alternate between flowing in one direction to flowing in the other direction (alternating current, AC). An AC voltage source produces an alternating current. These are usually sinusoidal waveforms, although other shapes, such as triangular or rectangular waveforms, are used in certain applications. In this section, we will only consider sinusoidal waveforms. An AC voltage and current can therefore be expressed as:

$$V = V_0 \cos \omega t$$
$$\text{and } I = I_0 \cos(\omega t - \phi)$$

where V_0 and I_0 are the maximum voltage and current, and ϕ is the phase difference between them. The amount by which the current leads or lags the voltage (i.e. the value of ϕ) depends on which components are present in the circuit. For a purely resistive circuit (no capacitors or inductors), the voltage and current are in phase:

$$I_R = \frac{V}{R} = I_0\cos(\omega t)$$
$$\therefore \phi_R = 0$$

For a capacitive circuit, voltage *lags* current by $\pi/2$:

$$I_C = C\frac{\mathrm{d}V}{\mathrm{d}t} = -CV_0\omega\sin(\omega t) = CV_0\omega\cos(\omega t + \pi/2)$$
$$\therefore \phi_C = -\frac{\pi}{2}$$

For an inductive circuit, voltage *leads* current by $\pi/2$:

$$I_L = \frac{1}{L}\int V\mathrm{d}t = \frac{V_0}{\omega L}\sin(\omega t) = \frac{V_0}{\omega L}\cos(\omega t - \pi/2)$$
$$\therefore \phi_L = \frac{\pi}{2}$$

Complex exponential form

Euler's formula states that:

$$\mathrm{e}^{\mathrm{i}\phi} = \cos\phi + \mathrm{i}\sin\phi$$

Therefore, the voltage and current of an AC source can be expressed as:

$$V = \mathrm{Re}\left\{V_0 e^{j\omega t}\right\} \equiv \mathrm{Re}\left\{\tilde{V}\right\}$$
$$\text{and } I = \mathrm{Re}\left\{I_0 e^{j(\omega t-\phi)}\right\} \equiv \mathrm{Re}\left\{\tilde{I}\right\}$$

In circuit theory, we use j to represent $\sqrt{-1}$ rather than i so as not to confuse it with current. Therefore, the complex numbers that represent AC voltage and current are:

$$\tilde{V} = V_0 e^{j\omega t}$$
$$\text{and } \tilde{I} = I_0 e^{j(\omega t-\phi)}$$

where the tilde indicates that the variable is complex. Of course, we could have equivalently taken the imaginary part of $\tilde{V}$ or $\tilde{I}$, giving a sine wave instead of a cosine wave, which corresponds to a shift in the initial phase of the wave. It's convenient to express voltage and current as complex exponential terms when dealing with AC sources, since these account for phase and magnitude in an implicit and algebraically-convenient manner. We then take the real or imaginary part in order to determine the physical amplitude or phase.

9.4.1 Complex impedances

From Ohm's Law, the voltage of an AC source is:

$$\tilde{V} = \tilde{I}\tilde{Z}$$

where $\tilde{Z}$ [Ω] is the *complex impedance* of the circuit. Using complex algebra, the magnitude of the voltage is then:

$$|\tilde{V}| = |\tilde{I}||\tilde{Z}|$$

and the phase is:

$$\mathrm{phase}(\tilde{V}) = \mathrm{phase}(\tilde{I}) + \mathrm{phase}(\tilde{Z})$$

By considering a series RLC circuit, we can determine expressions for the complex impedance of a resistor, a capacitor, and an inductor:

$$\tilde{Z}_R = R$$
$$\tilde{Z}_C = \frac{1}{j\omega C} = -\frac{j}{\omega C}$$
$$\tilde{Z}_L = j\omega L$$

(Eqs. 9.33–9.35 and Derivation 9.5). We see that the impedance of the resistor is purely real; therefore, there is no phase difference between V and I: $\phi_R = 0$. The impedance of a capacitor and inductor is purely imaginary. $\tilde{Z}_C$ points along −j; therefore, the voltage across a capacitor *lags* the current through it by $\pi/2$: $\therefore\ \phi_C = -\pi/2$. $\tilde{Z}_L$ points along +j; therefore, the voltage across an inductor *leads* the current through it by $\pi/2$: $\therefore\ \phi_L = \pi/2$.

Complex impedances (Derivation 9.5):

$$\tilde{Z}_R = R \tag{9.33}$$
$$\tilde{Z}_C = \frac{1}{j\omega C} = -\frac{j}{\omega C} \tag{9.34}$$
$$\tilde{Z}_L = j\omega L \tag{9.35}$$

Complex impedances add in the same manner as resistors: impedances in series add linearly, and the reciprocal of impedances in parallel add linearly:

$$\tilde{Z} = \sum_i \tilde{Z}_i \quad \text{in series}$$
$$\text{and } \frac{1}{\tilde{Z}} = \sum_i \frac{1}{\tilde{Z}_i} \quad \text{in parallel}$$

9.4.2 Electronic filters

Electric signals often have a range of both high- and low-frequency AC components. *Electronic filters* are electric circuits that are designed to remove unwanted frequency components from a signal, and to "pass" wanted components. Two common examples are a *low-pass filter*, which allows low-frequency components and removes high-frequency ones, and a *high-pass filter*, which allows high-frequency components and removes low-frequency ones from the signal.

Low-pass filter

A simple low-pass filter can be designed using a series RC circuit. Taking the output voltage over the capacitor, the ratio of output voltage to input voltage tends to 1 for low-frequency components and to 0 for high-frequency components (Eq. 9.36 and Derivation 9.6). The turn-over frequency is:

$$\omega_0 = \frac{1}{RC}$$

A high-pass filter is obtained by taking the output voltage over the resistor rather than the capacitor.

Derivation 9.5: Derivation of complex impedances (Eqs. 9.33–9.35)

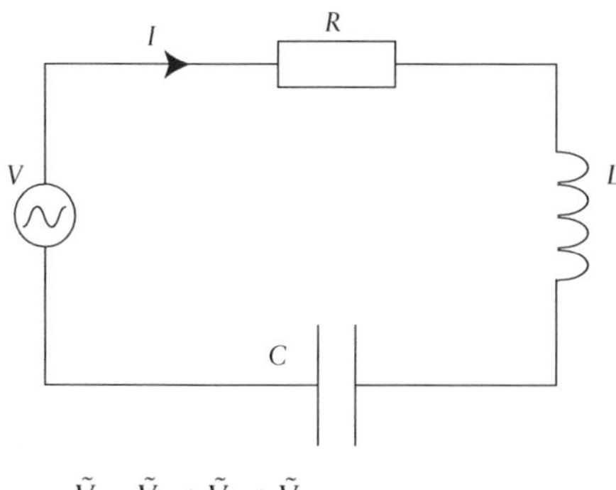

$$\tilde{V} = \tilde{V}_R + \tilde{V}_L + \tilde{V}_C$$
$$= \tilde{I}R + L\frac{d\tilde{I}}{dt} + \frac{1}{C}\int \tilde{I}dt$$
$$V_0 e^{j\omega t} = RI_0 e^{j(\omega t-\phi)} + LI_0 \frac{d}{dt}\left(e^{j(\omega t-\phi)}\right) + \frac{I_0}{C}\int e^{j(\omega t-\phi)}dt$$
$$= RI_0 e^{j(\omega t-\phi)} + j\omega LI_0 e^{j(\omega t-\phi)} + \frac{I_0}{j\omega C}e^{j(\omega t-\phi)}$$
$$\therefore \tilde{V} = \tilde{I}\left(R + j\omega L + \frac{1}{j\omega C}\right)$$
$$\equiv \tilde{I}\left(\tilde{Z}_R + \tilde{Z}_L + \tilde{Z}_C\right)$$

where $\tilde{Z}_R = R$

$$\tilde{Z}_L = j\omega L$$
$$\tilde{Z}_C = \frac{1}{j\omega C}$$

- Starting point: consider a series RLC circuit with an AC voltage source. The voltage and current produced by the source are:

$$\tilde{V} = V_0 e^{j\omega t}$$
and $\tilde{I} = I_0 e^{j(\omega t-\phi)}$

where ϕ is the phase lag between the voltage and the current.

- The total voltage equals the sum of voltage over each component:

$$\tilde{V} = \tilde{V}_R + \tilde{V}_L + \tilde{V}_C$$
where $\tilde{V}_R = \tilde{I}R$
$$\tilde{V}_L = L\frac{d\tilde{I}}{dt}$$
and $\tilde{V}_C = \frac{\tilde{Q}}{C} = \frac{1}{C}\int \tilde{I}dt$

- Differentiating or integrating the current, $\tilde{I}$, we find expressions for the complex impedances $\tilde{Z}_R$, $\tilde{Z}_L$, and $\tilde{Z}_C$.

Low-pass filter (series RC circuit) (Derivation 9.6):

$$\left|\frac{V_{out}}{V_{in}}\right| = \frac{1}{\sqrt{1+(\omega/\omega_0)^2}} \quad (9.36)$$
$$\to 1 \text{ for } \omega \ll \omega_0$$
$$\to 0 \text{ for } \omega \gg \omega_0$$

where $\omega_0 = \frac{1}{RC}$

High-pass filter

A simple high-pass filter can be designed using a series RL circuit. Taking the output voltage over the inductor, the ratio of output voltage to input voltage tends to 0 for low-frequency components and to 1 for high-frequency components (Eq. 9.37 and Derivation 9.7).

In this case, the turn-over frequency is:

$$\omega_0 = \frac{R}{L}$$

A low-pass filter is obtained by taking the output voltage over the resistor rather than the inductor.

High-pass filter (series RL circuit) (Derivation 9.7):

$$\left|\frac{V_{out}}{V_{in}}\right| = \frac{(\omega/\omega_0)}{\sqrt{1+(\omega/\omega_0)^2}} \quad (9.37)$$
$$\to 0 \text{ for } \omega \ll \omega_0$$
$$\to 1 \text{ for } \omega \gg \omega_0$$

where $\omega_0 = \frac{R}{L}$

Derivation 9.6: Derivation of the low-pass filter voltage output (Eq. 9.36)

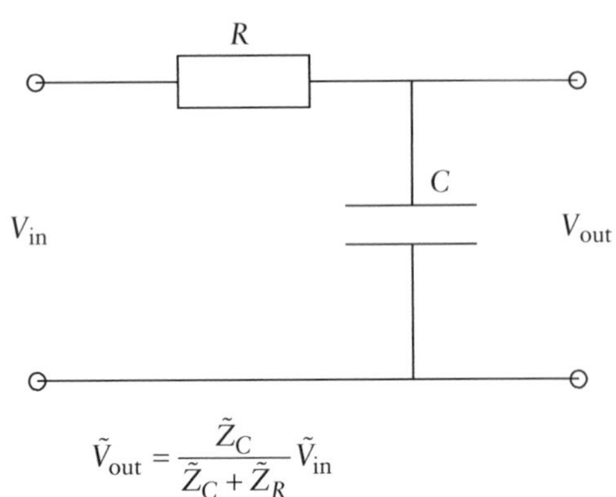

$$\tilde{V}_{out} = \frac{\tilde{Z}_C}{\tilde{Z}_C + \tilde{Z}_R}\tilde{V}_{in}$$

$$\therefore \frac{\tilde{V}_{out}}{\tilde{V}_{in}} = \frac{-j/\omega C}{-j/\omega C + R} = \frac{-j}{-j+\omega RC} = \frac{1-j\omega RC}{1+(\omega RC)^2} = \frac{1-j(\omega/\omega_0)}{1+(\omega/\omega_0)^2}$$

$$\text{where } \omega_0 = \frac{1}{RC}$$

Therefore, the magnitude and phase of output voltage to input voltage are:

$$\left|\frac{V_{out}}{V_{in}}\right| = \sqrt{\left(\text{Re}\left\{\frac{\tilde{V}_{out}}{\tilde{V}_{in}}\right\}\right)^2 + \left(\text{Im}\left\{\frac{\tilde{V}_{out}}{\tilde{V}_{in}}\right\}\right)^2} = \frac{\sqrt{1+(\omega/\omega_0)^2}}{1+(\omega/\omega_0)^2} = \frac{1}{\sqrt{1+(\omega/\omega_0)^2}}$$

$$\text{and } \phi = \tan^{-1}\left(\text{Im}\{\tilde{V}_{out}/\tilde{V}_{in}\}/\text{Re}\{\tilde{V}_{out}/\tilde{V}_{in}\}\right) = -\tan^{-1}(\omega/\omega_0)$$

- Starting point: using a potential divider, the output voltage across the capacitor in a series RC circuit is:

$$\tilde{V}_{out} = \frac{\tilde{Z}_C}{\tilde{Z}_C + \tilde{Z}_R}\tilde{V}_{in}$$

$$\text{where } \tilde{Z}_C = -\frac{j}{\omega C}$$

$$\text{and } \tilde{Z}_R = R$$

- Rearranging, we find that the magnitude of the ratio of output voltage to input voltage is:

$$\left|\frac{V_{out}}{V_{in}}\right| = \frac{1}{\sqrt{1+(\omega/\omega_0)^2}}$$

$$\text{where } \omega_0 = \frac{1}{RC}$$

- In the limit $\omega \ll \omega_0$:

$$\left|\frac{V_{out}}{V_{in}}\right| \to 1$$

and in the limit $\omega \gg \omega_0$:

$$\left|\frac{V_{out}}{V_{in}}\right| \to 0$$

Therefore, the circuit produces an output voltage for low-frequency voltage components, but not for high-frequency components ⇒ low-pass filter.

9.4.3 Power dissipation

The electric power of an AC circuit is given by the product of the real (or imaginary) parts of $\tilde{V}$ and $\tilde{I}$:

$$P = \text{Re}\{\tilde{V}\}\text{Re}\{\tilde{I}\}$$

By substituting in for $\tilde{V}$ and $\tilde{I}$ and using trigonometric identities, as shown in Derivation 9.8, we find that the mean power dissipated per cycle depends on the phase between V and I:

$$\langle P\rangle = \frac{V_0 I_0}{2}\cos\phi$$

(Eq. 9.38 and Derivation 9.8). For a purely resistive circuit, $\phi = 0 \therefore \langle P\rangle = V_0 I_0/2$. For a purely capacitive or inductive circuit, $\phi = \pm\pi/2 \therefore \langle P\rangle = 0$. Therefore, power is only dissipated in resistors, since the work done to take a capacitor or an inductor through one complete cycle is zero (for example, work is done *on* a capacitor while it is charging and *by* a capacitor while it is discharging). Therefore, since the resistive part of the

Derivation 9.7: Derivation of the high-pass filter voltage output (Eq. 9.37)

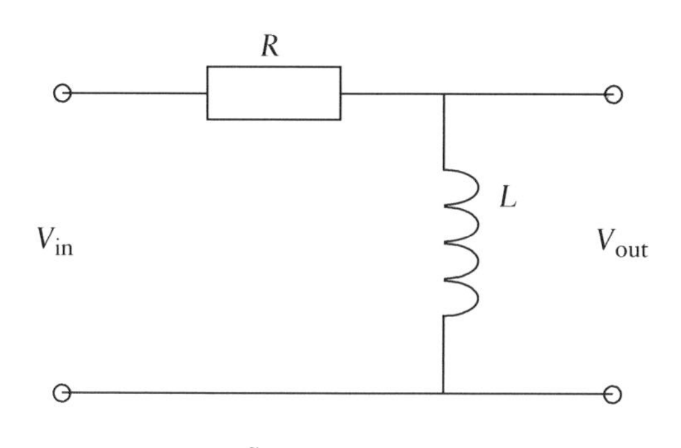

$$\tilde{V}_{out} = \frac{\tilde{Z}_L}{\tilde{Z}_R + \tilde{Z}_L}\tilde{V}_{in}$$

$$\therefore \frac{\tilde{V}_{out}}{\tilde{V}_{in}} = \frac{j\omega L}{R + j\omega L} = \frac{j\omega L / R}{1 + j\omega L / R} = \frac{(\omega L / R)^2 + j(\omega L / R)}{1 + (\omega L / R)^2} = \frac{(\omega / \omega_0)^2 + j(\omega / \omega_0)}{1 + (\omega / \omega_0)^2}$$

where $\omega_0 = \frac{R}{L}$

Therefore, the magnitude and phase of output voltage to input voltage are:

$$\left|\frac{V_{out}}{V_{in}}\right| = \sqrt{\left(\mathrm{Re}\left\{\frac{\tilde{V}_{out}}{\tilde{V}_{in}}\right\}\right)^2 + \left(\mathrm{Im}\left\{\frac{\tilde{V}_{out}}{\tilde{V}_{in}}\right\}\right)^2} = \frac{\sqrt{(\omega/\omega_0)^4 + (\omega/\omega_0)^2}}{1 + (\omega/\omega_0)^2} = \frac{(\omega/\omega_0)}{\sqrt{1 + (\omega/\omega_0)^2}}$$

and $\phi = \tan^{-1}\left(\mathrm{Im}\{\tilde{V}_{out}/\tilde{V}_{in}\} / \mathrm{Re}\{\tilde{V}_{out}/\tilde{V}_{in}\}\right) = \tan^{-1}(\omega_0/\omega)$

- Starting point: using a potential divider, the output voltage across the inductor in an RL circuit is:

$$\tilde{V}_{out} = \frac{\tilde{Z}_L}{\tilde{Z}_R + \tilde{Z}_L}\tilde{V}_{in}$$

where $\tilde{Z}_L = j\omega L$

and $\tilde{Z}_R = R$

- Rearranging, we find that the magnitude of the ratio of output voltage to input voltage is:

$$\left|\frac{V_{out}}{V_{in}}\right| = \frac{(\omega/\omega_0)}{\sqrt{1 + (\omega/\omega_0)^2}}$$

where $\omega_0 = \frac{R}{L}$

- In the limit $\omega \ll \omega_0$:

$$\left|\frac{V_{out}}{V_{in}}\right| \to 0$$

and in the limit $\omega \gg \omega_0$:

$$\left|\frac{V_{out}}{V_{in}}\right| \to 1$$

Therefore, the circuit produces an output voltage for high-frequency voltage components but not for low-frequency components ⇒ high-pass filter.

complex impedance corresponds to the real part, the average power dissipated per cycle is:

$$\langle P \rangle = \frac{1}{2} I_0^2 \,\mathrm{Re}\{\tilde{Z}\}$$

Furthermore, by manipulating Eq. 9.38, we find that the power can be expressed in terms of the complex quantities $\tilde{V} = V_0 e^{j\omega t}$ and $\tilde{I}^* = I_0 e^{-j(\omega t - \phi)}$, where $\tilde{I}^*$ is the complex conjugate of $\tilde{I}$:

$$\langle P \rangle = \frac{V_0 I_0}{2}\cos\phi = \frac{1}{2}\mathrm{Re}\{V_0 I_0 e^{j\phi}\} = \frac{1}{2}\mathrm{Re}\{\tilde{V}\tilde{I}^*\}$$

Derivation 9.8: Derivation of AC power dissipation (Eq. 9.38)

$$
\begin{aligned}
\langle P\rangle &= \frac{1}{T}\int_0^T P\,\mathrm{d}t \\
&= \frac{V_0 I_0}{T}\int_0^T \cos(\omega t)\cos(\omega t-\phi)\,\mathrm{d}t \\
&= \frac{V_0 I_0}{T}\int_0^T \cos^2(\omega t)\cos\phi + \cos(\omega t)\sin(\omega t)\sin\phi\,\mathrm{d}t \\
&= \frac{V_0 I_0}{2T}\int_0^T \left[(1+2\cos(2\omega t))\cos\phi + \sin(2\omega t)\sin\phi\right]\mathrm{d}t \\
&= \frac{V_0 I_0}{2T}\int_0^T \cos\phi\,\mathrm{d}t \\
&= \frac{V_0 I_0}{2}\cos\phi \\
&= V_{\mathrm{rms}} I_{\mathrm{rms}}\cos\phi
\end{aligned}
$$

- Starting point: the electrical power of an AC circuit is:

$$
\begin{aligned}
P &= \mathrm{Re}\{\tilde{V}\}\mathrm{Re}\{\tilde{I}\} \\
&= V_0 I_0 \cos(\omega t)\cos(\omega t-\phi)
\end{aligned}
$$

for $\tilde{V} = V_0 \mathrm{e}^{\mathrm{j}\omega t}$

and $\tilde{I} = I_0 \mathrm{e}^{\mathrm{j}(\omega t-\phi)}$

where the mean power dissipated per cycle of period T is:

$$\langle P\rangle = \frac{1}{T}\int_0^T P\,\mathrm{d}t$$

- Use the trigonometric identities:

$$\cos(A-B) = \cos A\cos B + \sin A\sin B$$

$$\cos^2 A = \frac{1}{2}(1+2\cos(2A))$$

and $\cos A\sin A = \frac{1}{2}\sin(2A)$

to rearrange the expression for $\langle P\rangle$.

- The integral of $\cos(2\omega t)$ and $\sin(2\omega t)$ over a complete cycle is zero.
- The rms voltage and current for sinusoidal V and I are:

$$V_{\mathrm{rms}} = \frac{V_0}{\sqrt{2}}$$

and $I_{\mathrm{rms}} = \frac{I_0}{\sqrt{2}}$

Power dissipation (Derivation 9.8):

$$
\begin{aligned}
\langle P\rangle &= \frac{V_0 I_0}{2}\cos\phi \qquad (9.38)\\
&\equiv \frac{1}{2} I_0^2\,\mathrm{Re}\{\tilde{Z}\} \\
&\equiv \frac{1}{2}\mathrm{Re}\{\tilde{V}\tilde{I}^*\}
\end{aligned}
$$

9.4.4 Series resonance and Q-value

Series resonance

Resonance in a series RLC circuit is defined as the frequency at which impedance is a minimum and therefore current through the circuit is a maximum. This occurs when there is zero phase between V and I, therefore the impedance is purely real (purely resistive). Hence, at resonance, the magnitudes of the capacitive and inductive impedances are equal, and they are π out of phase with each other, such that contributions from these components cancel. Therefore, at resonance:

$$\omega L = \frac{1}{\omega C}$$

Therefore, the resonant frequency is:

$$\omega_0 = \frac{1}{\sqrt{LC}}$$

This corresponds to the natural frequency of a series RLC circuit, which we derived in Derivation 9.4.

Quality factor

Resonant circuits are characterized by their *quality factor*, Q. This is a dimensionless quantity that is equal to the ratio of resonant frequency, ω_0, to the range of frequencies either side of the resonant frequency between which the magnitude of the current is greater than the root mean square (rms) current. This spread of frequencies, $\Delta\omega$, is called the *bandwidth* of the circuit (Fig. 9.6). It corresponds to the range of frequencies between which the output signal is significant. The Q-value is therefore defined as:

$$Q = \frac{\omega_0}{\Delta\omega} \tag{9.39}$$

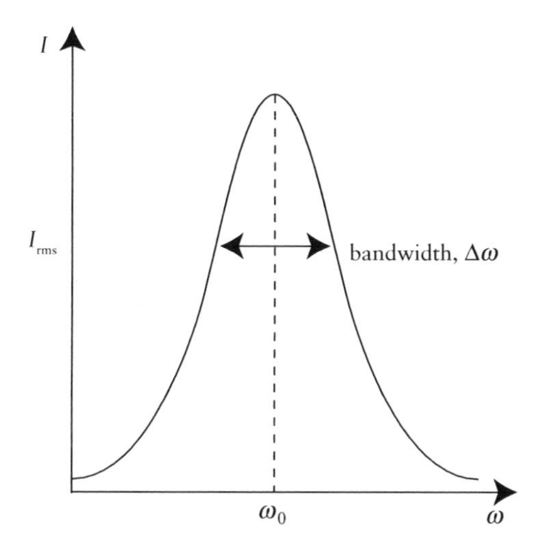

Figure 9.6: Frequency dependence of AC current through a series RLC circuit. The current is maximum at the *resonant frequency*, ω_0 of the circuit.

Sharply peaked maxima therefore have a high Q-value, whereas flatter resonances have a lower Q-value. Another definition of Q, which is equivalent to $\omega_0/\Delta\omega$, is the energy stored in the system compared to the energy lost per cycle at resonance:

$$\begin{aligned} Q &= 2\pi\left(\frac{\text{energy stored in the circuit}}{\text{energy loss per cycle at resonance}}\right) \\ &= 2\pi\frac{U}{\Delta U} \end{aligned}$$

Using this definition, we find that Q is inversely proportional to the resistance of the circuit, R:

$$Q = \frac{\omega_0 L}{R} = \frac{1}{R}\sqrt{\frac{L}{C}}$$

(Eq. 9.40 and Derivation 9.9). Comparing this to Eq. 9.39, we see that the bandwidth of the signal is:

$$\Delta\omega = \frac{R}{L}$$

Series RLC Q-value (Derivation 9.9):

$$\begin{aligned} Q &= \frac{\omega_0 L}{R} \\ &= \frac{1}{R}\sqrt{\frac{L}{C}} \end{aligned} \tag{9.40}$$

$$\text{where } \omega_0 = \frac{1}{\sqrt{LC}}$$

Derivation 9.9: Derivation of the Q-value of a series RLC circuit (Eq. 9.40)

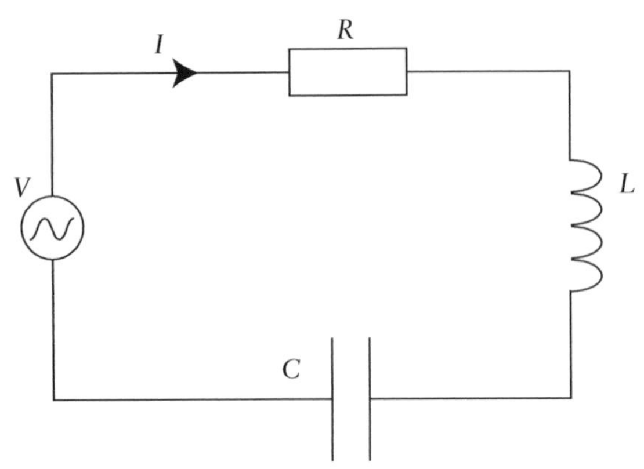

Energy stored in the system:

$$U = \frac{1}{2}CV_C^2 + \frac{1}{2}LI_L^2$$

$$= \frac{1}{2}C\frac{LI_0^2}{C}\sin^2(\omega t - \phi) + \frac{1}{2}LI_0^2\cos^2(\omega t - \phi)$$

$$= \frac{1}{2}LI_0^2$$

Average power loss per cycle:

$$\langle \Delta P \rangle = \frac{1}{2}I_0 V_0 \underbrace{\cos\phi}_{=1}$$

$$= \frac{1}{2}I_0 V_0$$

$$= \frac{1}{2}I_0^2 R$$

Energy loss per cycle at resonance:

$$\Delta U = \frac{2\pi}{\omega_0}\langle \Delta P \rangle$$

$$= \frac{\pi}{\omega_0}I_0^2 R$$

$\Rightarrow \therefore$ the Q-value is:

$$Q = 2\pi\frac{U}{\Delta U}$$

$$= \frac{\omega_0 L}{R}$$

$$= \frac{1}{R}\sqrt{\frac{L}{C}} \quad \text{for } \omega_0 = 1/\sqrt{LC}$$

- Starting point: the energy stored in an RLC circuit is stored in the capacitor and the inductor:

$$U = \frac{1}{2}CV_C^2 + \frac{1}{2}LI_L^2$$

- The voltage across the capacitor is:

$$V_C = \text{Re}\{\tilde{I}\tilde{Z}_C\}$$

$$= \text{Re}\left\{\frac{I_0 e^{j(\omega t - \phi)}}{j\omega C}\right\}$$

$$= \frac{I_0}{\omega C}\sin(\omega t - \phi)$$

- For a series RLC circuit, the resonant frequency is:

$$\omega_0 = \frac{1}{\sqrt{LC}}$$

$$\therefore V_C = I_0\sqrt{\frac{L}{C}}\sin(\omega t - \phi) \text{ at resonance.}$$

- The current through the inductor is:

$$I_L = \text{Re}\{\tilde{I}\}$$

$$= I_0\cos(\omega t - \phi)$$

- Use the expressions for V_C and I_L in the expression for the total energy. We find that the stored energy is constant and alternates between capacitor energy and inductor energy at resonance:

$$U = \frac{1}{2}LI_0^2 = \frac{1}{2}CV_0^2$$

- At resonance, the impedance is purely real $\therefore$ $\tilde{Z} = R$ and the phase lag between the voltage and current is zero: $\phi = 0$. Therefore, the average power loss per cycle is $\langle \Delta P \rangle = I_0^2 R/2$.

Chapter 10
Electromagnetic induction

Chapters 7 and 8 described the properties of electric and magnetic fields that are constant in time. In this chapter, we will start to look at the properties of time-varying fields. We find that a changing magnetic field produces an electric field, and a changing electric field produces a magnetic field. These phenomena constitute the subject of *electromagnetic induction*. The induction of an electric field by a changing magnetic field was discovered by Faraday in the 1830s, and the induction of a magnetic field by a changing electric field was discovered several decades later by Maxwell. The properties of electromagnetic induction are embodied in two of Maxwell's equations: Faraday's Law (curl of $\vec{E}$) and Ampère's Law with Maxwell's correction (curl of $\vec{B}$). In this chapter, we will derive these two equations, to arrive finally at Maxwell's four equations.

10.1 Faraday's Law

In 1831, Faraday performed a series of experiments that involved changing the magnetic flux through a closed loop of wire (for example, by moving a loop of wire through the magnetic field of a nearby magnet). Doing this, he found that a current flowed through the wire. This is an example of electromagnetic induction: the changing magnetic flux through the wire loop induces current to flow through the wire.

Electromotive force

For a current to flow through a wire, a potential difference must be applied across two ends of the wire. The work done per unit charge required to set up such a potential difference is referred to as the *electromotive force* (emf), ξ [V]. Said another way, the emf is a source of energy that tends to drive current flow through a circuit. The term "electromotive force" is misleading since it is not a force, but is an energy per unit charge, and therefore has units of volts (recall $V = W/q$).[1]

Sources of emf use non-electrostatic energy to separate positive and negative charge within the source, thus creating a potential difference across two open-circuit terminals that, when attached to an electrical circuit to close the loop, drives current flow around the circuit. Some common sources of emf include photovoltaic cells, which use solar energy to separate charge, thermoelectric devices, which use thermal energy to separate charge, and of course batteries, which use chemical energy (from chemical reactions) to separate charge. Sources of emf therefore store potential energy.

For a source to generate an emf ξ across its open-circuit terminals requires an amount of work W_ξ to be done for each unit of charge separation, q, where:

$$\xi = \frac{W_\xi}{q}$$

When attached to an electrical circuit, the source can then deliver an electric energy U_{emf} to a charge in the circuit, where, for an ideal emf source:[2]

$$U_{\text{emf}} = W_\xi$$

The charge then loses this energy in one loop of the circuit, doing work in the process (for example, to power a light bulb).

If we define $W_\xi = \int \vec{F}_\xi \cdot d\vec{l}$, where $\vec{F}_\xi$ is an arbitrary force against which work was done by the source to generate the emf, then in terms of $\vec{F}_\xi$, we have:

$$\begin{aligned}\xi &= \int_{\substack{\text{source}\\\text{terminals}}} \frac{\vec{F}_\xi}{q} \cdot d\vec{l} \\ &= \oint_L \frac{\vec{F}_\xi}{q} \cdot d\vec{l} \qquad (10.1)\end{aligned}$$

[1] For this reason, the term *electromotance* is sometimes employed instead.

[2] In a closed circuit, the potential difference across the terminals of a source of emf, V_{source}, is less than that across its open-circuit terminals, ξ, due to internal resistance of the source. Therefore, the energy delivered to a charge, $U_{\text{emf}} = qV_{\text{source}}$, is less than $W_\xi = q\xi$. An ideal emf source has negligible internal resistance $\therefore U_{\text{emf}} = W$.

We can change the integration limits to a closed loop since $\vec{F}_\xi = 0$ outside the source of the emf.

Motional emf and magnetic flux

What generates the emf that drives current around the loop of wire that Faraday moved through a magnetic field? To answer this question, consider a charge q in a wire. If the wire is moved through a magnetic field $\vec{B}$ at a velocity $\vec{v}$, then q experiences a force:

$$\vec{F}_B = q\vec{v} \times \vec{B}$$

If we take the directions of $\vec{v}$, $\vec{B}$, and the wire to be perpendicular to each another, as shown in Fig. 10.1, then the direction of $\vec{F}_B$ is along the wire. Therefore, current flows through the wire due to $\vec{F}_B$. An emf generated in this way (i.e. by a magnetic force acting on charges in a moving wire) is referred to as *motional emf.* Using $\vec{F}_\xi = q\vec{v}\times\vec{B}$ in Eq. 10.1, the motional emf is given by:

$$\xi_B = \oint_L (\vec{v} \times \vec{B}) \cdot d\vec{l} \tag{10.2}$$

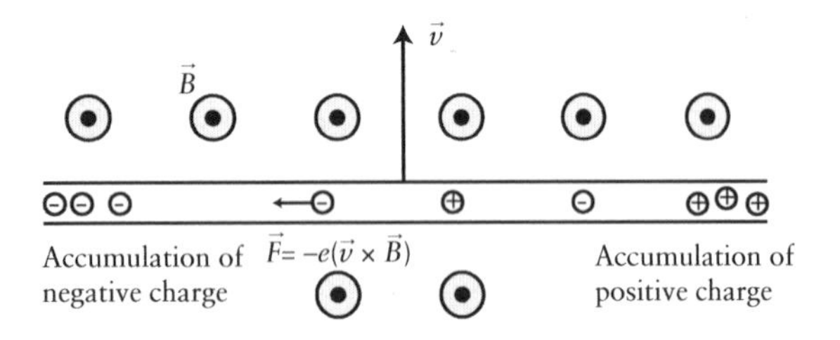

Figure 10.1: Induction of an electric field by a wire moving relative to a B-field.

Now let's go back to Faraday's experiment. Consider an element of wire $d\vec{l}$ moving at velocity $\vec{v}$ through a magnetic field $\vec{B}$. The area swept out by that element in a time dt is $d\vec{s} = \vec{v}dt \times d\vec{l}$. If the wire sweeps out a total area ΔS, then the magnetic flux through that area is:

$$\Phi_B = \int_{\Delta S} \vec{B} \cdot d\vec{s} = \int_{\Delta S} \vec{B} \cdot (\vec{v}dt \times d\vec{l})$$

Therefore, the rate of change of flux is:

$$\therefore \frac{d\Phi_B}{dt} = \int_L \vec{B} \cdot (\vec{v} \times d\vec{l}) = \int_L (\vec{B} \times \vec{v}) \cdot d\vec{l} = -\int_L (\vec{v} \times \vec{B}) \cdot d\vec{l}$$

Integrating over the closed loop of wire:

$$\frac{d\Phi_B}{dt} = -\underbrace{\oint_L (\vec{v} \times \vec{B}) \cdot d\vec{l}}_{\text{motional emf}} \tag{10.3}$$

We recognize the integral on the right-hand side as motional emf (Eq. 10.2):

$$\Rightarrow \therefore \frac{d\Phi_B}{dt} = -\xi_B \tag{10.4}$$

Therefore, Faraday found that magnetic flux cut by a moving wire gives rise to a motional emf, resulting in a current to flow through the wire.

Motional emf versus induced emf

In the example described above, $d\Phi_B/dt$ was equal to the rate of flux *cut* by a wire moving through a magnetic field. However, a $d\Phi_B/dt$ can also be generated by a stationary loop of wire in the vicinity of a changing magnetic field strength. This would lead to a changing Φ_B through the circuit, which, from Eq. 10.4, would be equal to an emf, ξ_B. However, in this case, the velocity of the wire is $\vec{v} = 0$; therefore, from Eq. 10.2, we would expect $\xi_B = 0$. How is this then compatible with Eq. 10.4?

The answer is that, although Eq. 10.4 holds true no matter how $d\Phi_B/dt$ is generated, the detailed mechanism for establishing ξ_B differs depending on the frame of reference of the observer. As an example, consider an open-ended strand of wire moving relative to a magnetic field, $\vec{B}$, where the relative velocity between them is $\vec{v}$, as illustrated in Fig. 10.1. Charges in the wire experience a force $\vec{F} = q\vec{v}\times\vec{B}$—they do not differentiate whether they are the ones moving or whether $\vec{B}$ is "moving" relative to them. The action of $\vec{F}$ on mobile electrons causes them to drift down one end of the wire, resulting in a net separation of positive and negative charge along the wire. This charge separation, in turn, results in a potential difference across the terminals, and hence an electric field is generated, or *induced*, across the wire. If the strand of wire is then attached to a closed electrical circuit, the induced electric field produces an emf, ξ, that drives current around the circuit.

This induced electric field is what generates an emf in the case of a stationary wire in the vicinity of a changing magnetic field. In this case, ξ is referred to as an *induced* emf, rather than a *motional* emf, and the change in flux $d\Phi_B/dt$ is referred to as a change in flux *linked* by the circuit, rather than a change in flux *cut*. The current that flows through a circuit under an induced emf is referred to as an *induced current.*

Despite the differences in nomenclature, the fundamental principles underlying motional emf and induced emf are the same.

ξ and $d\Phi_B/dt$ are thus related in the same way as Eq. 10.4, by:

$$\xi = -\frac{d\Phi_B}{dt}$$

(Eq. 10.5). This equation is known as *Faraday's Flux Law*. For a coil of wire consisting of N identical turns, each with flux Φ_B through them, then the total flux, Λ, is $\Lambda = N\Phi_B$, and the induced/motional emf is N times greater than for a single loop (Eq. 10.6).

Faraday's Flux Law (see text for derivation):

$$\text{Single loop}: \quad \xi = -\frac{d\Phi_B}{dt} \qquad (10.5)$$

$$N\text{-turn coil}: \quad \xi_N = -\frac{d\Lambda}{dt} \qquad (10.6)$$

$$\text{where } \Lambda = N\Phi_B$$

Magnitude of the induced electric field

The magnetic force on a charge, resulting from a relative motion between charges in a wire and an external magnetic field, acts in a direction to separate positive and negative charges. In contrast, the electric field induced as a consequence of this charge separation produces an electric force that acts in the opposite direction (refer to Fig. 10.1). These opposing forces rapidly establish a charge equilibrium in the wire, such that the electric and magnetic forces balance:

$$\vec{F}_E = -\vec{F}_B$$
$$\therefore q\vec{E}_\xi = -q\vec{v} \times \vec{B}$$

Therefore, the strength of the induced electric field is:

$$|\vec{E}_\xi| = |\vec{v}||\vec{B}|\sin\theta$$

where θ is the angle between $\vec{v}$ and $\vec{B}$.

Lenz's Law

Lenz's Law describes the direction in which induced current flows. It states that the induced current flows in the direction to *oppose* the change that caused it. For example, if the magnetic flux through a circuit decreases, then current flows in a direction to *increase* the flux, and vice versa. Lenz's Law is often introduced as a separate law to Faraday's Flux Law, but it is, in fact, just a statement of the minus sign in Eq. 10.5.

Faraday's Law in terms of $\vec{E}$ and $\vec{B}$

With a few straightforward manipulations, we can express Faraday's Flux Law in terms of $\vec{E}$ and $\vec{B}$. Invoking Stokes's theorem, the induced emf in terms of an induced electric field $\vec{E}$ is:

$$\xi = \oint_L \vec{E} \cdot d\vec{l} \qquad (10.7)$$

$$= \int_S \nabla \times \vec{E} \cdot d\vec{s} \qquad (10.8)$$

And from Faraday's Flux Law, the induced emf in terms of $\vec{B}$ is:

$$\xi = -\frac{d\Phi_B}{dt} = -\int_S \frac{\partial \vec{B}}{\partial t} \cdot d\vec{s} \qquad (10.9)$$

Equating Eq. 10.9 to Eqs. 10.7 and 10.8, in turn, we recover an integral and differential form of Faraday's Law, respectively, in terms of $\vec{E}$ and $\vec{B}$ (Eqs. 10.10 and 10.11). Written in this way, Faraday's Law is another of Maxwell's equations. Faraday's Law states that a changing magnetic field induces an electric field. For zero or constant magnetic field, we find $\nabla \times \vec{E} = 0$, which is consistent with the electrostatic case.

Faraday's Law (see text for derivation):

Integral form:

$$\oint_L \vec{E} \cdot d\vec{l} = -\int_S \frac{\partial \vec{B}}{\partial t} \cdot d\vec{s} \qquad (10.10)$$

Differential form:

$$\nabla \times \vec{E} = -\frac{\partial \vec{B}}{\partial t} \qquad (10.11)$$

10.2 Inductors and inductance

Now we'll look at an electrical device that is based on Faraday's Flux Law—the *inductor*. Inductors usually consist of a coil of wire, but other configurations, such as parallel wires or coaxial cables, are common too. If the current through an inductor changes, then so does the magnetic flux through the wire configuration, which, from Faraday's Flux Law, induces an emf.

The emf, in turn, induces current to flow through the inductor. From Lenz's Law, the induced current flows in a direction to oppose the initial change in current. For example, an initial decrease in current leads to a change in flux that induces a current to flow in the direction to *increase* it again.

Inductors are thought of as the magnetic equivalent of the electric capacitor, in that they store magnetic potential energy (see Section 10.2.3). They are characterized by their *self-inductance*, L [H], which is a measure of their capability to induce an emf in response to a change of current. In systems consisting of pairs of inductors, a change in current in one may induce an emf in the other, which gives rise to *mutual inductance*, M [H]. This is a measure of the capability of one inductor to induce an emf in another. Below, we will look at each in turn.

10.2.1 Self-inductance

The *self-inductance*, L [H], of an inductor is defined as the ratio of the magnetic flux Φ linking the inductor, to the current I flowing through it:

$$L = \frac{\Phi}{I}$$

(Eq. 10.12). The value of L depends on the geometry of the inductor. Example calculations of the self-inductance of some common inductors are provided in Worked example 10.1. For inductors whose inductance is current-dependent, the *differential* or *small signal* inductance, L_d, is defined by the derivative:

$$L_d = \frac{d\Phi}{dI}$$

(Eq. 10.13). $L = L_d$ provided the inductance (the inductor geometry) is current independent.

Back emf

A change in current through an inductor leads to a change in magnetic flux linked, which induces an emf. This induced emf is called the *back emf*, ξ_L [V], since from Lenz's Law, it acts in the direction that opposes the change. From Faraday's Flux Law, the back emf is given by:

$$\begin{aligned}\xi_L &= -\frac{d\Phi}{dt} \\ &= -\frac{d\Phi}{dI}\frac{dI}{dt} \\ &= -L\frac{dI}{dt}\end{aligned}$$

Therefore, ξ_L is proportional to the rate of change of current, where the constant of proportionality is the self-inductance (Eq. 10.14).

Self-inductance and back emf:

Self-inductance:

$$L \stackrel{\text{def}}{=} \frac{\Phi}{I} \tag{10.12}$$

Differential self-inductance:

$$L_d \stackrel{\text{def}}{=} \frac{d\Phi}{dI} \tag{10.13}$$

Back emf:

$$\xi_L = -L\frac{dI}{dt} \tag{10.14}$$

10.2.2 Mutual inductance

For two inductors i and j, the magnetic flux of $\vec{B}_i$ linked by j is:

$$\Phi_j = \int_{S_j} \vec{B}_i \cdot d\vec{s}_j$$

Similarly, $\Phi_i = \int_{S_i} \vec{B}_j \cdot d\vec{s}_i$

A change in $\vec{B}_i$ therefore leads to a change in Φ_j, which, in turn, induces an emf and current in j, where:

$$\xi_j = -\frac{d\Phi_j}{dt}$$

The current induced in j then changes $\vec{B}_j$, which, in turn, leads to a change in Φ_i, thus inducing an emf and current in i, where:

$$\xi_i = -\frac{d\Phi_i}{dt}$$

This phenomenon is, not surprisingly, called *mutual induction*. The *mutual inductance*, M [H], is defined in an analogous way to self-inductance, L, as the ratio of flux linked in one inductor to the current in the other inductor:

$$M = \frac{\Phi_j}{I_i}$$

Through reciprocity, the mutual inductance takes the same value whether we calculate the ratio of Φ_j to I_i, or swap the subscripts:

$$\therefore M = \frac{\Phi_j}{I_i} = \frac{\Phi_i}{I_j}$$

(Eq. 10.15). The value of M depends on the geometries and separation of the two inductors. For example, consider two coaxial solenoids: a large solenoid of length l_1 with $n_1 = N_1/l_1$ turns per unit length, and a smaller solenoid within the larger one of length l_2 with $n_2 = N_2/l_2$ turns per unit length. Assume the magnetic field of the large solenoid is constant over the dimensions of the smaller solenoid, where the magnitude of the field is $|\vec{B}_1| = \mu_0 I_1 n_1$ (refer to Worked example 8.3). The flux of $\vec{B}_1$ through the smaller coil is then:

$$\Phi_2 = \int \vec{B}_1 \cdot d\vec{s}_2$$
$$= \left|\vec{B}_1\right| N_2 A_2$$
$$= \mu_0 I_1 n_1 n_2 l_2 A_2$$

where A_2 is the cross-sectional area of the smaller solenoid. Therefore, the mutual inductance of the coaxial solenoids is:

$$M = \frac{\Phi_2}{I_1} = \mu_0 n_1 n_2 l_2 A_2$$

This expression has a similar form to the self-inductance of a single solenoid, $L = \mu_0 n^2 l A$, calculated in Worked example 10.1.

WORKED EXAMPLE 10.1: Determining the self-inductance of some common configurations from their magnetic fields

a) N-turn solenoid

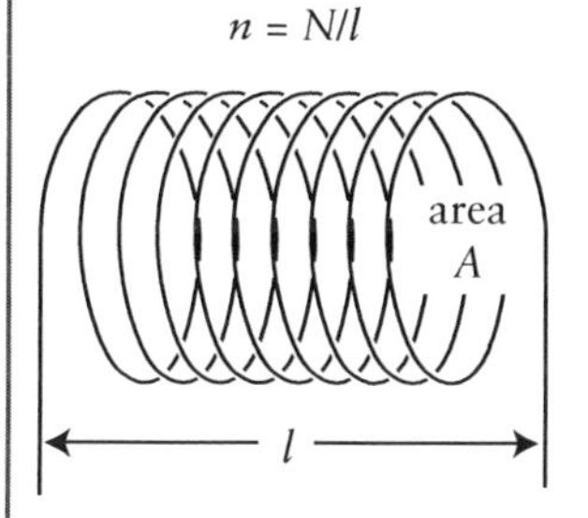

length l
area A
n turns per unit length

$$\left|\vec{B}\right| = \mu_0 I n$$
$$\Phi = \int \vec{B} \cdot d\vec{s}$$
$$= NA\left|\vec{B}\right|$$
$$= \mu_0 A n^2 l I$$
$$\therefore L = \frac{\Phi}{I}$$
$$= \mu_0 A n^2 l$$
$$= \mu_0 \pi r^2 n^2 l$$

for circular solenoid

b) Parallel wires in the same circuit

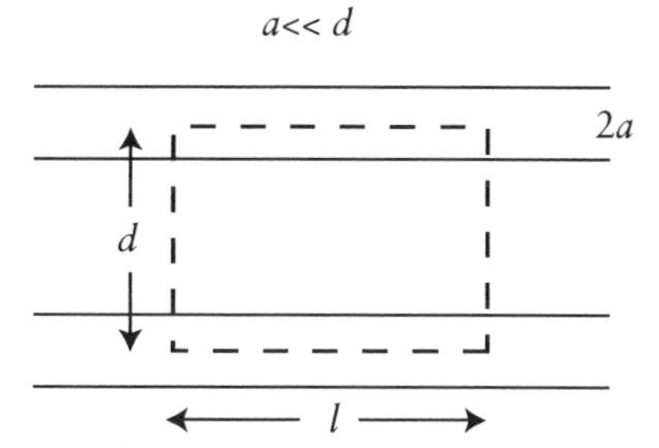

length l
wire radius a
wire separation $(d–2a)$

$$\left|\vec{B}\right| = \frac{\mu_0 I}{2\pi r}$$
$$\Phi = 2\int_a^{d-a} \vec{B} \cdot l d\vec{r}$$
$$= 2\int_a^{d-a} \frac{\mu_0 I l}{2\pi r} dr$$
$$\simeq \frac{\mu_0 I l}{\pi} \ln(d/a)$$

for $a \ll d$

$$\therefore L = \frac{\Phi}{I}$$
$$= \frac{\mu_0 l}{\pi} \ln(d/a)$$

c) Coaxial cable

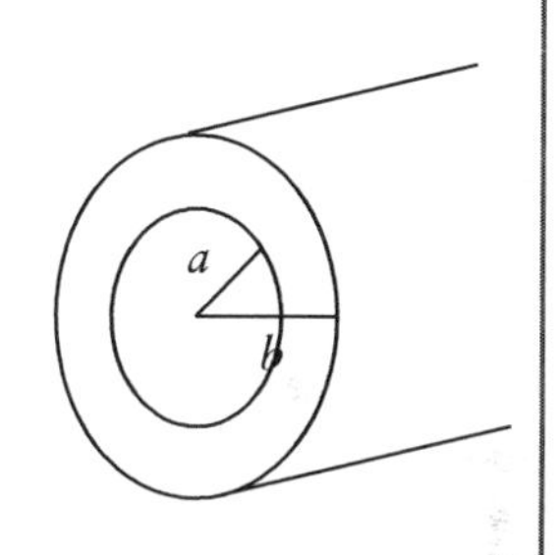

length l
inner radius a
outer radius b

$$\left|\vec{B}\right| = \frac{\mu_0 I}{2\pi r}$$
$$\Phi = \int_a^b \vec{B} \cdot l d\vec{r}$$
$$= \int_a^b \frac{\mu_0 I l}{2\pi r} dr$$
$$= \frac{\mu_0 I l}{2\pi} \ln(b/a)$$
$$\therefore L = \frac{\Phi}{I}$$
$$= \frac{\mu_0 l}{2\pi} \ln(b/a)$$

Refer to Worked example 8.3 for determining the magnetic field of each configuration from Ampère's Law.

When calculating mutual inductances, it's useful to consider whether it's more straightforward to determine $M = \Phi_1/I_2$ or $M = \Phi_2/I_1$; this usually corresponds to the configuration for which $\vec{B}$ is uniform over the relevant area for determining Φ.

For inductors whose inductance is current-dependent, the *differential* or *small signal* inductance, M_d, is defined by the derivative:

$$M_d = \frac{d\Phi_j}{dI_i} = \frac{d\Phi_i}{dI_j}$$

(Eq. 10.16). $M = M_d$ provided the inductance (the inductor geometry) is current independent.

Induced emf

Expressing the induced emf in terms of M, we have:

$$\begin{aligned}\xi_j &= -\frac{d\Phi_j}{dt} \\ &= -\frac{d\Phi_j}{dI_i}\frac{dI_i}{dt} \\ &= -M\frac{dI_i}{dt}\end{aligned}$$

$$\text{Similarly, } \xi_i = -M\frac{dI_j}{dt}$$

(Eqs. 10.17 and 10.18). These principles provide the basis of how electrical transformers work: a changing current in the primary coil results in a changing magnetic flux in the transformer's core, thus changing the magnetic flux through the secondary coil, which, in turn, induces an emf in the secondary coil. Therefore, changing the current through a circuit influences current flow through nearby circuits without there being any direct electrical connection between them.

Mutual inductance and induced emf:

Mutual inductance:

$$M \overset{\text{def}}{=} \frac{\Phi_j}{I_i} = \frac{\Phi_i}{I_j} \quad (10.15)$$

Differential mutual inductance:

$$M_d \overset{\text{def}}{=} \frac{d\Phi_j}{dI_i} = \frac{d\Phi_i}{dI_j} \quad (10.16)$$

where the flux linked in one inductor due to the other is:

$$\Phi_j = \int_{S_j} \vec{B}_i \cdot d\vec{s}_j$$

$$\text{or } \Phi_i = \int_{S_i} \vec{B}_j \cdot d\vec{s}_i$$

Induced emf:

$$\xi_j = -M\frac{dI_i}{dt} \quad (10.17)$$

$$\text{and } \xi_i = -M\frac{dI_j}{dt} \quad (10.18)$$

The subscripts represent inductors 1 and 2, where $i \neq j$.

10.2.3 Potential energy of an inductor

To get a current flowing through an inductor, work must be done against the back emf. (From Lenz's Law, the back emf is induced in a direction to *decrease* the current.) This work is stored as potential energy in the magnetic field of the inductor and can be recovered if the current is reduced slowly to zero. The potential energy stored in an inductor, U_L, is given by:

$$U_L = \frac{1}{2}I\Phi$$

(Eq. 10.19 and Derivation 10.1). Eq. 10.19 has an equivalent form to the potential energy of a capacitor ($U_C = QV/2$, Eq. 7.47).

For a set of N inductors, the potential energy is the sum of energies due to the self-inductance of each inductor plus the mutual inductance between pairs of inductors (Eq. 10.20 and Derivation 10.2).

Potential energy of an inductor (Derivations 10.1 and 10.2):

One inductor:

$$U_L = \frac{1}{2}I\Phi, \frac{1}{2}LI^2, \text{ or } \frac{\Phi^2}{2L} \quad (10.19)$$

N inductors:

$$U_L^N = \frac{1}{2}\sum_{i=1}^{N} L_i I_i^2 + \sum_{i=1}^{N}\sum_{\substack{j=1 \\ j>i}}^{N} M_{ij} I_i I_j \quad (10.20)$$

10.2.4 Magnetic energy density

By expressing Φ in terms of the magnetic vector potential, $\vec{A}$, where $\vec{B} = \nabla \times \vec{A}$, as shown in Derivation 10.3, we can express U_L in terms of the magnetic field, $\vec{B}$, of a system:

$$U_L \equiv U_B = \frac{1}{2\mu_0} \int_{\substack{\text{all}\\\text{space}}} \left|\vec{B}\right|^2 \mathrm{d}\tau$$

(Eq. 10.22 and Derivation 10.3). Written in this way, the potential energy is usually referred to as *magnetic energy*, U_B [J]. Magnetic energy is thus stored in the magnetic field of a system. It has an analogous form to the expression for electric energy (refer to Eq. 7.36):

$$U_E = \frac{1}{2}\varepsilon_0 \int_{\substack{\text{all}\\\text{space}}} \left|\vec{E}\right|^2 \mathrm{d}\tau$$

Derivation 10.1: Derivation of the energy stored in an inductor (Eq. 10.19)

$$\begin{aligned} U_L &= \int P \mathrm{d}t \\ &= \int I'V' \mathrm{d}t \\ &= L\int I' \frac{\mathrm{d}I'}{\mathrm{d}t}\mathrm{d}t \\ &= L\int_0^I I' \mathrm{d}I' \\ &= \frac{1}{2}LI^2 \\ &\equiv \frac{1}{2}I\Phi \\ &\equiv \frac{\Phi^2}{2L} \end{aligned}$$

The primed variables are to distinguish them from the integration limits.

- Starting point: assuming L is constant in time, differentiating the expression $\Phi = LI$ with respect to time, we have:

$$\frac{\mathrm{d}\Phi}{\mathrm{d}t} = L\frac{\mathrm{d}I}{\mathrm{d}t}$$

- From the Flux Law:

$$\begin{aligned} \frac{\mathrm{d}\Phi}{\mathrm{d}t} &= -\xi = V \\ \therefore V &= L\frac{\mathrm{d}I}{\mathrm{d}t} \end{aligned} \qquad (10.21)$$

where ξ is the emf induced in an inductor and hence V is the voltage across an inductor in a circuit.

- The energy stored in an inductor is given by the time integral over the instantaneous power, $P = IV = I\mathrm{d}I/\mathrm{d}t$. Integrating, we find that the energy stored in an inductor is:

$$U_L = \frac{1}{2}LI^2$$

Use the expression for inductance ($L = \Phi/I$) to obtain different expressions for U_L.

Derivation 10.2: Derivation of the energy stored in N inductors (Eq. 10.20)

$$\begin{aligned} U_L^N &= \frac{1}{2}\sum_{i=1}^{N} I_i\Phi_i \\ &= \frac{1}{2}\sum_{i=1}^{N} I_i \left(L_i I_i + \sum_{\substack{j=1\\j>i}}^{N} M_{ij}I_j \right) \\ &= \frac{1}{2}\sum_{i=1}^{N} L_i I_i^2 + \sum_{i=1}^{N}\sum_{\substack{j=1\\j>i}}^{N} M_{ij}I_i I_j \end{aligned}$$

- Starting point: total flux linking the ith inductor is from a combination of its self-inductance and the sum of mutual inductances with the other components:

$$\Phi_i = L_i I_i + \sum_{\substack{j=1\\j>i}}^{N} M_{ij}I_j$$

Since $M_{ij} = M_{ji}$, the limits in the sum require that $j > i$ to avoid double counting.

- The total energy of N inductors is the sum of the energy of the individual inductors, where $U_L = I\Phi/2$ for one inductor. Use the expression for Φ_i above in U_L^N.

Derivation 10.3: Derivation of the energy stored in a magnetic field (Eq. 10.22)

$$
\begin{aligned}
U_L &\equiv U_B \\
&= \frac{1}{2} I\Phi \\
&= \frac{1}{2} I \oint_L \vec{A} \cdot d\vec{l} \\
&= \frac{1}{2} \oint_L \vec{A} \cdot \vec{I} dl \\
&= \frac{1}{2} \int_V \vec{A} \cdot \vec{J} d\tau \\
&= \frac{1}{2\mu_0} \int_V \vec{A} \cdot \left(\nabla \times \vec{B}\right) d\tau \\
&= \frac{1}{2\mu_0} \int_V \left(\vec{B} \cdot \vec{B} - \nabla \cdot \left(\vec{A} \times \vec{B}\right)\right) d\tau \\
&= \frac{1}{2\mu_0} \left(\int_V \left|\vec{B}\right|^2 d\tau - \oint_S \left(\vec{A} \times \vec{B}\right) \cdot d\vec{s} \right) \\
&= \frac{1}{2\mu_0} \int_{\substack{\text{all} \\ \text{space}}} \left|\vec{B}\right|^2 d\tau
\end{aligned}
$$

- Starting point: in terms of the magnetic vector potential, $\vec{A}$, where $\vec{B} = \nabla \times \vec{A}$, and invoking Stokes's theorem, the magnetic flux is:

$$
\begin{aligned}
\Phi &= \int_S \vec{B} \cdot d\vec{s} \\
&= \int_S \left(\nabla \times \vec{A}\right) \cdot d\vec{s} \\
&= \oint_L \vec{A} \cdot d\vec{l}
\end{aligned}
$$

- Using this in the expression for $U_L (U_L = I\Phi/2)$, and letting the subscript $L \to B$ to stand for magnetic energy, we can express the magnetic energy in terms of $\vec{A}$ and $\vec{J}$:

$$U_B = \frac{1}{2} \int_V \vec{A} \cdot \vec{J} d\tau$$

This expression is the magnetic equivalent of the electric energy density:

$$U_E = \frac{1}{2} \int V \rho d\tau$$

(refer to Eq. 7.35).

- Use Ampère's Law ($\nabla \times \vec{B} = \mu_0 \vec{J}$) to eliminate $\vec{J}$, along with the product rule:

$$
\begin{aligned}
\nabla \cdot (\vec{A} \times \vec{B}) &= \vec{B} \cdot \left(\nabla \times \vec{A}\right) - \vec{A} \cdot \left(\nabla \times \vec{B}\right) \\
&= \vec{B} \cdot \vec{B} - \vec{A} \cdot \left(\nabla \times \vec{B}\right)
\end{aligned}
$$

- Invoke the divergence theorem on the integral over $\nabla \cdot (\vec{A} \times \vec{B})$, and let $\vec{B} \cdot \vec{B} = |\vec{B}|^2$.
- When integrating over all space, the volume integral dominates over the surface integral.

Dividing U_B by volume, we get the *magnetic energy density*, u_B [J m^{-3}] (Eq. 10.23). Again, this is analogous to the electric energy density (refer to Eq. 7.37):

$$u_E = \frac{1}{2} \varepsilon_0 \left|\vec{E}\right|^2$$

Energy stored in a magnetic field (Derivation 10.3):

Total energy:

$$U_B = \frac{1}{2\mu_0} \int_{\substack{\text{all} \\ \text{space}}} \left|\vec{B}\right|^2 d\tau \qquad (10.22)$$

Energy density (energy per unit volume):

$$\therefore u_B = \frac{1}{2\mu_0} \left|\vec{B}\right|^2 \qquad (10.23)$$

10.3 The displacement current

Faraday's Law states that a changing magnetic field induces an electric field. Similarly, Maxwell deduced that a changing electric field induces a *magnetic* field, as we shall see through the following argument.

The integral form of Ampère's Law is:

$$\oint_L \vec{B} \cdot d\vec{l} = \mu_0 I_{enc}$$

We can deduce that this "static" form of Ampère's Law is flawed by considering the charging of a parallel-plate capacitor in a closed electrical circuit. A closed path around the current-carrying wire placed just outside a capacitor plate encloses some current, I. However, the same closed path placed just below the plate encloses zero current; thus, from Ampère's Law, we would expect $\vec{B}$ to suddenly drop to zero within a very small displacement. However, $\vec{B}$ is a continuous function in space, therefore it cannot drop to zero in a small displacement.

Therefore, Maxwell realized that an additional term must be added to Ampère's Law in order to prevent a discontinuity in $\vec{B}$ in such instances. This term is called the *displacement current*, I_D[A], or the *displacement current density*, $\vec{J}_D$[Am^{-2}], where I_D and $\vec{J}_D$ are related by:

$$I_D = \int_S \vec{J}_D \cdot d\vec{s}$$

We can determine an expression for $\vec{J}_D$ by considering the differential form of Ampère's Law:

$$\nabla \times \vec{B} = \mu_0 \vec{J}$$

Taking the divergence of both sides, we have:

$$\nabla \cdot (\nabla \times \vec{B}) = \mu_0 \nabla \cdot \vec{J}$$

Mathematically, we require that $\nabla \cdot \vec{J} = 0$, since the divergence of a curl is equal to zero. From the continuity equation (refer to Section 9.1.1), this term is only equal to zero when rate of change of charge density, ρ, is equal to zero:

$$\nabla \cdot \vec{J} = -\frac{\partial \rho}{\partial t} = 0$$

i.e. only in the magnetostatic case. However, for systems in which $\partial\rho/\partial t \neq 0$, such as for the capacitor-plate example described above, we find that the term $\partial\rho/\partial t$ corresponds to the displacement current term. Using the continuity equation together with Gauss's Law, $\vec{J}_D$ is found to be proportional to the rate of change of electric field, $\vec{E}$, created by ρ:

$$\vec{J}_D = \varepsilon_0 \frac{\partial \vec{E}}{\partial t}$$

$$\text{and } I_D = \int_S \vec{J}_D \cdot d\vec{s} = \varepsilon_0 \frac{\partial \Phi_E}{\partial t}$$

(Eqs. 10.24 and 10.25 and Derivation 10.4). Φ_E is the electric flux through the surface S, and is defined by:

$$\Phi_E = \int_S \vec{E} \cdot d\vec{s}$$

Derivation 10.4: Derivation of the displacement current and Ampère's Law (Eqs. 10.24 and 10.26)

$$\nabla \cdot \vec{J} = -\frac{\partial \rho}{\partial t}$$

$$= -\frac{\partial}{\partial t}\varepsilon_0(\nabla \cdot \vec{E})$$

$$= -\nabla \left(\varepsilon_0 \frac{\partial \vec{E}}{\partial t}\right)$$

$$\therefore \nabla \cdot \left(\vec{J} + \varepsilon_0 \frac{\partial \vec{E}}{\partial t}\right) = 0$$

Define the total current as:

$$\vec{J}_{tot} = \vec{J} + \varepsilon_0 \frac{\partial \vec{E}}{\partial t}$$

$$\therefore \nabla \cdot \vec{J}_{tot} = 0$$

Using this current in Ampère's Law:

$$\nabla \times \vec{B} = \mu_0 \vec{J}_{tot}$$

$$= \mu_0 \vec{J} + \mu_0 \varepsilon_0 \frac{\partial \vec{E}}{\partial t}$$

- Starting point: the continuity equation shows that current density $\vec{J}$ and charge density ρ are related by:

  $$\nabla \cdot \vec{J} = -\frac{\partial \rho}{\partial t}$$

  (refer to Section 9.1.1 for derivation).
- Use Gauss's Law, $\nabla \cdot \vec{E} = \rho/\varepsilon_0$, to eliminate ρ from the continuity equation, and rearrange.
- For zero divergence, define the total current as:

  $$\vec{J}_{tot} = \vec{J} + \varepsilon_0 \frac{\partial \vec{E}}{\partial t} \equiv \vec{J}_C + \vec{J}_D$$

  $$\text{where } \vec{J}_D = \varepsilon_0 \frac{\partial \vec{E}}{\partial t}$$

  Therefore, the total current comprises two components: the conduction current due to flow of mobile charge, $\vec{J}$ (or $\vec{J}_C$), and the displacement current due to a changing electric field, $\vec{J}_D$.
- Substitute $\vec{J}_{tot}$ into Ampère's Law. Written in this way, $\nabla \cdot \nabla \times \vec{B} = \mu_0 \nabla \cdot \vec{J}_{tot} = 0$, as required mathematically.

Displacement current (Derivation 10.4):

$$\vec{J}_D = \varepsilon_0 \frac{\partial \vec{E}}{\partial t} \quad (10.24)$$

$$\text{and } I_D = \varepsilon_0 \frac{\partial \Phi_E}{\partial t} \quad (10.25)$$

The displacement current is distinguished from the *conduction current*, $\vec{J}$ (or $\vec{J}_C$), which is current that corresponds to the flow of mobile charge (which we have used exclusively up until now).

Ampère's Law with Maxwell's correction

In Derivation 10.4, we saw that in order for the divergence of the curl of $\vec{B}$ in Ampère's Law to equal zero, we needed to define the total current as the sum of conduction and displacement currents:

$$\vec{J}_{\text{tot}} = \vec{J} + \varepsilon_0 \frac{\partial \vec{E}}{\partial t}$$

Using this in the differential form of Ampère's Law, we have:

$$\begin{aligned} \nabla \times \vec{B} &= \mu_0 \vec{J}_{\text{tot}} \\ &= \mu_0 \vec{J} + \mu_0 \varepsilon_0 \frac{\partial \vec{E}}{\partial t} \end{aligned}$$

This is *Ampere's Law with Maxwell's correction* (Eq. 10.26). Integrating both sides over area and invoking Stokes's theorem, we find:

$$\begin{aligned} \int_S \nabla \times \vec{B} \cdot d\vec{s} &= \mu_0 \int_S \vec{J} \cdot d\vec{s} + \mu_0 \varepsilon_0 \int_S \frac{\partial \vec{E}}{\partial t} \cdot d\vec{s} \\ &= \mu_0 I_{\text{enc}} + \mu_0 \varepsilon_0 \frac{\partial \Phi_E}{\partial t} \end{aligned}$$

$$\text{and } \int_S \nabla \times \vec{B} \cdot d\vec{s} = \oint_L \vec{B} \cdot d\vec{l}$$

Equating these expressions gives us Ampère's Law with Maxwell's correction in integral form (Eq. 10.27). Therefore, a changing electric field induces a current (the displacement current), which, in turn, produces a magnetic field. Similarly, Faraday's Law states that a changing magnetic field induces an electric field. This is the basis of the propagation of *electromagnetic waves*, covered in Chapter 12.

Ampère's Law with Maxwell's correction (Derivation 10.4):

Differential form:

$$\nabla \times \vec{B} = \mu_0 \vec{J} + \mu_0 \varepsilon_0 \frac{\partial \vec{E}}{\partial t} \quad (10.26)$$

Integral form:

$$\oint_L \vec{B} \cdot d\vec{l} = \mu_0 I_{\text{enc}} + \mu_0 \varepsilon_0 \frac{\partial \Phi_E}{\partial t} \quad (10.27)$$

10.4 Maxwell's equations, finally

We are finally ready to collect together Maxwell's equations, which we've derived throughout the previous few chapters. Maxwell's four equations express the divergence and curl of $\vec{E}$ and $\vec{B}$:

- **Gauss's Law:**
  the divergence of $\vec{E}$ (Eq. 10.28).
- **Gauss's Law for magnetism:**
  the divergence of $\vec{B}$ (Eq. 10.29).
- **Faraday's Law:**
  the curl of $\vec{E}$ (Eq. 10.30).
- **Ampère's Law:**
  the curl of $\vec{B}$ (Eq. 10.31).

They each have a differential form and an integral form, and we can convert between them by invoking either the divergence theorem (for Eqs. 10.28 and 10.29) or Stokes's theorem (for Eqs. 10.30 and 10.31).

Maxwell's equations illustrate that electric fields are produced by electric charge or by a changing magnetic field (Eqs. 10.28 and 10.30), whereas magnetic fields are produced by electric current or by a changing electric field (Eqs. 10.29 and 10.31). The asymmetry between the equations for $\vec{E}$ and $\vec{B}$ arises because there are no magnetic monopoles, and therefore no sources or sinks of magnetic field, whereas electric monopoles (for example, protons and electrons) do exist in nature. Maxwell's equations, together with the Lorentz Force Law ($\vec{F} = q(\vec{E} + \vec{v} \times \vec{B})$, Eq. 8.14), essentially encapsulate all of electromagnetism. In the following chapters, we will look at how to apply Maxwell's equations in order to understand the properties of electromagnetic fields in matter (Chapter 11), and the properties of electromagnetic waves (Chapter 12).

Maxwell's equations:

Gauss's Law: $\nabla \cdot \vec{E} = \dfrac{\rho}{\varepsilon_0}$ $\qquad \oint_S \vec{E} \cdot d\vec{s} = \dfrac{Q_{\text{enc}}}{\varepsilon_0}$ (10.28)

Gauss's Law for magnetism: $\nabla \cdot \vec{B} = 0$ $\qquad \oint_S \vec{B} \cdot d\vec{s} = 0$ (10.29)

Faraday's Law: $\nabla \times \vec{E} = -\dfrac{\partial \vec{B}}{\partial t}$ $\qquad \oint_L \vec{E} \cdot d\vec{l} = -\dfrac{\partial \Phi_B}{\partial t}$ (10.30)

Ampère's Law: $\nabla \times \vec{B} = \mu_0 \vec{J} + \mu_0 \varepsilon_0 \dfrac{\partial \vec{E}}{\partial t}$ $\qquad \oint_L \vec{B} \cdot d\vec{l} = \mu_0 I_{\text{enc}} + \mu_0 \varepsilon_0 \dfrac{\partial \Phi_E}{\partial t}$ (10.31)

Chapter 11
Electromagnetic fields in matter

In this chapter, we look at the properties of electromagnetic fields in matter. Matter is made up of charged particles: positively charged nuclei and negatively charged electrons. These are influenced by an applied electric and magnetic field in such a way that the material itself produces its own electric or magnetic field. A material that produces its own electric field is said to be *polarized* and one that produces its own magnetic field is said to be *magnetized*. In this chapter, we'll look at some of the properties of polarization and magnetization.

11.1 Electric fields in matter

Materials are often classified as either conductors or insulators, depending on their resistance to the flow of electric current through them.[1] Conductors have mobile electrons that are free to drift in the presence of an electric field, and therefore conductors have a low resistance to current flow (Section 7.7). In contrast, insulators have electrons that are tightly bound to the nucleus, and therefore insulators have a high resistance to current flow. Electrical insulators are often called *dielectrics*. Atoms in dielectrics are depicted as a positive nucleus surrounded by a negative electron cloud (Fig. 11.1).

HILS materials

In this section, we will deal solely with so-called *HILS* materials, which are materials whose properties are Homogeneous (the properties do not vary with position), Isotropic (the properties do not vary with orientation), Linear (the superposition principle holds for $\vec{E}$ and $\vec{B}$), and Stationary (the properties do not vary with time).

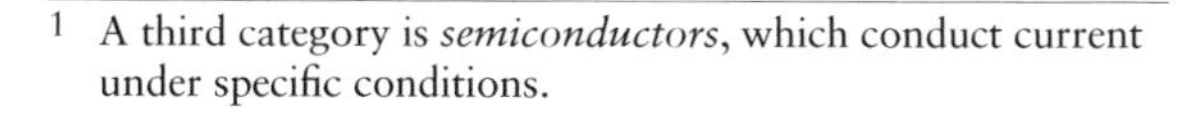
[1] A third category is *semiconductors*, which conduct current under specific conditions.

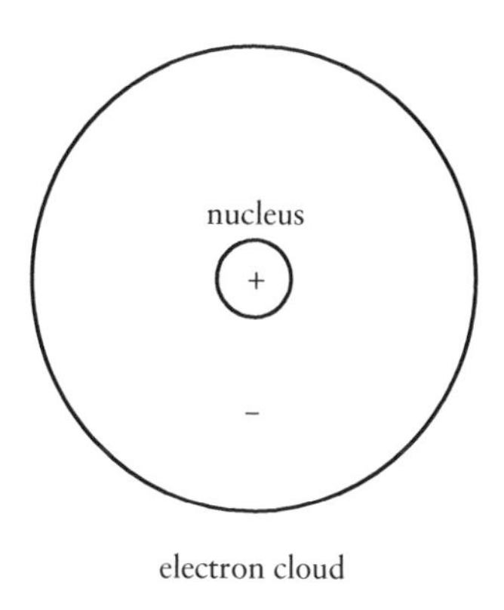

Figure 11.1: Nucleus and electron cloud.

11.1 1 Polarization

When placed in an external electric field, a dielectric becomes *polarized*, which means it acquires an internal electric field. Polarization is the separation of positive and negative charges within an atom to create an electric dipole. The net alignment of these atomic dipoles produce an overall charge separation within the dielectric, which produces an electric field. For example, consider first the atom illustrated in Fig. 11.1. When placed in an external electric field, the positive nucleus and negative electron cloud move a minute amount in opposite directions, causing them to become off-centered. The atom therefore acquires a dipole moment, $\vec{p} = q\vec{l}$, where $\vec{l}$ represents the separation of the nucleus from the center of the electron cloud (Fig.11.2a). This is called *electronic* polarization. The net alignment of the dipole moments from all the atoms produces an overall *polarization*, $\vec{P}$ [Cm^{-2}], of the material, where $\vec{P}$ is the electric dipole moment per unit volume. Therefore, for a material with n dipoles per unit volume, the polarization is defined as $\vec{P} = n\vec{p}$ (Eq. 11.1).

Orientational and atomic polarization

As well as electronic polarization, the two other main types of polarization are *orientational* and *atomic* polarization. Orientational polarization occurs in dielectrics that are composed of polar molecules (ones that have a permanent dipole moment). In the absence of an external electric field, the dipole moments are randomly aligned, and the material is not polarized. An external electric field causes the polar molecules to align, thus polarizing the material (Fig. 11.2b). Atomic polarization occurs in dielectrics that are composed of ions of different signs, such as NaCl; an external electric field causes the positive and negative ions to separate (Fig.11.2c).

Electric susceptibility

The *electric susceptibility*, χ_e, is a measure of the readiness at which a material polarizes in an external electric field. In linear and isotropic dielectrics, the polarization, $\vec{P}$, is proportional to the external electric field, $\vec{E}$:

$$\vec{P} \propto \varepsilon_0 \vec{E}$$

By including the constant ε_0, both right- and left-hand sides have the same dimensions of Cm^{-2}. The dimensionless proportionality constant is the electric susceptibility:

$$\therefore \vec{P} = \chi_e \varepsilon_0 \vec{E}$$

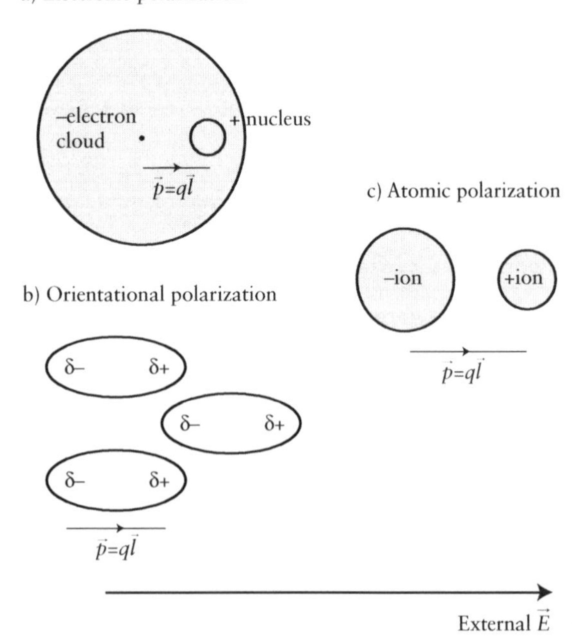

Figure 11.2: Polarization of dielectrics. $\vec{p}$ is the induced dipole moment.

(Eq. 11.2) The higher χ_e, the greater $|\vec{P}|$ is for a given external electric field. For example, materials with polar molecules (like water) polarize more readily than those with non-polar molecules, and therefore have a higher susceptibility.

Electric polarization:

In terms of atomic electric dipole moment, $\vec{p}$:

$$\vec{P} = n\vec{p} \tag{11.1}$$

where $\vec{p} = q\vec{l}$

In terms of the external electric field, $\vec{E}$:

$$\vec{P} = \chi_e \varepsilon_0 \vec{E} \tag{11.2}$$

where χ_e is the electric susceptibility.

11.1.2 Free and bound charge

Charge that is associated with polarization of dielectrics is called *bound charge*, q_b [C], since it is effectively "bound" to the atom. The charge per unit area is the *bound surface charge density*, σ_b [Cm^{-2}], where:

$$\sigma_b = \frac{dq_b}{ds}$$

and the charge per unit volume is the *bound volume charge density*, ρ_b [$\mathrm{C\,m}^{-3}$], where:

$$\rho_b = \frac{dq_b}{d\tau}$$

If a material is not polarized, then the bound charge density equals zero. All other charge (i.e. charge that is not associated with polarization) is called *free charge*, q_f [C], σ_f [Cm^{-2}], or ρ_f [Cm^{-3}]. For example, ions embedded in a dielectric constitute free charge, as do mobile electrons in a conductor or ions in a solution. These charges are present, even in the absence of polarization. The total charge is then the sum of free and bound charges:

$$q = q_f + q_b$$

Bound charge densities

The electric potential of an ideal electric dipole[2] is:

$$V(\vec{r}) = \frac{\vec{p} \cdot \hat{r}}{4\pi\varepsilon_0 r^2}$$

[2] Recall from Section 7.5 that an ideal dipole is one in which the separation vector, $\vec{l}$, is much less than the position vector, $\vec{r}$: $|\vec{l}| \ll |\vec{r}|$. Since the dipoles induced in materials are of atomic dimensions, this approximation is valid for most observations.

(refer to Section 7.5). This expression can be separated into the potential produced by a bound surface charge density, σ_b, and a bound volume charge density ρ_b (Derivation 11.1). Doing this, we find expressions for σ_b and ρ_b in terms of $\vec{P}$: the surface charge density is equal to the component of polarization perpendicular to the surface (Eq. 11.3 and Derivation 11.1), whereas the volume charge density is equal to minus the divergence of the polarization[3] (Eq. 11.4 and Derivation 11.1).

Bound charge densities (Derivation 11.1):

$$\sigma_b = \vec{P}\cdot\hat{n} \quad (11.3)$$

$$\rho_b = -\nabla\cdot\vec{P} \quad (11.4)$$

where $\hat{n}$ is a unit vector perpendicular to the surface.

Polarization current

Polarization current is the "flow" of bound charge. A polarization current flows if the polarization, $\vec{P}$, changes with time. Using the expression for the bound surface charge density from Eq. 11.3, we find that the *polarization current density*, $\vec{J}_p$ [A m^{-2}], equals the rate of change of polarization (Eq. 11.5 and Derivation 11.2).

Polarization current density (Derivation 11.2):

$$\vec{J}_p = \frac{d\vec{P}}{dt} \quad (11.5)$$

11.1.3 The electric displacement field

Letting the total charge density equal the sum of free and bound charge densities, Gauss's Law can be written as:

$$\nabla\cdot\vec{E} = \frac{\rho}{\varepsilon_0} = \frac{\rho_f + \rho_b}{\varepsilon_0}$$

Using the expression for ρ_b from Eq. 11.4 and rearranging (the div operator is linear), we have:

$$\nabla\cdot\vec{E} = \frac{\rho_f - \nabla\cdot\vec{P}}{\varepsilon_0}$$

$$\therefore \nabla\cdot\underbrace{\left(\varepsilon_0\vec{E}+\vec{P}\right)}_{\vec{D}} = \rho_f$$

Thus, doing this, we can define another vector field, called the *electric displacement field*, $\vec{D}$ [Cm^{-2}], which is a combination of the applied electric field and the dielectric polarization:

$$\vec{D} \overset{\text{def}}{=} \varepsilon_0\vec{E}+\vec{P}$$

(Eq. 11.8).

Gauss's Law in dielectrics

Following on from the above discussion, Gauss's Law in dielectrics equates the divergence of the electric displacement field to the free charge density (Eq. 11.6 and Derivation 11.3). Integrating over volume and invoking the divergence theorem, we can express Gauss's Law in integral form; this equates the flux of $\vec{D}$ through a closed surface S to the free charge enclosed within that surface (Eq. 11.7 and Derivation 11.3).

Gauss's Law in dielectrics (Derivation 11.3):

Differential form:

$$\nabla\cdot\vec{D} = \rho_f \quad (11.6)$$

Integral form:

$$\oint_S \vec{D}\cdot d\vec{s} = (q_f)_{enc} \quad (11.7)$$

where the electric displacement field is:

$$\vec{D} \overset{\text{def}}{=} \varepsilon_0\vec{E}+\vec{P} \quad (11.8)$$

$\vec{D}$ is a construct that combines $\vec{E}$ and $\vec{P}$ in such a way that we do not have to deal explicitly with the effects of polarization. Therefore, $\vec{D}$ simplifies expressions and calculations when dielectrics are involved. Since field lines of $\vec{D}$ begin and end on free charges, the value of $\vec{D}$ is unaltered by the polarization of a dielectric. On the other hand, field lines of $\vec{E}$ begin and end on both free and polarization charges, which means that the value of $\vec{E}$ changes depending on the polarization of a material. Polarized dielectrics set up an internal electric field that

[3] The minus sign in the expression for ρ_b is because $\vec{P}$ is a vector that points from negative to positive charge; therefore, a positive outward polarization is produced by an accumulation of negative bound charge.

Derivation 11.1: Derivation of bound current densities (Eqs. 11.3 and 11.4)

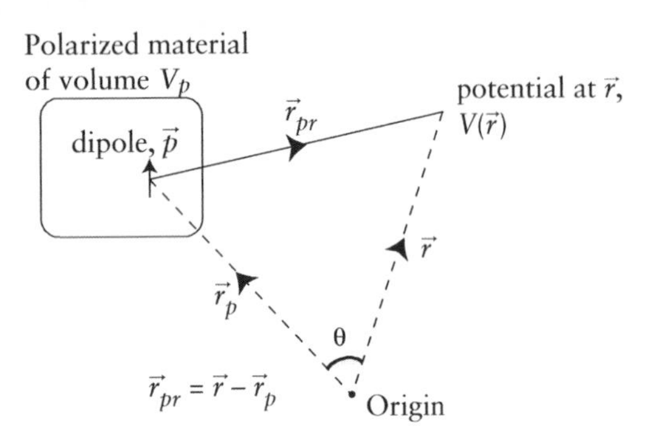

$$
\begin{aligned}
V(\vec{r}) &= N\frac{\vec{p}\cdot\hat{r}_{pr}}{4\pi\varepsilon_0 r_{pr}^2} \\
&= \int_{V_p} \frac{\vec{P}\cdot\hat{r}_{pr}}{4\pi\varepsilon_0 r_{pr}^2}\,\mathrm{d}\tau_p \\
&= \frac{1}{4\pi\varepsilon_0}\int_{V_p} \vec{P}\cdot\nabla_p\left(\frac{1}{r_{pr}}\right)\mathrm{d}\tau_p \\
&= \frac{1}{4\pi\varepsilon_0}\int_{V_p}\left(\nabla_p\cdot\left(\frac{\vec{P}}{r_{pr}}\right) - \frac{\nabla_p\cdot\vec{P}}{r_{pr}}\right)\mathrm{d}\tau_p \\
&= \frac{1}{4\pi\varepsilon_0}\left(\oint_{S_p}\frac{\vec{P}}{r_{pr}}\cdot\mathrm{d}\vec{s}_p - \int_{V_p}\frac{1}{r_{pr}}\nabla_p\cdot\vec{P}\,\mathrm{d}\tau_p\right) \\
&= \frac{1}{4\pi\varepsilon_0}\left(\oint_{S_p}\frac{1}{r_{pr}}\underbrace{\vec{P}\cdot\hat{n}}_{\text{surface charge, }\sigma_b}\,\mathrm{d}s_p - \int_{V_p}\frac{1}{r_{pr}}\underbrace{\nabla_p\cdot\vec{P}}_{\text{volume charge, }\rho_b}\,\mathrm{d}\tau_p\right) \\
&\equiv \frac{1}{4\pi\varepsilon_0}\left(\oint_{S_p}\frac{\sigma_b}{r_{pr}}\mathrm{d}s_p + \int_{V_p}\frac{\rho_b}{r_{pr}}\mathrm{d}\tau_p\right)
\end{aligned}
$$

$$\therefore\ \sigma_b = \vec{P}\cdot\hat{n}$$
$$\text{and}\quad \rho_b = -\nabla\cdot\vec{P}$$

The subscript p refers to coordinates over which the material is polarized (i.e. where dipoles are present). Therefore, V_p and S_p are volume and surface area of polarized matter, respectively.

- Starting point: the electric potential of a polarized dielectric with N identical electric dipoles is N times that of a single electric dipole:

$$V(\vec{r}) = N\frac{\vec{p}\cdot\hat{r}_{pr}}{4\pi\varepsilon_0 r_{pr}^2}$$

where, for a dipole located at $\vec{r}_p$, the separation vector, $\vec{r}_{pr}$, is defined as $\vec{r}_{pr} = \vec{r} - \vec{r}_p$ (refer to figure).
- The dipole number density is $n = N/V_p$, where V_p is the volume of polarized material (not to be confused with the electric potential). Therefore:

$$N\vec{p} = \vec{P}V_p \equiv \int_{V_p}\vec{P}\,\mathrm{d}\tau$$

where the polarization of the material is $\vec{P} = n\vec{p}$.
- Differentiating with respect to r_p, we have:

$$\nabla_p\left(\frac{1}{r_{pr}}\right) = \frac{\hat{r}_{pr}}{r_{pr}^2}$$

Hint: let $r_{pr} = |\vec{r}_{pr}|$
$= ((\vec{r}-\vec{r}_p)\cdot(\vec{r}-\vec{r}_p))^{1/2}$
$= (r^2 + r_p^2 - 2rr_p\cos\theta)^{1/2}$
and differentiate the reciprocal with respect to r_p using the Del operator in spherical polar coordinates.
- Use the identity:

$$\nabla\cdot\left(f\vec{G}\right) = \vec{G}\cdot(\nabla f) + f\left(\nabla\cdot\vec{G}\right)$$

to separate the integral into two terms.
- Invoke the divergence theorem to express $\nabla_p\cdot(\vec{P}/r_{pr})$ as a surface integral, and let $\mathrm{d}\vec{s}_p = \hat{n}\,\mathrm{d}s_p$, where $\hat{n}$ is normal to the surface.
- The potential can therefore be expressed as the superposition of a surface charge density term and a volume charge density term.

Derivation 11.2: Derivation of polarization current density (Eq. 11.5)

$$\vec{J}_p = \frac{\mathrm{d}I_p}{\mathrm{d}s_\perp}$$
$$= \frac{\mathrm{d}}{\mathrm{d}s_\perp}\left(\frac{\mathrm{d}q_\mathrm{b}}{\mathrm{d}t}\right)$$
$$= \frac{\mathrm{d}}{\mathrm{d}s_\perp}\left(\frac{\sigma_\mathrm{b}\mathrm{d}s}{\mathrm{d}t}\right)$$
$$= \frac{\mathrm{d}}{\mathrm{d}s_\perp}\left(\frac{\vec{P}\cdot\mathrm{d}\vec{s}}{\mathrm{d}t}\right)$$
$$= \frac{\mathrm{d}}{\mathrm{d}s_\perp}\left(\frac{\vec{P}\mathrm{d}s_\perp}{\mathrm{d}t}\right)$$
$$= \frac{\mathrm{d}\vec{P}}{\mathrm{d}t}$$

- Starting point: the polarization current density, $\vec{J}_p$, is the polarization current I_p per unit area-perpendicular-to-the-flow, $s_\perp$, where I_p is the rate of change of bound charge:

$$I_p = \frac{\mathrm{d}q_\mathrm{b}}{\mathrm{d}t}$$

- Expressing dq_b as a surface charge density, and using:

$$\sigma_\mathrm{b} = \vec{P}\cdot\hat{n}$$

(Eq. 11.3), we find that the polarization current density equals the rate of change of polarization.

Derivation 11.3: Derivation of Gauss's Law in dielectrics (Eqs. 11.6 and 11.7)

Differential form:

$$\nabla\cdot\vec{E} = \frac{\rho}{\varepsilon_0}$$
$$= \frac{\rho_\mathrm{f}+\rho_\mathrm{b}}{\varepsilon_0}$$
$$= \frac{1}{\varepsilon_0}(\rho_\mathrm{f} - \nabla\cdot\vec{P})$$
$$\therefore \nabla\cdot(\varepsilon_0\vec{E}+\vec{P}) = \rho_\mathrm{f}$$
$$\therefore \nabla\cdot\vec{D} = \rho_\mathrm{f}$$

Integral form:

$$\int_V \nabla\cdot\vec{D}\,\mathrm{d}\tau = \int_V \rho_\mathrm{f}\,\mathrm{d}\tau$$
$$= q_f$$
$$\text{and } \int_V \nabla\cdot\vec{D}\,\mathrm{d}\tau = \oint_S \vec{D}\cdot\mathrm{d}\vec{s}$$
$$\therefore \oint_S \vec{D}\cdot\mathrm{d}\vec{s} = (q_\mathrm{f})_\mathrm{enc}$$

- Starting point: let the total charge density in Gauss's Law equal the sum of free and bound charge densities, and use:

$$\rho_\mathrm{b} = -\nabla\cdot\vec{P}$$

to eliminate ρ_b.
- Divergence is a linear operation, which means we can collect the divergence terms together.
- Define the electric displacement field as:

$$\vec{D} = \varepsilon_0\vec{E}+\vec{P}$$

to express Gauss's Law in dielectrics as the divergence of $\vec{D}$.
- The integral form is found by integrating the differential form over volume and invoking the divergence theorem.

opposes the external field (Fig. 11.3). In free space and in materials where $\vec{P}=0$, the electric displacement and electric field are equivalent: $\vec{D}=\varepsilon_0\vec{E}$.

Permittivity

For a general HILS[4] material, the electric displacement, $\vec{D}$, and electric field, $\vec{E}$, are proportional to one another and are related by:

$$\vec{D}=\varepsilon\vec{E}$$

(Eq. 11.9), where ε is the *permittivity* [F m^{-2}] of the material. Equating this expression for $\vec{D}$ to Eq. 11.8, and using $\vec{P}=\chi_e\varepsilon_0\vec{E}$ to eliminate $\vec{P}$, we find an expression for ε in terms of the electric susceptibility:

$$\begin{aligned}\varepsilon\vec{E} &= \varepsilon_0\vec{E}+\vec{P}\\ &= \varepsilon_0\vec{E}+\chi_e\varepsilon_0\vec{E}\\ &= (1+\chi_e)\varepsilon_0\vec{E}\\ \therefore \varepsilon &= (1+\chi_e)\varepsilon_0\end{aligned}$$

(Eq. 11.10). Therefore, the permittivity of a dielectric is a measure of how readily it polarizes in response to an external electric field. The *relative permittivity*, ε_r, is then defined as the ratio of ε to ε_0:

$$\varepsilon_r=\frac{\varepsilon}{\varepsilon_0}$$

(Eq. 11.11).

Electric displacement:

$$\vec{D}=\varepsilon\vec{E}=\varepsilon_0\varepsilon_r\vec{E} \qquad (11.9)$$

Permittivity:

$$\varepsilon=(1+\chi_e)\varepsilon_0 \qquad (11.10)$$

Relative permittivity ($\varepsilon_r=\varepsilon/\varepsilon_0$):

$$\varepsilon_r=1+\chi_e \qquad (11.11)$$

As we shall see in later examples, the only difference between electric properties in free space and in dielectrics is that ε_0 is replaced by $\varepsilon=\varepsilon_0\varepsilon_r$. For most dielectrics, ε_r is typically between 3 and 10. For example, $\varepsilon_r\simeq 3$ for paper, 5 for rubber, and 10 for graphite. This means that graphite acquires a greater polarization than paper in the same external electric field. The relative permittivity of distilled water has a high value of $\varepsilon_r\simeq 80$, which means that water is readily polarized. This is due to the combination of it having polar molecules and it being a liquid. Therefore, the permanent dipoles readily align in an external electric field to produce a large polarization. For a vacuum and (to a good approximation) air, $\varepsilon_r=1$.

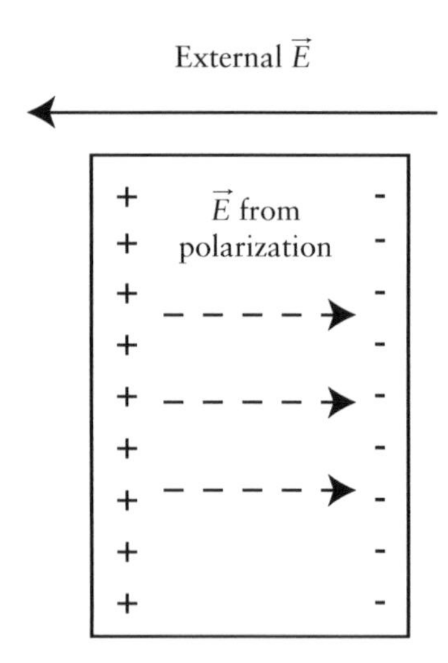

Figure 11.3: The internal electric field set up by polarized dielectric opposes the external field.

11.1.4 Electric energy stored in dielectrics

The electric energy density stored in an electric field in free space is:

$$u_E=\frac{1}{2}\varepsilon_0\left|\vec{E}\right|^2$$

(Section 7.4.2). In a dielectric, the energy density is a factor ε_r times that in free space:

$$(u_E)_{\text{dielectric}}=\varepsilon_r\,(u_E)_{\text{free space}}$$

(Eqs. 11.12 and 11.13 and Derivation 11.4).

Electric energy in dielectrics (Derivation 11.4):

Total energy:

$$\begin{aligned}U_E &= \frac{1}{2}\varepsilon_0\varepsilon_r\int_{\substack{\text{all}\\\text{space}}}\left|\vec{E}\right|^2 d\tau\\ &\equiv \frac{1}{2}\int_{\substack{\text{all}\\\text{space}}}\vec{D}\cdot\vec{E}\,d\tau \qquad (11.12)\end{aligned}$$

Energy density (energy per unit volume):

$$\begin{aligned}\therefore u_E &= \frac{1}{2}\varepsilon_0\varepsilon_r\left|\vec{E}\right|^2\\ &\equiv \frac{1}{2}\vec{D}\cdot\vec{E} \qquad (11.13)\end{aligned}$$

[4] Homogeneous, isotropic, linear, and stationary.

11.1.5 Capacitors and dielectrics

If a dielectric is placed between the conducting plates of a charged capacitor, then the electric field between the plates is *reduced* and the capacitance is *increased*. The reason is as follows. Let's call the electric field produced by the capacitor in the absence of a dielectric $\vec{E}_{fs}$, where "fs" stands for free space. A dielectric between the conducting plates of a capacitor becomes polarized by $\vec{E}_{fs}$ in such a way that the electric field produced in the dielectric by bound charges *opposes* $\vec{E}_{fs}$ (Fig.11.4). Therefore, the resultant electric field with a dielectric in a capacitor ($\vec{E}_{di}$), which is the sum of $\vec{E}_{fs}$ plus the opposing field produced by bound charges, is reduced. This means that the potential difference V between the plates is also reduced. Therefore, since $C = Q/V$, and since the charge Q on the plates does not change, then the capacitance increases. The electric field between the plates is reduced by a factor ε_r, whereas the capacitance is increased by a factor ε_r (Eqs. 11.14 and 11.15 and Derivation 11.5).

Capacitors with dielectrics (Derivation 11.5):

$$\vec{E}_{di} = \frac{\vec{E}_{fs}}{\varepsilon_r} \tag{11.14}$$

$$C_{di} = \varepsilon_r C_{fs} \tag{11.15}$$

di: dielectric, fs: free space.

Placing a dielectric in a capacitor has practical applications, since a higher capacitance can be achieved at a given plate separation. The reduced electric field between plates also reduces the possibility of shorting out the circuit by sparking.

Derivation 11.4: Derivation of the electric energy in matter (Eqs. 11.12 and 11.13)

$$\begin{aligned}
U_E &= \frac{1}{2}\int V \,\mathrm{d}q_f \\
&= \frac{1}{2}\int_{V_p} \rho_f V \,\mathrm{d}\tau \\
&= \frac{1}{2}\int_{V_p} (\nabla\cdot\vec{D})V \,\mathrm{d}\tau \\
&= \frac{1}{2}\left(\oint_{S_p} V\vec{D}\cdot\mathrm{d}\vec{s} - \int_{V_p}\vec{D}\cdot(\nabla V)\mathrm{d}\tau\right) \\
&= \frac{1}{2}\left(\oint_{S_p} V\vec{D}\cdot\mathrm{d}\vec{s} + \int_{V_p}\vec{D}\cdot\vec{E}\mathrm{d}\tau\right) \\
&= \frac{1}{2}\int_{\substack{\text{all}\\ \text{space}}} \vec{D}\cdot\vec{E}\mathrm{d}\tau \\
&\equiv \frac{1}{2}\varepsilon_0\varepsilon_r \int_{\substack{\text{all}\\ \text{space}}} \left|\vec{E}\right|^2 \mathrm{d}\tau
\end{aligned}$$

$$\begin{aligned}
\therefore u_E &= \frac{1}{2}\vec{D}\cdot\vec{E} \\
&\equiv \frac{1}{2}\varepsilon_0\varepsilon_r\left|\vec{E}\right|^2
\end{aligned}$$

V_p and S_p correspond to volume and surface area of polarized matter, respectively, so as not to confuse V_p with electric potential, V.

- Starting point: the electric energy stored in a distribution of free charge, q_f, is:

$$U_E = \frac{1}{2}\int V \,\mathrm{d}q_f$$

(refer to Eq. 7.35). Use $\mathrm{d}q_f = \rho_f \mathrm{d}\tau$ to express U_E as a volume integral.
- Use Gauss's Law in dielectrics ($\rho_f = \nabla\cdot\vec{D}$) to eliminate ρ_f.
- Integrate by parts using the vector identity:

$$\int_V (\nabla\cdot\vec{F})\phi \mathrm{d}\tau = \oint_S \phi\vec{F}\cdot\mathrm{d}\vec{s} - \int_V \vec{F}\cdot(\nabla\phi)\mathrm{d}\tau$$

- Use $\nabla V = -\vec{E}$ to eliminate ∇V.
- When integrating over all space, the volume integral dominates over the surface integral.
- Using $\vec{D} = \varepsilon_0\varepsilon_r\vec{E}$, we see that the electric energy in matter is ε_r times the energy density in free space (refer to Eq. 7.36).
- The energy density, u_E, is the electric energy per unit volume.

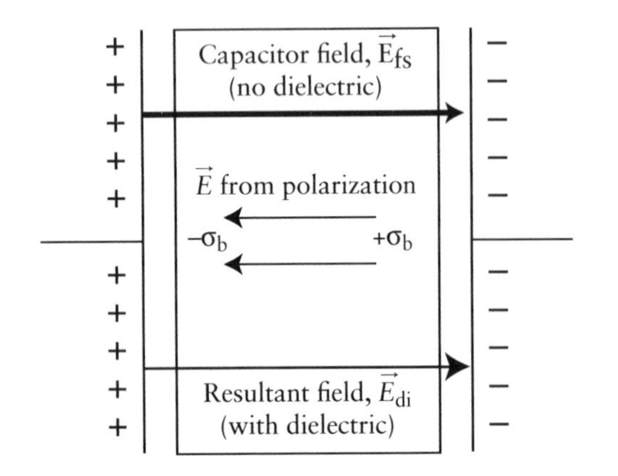

Figure 11.4: Electric field of a capacitor with a dielectric.

11.2 Magnetic fields in matter

A material can become *magnetized* when placed in an external magnetic field, which means it produces its own magnetic field. *All* materials, to a greater or lesser extent, can be magnetized by an external magnetic field.[5] The nature and strength of magnetization depends on the atomic properties of the material. Three common types of magnetization are diamagnetism, paramagnetism, and ferromagnetism, although others exist (such as antiferromagnetism, ferrimagnetism, metamagnetism, to name a few). The differences between diamagnetism, paramagnetism, and ferromagnetism are described briefly below.

11.2.1 Magnetization

In dielectric materials, a separation of positive and negative charge within an atom gives rise to an atomic electric dipole moment. The net alignment of these atomic dipoles in an external electric field produces a net polarization. Analogously, atoms contain orbiting electrons that act as small current loops, which thereby give rise to an atomic *magnetic* dipole moment, $\vec{m}$. An external magnetic field tends to align the dipole moments, such that the material acquires an overall *magnetization*, $\vec{M}$ [A m^{-1}]. For a material with n magnetic dipole moments per unit volume, magnetization is defined as $\vec{M} = n\vec{m}$ (Eq. 11.16).

Magnetization:

In terms of atomic magnetic dipole moment, $\vec{m}$:

$$\vec{M} = n\vec{m} \tag{11.16}$$

where $\vec{m} = I\vec{s}$

[5] In contrast, only dielectrics (electrical insulators), but not conductors, can be polarized by an external electric field.

Diamagnetism

Diamagnetism is the weakest type of magnetism. It occurs in *all* materials when placed in an external magnetic field, even those we generally think of as non-magnetic, such as wood, plastic, or water. Diamagnetism is explained by Faraday's Law of Induction: the application of an external magnetic field induces atomic magnetic dipole moments, which align in a direction to generate a magnetic field to oppose the change. Therefore, the magnetization produced in diamagnets is anti-parallel to the applied field (Fig. 11.5a).

Paramagnetism

Paramagnetism occurs in materials that have unpaired electrons, such as aluminum or sodium, and thus have permanent magnetic dipole moments due to the spin of the unpaired electron. In the absence of an external field, the permanent dipole moments point in random directions because of thermal motion, and so the material has no net magnetism. An external field aligns the dipole moments with the field, resulting in a magnetization that is parallel to the external field (Fig. 11.5b). Paramagnetism overwhelms the much weaker diamagnetic effects.

Ferromagnetism

Ferromagnetism occurs only in some materials, such as iron, nickel, and cobalt, and produces the strongest magnetic fields. Like paramagnetic materials, ferromagnets have permanent atomic magnetic dipole moments from unpaired electron spins. The special property that sets them apart is the presence of so-called *domains*, which are regions within the material where the magnetic dipole moments are aligned, even in the absence of an applied magnetic field. These domains are randomly oriented in the absence of an external field, thus the material has no net magnetism. The effect of an external field is to align the domains, producing a strong magnetization parallel to the external field (Fig. 11.5c). When the applied field is removed, the domains remain aligned. Therefore, ferromagnets retain their magnetization, thus producing permanent magnets. Heat, mechanical impacts, or reversed external fields can reduce the magnetization.

11.2.2 Free and bound currents

Current that is associated with magnetization is called *bound current*, $\vec{I}_b$ [A], since it is effectively "bound" to the atom. Current flowing across a surface, per unit width-perpendicular-to-the-flow, is the *bound surface current density*, $\vec{K}_b$ [A m^{-1}], where:

$$\vec{K}_b = \frac{d\vec{I}_b}{dl_\perp}$$

Derivation 11.5: Derivation of capacitors with dielectrics (Eqs. 11.14 and 11.15)

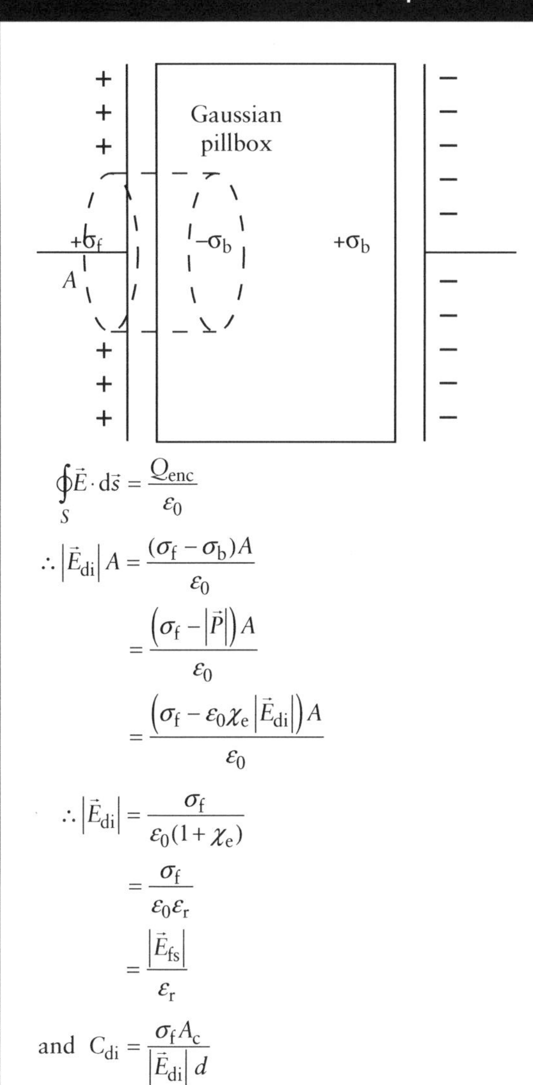

$$\oint_S \vec{E}\cdot d\vec{s} = \frac{Q_{enc}}{\varepsilon_0}$$

$$\therefore \left|\vec{E}_{di}\right| A = \frac{(\sigma_f - \sigma_b)A}{\varepsilon_0} = \frac{\left(\sigma_f - \left|\vec{P}\right|\right)A}{\varepsilon_0} = \frac{\left(\sigma_f - \varepsilon_0\chi_e\left|\vec{E}_{di}\right|\right)A}{\varepsilon_0}$$

$$\therefore \left|\vec{E}_{di}\right| = \frac{\sigma_f}{\varepsilon_0(1+\chi_e)} = \frac{\sigma_f}{\varepsilon_0\varepsilon_r} = \frac{\left|\vec{E}_{fs}\right|}{\varepsilon_r}$$

$$\text{and } C_{di} = \frac{\sigma_f A_c}{\left|\vec{E}_{di}\right| d} = \frac{\varepsilon_0\varepsilon_r A_c}{d} = \varepsilon_r C_{fs}$$

- Starting point: Gauss's Law in integral form is:

$$\oint_S \vec{E}\cdot d\vec{s} = \frac{Q_{enc}}{\varepsilon_0}$$

- (Refer to Fig. 11.4). Use the following assumptions in Gauss's Law: (i) The electric field in the dielectric material, $\vec{E}_{di}$, is parallel to $d\vec{s}$, which means we can drop the dot product, (ii) $|\vec{E}_{di}|$ is uniform over the area, A, of a Gaussian pillbox, which means that we can take $\vec{E}_{di}$ out of the surface integral, and (iii) $Q_{enc} = (\sigma_f - \sigma_b)A$, where σ_f is the free surface charge density on the capacitor plate and σ_b is the bound charge in the dielectric.
- The bound charge surface density is (refer to Eq. 11.3)

$$\sigma_b = \vec{P}\cdot\hat{n} = \left|\vec{P}\right|$$

since $\vec{P}$ and $\hat{n}$ are parallel.
- Use $\vec{P} = \varepsilon_0\,\chi_e\,\vec{E}_{di}$ to eliminate $|\vec{P}|$.
- Rearrange and use $\varepsilon_r = 1 + \chi_e$ to show that the electric field in the dielectric is a factor ε_r less than that in free space (from Worked example 7.2, $|\vec{E}_{fs}| = \sigma_f/\varepsilon_0$).
- For a parallel-plate capacitor, the capacitance is:

$$C = \frac{Q}{V} = \frac{\sigma A_c}{\left|\vec{E}\right| d}$$

where σ is the surface charge on the plates, A_c is the area of the plates, $\vec{E}$ is the electric field between the plates and d is the distance between them.
- Substituting in for $|\vec{E}_{di}|$, we find that the capacitance in the presence of the dielectric is a factor ε_r greater than the capacitance in free space (from Worked example 7.5, $C_{fs} = \varepsilon_0 A_c/d$).

and current flowing through a volume, per unit area-perpendicular-to-the-flow, is the *bound volume current density*, $\vec{J}_b$ [A m^{-2}], where:

$$\vec{J}_b = \frac{d\vec{I}_b}{ds_\perp}$$

If a material is not magnetized, then the bound current equals zero. Current that is not associated with magnetization is called *free current*, $\vec{I}_f$ [A], $\vec{K}_f$ [A m^{-1}], or $\vec{J}_f$ [A m^{-2}]. For example, conduction electrons flowing through a wire constitute a free current. The total current is then the sum of free and bound current:

$$\vec{I} = \vec{I}_f + \vec{I}_b$$

Bound current densities

The magnetic vector potential of an ideal magnetic dipole[6] is:

6 An ideal dipole is one in which the dimensions of dipole area, $\vec{s}$, is much less than the position vector, $\vec{r}$. Since magnetic dipoles in materials are on the order of atomic dimensions, this approximation is valid for most observations.

$$\vec{A}(\vec{r}) = \frac{\mu_0 \vec{m} \times \hat{r}}{4\pi r^2}$$

(refer to Section 8.5). This expression can be separated into the potential produced by a bound surface current density, $\vec{K}_b$, and a bound volume current density, $\vec{J}_b$ (Derivation 11.6). Doing this, we find expressions for $\vec{K}_b$ and $\vec{J}_b$ in terms of $\vec{M}$: the surface current density is equal to the component of magnetization along the surface (Eq. 11.17 and Derivation 11.6), whereas the volume current density is equal to the curl of the magnetization (Eq. 11.18 and Derivation 11.6). Notice the similarities between these expressions and their electric counterparts (Eqs. 11.3 and 11.4).

Bound current densities (Derivation 11.6):

$$\vec{K}_b = \vec{M} \times \hat{n} \qquad (11.17)$$

$$\vec{J}_b = \nabla \times \vec{M} \qquad (11.18)$$

where $\hat{n}$ is a unit vector perpendicular to the surface.

11.2.3 The auxiliary magnetic field

Letting the total current density equal the sum of free and bound current densities, Ampère's Law (in the absence of a changing electric field) can be written as:

$$\nabla \times \vec{B} = \mu_0 \vec{J}$$
$$= \mu_0 \left(\vec{J}_f + \vec{J}_b\right)$$

Using the expression for $\vec{J}_b$ from Eq. 11.18 and rearranging (the curl operator is linear), we have:

$$\nabla \times \vec{B} = \mu_0 \left(\vec{J}_f + \nabla \times \vec{M}\right)$$

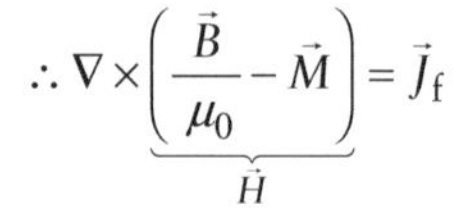

$$\therefore \nabla \times \underbrace{\left(\frac{\vec{B}}{\mu_0} - \vec{M}\right)}_{\vec{H}} = \vec{J}_f$$

Thus, doing this, we can define another vector field, called the *auxiliary magnetic field*, $\vec{H}$ [A m^{-1}], which is a combination of the applied magnetic field, $\vec{B}$, and the magnetization, $\vec{M}$, which it gives rise to (Eq. 11.21):

$$\vec{H} \stackrel{\text{def}}{=} \frac{\vec{B}}{\mu_0} - \vec{M}$$

Ampère's Law in magnetic media

Following on from the above discussion, Ampère's Law in magnetic media equates the curl of the auxiliary magnetic field to the free current density:

$$\nabla \times \vec{H} = \vec{J}_f$$

In the presence of a time-varying electric field, we have two additional current terms: the displacement current, $\vec{J}_D$, and the polarization current, $\vec{J}_p$. In this case, we find that the curl of $\vec{H}$ is equal to the sum of the free current density plus the rate of change of the electric displacement field (Eq. 11.19 and Derivation 11.7). Integrating over a surface and invoking Stokes's theorem, we can express Ampère's Law in integral form; this equates the closed line integral of $\vec{H}$ to the free current flowing through the loop plus the rate of change of electric displacement flux, Φ_D, through the surface enclosed by the loop (Eq. 11.20 and Derivation 11.7). $\vec{H}$ is a construct that combines $\vec{B}$ and $\vec{M}$ in such a way that we do not have to deal explicitly with the effects of magnetization. Similarly, polarization is incorporated in $\vec{D}$, which means we do not have to deal explicitly with the effects of polarization either.

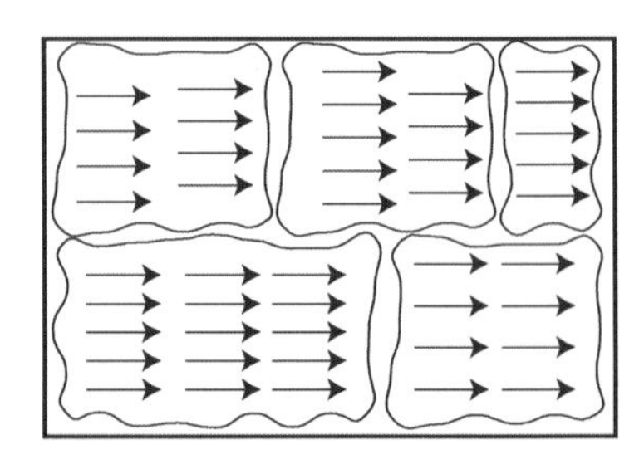

Figure 11.5: a) Diamagnetism: alignment of induced dipole moments anti-parallel to the applied field, b) paramagnetism: alignment of permanent dipole moments parallel to the applied field, and c) ferromagnetism: alignment of domains of permanent dipole moments parallel to applied field.

Derivation 11.6: Derivation of bound current densities (Eqs. 11.17 and 11.18)

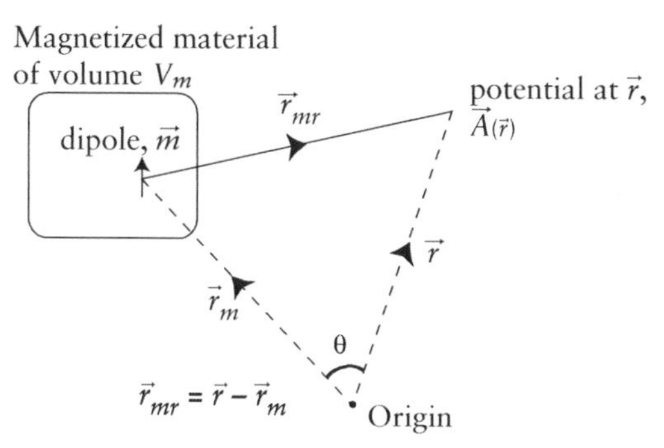

$$\vec{A}(\vec{r}) = N\frac{\mu_0 \vec{m}\times\hat{r}_{mr}}{4\pi r_{mr}^2}$$

$$= \int_{V_m} \frac{\mu_0 \vec{M}\times\hat{r}_{mr}}{4\pi r_{mr}^2}\,\mathrm{d}\tau_m$$

$$= \frac{\mu_0}{4\pi}\int_{V_m} \vec{M}\times\nabla_m\left(\frac{1}{r_{mr}}\right)\mathrm{d}\tau_m$$

$$= \frac{\mu_0}{4\pi}\int_{V_m}\left(\frac{\nabla_m\times\vec{M}}{r_{mr}} - \nabla_m\times\left(\frac{\vec{M}}{r_{mr}}\right)\right)\mathrm{d}\tau_m$$

$$= \frac{\mu_0}{4\pi}\left(\int_{V_m}\frac{\nabla_m\times\vec{M}}{r_{mr}}\mathrm{d}\tau_m + \oint_{S_m}\frac{\vec{M}}{r_{mr}}\times\mathrm{d}\vec{s}_m\right)$$

$$= \frac{\mu_0}{4\pi}\left(\int_{V_m}\frac{1}{r_{mr}}\underbrace{\nabla_m\times\vec{M}}_{\text{volume current, }\vec{J}_b}\,\mathrm{d}\tau_m + \oint_{S_m}\frac{1}{r_{mr}}\underbrace{\vec{M}\times\hat{n}}_{\text{surface current, }\vec{K}_b}\,\mathrm{d}s_m\right)$$

$$\equiv \frac{\mu_0}{4\pi}\left(\int_{V_m}\frac{\vec{J}_b}{r_{mr}}\mathrm{d}\tau_m + \oint_{S_m}\frac{\vec{K}_b}{r_{mr}}\mathrm{d}s_m\right)$$

$$\therefore \vec{K}_b = \vec{M}\times\hat{n}$$

$$\text{and } \vec{J}_b = \nabla\times\vec{M}$$

The subscript m refers to coordinates over which the material is magnetized (i.e. where magnetic dipoles are present). Therefore, V_m and S_m are volume and surface area of magnetized matter, respectively.

- Starting point: the magnetic vector potential of a magnetized material with N identical magnetic dipole moments is N times that of a single magnetic dipole:

$$\vec{A}(\vec{r}) = N\frac{\mu_0 \vec{m}\times\hat{r}_{mr}}{4\pi r_{mr}^2}$$

where, for a dipole located at $\vec{r}_m$, the separation vector $\vec{r}_{mr}$ is $\vec{r}_{mr} = \vec{r} - \vec{r}_m$ (refer to figure).

- The dipole number density is $n = N/V_m$, where V_m is the volume of magnetized material. Therefore:

$$N\vec{m} = \vec{M}V_m \equiv \int_{V_m}\vec{M}\mathrm{d}\tau$$

where the magnetization is $\vec{M} = n\vec{m}$.

- Differentiating with respect to r_m, we have:

$$\nabla_m\left(\frac{1}{r_{mr}}\right) = \frac{\hat{r}_{mr}}{r_{mr}^2}$$

Hint: let $r_{mr} = |\vec{r}_{mr}| = ((\vec{r}-\vec{r}_m)\cdot(\vec{r}-\vec{r}_m))^{1/2} = (r^2 + r_m^2 - 2rr_m\cos\theta)^{1/2}$ and differentiate with respect to r_m the reciprocal using the Del operator in spherical polar coordinates.

- Use the identities:

$$\nabla\times\left(f\vec{G}\right) = f\left(\nabla\times\vec{G}\right) - \vec{G}\times(\nabla f)$$

$$\text{and } \int_V\left(\nabla\times\vec{G}\right)\mathrm{d}\tau = -\int_S\vec{G}\times\mathrm{d}\vec{s}$$

to separate the integral into two terms.

- Let $\mathrm{d}\vec{s}_m = \hat{n}\mathrm{d}s_m$ where $\hat{n}$ is normal to the surface.
- The vector potential can therefore be expressed as the superposition of a surface current density term and a volume current density term.

Ampère's Law in magnetic media (Derivation 11.7):

Differential form:

$$\nabla \times \vec{H} = \vec{J}_\mathrm{f} + \frac{\partial \vec{D}}{\partial t} \tag{11.19}$$

Integral form:

$$\oint_L \vec{H} \cdot \mathrm{d}\vec{l} = (I_\mathrm{f})_\mathrm{enc} + \frac{\partial \Phi_D}{\partial t} \tag{11.20}$$

where the auxiliary magnetic field is:

$$\vec{H} \stackrel{\mathrm{def}}{=} \frac{\vec{B}}{\mu_0} - \vec{M} \tag{11.21}$$

Sources and sinks of $\vec{H}$

Gauss's Law for magnetism states that the divergence of $\vec{B}$ is $\nabla \cdot \vec{B} = 0$, implying that there are no sources or sinks of magnetic field. However, taking the divergence of Eq. 11.21, the divergence of $\vec{H}$ is:

$$\nabla \cdot \vec{H} = -\nabla \cdot \vec{M}$$

This means that sources of $\vec{H}$ are sinks of $\vec{M}$, i.e. the disappearance of $\vec{M}$ gives rise to $\vec{H}$, and vice versa. We can illustrate these characteristics by considering an iron-cored toroidal coil with a small gap (Fig. 11.6). The current through the toroid produces a $\vec{B}$-field that is constant throughout its core. This $\vec{B}$-field produces a magnetization, $\vec{M}$, that is also constant throughout the core, but falls to zero in the small air gap since there is no material in the gap. The disappearance of $\vec{M}$ thus gives rise to a large $\vec{H}$ in the air gap. The large magnetization of the iron produces a far stronger magnetic field in the air gap than a corresponding air-core toroid operating at the same current would.

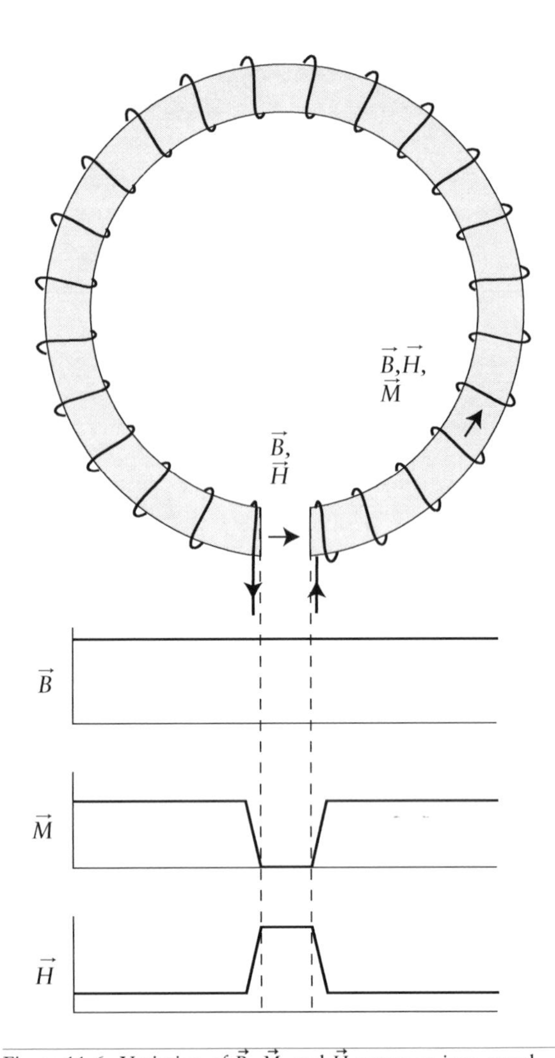

Figure 11.6: Variation of $\vec{B}$, $\vec{M}$, and $\vec{H}$ across an iron-cored toroidal coil with an air gap.

Susceptibility and permeability

In linear and isotropic dielectrics, polarization is related to electric field by the electric susceptibility, χ_e:$\vec{P} = \chi_\mathrm{e}\,\varepsilon_0\,\vec{E}$. In linear and isotropic magnetic materials, magnetization is related to the *auxiliary* magnetic field by the *magnetic* susceptibility, χ_m:

$$\vec{M} = \chi_\mathrm{m}\vec{H}$$

(Eq. 11.22). (This equation does not hold for ferromagnets.) χ_m is a dimensionless parameter, and is a measure of how readily a material becomes magnetized in response to an applied magnetic field. χ_m is negative for diamagnets since the magnetization opposes the applied magnetic field (Faraday and Lenz's Law for an induced magnetic dipole moment), and positive for paramagnets and ferromagnets since the magnetization is parallel to the applied field (the torque on a permanent dipole in a magnetic field tends to align the dipole with the field). Using $\vec{M} = \chi_\mathrm{m}\,\vec{H}$, we find that $\vec{B}$ is proportional to $\vec{H}$:

$$\begin{aligned}
\vec{H} &= \frac{\vec{B}}{\mu_0} - \vec{M} \\
\therefore\ \vec{B} &= \mu_0\left(\vec{H} + \vec{M}\right) \\
&= \mu_0\left(\vec{H} + \chi_\mathrm{m}\vec{H}\right) \\
&= \underbrace{(1+\chi_\mathrm{m})\,\mu_0}_{=\mu}\,\vec{H}
\end{aligned}$$

Derivation 11.7: Derivation of Ampère's Law in a magnetic material (Eqs. 11.19 and 11.20)

Differential form:

$$\nabla\times\vec{B}=\mu_0\left(\vec{J}+\varepsilon_0\frac{\partial\vec{E}}{\partial t}\right)$$
$$=\mu_0\left(\vec{J}_\text{f}+\vec{J}_p+\vec{J}_\text{b}+\varepsilon_0\frac{\partial\vec{E}}{\partial t}\right)$$
$$=\mu_0\left(\vec{J}_\text{f}+\frac{\partial\vec{P}}{\partial t}+\nabla\times\vec{M}+\varepsilon_0\frac{\partial\vec{E}}{\partial t}\right)$$
$$=\mu_0\left(\vec{J}_\text{f}+\frac{\partial(\vec{P}+\varepsilon_0\vec{E})}{\partial t}+\nabla\times\vec{M}\right)$$
$$=\mu_0\left(\vec{J}_\text{f}+\frac{\partial\vec{D}}{\partial t}+\nabla\times\vec{M}\right)$$
$$\therefore\nabla\times\left(\frac{\vec{B}}{\mu_0}-\vec{M}\right)=\vec{J}_\text{f}+\frac{\partial\vec{D}}{\partial t}$$
$$\therefore\nabla\times\vec{H}=\vec{J}_\text{f}+\frac{\partial\vec{D}}{\partial t}$$

Integral form:

$$\int_S\left(\nabla\times\vec{H}\right)\cdot d\vec{s}=\int_S\left(\vec{J}_\text{f}+\frac{\partial\vec{D}}{\partial t}\right)\cdot d\vec{s}$$
$$=(I_\text{f})_\text{enc}+\frac{\partial\Phi_D}{\partial t}$$
and $$\int_S\left(\nabla\times\vec{H}\right)\cdot d\vec{s}=\oint_L\vec{H}\cdot d\vec{l}$$
$$\therefore\oint_L\vec{H}\cdot d\vec{l}=(I_\text{f})_\text{enc}+\frac{\partial\Phi_D}{\partial t}$$

- Starting point: in the absence of magnetic material, Ampère's Law in differential form is:

$$\nabla\times\vec{B}=\mu_0\left(\vec{J}+\varepsilon_0\frac{\partial\vec{E}}{\partial t}\right)$$

(refer to Section 10.3).
- Let the total current density in Ampère's Law equal the sum of free, bound, and polarization current densities:

$$J=\vec{J}_\text{f}+\vec{J}_p+\vec{J}_\text{b}$$
where $\vec{J}_p=\dfrac{\partial\vec{P}}{\partial t}$

and $\vec{J}_\text{b}=\nabla\times\vec{M}$
- The definition of the electric displacement field is $\vec{D}=\vec{P}+\varepsilon_0\vec{E}$.
- Curl is a linear operation, which means we can collect the curl terms together.
- Define the auxiliary magnetic field as:

$$\vec{H}=\frac{\vec{B}}{\mu_0}-\vec{M}$$

to express Ampère's Law in magnetic media as the curl of $\vec{H}$.
- The integral form is then found by integrating the differential form over a surface and invoking Stokes's theorem. Let Φ_D represent the flux of the electric displacement field: $\Phi_D=\int_S\vec{D}\cdot d\vec{s}$.

The proportionality constant is the *permeability*, μ [H m^{-1}] (Eqs. 11.23 and 11.24). The *relative permeability*, μ_r, is then the ratio of μ to the permeability of free space, μ_0 : $\mu_\text{r}=\mu/\mu_0$ (Eq. 11.25). For diamagnets and many paramagnets, $\mu_\text{r}\simeq 1$ since magnetization is a weak effect. (For diamagnets, $\mu_\text{r}<1$ and $\chi_\text{m}<0$, and for paramagnets, $\mu_\text{r}>1$ and $\chi_\text{m}>0$ since their magnetization points in opposite directions.) In contrast, ferromagnets have large values of μ_r, ranging between about 10^2 to 10^5 in common materials.

Susceptibility:

$$\vec{M}=\chi_\text{m}\vec{H} \tag{11.22}$$
and $$\vec{B}=\mu\vec{H}$$
$$\equiv\mu_0\,\mu_\text{r}\vec{H} \tag{11.23}$$

where χ_m is the magnetic susceptibility.

Permeability:

$$\mu=(1+\chi_\text{m})\mu_0 \tag{11.24}$$

Relative permeability ($\mu_\text{r}=\mu/\mu_0$):
$$\mu_\text{r}=1+\chi_\text{m} \tag{11.25}$$

11.2.4 Magnetic energy stored in matter

The magnetic energy density stored in a magnetic field in free space is:

$$u_B = \frac{1}{2\mu_0}\left|\vec{B}\right|^2$$

(Section 10.2.4). In magnetic materials, energy density is a factor μ_r times less than in free space:

$$(u_B)_{\text{matter}} = \frac{(u_B)_{\text{free space}}}{\mu_r}$$

(Eqs. 11.26 and 11.27 and Derivation 11.8). Again, notice the similarity between these expressions for U_B and their electric counterparts in Eqs. 11.12 and 11.13.

Magnetic energy in matter (Derivation 11.8):

Total energy:

$$U_B = \frac{1}{2\mu_0\mu_r}\int_{\substack{\text{all}\\\text{space}}} \left|\vec{B}\right|^2 d\tau$$

$$\equiv \frac{1}{2}\int_{\substack{\text{all}\\\text{space}}} \vec{H}\cdot\vec{B}\,d\tau \qquad (11.26)$$

Energy density (energy per unit volume):

$$\therefore u_B = \frac{1}{2\mu_0\mu_r}\left|\vec{B}\right|^2$$

$$\equiv \frac{1}{2}\vec{H}\cdot\vec{B} \qquad (11.27)$$

11.3 Maxwell's equations in matter

Maxwell's equations in matter are usually expressed in terms of all four variables, $\vec{E}$, $\vec{D}$, $\vec{B}$, and $\vec{H}$. There is one equation for each variable, in terms of either its divergence or its curl. The differential and integral forms are summarized in Eqs. 11.28–11.31. Expressions for $\vec{D}$, $\vec{H}$, ρ, and $\vec{J}$ are summarized in Eqs. 11.32–11.37.

11.3.1 Boundary conditions

Maxwell's equations provide solutions for electric and magnetic fields in *any* material. The solutions generally depend on the material's relative permittivity and permeability, ε_r and μ_r. At the boundaries between two materials, there is a discontinuity in the field. Thus, at boundaries, we need to match the solutions on either side of the boundary using standard *boundary conditions*. Each variable $\vec{E}$, $\vec{D}$, $\vec{B}$, and $\vec{H}$ obeys a different boundary condition. These conditions are easily derived using the integral forms of Maxwell's equations in matter, by performing either a closed surface integral or closed line integral across a boundary and equating the integral to the enclosed charge or current (Derivations 11.9 and 11.10). Doing this, we find that in dielectric materials, the difference between the components of $\vec{D}$ *perpendicular* to the boundary equals the free surface charge density at the boundary, σ_f (Eq. 11.38), whereas the difference between the components of $\vec{E}$ *parallel* to the boundary equals zero (Eq. 11.39). For magnetic materials, the difference between the components of $\vec{H}$ *parallel* to the boundary equals the magnitude of free surface current density at the boundary, $|\vec{K}_f|$ (Eq. 11.40), whereas the difference between the components of $\vec{B}$ *perpendicular* to the boundary equals zero (Eq. 11.41). An example of using Laplace's equation and boundary conditions to solve the electric potential of a dielectric rod placed in an external electric field is provided in Worked example 11.1.

Boundary conditions (Derivations 11.9 and 11.10):

For electric fields:

$$\left|\vec{D}\right|^{\perp}_{\text{out}} - \left|\vec{D}\right|^{\perp}_{\text{in}} = \sigma_f \qquad (11.38)$$

$$\left|\vec{E}\right|^{\parallel}_{\text{out}} - \left|\vec{E}\right|^{\parallel}_{\text{in}} = 0 \qquad (11.39)$$

For magnetic fields:

$$\left|\vec{H}\right|^{\parallel}_{\text{out}} - \left|\vec{H}\right|^{\parallel}_{\text{in}} = \left|\vec{K}_f\right| \qquad (11.40)$$

$$\left|\vec{B}\right|^{\perp}_{\text{out}} - \left|\vec{B}\right|^{\perp}_{\text{in}} = 0 \qquad (11.41)$$

Derivation 11.8: Derivation of the energy stored in a magnetic field (Eqs. 11.26 and 11.27)

$$
\begin{aligned}
U_B &= \frac{1}{2} I_f \Phi_B \\
&= \frac{1}{2} I_f \oint_L \vec{A} \cdot d\vec{l} \\
&= \frac{1}{2} \oint_L \vec{A} \cdot \vec{I}_f dl \\
&= \frac{1}{2} \int_V \vec{A} \cdot \vec{J}_f d\tau \\
&= \frac{1}{2} \int_V \vec{A} \cdot \left(\nabla \times \vec{H}\right) d\tau \\
&= \frac{1}{2} \int_V \left(\vec{H} \cdot \vec{B} - \nabla \cdot \left(\vec{A} \times \vec{H}\right)\right) d\tau \\
&= \frac{1}{2} \left(\int_V \vec{H} \cdot \vec{B} d\tau - \oint_S \left(\vec{A} \times \vec{H}\right) \cdot d\vec{s} \right) \\
&= \frac{1}{2} \int_{\substack{\text{all} \\ \text{space}}} \vec{H} \cdot \vec{B} d\tau \\
&\equiv \frac{1}{2\mu_0 \mu_r} \int_{\substack{\text{all} \\ \text{space}}} \left|\vec{B}\right|^2 d\tau
\end{aligned}
$$

$$
\begin{aligned}
\therefore u_B &= \frac{1}{2} \vec{H} \cdot \vec{B} \\
&\equiv \frac{1}{2\mu_0 \mu_r} \left|\vec{B}\right|^2
\end{aligned}
$$

- Starting point: the magnetic energy stored in a magnetic flux Φ_B produced by a free current I_f is:

$$U_B = \frac{1}{2} I_f \Phi_B$$

- In terms of the magnetic vector potential, $\vec{A}$, where ($\vec{B} = \nabla \times \vec{A}$), and invoking Stokes's theorem, the magnetic flux is:

$$
\begin{aligned}
\Phi_B &= \int_S \vec{B} \cdot d\vec{s} \\
&= \int_S \left(\nabla \times \vec{A}\right) \cdot d\vec{s} \\
&= \oint_L \vec{A} \cdot d\vec{l}
\end{aligned}
$$

- Using this in the expression for U_B, we can express the magnetic energy in terms of $\vec{A}$ and $\vec{J}_f$:

$$U_B = \frac{1}{2} \int_V \vec{A} \cdot \vec{J}_f d\tau$$

This expression is the magnetic equivalent of the electric energy density of a free charge distribution, ρ_f:

$$U_E = \frac{1}{2} \int V \rho_f d\tau$$

- Use Ampère's Law for magnetic materials ($\nabla \times \vec{H} = \vec{J}_f$) to eliminate $\vec{J}_f$, along with the product rule:

$$
\begin{aligned}
\nabla \cdot \left(\vec{A} \times \vec{H}\right) &= \vec{H} \cdot \left(\nabla \times \vec{A}\right) - \vec{A} \cdot \left(\nabla \times \vec{H}\right) \\
&= \vec{H} \cdot \vec{B} - \vec{A} \cdot \left(\nabla \times \vec{H}\right)
\end{aligned}
$$

- Invoke the divergence theorem on the integral over $\nabla \cdot (\vec{A} \times \vec{H})$.
- When integrating over all space, the volume integral dominates over the surface integral.
- Using $\vec{B} = \mu_0 \mu_r \vec{H}$, we see that the magnetic energy in matter is μ_r times less than the energy density in free space (refer to Eq. 10.22).
- The energy density, u_B, is the magnetic energy per unit volume.

Maxwell's equations in matter:

$$\text{Gauss's Law: } \nabla\cdot\vec{D}=\rho_{\mathrm{f}} \qquad \oint_S \vec{D}\cdot d\vec{s}=(q_{\mathrm{f}})_{\mathrm{enc}} \tag{11.28}$$

$$\text{Gauss's Law for magnetism: } \nabla\cdot\vec{B}=0 \qquad \oint_S \vec{B}\cdot d\vec{s}=0 \tag{11.29}$$

$$\text{Faraday's Law: } \nabla\times\vec{E}=-\frac{\partial\vec{B}}{\partial t} \qquad \oint_L \vec{E}\cdot d\vec{l}=-\frac{\partial\Phi_B}{\partial t} \tag{11.30}$$

$$\text{Ampère's Law: } \nabla\times\vec{H}=\vec{J}_{\mathrm{f}}+\frac{\partial\vec{D}}{\partial t} \qquad \oint_L \vec{H}\cdot d\vec{l}=(I_{\mathrm{f}})_{\mathrm{enc}}+\frac{\partial\Phi_D}{\partial t} \tag{11.31}$$

where:

$$\text{Electric displacement field: } \vec{D}=\varepsilon_0\vec{E}+\vec{P} \text{ (always)} \tag{11.32}$$

$$=\varepsilon_0\varepsilon_{\mathrm{r}}\vec{E} \text{ (in linear and isotropic materials)} \tag{11.33}$$

$$\equiv\varepsilon\vec{E}$$

$$\text{Auxiliary magnetic field: } \vec{H}=\frac{\vec{B}}{\mu_0}-\vec{M} \text{ (always)} \tag{11.34}$$

$$=\frac{\vec{B}}{\mu_0\mu_{\mathrm{r}}} \text{ (in linear and isotropic materials)} \tag{11.35}$$

$$\equiv\frac{\vec{B}}{\mu}$$

$$\text{Total charge density: } \rho=\rho_{\mathrm{f}}+\rho_{\mathrm{b}}$$

$$=\nabla\cdot\vec{D}-\nabla\cdot\vec{P} \tag{11.36}$$

$$\text{Total current density: } \vec{J}=\vec{J}_{\mathrm{f}}+\vec{J}_{\mathrm{b}}+\vec{J}_p$$

$$=\nabla\times\vec{H}-\frac{\partial\vec{D}}{\partial t}+\nabla\times\vec{M}+\frac{\partial\vec{P}}{\partial t} \tag{11.37}$$

Derivation 11.9: Derivation of boundary conditions for electric fields (Eqs. 11.38 and 11.39)

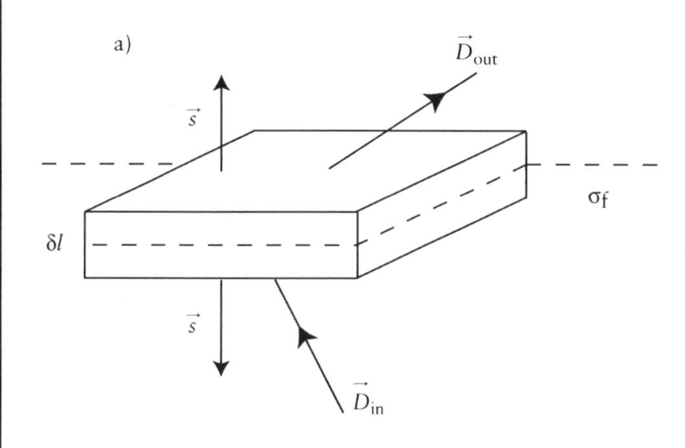

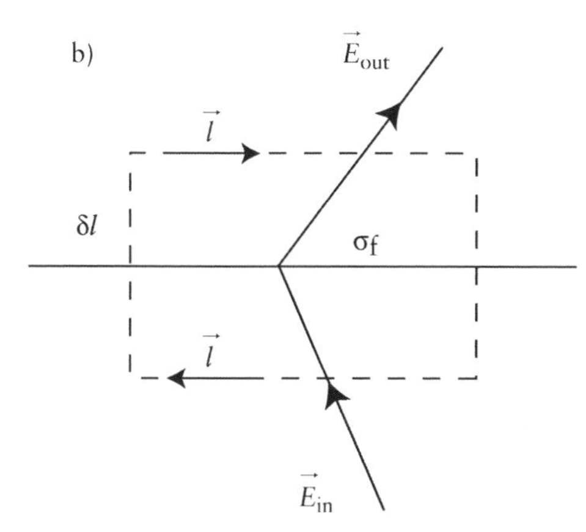

Electric displacement field, $\vec{D}$:

$$\oint_S \vec{D}\cdot d\vec{s} = (q_f)_{enc}$$

and $$\oint_S \vec{D}\cdot d\vec{s} = \vec{D}_{out}\cdot\vec{s} - \vec{D}_{in}\cdot\vec{s}$$

$$= \left|\vec{D}\right|^{\perp}_{out}\left|\vec{s}\right| - \left|\vec{D}\right|^{\perp}_{in}\left|\vec{s}\right|$$

$$\therefore \left|\vec{D}\right|^{\perp}_{out} - \left|\vec{D}\right|^{\perp}_{in} = \frac{(q_f)_{enc}}{\left|\vec{s}\right|}$$

$$= \sigma_f$$

Electric field, $\vec{E}$:

$$\oint_L \vec{E}\cdot d\vec{l} = 0$$

and $$\oint_L \vec{E}\cdot d\vec{l} = \vec{E}_{out}\cdot\vec{l} - \vec{E}_{in}\cdot\vec{l}$$

$$= \left|\vec{E}\right|^{\parallel}_{out}\left|\vec{l}\right| - \left|\vec{E}\right|^{\parallel}_{in}\left|\vec{l}\right|$$

$$\therefore \left|\vec{E}\right|^{\parallel}_{out} - \left|\vec{E}\right|^{\parallel}_{in} = 0$$

Electric displacement field, $\vec{D}$:

- The integral form of Gauss's Law in dielectrics is:

$$\oint_S \vec{D}\cdot d\vec{s} = (q_f)_{enc}$$

  where S represents a Gaussian pillbox of area $|\vec{s}|$ (refer to diagram a).
- The contribution from the sides equals zero as $\delta l \to 0$; therefore, Gauss's Law includes only the contributions from the surfaces $\vec{s}$.
- Express the dot product in terms of the components of $\vec{D}$ perpendicular to the boundary (parallel or anti-parallel to $\vec{s}$), $|\vec{D}|^{\perp}$.

Electric field, $\vec{E}$:

- The integral form of Faraday's Law for a constant magnetic field is:

$$\oint_L \vec{E}\cdot d\vec{l} = 0$$

  where L represents a rectangular loop of length $|\vec{l}|$ (refer to diagram b).
- The contribution from the sides equals zero as $\delta l \to 0$; therefore Faraday's Law includes only the contributions from the sides $\vec{l}$.
- Express the dot product in terms of the components of $\vec{E}$ parallel to the boundary (parallel or anti-parallel to $\vec{l}$), $|\vec{E}|^{\parallel}$.

Derivation 11.10: Derivation of boundary conditions for magnetic fields (Eqs. 11.40 and 11.41)

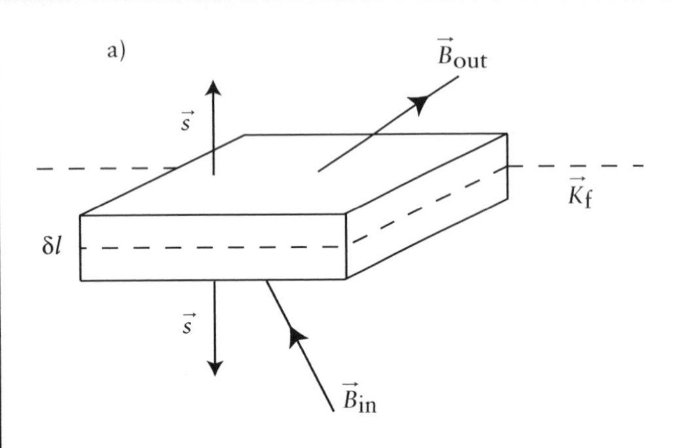

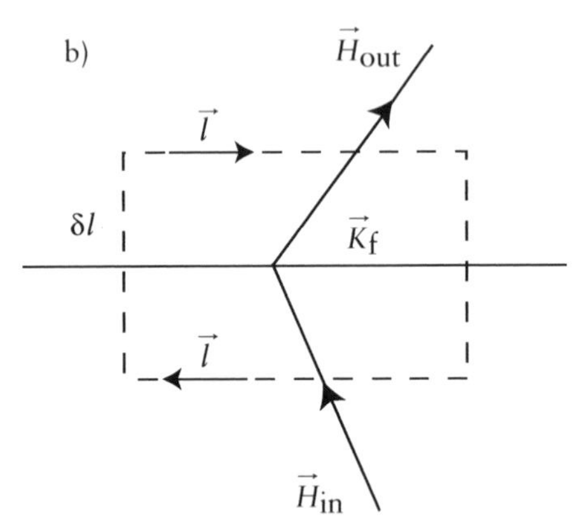

Magnetic field, $\vec{B}$:

$$\oint_S \vec{B} \cdot d\vec{s} = 0$$

and $\oint_S \vec{B} \cdot d\vec{s} = \vec{B}_{out} \cdot \vec{s} - \vec{B}_{in} \cdot \vec{s}$

$$= \left|\vec{B}\right|_{out}^{\perp} |\vec{s}| - \left|\vec{B}\right|_{in}^{\perp} |\vec{s}|$$

$$\therefore \left|\vec{B}\right|_{out}^{\perp} - \left|\vec{B}\right|_{in}^{\perp} = 0$$

Auxiliary magnetic field, $\vec{H}$:

$$\oint_L \vec{H} \cdot d\vec{l} = (I_f)_{enc}$$

and $\oint_L \vec{H} \cdot d\vec{l} = \vec{H}_{out} \cdot \vec{l} - \vec{H}_{in} \cdot \vec{l}$

$$= \left|\vec{H}\right|_{out}^{\parallel} \left|\vec{l}\right| - \left|\vec{H}\right|_{in}^{\parallel} \left|\vec{l}\right|$$

$$\therefore \left|\vec{H}\right|_{out}^{\parallel} - \left|\vec{H}\right|_{in}^{\parallel} = \frac{(I_f)_{enc}}{\left|\vec{l}\right|}$$

$$= \left|\vec{K}\right|_f$$

Magnetic field, $\vec{B}$:

- The integral form of Gauss's Law for magnetism in a magnetic material is:

  $$\oint_S \vec{B} \cdot d\vec{s} = 0$$

  where S represents a Gaussian pillbox of area $|\vec{s}|$ (refer to diagram a).
- The contribution from the sides equals zero as $\delta l \to 0$; therefore, Gauss's Law includes only the contributions from the surfaces $\vec{s}$.
- Express the dot product in terms of the components of $\vec{B}$ perpendicular to the boundary (parallel or anti-parallel to $\vec{s}$), $|\vec{B}|^{\perp}$.

Auxiliary magnetic field, $\vec{H}$:

- The integral form of Ampère's Law for a constant electric field in magnetic media is:

  $$\oint_L \vec{H} \cdot d\vec{l} = (I_f)_{enc}$$

  where L represents a rectangular loop of length $|\vec{l}|$ (refer to diagram b).
- The contribution from the sides equals zero as $\delta l \to 0$; therefore, Ampère's Law includes only the contributions from the sides $|\vec{l}|$.
- Express the dot product in terms of the components of $\vec{H}$ parallel to the boundary (parallel or anti-parallel to $\vec{l}$), $|\vec{H}|^{\parallel}$.

WORKED EXAMPLE 11.1: Using Laplace's equation to determine the electric potential of a dielectric in an external electric field

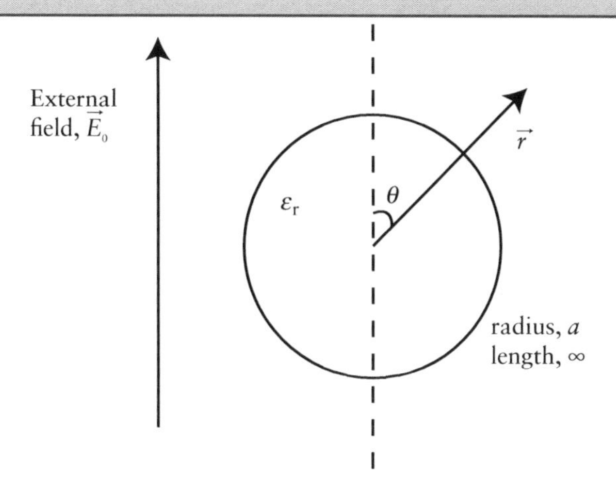

Consider an uncharged dielectric rod of relative permittivity ε_r, radius a, and infinite length, placed in a uniform electric field $\vec{E}_0$ such that the rod axis is perpendicular to the field (refer to figure). $\rho = 0$ outside the rod since there are no free charges present, and $\rho = 0$ inside the rod since positive and negative polarization charges cancel out. Therefore, in these regions, Laplace's equation is satisfied:

$$\nabla^2 V_{\text{in}} = 0 \quad \text{and} \quad \nabla^2 V_{\text{out}} = 0$$

However, on the boundaries $\rho \neq 0$ due to the accumulation of bound charge; therefore, the potential satisfies Poisson's equation on the boundary:

$$\nabla^2 V_{(r=a)} = -\frac{\rho}{\varepsilon_0}$$

To determine the electric potential of this system, we do not have to directly solve Poisson's equation—we can solve Laplace's equation for V_{in} and V_{out} and use boundary conditions to match the solutions at $r = a$, as outlined below.

General solutions of V_{in} and V_{out}

In cylindrical polar coordinates in two dimensions, Laplace's equation is:

$$r\frac{\partial}{\partial r}\left(r\frac{\partial V}{\partial r}\right) + \frac{\partial^2 V}{\partial \theta^2} = 0$$

It's straightforward to show by substitution that:

$$V_{\text{out}}(r,\theta) = A_1 r\cos\theta + \frac{B_1\cos\theta}{r}$$

$$\text{and} \quad V_{\text{in}}(r,\theta) = A_2 r\cos\theta$$

are possible solutions of Laplace's equation, where A_1, B_1, and A_2 are constants of integration to be set by boundary conditions.

Boundary conditions

Applying boundary conditions, we find expressions for A_1, B_1, and A_2.

1. At radial distances far from the rod ($r \gg a$), the potential tends to the potential of a uniform electric field:

$$V_{\text{out}}(r \gg a, \theta) = -E_0 r\cos\theta$$
$$\therefore A_1 = -E_0$$

2. The electric potential must be continuous at the boundary $\therefore V_{\text{out}} = V_{\text{in}}$ at $r = a$:

$$\therefore -E_0 a\cos\theta + \frac{B_1}{a}\cos\theta = A_2 a\cos\theta$$
$$\therefore \frac{B_1}{a^2} - E_0 = A_2 \tag{11.42}$$

3. The perpendicular components of the electric displacement field satisfy:

$$\left|\vec{D}\right|^{\perp}_{\text{out}} - \left|\vec{D}\right|^{\perp}_{\text{in}} = \sigma_f$$

(refer to Derivation 11.9). In this example, there are no free charges present $\therefore \sigma_f = 0$:

$$\therefore \left|\vec{D}\right|^{\perp}_{\text{out}} = \left|\vec{D}\right|^{\perp}_{\text{in}}$$
$$\therefore \varepsilon_0\left|\vec{E}\right|_{\text{out}}\hat{r} = \varepsilon_r\varepsilon_0\left|\vec{E}\right|_{\text{in}}\hat{r}$$

The radial component of $\vec{E} = -\nabla V$ is $\partial V/\partial r$, therefore in terms of the electric potential, the boundary condition is:

$$\therefore \left(\frac{\partial V_{\text{out}}}{\partial r}\right)_{r=a} = \varepsilon_r\left(\frac{\partial V_{\text{in}}}{\partial r}\right)_{r=a}$$
$$\therefore -E_0\cos\theta - \frac{B_1}{a^2}\cos\theta = \varepsilon_r A_2\cos\theta$$
$$\therefore -\frac{B_1}{a^2} - E_0 = \varepsilon_r A_2 \tag{11.43}$$

WORKED EXAMPLE 11.1: Using Laplace's equation to determine the electric potential of a dielectric in an external electric field (cont.)

Electric potential

Using Eqs. 11.42 and 11.43 to solve for B_1 and A_2, we find:

$$B_1 = -E_0 a^2 \left(\frac{1-\varepsilon_r}{1+\varepsilon_r}\right)$$

and $A_2 = -E_0 \left(\frac{2}{1+\varepsilon_r}\right)$

Therefore, the electric potential outside and inside the rod are:

$$V_{out}(r,\theta) = A_1 r\cos\theta + \frac{B_1\cos\theta}{r}$$

$$= -E_0\left(r + \frac{a^2}{r}\left(\frac{1-\varepsilon_r}{1+\varepsilon_r}\right)\right)\cos\theta$$

and $V_{in}(r,\theta) = A_2 r\cos\theta$

$$= -E_0\left(\frac{2}{1+\varepsilon_r}\right) r\cos\theta$$

Electric field inside rod

Letting $z = r\cos\theta$, we see that the net electric field inside the rod is uniform along the $+z$-direction:

$$V_{in} = -E_0\left(\frac{2}{1+\varepsilon_r}\right) z$$

$$\therefore \vec{E}_{in} = -\nabla V_{in}$$

$$= \frac{\partial}{\partial z}\left(E_0\left(\frac{2}{1+\varepsilon_r}\right) z\right)\hat{z}$$

$$= E_0\left(\frac{2}{1+\varepsilon_r}\right)\hat{z}$$

$$= \left(\frac{2}{1+\varepsilon_r}\right)\vec{E}_0$$

Therefore, the resultant field is parallel to the applied field, with an amplitude scale factor of $2/(1 + \varepsilon_r)$.

Chapter 12
Electromagnetic waves

Incorporated in Maxwell's equations is the existence of electromagnetic waves. A wave is a disturbance that propagates in space and time; an electromagnetic wave consists of a propagating electric and magnetic field. We are able to see qualitatively how electromagnetic induction gives rise to self-propagating electromagnetic waves. From Faraday's Law, a changing magnetic field induces a changing electric field, and from Ampère's Law, a changing electric field induces a changing magnetic field. This, in turn, induces a changing electric field, and so on. We're of course very familiar with electromagnetic waves—light, radio waves, and microwaves, for example, are all electromagnetic waves. The type of electromagnetic wave depends on its frequency. The spectrum of waves of different frequency make up the *electromagnetic spectrum*, which is shown in Fig. 12.1. In this chapter, we will first look at the properties of electromagnetic waves in free space (a vacuum), followed by their properties in dielectric (non-conducting) matter.

12.1 Electromagnetic waves in free space

Free space (a theoretically perfect vacuum) contains no matter. Therefore, the charge and current densities in free space equal zero:

Charge density: $\rho = 0$

Current density: $\vec{J} = 0$

Furthermore, by definition, the relative permittivity, ε_r, and permeability, μ_r, in free space both equal one:

Relative permittivity: $\varepsilon_r = 1$

Relative permeability: $\mu_r = 1$

Using these properties in Maxwell's equations (refer to Eqs. 11.28–11.31), we find that the divergence of $\vec{E}$ and $\vec{B}$ in free space are equal to zero (Eqs. 12.1 and 12.2). This makes intuitive sense—there are no sources or sinks of $\vec{E}$ and $\vec{B}$ (charge and current) in free space.[1] However, the curl of $\vec{E}$ and $\vec{B}$ in free space do not equal zero, but instead depend on the rate of change of $\vec{B}$ or $\vec{E}$, respectively (Eqs. 12.3 and 12.4). We shall see next how these equations describe self-propagating, non-attenuating electromagnetic waves.

Maxwell's equations in free space:
($\rho = 0, \vec{J} = 0, \varepsilon_r = 1, \mu_r = 1$):

Gauss's Law:

$$\nabla \cdot \vec{E} = 0 \quad (12.1)$$

Gauss's Law for magnetism:

$$\nabla \cdot \vec{B} = 0 \quad (12.2)$$

Faraday's Law:

$$\nabla \times \vec{E} = -\frac{\partial \vec{B}}{\partial t} \quad (12.3)$$

Ampère's Law:

$$\nabla \times \vec{B} = \varepsilon_0 \mu_0 \frac{\partial \vec{E}}{\partial t} \quad (12.4)$$

12.1.1 The wave equations

A wave is a disturbance that propagates in space and time. An arbitrary function $f(x, t)$ represents a wave if it satisfies the *wave equation*:

[1] $\nabla \cdot \vec{B} = 0$ always since magnetic monopoles do not exist.

$$\frac{\partial^2 f}{\partial x^2} = \frac{1}{v^2}\frac{\partial^2 f}{\partial t^2} \quad \text{in } 1D$$

$$\text{or } \nabla^2 f = \frac{1}{v^2}\frac{\partial^2 f}{\partial t^2} \quad \text{in } 3D$$

(refer to Section 13.2). By decoupling Faraday's and Ampère's Laws, we find that both $\vec{E}$ and $\vec{B}$ satisfy a wave equation (Eqs. 12.5 and 12.6 and Derivation 12.1). We can read off their velocity as:

$$\frac{1}{v^2} = \varepsilon_0 \mu_0$$

$$\therefore v = \frac{1}{\sqrt{\varepsilon_0 \mu_0}}$$

(more on the significance of this velocity below).

Wave equation in free space (Derivation 12.1):

Electric field:

$$\nabla^2 \vec{E} = \varepsilon_0 \mu_0 \frac{\partial^2 \vec{E}}{\partial t^2} \tag{12.5}$$

Magnetic field:

$$\nabla^2 \vec{B} = \varepsilon_0 \mu_0 \frac{\partial^2 \vec{B}}{\partial t^2} \tag{12.6}$$

We can see qualitatively from Faraday's and Ampère's Law that electromagnetic waves are self-propagating: a changing magnetic field induces an electric field (Faraday's Law) and a changing electric field induces a magnetic field (Ampère's Law). Thus, the fields are mutually inducing. Since the divergence of both $\vec{E}$ and $\vec{B}$ are zero in free space, there are no other sources or sinks of the fields. Therefore, in theory, electromagnetic waves can propagate indefinitely through free space without attenuation.

12.1.2 Wave properties

It's straightforward to show that sinusoidal plane-wave solutions satisfy the wave equations for $\vec{E}$ and $\vec{B}$:

$$\vec{E}(\vec{r},t) = \vec{E}_0 \sin(\vec{k}\cdot\vec{r} - \omega t)$$

$$\text{or } \vec{E}(\vec{r},t) = \vec{E}_0 \sin(\vec{k}\cdot\vec{r} - \omega t + \phi)$$

and similarly for $\vec{B}$ (Eqs. 12.7 and 12.9), where ϕ is an arbitrary phase shift and $\vec{E}_0$ is the maximum amplitude. The phase of the wave has a space component, $\vec{k}\cdot\vec{r}$, and a time component, ωt. The *wavevector*, $\vec{k}$ [m^{-1}], describes the direction of wave propagation ($\hat{k}$) and the wavelength ($\lambda = 1/|\vec{k}|$). The *angular frequency*, ω [s^{-1}], describes the time period ($T = 2\pi/\omega$).

Complex exponential form

Throughout this chapter, we will use the *complex exponential form* of solution for $\vec{E}$ and $\vec{B}$, since they are more tractable than sinusoidal solutions (Eqs. 12.8 and 12.10). The tildes in Eqs. 12.8 and 12.10 denote complex solutions, but the functions that describe the physical waves correspond to the real or imaginary parts of $\tilde{\vec{E}}$ and $\tilde{\vec{B}}$ (thus giving sin or cos solutions). It is common to omit the tildes, therefore we have omitted the tildes in the rest of the chapter.

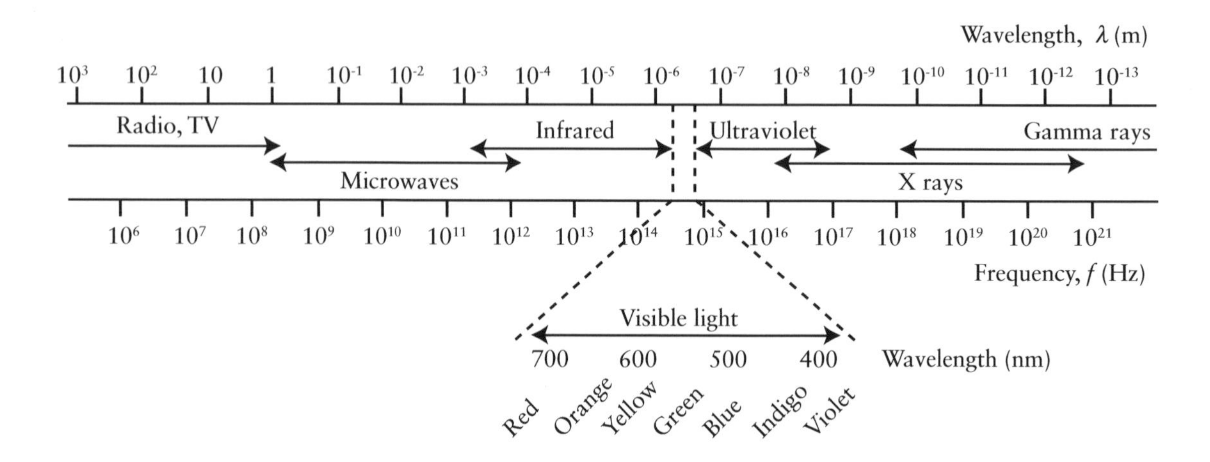

Figure 12.1: The electromagnetic spectrum

Derivation 12.1: Derivation of the wave equations in free space (Eqs. 12.5 and 12.6)

$$\nabla\times\vec{E}=-\frac{\partial\vec{B}}{\partial t}$$

$$\therefore \nabla\times\nabla\times\vec{E}=-\nabla\times\frac{\partial\vec{B}}{\partial t}$$

$$=-\frac{\partial}{\partial t}(\nabla\times\vec{B})$$

$$=-\varepsilon_0\mu_0\frac{\partial^2\vec{E}}{\partial t^2}$$

$$\text{and } \nabla\times\nabla\times\vec{E}=\nabla(\nabla\cdot\vec{E})-\nabla^2\vec{E}$$

$$=-\nabla^2\vec{E}$$

$$\therefore \nabla^2\vec{E}=\varepsilon_0\mu_0\frac{\partial^2\vec{E}}{\partial t^2}$$

Similarly, using Ampère's Law:

$$\nabla^2\vec{B}=\varepsilon_0\mu_0\frac{\partial^2\vec{B}}{\partial t^2}$$

- Starting point: Faraday's and Ampère's Laws in free space are:

  $$\nabla\times\vec{E}=-\frac{\partial\vec{B}}{\partial t}$$

  $$\text{and } \nabla\times\vec{B}=\varepsilon_0\mu_0\frac{\partial\vec{E}}{\partial t}$$

  Take the curl of Faraday's Law, and use Ampère's Law to express $\nabla\times\vec{B}$ in terms of $\vec{E}$. (The curl operator does not act on the time derivative.)
- The vector identity for the curl of the curl of a vector is:

  $$\nabla\times\nabla\times\vec{E}=\nabla(\nabla\cdot\vec{E})-\nabla^2\vec{E}$$

  where $\nabla\cdot\vec{E}=0$ in free space.
- Equate the two expressions for $\nabla\times\nabla\times\vec{E}$. Doing this, we find that $\vec{E}$ satisfies a second-order partial differential equation, which is a three-dimensional wave equation.
- We can use a similar method (take the curl of Ampère's Law and eliminate $\nabla\times\vec{E}$ using Faraday's Law) to determine an identical wave equation for $\vec{B}$.

Plane-wave solutions to the three-dimensional wave equation:

Electric field:

$$\vec{E}(\vec{r},t)=\vec{E}_0\sin(\vec{k}\cdot\vec{r}-\omega t) \quad (12.7)$$

$$\text{or } \tilde{\vec{E}}(\vec{r},t)=\tilde{\vec{E}}_0 e^{i(\vec{k}\cdot\vec{r}-\omega t)} \quad (12.8)$$

Magnetic field:

$$\vec{B}(\vec{r},t)=\vec{B}_0\sin(\vec{k}\cdot\vec{r}-\omega t) \quad (12.9)$$

$$\text{or } \tilde{\vec{B}}(\vec{r},t)=\tilde{\vec{B}}_0 e^{i(\vec{k}\cdot\vec{r}-\omega t)} \quad (12.10)$$

where

$$\vec{r}=x\hat{x}+y\hat{y}+z\hat{z}$$

$$\vec{k}=k_x\hat{x}+k_y\hat{y}+k_z\hat{z}$$

$$\therefore \vec{k}\cdot\vec{r}=k_x x+k_y y+k_z z$$

Plane-wave derivatives (Derivation 12.2):

For a plane-wave solution $\vec{E}=\vec{E}_0 e^{i(\vec{k}\cdot\vec{r}-\omega t)}$
Space derivatives:

$$\nabla\rightarrow i\vec{k} \quad (12.11)$$

$$\therefore \nabla\cdot\vec{E}=i\vec{k}\cdot\vec{E}$$

$$\nabla\times\vec{E}=i\vec{k}\times\vec{E}$$

$$\nabla^2\vec{E}=-\left|\vec{k}\right|^2\vec{E}$$

Time derivatives:

$$\frac{\partial}{\partial t}\rightarrow -i\omega \quad (12.12)$$

$$\therefore \frac{\partial\vec{E}}{\partial t}=-i\omega\vec{E}$$

$$\frac{\partial^2\vec{E}}{\partial t^2}=-\omega^2\vec{E}$$

Space and time derivatives

It is useful at this point to determine the space and time derivatives of the plane-wave solutions in Eqs. 12.8 and 12.10 since we use them frequently. We find that the ∇ operator multiplies the solution by $i\vec{k}$, and the $\partial/\partial t$ operator multiplies it by $-i\omega$ (Eqs. 12.11 and 12.12, Derivation 12.2).

Wave velocity

Substituting the plane-wave solutions for $\vec{E}$ into its wave equation, we have:

$$\nabla^2 \vec{E} = \varepsilon_0 \mu_0 \frac{\partial^2 \vec{E}}{\partial t^2}$$

$$\therefore -\left|\vec{k}\right|^2 \vec{E} = -\varepsilon_0 \mu_0 \omega^2 \vec{E}$$

$$\therefore \left|\vec{k}\right|^2 = \varepsilon_0 \mu_0 \omega^2$$

The velocity of propagation of a wave is defined as $v = \omega/k$. Therefore, the velocity of electromagnetic waves in free space is:

$$\begin{aligned} v &= \frac{\omega}{\left|\vec{k}\right|} \\ &= \frac{1}{\sqrt{\varepsilon_0 \mu_0}} \\ &= \frac{1}{\sqrt{8.85 \cdot 10^{-12} \cdot 4\pi \cdot 10^{-7}}} \\ &\simeq 3 \cdot 10^8 \text{ ms}^{-1} \end{aligned}$$

We all recognize this value as the *speed of light* in a vacuum, c [ms^{-1}] (Eq. 12.13). Electromagnetic waves of *all* frequencies travel at this velocity in free space (and, to a very good approximation, in air).

Wave velocity in free space:

$$\begin{aligned} c &= \frac{1}{\sqrt{\varepsilon_0 \mu_0}} \\ &\simeq 3 \cdot 10^8 \text{ ms}^{-1} \end{aligned} \qquad (12.13)$$

Properties of electromagnetic waves

Let's now use the plane-wave solutions for $\vec{E}$ and $\vec{B}$ in Maxwell's equations to determine additional properties of electromagnetic waves (which are summarized in Fig. 12.2):

1. From Gauss's Law for $\vec{E}$ and $\vec{B}$:

$$\nabla \cdot \vec{E} = 0 \rightarrow i\vec{k} \cdot \vec{E} = 0$$

$$\nabla \cdot \vec{B} = 0 \rightarrow i\vec{k} \cdot \vec{B} = 0$$

$\therefore$ $\vec{E}$ and $\vec{B}$ are perpendicular to the direction of wave propagation, $\vec{k} \Rightarrow$ electromagnetic waves are *transverse* waves (Eqs. 12.14 and 12.15).

2. From Faraday's Law:

$$\nabla \times \vec{E} = -\frac{\partial \vec{B}}{\partial t} \rightarrow i\vec{k} \times \vec{E} = i\omega \vec{B}$$

$$\left|\vec{k}\right| \hat{k} \times \vec{E} = \omega \vec{B}$$

$$\therefore \hat{k} \times \vec{E} = c\vec{B}$$

and from Ampère's Law:

$$\nabla \times \vec{B} = \frac{1}{c^2}\frac{\partial \vec{E}}{\partial t} \rightarrow i\vec{k} \times \vec{B} = -\frac{i\omega}{c^2}\vec{E}$$

$$\left|\vec{k}\right| \hat{k} \times \vec{B} = -\frac{\omega}{c^2}\vec{E}$$

$$\therefore \hat{k} \times \vec{B} = -\frac{1}{c}\vec{E}$$

$\therefore$ $\vec{E}$ and $\vec{B}$ are *orthogonal* to one another $\Rightarrow$ $\vec{E}$ and $\vec{B}$ are therefore linearly independent (Eqs. 12.16 and 12.17).

3. The amplitude of $\vec{E}$ is numerically a factor c times greater than the amplitude of $\vec{B}$ (Eq. 12.18). However, as we shall see in Section 12.2, the component of energy density from the electric field equals that from the magnetic field. It is this unique ratio that allows a wave to propagate: a changing electric field makes just enough changing magnetic field to make just enough changing electric field again, and so on.

Properties of electromagnetic waves (refer to Fig. 12.2):

For a wave traveling along $\vec{k}$:
$\vec{E}$ and $\vec{B}$ components are transverse:

$$\hat{k} \cdot \vec{E} = 0 \qquad (12.14)$$

$$\hat{k} \cdot \vec{B} = 0 \qquad (12.15)$$

$\vec{E}$ and $\vec{B}$ are orthogonal to each other:

$$\hat{k} \times \vec{E} \propto \vec{B} \qquad (12.16)$$

$$\hat{k} \times \vec{B} \propto -\vec{E} \qquad (12.17)$$

The ratio of the amplitudes of $\vec{E}$ and $\vec{B}$ is c:

$$\hat{k} \times \vec{E} = c\vec{B}$$

$$\therefore \left|\vec{E}\right| = c\left|\vec{B}\right| \qquad (12.18)$$

Polarization

By convention, the direction of $\vec{E}$ specifies the *polarization* of an electromagnetic wave. If $\vec{E}$ remains in a single plane, for example, in the x–z plane, as illustrated in Fig. 12.2, then the wave is said to be *plane* polarized (or linearly polarized). *Circular* polarization is achieved by the superposition of two orthogonal plane-polarized waves with a $\pi/2$ phase difference between them. This produces a resultant $\vec{E}$-field that rotates around the axis of propagation.

Derivation 12.2: Derivation of the space and time derivatives of plane waves (Eqs. 12.11 and 12.12)

Space derivative:
(recall $\vec{k} \cdot \vec{r} = k_x x + k_y y + k_z z$)

$$\nabla \cdot \vec{E} = \left(\frac{\partial}{\partial x}\hat{x} + \frac{\partial}{\partial y}\hat{y} + \frac{\partial}{\partial z}\hat{z} \right) \cdot \vec{E}_0 e^{i(\vec{k}\cdot\vec{r}-\omega t)}$$
$$= i\left(k_x\hat{x} + k_y\hat{y} + k_z\hat{z}\right) \cdot \vec{E}_0 e^{i(\vec{k}\cdot\vec{r}-\omega t)}$$
$$= i\vec{k} \cdot \vec{E}_0 e^{i(\vec{k}\cdot\vec{r}-\omega t)}$$
$$\therefore \nabla^2 \vec{E} = (i\vec{k} \cdot i\vec{k})\vec{E}$$
$$= -\left|\vec{k}\right|^2 \vec{E}$$

The ∇ operator multiplies the solution by $i\vec{k}$:

$$\nabla \to i\vec{k}$$

Time derivative:

$$\frac{\partial}{\partial t}\vec{E} = \frac{\partial}{\partial t}\vec{E}_0 e^{i(\vec{k}\cdot\vec{r}-\omega t)}$$
$$= -i\omega \vec{E}_0 e^{i(\vec{k}\cdot\vec{r}-\omega t)}$$
$$= -i\omega\vec{E}$$
$$\therefore \frac{\partial^2}{\partial t^2}\vec{E} = (-i\omega)^2 \vec{E}$$
$$= -\omega^2\vec{E}$$

The $\frac{\partial}{\partial t}$ operator multiplies the solution by $-i\omega$:

$$\frac{\partial}{\partial t} \to -i\omega$$

Characteristic impedance

The ratio of the amplitudes of electric field, $\vec{E}$ [V m^{-1}], to the auxiliary magnetic field, $\vec{H}$ [Am^{-1}], has units of resistance:

$$\frac{\left[\text{Vm}^{-1}\right]}{\left[\text{Am}^{-1}\right]} \equiv [\Omega]$$

This ratio represents the *characteristic impedance*, $Z\,[\Omega]$:

$$Z \stackrel{\text{def}}{=} \frac{|\vec{E}|}{|\vec{H}|}$$

(Eq. 12.19). Z is a property of the medium through which a wave is traveling. In free space: $|\vec{E}| = c|\vec{B}|$ and $|\vec{B}| = \mu_0|\vec{H}|$. Therefore, the *impedance of free space*, Z_0, is a constant that depends only on μ_0 and ε_0 (Eq. 12.20).

Characteristic impedance:

$$Z = \frac{|\vec{E}|}{|\vec{H}|} \qquad (12.19)$$

Impedance of free space:

$$Z_0 = c\,\mu_0$$
$$= \sqrt{\frac{\mu_0}{\varepsilon_0}}$$
$$\simeq 377\ \Omega \qquad (12.20)$$

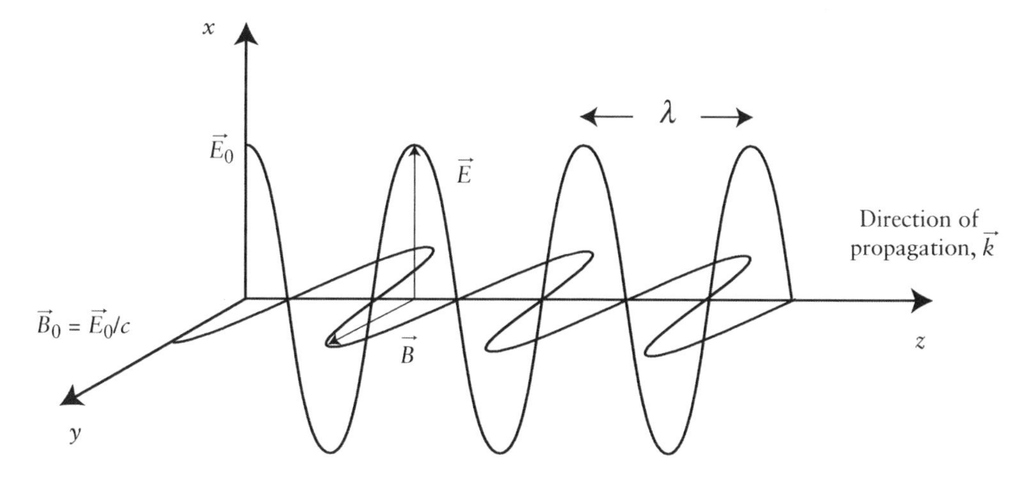

Figure 12.2: An electromagnetic wave. $\vec{E}$ and $\vec{B}$ are orthogonal to each other, and to the direction of propagation. $\vec{E}$ and $\vec{B}$ have the same frequency and wavelength, λ. Although the amplitude of $\vec{E}$ is numerically c times the amplitude of $\vec{B}$, the energy density stored in either component is equal.

12.2 Energy and momentum

Electromagnetic waves carry both energy and momentum and therefore convey power away from their source.

12.2.1 Electromagnetic energy density

The electric and magnetic energy densities stored in an electric or magnetic field, respectively, are:

$$u_E = \frac{1}{2}\varepsilon_0 \left|\vec{E}\right|^2$$

and $$u_B = \frac{1}{2\mu_0}\left|\vec{B}\right|^2$$

(refer to Sections 7.4 and 10.2). The total energy density carried by an electromagnetic wave is the sum of these two contributions (Eq. 12.21). Using $\left|\vec{E}\right| = c\left|\vec{B}\right| = \left|\vec{B}\right| / \sqrt{\varepsilon_0\mu_0}$, we find that the contributions from the electric and magnetic components are equal:

$$\begin{aligned} u_E &= \frac{1}{2}\varepsilon_0\left|\vec{E}\right|^2 \\ &= \frac{\varepsilon_0}{2\varepsilon_0\mu_0}\left|\vec{B}\right|^2 \\ &= \frac{1}{2\mu_0}\left|\vec{B}\right|^2 \\ &= u_B \end{aligned}$$

Electromagnetic energy density in a wave:

$$\begin{aligned} u_{EM} &= u_E + u_B \\ &= \frac{1}{2}\varepsilon_0\left|\vec{E}\right|^2 + \frac{1}{2\mu_0}\left|\vec{B}\right|^2 \end{aligned} \qquad (12.21)$$

where $u_E = u_B$.

12.2.2 The Poynting vector

The *Poynting vector*, $\vec{S}$ [Wm^{-2}], defines the power flux (power per unit area) carried by a wave. An expression for the Poynting vector in terms of $\vec{E}$ and $\vec{B}$ is found by calculating the rate of change of the wave's energy density, du_{EM}/dt, as shown in Derivation 12.3, and is found to be proportional to the cross product of $\vec{E}$ and $\vec{B}$:

$$\vec{S} = \frac{1}{\mu_0}(\vec{E}\times\vec{B})$$

(Eqs. 12.22 and 12.23 and Derivation 12.3). The direction of $\vec{S}$ is perpendicular to both $\vec{E}$ and $\vec{B}$, and hence is along the direction of wave propagation.[2] The element of power, dP, transmitted through an element of area $d\vec{s}$ corresponds to the component of $\vec{S}$ along $d\vec{s}$:

[2] This is only true in a HILS material. It's not true in anisotropic materials, such as birefringent materials (where refractive index depends on direction).

$$dP = \vec{S}\cdot d\vec{s}$$

Therefore, the total power carried through a closed surface, S, by a wave is given by the integral $P = \int dP$ (Eq. 12.24).

Poynting vector (Derivation 12.3):

$$\vec{S} = \frac{1}{\mu_0}(\vec{E}\times\vec{B}) \qquad (12.22)$$

$$\equiv \vec{E}\times\vec{H} \qquad (12.23)$$

Power carried out of a closed surface, S:

$$P = \oint_S \vec{S}\cdot d\vec{s} \qquad (12.24)$$

Calculating the Poynting vector

We can determine the magnitude and direction of the Poynting vector for electromagnetic waves traveling through electric circuits with a given current, I, and potential, V (Worked example 12.1). The direction of the Poynting vector is found from the cross product of $\vec{E}$ and $\vec{B}$, as illustrated in the diagrams in Worked example 12.1. From the examples shown, we find that the magnitude of $\vec{S}$ is equal to the power, $P = VI$, divided by the area perpendicular to the direction of power flow (orthogonal to $\vec{S}$).

Radiation pressure

Electromagnetic waves exert *radiation pressure*, $\vec{R}$ [Pa], on a surface. Radiation pressure arises due to the change in momentum,[3] $\vec{p}$, of a photon against a surface. From Newton's Second Law, force is related to momentum change by $\vec{F} = d\vec{p}/dt$, and the momentum of a photon is related to its energy, U, by $\vec{p} = U\hat{n}/c$, where $\hat{n}$ is a unit vector pointing along $\vec{p}$. Therefore, the radiation pressure is:

$$\begin{aligned} \vec{R} &= \frac{\vec{F}}{s_\perp} \\ &= \frac{1}{s_\perp}\frac{d\vec{p}}{dt} \\ &= \frac{1}{s_\perp c}\frac{dU}{dt}\hat{n} \\ &= \frac{\vec{S}}{c} \end{aligned}$$

where $s_\perp$ is a perpendicular component of area (Eq. 12.25). Therefore, the radiation pressure is related to the Poynting vector by a factor of c. If an incident wave is reflected off a surface, rather than absorbed by it, then the radiation pressure is $\vec{R} = 2\vec{S}/c$ to account for the pressure of both incoming and reflected waves.

[3] Momentum, $\vec{p}$, in this section is not to be confused with an electric dipole moment, also denoted by $\vec{p}$.

Derivation 12.3: Derivation of the Poynting vector (Eqs. 12.22–12.24)

$$\begin{aligned}
P &= -\int_V \frac{\partial u_{EM}}{\partial t} d\tau \\
&= -\int_V \frac{\partial}{\partial t}\left(\frac{1}{2}\varepsilon_0 |\vec{E}|^2 + \frac{1}{2\mu_0}|\vec{B}|^2\right) d\tau \\
&= -\int_V \left(\varepsilon_0 \vec{E}\cdot\frac{\partial \vec{E}}{\partial t} + \frac{1}{\mu_0}\vec{B}\cdot\frac{\partial \vec{B}}{\partial t}\right) d\tau \\
&= -\int_V \left(\frac{1}{\mu_0}\vec{E}\cdot(\nabla\times\vec{B}) - \frac{1}{\mu_0}\vec{B}\cdot(\nabla\times\vec{E})\right) d\tau \\
&= \frac{1}{\mu_0}\int_V \nabla\cdot(\vec{E}\times\vec{B}) d\tau \\
&= \frac{1}{\mu_0}\oint_S (\vec{E}\times\vec{B})\cdot d\vec{s} \\
&\equiv \oint_S \vec{S}\cdot d\vec{s}
\end{aligned}$$

$$\text{where } \vec{S} = \frac{1}{\mu_0}\vec{E}\times\vec{B} \equiv \vec{E}\times\vec{H}$$

- Starting point: the power carried by an electromagnetic wave in free space, where there are no energy losses, equals the rate of work done:

$$P = \frac{\partial W}{\partial t} = -\frac{\partial U_{EM}}{\partial t} = -\int_V \frac{\partial u_{EM}}{\partial t} d\tau$$

where the wave's energy density is the sum of electric and magnetic components:

$$u_{EM} = \frac{1}{2}\varepsilon_0 |\vec{E}|^2 + \frac{1}{2\mu_0}|\vec{B}|^2$$

- Let $|\vec{E}|^2 = \vec{E}\cdot\vec{E}$ and differentiate. Similarly for $\vec{B}$.
- Use Ampère's Law and Faraday's Law in free space to eliminate the time derivatives:

$$\text{Ampère}: \ \nabla\times\vec{B} = \varepsilon_0 \mu_0 \frac{\partial \vec{E}}{\partial t}$$

$$\text{Faraday}: \ \nabla\times\vec{E} = -\frac{\partial \vec{B}}{\partial t}$$

- Use the vector identity:

$$\vec{G}\cdot(\nabla\times\vec{F}) - \vec{F}\cdot(\nabla\times\vec{G}) = \nabla\cdot(\vec{F}\times\vec{G})$$

to simplify the integral.
- Invoke the divergence theorem to convert the volume integral into a closed surface integral, where S is the surface that encloses a volume V.
- The Poynting vector, $\vec{S}$, is defined as the power per unit area (the power flux) carried by a wave.

Radiation pressure:

$$\vec{R} = \frac{\vec{S}}{c} \qquad (12.25)$$

For a wave reflected off a surface, the radiation pressure is doubled ($\vec{R}_r = 2\vec{S}/c$).

12.3 Electromagnetic waves in non-conducting matter

In non-conducting matter (≡ insulators or dielectric materials), the free charge and free current densities equal zero:

$$\text{Free charge density}: \rho_f = 0$$

$$\text{Free current density}: \vec{J}_f = 0$$

The permittivity, ε, and permeability, μ, of a dielectric material describe how readily it is polarized or magnetized by an applied electric or magnetic field. Therefore, we have:

$$\text{Permittivity}: \varepsilon = \varepsilon_0 \varepsilon_r$$

$$\text{Permeability}: \mu = \mu_0 \mu_r$$

where ε_r and μ_r are the material's relative permittivity and relative permeability, respectively. As we shall see in this section, these parameters characterize the propagation properties of electromagnetic waves in dielectrics.

12.3.1 Maxwell's equations

In an HILS material, the electric displacement field $\vec{D}$ and auxiliary magnetic field $\vec{H}$ are related to the electric and magnetic fields by:

$$\vec{D} = \varepsilon\vec{E} \quad \text{and} \quad \vec{H} = \frac{\vec{B}}{\mu}$$

Therefore, using the expressions listed above, it's straightforward to write Maxwell's equations in non-conducting matter (Eqs. 12.26–12.29). We find that Maxwell's equations in non-conducting matter are almost identical to those in free space, with the only differences being that ε_0 is replaced by ε, and μ_0 by μ (compare Eqs. 12.26–12.29 with Eqs. 12.1–12.4).

WORKED EXAMPLE 12.1: Calculating the Poynting vector of electromagnetic waves traveling through a circuit with given V and I (Eq. 12.22)

a) Straight conducting wire

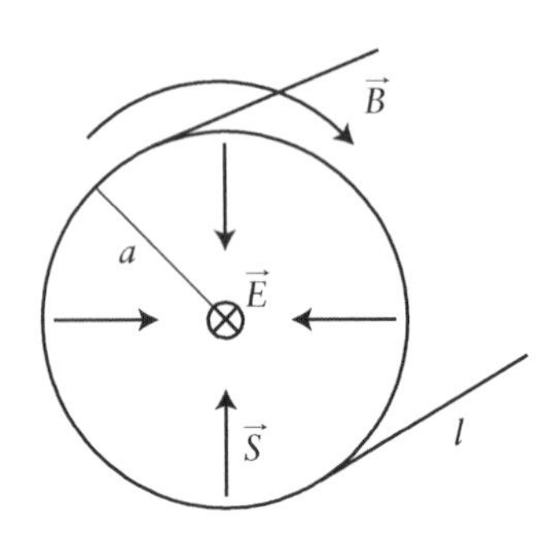

wire radius a
length l

$$|\vec{E}| = \frac{V}{l}$$
$$|\vec{B}| = \frac{\mu_0 I}{2\pi a}$$
$$\therefore |\vec{S}| = \frac{|\vec{E}||\vec{B}|}{\mu_0} = \frac{VI}{2\pi a l}$$

b) Circular plate capacitor

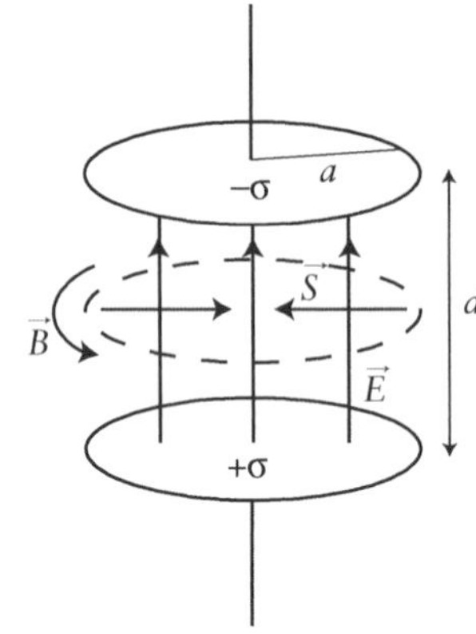

plate radius a
plate separation d

$$|\vec{E}| = \frac{V}{d}$$
$$|\vec{B}| = \frac{\mu_0 I_D}{2\pi a}$$
$$\therefore |\vec{S}| = \frac{|\vec{E}||\vec{B}|}{\mu_0} = \frac{VI_D}{2\pi a d}$$

c) Coaxial cable

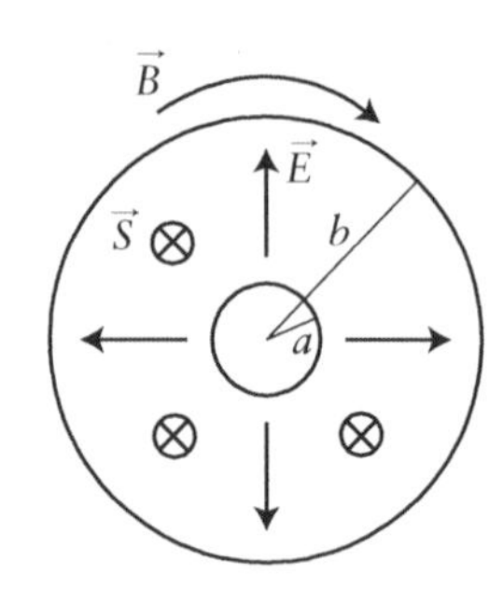

inner radius a
outer radius b

$$|\vec{E}| = \frac{V}{r\ln(b/a)}$$
$$|\vec{B}| = \frac{\mu_0 I}{2\pi r}$$
$$\therefore |\vec{S}| = \frac{|\vec{E}||\vec{B}|}{\mu_0} = \frac{VI}{2\pi r^2 \ln(b/a)}$$

Refer to Worked example 7.5 for determining $|\vec{E}|$ *and Worked example 8.4 for determining* $|\vec{B}|$.

Maxwell's equations in non-conducting matter:

($\rho_f = 0, \vec{J}_f = 0, \varepsilon = \varepsilon_0\varepsilon_r, \mu = \mu_0\mu_r$):

Gauss's Law:

$$\nabla \cdot \vec{E} = 0 \tag{12.26}$$

Gauss's Law for magnetism:

$$\nabla \cdot \vec{B} = 0 \tag{12.27}$$

Faraday's Law:

$$\nabla \times \vec{E} = -\frac{\partial \vec{B}}{\partial t} \tag{12.28}$$

Ampère's Law:

$$\nabla \times \vec{B} = \varepsilon\mu\frac{\partial \vec{E}}{\partial t} \tag{12.29}$$

12.3.2 The wave equations

We can use an identical treatment as we did in free space—decouple Faraday's and Ampère's Laws by taking their curl—to determine wave equations for $\vec{E}$ and $\vec{B}$ in non-conducting matter (refer to Derivation 12.1, using ε and μ in place of ε_0 and μ_0). The wave equations are provided in Eqs. 12.30 and 12.31 (compare these with Eqs. 12.5 and 12.6). The solutions for $\vec{E}$ and $\vec{B}$ are therefore plane-wave solutions:

$$\vec{E}(\vec{r},t) = \vec{E}_0 \sin(\vec{k}\cdot\vec{r} - \omega t)$$
$$\text{or } \tilde{\vec{E}}(\vec{r},t) = \tilde{\vec{E}}_0 e^{i(\vec{k}\cdot\vec{r} - \omega t)}$$

and similarly for $\vec{B}$ (refer to Section 12.1.2 and Eqs. 12.7–12.10).

Wave equations in non-conducting matter (refer to Derivation 12.1):

Electric field:

$$\nabla^2 \vec{E} = \varepsilon\mu \frac{\partial^2 \vec{E}}{\partial t^2} \quad (12.30)$$

Magnetic field:

$$\nabla^2 \vec{B} = \varepsilon\mu \frac{\partial^2 \vec{B}}{\partial t^2} \quad (12.31)$$

Due to the similarity between the wave equations in free space and in non-conducting matter, we can anticipate that the properties of electromagnetic waves in matter are identical to those in free space, but with ε_0 and μ_0 scaled by ε_r and μ_r, as we shall see below.

12.3.3 Velocity and refractive index

Let's first determine the velocity of electromagnetic waves in non-conducting matter. Using plane-wave solutions in the wave equations, along with $v = \omega / |\vec{k}|$, we find that the velocity of electromagnetic waves in matter is:

$$v = \frac{1}{\sqrt{\varepsilon\mu}} = \frac{c}{\sqrt{\varepsilon_r \mu_r}}$$

(Eq. 12.32). The product $\varepsilon_r\mu_r$ is always greater than 1, therefore the speed of waves in matter is slower than in a vacuum.

Velocity and refractive index:

Wave velocity:

$$v = \frac{1}{\sqrt{\varepsilon\mu}} = \frac{c}{\sqrt{\varepsilon_r \mu_r}} \quad (12.32)$$

Refractive index:

$$n = \frac{c}{v} \quad (12.33)$$

$$= \sqrt{\varepsilon_r \mu_r} \quad (12.34)$$

$$\simeq \sqrt{\varepsilon_r} \quad \text{for } \mu_r \simeq 1 \quad (12.35)$$

Refractive index

The ratio of wave velocity (strictly speaking, wave speed) in free space to that in matter is called the *refractive index*, n (Eq. 12.33). The refractive index depends only on ε_r and μ_r (Eq. 12.34). Most non-conducting materials are only weakly magnetic and therefore have $\mu_r \simeq 1$ (ferromagnets have high μ_r). Therefore, the refractive index is often approximated to the square root of ε_r (Eq. 12.35).

Dispersion

An electromagnetic wave traveling through a non-conducting material interacts with the material's electrons, which alters the wave's properties, such as its speed. Let's assume that the material is non-magnetic, such that $\mu_r \simeq 1$. The oscillating electric field in the wave exerts an oscillating force on bound electrons ($\vec{F} = q\vec{E}$ for a charge q), which results in an oscillating displacement of the electrons from their equilibrium position. This, in turn, gives rise to oscillating atomic electric dipole moments, and hence a net polarization of the material.

Now let's express the concepts in the above paragraph mathematically. The electric field in a wave varies sinusoidally in time, say $\vec{E} = \vec{E}_0 \cos\omega t$. Therefore, for a plane-polarized wave of frequency ω, an electron experiences an oscillating driving force of:

$$\vec{F}_D = q\vec{E} = q\vec{E}_0 \cos\omega t$$

(for an electron, $q = -e$, but we will use the variable q in this discussion for generality). This driving force displaces the electron from its equilibrium position. For a displacement $\vec{x}$, the restoring force on the electron obeys Hooke's Law:

$$\vec{F}_R = -k\vec{x} = -m\omega_0^2\vec{x}$$

where k is the spring constant, ω_0 is the natural frequency of the system, and m is the electron mass (refer to Section 2.3.1 for why $k = m\,\omega_0^2$). Assuming that there are no damping forces that act on the bound electron, then, using Newton's Second Law, the forces acting on the electron are:

$$\vec{F}_D + \vec{F}_R = m\frac{d^2\vec{x}}{dt^2}$$

Rearranging and using the expressions above for $\vec{F}_D$ and $\vec{F}_R$, the equation of motion of the electron is:

$$m\frac{d^2\vec{x}}{dt^2} + m\omega_0^2\vec{x} = q\vec{E}_0 \cos\omega t \quad (12.36)$$

This differential equation describes driven, simple harmonic motion (refer to Section 2.3.3). We can solve this equation for $\vec{x}$ and use the solution to determine an expression for the refractive index of the material (Eq. 12.37 and Derivation 12.4).

The refractive index is a function of a wave's angular frequency, ω. Therefore, since $v = c/n$, the velocity of a wave through a material is a function of its frequency. The phenomenon of different frequencies traveling at different speeds through a material is called *dispersion*. Therefore, the different frequencies that make up a non-monochromatic wave will separate (or *disperse*) as they propagate through a dispersive material. For example, white light propagating through a triangular glass prism separates into its different color components and produces a rainbow.

Dispersion relation (Derivation 12.4):

(refractive index)

$$n(\omega) \simeq 1 + \frac{n_q q^2}{2m\varepsilon_0\left(\omega_0^2 - \omega^2\right)} \qquad (12.37)$$

12.3.4 Characteristic impedance

The *characteristic impedance* of a material, $Z\,[\Omega]$, is defined as the ratio of $|\vec{E}|$ to $|\vec{H}|$. Using $|\vec{E}| = v|\vec{B}|$ and $|\vec{B}| = \mu|\vec{H}|$, we have:

$$\begin{aligned} Z &= \frac{|\vec{E}|}{|\vec{H}|} \\ &= \mu v \\ &= \sqrt{\frac{\mu}{\varepsilon}} \end{aligned}$$

where $v = 1/\sqrt{\varepsilon\mu}$. Comparing this to Eq. 12.20 for the impedance of free space, Z_0, we see that the characteristic impedance of a material is related to Z_0 by replacing ε_0 for ε and μ_0 for μ, as expected (Eq. 12.38).

Characteristic impedance:

$$\begin{aligned} Z &= \mu v \\ &= \sqrt{\frac{\mu}{\varepsilon}} \\ &= \sqrt{\frac{\mu_r}{\varepsilon_r}} Z_0 \end{aligned} \qquad (12.38)$$

12.3.5 Energy density and Poynting vector

The energy density, u_{EM}, and Poynting vector, $\vec{S}$, of electromagnetic waves in non-conducting media are similar to their free-space counterpart, with ε_0 replaced by ε and μ_0 by μ (Eqs. 12.39 and 12.40, respectively).

Energy density and Poynting vector (refer to Derivation 12.3 for expression for $\vec{S}$):

Energy density:

$$\begin{aligned} u_{EM} &= \frac{1}{2}\varepsilon\left|\vec{E}\right|^2 + \frac{1}{2\mu}\left|\vec{B}\right|^2 \\ &= \frac{1}{2}\vec{E}\cdot\vec{D} + \frac{1}{2}\vec{B}\cdot\vec{H} \end{aligned} \qquad (12.39)$$

Poynting vector:

$$\begin{aligned} \vec{S} &= \frac{1}{\mu}\vec{E}\times\vec{B} \\ &= \vec{E}\times\vec{H} \end{aligned} \qquad (12.40)$$

where $\vec{D} = \varepsilon\vec{E}$ and $\vec{H} = \vec{B}/\mu$ for a HILS material.

Energy dissipation in conducting matter

In a HILS material, where $\vec{B} = \mu\vec{H}$, the Poynting vector is:

$$\vec{S} = \vec{E}\times\vec{H}$$

Although this result was derived for non-conducting matter, it also holds for *conducting* matter, i.e. for $\vec{J}_f \neq 0$. However, in the latter case, the Poynting vector takes into account power dissipation (which is zero if $\vec{J}_f = 0$) as well as the rate of change of electric and magnetic energy. This is most clearly seen by performing Derivation 12.3 "in reverse."

Using vector identity $\nabla\cdot(\vec{F}\times\vec{G}) = \vec{G}\cdot(\nabla\times\vec{F}) - \vec{F}\cdot(\nabla\times\vec{G})$, the divergence of the Poynting vector is:

$$\begin{aligned} \nabla\cdot\vec{S} &= \nabla\cdot\left(\vec{E}\times\vec{H}\right) \\ &= \vec{H}\cdot\left(\nabla\times\vec{E}\right) - \vec{E}\cdot\left(\nabla\times\vec{H}\right) \end{aligned}$$

Eliminating the curl terms using Faraday's and Ampère's Laws in the case where $\vec{J}_f \neq 0$ (refer to Eqs. 11.30 and 11.31), and rearranging, we find:

Derivation 12.4: Derivation of the dispersion relation (Eq. 12.37)

$$m\frac{d^2\vec{x}}{dt^2}+m\omega_0^2\vec{x}=q\vec{E}_0\cos\omega t$$

Try $\vec{x}=\vec{x}_0\cos\omega t$:

$$-m\omega^2\vec{x}_0\cos\omega t+m\omega_0^2\vec{x}_0\cos\omega t=q\vec{E}_0\cos\omega t$$

$$\therefore\ \vec{x}_0=\frac{q\vec{E}_0}{m(\omega_0^2-\omega^2)}$$

$$\vec{P}_0=n_q q\vec{x}_0=\frac{n_q q^2\vec{E}_0}{m\left(\omega_0^2-\omega^2\right)}$$

$$\text{and}\ \vec{P}_0=\chi_e\varepsilon_0\vec{E}_0=(\varepsilon_r-1)\varepsilon_0\vec{E}_0$$

$$\therefore\ \varepsilon_r=1+\frac{n_q q^2}{m\varepsilon_0\left(\omega_0^2-\omega^2\right)}$$

$$n\simeq\sqrt{\varepsilon_r}=\sqrt{1+\frac{n_q q^2}{m\varepsilon_0\left(\omega_0^2-\omega^2\right)}}\simeq 1+\frac{n_q q^2}{2m\varepsilon_0\left(\omega_0^2-\omega^2\right)}$$

- Starting point: the equation of motion of a charge q of mass m oscillating in the electric field of an electromagnetic wave of frequency ω is:

$$m\frac{d^2\vec{x}}{dt^2}+m\omega_0^2\vec{x}=q\vec{E}_0\cos\omega t$$

(refer to text).
- Try an oscillatory solution, $\vec{x}=\vec{x}_0\cos\omega t$, and rearrange to find an expression for the displacement amplitude, $\vec{x}_0$.
- The polarization amplitude is $\vec{P}_0=n_q\vec{p}_0=n_q q\vec{x}_0$, where n_q is the charge density, and $\vec{p}_0=q\vec{x}_0$ is the electric dipole amplitude.
- The polarization and electric susceptibility for a HILS material are defined as:

$$\vec{P}=\chi_e\varepsilon_0\vec{E}$$
$$\text{and}\ \chi_e=\varepsilon_r-1$$

Use these, along with the expression for $\vec{P}_0$ above, to obtain an expression for the relative permittivity, ε_r.
- Refractive index is $n\simeq\sqrt{\varepsilon_r}$ (Eq. 12.35). Use a binomial expansion to approximate the expression for n.

$$\nabla\cdot\vec{S}=-\mu\vec{H}\cdot\left(\frac{\partial\vec{H}}{\partial t}\right)-\vec{E}\cdot\left(\vec{J}_f+\varepsilon\frac{\partial\vec{E}}{\partial t}\right)=-\frac{\partial}{\partial t}\underbrace{\left(\frac{\varepsilon}{2}|\vec{E}|^2+\frac{\mu}{2}|\vec{H}|^2\right)}_{\text{energy density}}-\underbrace{\vec{E}\cdot\vec{J}_f}_{\substack{\text{dissipated}\\ \text{power}}}$$

We recognize the first term as the rate of change of energy density—or the power density—of the wave. The second term equals the dissipated power density:[4]

$$P=VI\sim\int\vec{E}\cdot d\vec{l}\int\vec{J}_f\cdot d\vec{s}\sim\int\vec{E}\cdot\vec{J}_f\,d\tau$$

[4] Strictly speaking, it is non-trivial to combine the integrals for I and V, but at least we get a sense that $\vec{E}\cdot\vec{J}$ represents power density.

The dissipated power manifests itself as heat, which is why electrical circuits heat up when current flows through them. Therefore, the Poynting vector is a statement of the conservation of energy: it equals the sum of power carried in the electric and magnetic components of a wave plus power losses from the wave to its surroundings. However, in non-conducting materials, such as free space or perfect dielectrics, $\vec{J}_f=0$; therefore, there is no loss of power from a propagating wave.

12.4 Reflection and transmission at a boundary

A wave that is incident on a boundary between two non-conducting media is partially reflected back into the medium through which the incident wave is traveling, and partially transmitted through the second medium. Below, we will look at the properties of reflected and transmitted waves. Throughout this section, we will consider only non-magnetic materials, therefore $\mu_r=1$.

12.4.1 Laws of reflection and refraction

Consider a plane wave incident on a boundary between two non-conducting media, as shown in Fig. 12.3. The refractive indices of the two media are n_1 and n_2. The wavevectors of the incident, reflected, and transmitted waves are $\vec{k}_i$, $\vec{k}_r$, and $\vec{k}_t$, respectively, and they make angles θ_i, θ_r, and θ_t with the normal to the boundary.

The waves are described by plane-wave solutions with electric-field components of the form:

$$\vec{E}_i = \vec{E}_{i_0} e^{(\vec{k}_i \cdot \vec{r} - \omega_i t)} \tag{12.41}$$

$$\vec{E}_r = \vec{E}_{r_0} e^{(\vec{k}_r \cdot \vec{r} - \omega_r t)} \tag{12.42}$$

$$\vec{E}_t = \vec{E}_{t_0} e^{(\vec{k}_t \cdot \vec{r} - \omega_t t)} \tag{12.43}$$

(and the same for $\vec{B}$). The incident electric-field component of the wave is the sum of those of the reflected and transmitted waves:

$$\vec{E}_{i_0} e^{(\vec{k}_i \cdot \vec{r} - \omega_i t)} = \vec{E}_{r_0} e^{(\vec{k}_r \cdot \vec{r} - \omega_r t)} + \vec{E}_{t_0} e^{(\vec{k}_t \cdot \vec{r} - \omega_t t)}$$

Since this equation must be valid for all times, and for all points along the plane of the interface, the temporal and spatial parts of the exponents, which represent the temporal and spatial phase of the wave, must independently be equal:

$$\therefore \omega_i t = \omega_r t = \omega_t t$$
$$\text{and } \vec{k}_i \cdot \vec{r} = \vec{k}_r \cdot \vec{r} = \vec{k}_t \cdot \vec{r}$$

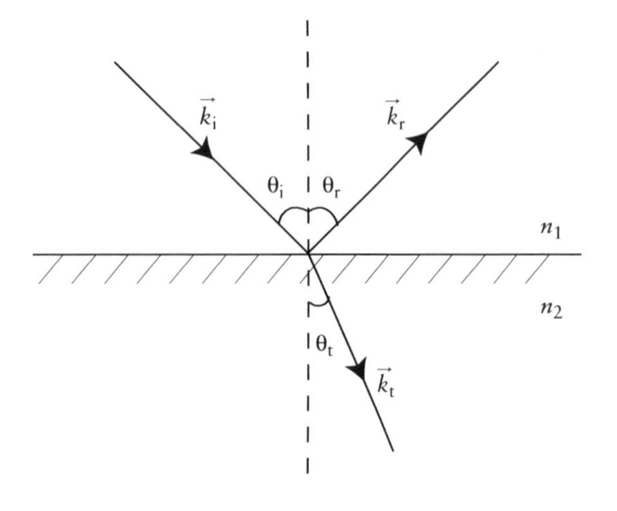

Figure 12.3: Reflection and transmission of a wave incident on a boundary between two non-conducting media.

From these expressions, we find the following wave properties:
(i) The frequencies of all three waves are equal:

$$\omega_i = \omega_r = \omega_t \equiv \omega$$

(ii) The law of reflection is obeyed:

$$\vec{k}_i \cdot \vec{r} = \vec{k}_r \cdot \vec{r}$$
$$\therefore |\vec{k}_i||\vec{r}|\cos(90 - \theta_i) = |\vec{k}_r||\vec{r}|\cos(90 - \theta_r)$$
$$\therefore |\vec{k}_i|\sin\theta_i = |\vec{k}_r|\sin\theta_r$$
$$\frac{n_1\omega}{c}\sin\theta_i = \frac{n_1\omega}{c}\sin\theta_r$$
$$\Rightarrow \therefore \theta_i = \theta_r$$

(iii) Snell's Law of refraction is obeyed:

$$\vec{k}_i \cdot \vec{r} = \vec{k}_t \cdot \vec{r}$$
$$\therefore |\vec{k}_i||\vec{r}|\cos(90 - \theta_i) = |\vec{k}_t||\vec{r}|\cos(90 - \theta_t)$$
$$\therefore |\vec{k}_i|\sin\theta_i = |\vec{k}_t|\sin\theta_t$$
$$\frac{n_1\omega}{c}\sin\theta_i = \frac{n_2\omega}{c}\sin\theta_t$$
$$\Rightarrow \therefore n_1\sin\theta_i = n_2\sin\theta_t$$

(show (ii) and (iii) using (i) and $v = c/n = \omega/|\vec{k}|$).

12.4.2 The Fresnel equations

If we let the amplitude of the incident wave be $|\vec{E}_i|$, then we can express the amplitudes of the reflected and transmitted waves as fractions of $|\vec{E}_i|$:

$$|\vec{E}_r| = r|\vec{E}_i|$$
$$\text{and } |\vec{E}_t| = t|\vec{E}_i|$$

where r and t are called the *amplitude coefficients* of the reflected and transmitted waves, respectively. The *Fresnel equations* are a set of equations that express r and t in terms of the refractive indices of the two media, n_1 and n_2, and the wave angles θ_i, θ_r, and θ_t.
To find expressions for r and t, we usually decompose the incident wave into two components: the components of $\vec{E}$ parallel and perpendicular to the plane of incidence. For example, if we define the plane of incidence as the plane of the paper (the plane in which the $\vec{k}$ vectors lie in Fig. 12.3), then the perpendicular component of $\vec{E}$ points out of or into the page, and the parallel component points along the page. We therefore define two sets of

amplitude coefficients: a set for $\vec{E}$ perpendicular to the plane of incidence, $r_\perp$ and $t_\perp$ (Eqs. 12.44 and 12.45, Derivation 12.5), and a set for $\vec{E}$ parallel to the plane of incidence, $r_\parallel$ and $t_\parallel$ (Eqs. 12.46 and 12.47, Derivation 12.5). Setting $\theta_i = 0°$ and $\theta_t = 0°$ in these equations gives the reflection and transmission amplitude coefficients for normal incidence (Eqs. 12.48 and 12.49).

Fresnel equations (Derivation 12.5):

$\vec{E}$ perpendicular to plane of incidence:

$$r_\perp = \frac{n_1 \cos\theta_i - n_2 \cos\theta_t}{n_1 \cos\theta_i + n_2 \cos\theta_t} \quad (12.44)$$

$$t_\perp = \frac{2n_1 \cos\theta_i}{n_1 \cos\theta_i + n_2 \cos\theta_t} \quad (12.45)$$

$\vec{E}$ parallel to plane of incidence:

$$r_\parallel = \frac{n_1 \cos\theta_t - n_2 \cos\theta_i}{n_1 \cos\theta_t + n_2 \cos\theta_i} \quad (12.46)$$

$$t_\parallel = \frac{2n_1 \cos\theta_i}{n_1 \cos\theta_t + n_2 \cos\theta_i} \quad (12.47)$$

Normal incidence ($\theta_i = \theta_t = 0°$):

$$r_n = \frac{n_1 - n_2}{n_1 + n_2} \quad (12.48)$$

$$t_n = \frac{2n_1}{n_1 + n_2} \quad (12.49)$$

The Fresnel equations give the *amplitude* coefficients of reflected and transmitted radiation; the reflected or transmitted *power* coefficients (reflectance, R, and transmittance, T, respectively) are related to the square of the amplitude coefficients:

$$R_\perp = r_\perp^2 \text{ and } R_\parallel = r_\parallel^2$$
$$T_\perp = 1 - R_\perp \text{ and } T_\parallel = 1 - R_\parallel$$

12.4.3 The Brewster angle

For the component of $\vec{E}$ parallel to the plane of incidence, there is an angle of incidence at which there is no reflected wave: $r_\parallel = 0$. The angle of incidence at which this occurs is called the *Brewster angle*, θ_B (Eq. 12.54, Derivation 12.6, and Fig. 12.4).

The Brewster angle (Derivation 12.6):

$$\tan\theta_B = \frac{n_2}{n_1} \quad (12.54)$$

Fig. 12.4 illustrates that the reflected ray of an incident ray at the Brewster angle is polarized along the plane of the boundary. This is one method of producing polarized light. For example, the Brewster angle for water is $\theta_B \simeq \tan^{-1}(1.33) \simeq 53°$, and for glass is $\theta_B \simeq \tan^{-1}(1.5) \simeq 56°$. Polarized sunglasses are designed to block the horizontal $\vec{E}$ component of light, which strongly reduces the glare from reflections off water and other highly reflective surfaces when the sun is low in the sky.

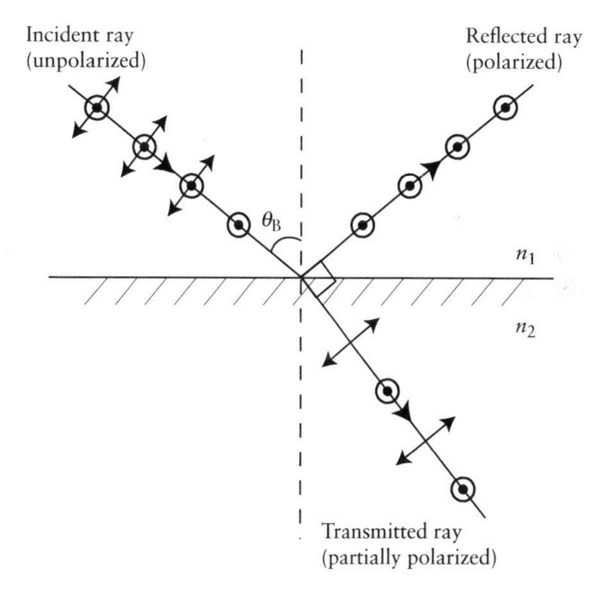

Figure 12.4: The Brewster angle. For unpolarized light incident on a surface at the Brewster angle, the electric-field component in the plane of incidence (the plane of the paper) is not reflected. This means that the reflected light is polarized parallel to the plane of the boundary interface (out of the paper). The transmitted ray is partially polarized.

Derivation 12.5: Derivation of the Fresnel equations (Eqs. 12.44–12.47)

$\vec{E}$ perpendicular to plane of incidence:

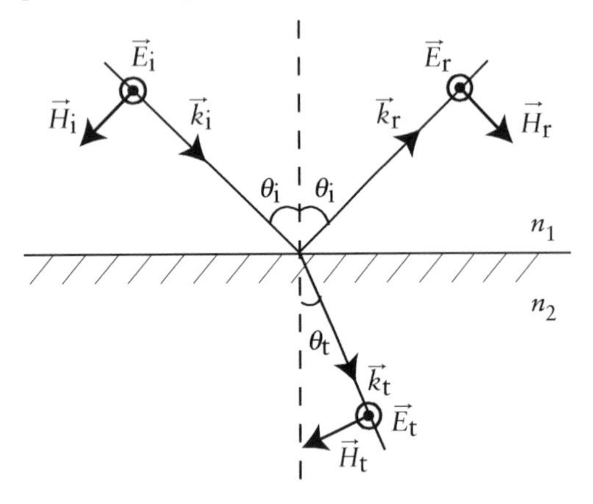

$\vec{E}$ parallel to plane of incidence:

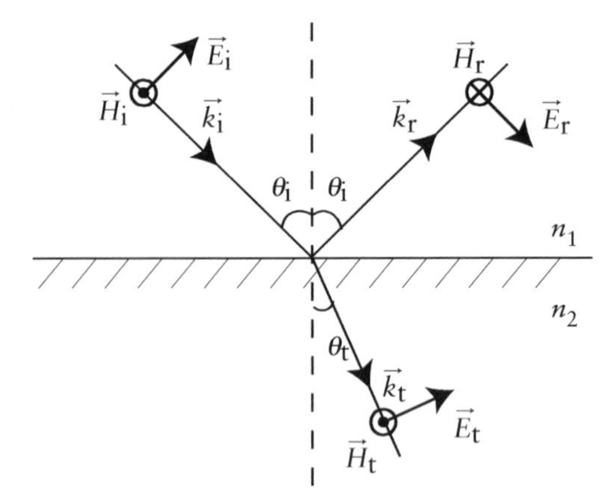

Boundary conditions (refer to Section 11.3.1):

$$\left|\vec{E}\right|_1^{\parallel} = \left|\vec{E}\right|_2^{\parallel}$$

and $\left|\vec{H}\right|_1^{\parallel} = \left|\vec{H}\right|_2^{\parallel}$

$|\vec{E}|$ $\perp$ to plane of incidence:

$$r_{\perp} = \frac{\left|\vec{E}_r\right|}{\left|\vec{E}_i\right|} = \frac{n_1 \cos\theta_i - n_2 \cos\theta_t}{n_1 \cos\theta_i + n_2 \cos\theta_t}$$

$$t_{\perp} = \frac{\left|\vec{E}_t\right|}{\left|\vec{E}_i\right|} = \frac{2n_1 \cos\theta_i}{n_1 \cos\theta_i + n_2 \cos\theta_t}$$

$|\vec{E}|_{\parallel}$ to plane of incidence:

$$r_{\parallel} = \frac{\left|\vec{E}_r\right|}{\left|\vec{E}_i\right|} = \frac{n_1 \cos\theta_t - n_2 \cos\theta_i}{n_1 \cos\theta_t + n_2 \cos\theta_i}$$

$$t_{\parallel} = \frac{\left|\vec{E}_t\right|}{\left|\vec{E}_i\right|} = \frac{2n_1 \cos\theta_i}{n_1 \cos\theta_t + n_2 \cos\theta_i}$$

- Starting point: let the amplitudes of the incident, reflected, and transmitted electric field be $|\vec{E}_i|$, $|\vec{E}_r|$, and $|\vec{E}_t|$, respectively, and of the auxiliary magnetic field be $|\vec{H}_i|, |\vec{H}_r|$, and $|\vec{H}_t|$.
- $|\vec{E}|$ and $|\vec{H}|$ are related to one another by:

$$\left|\vec{E}\right| = v\left|\vec{B}\right| = \frac{c\mu_r\mu_0}{n}\left|\vec{H}\right| = \frac{c\mu_0}{n}\left|\vec{H}\right|$$

where $\mu_r = 1$ for non-conducting media.

- $|\vec{E}|$ $\perp$ to the plane of incidence: from the boundary conditions and the diagram, the wave amplitudes are related by:

$$\left|\vec{E}_i\right| + \left|\vec{E}_r\right| = \left|\vec{E}_t\right|$$

$$\text{and } \left|\vec{H}_i\right|\cos\theta_i - \left|\vec{H}_r\right|\cos\theta_i = \left|\vec{H}_t\right|\cos\theta_t \quad (12.50)$$

$$\therefore n_1\left|\vec{E}_i\right|\cos\theta_i - n_1\left|\vec{E}_r\right|\cos\theta_i = n_2\left|\vec{E}_t\right|\cos\theta_t \quad (12.51)$$

Eliminate $|\vec{E}_t|$ from Eq. 12.51 using Eq. 12.50 and rearrange to find $r_{\perp}$; eliminate $|\vec{E}_r|$ from Eq. 12.51 using Eq. 12.50 and rearrange to find $t_{\perp}$.

- $|\vec{E}|_{\parallel}$ to the plane of incidence: using the boundary conditions and the diagram, the wave amplitudes are related by:

$$\left|\vec{E}_i\right|\cos\theta_i + \left|\vec{E}_r\right|\cos\theta_i = \left|\vec{E}_t\right|\cos\theta_t \quad (12.52)$$

$$\text{and } \left|\vec{H}_i\right| - \left|\vec{H}_r\right| = \left|\vec{H}_t\right|$$

$$\therefore n_1\left|\vec{E}_i\right| - n_1\left|\vec{E}_r\right| = n_2\left|\vec{E}_t\right| \quad (12.53)$$

Eliminate $|\vec{E}_t|$ from Eq. 12.52 using Eq. 12.53 and rearrange to find $r_{\parallel}$; eliminate $|\vec{E}_r|$ from Eq. 12.52 using Eq. 12.53 and rearrange to find $t_{\parallel}$.

Derivation 12.6: Derivation of the Brewster angle (Eq. 12.54)

- Starting point: the Fresnel equation for $r_{\parallel}$ is:

$$r_{\parallel} = \frac{n_1 \cos\theta_t - n_2 \cos\theta_i}{n_1 \cos\theta_t + n_2 \cos\theta_i}$$

- There is no reflected parallel component of $\vec{E}$ when $r_{\parallel} = 0$:

$$\therefore n_1 \cos\theta_t - n_2 \cos\theta_i = 0$$

- Use Snell's Law ($n_1 \sin\theta_i = n_2 \sin\theta_t$) to eliminate n_2/n_1.
- Use the trigonometric identities:

$$\sin\theta\cos\theta = \frac{1}{2}\sin 2\theta$$

and $\sin 2\theta_1 - \sin 2\theta_2 = \sin(\theta_1 - \theta_2)\cos(\theta_1 + \theta_2)$

to manipulate the expression for zero reflection.
- For zero reflection, we find that either the refractive indices are equal ($\theta_i = \theta_t$), or the sum of incident and transmitted angles is $\theta_i + \theta_t = \pi/2$.
- Use the latter condition in Snell's Law to eliminate θ_t. The angle of incidence at which there is zero reflection is the Brewster angle, θ_B.

$$\cos\theta_t - \frac{n_2}{n_1}\cos\theta_i = 0$$

$$\sin\theta_t \cos\theta_t - \sin\theta_i \cos\theta_i = 0$$

$$\frac{1}{2}\sin 2\theta_t - \frac{1}{2}\sin 2\theta_i = 0$$

$$\sin(\theta_t - \theta_i)\cos(\theta_t + \theta_i) = 0$$

For this to be true:

either $\theta_t - \theta_i = 0$

or $\theta_t + \theta_i = \dfrac{\pi}{2}$

$\therefore$ Using the latter in Snell's Law (let $\theta_i = \theta_B$):

$$n_1 \sin\theta_B = n_2 \sin\theta_t$$

$$\therefore n_1 \sin\theta_B = n_2 \sin\left(\frac{\pi}{2} - \theta_B\right) = n_2 \cos\theta_B$$

$$\therefore \tan\theta_B = \frac{n_2}{n_1}$$

Part III
Waves and Optics

Introduction

Optics is the study of the behavior and properties of light, and its interaction with matter. Since light is an electromagnetic wave, optics is usually taught alongside or after a course on *wave motion* (Chapter 13). When light interacts with matter on scales much larger than its wavelength (>>10 μm), the wave nature of light is not apparent. The study of optics on these scales is called *geometrical optics* (Chapter 14). In contrast, when light interacts with matter on scales on the order of its wavelength (~10 μm), then we encounter wave phenomena such as diffraction and interference. The study of optics on these scales is called *wave optics* (Chapter 15).

Practical applications

Practical applications of optics include optical instruments for imaging, such as telescopes, microscopes, cameras, and optical fibers (geometrical optics); optical instruments for spectroscopy, such as the diffraction grating and the Michelson–Morley and Fabry–Perot interferometers (wave optics); and optometry (the study of eyes and vision).

My two cents

Optics, not surprisingly, is a subject best understood and learnt through visual examples. Of course, we all have a good conceptual grasp of reflection and refraction since we encounter these phenomena daily. However, we are less familiar with diffraction and interference. As a consequence, the simple questions of "what do we see and where do we see it?" are in fact not so simple. The best way around this problem is to do some laboratory practicals on optics—for example using a diffraction grating to determine the wavelength of an unknown source of light. The necessary instruments obviously have to be provided by the university, but hopefully you can get your hands on them; I found that a few experiments went a long way for my understanding of optical wave phenomena.

Recommended books

- Born, M. and Wolf, E., 1999, *Principles of Optics: Electromagnetic Theory of Propagation, Interference and Diffraction of Light*, 7th edition. Cambridge University Press.
- Hecht, E., 2002, *Optics*, 4th edition. Addison Wesley.
- Hecht, E., 1974, *Schaum's Outline of Optics*. McGraw-Hill.

I personally had trouble finding an optics book that I really liked. I used *Optics* by Hecht for my undergraduate work, and while it's a thorough and clearly written book (with a lot of good diagrams), it gave more detail than the level of an introductory course, and as a consequence was not always useful. *Schaum's Outline of Optics*, by the same author, provides a lot of worked examples, at a more appropriate level for an introductory course. I have never used the book by Born and Wolf, but have been told it is a good book—also very thorough.

Chapter 13
Wave motion

13.1 Types of wave

In its broadest terms, a wave is defined as a disturbance that propagates through space and time, at a velocity v [$\mathrm{ms^{-1}}$] Waves transfer energy as they propagate, but do not transfer mass. In classical physics,[1] waves are classified as either *mechanical waves* (e.g. sound waves, water waves, seismic waves, waves on a string) or *electromagnetic waves* (e.g. radio waves, light, gamma rays).

Mechanical waves

Mechanical waves require a material medium for propagation, such as air, water, or wood. These waves are transmitted through the medium by vibrating particles, but the particles themselves do not propagate. Mechanical waves are classified as either *transverse* or *longitudinal* waves, depending on the direction of particle vibration relative to the direction of wave propagation. In transverse waves, such as waves on a string, particle displacement is perpendicular to the direction of wave propagation; in longitudinal waves, such as sound waves, particle displacement is parallel to the direction of propagation (Fig. 13.1). Some waves involve a combination of both longitudinal and transverse motion. For example, surface water waves involve circular motion of water particles and surface seismic waves (called Rayleigh waves) involve elliptical particle motion.

Electromagnetic waves

In contrast to mechanical waves, electromagnetic waves can travel through either a material or a vacuum. Electromagnetic waves are propagating electric and magnetic fields, and their properties are encapsulated in Maxwell's equations (refer to Chapter 12). Electromagnetic waves are transverse waves; the electric and magnetic field components are perpendicular to one another, and to the direction of wave propagation (Fig. 13.2).

The continuous spectrum of electromagnetic waves of different wavelength and frequency is called the *electromagnetic spectrum* (Fig. 13.3). At the long-wavelength end of the spectrum (wavelengths on the order of centimeters to several meters) are radio waves, TV waves, and microwaves; at the short-wavelength end of the spectrum (wavelengths on the order of atomic dimensions) are X-rays and gamma rays. In between these two extremes is the visible spectrum (wavelengths on the order of nanometers), with infrared waves just beyond the red end of the visible spectrum and ultraviolet waves just beyond the violet end of the spectrum. All electromagnetic waves travel at the same speed in a vacuum, $c \simeq 3 \cdot 10^8\,\mathrm{m\,s^{-1}}$. The wavelength, λ[m], and frequency, f [Hz], of an electromagnetic wave in a vacuum are related to one another by $\lambda = c/f$.

13.2 The wave equation

The *wave equation* is a second-order partial differential equation that describes how a wave propagates in space and time. A function $f(x, t)$ represents a wave traveling at velocity v if it satisfies the wave equation:

$$\frac{\partial^2 f}{\partial x^2} = \frac{1}{v^2}\frac{\partial^2 f}{\partial t^2}$$

(Eq. 13.1 and Derivation 13.1). This equation holds for both mechanical waves (transverse and longitudinal) and electromagnetic waves. For mechanical waves, it is essentially a statement of Newton's Second Law—it is derived by a consideration of the forces acting on a segment of the medium though which the wave is propagating, as shown in Derivation 13.1. For electromagnetic waves, it is derived from Maxwell's equations (see Chapter 12).

[1] In quantum physics, we refer to *matter waves*, which are neither mechanical nor electromagnetic waves. These represent the wave associated with a quantum particle. See Chapter 16 for more detail.

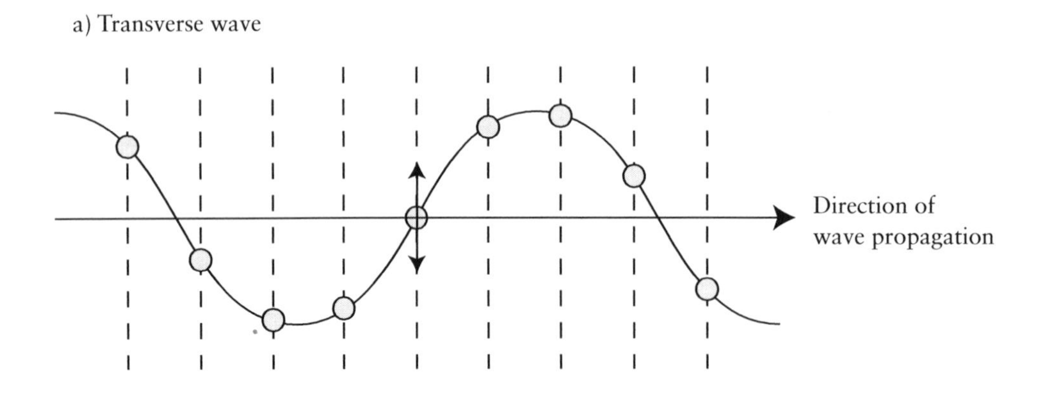

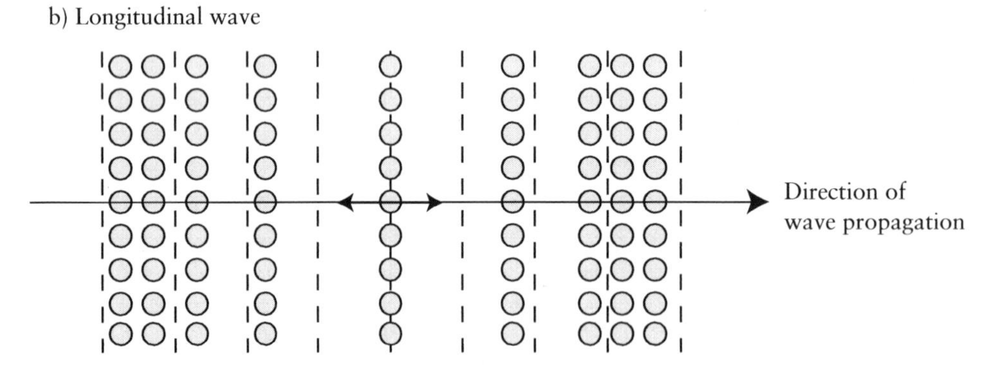

Figure 13.1: a) A transverse wave and b) a longitudinal wave. Particle displacement is perpendicular to the direction of wave propagation in a transverse wave and parallel to the direction of propagation in a longitudinal wave.

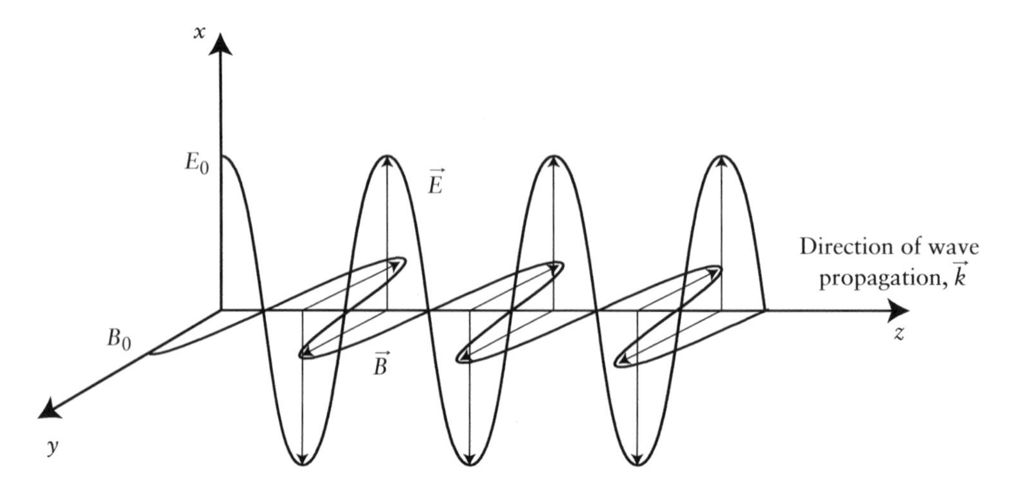

Figure 13.2: An electromagnetic wave. The electric field, $\vec{E}$, and magnetic field, $\vec{B}$, oscillate perpendicular to one another, and both components are perpendicular to the direction of wave propagation.

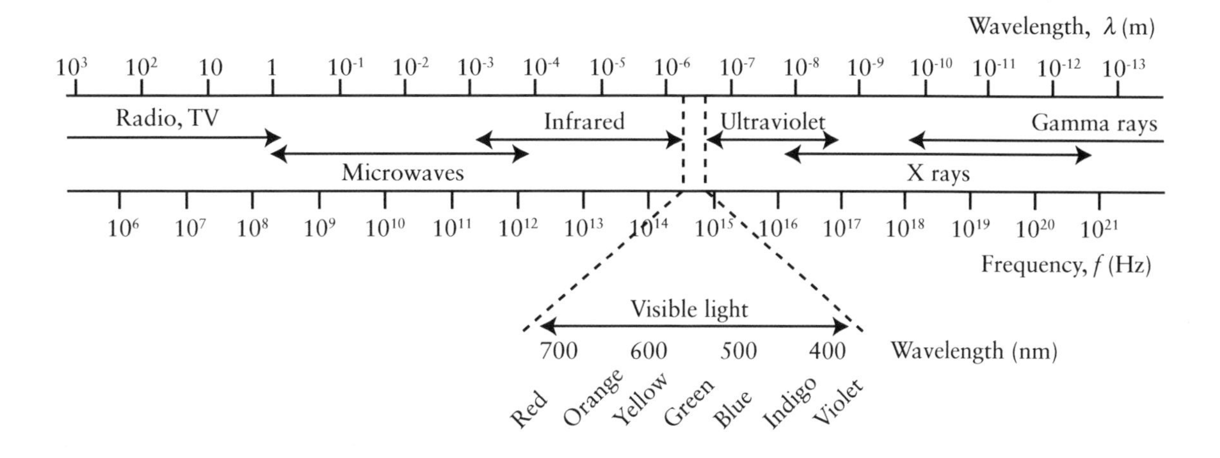

Figure 13.3: The electromagnetic spectrum.

Wave equation (Derivation 13.1):

$$\frac{\partial^2 f}{\partial x^2} = \frac{1}{v^2}\frac{\partial^2 f}{\partial t^2} \qquad (13.1)$$

where the general solution is $f = f(x \pm vt)$.

General solution of the wave equation

It's straightforward to show that any function of the form $f(x - vt)$ or $f(x + vt)$ satisfies the wave equation. If we let $y = x \pm vt$, then:

$$\frac{\partial f}{\partial x} = \frac{\partial f}{\partial y}\frac{\partial y}{\partial x} = \frac{\partial f}{\partial y}$$

$$\therefore \frac{\partial^2 f}{\partial x^2} = \frac{\partial}{\partial x}\left(\frac{\partial f}{\partial y}\right) = \frac{\partial}{\partial y}\left(\frac{\partial f}{\partial x}\right) = \frac{\partial^2 f}{\partial y^2}$$

and $$\frac{\partial f}{\partial t} = \frac{\partial f}{\partial y}\frac{\partial y}{\partial t} = \pm v\frac{\partial f}{\partial y}$$

$$\therefore \frac{\partial^2 f}{\partial t^2} = \pm v\frac{\partial}{\partial t}\left(\frac{\partial f}{\partial y}\right) = \pm v\frac{\partial}{\partial y}\left(\frac{\partial f}{\partial t}\right) = v^2\frac{\partial^2 f}{\partial y^2}$$

$$\Rightarrow \therefore \frac{\partial^2 f}{\partial x^2} = \frac{1}{v^2}\frac{\partial^2 f}{\partial t^2}$$

Therefore, any function where the space and time variables are related by $x \pm vt$ represents a wave.

We can understand this statement conceptually if we consider an arbitrary function of space and time, say $g(x, t)$, at two points in time, say time 0 and t. At a time $t = 0$, the function $g(x, 0)$ describes the shape of the wave at that snapshot in time. A wave traveling at constant velocity v propagates a distance $\Delta x = vt$ in a time t. For a wave of constant shape, the function at a time t looks like the function at $t = 0$ if:

$$g(x - vt, t) = g(x, 0)$$

(refer to Fig. 13.4). Since this holds true at all times, *any* function that depends on x and t in the combination $x \pm vt$ represents a wave traveling at velocity v. A function $f(x - vt)$ represents a wave traveling in the $+x$-direction, and a function $f(x + vt)$ represents a wave traveling in the $-x$-direction.

Derivation 13.1: Derivation of the wave equation for transverse waves on a string (Eq. 13.1)

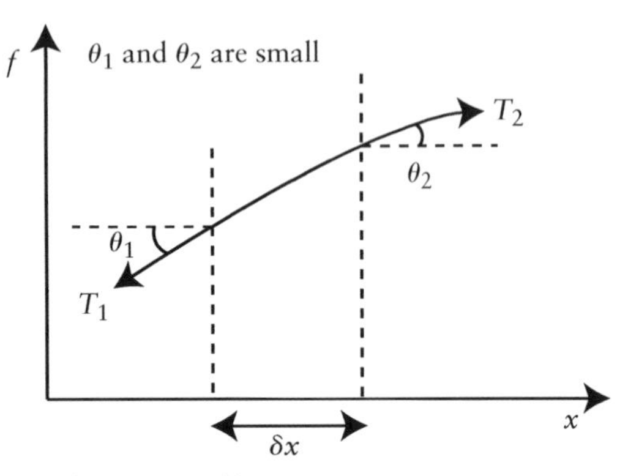

Resolving vertically:

$$F = T\sin\theta_2 - T\sin\theta_1$$
$$\simeq T\tan\theta_2 - T\tan\theta_1$$
$$= T\left(\left(\frac{\partial f}{\partial x}\right)_2 - \left(\frac{\partial f}{\partial x}\right)_1\right)$$
$$\simeq T\left(\left(\frac{\partial f}{\partial x}\right)_1 + \frac{\partial}{\partial x}\left(\frac{\partial f}{\partial x}\right)\delta x - \left(\frac{\partial f}{\partial x}\right)_1\right)$$
$$= T\left(\frac{\partial^2 f}{\partial x^2}\right)\delta x$$

From Newton's Second Law:

$$F = ma$$
$$= \rho\delta x\left(\frac{\partial^2 f}{\partial t^2}\right)$$

Therefore, equating forces:

$$T\left(\frac{\partial^2 f}{\partial x^2}\right)\delta x = \rho\delta x\left(\frac{\partial^2 f}{\partial t^2}\right)$$
$$\therefore \left(\frac{\partial^2 f}{\partial x^2}\right) = \underbrace{\frac{\rho}{T}}_{=1/v^2}\left(\frac{\partial^2 f}{\partial t^2}\right)$$
$$\left(\frac{\partial^2 f}{\partial x^2}\right) = \frac{1}{v^2}\left(\frac{\partial^2 f}{\partial t^2}\right)$$

- Starting point: a string under tension, T, supports transverse waves. Consider a segment of string of length δx with tension T_1 and T_2 acting either side (refer to diagram).
- Resolving horizontally, the lateral force on the segment of string is $T_1\cos\theta_1 = T_2\cos\theta_2$. For small angles, $\cos\theta \simeq 1$:

$$\therefore T_1 \simeq T_2 = T$$

- Resolving vertically, the net force on the string is $F = T\sin\theta_2 - T\sin\theta_1$. In the small-angle approximation, $\sin\theta \simeq \tan\theta = (\partial f/\partial x)$.

- For a small segment of string, δx, we can approximate $(\partial f/\partial x)_2$ by a Taylor expansion about $(\partial f/\partial x)_1$:

$$\therefore \left(\frac{\partial f}{\partial x}\right)_2 \simeq \left(\frac{\partial f}{\partial x}\right)_1 + \frac{\partial}{\partial x}\left(\frac{\partial f}{\partial x}\right)\delta x$$

- From Newton's Second Law, the net force on a segment of string is $F = ma = (\rho\delta x)a$, where ρ is the mass per unit length of the string and a is the acceleration of the string.
- Equating forces, we see that the vertical displacement of the string is governed by a second-order partial differential equation:

$$\left(\frac{\partial^2 f}{\partial x^2}\right) = \frac{\rho}{T}\left(\frac{\partial^2 f}{\partial t^2}\right)$$

This has the standard form of a wave equation that describes the propagation of a wave traveling at velocity:

$$v = \sqrt{\frac{T}{\rho}}$$

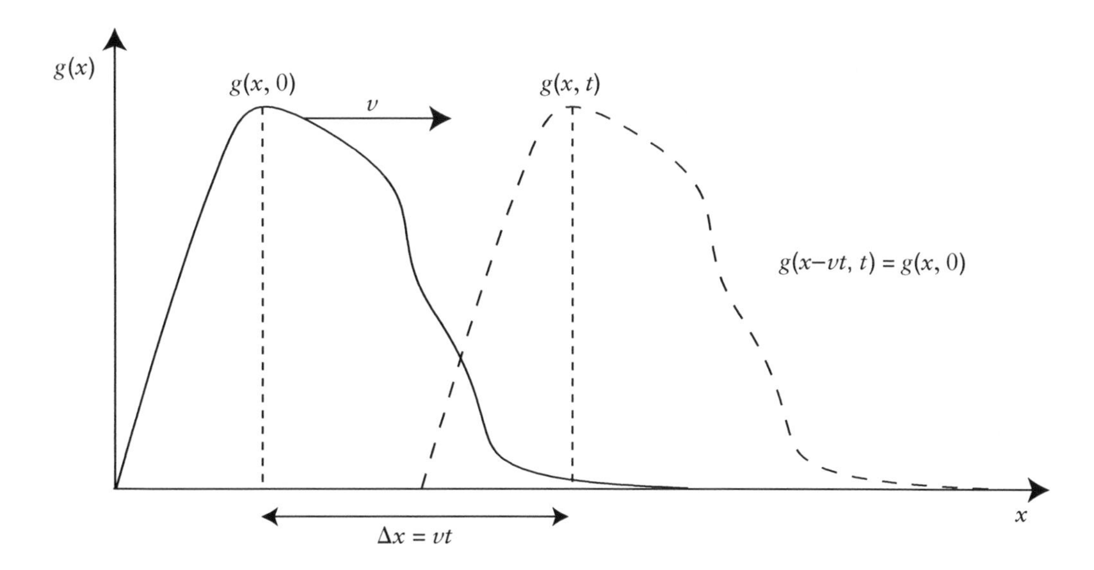

Figure 13.4: Graphical representation of a wave. Any function of space and time where the variables are related by $x \pm vt$ represents a wave.

13.2.1 Harmonic waves

It is straightforward to see by direct substitution into Eq. 13.1 that sinusoidal functions of the form:

$$f(x,t) = A\sin(x \pm vt)$$
$$\text{or } g(x,t) = B\cos(x \pm vt)$$

satisfy the wave equation, where A and B are arbitrary constants. These are mathematically the most basic wave, and are referred to as *harmonic* waves or *sinusoidal* waves.

Wavenumber and angular frequency

If we multiply the argument of the sine function by a constant k, we find:

$$\begin{aligned} f(x,t) &= A\sin(k(x \pm vt)) \\ &= A\sin(kx \pm kvt) \\ &\equiv A\sin(kx \pm \omega t) \end{aligned}$$

(Eq. 13.2). This is the standard way of expressing a harmonic wave. The *wavenumber*, k [rad m^{-1}], is inversely proportional to the *wavelength*, λ [m], and is defined as:

$$k = \frac{2\pi}{\lambda}$$

(Eq. 13.3). The product kx therefore has units of radians, and it represents the spatial phase of the wave at a position x: $\phi_x = kx = 2\pi x/\lambda$. The *angular frequency*, ω [rad s^{-1}], is proportional to the *frequency*, f [s^{-1}], and is defined as:

$$\omega = 2\pi f$$

(Eq. 13.4). The product ωt therefore also has units of radians, and it represents the temporal phase of the wave at a time t: $\phi_t = \omega t = 2\pi t/T$, where $T = 1/f$ corresponds to the time of one period (Eq. 13.5).

We saw above that angular frequency is related to the wavenumber by $\omega = kv$. Therefore, the wave velocity is:

$$v = \frac{\omega}{k} = f\lambda$$

(Eq. 13.6). These wave properties are illustrated in Fig. 13.5.

Harmonic waves:

$$f(x,t) = A\sin(\underbrace{kx \pm \omega t + \phi}_{\text{phase}}) \quad (13.2)$$

where ϕ is an arbitrary initial phase, and:

$$\text{Wavenumber: } k = \frac{2\pi}{\lambda} \quad (13.3)$$

$$\text{Angular frequency: } \omega = 2\pi f \quad (13.4)$$

$$\text{Time period: } T = \frac{1}{f} = \frac{2\pi}{\omega} \quad (13.5)$$

$$\text{Velocity: } v = f\lambda = \frac{\omega}{k} \quad (13.6)$$

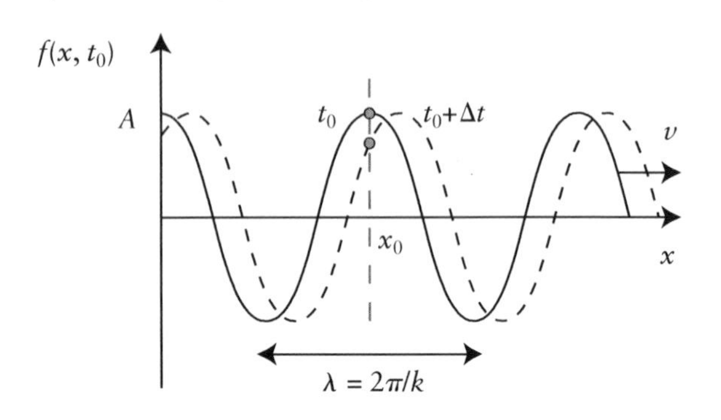

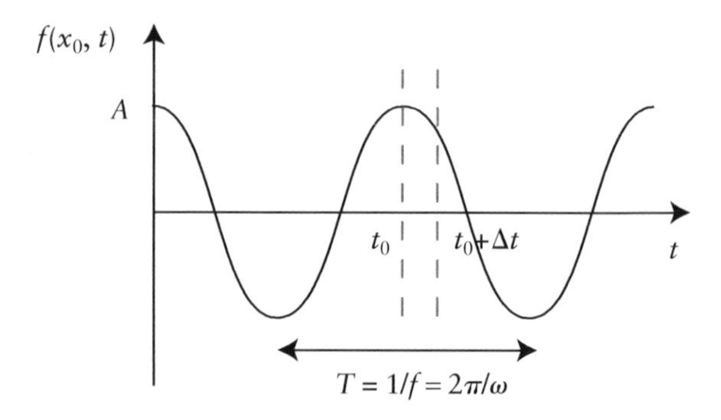

Figure 13.5: Propagation of a harmonic wave. a) Variation of the shape of the wave with distance at a snapshot in time, t_0. A time Δt later, the wave has traveled to the right. b) Variation of the amplitude of oscillation at a point x_0 with time. The displacements at t_0 and $t_0 + \Delta t$ correspond to the two points marked in a).

13.3 Superposition of waves

The wave equation is a so-called linear differential equation. This means that if $f_1(x, t)$ and $f_2(x, t)$ both satisfy the wave equation, then so does their sum, $g(x, t) = f_1(x, t) + f_2(x, t)$:

$$\begin{aligned}\frac{\partial^2 g}{\partial x^2} &= \frac{\partial^2 (f_1 + f_2)}{\partial x^2} \\ &= \frac{\partial^2 f_1}{\partial x^2} + \frac{\partial^2 f_2}{\partial x^2} \\ &= \frac{1}{v^2}\frac{\partial^2 f_1}{\partial t^2} + \frac{1}{v^2}\frac{\partial^2 f_2}{\partial t^2} \\ &= \frac{1}{v^2}\frac{\partial^2 (f_1 + f_2)}{\partial t^2} \\ &= \frac{1}{v^2}\frac{\partial^2 g}{\partial t^2}\end{aligned}$$

Therefore, the net amplitude of two waves arriving at the same point and same time equals the sum of their individual amplitudes. This statement is called the *principle of superposition*, and it applies to many physical systems. When waves sum together to produce a regular new wave pattern, they are said to exhibit *interference*.

13.3.1 Interference of traveling waves

Let's first look at the interference of two waves that have the same amplitude, wavelength, and frequency, traveling in the same direction with a constant phase difference ϕ between them:

$$f_1(x,t) = A\sin(kx - \omega t)$$

$$f_2(x,t) = A\sin(kx - \omega t + \phi)$$

The resultant wave is given by the sum of the individual waves. Using the trigonometric identity for the sum of two sine functions:

$$\sin(x) + \sin(y) = 2\cos\left(\frac{x-y}{2}\right)\sin\left(\frac{x+y}{2}\right)$$

the resultant wave is then:

$$\begin{aligned}f(x,t) &= f_1(x,t) + f_2(x,t) \\ &= A\sin(kx - \omega t) + A\sin(kx - \omega t + \phi) \\ &= \underbrace{2A\cos\left(\frac{\phi}{2}\right)}_{\text{constant new amplitude}}\underbrace{\sin\left(kx - \omega t + \frac{\phi}{2}\right)}_{\text{traveling wave}}\end{aligned}$$

This represents a traveling wave with the same wavelength and frequency as the initial waves, but with a different amplitude. The amplitude depends on the phase, ϕ, between the two waves. If the two waves are in phase with each other (meaning their peaks line up), then $\phi = 0$ and the new amplitude is $2A$ (double the initial amplitudes). In this case, the waves are said to exhibit *constructive interference* (Fig. 13.6a). If the two waves are initially half a wavelength out of phase, then $\phi = \pi$, and the resultant amplitude is zero. This is called *destructive interference*, and the waves cancel each other out (Fig. 13.6b). For intermediate values of ϕ, the amplitude of the resultant wave has a value between zero and $2A$.

a) Constructive interference

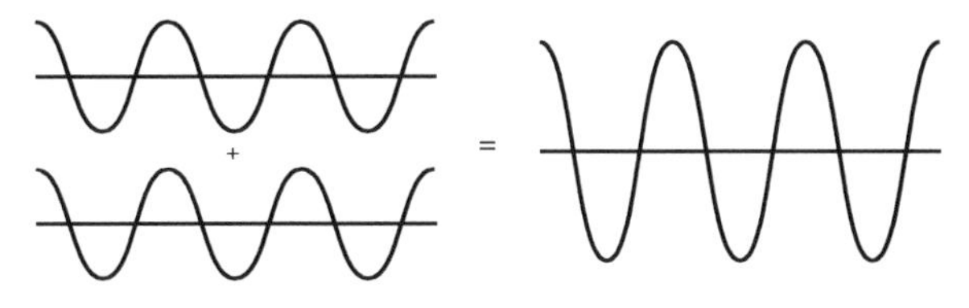

b) Destructive interference

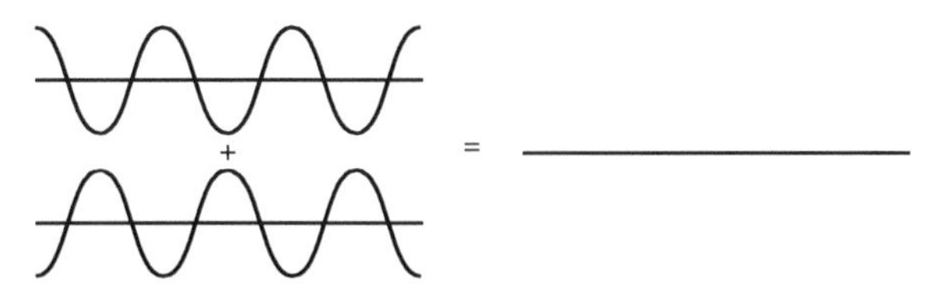

Figure 13.6: a) Constructive interference of waves that are in phase with each other and b) destructive interference of waves that are half a wavelength out of phase with each other.

13.3.2 Standing waves

Next, let's consider the interference of two waves of equal amplitude, wavelength, and frequency, but this time traveling in *opposite* directions:

$$f_1(x,t) = A\sin\left(kx - \omega t\right)$$

$$f_2(x,t) = A\sin\left(kx + \omega t\right)$$

Using the trigonometric identity for the sum of two sine functions, the superposition of these waves is:

$$\begin{aligned} f(x,t) &= f_1(x,t) + f_2(x,t) \\ &= A\sin\left(kx - \omega t\right) + A\sin\left(kx + \omega t\right) \\ &= \underbrace{2A\cos\left(\omega t\right)}_{\text{time-varying amplitude}} \underbrace{\sin\left(kx\right)}_{\text{standing wave}} \end{aligned}$$

This is not a function of $kx - \omega t$, which means it is not a traveling wave. The spatial dependence of the wave is $\sin(kx)$, and this does not vary with time. The amplitude of the wave, however, *does* vary with time, and is given by $2A\cos(\omega t)$. This type of wave is called a *standing* or *stationary* wave, since it does not propagate through space. There are some points of $f(x, t)$ that always have zero amplitude, for example when $kx = 0, \pi, 2\pi$, etc. These points are called *nodes*. In between nodes, the amplitude of the standing wave varies from zero when $\cos(\omega t) = 0$, to a maximum of $2A$ when $\cos(\omega t) = 1$. The points of maximum amplitude, which fall mid-way between the nodes, are called *anti-nodes*. Fig. 13.7 illustrates the change in the amplitude of a standing wave with time.

13.3.3 Beating waves

Now let's add two waves traveling in the same direction, with the same amplitudes, but different wavenumbers, k_1 and k_2, and different frequencies, ω_1 and ω_2. Let the average wavenumber and frequency be k and ω respectively, and each wave have a small offset from the average of either $\pm\delta k$ or $\pm\delta\omega$. Therefore, the two wavenumbers are:

$$k_1 = k + \delta k$$

and $k_2 = k - \delta k$

and the angular frequencies are:

$$\omega_1 = \omega + \delta\omega$$

and $\omega_2 = \omega - \delta\omega$

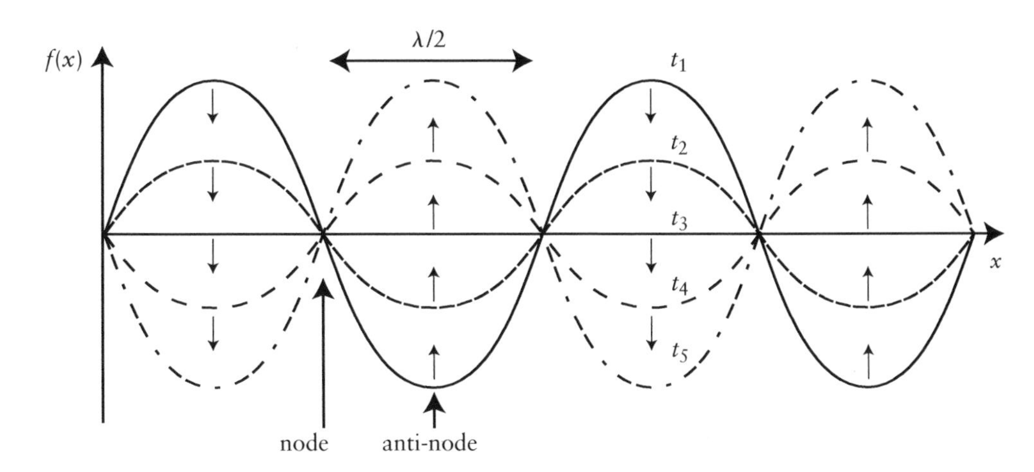

Figure 13.7: A standing wave. The amplitude of the wave is always zero at nodes and reaches a maximum at anti-nodes.

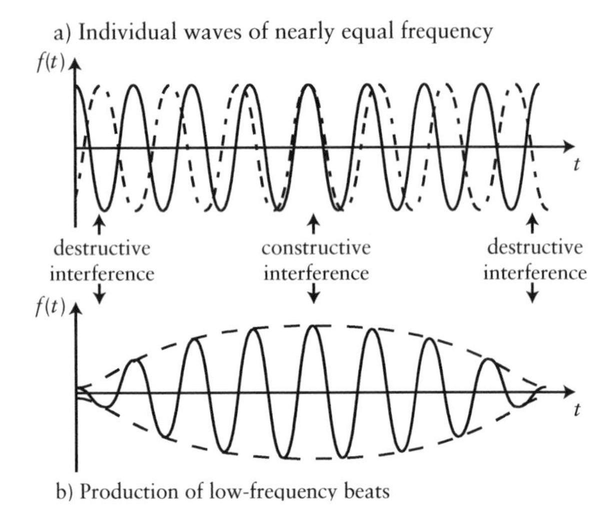

Figure 13.8: The interference of two waves of nearly equal frequencies and similar amplitudes traveling in the same direction produces beats at a lower frequency.

If we now superimpose the two waves, we find:

$$\begin{aligned} f(x,t) &= A\sin(k_1x-\omega_1t)+A\sin(k_2x-\omega_2t) \\ &= A\sin((k+\delta k)x-(\omega+\delta\omega)t) \\ &\quad + A\sin((k-\delta k)x-(\omega-\delta\omega)t) \\ &= \underbrace{2A\cos(\delta kx-\delta\omega t)}_{\text{envelope}}\ \underbrace{\sin(kx-\omega t)}_{\text{traveling wave}} \end{aligned} \tag{13.7}$$

This represents a traveling wave of wavenumber k and frequency ω, whose envelope is modulated by a long-wavelength traveling wave of wavenumber δk and frequency $\delta\omega$. This superposition therefore produces *beating* waves, whose amplitudes are modulated by a harmonic envelope, as illustrated in Fig. 13.8.

Properties of beating waves

From Eq. 13.7, it is straightforward to determine the properties of the traveling wave and its envelope. The wavenumber and frequency of the traveling wave are the average of those of the individual waves:

$$k = \frac{k_1+k_2}{2}$$

$$\text{and}\quad \omega = \frac{\omega_1+\omega_2}{2}$$

whereas those of the *envelope* are half the difference of those of the individual waves:

$$\delta k = \frac{k_1-k_2}{2}$$

$$\text{and}\quad \delta\omega = \frac{\omega_1-\omega_2}{2}$$

The beat frequency is double the envelope frequency since there are two beats per envelope wavelength:

$$\begin{aligned} \therefore \omega_B &= 2\delta\omega \\ &= \omega_1-\omega_2 \end{aligned}$$

(refer to Fig. 13.8). Since $\delta k << k$ and $\delta\omega << \omega$, the wavelength and time period of the envelope are much greater than those of the traveling wave.

Phase and group velocity

The velocity of the resultant traveling wave is $v=\omega/k$. This is called the *phase velocity*, v_p, since it corresponds to the velocity of any given phase of the wave, for example, a trough or a crest (Eq. 13.8). The velocity of the envelope is $v=\delta\omega/\delta k$. The envelope velocity is called the *group velocity*, v_g, and is defined by the derivative:

$$v_g = \frac{d\omega}{dk}$$

(Eq. 13.9). The group velocity is often thought of as the velocity at which energy or information is conveyed along a wave.

Wave velocities:

$$\text{Phase velocity:}\quad v_p = \frac{\omega}{k} \tag{13.8}$$

$$\text{Group velocity:}\quad v_g = \frac{d\omega}{dk} \tag{13.9}$$

Dispersion

Angular frequency is a function of wavenumber: $\omega=\omega(k)$. This function is called a *dispersion* relation, and from it we can determine both the phase and group velocity of a wave. For example, $\omega=vk$ corresponds to a dispersion relation. If v is constant, then:

$$v_p = \frac{\omega}{k} = v$$

$$\text{and}\quad v_g = \frac{d\omega}{dk} = v$$

i.e. the phase and group velocities are equal. However, if v is not constant, then the phase and group velocities are not constant:

$$v_p = \frac{\omega}{k} = v$$

and $$v_g = \frac{d\omega}{dk} = v + k\frac{dv}{dk}$$

In general, $v_g < v_p$. For example, v is not constant for white light propagating through a so-called *dispersive medium*, such as a prism; the different frequency components in white light travel at different velocities from one another. Therefore, the envelope of the wave becomes distorted as it propagates. This results in a spreading out (or *dispersion*) of the wave into its different frequency components, which is why a prism produces a rainbow when white light is shone through it. A vacuum however is a non-dispersive medium, therefore electromagnetic waves of any frequency propagate at $v_p = v_g = c$ through a vacuum, and there is no dispersion.

13.4 The Doppler effect

Whenever there is relative motion between a wave source and an observer, the apparent frequency of the wave to the observer is different from the actual frequency of emitted waves. This change in observed frequency is called the *Doppler effect*. If the relative motion brings the source towards the observer, the apparent frequency is higher than the emitted frequency, $f' > f$, where the emitted wave frequency is $f = v/\lambda$ (Eqs. 13.10 and 13.12, Derivation 13.2). Conversely, if the relative motion takes the source away from the observer, the apparent frequency is lower than the emitted frequency, $f' < f$ (Eqs. 13.11 and 13.13, Derivation 13.2). For example, sound waves emitted by a siren from an approaching ambulance have a higher pitch (higher frequency) than one that is retreating.

Shock waves

If the velocity of a moving source is greater than the wave velocity, then the peaks of each subsequently emitted wave constructively interfere with those of the previously emitted ones. This results in a *shock wave*, which propagates out at the velocity of the wave (Fig. 13.9). For example, when a moving source of sound travels at a speed greater than the speed of sound, such as a Concorde, the shock wave corresponds to a loud sound called a supersonic boom.

Doppler effect (Derivation 13.2):

Moving source (source velocity v_S):

$$\text{Towards O: } f' = \frac{f}{(1 - v_s / v)} \quad (13.10)$$

$$\text{Away from O: } f' = \frac{f}{(1 + v_s / v)} \quad (13.11)$$

Moving observer (observer velocity v_O):

$$\text{Towards S: } f' = f(1 + v_o / v) \quad (13.12)$$

$$\text{Away from S: } f' = f(1 - v_o / v) \quad (13.13)$$

where O is the observer and S is the source.

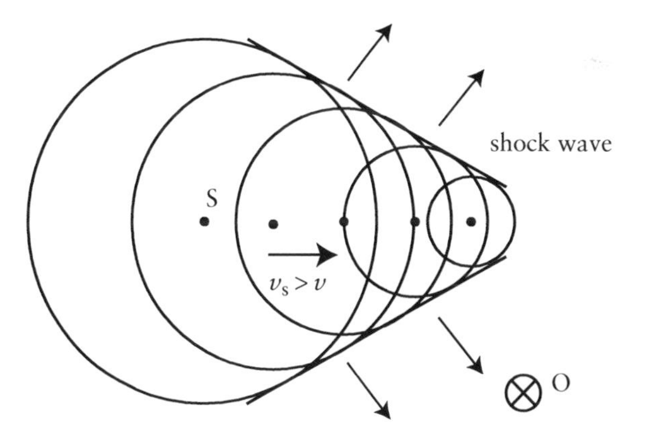

Figure 13.9: Constructive interference of waves emitted from a source traveling at $v_S > v$ results in a propagating shock wave.

Derivation 13.2: Derivation of the Doppler effect (Eqs. 13.10–13.13)

a) Moving source:

Waves emitted by moving source S, at frequency f

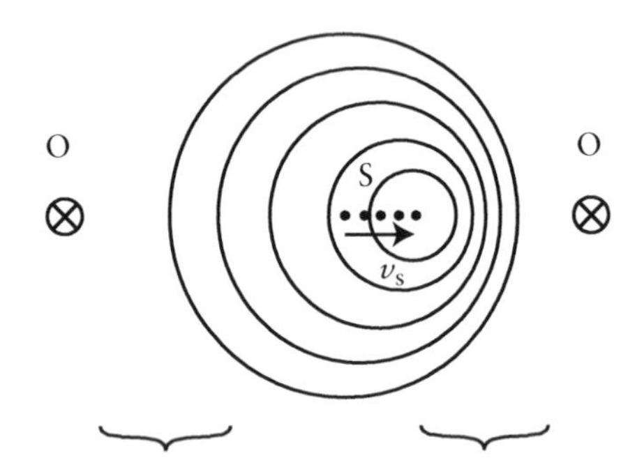

λ increased by an amount $\Delta s = v_S/f$: $\lambda' = \lambda + \Delta s$

λ reduced by an amount $\Delta s = v_S/f$: $\lambda' = \lambda - \Delta s$

S away from O:

$$\lambda' = \lambda + \Delta s$$
$$\frac{v}{f'} = \frac{v}{f} + \frac{v_S}{f}$$
$$\therefore f' = \frac{f}{(1 + v_S / v)}$$

S towards O:

$$\lambda' = \lambda - \Delta s$$
$$\frac{v}{f'} = \frac{v}{f} - \frac{v_S}{f}$$
$$\therefore f' = \frac{f}{(1 - v_S / v)}$$

b) Moving observer:

Waves emitted by stationary source S, at frequency f

O O S v_O v_O

apparent wave velocity decreased by an amount v_O: $v' = v - v_O$

apparent wave velocity increased by an amount v_O: $v' = v + v_O$

O away from S:

$$v' = v - v_O$$
$$\lambda f' = v(1 - v_O / v)$$
$$\therefore f' = f(1 - v_O / v)$$

O towards S:

$$v' = v + v_O$$
$$\lambda f' = v(1 + v_O / v)$$
$$\therefore f' = f(1 + v_O / v)$$

Chapter 14
Geometrical optics

Geometrical optics describes the behavior of light when it interacts with matter on scales much larger than the wavelength of light (>>10 µm). On these scales, wave phenomena such as interference are not apparent. In geometrical optics, we represent light schematically by light rays, which are straight lines that point in the direction of propagation of the wave. Light rays change direction when they are reflected or refracted at a surface. We can determine the change in direction at an interface using simple geometrical constructions. Understanding the laws of reflection and refraction is important in the design of optical instruments used for imaging, such as microscopes and telescopes.

14.1 Fermat's principle

Refractive index and optical path length

The speed of light in a vacuum is $c \simeq 3 \cdot 10^8\ \text{ms}^{-1}$. In a material medium, wave speed is less than c due to interactions between light and atoms in the medium. The ratio of the speed of light in a vacuum to that in matter is called the *refractive index*, n (Eq. 14.1). Since $v < c$ for all media, the refractive index is always greater than 1. The refractive index of glass is $n \simeq 1.5$, for water, $n \simeq 1.33$, and for air, to a good approximation, $n \simeq 1$. Although the velocity of light decreases when it travels through a medium, its *frequency* remains constant: $\omega_v = \omega_c$. Therefore, the wavelength of light is shorter in a medium relative to a vacuum: $\omega_v = \omega_c \rightarrow vk_v = ck_c$ $\therefore \lambda_v = \lambda_c/n$.

The *optical path length*, OPL [m], corresponds to the distance in a vacuum that has the same number of wavelengths as the distance x traveled by the same wave in a medium of refractive index n:

$$\text{OPL} = nx$$

(Eq. 14.2). The total optical path length of a wave passing through several media of different refractive indices is the sum of individual optical path lengths:

$$\text{OPL} = n_1x_1 + n_2x_2 + n_3x_3 \ldots = \sum_{i=1}^{N} n_i x_i$$

Where x_i corresponds to the distance travelled in the ith medium.

> **Refractive index and optical path length:**
>
> Refractive index:
>
> $$n = \frac{c}{v} \qquad (14.1)$$
>
> Optical path length:
>
> $$\text{OPL} = nx \qquad (14.2)$$

Fermat's principle of least time

Fermat's principle of least time states that "*The path traveled by light through a vacuum or a medium corresponds to the path that is traversed in the least time.*" This principle can be used to determine path traveled by light upon reflection or refraction.

The time in which light travels a distance x in a medium of refractive index n is proportional to the optical path length:

$$t = \frac{x}{v} = \frac{nx}{c} = \frac{\text{OPL}}{c}$$

Therefore, the path of least time corresponds to the shortest optical path length between two points. Notice also from this definition that the optical path length corresponds to the distance that would be traveled by light in a vacuum in the same amount of time that it travels a distance x in a medium. The two lengths, OPL and x, therefore correspond to the same temporal and spatial phase change as the wave advances.

Below, we will use Fermat's principle to deduce the laws of reflection and refraction.

14.2 Reflection and refraction

Light rays

In geometrical optics, we use *light rays* to represent the direction of straight-line streams of light. Light rays are mathematical constructs used to visualize the progression of light, and are not physical entities.

The direction of a light ray changes when light is incident on the surface of a material, such as a mirror or a lens. Light is either *reflected* by, *absorbed* by, or *transmitted* through a material. The change of direction on transmission is called refraction.

14.2.1 Laws of reflection and refraction

We can use Fermat's principle to determine the change in direction of a light ray on reflection or refraction at a surface. The *law of reflection* states that the angle of reflection, θ_r, is equal to the incident angle, θ_i (Eq. 14.3 and Derivation 14.1).

At a boundary between two media of refractive indices n_1 and n_2, a light ray is refracted by an amount that depends on n_1 and n_2. For light incident at an angle θ_1 relative to the normal, the refracted angle relative to the normal is:

$$\sin\theta_2 = \frac{n_1}{n_2}\sin\theta_1$$

(Eq. 14.4 and Derivation 14.2). This is known as *Snell's Law of refraction*. Incident light that makes only a small angle with the normal is called *a paraxial ray*. Therefore, in the *paraxial approximation*, we have $\sin\theta \simeq \theta$, and Snell's Law is given by Eq. 14.5.

Laws of reflection and refraction (Derivations 14.1 and 14.2):

Law of reflection:

$$\theta_i = \theta_r \tag{14.3}$$

Snell's Law of refraction:

$$n_1 \sin\theta_1 = n_2 \sin\theta_2 \tag{14.4}$$

Paraxial approximation, $\sin\theta \simeq \theta$:

$$n_1\theta_1 \simeq n_2\theta_2 \tag{14.5}$$

Derivation 14.1: Derivation of the law of reflection (Eq. 14.3)

$$\begin{aligned}
\text{POQ} &= n_1 L_i + n_1 L_r \\
&= n_1\sqrt{(h_i^2 + x^2)} + n_1\sqrt{(h_r^2 + (l-x)^2)} \\
\frac{d(\text{POQ})}{dx} &= \frac{n_1 x}{\sqrt{(h_i^2 + x^2)}} - \frac{n_1(l-x)}{\sqrt{(h_r^2 + (l-x)^2)}} \\
&= n_1 \sin\theta_i - n_1 \sin\theta_r \\
&= 0 \text{ when POQ is a minimum} \\
\therefore n_1 \sin\theta_i &= n_1 \sin\theta_r \\
\therefore \theta_i &= \theta_r
\end{aligned}$$

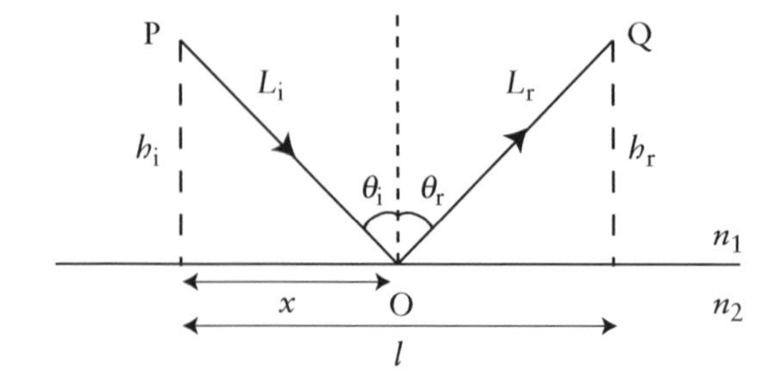

- Starting point: the total optical path length between points P and Q is the sum of optical path lengths between PO and OQ.
- Minimize the optical path length, POQ, with respect to x.
- The optical path length is a minimum when the angle of incidence equals the angle of reflection.

Derivation 14.2: Derivation of Snell's Law of refraction (Eq. 14.4)

$$\mathrm{POQ} = n_1 L_1 + n_2 L_2$$
$$= n_1\sqrt{(h_1^2 + x^2)} + \sqrt{(h_2^2 + (l-x)^2)}$$
$$\frac{\mathrm{d(POQ)}}{\mathrm{d}x} = \frac{n_1 x}{\sqrt{(h_1^2 + x^2)}} - \frac{n_2(l-x)}{\sqrt{(h_2^2 + (l-x)^2)}}$$
$$= n_1 \sin\theta_1 - n_2 \sin\theta_2$$
$$= 0 \text{ when POQ is a minimum}$$
$$\therefore n_1 \sin\theta_1 = n_2 \sin\theta_2$$

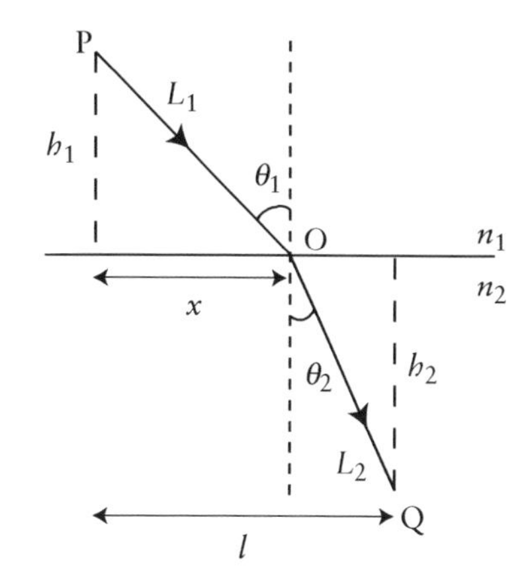

- Starting point: the total optical path length between points P and Q is the sum of optical path lengths between PO and OQ.
- Minimize the optical path length, POQ, with respect to x.
- We recover Snell's Law of refraction when the optical path length is a minimum.

Specular and diffuse reflection

Reflection is classified as either *specular* or *diffuse*. Specular reflection occurs on smooth surfaces, such as glass and mirrors, where parallel light is reflected in the same direction (Fig. 14.1a). In contrast, diffuse reflection occurs on rough surfaces, where parallel light is reflected in all directions (Fig. 14.1b). Specular reflection makes objects appear shiny and enables us to see other objects "in the reflection" of the reflecting surface, whereas diffuse reflection allows us to identify a rougher texture.

14.2.2 Total internal reflection

When light travels from a dense to a less dense material, such that $n_2 < n_1$ in Snell's Law, then light rays are refracted *away* from the normal ($\theta_2 > \theta_1$). The angle of incidence at which light is refracted along the boundary between the two media, i.e. when $\theta_2 = 90°$, is called the *critical angle*, θ_c (Eq. 14.6 and Fig. 14.2). At angles of incidence greater than θ_c, all the light is reflected back into the medium of the incident ray. This is called *total internal reflection*, and has many practical applications, for example, in binoculars, endoscopes, and optical fibers.

Critical angle:

$$\sin\theta_c = \frac{n_2}{n_1} \qquad (14.6)$$

a) Specular reflection

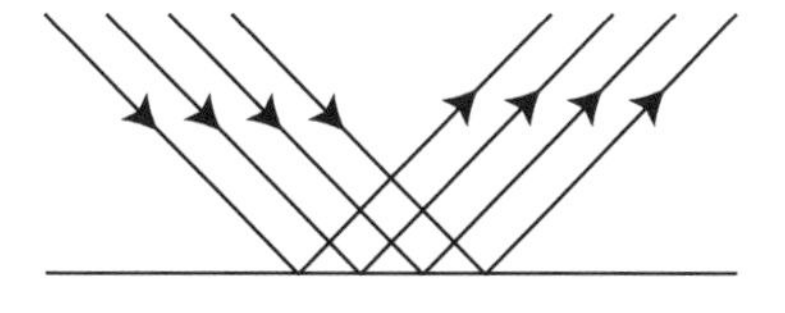

b) Diffuse reflection

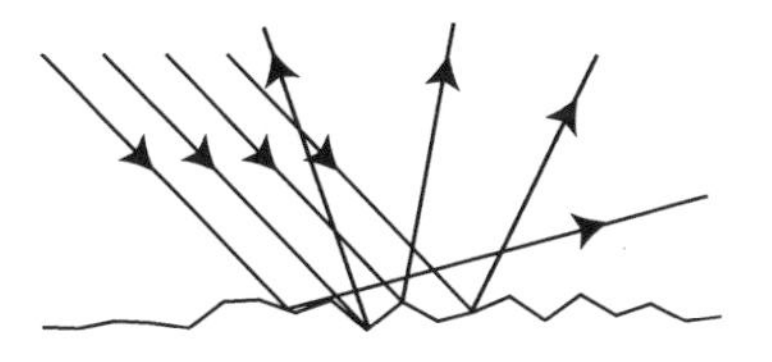

Figure 14.1: a) Specular reflection and b) diffuse reflection.

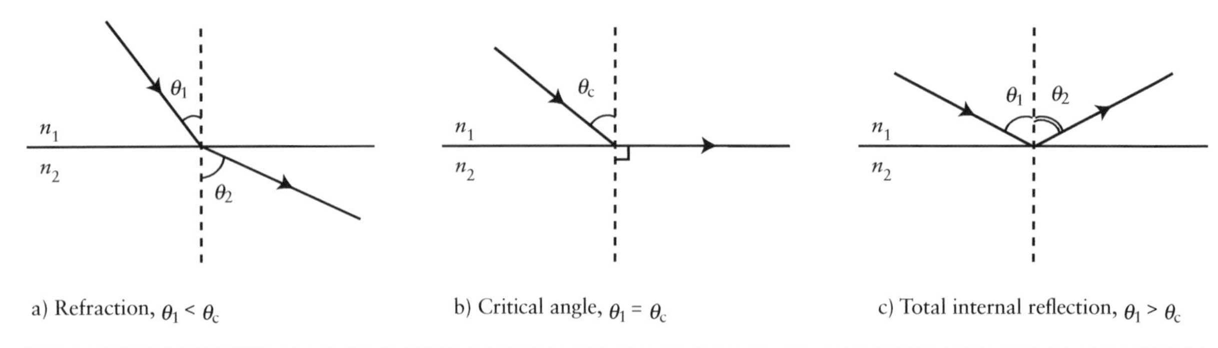

Figure 14.2: a) Refraction, b) critical angle, and c) total internal reflection at a boundary where $n_2 < n_1$.

Optical fibers

Optical fibers use total internal reflection to transmit light down them without any losses. They are usually made of a glass or plastic core and are surrounded by a cladding that has a lower refractive index than the core. The purpose of the cladding is to protect the fiber core from scratches, which could lead to losses of light. The maximum angle at which a ray incident on an optical fiber undergoes total internal reflection depends on the refractive indices of the core, n_{core}, the cladding, n_{clad}, and of the material surrounding the fiber, n_0 (Eq. 14.7 and Derivation 14.3).

Maximum angle of incidence on an optical fiber (Derivation 14.3):

$$\sin\theta_{\max} = \frac{\sqrt{n_{\text{core}}^2 - n_{\text{clad}}^2}}{n_0} \qquad (14.7)$$

where $n_{\text{core}} > n_{\text{clad}}$

14.3 Ray diagrams and images

We can use *ray diagrams* to determine where an image forms after light from an object passes through an optical instrument involving reflection off mirrors or refraction through lenses. Fig. 14.3 shows a typical ray diagram for a thin convex lens. The lens is placed perpendicular to the *principal axis*. The lens has two *focal points*, which are a distance f either side of the lens. The object is placed a distance $u > f$ from the center of the lens. Light emanates from the object in all directions, however, to determine the nature and position of the image, we can use three "landmark" rays (refer to Fig. 14.3):

1. The ray parallel to the principal axis, which emerges through the focal point.
2. The ray passing through the focal point, which emerges parallel to the principal axis.
3. The ray passing through the point of intersection between the principal axis and the lens, which emerges undeviated.

The image is then formed at the point at which any two of these light rays intersect, at a distance v from the center of the lens. We can use a similar ray diagram for image construction of a spherical mirror rather than a convex lens, except that the undeviated ray corresponds to the ray that passes through the center of curvature of the mirror rather than through the point of intersection between the principal axis and the mirror.

14.3.1 Real and virtual images

Using the ray construction described above, we can determine the nature and position of the images formed by single mirrors and lenses. Fig. 14.4 shows the image construction of a convex and concave mirror and lens, for an object placed at a distance greater than the focal length of the component in each case.[1] For a concave mirror and a convex lens, the reflected or emerging light rays are *convergent*—they come together—and an image forms at the point at which they cross (refer to Fig. 14.4). In this case, the image is said to be a *real* image, since it is formed by the intersection of real light rays. In the case of a con*vex* mirror and a con*cave* lens, however, the light rays are *divergent*, which means that they never cross. Now,

[1] The image can change dramatically depending on the object location. For example, an object inside of the focal length of a convex lens leads to a virtual, upright, and magnified image (see Section 14.3.5).

Derivation 14.3: Derivation of the maximum optical fiber angle (Eq. 14.7)

$$
\begin{aligned}
n_0 \sin\theta_{max} &= n_{core} \sin\theta_r \\
&= n_{core} \cos\theta_c \\
&= n_{core}\sqrt{(1-\sin^2\theta_c)} \\
&= n_{core}\sqrt{1-(n_{clad}/n_{core})^2} \\
&= \sqrt{n_{core}^2 - n_{clad}^2}
\end{aligned}
$$

$$
\therefore \sin\theta_{max} = \frac{\sqrt{n_{core}^2 - n_{clad}^2}}{n_0}
$$

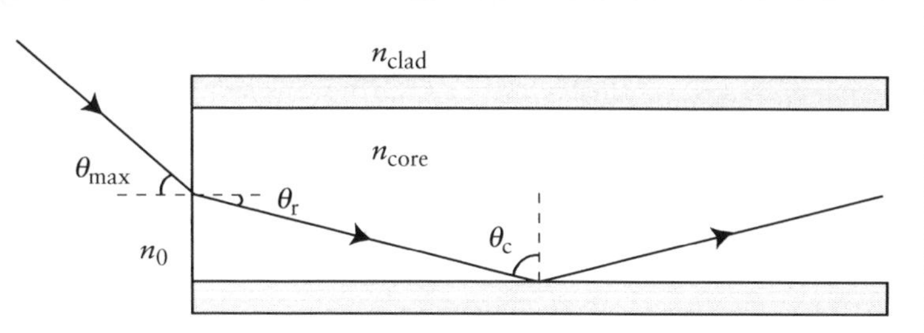

- Starting point: use Snell's Law of refraction at the boundary between the core and its surroundings.
- Use geometry to express θ_r in terms of the critical angle θ_c (figure not drawn to scale).
- Use the definition of the critical angle, $\sin\theta_c = n_2/n_1$, to eliminate θ_c, and rearrange.

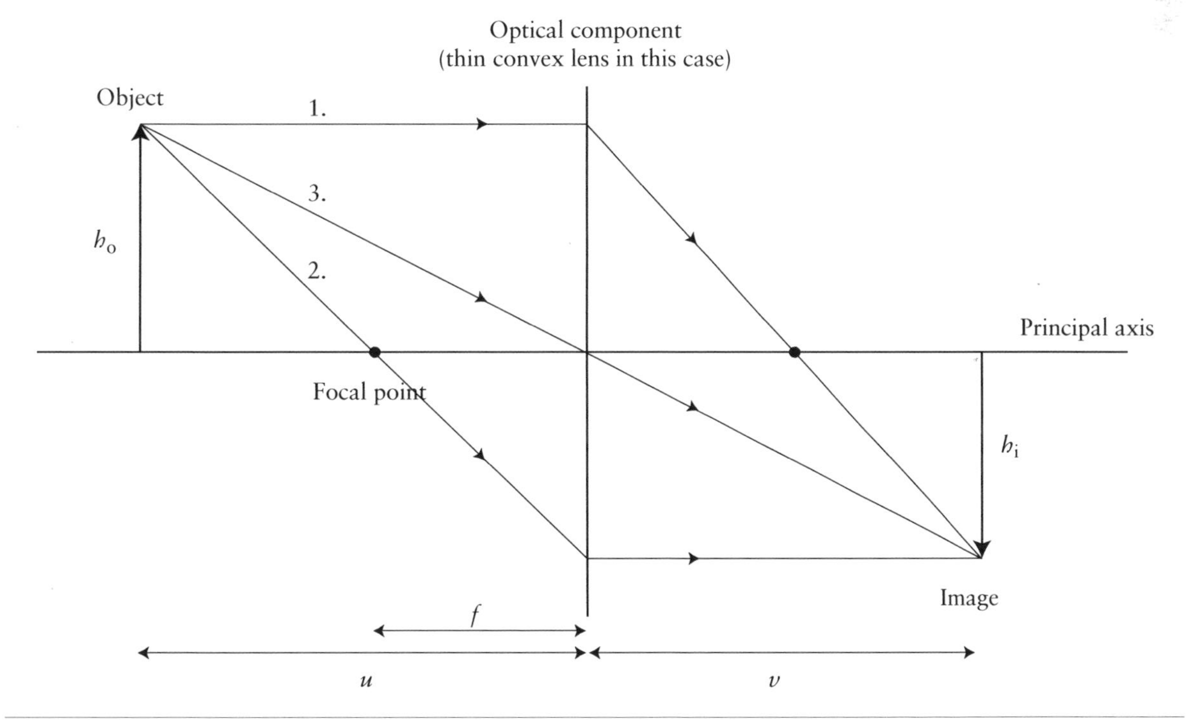

Figure 14.3: Ray diagram and three light rays used for image construction (refer to text).

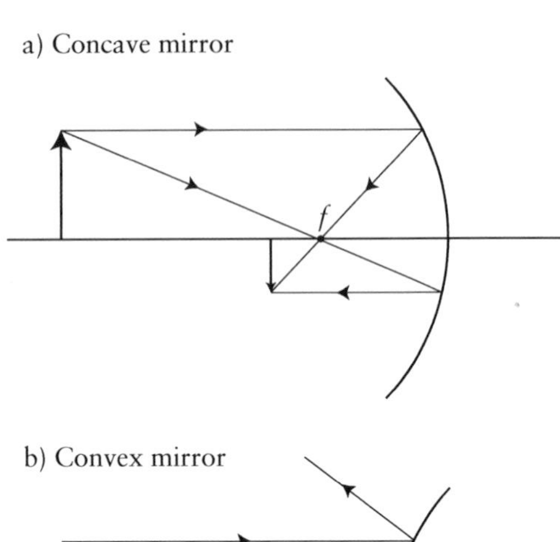

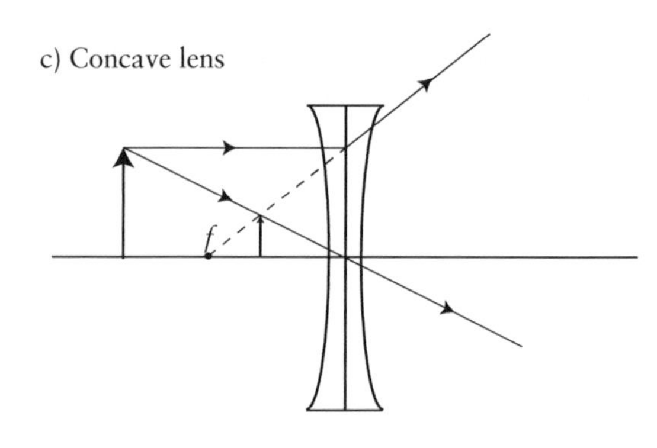

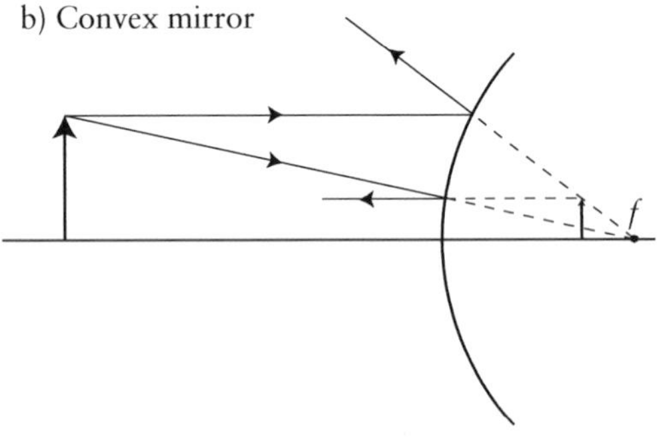

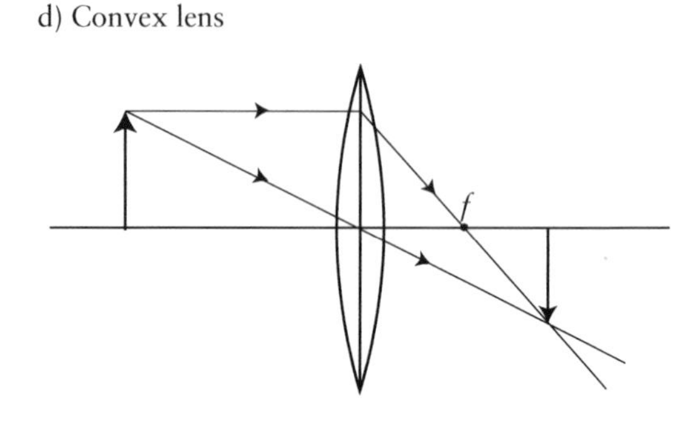

Figure 14.4: Ray diagrams and image construction of a concave and convex mirror and lens. All objects are placed at a distance greater than the focal length, f, of the optical component. In this case, the image is a) real and inverted for a concave mirror, b) virtual and upright for a convex mirror, c) virtual and upright for a concave lens, and d) real and inverted for a convex lens.

a *virtual* image is formed at the point at which the light rays *appear* to be coming from. We can "see" both real and virtual images. Our brain interprets light rays as traveling in straight lines, therefore it interprets diverging light rays as coming from the point at which a virtual image is formed. Real images can be perceived by focusing them on a screen. For example, cameras use convex lenses to focus light onto a photographic plate, hence producing a real image on the plate. We can then see the image by looking at the photographic plate.

14.3.2 Thin lens equation

The relationship between focal length f, object distance u and image distance v, for a thin convex lens, is given by the *thin lens equation*:

$$\frac{1}{f}=\frac{1}{u}+\frac{1}{v}$$

(Eq. 14.8 and Derivation 14.4). From this equation, we find that for a real image to form, the minimum distance between the object and image is equal to four times the focal length:

$$u+v \geqslant 4f$$

(Eq. 14.9 and Derivation 14.5)

Thin lens equation (Derivations 14.4 and 14.5):

$$\frac{1}{f}=\frac{1}{u}+\frac{1}{v} \tag{14.8}$$

Minimum distance between object and image:

$$u+v \geqslant 4f \tag{14.9}$$

Sign convention

The sign convention for the thin lens equation is that *"real is positive"*. This means that real object and image distances are $+u$ and $+v$, respectively, whereas virtual object and image distances are $-u$ and $-v$, respectively. Similarly, the focal length of a concave mirror and

Derivation 14.4: Derivation of the thin lens equation (Eq. 14.8)

From θ_1: $\frac{h_o}{h_i} = \frac{u}{v}$

From θ_2: $\frac{h_o}{h_i} = \frac{f}{v-f}$

$$\therefore \frac{u}{v} = \frac{f}{v-f}$$

$$\therefore \frac{1}{f} = \frac{1}{u} + \frac{1}{v}$$

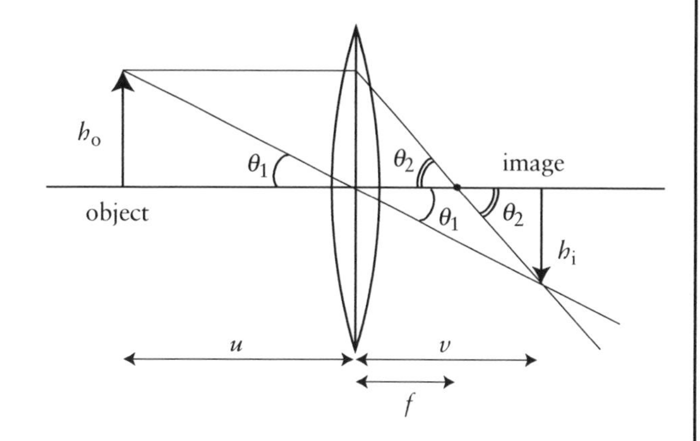

- Starting point: we can determine two expressions for similar triangles: one expression using angle θ_1 and one using θ_2.
- Equate these expressions and rearrange to determine the thin lens equation.

Derivation 14.5: Derivation of the minimum distance between object and image (Eq. 14.9)

$$\frac{1}{f} = \frac{1}{u} + \frac{1}{v}$$

$$= \frac{1}{u} + \frac{1}{(L-u)}$$

$$= \frac{L}{u(L-u)}$$

$$\therefore u^2 - uL + Lf = 0$$

$$\therefore u = \frac{L \pm \sqrt{L^2 - 4Lf}}{2}$$

where $L^2 - 4Lf \geqslant 0$

$$\therefore u + v \geqslant 4f$$

- Starting point: let the distance between an object and an image be $L = u + v$, and use this to eliminate v from the thin lens equation.
- The solutions to the quadratic in u are only real and different if:

$$L^2 - 4Lf \geqslant 0$$

$$\therefore L \geqslant 4f$$

$$\therefore u + v \geqslant 4f$$

convex lens, which form real images, is $+f$, whereas those for a convex mirror and concave lens, which form virtual images, is $-f$.

14.3.3 The spherical mirror equation

We can determine a similar equation to the thin lens equation for spherical mirrors, which relates image and object distances to the radius of curvature, R, of a spherical mirror (Eq. 14.10 and Derivation 14.6). From this, we see that the radius of curvature of a spherical mirror is twice its focal length (Eq. 14.11). The mirror equation only holds when the paraxial approximation is valid (i.e. when incident light makes only a small angle with the normal such that $\sin\theta \simeq \tan\theta \simeq \theta$).

Derivation 14.6: Derivation of the spherical mirror equation for paraxial rays (Eqs. 14.10 and 14.11)

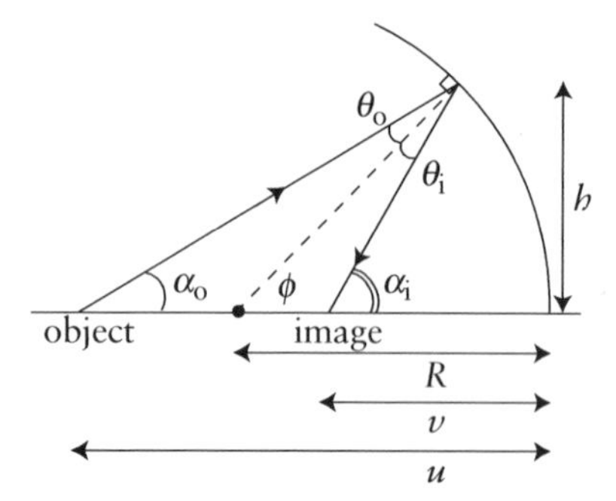

- Starting point: in the paraxial approximation, the angles are constrained to be very small, ∴ $\tan\theta \simeq \theta$ (image not drawn to scale, for clarity). Using the paraxial approximation, the angles in the diagram are therefore:

$$\alpha_o \simeq \frac{h}{u},\ \alpha_i \simeq \frac{h}{v},\ \text{and}\ \phi \simeq \frac{h}{R}$$

$$\theta_o = \phi - \alpha_o = h\left(\frac{1}{R} - \frac{1}{u}\right)$$

and

$$\theta_i = \alpha_i - \phi = h\left(\frac{1}{v} - \frac{1}{R}\right)$$

From the law of reflection:

$$\theta_o = \theta_i$$

$$\therefore h\left(\frac{1}{R} - \frac{1}{u}\right) = h\left(\frac{1}{v} - \frac{1}{R}\right)$$

$$\therefore \frac{1}{u} + \frac{1}{v} = \frac{2}{R}$$

∴ For a spherical mirror:

$$f = \frac{R}{2}$$

Spherical mirror equation (Derivation 14.6):

$$\frac{1}{u} + \frac{1}{v} = \frac{2}{R} \qquad (14.10)$$

Focal length of a spherical mirror:

$$f = \frac{R}{2} \qquad (14.11)$$

Lens maker's formula for a thin lens (Derivation 14.7):

$$\frac{1}{f} \simeq \left(\frac{n_{\text{lens}}}{n_0} - 1\right)\left(\frac{1}{R_1} - \frac{1}{R_2}\right) \qquad (14.12)$$

where n_0 is the refractive index of the surrounding medium.

14.3.4 Lens maker's formula

The focal length of a thin lens is related to the radii of curvature of either surface by the *lens maker's formula* (Eq. 14.12 and Derivation 14.7). Like the spherical mirror equation, this equation only holds when the paraxial approximation is valid. Additionally, the lens maker's formula applies only to *thin* lenses, which are defined as having a thickness, d, that is much less than the radii of curvature of either refracting surface; $d \ll R_1$ and $d \ll R_2$.

14.3.5 Magnification

The *magnification* of an image is the ratio of image size to object size. *Linear* magnification (M_L) is the ratio of linear dimensions. For example, magnification could be determined from the ratio of the image height to object height, or the ratio of image distance to object distance:

$$M_L = \frac{h_i}{h_o} \equiv \frac{v}{u}$$

Derivation 14.7: Derivation of the lens maker's formula (Eq. 14.12)

$$\theta_o = \phi + \alpha_o$$
$$= h\left(\frac{1}{R} + \frac{1}{u}\right)$$

and $\theta_i = \phi - \alpha_i$

$$= h\left(\frac{1}{R} - \frac{1}{v}\right)$$

Snell's Law in the paraxial approximation:

$$n_1\theta_o = n_2\theta_i$$
$$\therefore n_1 h\left(\frac{1}{R} + \frac{1}{u}\right) = n_2 h\left(\frac{1}{R} - \frac{1}{v}\right)$$
$$\therefore \frac{n_1}{u} + \frac{n_2}{v} = \frac{n_2 - n_1}{R}$$

For two spherical surfaces:

$$1^{\text{st}} \text{ surface: } \frac{n_0}{u} + \frac{n_{\text{lens}}}{s'} = \frac{n_{\text{lens}} - n_0}{R_1}$$

$$2^{\text{nd}} \text{ surface: } \frac{n_{\text{lens}}}{(-s')} + \frac{n_0}{v} = \frac{n_0 - n_{\text{lens}}}{(-R_2)}$$

Combining these equations:

$$\frac{n_0}{u} + \frac{n_0}{v} = (n_{\text{lens}} - n_0)\left(\frac{1}{R_1} - \frac{1}{R_2}\right)$$
$$\therefore \frac{1}{f} = \frac{1}{u} + \frac{1}{v}$$
$$= \left(\frac{n_{\text{lens}}}{n_0} - 1\right)\left(\frac{1}{R_1} - \frac{1}{R_2}\right)$$

where n_{lens} is the refractive index of the lens and n_0 is the refractive index of the surrounding medium.

Single spherical surface of a thin lens:

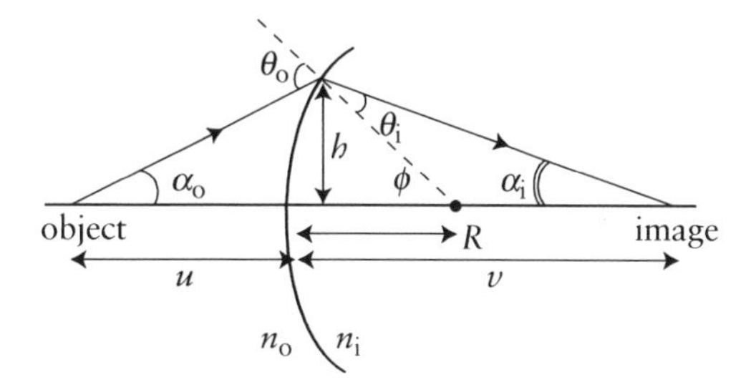

- Starting point: in the paraxial approximation, the angles are constrained to be very small $\therefore \sin\theta \simeq \tan\theta \simeq \theta$ (image not drawn to scale, for clarity). Using the paraxial approximation, the angles in the diagram are therefore:

$$\alpha_o \simeq \frac{h}{u}, \quad \alpha_i \simeq \frac{h}{v} \quad \text{and} \quad \phi \simeq \frac{h}{R}$$

- Use these angles to find an expression for Snell's Law in the paraxial approximation:

$$n_1\theta_o = n_2\theta_i$$

- A thin lens consists of two spherical surfaces, where refraction at either surface obeys Snell's Law in the paraxial approximation. Let n_{lens} be the refractive index of the lens and n_0 is the refractive index of the surrounding medium. Therefore, at the first surface (surrounding medium to lens), we have $n_1 \to n_0$ and $n_2 \to n_{\text{lens}}$, and at the second surface (lens to surrounding medium), we have $n_1 \to n_{\text{lens}}$ and $n_2 \to n_0$.
- Let the magnitudes of the radii of curvature of either surface be R_1 and R_2. Since the second surface is a concave lens, then using the real is positive sign convention, $R_2 \to -R_2$ for the second surface.
- A thin lens is defined as $R_1 << d$ and $R_2 << d$, where d is the lens thickness.
- Let the intermediate image formed by the first surface be at a distance $s' - d \simeq s'$. This image then acts as the virtual object for the second surface $(u \to -s')$.

(Eq. 14.13, refer to Fig. 14.3). The sign associated with magnification tells us whether the image is upright (M_L is negative) or inverted (M_L is positive).

Magnifying glass

If the object distance is less than the focal length of a thin convex lens, then from the thin lens equation, the image distance is $|v| > |u|$, and therefore the image is enlarged ($|M_L| > 1$). This is the principle of a *magnifying glass*, which produces a virtual, upright, enlarged image (Fig. 14.5). If the object distance is $u = f$, then a virtual image is formed at $v = \infty$ (parallel rays). In this case, it is not useful to measure linear magnification, since $M_L \to \infty$ as $v \to \infty$, and so we must use angular magnification instead.

Angular magnification

Another way to measure magnification, often used in telescopes for astronomy, is *angular* magnification. The eye perceives the size of an object by the size of the image on the retina, which is proportional to the angle subtended by the object, α (refer to Fig. 14.6a). α is usually specified as the angle subtended by an object when it is placed at the *near point*, D, of the eye. This corresponds to the closest distance from the eye at which an image can be comfortably focused, and corresponds to ~25 cm for humans. An object of height h_o then has an angular size of $\alpha \simeq h_o / D$. An optical instrument that produces an image that subtends a larger visual angle than α, such as a magnifying glass, produces an angular magnification of:

$$M_\theta = \frac{\beta}{\alpha}$$

(Eq. 14.14) where β is the angle subtended at the eye by the magnified image (refer to Fig. 14.6b). An image of height h_i at the near point then has an angular size of $\beta \simeq h_i / D$. Therefore, we see that angular and linear magnification are equivalent:

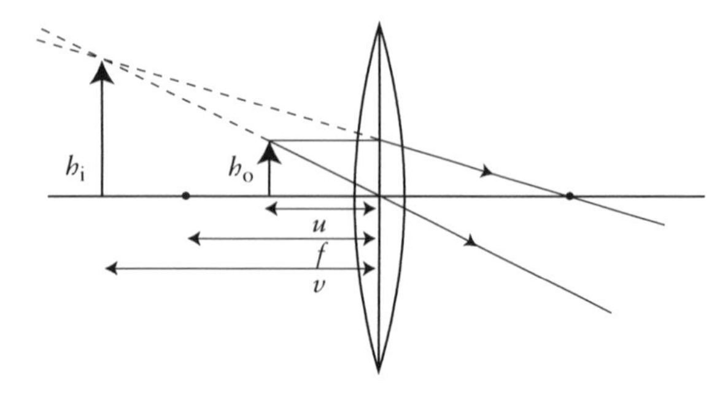

Figure 14.5: Ray diagram of a magnifying glass. The image of an object placed inside the focal length of a convex lens is virtual, magnified, and upright.

$$M_\theta = \frac{\beta}{\alpha} \simeq \frac{h_i / D}{h_o / D} = M_L$$

Magnification (refer to Figs. 14.3 and 14.6):

Linear magnification:

$$M_L = \frac{h_i}{h_o} \equiv \frac{v}{u} \qquad (14.13)$$

Angular magnification:

$$M_\theta = \frac{\beta}{\alpha} \qquad (14.14)$$

Magnification of an image at $v = \infty$

From the triangles in Fig. 14.6, we see that $\alpha \simeq h_o / D$ and $\beta \simeq h_o / u$. Therefore, the magnification is:

$$M_\theta = \frac{\beta}{\alpha} = \frac{D}{u}$$

If the object distance is $u = f$, a virtual image is formed at $v = \infty$ (parallel rays). In this case, the magnification is:

$$M_\theta = \frac{\beta}{\alpha} = \frac{D}{f}$$

This image has a smaller magnification than if $u < f$, but is has a more restful effect on the eye since the eye muscles are more relaxed.

a) Unaided eye

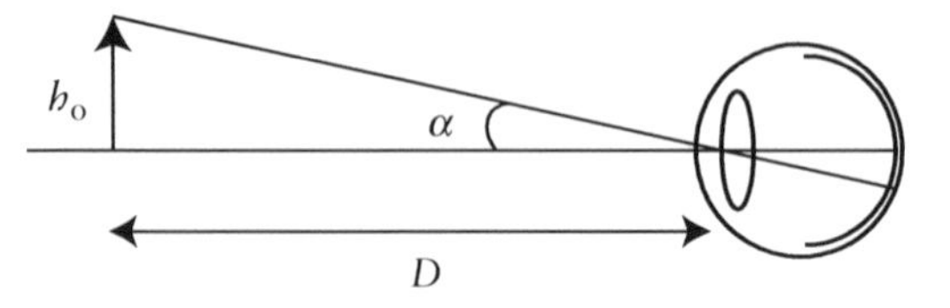

b) Magnifying lens

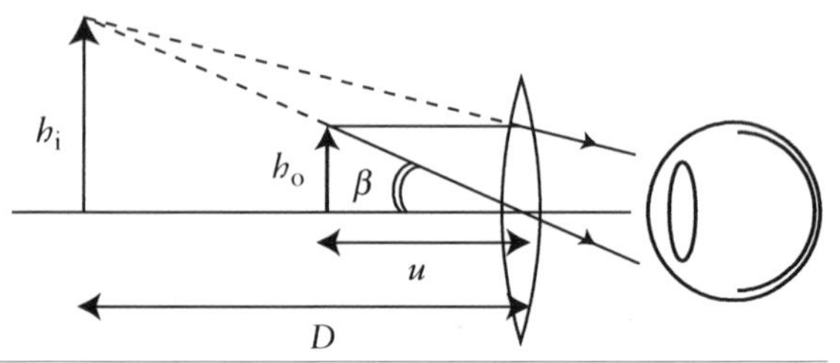

Figure 14.6: Ray diagram of angular magnification. An object at the near point, D, subtends an angle α at the unaided eye. The action of a magnifying glass is to increase the angle subtended at the eye to an angle $\beta > \alpha$.

14.3.6 Optical instruments for imaging

Optical instruments, such as telescopes and microscopes, often use a combination of two or more thin lenses in order to achieve the desired image magnification. The compound microscope and astronomical telescope employ two thin convex lenses: an *objective* lens (the first lens) and an *eyepiece* lens (the second lens) (Fig. 14.7). Light from an object is focused by the objective lens to produce an intermediate image, which then acts as the object for the eyepiece lens. In the case of a compound microscope, the lenses are arranged such that the intermediate image forms inside the focal length of the eyepiece lens. This arrangement produces a magnified, inverted, virtual image at the near point (Fig. 14.7a). In the case of an astronomical telescope, the lenses are arranged such that the intermediate image forms at the focal length of the eyepiece lens. The eyepiece therefore produces a magnified, inverted, virtual image at ∞ (Fig. 14.7b). In this case, magnification is less than that of the near point image, but viewing is more relaxed on the eye. The angular magnification of an astronomical telescope is the ratio of focal lengths of the objective, f_o, and eyepiece, f_e, lenses:

$$M_\theta = \frac{\beta}{\alpha} = \frac{h / f_e}{h / f_o} = \frac{f_o}{f_e}$$

(refer to Fig. 14.7b).

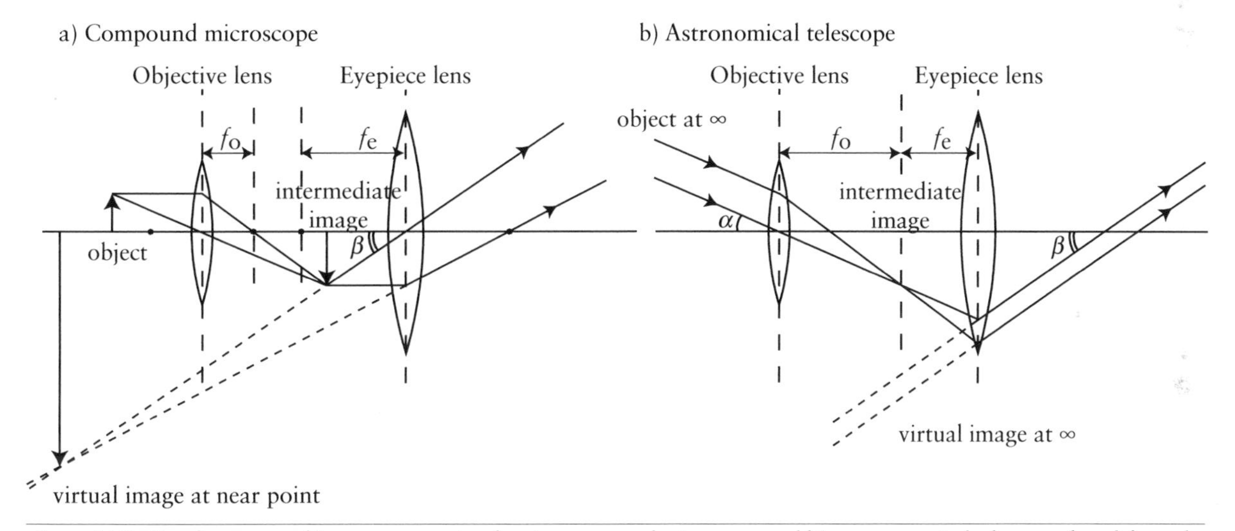

Figure 14.7: Ray diagrams and image construction for a) a compound microscope and b) an astronomical telescope. f_o and f_e are the focal lengths of the objective and eyepiece lenses, respectively.

Chapter 15
Wave optics

Whereas geometrical optics describes the properties of light when it interacts with matter on scales much larger than its wavelength, wave optics describes its properties when it interacts with matter on scales on the order of the wavelength of light (~10 μm). On these scales, wave phenomena such as diffraction and interference are apparent. Diffraction is the spreading out of light around small objects or through apertures whose dimensions are comparable with the light's wavelength; interference is the superposition of coherent light from one or more sources of light. We are familiar with interference in our day-to-day lives. For example, the colorful patterns on the surface of bubbles of soap are due to interference of white light. In this case, the thickness of the soap bubble is on the order of λ. Understanding the properties of diffraction and interference is important for the design of optical instruments used for spectroscopy, such as the diffraction grating, from which we can determine properties of an unknown source of light. For example, light emitted from atoms can be analyzed using spectroscopy instruments to determine details of their internal structure.

15.1 Huygen's principle

Light rays and wavefronts

In geometrical optics, we are not concerned with the wave nature of light. Therefore, we could represent straight-line streams of light by light rays, and use Fermat's principle to determine the laws of reflection and refraction of a light ray incident on a surface.

In wave optics, we *are* concerned with the wave nature of light, in which case we use a series of *wavefronts* to represent light. A wavefront is a line that connects points on a wave that have the same phase, for example crests or troughs. The distance between wavefronts corresponds to the wavelength, and the direction of propagation is perpendicular to the wavefront (Fig. 15.1). Thus, light rays are perpendicular to wavefronts.

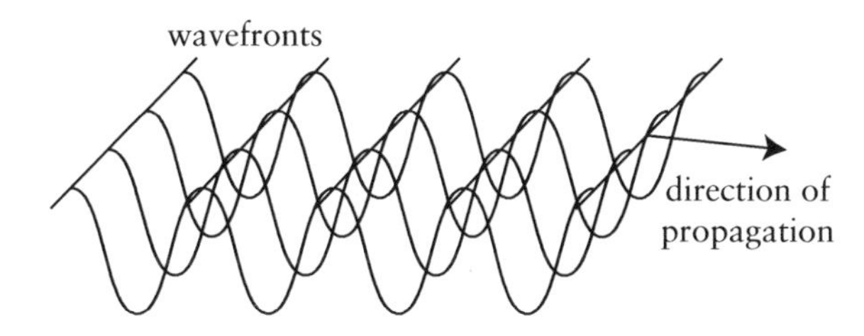

Figure 15.1: Wavefronts are lines that connect points on a wave that have the same phase. The direction of propagation is perpendicular to the wavefronts.

Huygen's principle

Huygen's principle is a construction from which the position of a new wavefront is determined from the previous one. It states that "*every point along a wavefront is a source of spherical secondary wavelets; the wavefront at some later time is the envelope of these wavelets*" (Fig. 15.2).

Huygen's principle is a theoretical construct from which we can determine the law of reflection and Snell's Law of refraction (Derivation 15.1). It is a statement of the superposition of waves—secondary wavelets constructively interfere at points where they are in phase with each other, and destructively interfere at points where they are half a wavelength out of phase with each other. The resultant wavefront therefore corresponds to the points of constructive interference.

A convenient way to determine the resultant wavefront from individual wavelets is to represent a wavelet as a *phasor*, described below, and to then determine the sum of all phasors.

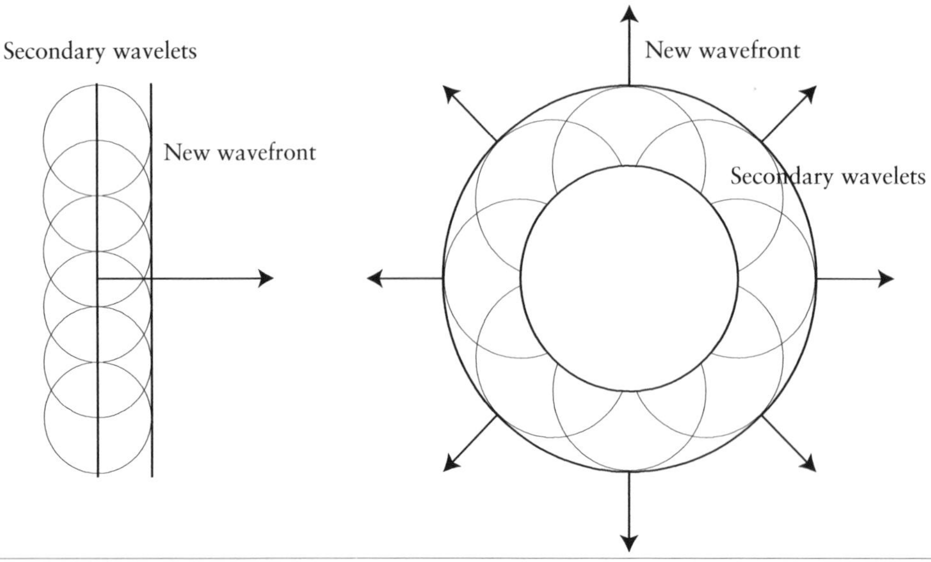

Figure 15.2: Huygen's principle applied to plane and circular wavefronts.

Derivation 15.1: Derivation of the laws of reflection and refraction using Huygen's principle

Consider two parallel light rays incident on a boundary. The incident wavefront AC reaches the boundary at point A. In the time the reflected or refracted light ray travels from A to B, the adjacent ray travels from C to D. If we consider a spherical secondary wavelet emanating from point A, the reflected or refracted wavefront BD forms when BD is tangential to the secondary wavelet. We can then determine the laws of reflection and refraction using geometrical argument.

Law of reflection:

Secondary wavelet emanating from point A

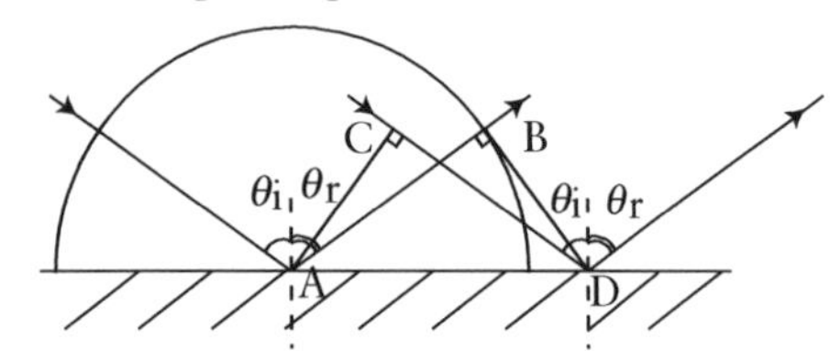

$$\sin\theta_i = \frac{CD}{AD}$$

$$\sin\theta_r = \frac{AB}{AD}$$

$$\therefore \frac{\sin\theta_i}{CD} = \frac{\sin\theta_r}{AB}$$

$$\frac{\sin\theta_i}{v_1 t} = \frac{\sin\theta_r}{v_1 t}$$

$$\therefore \theta_i = \theta_r$$

Snell's Law of refraction:

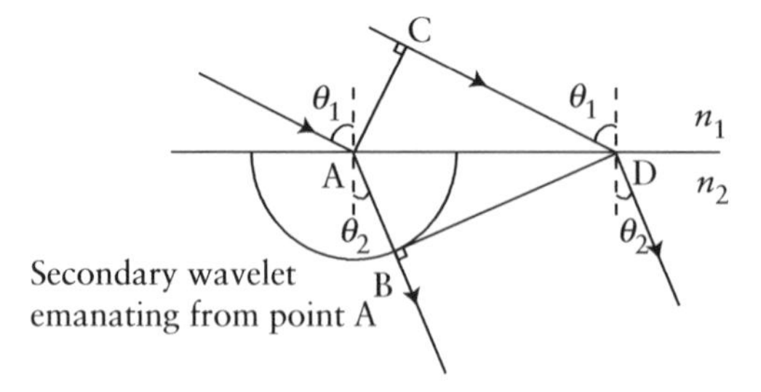

$$\sin\theta_1 = \frac{CD}{AD}$$

$$\sin\theta_2 = \frac{AB}{AD}$$

$$\therefore \frac{\sin\theta_1}{CD} = \frac{\sin\theta_2}{AB}$$

$$\frac{\sin\theta_1}{v_1 t} = \frac{\sin\theta_2}{v_2 t}$$

$$\frac{n_1\sin\theta_1}{ct} = \frac{n_2\sin\theta_2}{ct}$$

$$\therefore n_1\sin\theta_1 = n_2\sin\theta_2$$

15.1.1 Phasor representation of a wave

Using Euler's formula ($Ae^{i\theta} = A\cos\theta + A i\sin\theta$), we see that we can express a harmonic wave as the real or imaginary part of a complex exponential:

$$A\sin(kx - \omega t + \phi) = \mathrm{Im}\left\{Ae^{i(kx-\omega t+\phi)}\right\}$$

$$A\cos(kx - \omega t + \phi) = \mathrm{Re}\left\{Ae^{i(kx-\omega t+\phi)}\right\}$$

If we absorb the arbitrary initial phase, ϕ, into the coefficient, then we can express a wave by:

$$\tilde{f}(x,t) = \tilde{A}e^{i(kx-\omega t)}$$
$$\equiv \tilde{A}e^{i\delta}$$

where the tilde indicates that the number or function is complex. It is straightforward to see by substitution that $\tilde{f}(x, t)$ satisfies the wave equation:

$$\frac{\partial^2 \tilde{f}}{\partial x^2} = \frac{1}{v^2}\frac{\partial^2 \tilde{f}}{\partial t^2}$$

where both the real and imaginary parts are solutions. It's therefore common to omit the tilde when expressing a wave in complex exponential form, and it is understood that the physical wave corresponds to the real or imaginary part of the complex function (Eq. 15.1).

Complex exponential representation (phasor):

$$f(x,t) = A\sin(kx \pm \omega t)$$
$$\equiv Ae^{i(kx\pm\omega t)}$$
$$\equiv Ae^{i\delta} \quad (15.1)$$

where $\delta = kx \pm \omega t$

Phasors

We can represent a complex number, $Ae^{i\delta}$, on an Argand diagram, where A is the distance of the complex number from the origin and δ is its angle with the positive real axis (Fig. 15.3). Similarly, we can represent a wave written in complex exponential form in this way; this representation of a wave is called a *phasor*. The magnitude A represents the wave amplitude, and the argument $\delta = kx - \omega t$, represents its phase (i.e. the progression of the wave as a fraction of a complete cycle, where a complete cycle corresponds to 2π radians). Since δ increases with time, the evolution of the phasor with time is an counter-clockwise rotation (refer to Fig. 15.3). It is convenient to use this representation for determining interference patterns of multiple secondary wavelets, as we shall see below.

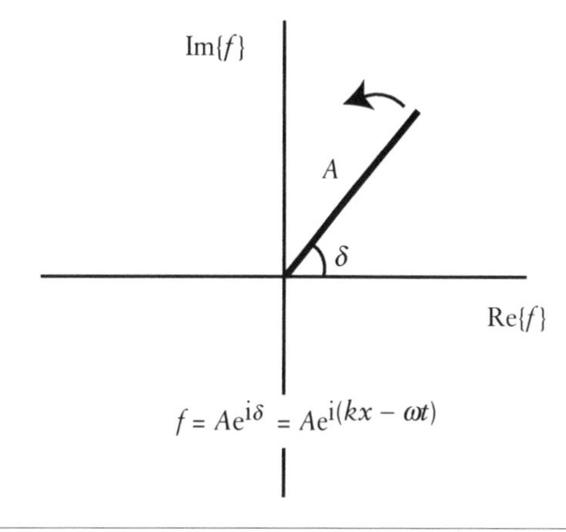

Figure 15.3: Phasor representation of a wave.

Addition of phasors

If we represent the nth Huygen wavelet of a wavefront as a phasor, $a_n e^{i\delta_n}$, then from Huygen's principle, the resultant wave is the superposition of individual wavelets:

$$U_{\text{resultant}} = \sum_n a_n e^{i\delta_n}$$
$$\equiv Ae^{i\delta}$$

In the following sections, we will use phasor addition and Huygen's principle to determine different intensity patterns that result from interfering wavelets.

15.2 Interference

Coherent waves

The phase, δ, of a wave corresponds to its progression around a complete cycle of 2π radians, and it varies with both space and time:

$$\delta = kx - \omega t$$

Two waves with a constant phase difference between them, $\Delta\delta$, are said to be *coherent*. Said another way, coherent waves have the same frequency, ω, but may be at a different progression around a complete cycle from one another.[1] In contrast, two waves for which $\Delta\delta$ is a function of time do not have a constant phase difference between them, and are said to be *in*coherent.

1 Spatially coherent waves also have the same value of k at all positions.

If two coherent waves meet at a particular position, they have the same phase difference between them at all times. For example, if two peaks meet at a given position, there will always be constructive interference at that point, whereas if two troughs meet at a given position, there will always be destructive interference at that point. Therefore, the superposition of coherent waves forms a spatial pattern of constructive and destructive interference that is constant in time; this type of pattern is called an *interference pattern*. Below, we will determine the interference patterns of various standard configurations using phasor addition and Huygen's principle.

15.2.1 Young's double-slit experiment

One example of wave interference is *Young's double-slit* experiment, which was named after Thomas Young who first conducted it. The experiment involves shining monochromatic, coherent light onto two thin slits of width a spaced a distance d apart, as shown in Fig. 15.4. For now, we treat the slit width as being much smaller than the slit separation: $a << d$.

According to Huygen's principle, each slit acts as a source of secondary wavelets. These wavelets emanate spherically from the slits, and they interfere where they overlap. This results in an *interference pattern* of alternate bright and dark fringes, which can be seen on a screen that is a perpendicular distance L from the slits (refer to Fig. 15.4). A bright fringe is produced where the wavelets constructively interfere (at positions where they are in phase with each other), and a dark fringe is produced where they destructively interfere (at positions where they are half a wavelength out of phase with each other).

Using phasor addition, we can determine the intensity pattern of interference fringes as we move across the screen (Eq. 15.2, Derivation 15.2, and Fig. 15.5).

Young's double-slit experiment (refer to Fig. 15.4) (Derivation 15.2):

Interference pattern:

$$I(\beta) = I_0 \cos^2\beta \quad (15.2)$$

where $\beta = \dfrac{kd\sin\theta}{2}$

For constructive interference:

$$d\sin\theta = n\lambda \quad (15.3)$$

where $n = 0, \pm1, \pm2, \ldots$

Fringe spacing:

$$\Delta x = \frac{L\lambda}{d} \quad (15.4)$$

The condition for constructive interference (bright fringes) is that the path difference between the two phasors must equal an integer number of wavelengths:

$$\underbrace{d\sin\theta}_{\text{path difference}} = n\lambda$$

(Eq. 15.3). This equation is only valid in the far field ($L >> d$). We arrive at the same result if we consider that $I(\beta)$ is a maximum ($I(\beta) = I_0$) when $\cos^2(\beta) = 1$, i.e. when $\beta = \dfrac{kd\sin\theta}{2} = n\pi$. Conversely, the condition for destructive interference (dark fringes) is that the path

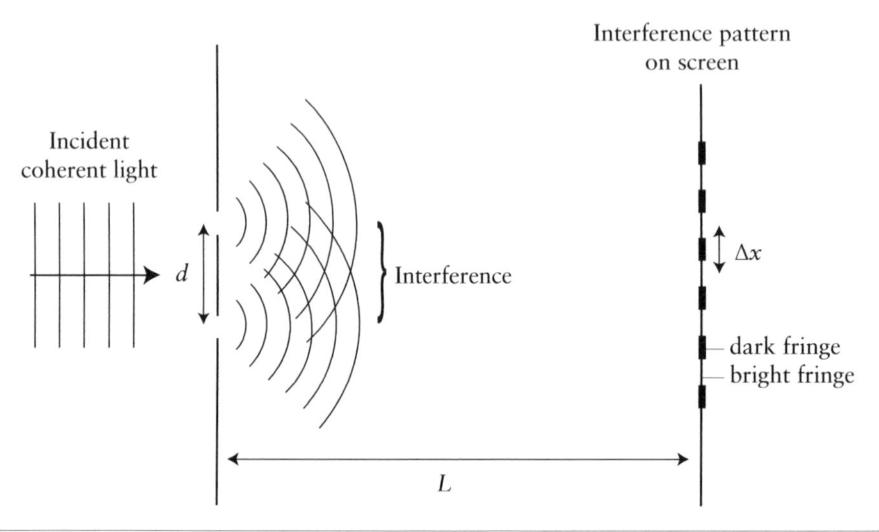

Figure 15.4: Experimental setup of Young's double-slit experiment.

Derivation 15.2: Derivation of Young's double-slit interference pattern (Eq. 15.2)

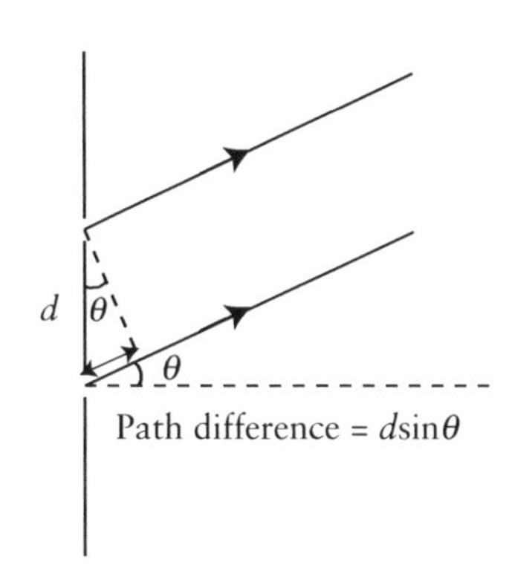

$$\begin{aligned} U(\theta) &= U_0 + U_0 e^{-i\delta} \\ &= U_0\left(1 + e^{-2i\beta}\right) \\ &= U_0 e^{-i\beta}\left(e^{i\beta} + e^{-i\beta}\right) \\ &= 2U_0 e^{-i\beta}\cos\beta \\ \therefore I(\theta) &= |U(\theta)|^2 \\ &= 4U_0^2\cos^2\beta \\ &\equiv I_0\cos^2\beta \end{aligned}$$

where $\beta = \dfrac{kd\sin\theta}{2}$

- Starting point: consider coherent light incident on two narrow slits separated a distance d. Secondary wavelets emanate radially from each slit and interfere with each other where they overlap. Let each wavelet have the same amplitude, U_0, and the relative phase between them be δ.
- The resultant phasor of the interfering wavelets is the sum of individual wavelet phasors: $U(\delta) = U_0 + U_0 e^{-i\delta}$.
- Consider two parallel rays at an angle θ to the normal. The phase difference between the two wavelets at this angle is $\delta = kx = kd\sin\theta = 2\beta$, where $\beta = kd\sin\theta/2$. The phase difference does not vary with time since the wavelets are coherent.
- Use the trigonometric identity:

  $e^{i\phi} + e^{-i\phi} = 2\cos\phi$

  to simplify the expression.
- The intensity distribution as a function of θ, $I(\theta)$, equals the square of the resultant phasor amplitude: $I(\theta) = |U(\theta)|^2$.

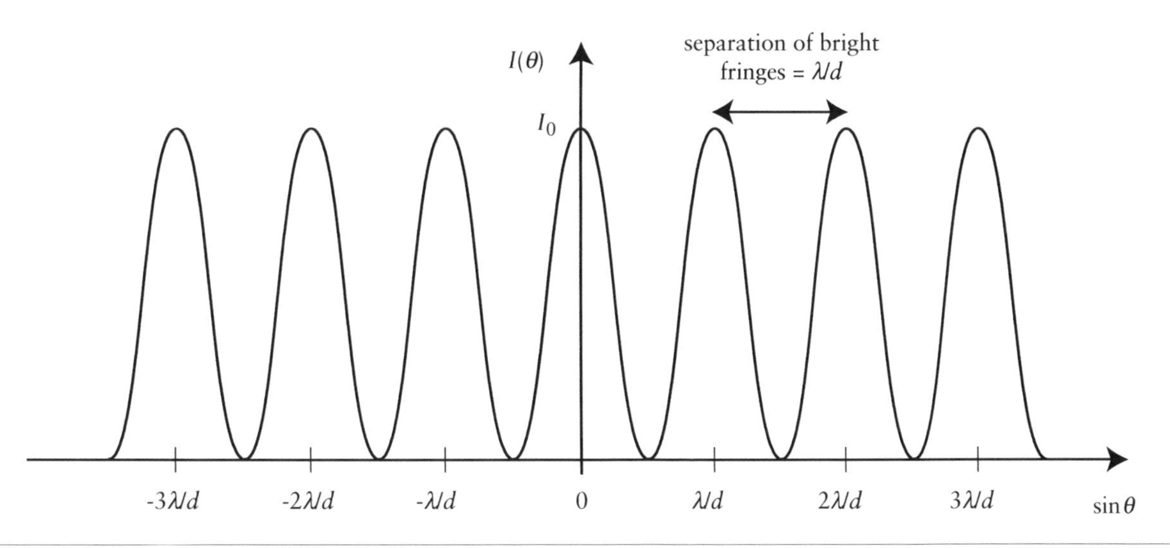

Figure 15.5: Double-slit interference pattern (refer to Eqs. 15.2 and 15.3).

difference between two phasors is half a wavelength since these correspond to regions at which their amplitudes cancel. Using the expression for the condition for constructive interference, the spacing between adjacent bright fringes is then:

$$\Delta x = L \sin\theta \\ = \frac{L\lambda}{d}$$

where L is the perpendicular distance between the slits and an imaging screen (refer to Fig. 15.4). Young's double-slit experiment is used in spectroscopy experiments to determine the wavelength of an unknown source, upon measuring d, L and Δx.

15.3 Diffraction

Diffraction is the spreading out of waves as a result of passing through an aperture or across an edge. The effects of diffraction are significant when the dimensions of the diffracting object are comparable with the wavelength of light. An illustration of diffraction through an aperture using Huygen's principle to construct wavefronts is shown in Fig. 15.6. The secondary Huygen wavelets on the edges of the aperture have no adjacent wavelets on one side to destructively interfere with, and so they continue to propagate spherically, which results in the wave spreading out. If the aperture is very narrow, such that only one secondary wavelet forms in it, then the diffracted wave emanates spherically.

Diffraction is usually accompanied by interference of the secondary wavelets that form in the plane of the diffracting object or aperture. The intensity pattern of the resultant interference pattern depends on the nature of the diffracting object. In this section, we will determine the interference pattern of a single aperture and of a grating with several adjacent slits.

15.3.1 Fraunhofer diffraction

We classify diffraction into two types: *Fraunhofer* or *far-field* diffraction and *Fresnel* or *near-field* diffraction. The two specifications for Fraunhofer diffraction are that (i) incoming light is parallel and (ii) the image plane is a large distance from the plane of the diffracting object. In contrast, Fresnel diffraction is the more general case where incident wavefronts are curved, and the image plane is nearby. Generally, Fraunhofer diffraction applies to slits and circular apertures, for example, diffraction in a telescope, whereas Fresnel diffraction applies to opaque edges and barriers, such as light diffracting around a pinhead.

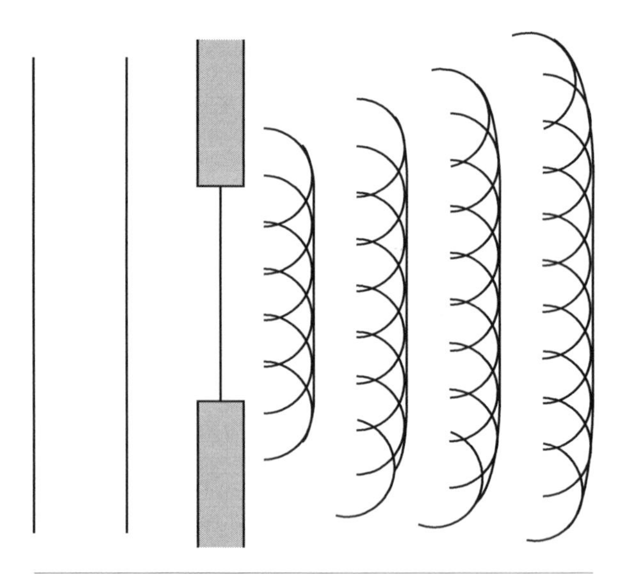

Figure 15.6: Diffraction using Huygen's principle.

Fresnel diffraction is more complex mathematically than Fraunhofer diffraction, so in an introductory optics course, we deal almost exclusively with Fraunhofer diffraction. The conditions for Fraunhofer diffraction are achieved experimentally by the setup shown in Fig. 15.7. Such an arrangement also ensures that coherent light is incident on the the aperture, which results in an interference pattern.

The Fresnel number

The *Fresnel number*, F, is a dimensionless quantity that indicates when the approximation for Fraunhofer diffraction is valid (Eq. 15.5 and Derivation 15.3). If $F \ll 1$, then the curvature of an incident wavefront is small, which means we can approximate to Fraunhofer diffraction. For larger values of F, such that $F \gtrsim 1$, the curvature of an incident wavefront is significant, and therefore we can no longer approximate to Fraunhofer diffraction. In the following sections, we will assume $F \ll 1$, such that the approximation for Fraunhofer diffraction is valid.

Fresnel number (Derivation 15.3):

$$F \simeq \frac{a^2}{8L\lambda} \qquad (15.5)$$

Fraunhofer diffraction: $F \ll 1$;
Fresnel diffraction: $F \gtrsim 1$.

15.3.2 Single slit

Light incident on a single slit of width a, where a is on the order of the wavelength of light, produces an

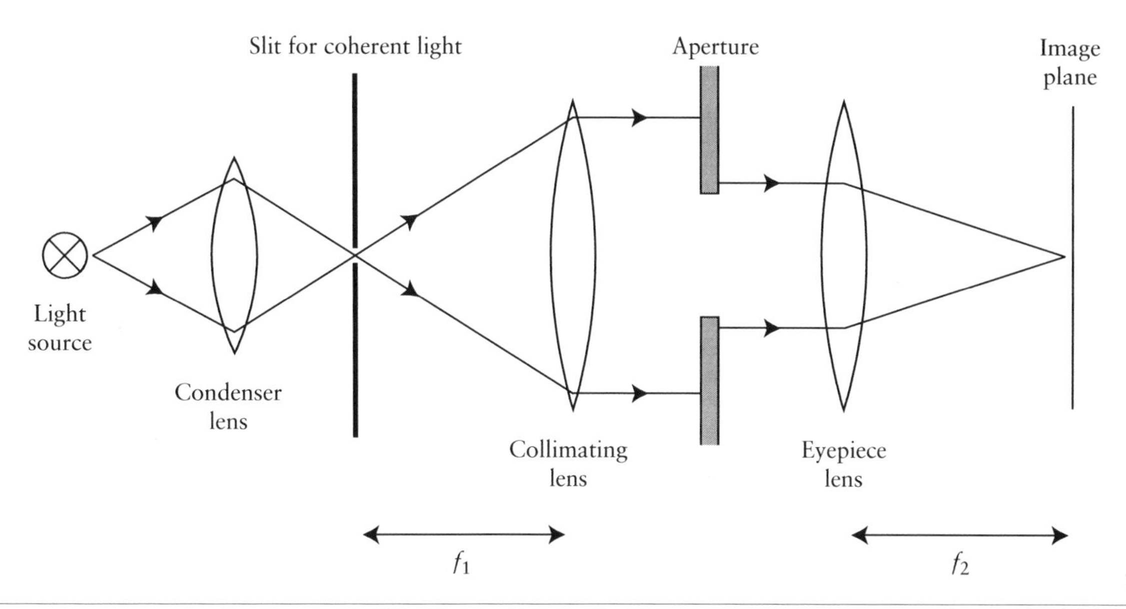

Figure 15.7: Experimental setup to achieve conditions required for Fraunhofer diffraction: (i) incoming light on diffracting aperture is parallel and (ii) the image plane is a large distance from the plane of the diffracting aperture. The condenser lens focuses light from a light source on a fine slit. Light diffracted through the slit is therefore coherent. If the slit is placed at the focal length of a collimating lens, then light passing through that lens emerges parallel to the principal axis. Light passing through the diffracting aperture is focused onto the image plane by an eyepiece lens.

Derivation 15.3: Derivation of the Fresnel number, *F* (Eq. 15.5)

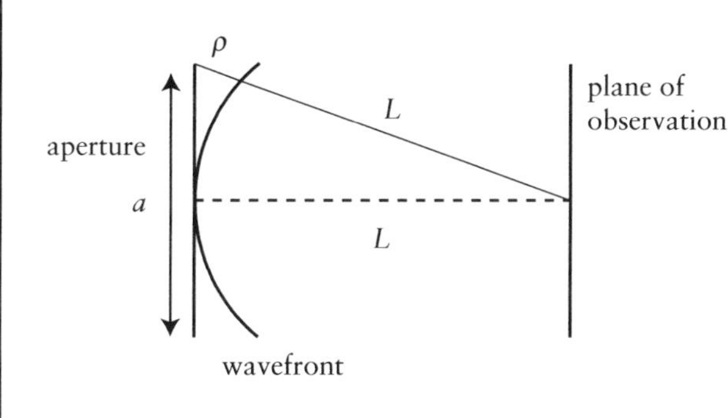

$$(L+\rho)^2 = \left(\frac{a}{2}\right)^2 + L^2$$

$$\text{and } (L+\rho)^2 = L^2 + \rho^2 + 2L\rho$$

$$\simeq L^2 + 2L\rho \quad \text{for } \rho \ll L$$

$$\therefore \rho \simeq \frac{a^2}{8L}$$

$$\therefore F = \frac{\rho}{\lambda} \simeq \frac{a^2}{8L\lambda}$$

- Starting point: consider a curved wavefront incident on an aperture. Let the curvature of the wavefront be defined by the parameter ρ, as shown in the figure. Assume that $\rho \ll L$, where L is the perpendicular distance between the aperture and plane of observation.
- Define the *Fresnel number*, F, as the ratio of curvature parameter ρ to the wavelength λ:

$$F = \frac{\rho}{\lambda}$$

- Using trigonometry, we find $\rho \simeq a^2 / 8L$, where a is the aperture width.
- For Fraunhofer diffraction: $\rho \ll \lambda \therefore F \ll 1$.
- For Fresnel diffraction: $\rho \gtrsim \lambda \therefore F \gtrsim 1$.

interference pattern of bright and dark fringes. We can find the intensity distribution of the interference fringes using phasor addition of interfering secondary Huygen wavelets that are produced in the slit (Eq. 15.6, Derivation 15.4, and Fig. 15.8). This intensity pattern is called the *single-slit diffraction pattern*.

Derivation 15.4: Derivation of the single-slit diffraction pattern (Eq. 15.6)

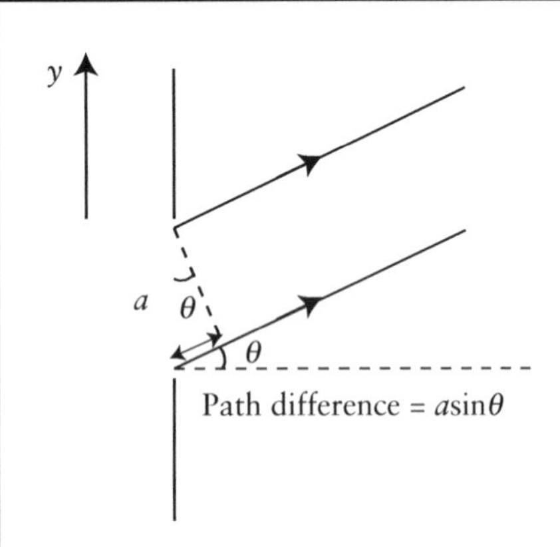

- Starting point: consider coherent light incident on a slit of width a. According to Huygen's principle, secondary wavelets are produced along the length of the slit. These interfere to produce a diffraction pattern.
- Let each wavelet be represented by a phasor $U_0 e^{-i\delta}$, where U_0 is the phasor amplitude and δ is the phase difference between two adjacent wavelets.
- The distance along the slit, y, is continuous, therefore δ is a continuous function of distance along the slit: $\delta = \delta(y)$.
- The resultant phasor of the interfering wavelets is the integral of individual wavelet phasors: $U(\delta) = \int U_0 e^{-i\delta(y)} dy$.
- Consider parallel rays at an angle θ to the normal. The phase difference between the two adjacent wavelets at this angle is $\delta(y) = ky \sin\theta = y\gamma$, where $\gamma = k\sin\theta$. The phase difference does not vary with time since the wavelets are coherent.
- Use the trigonometric identity:

$$e^{i\phi} - e^{-i\phi} = 2i\sin\phi$$

to simplify the expression.

- Let $\alpha = ka\sin\theta/2$ $(\equiv a\gamma/2)$.
- The intensity distribution as a function of θ, $I(\theta)$, equals the square of the resultant phasor amplitude: $I(\theta) = |U(\theta)|^2$.

$$U(\theta) = \int_0^a U_0 e^{-i\delta(y)} dy$$

$$= \int_0^a U_0 e^{-iy\gamma} dy$$

$$= U_0 \frac{\left(e^{-ia\gamma} - 1\right)}{-i\gamma}$$

$$= U_0 e^{-ia\gamma/2} \frac{\left(e^{-ia\gamma/2} - e^{ia\gamma/2}\right)}{-i\gamma}$$

$$= U_0 e^{-ia\gamma/2} \frac{2i\sin(a\gamma/2)}{i\gamma}$$

$$= U_0 a e^{-ia\gamma/2} \frac{\sin(a\gamma/2)}{(a\gamma/2)}$$

$$= U_0 a e^{-i\alpha} \frac{\sin\alpha}{\alpha}$$

$$\therefore I(\theta) = |U(\theta)|^2$$

$$= (U_0 a)^2 \left(\frac{\sin\alpha}{\alpha}\right)^2$$

$$\equiv I_0 \left(\frac{\sin\alpha}{\alpha}\right)^2$$

$$\text{where } \alpha = \frac{ka\sin\theta}{2}$$

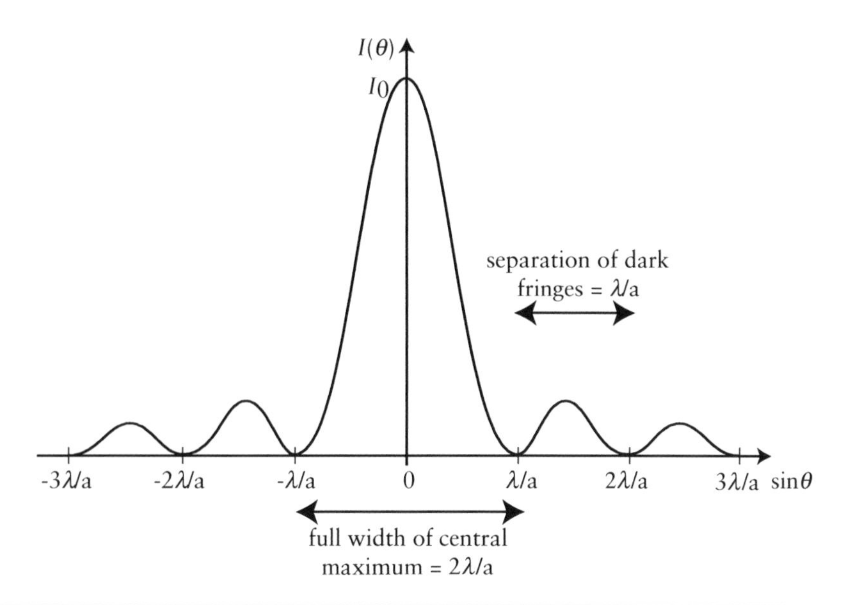

Figure 15.8: Single-slit diffraction pattern (refer to Eqs. 15.6–15.8).

Single slit interference pattern (Derivation 15.4):

$$I(\alpha) = I_0\left(\frac{\sin\alpha}{\alpha}\right)^2 \tag{15.6}$$

where $\alpha = \dfrac{ka\sin\theta}{2}$

For destructive interference:

$$a\sin\theta = m\lambda \tag{15.7}$$

where $m = \pm 1, \pm 2, \ldots$

Width of central maximum:

$$\Delta y = \frac{2L\lambda}{a} \tag{15.8}$$

At $\theta = 0$, the phase difference between all adjacent wavelets is zero; therefore, they constructively interfere to produce a central bright fringe of intensity I_0. The subsequent bright fringes are secondary maxima, with intensities $< I_0$, since not all wavelets are in phase (some wavelets destructively interfere). The dark fringes correspond to regions of net destructive interference between all wavelets, which occurs when the phasors "wrap around" such that their sum forms a closed circle ($\therefore$ resultant amplitude $= 0$). This corresponds to angles at which the path difference between two phasors at either extreme of the slit equals an integer number of wavelengths. Therefore, the condition for destructive interference is:

$$\underbrace{a\sin\theta}_{\text{path difference}} = m\lambda$$

where $m = \pm 1, \pm 2, \ldots$

(Eq. 15.7). We arrive at the same result if we consider that $I(a) = 0$ when $\sin^2 a = 0$, i.e. when $a = \dfrac{ka\sin\theta}{2} = m\pi$. The fringe at $m = 0$ corresponds to the bright central maximum. This is, of course, in agreement with the expression for $I(a)$ in Eq. 15.6 since $\sin\theta \simeq \theta$ as $\theta \to 0$:

$$\therefore I(\alpha \to 0) \simeq I_0\left(\frac{\alpha}{\alpha}\right)^2 = I_0$$

Using the condition for destructive interference (Eq. 15.7), the width of the central maximum is:

$$\Delta y = 2L\sin\theta$$
$$= \frac{2L\lambda}{a}$$

(Eq. 15.8). The width of the central maximum is double that of the secondary maxima (refer to Fig. 15.8).

15.3.3 Circular apertures

The diffraction pattern of a circular aperture consists of a central bright spot, with concentric dimmer rings around it that correspond to secondary maxima (Fig. 15.9). This intensity distribution is called an *Airy pattern*, and the central bright spot is the *Airy disk*; the Airy pattern has a similar functional dependence to the single-slit intensity distribution, but in this case, the first minimum occurs when:

$$\sin\theta = 1.22\frac{\lambda}{a}$$

rather than when $\sin\theta = \lambda / a$, which was the case for a linear aperture.

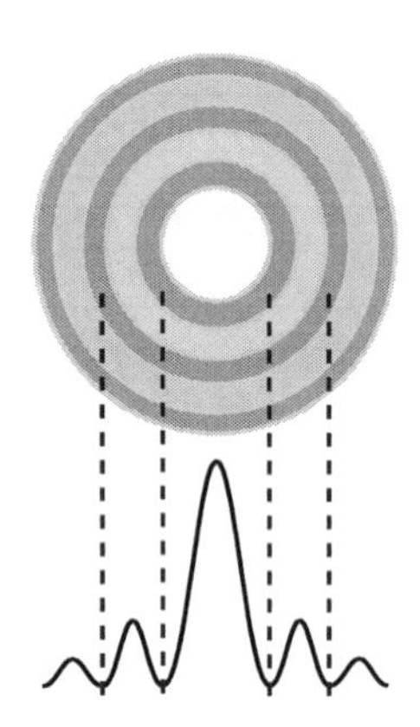

Figure 15.9: Airy pattern: the diffraction pattern of a spot of light formed by a circular aperture. The central bright maxima is called the Airy disk.

Resolution limit

Diffraction through a circular lens means that the image of a point source of light is not a bright dot, but is spread out in an Airy pattern. This phenomenon occurs in optical devices such as telescopes and cameras, and it limits their ability to resolve two small nearby objects, such as two nearby stars. Diffraction therefore imposes a *resolution limit* on an optical device, which corresponds to the minimum angular separation between two objects below which they cannot be discerned as two separate bodies. The limit at which two objects are *just* resolved is known as the *Rayleigh criterion*. This criterion is that the minimum angular separation between two objects, θ_{min}, corresponds to the separation when the central maximum of one Airy pattern coincides with the first minimum of the other:

$$\therefore \sin(\theta_{min}) = 1.22\frac{\lambda}{a}$$

(Eq. 15.9 and Fig. 15.10). The linear resolution of an optical device depends on the focal length of the diffracting lens. The minimum separation between two just-resolved objects is:

$$\Delta x_{min} = f\sin(\theta_{min}) = 1.22\frac{f\lambda}{a}$$

(Eq. 15.10). As a rule of thumb, we say that for far-away objects, $\sin\theta \simeq \theta$; therefore, the resolution limit is:

$$\theta_{min} \simeq \frac{\lambda}{a}$$

(Eqs. 15.11 and 15.12).

Resolution limit:

Rayleigh criterion:

$$\sin(\theta_{min}) = 1.22\frac{\lambda}{a} \qquad (15.9)$$

Linear resolution:

$$\Delta x_{min} = f\sin(\theta_{min}) = 1.22\frac{f\lambda}{a} \qquad (15.10)$$

For far-away objects, $\sin\theta \simeq \theta$:

$$\therefore \theta_{min} \simeq \frac{\lambda}{a} \qquad (15.11)$$

$$\text{and } \Delta x_{min} \simeq \frac{f\lambda}{a} \qquad (15.12)$$

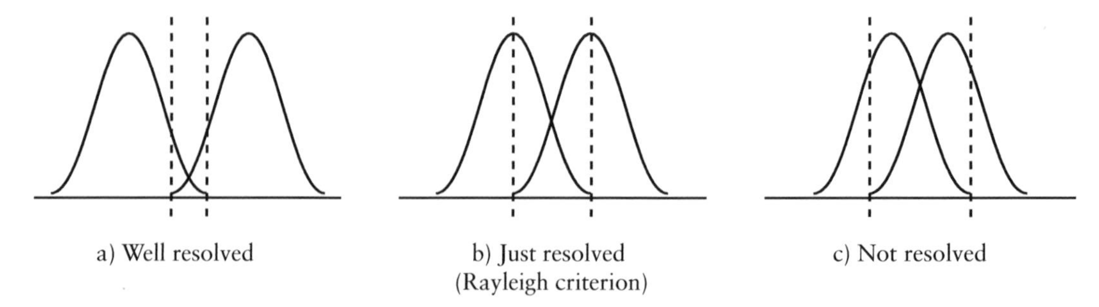

Figure 15.10: The Rayleigh criterion and the resolution limit. Two point sources are a) well resolved if their diffraction patterns have little overlap, b) just resolved if the peak of one diffraction pattern coincides with the first minimum of the other (Rayleigh criterion), and c) not resolved if the angular separation is less than the Rayleigh criterion.

15.4 The diffraction grating

A *diffraction grating* is an optical device that consists of multiple thin slits. Light incident on a diffraction grating produces an interference pattern similar to Young's double-slit interference pattern, but the interference fringes have a larger separation between them. We determine the intensity distribution in a similar way as we did for Young's double-slit experiment or a single aperture: using phasor addition of interfering secondary Huygen wavelets that are produced in each slit of the grating. Doing this, we see that the interference distribution of a diffraction grating depends on the number of slits, N, in the grating (Eq. 15.13, Derivation 15.5, and Fig. 15.11).

The bright maxima correspond to angles at which the path difference between two adjacent phasors is an integral number of wavelengths; at these angles, the wavelets from each slit are in phase with each other and they constructively interfere:

$$\underbrace{d\sin\theta}_{\text{path difference}} = n\lambda$$

(Eq. 15.14). The dark fringes correspond to regions of net destructive interference between all wavelets. At these points, the intensity pattern is $I(\beta) = 0$. From Eq. 15.13, this occurs when:

$$\sin(N\beta) = \sin(m\pi)$$

$$\therefore \frac{Nkd\sin\theta}{2} = m\pi$$

$$\therefore d\sin\theta = \frac{2\pi m}{kN} = \frac{m\lambda}{N}$$

where $m = 1, 2, 3,\ldots, (N-1)$. (The points at which $m = 0$ or N correspond to the condition for constructive interference, so these are the maxima.) Therefore, there are $N - 1$ minima between two maxima (refer to Fig. 15.11). Additionally, from the condition for destructive interference, we see that the width of the maxima is proportional to $1 / N$. Therefore, a change in N for a constant slit spacing d has the following effects on the interference pattern of a diffraction grating: for larger N, (i) the number of minima between two maxima, $\#_{min}$, is greater ($\#_{min} = N-1$), (ii) the width of the maxima, Δ_{max}, are narrower ($\Delta_{max} \propto 1 / N$) (but are in the same positions for all N since the condition for constructive interference does not depend on N), and (iii) the maximum intensity, I_{max}, is reduced ($I_{max} \propto 1 / N^2$).

Diffraction grating (Derivation 15.5):

Intensity pattern:

$$I(\beta) = \frac{I_0}{N^2}\left(\frac{\sin(N\beta)}{\sin\beta}\right)^2 \qquad (15.13)$$

$$\text{where } \beta = \frac{kd\sin\theta}{2}$$

For constructive interference:

$$d\sin\theta = n\lambda \qquad (15.14)$$

where $n = 0, \pm1, \pm2,\ldots$

For destructive interference:

$$d\sin\theta = \frac{m\lambda}{N} \qquad (15.15)$$

where $m = 1, 2,\ldots, (N - 1)$.

Finite-width slits

If the slits are not of negligible width, but have a finite width a, then secondary wavelets *within* a slit interfere, as well as wavelets *between* slits. The resulting intensity pattern is a combination of the multiple-slit diffraction grating pattern (Eq. 15.13), with an intensity envelope characteristic of the single-slit diffraction pattern (Eq. 15.6) (Fig. 15.12):

$$I(\beta) = \frac{I_0}{N^2}\left(\frac{\sin(N\beta)}{\sin\beta}\right)^2\left(\frac{\sin\alpha}{\alpha}\right)^2$$

$$\text{where } \beta = \frac{kd\sin\theta}{2}$$

$$\text{and } \alpha = \frac{ka\sin\theta}{2}$$

Double-slit pattern

We recover the double-slit interference pattern by setting $N = 2$ in Eq. 15.13 and using the double-angle formula $\sin 2\phi = 2\sin\phi\cos\phi$:

$$I_2(\beta) = I_0\left(\frac{\sin(2\beta)}{2\sin\beta}\right)^2$$

$$= I_0(\cos\beta)^2$$

$$= I_0\cos^2\beta$$

Derivation 15.5: Derivation of the diffraction grating pattern (Eq. 15.13)

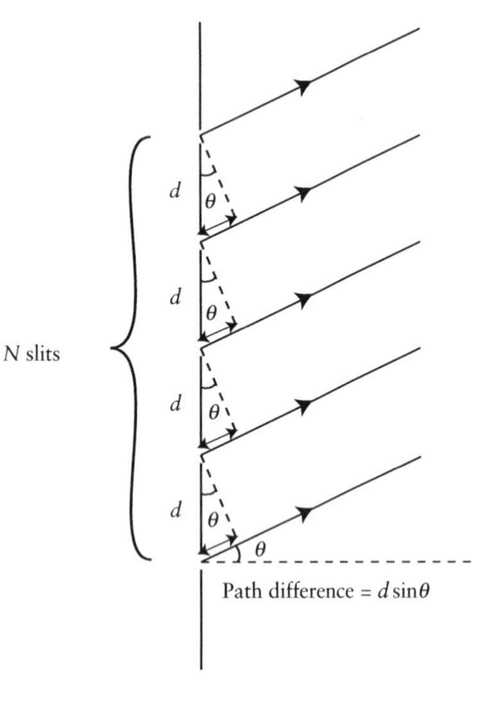

- Starting point: consider coherent light incident on a diffraction grating with N slits, each a distance d apart. According to Huygen's principle, secondary wavelets emanate radially from each slit and interfere with each other where they overlap.
- Let each wavelet be represented by a phasor $U_0 e^{-i\delta}$, where U_0 is the phasor amplitude and δ is the phase difference between two wavelets in adjacent slits.
- The resultant phasor of the interfering wavelets is the sum of individual wavelet phasors:

$$U(\delta) = \sum_{n=0}^{N-1} U_0 e^{-in\delta}.$$

- Consider parallel rays at an angle θ to the normal. The phase difference between the two wavelets in adjacent slits at this angle is $\delta = kd \sin\theta = 2\beta$, where $\beta = kd \sin\theta / 2$. The phase difference does not vary with time since the wavelets are coherent.
- The sum of wavelet phasors corresponds to a geometric progression with first term $a = U_0$ and ratio $r = e^{-2i\beta}$. The sum of a finite geometric progression is:

$$\sum_{p=0}^{N} ar^p = \frac{a\left(1-r^{N+1}\right)}{(1-r)}$$

- Use the trigonometric identity:

$$e^{i\phi} - e^{-i\phi} = 2i\sin\phi$$

to simplify the expression.

- The intensity distribution as a function of θ, $I(\theta)$, equals the square of the resultant phasor amplitude: $I(\theta) = |U(\theta)|^2$.
- Normalization: in the small-angle approximation, $\sin\theta \simeq \theta$. Therefore, the maximum intensity is $I_0 = I(\theta \to 0) = U_0^2 N^2$. Since the maximum intensity scales inversely with N^2, let the coefficient be $U_0^2 \to I_0 / N^2$.

$$\begin{aligned}
U(\theta) &= U_0 + U_0 e^{-i\delta} + U_0 e^{-i2\delta} + \ldots + U_0 e^{-i(N-1)\delta} \\
&= \sum_{n=0}^{N-1} U_0 e^{-in\delta} \\
&= \sum_{n=0}^{N-1} U_0 e^{-2in\beta} \\
&= U_0 \left(\frac{1 - e^{-2iN\beta}}{1 - e^{-2i\beta}} \right) \\
&= U_0 \frac{e^{-iN\beta}}{e^{-i\beta}} \left(\frac{e^{iN\beta} - e^{-iN\beta}}{e^{i\beta} - e^{-i\beta}} \right) \\
&= U_0 \frac{e^{-iN\beta}}{e^{-i\beta}} \left(\frac{\sin(N\beta)}{\sin\beta} \right)
\end{aligned}$$

$$\begin{aligned}
\therefore I(\theta) &= |U(\theta)|^2 \\
&= U_0^2 \left(\frac{\sin(N\beta)}{\sin\beta} \right)^2 \\
&= \frac{I_0}{N^2} \left(\frac{\sin(N\beta)}{\sin\beta} \right)^2
\end{aligned}$$

where $\beta = \dfrac{kd\sin\theta}{2}$

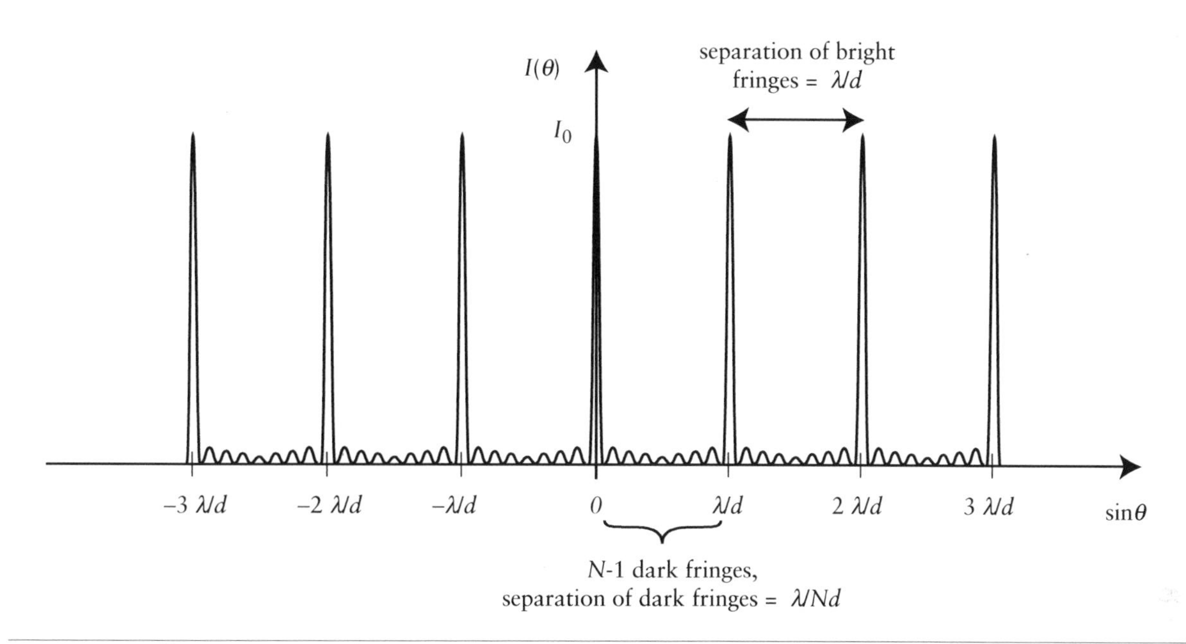

Figure 15.11: Diffraction grating interference pattern (refer to Eqs. 15.13–15.15).

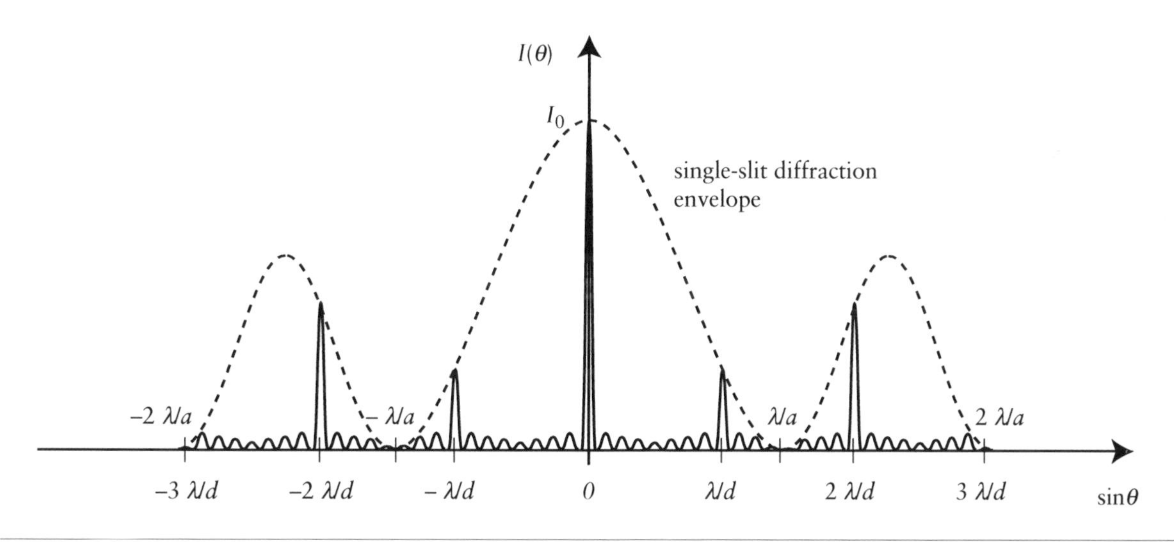

Figure 15.12: Diffraction grating interference pattern with finite slit width, a. The intensity of the interference pattern is modulated by a single-slit diffraction envelope, $(\sin\alpha / \alpha)^2$, where $\alpha = ka \sin\theta / 2$.

Single-slit pattern

As θ decreases in Eq. 15.13, $\sin\beta \rightarrow \beta$ before $\sin(N\beta) \rightarrow (N\beta)$. In this limit, the intensity pattern resembles the single-slit diffraction pattern:

$$I(\beta) \rightarrow I_0 \left(\frac{\sin(N\beta)}{(N\beta)} \right)^2$$

where the equivalent slit width a is equal to Nd.

Angular dispersion

The positions of the diffraction maxima depend on the wavelength of incident light:

$$d \sin\theta = n\lambda$$

Therefore, if non-monochromatic light, for example white light, is incident on a grating, the diffraction maxima are separated into different wavelength components. This provides a way of separating and identifying different wavelength components in an unknown sample of light.

The *angular dispersion*, D, of a diffraction grating is defined as:

$$D = \frac{d\theta}{d\lambda}$$

Differentiating the condition for constructive interference (Eq. 15.14), we have:

$$d\cos\theta \frac{d\theta}{d\lambda} = n$$

Therefore, the angular dispersion of a diffraction grating is:

$$\therefore \frac{d\theta}{d\lambda} = \frac{n}{d\cos\theta} = \frac{\tan\theta}{\lambda}$$

(Eq. 15.16).

Resolving power

The angular separation between spectral components of different wavelength tells us whether two fringes of different wavelength can be resolved. Recall that the Rayleigh criterion states that two images are just resolved if the maximum of one diffraction pattern coincides with the first minimum of the other.

Differentiating the condition for destructive interference (Eq. 15.15), we have:

$$d\cos\theta \frac{d\theta}{dm} = \frac{\lambda}{N}$$

The angular separation between the principal maximum and first minimum corresponds to $dm = 1$, therefore the angular separation for the Rayleigh criterion is:

$$\Delta\theta_{\text{Rayleigh}} = \frac{\lambda}{Nd\cos\theta}$$

In addition, from Eq. 15.16, the angular separation of two spectral components with a wavelength difference of $\Delta\lambda$ is:

$$\Delta\theta_{\Delta\lambda} = \frac{\tan\theta}{\lambda}\Delta\lambda = \frac{n}{d\cos\theta}\Delta\lambda$$

Therefore, using the Rayleigh criterion, these two components are resolved if:

$$\begin{aligned} \Delta\theta_{\Delta\lambda} &\geqslant \Delta\theta_{\text{Rayleigh}} \\ \frac{n}{d\cos\theta}\Delta\lambda &\geqslant \frac{\lambda}{Nd\cos\theta} \\ \Rightarrow \therefore nN &\geqslant \frac{\lambda}{\Delta\lambda} \end{aligned}$$

(refer to Fig. 15.10). Said another way, the nth maximum of two spectral lines of wavelength λ and $\lambda + \Delta\lambda$ are resolved if the above condition is satisfied.

The *resolving power*, R, of a diffraction grating is defined as:

$$R = \frac{\lambda}{\Delta\lambda}$$

where $\Delta\lambda$ corresponds to the smallest resolvable wavelength difference. This is a parameter that defines the performance of a diffraction grating. Therefore, from the Rayleigh criterion, we see that the resolving power of a diffraction grating depends on the number of slits, N, and the maximum diffraction order observed, n: $R = nN$ (Eq. 15.17).

Grating dispersion and resolving power:

Angular dispersion:

$$D = \frac{d\theta}{d\lambda} = \frac{\tan\theta}{\lambda} \quad (15.16)$$

Resolving power:

$$R = \frac{\lambda}{\Delta\lambda} = nN \quad (15.17)$$

Part IV
Quantum Physics

Introduction

Quantum physics is the study of the behavior and properties of matter and radiation on atomic and sub-atomic scales ($\lesssim 10^{-10}$ m). On these very small, so-called "quantum" scales, our familiar classical laws break down, and matter behaves in strange and counterintuitive ways. A major breakthrough of quantum physics was the discovery that waves have a particle-like nature and particles have a wave-like nature; wave-like particles are referred to as *matter waves* (Chapter 16). The behavior of matter waves is described by the *Schrödinger equation*, which is the quantum equivalent of Newton's Classical Laws of Motion (Chapter 17). The properties of matter waves are probabilistic. In contrast, the properties of moving bodies in classical or relativistic physics are deterministic. In this way, experimental measurements of quantum systems are unpredictable—one can only determine the *probability* of a given outcome (Chapter 18). The final two chapters of this part are concerned with applying quantum physics to atomic electron states. First, we cover angular momentum properties of atomic electrons in general (Chapter 19), and then apply the theory to the hydrogen atom, which is the best understood quantum system (Chapter 20).

Practical applications

Practical applications of quantum mechanics include understanding atomic structure and atomic spectra, nuclear and particle physics, and properties of materials, such as semiconductors, paramagnets, and metals.

My two cents

Quantum mechanics is an abstract and difficult subject. The quantum mechanical world is contradictory to common sense, and we have no day-to-day experiences with which to make analogies. To understand the concepts requires a mathematical formalism—linear algebra—which itself is also very abstract. An introductory undergraduate course, therefore, consists mainly of setting the foundation with some mathematically tractable examples, on which we can then build in order to understand atomic structure (atomic physics) and properties of materials (condensed matter physics). However, the danger with solving "simple" systems—such as the infinite square potential well, which can be solved easily using standard differential equations—is that we can quickly lose sight of what the solutions are *telling* us about the quantum world. Furthermore, in solving the infinite square potential well it is often not emphasized how the solutions ("standing waves in a rectangle") apply to a more physically relevant system. So my advice would be, even if you've solved the infinite square well countless times, take some time to be sure you understand the implications of what the solutions tell us about a particle's state, since many of the concepts that arise in this straightforward example apply to more complex atomic systems.

Recommended books

- Eisberg, R. and Resnick, R., 1985, *Quantum Physics of Atoms, Molecules, Solids, Nuclei, and Particles*, 2nd edition. Wiley.
- Gasiorowicz, S., 2003, *Quantum Physics*, 3rd edition. Wiley.
- Griffiths, D. J., 2004, *Introduction to Quantum Mechanics*, 2nd edition. Benjamin Cummings.
- Rae, A., 2002, *Quantum Mechanics*, 4th edition. Institute of Physics Publishing.
- Shankar, R., 2008, *Principles of Quantum Mechanics*, 2nd edition. Springer.

The book by Eisberg and Resnick is a thorough, descriptive book that covers a whole range of topics covered in an undergraduate physics course: introductory quantum physics, atomic physics, statistical mechanics, condensed matter physics, and nuclear and particle physics. The other books focus just on quantum physics; Gasiorowicz, Griffiths and Rae cover the material at a level that's appropriate for an introductory course, while Shankar's book is more advanced. In my opinion, Rae's book is particularly clear, and provides a good level of discussion and description.

Chapter 16
Background to quantum mechanics

Quantum physics is the study of matter and radiation (electromagnetic waves) on atomic and sub-atomic scales. On these scales, matter and radiation do not behave as we would expect from classical theory. In the early 1900s, through a series of breakthrough experiments, the discovery of the wave–particle duality, whereby waves have a particle-like nature and particles have a wave-like nature, lead to the emergence of quantum mechanics. This chapter provides an overview of the background and postulates of quantum mechanics.

16.1 Photons and matter waves

Some of the key experiments that lead to the development of quantum theory involved measuring the properties of radiation emitted from matter, and their results were incompatible with classical predictions. In particular, through the incompatibility of black-body radiation and the photoelectric effect with classical expectations (see below), it was postulated that electromagnetic radiation is not a continuous wave, but comes in discrete "packets" of energy, called photons. The energy of a photon of *frequency*[1] ν was found to be $E = h\nu$, where h is a constant (see Section 16.2). Not only did these observations imply that waves have a particle-like nature, but they also implied that the energy of radiation is *quantized* into discrete values, where one "quantum of energy" is equal to $h\nu$ [J]. A brief overview of thermal radiation and the photoelectric effect are summarized below.[2]

Thermal radiation

A *black body* is an object that is a perfect absorber and perfect emitter of all frequencies of radiation. According to classical theory, the intensity of radiation emitted by a black body was predicted to be proportional to the square of its frequency: $I \propto \nu^2$ (this is the *Rayleigh–Jeans* Law). If this were the case, then the total energy emitted by a black body would tend to infinity due to its high-frequency (ultraviolet) components. This problem was referred to as the *ultraviolet catastrophe*, since it is physically impossible for an object to emit infinite power. This catastrophe was resolved by Planck and Einstein, who postulated that radiation was, in fact, emitted in discrete packets—photons—where the energy of a photon is $E = h\nu$. This being the case, high-frequency photons would be emitted in larger packets of energy. Therefore, the intensity of high-frequency radiation emitted by a black body would be less than its classical prediction, and would eventually tend to zero at very high frequencies. The intensity spectrum of radiation emitted by a black body is given by *Planck's Law*, which falls to zero at high-frequencies (Fig. 16.1, refer to Section 24.6.4 for more detail). The Rayleigh–Jeans Law is an approximation to Planck's Law in the low-frequency limit.

The photoelectric effect

The photoelectric effect is the phenomenon whereby electrons are emitted from a metal surface as a result of atomic absorption of electromagnetic waves (light), which provide the energy necessary to remove an electron from the surface.

The minimum energy required to remove an electron is called the *work function*, ϕ, which is measured in *electron volts* [eV].[3] The energy of an emitted electron is then:

$$E_e = \phi + T_e$$

1 It is also common to use the variable f to represent frequency.

2 Most of you should already be familiar with these experiments—if not, take the time to read about them now. For more detail, the reader is referred to the book by Eisberg and Resnick.

3 The electron volt is a unit of energy commonly used in relation to quantum phenomenon, where energies reach very small values when measured using the SI unit of Joules. It is defined as "the kinetic energy gained by an electron by accelerating it through a potential difference of 1 V". Therefore, the energy in J of 1 eV is $E = e \cdot V = 1.602 \cdot 10^{-19}$ J, where e is the electron charge ($e = 1.602 \cdot 10^{-19}$ C) and V is the potential difference ($V = 1$ V, where $1\ \mathrm{V} \equiv 1\ \mathrm{J\ C^{-1}}$).

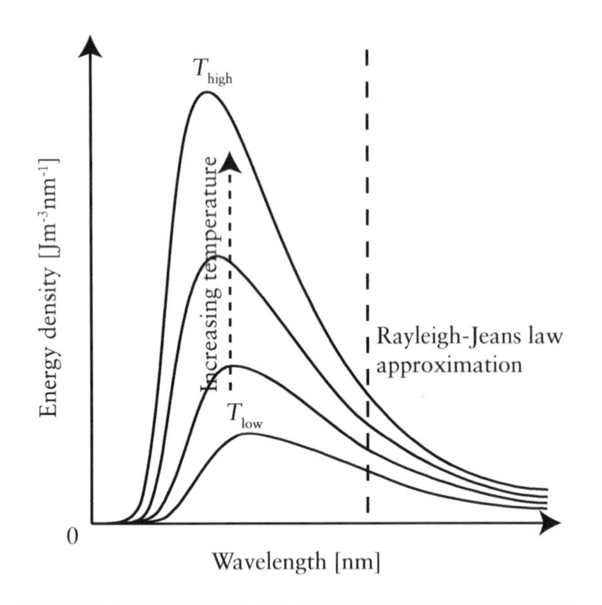

Figure 16.1: Planck's Law: emission spectrum of a black body. Planck's Law approximates to the Rayleigh–Jeans Law in the low-frequency (high-wavelength) limit.

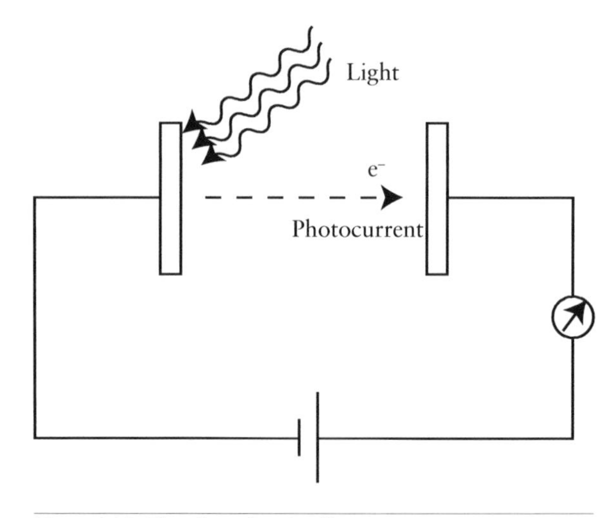

Figure 16.2: The photoelectric effect.

where T_e is the electron's kinetic energy. The classic experiment involved shining light on a metal cathode to liberate electrons, and to measure the current produced by emitted electrons, e$^-$ (i.e. the photocurrent) (Fig. 16.2). If light were a continuous wave, the classical hypothesis was that light of *any* frequency at a high enough intensity would eventually bring enough energy to the cathode to liberate electrons.

What was in fact observed was that there was a threshold frequency of light below which no electrons were emitted, even at high light intensities. This observation can be explained by the fact that light is not a continuous wave, but consists of discrete photons, where one photon carries an energy of magnitude $E = h\nu$. Therefore, a photon with energy $h\nu < \phi$ has insufficient energy to liberate an electron from the metal surface, no matter what the intensity of incident light is.

Matter waves

Shortly after the existence of photons (particle-like waves) was postulated, it was also discovered that electrons and other small particles (neutrons and helium, for example) have a wave-like nature. For example, the wave-like nature of electrons was demonstrated in the *Davisson–Germer* experiment, which was essentially a diffraction experiment for electrons. Electrons were "boiled off" a heated filament, and accelerated through a vacuum onto a nickel crystal, off which they were scattered (Fig. 16.3). Using an electron detector, it was found that the intensity of scattered electrons was not uniform in all directions, as would be predicted by classical theory, but had local peaks and troughs. The scattering pattern resembled a diffraction pattern with peaks corresponding to regions of constructive interference and troughs to regions of

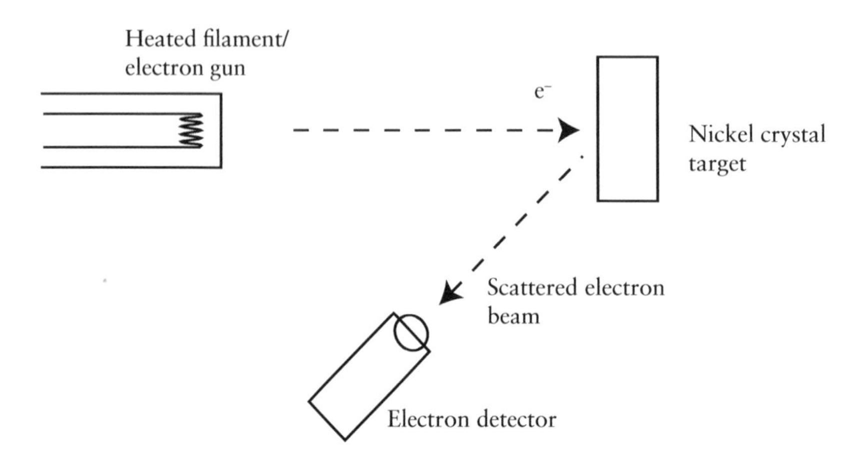

Figure 16.3: The Davisson–Germer experiment.

destructive interference. These observations implied that particles could behave as waves. Wave-like particles are referred to as *matter waves*. Throughout the next few chapters, we will introduce some of the strange and counterintuitive properties of matter waves.

16.2 Wave–particle duality

A principal breakthrough of quantum theory was the discovery that waves can behave as particles, and particles can behave as waves. This phenomenon is known as the *wave–particle duality*.

In classical physics, we are familiar with describing waves by their frequency, ω [s^{-1}], and wavenumber, k [m^{-1}] (or wavelength, $\lambda = 2\pi/k$), and with describing particles of mass m by their energy, E [J], and momentum, p [kg ms^{-2}]. Therefore, what the wave–particle duality implies in theoretical terms is that all particles have an associated frequency and wavenumber, and all waves have an associated energy and momentum.

The Planck and de Broglie relations

Max Planck discovered that E and ω are related to one another (for both particles and waves) via a seemingly simple relation, known as the *Planck relation*. This relation shows that E is proportional to ω (or to linear frequency, ν, where $\omega = 2\pi\nu$):

$$E = \hbar\omega$$
$$\text{or } E = h\nu$$

(Eq. 16.1). The proportionality constant in the top expression is the *reduced Planck's constant*, $\hbar$ [$1.05 \cdot 10^{-34}$ J s]. This is related to *Planck's constant*, h [$6.63 \cdot 10^{-34}$ J s] by a factor of 2π :

$$\hbar = \frac{h}{2\pi}$$

(Eq. 16.3). The small order of magnitude of h and $\hbar$ is significant in that it "sets the scale" at which quantum phenomenon manifest themselves (i.e. atomic scales, as we shall see below).

Around the same time, de Broglie discovered that p and k (or $\lambda = 2\pi/k$) are also proportional:

$$p = \hbar k$$
$$\text{or } p = \frac{h}{\lambda}$$

(Eq. 16.2). This is known as the *de Broglie relation* or the *de Broglie–Einstein relation*. The wavelength of a matter wave is called the *de Broglie wavelength*, $\lambda_B = 2\pi/k$. Therefore, the momentum of a matter wave is $p = h / \lambda_B$.

Wave–particle duality:

Planck's relation:

$$\begin{aligned} E &= \hbar\omega \\ &\equiv h\nu \end{aligned} \qquad (16.1)$$

de Broglie–Einstein relation:

$$\begin{aligned} p &= \hbar k \\ &\equiv \frac{h}{\lambda} \end{aligned} \qquad (16.2)$$

where the reduced Planck's constant is:

$$\hbar = \frac{h}{2\pi} \qquad (16.3)$$

Quantum scales

On what scale is the wave nature of matter apparent? We know from optics that wave phenomena (such as diffraction and interference) are apparent when the wavelength is of the same order of magnitude as the diffracting object or interference slit. Similarly, wave-like quantum phenomena are apparent when the de Broglie wavelength of a particle is comparable to its dimensions, or to its confinement in space.

It turns out that this corresponds to atomic or subatomic scales (i.e. $\lesssim 10^{-10}$ m). It's straightforward to see this if we consider an electron in a hydrogen atom. The mass of an electron is $m_e \approx 9.1 \cdot 10^{-31} \sim 10^{-30}$ kg. If we approximate its velocity to $v \sim 0.01c \sim 10^6$ ms^{-1}, then its momentum is:

$$\begin{aligned} p_e &= m_e v \\ &\sim 10^{-30} \cdot 10^6 \\ &\sim 10^{-24} \text{ kgms}^{-1} \end{aligned}$$

From Eq. 16.2, its de Broglie wavelength is then:

$$\begin{aligned} \lambda_e &= \frac{h}{p_e} \\ &\sim \frac{10^{-34}}{10^{-24}} \\ &\sim 10^{-10} \text{ m} \end{aligned}$$

This is the same order of magnitude as the dimensions of an atom—thus, quantum scales correspond to atomic scales. Quantum scales are set by Planck's constant, h, which is equivalent to relativistic effects being apparent when objects travel at speeds approaching the speed of light, c.

In contrast, we don't see quantum phenomena on "every-day" classical scales since the de Broglie wavelengths of classical objects are negligible compared to their dimension. For example, consider a bouncing ball with momentum $p = 10$ kg m s^{-1}. From Eq. 16.2, its de Broglie wavelength is therefore:

$$\begin{aligned}\lambda_B &= \frac{h}{p} \\ &\sim \frac{10^{-34}}{10} \\ &\sim 10^{-35}\ \text{m} \ll r_{\text{ball}}\end{aligned}$$

This is obviously far smaller than the ball's dimensions ($r_{\text{ball}} \sim 0.1$–1 m). Therefore, in this regime, we use a classical treatment.

16.3 Postulates of quantum mechanics

Quantum mechanics is based on a few postulates, which cannot be proved or deduced. However, the postulates are in very good agreement with experiment. Therefore, as long as they are not disproved by experiment, they are considered as axioms (i.e. non-provable, true statements). The four postulates are: the wavefunction postulate, the operator postulate, the measurement postulate, and the time-evolution postulate. These are discussed in order in this section and summarized in Summary box 16.1.

16.3.1 Wavefunction postulate

On quantum scales, we cannot treat particles of mass m as hard spheres located at position x at time t since this does not incorporate their wave nature. Instead, we represent particles by a function that varies in space and time: this function is called a *wavefunction*, $\psi(x,t)$. Wavefunctions do not represent a disturbance of matter itself—they cannot be interpreted in the same way as the "classical waves" that we are familiar with. In fact, we shall see in this chapter that wavefunctions are *complex* functions. Therefore, they cannot represent anything "real" by their nature. Instead, they should be regarded as abstract, mathematical *tools*. Throughout this chapter, we shall see how to use and interpret this tool in a way that is physically meaningful.

The *wavefunction postulate* states that:

Associated with any quantum mechanical particle is a wavefunction, $\psi(x,t)$, that determines everything that can be known about the system.

Summary box 16.1: Postulates of quantum mechanics

1. **Wavefunction postulate:** associated with any quantum mechanical particle is a wavefunction, $\psi(x,t)$, that determines everything that can be known about the system. The probability of finding a particle between points x and $x + \mathrm{d}x$ at time t is given by:

$$P(x,t)\mathrm{d}x = \psi^*(x,t)\psi(x,t)\mathrm{d}x = |\psi(x,t)|^2\,\mathrm{d}x$$

2. **Operator postulate:** associated with every physical observable, q, there is a mathematical operator, $\hat{Q}$, which is used in conjunction with the wavefunction. The eigenvalue equation for a physical observable q is:

$$\hat{Q}\psi(x,t) = q\psi(x,t)$$

where q is the eigenvalue for $\hat{Q}$.

3. **Measurement postulate:** measurement of a physical observable q corresponds to the expectation value of the corresponding operator, $\hat{Q}$:

$$\langle \hat{Q} \rangle = \int_{-\infty}^{\infty} \psi^*(x,t)\hat{Q}\psi(x,t)\mathrm{d}x$$

4. **Time-evolution postulate:** the time evolution of the wavefunction is governed by the Schrödinger equation:

$$\hat{H}\psi(x,t) = \mathrm{i}\hbar\frac{\partial\psi(x,t)}{\partial t}$$

For example, consider a sub-atomic particle such as an electron. We can define the *state* of that electron by variables such as its energy, its momentum, its position, etc. It turns out that *all* these variables that define the electron's state are contained within its wavefunction. There are elegant mathematical manipulations that then allow us to "extract" the relevant information from it, as we shall see throughout this chapter.

Wavefunction and probability

The second part of the wavefunction postulate provides us with a somewhat more conceptual interpretation of what a wavefunction represents. It states that:

The probability of a particle being found between positions x and $x + \mathrm{d}x$ is:

$$P(x,t)\mathrm{d}x = |\psi(x,t)|^2\,\mathrm{d}x \equiv \psi^*(x,t)\psi(x,t)\mathrm{d}x$$

where $\psi^*(x,t)$ is the complex conjugate of $\psi\,(x,t)$. Therefore, the product $\psi^*(x,t)\,\psi(x,t)$ is the *probability density function*[4] (the probability per unit distance) of a quantum mechanical particle (Eq. 16.4). The probability that a particle is located between two points, a and b, at a time t is therefore found by integrating the probability density function over that interval:

$$P_{ab}(t) = \int_a^b |\psi(x,t)|^2\,\mathrm{d}x$$

(Eq. 16.5 and Fig. 16.4).

> **Wavefunction and probability:**
>
> Probability density function:
>
> $$P(x,t) = |\psi(x,t)|^2 = \psi^*(x,t)\psi(x,t) \qquad (16.4)$$
>
> Probability that a particle is located between a and b at time t:
>
> $$P_{ab}(t) = \int_a^b |\psi(x,t)|^2\,\mathrm{d}x \qquad (16.5)$$

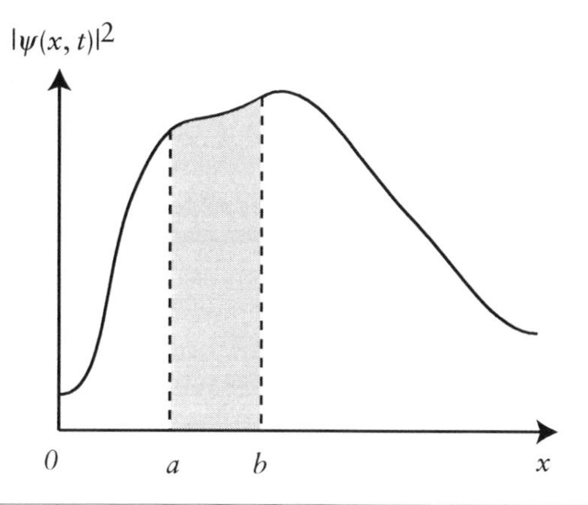

Figure 16.4: The probability that a particle is located between a and b is the integral of the probability density function, $|\psi\,(x,t)|^2$, over that interval.

Constraints on the wavefunction

There are a couple of additional constraints that $\psi\,(x,t)$ must satisfy in order to accurately describe the wavefunction of a particle. First, the probability of finding the particle *somewhere* must be one; therefore, the wavefunction must be normalized such that the probability integrated over all space equals one (Eq. 16.6). Second, both $\psi\,(x,t)$ and its slope $(\partial\psi/\partial x)$ must be a continuous function of x. Therefore, at a boundary $x = a$, the wavefunction and its derivative on either side of the boundary must be equal at $x = a$ (Eq. 16.7).

> **Constraints on the wavefunction:**
>
> Normalization condition:
>
> $$P_{\text{total}} = \int_{-\infty}^{\infty} |\psi(x,t)|^2\,\mathrm{d}x = 1 \qquad (16.6)$$
>
> Boundary conditions (boundary at $x = a$):
>
> $$\psi_1(a,t) = \psi_2(a,t)$$
>
> $$\text{and} \quad \left(\frac{\partial\psi_1}{\partial x}\right)_a = \left(\frac{\partial\psi_2}{\partial x}\right)_a \qquad (16.7)$$

Are plane waves wavefunctions?

Let's consider a wavefunction that has the form of a plane wave of frequency ω and wavevector k: $\psi\,(x,t) = A\mathrm{e}^{\mathrm{i}(kx-\omega t)}$. Does this represent a plausible wavefunction? The answer is both yes and no—we need further specification. *Infinite* plane waves do *not* represent wavefunctions because they are not normalizable:

$$\int_{-\infty}^{\infty} \left|A\mathrm{e}^{\mathrm{i}(kx-\omega t)}\right|^2 \mathrm{d}x = \int_{-\infty}^{\infty} A^2\mathrm{d}x \to \infty$$

However, they *are* normalizable if they are confined in a finite space, for example in a potential well or in experimental apparatus:

[4] In this vein, $\psi\,(x,t)$ is sometimes referred to as the *probability amplitude*.

$$\int_a^b \left|Ae^{i(kx-\omega t)}\right|^2 dx = A^2(b-a)$$

Therefore, localized plane waves represent wavefunctions provided they satisfy the boundary conditions at $x = a$ and $x = b$. This makes intuitive sense—we would imagine that a particle that is represented by a given wavefunction must be localized in space.

The assumption that a plane wave represents a wavefunction (provided it is localized) is a pretty reasonable assumption: a wave of *any* shape can be constructed from a linear superposition of plane waves of different frequencies and amplitudes (this is the essence of Fourier analysis). Therefore, a plane wave is the "building block" of all wavefunctions, from which waves of other shapes, such as Gaussian-shaped waves, can be constructed. We shall encounter examples of plane-wave solutions throughout this section.

16.3.2 Operator postulate

A mathematical *operator* is an object that acts on and modifies another mathematical object. For example, a *differential operator* is a differential function that returns the derivative of another function ($y = df/dx$ is an operation that returns the x derivative of a function f).

The *operator postulate* states that:

Associated with every physical observable, q, there is a mathematical operator, $\hat{Q}$, which is used in conjunction with the wavefunction.

A large part of the development of quantum mechanics was, in fact, the establishment of the operators associated with the parameters needed to describe a system, such as the energy operator or the momentum operator. We'll determine expressions for some of the most common operators below.

Space and time derivatives of ψ

Let's consider a wavefunction that has the form of a plane wave of frequency ω and wavevector k: $\psi(x,t) = Ae^{i(kx-\omega t)}$, normalized such that A is constant. The first space and time derivatives of $\psi(x,t)$ are then:

$$\begin{aligned}\frac{\partial}{\partial x}\psi(x,t) &= \frac{\partial}{\partial x}Ae^{i(kx-\omega t)}\\ &= ikAe^{i(kx-\omega t)}\\ &= ik\psi(x,t)\end{aligned}$$

$$\text{and}\quad \begin{aligned}\frac{\partial}{\partial t}\psi(x,t) &= \frac{\partial}{\partial t}Ae^{i(kx-\omega t)}\\ &= -i\omega Ae^{i(kx-\omega t)}\\ &= -i\omega\psi(x,t)\end{aligned}$$

Therefore, the operation of these derivatives on a plane-wave wavefunction is to multiply the wavefunction by ik or $-i\omega$ respectively:

$$\frac{\partial}{\partial x} \to ik$$

$$\text{and}\quad \frac{\partial}{\partial t} \to -i\omega$$

Rearranging, we are provided with the wavevector operator, denoted $\hat{k}$, and the angular frequency operator, $\hat{\omega}$, both of which are differential operators (Eqs. 16.8 and 16.9).

Wavevector and frequency operators:

$$\text{Wavevector:}\quad \hat{k} = -i\frac{\partial}{\partial x} \qquad (16.8)$$

$$\text{Frequency:}\quad \hat{\omega} = i\frac{\partial}{\partial t} \qquad (16.9)$$

Eigenvalues and operators

The second part of the operator postulate states that:

The eigenvalue equation[5] for a physical observable q is:

$$\hat{Q}\psi(x,t) = q\psi(x,t)$$

where q is the eigenvalue for $\hat{Q}$.

Let's see what this means in terms of $\hat{k}$ as an example. Acting on a wavefunction with $\hat{k}$, we find:

$$\begin{aligned}\hat{k}\psi &= -i\frac{\partial}{\partial x}\psi\\ &= -i(ik)\psi\\ &= k\psi\end{aligned}$$

This equation, $\hat{k}\psi = k\psi$, is the eigenvalue equation for $\hat{k}$. Therefore, the operation indeed returns the value of the wavevector, k (its eigenvalue). We will see in the next section that the Schrödinger equation is essentially the eigenvalue equation for the total energy operator (which is called the Hamiltonian, $\hat{H}$).

Energy and momentum operators

Using Eq. 16.8 for the expression for $\hat{k}$, it is straightforward to find expressions for the momentum operator, $\hat{p}$, and the kinetic energy operator, $\hat{T}$:

[5] Every operator has a given set of functions on which the operation is to multiply the function by a scalar quantity: $\hat{A}f(x) = \lambda f(x)$. The equation that represents such a transformation is the *eigenvalue equation* of a given operator. The functions that satisfy the eigenvalue equation are called *eigenfunctions* and the scale factor λ is the *eigenvalue*.

$$\hat{p} = \hbar\hat{k}$$
$$= -i\hbar\frac{\partial}{\partial x}$$

and
$$\hat{T} = \frac{\hat{p}^2}{2m}$$
$$= \left(-i\hbar\frac{\partial}{\partial x}\right)^2\frac{1}{2m}$$
$$= -\frac{\hbar^2}{2m}\frac{\partial^2}{\partial x^2}$$

(Eqs. 16.11 and 16.12).

The *Hamiltonian operator*, $\hat{H}$, is the operator that represents a particle's *total* energy (kinetic energy plus potential energy). Using the above expression for $\hat{T}$, we have:

$$\hat{H} = \hat{T} + \hat{V}$$
$$\equiv \frac{\hat{p}^2}{2m} + \hat{V}$$
$$\equiv \frac{\hbar^2\hat{k}^2}{2m} + \hat{V}$$
$$= -\frac{\hbar^2}{2m}\frac{\partial^2}{\partial x^2} + V(x)$$

(Eq. 16.10). We let the *potential energy operator*, $\hat{V}$, be an unknown function of space, $V(x)$, and assume it is constant in time (Eq. 16.13).

The Hamiltonian:

$$\hat{H} = \underbrace{-\frac{\hbar^2}{2m}\frac{\partial^2}{\partial x^2}}_{\hat{T}} + \underbrace{V(x)}_{\hat{V}} \quad (16.10)$$

Energy and momentum operators:

$$\text{Momentum: } \hat{p} = -i\hbar\frac{\partial}{\partial x} \quad (16.11)$$

$$\text{Kinetic energy: } \hat{T} = -\frac{\hbar^2}{2m}\frac{\partial^2}{\partial x^2} \quad (16.12)$$

$$\text{Potential energy: } \hat{V} = V(x) \quad (16.13)$$

16.3.3 Measurement postulate

Measurements on quantum mechanical systems are probabilistic, not deterministic. In probability theory, the most likely outcome of a given variable is given by the *expectation value*[6] of that variable. The *measurement postulate* states that:

Measurement of a physical observable q corresponds to the expectation value of the corresponding operator, $\hat{Q}$:

$$\langle\hat{Q}\rangle \stackrel{\text{def}}{=} \int_{-\infty}^{\infty}\psi^*(x,t)\hat{Q}\psi(x,t)\mathrm{d}x$$

(Eq. 16.14). For example, the expectation value of a particle's position is found by the expectation value of $\hat{x} = x$:

$$\langle\hat{x}\rangle = \int_{-\infty}^{\infty}\psi^*(x,t)x\psi(x,t)\mathrm{d}x$$
$$= \int_{-\infty}^{\infty}x\underbrace{|\psi(x,t)|^2}_{\text{Probability density}}\mathrm{d}x$$

(Eq. 16.15). This corresponds to the position at which a particle is most likely to be located.

Expectation values:

Expectation value of quantity $\hat{Q}$:

$$\langle\hat{Q}\rangle = \int_{-\infty}^{\infty}\psi^*(x,t)\hat{Q}\psi(x,t)\mathrm{d}x \quad (16.14)$$

Expectation value of position:

$$\langle\hat{x}\rangle = \int_{-\infty}^{\infty}x|\psi(x,t)|^2\mathrm{d}x \quad (16.15)$$

16.3.4 Time-evolution postulate

Wavefunctions are functions of time. Therefore, the probability of a particle being between x and $x + \mathrm{d}x$ evolves with time. The *time-evolution postulate* states that:

The time evolution of the wavefunction is governed by the Schrödinger equation:

$$\hat{H}\psi(x,t) = i\hbar\frac{\partial\psi(x,t)}{\partial t}$$

This is a complex partial differential wave equation. Although it may initially look daunting, its interpretation is conceptually straightforward, as we shall see below.

Deriving the Schrödinger equation

The total energy of a quantum mechanical particle is $\hat{E} = \hbar\hat{\omega}$. We have already found an expression for the frequency operator (see Eq. 16.9); therefore, acting on a wavefunction with the Hamiltonian, we have:

6 The expectation value, or mean, of an arbitrary quantity, X, which takes a value x with a probability density function $P(x)$ is defined as: $\langle X\rangle = \int xP(x)\mathrm{d}x$.

$$
\begin{aligned}
\hat{H}\psi(x,t) &= \hat{E}\psi(x,t) \\
&= \hbar\hat{\omega}\psi(x,t) \\
&= i\hbar\frac{\partial\psi(x,t)}{\partial t} \\
\therefore \hat{H}\psi(x,t) &= i\hbar\frac{\partial\psi(x,t)}{\partial t}
\end{aligned}
$$

This is the *Time-dependent Schrödinger equation*, or TDSE. We can express $\hat{H}$ in terms of the kinetic and potential energy (see Eq. 16.10). Doing this, we see that the TDSE is a wave equation that describes the behavior of a particle of mass m, moving under the influence of a potential energy function, $V(x)$ (Eq. 16.16).

Time-dependent Schrödinger equation (TDSE):

$$
\underbrace{\left(-\frac{\hbar^2}{2m}\frac{\partial^2}{\partial x^2} + V(x)\right)}_{\text{Hamiltonian, } \hat{H}}\psi(x,t) = i\hbar\frac{\partial\psi(x,t)}{\partial t} \tag{16.16}
$$

Solutions of the wavefunctions that satisfy the Schrödinger equation determine everything that can be known about a given system. Therefore, the Schrödinger equation is a very powerful and elegant equation. In the following chapter, we will solve the Schrödinger equation for various simple potential energy functions, $V(x)$.

Chapter 17
The Schrödinger equation

Newton's Classical Laws of Motion describe the time evolution of a classical body as a result of external forces acting on it. The Schrödinger equation is the quantum analog of Newton's Laws. It describes the time evolution of a particle a result of its energy (recall that force and energy/work are essentially two versions of the same thing). The Schrödinger equation describes the properties of particles on atomic and sub-atomic scales, and it approximates to Newton's Laws in the macroscopic limit (much like how Einstein's Theory of Special Relativity approximates to Newton's Laws in the low-velocity limit). In this chapter, we will solve the Schrödinger equation for particles in various potential energy wells. In doing this, we find that sub-atomic particles behave quite differently than we would expect from classical motion.

17.1 Separating the Schrödinger equation

We saw in the previous chapter that the time evolution of a wavefunction is governed by the TDSE (time-dependent Schrödinger equation):

$$\underbrace{\left(-\frac{\hbar^2}{2m}\frac{\partial^2}{\partial x^2}+V(x)\right)}_{\text{Hamiltonian, }\hat{H}}\psi(x,t)=\mathrm{i}\hbar\frac{\partial\psi(x,t)}{\partial t} \tag{17.1}$$

It's straightforward to separate this equation into two parts:[1] a spatial part (equation in x only) and a temporal part (equation in t only). Expressing the wavefunction as the product of a function in space, $\phi(x)$, and a function in time, $T(x)$, we have:

$$\psi(x,t)=\phi(x)T(t)$$

Using this function in Eq. 17.1, we find that the TDSE reduces to two ordinary differential equations, one that depends only on the variable x:

$$\underbrace{\left(-\frac{\hbar^2}{2m}\frac{\mathrm{d}^2}{\mathrm{d}x^2}+V(x)\right)}_{\text{Hamiltonian}}\phi(x)=E\phi(x) \tag{17.2}$$

and one that depends only on t:

$$\mathrm{i}\hbar\frac{\mathrm{d}T(t)}{\mathrm{d}t}=ET(t) \tag{17.3}$$

(Derivation 17.1). The variable E is a constant that equals the total energy of the system. Splitting up the Schrödinger equation in this way makes it easier to solve, and it allows us to distinguish a particle's spatial behavior from its time evolution.

Eigenvalue equations

Eqs. 17.2 and 17.3 are energy eigenvalue equations. The differential operator in the x-equation corresponds to the Hamiltonian (kinetic plus potential energy), and that in the t-equation corresponds to $\hat{E}=\hbar\hat{\omega}$, where $\hat{\omega}=\mathrm{i}\,\mathrm{d}/\mathrm{d}t$ (refer to Eq. 16.9). Dividing the t-equation by $\hbar$, we therefore recover the angular frequency eigenvalue equation:

$$x\text{-equation: } \hat{H}\phi(x)=E\phi(x)$$

$$t\text{-equation: } \hat{\omega}T(t)=\frac{E}{\hbar}T(t)\equiv\omega T(t)$$

The x-equation is known as the *Time-independent Schrödinger equation*, or the TISE (Eq. 17.4). In Section 17.2, we'll look at examples of solving the TISE for various one-dimensional potential functions, $V(x)$. For a constant potential, $V(x)=V_0$, the solutions are plane-wave solutions (Eq. 17.5 and Derivation 17.2).

Time-independent Schrödinger equation (TISE) (Derivations 17.1 and 17.2):

$$\underbrace{\left(-\frac{\hbar^2}{2m}\frac{\mathrm{d}^2}{\mathrm{d}x^2}+V(x)\right)}_{\text{Hamiltonian, }\hat{H}}\phi(x)=E\phi(x) \tag{17.4}$$

[1] This method is called the *separation of variables*, which is a standard method used to solve separable partial differential equations.

Derivation 17.1: Separation of variables: the time-dependent Schrödinger equation (TDSE)

1. **Partial differential equation:** the TDSE, with a time-independent potential energy function, $V(x)$, is:

$$\left(-\frac{\hbar^2}{2m}\frac{\partial^2}{\partial x^2}+V(x)\right)\psi(x,t)=\mathrm{i}\hbar\frac{\partial\psi(x,t)}{\partial t}$$

2. **Separation function:** separate the spatial part from the temporal part by letting $\psi(x, t) = \phi(x)T(t)$:

$$\left(-T(t)\frac{\hbar^2}{2m}\frac{\partial^2\phi(x)}{\partial x^2}+V(x)\phi(x)T(t)\right)=\phi(x)\mathrm{i}\hbar\frac{\partial T(t)}{\partial t}$$

3. **Separation constant:** divide by $\phi(x)T(t)$ and rearrange to make one side a function of x only and the other side a function of t only:

$$\underbrace{\frac{1}{\phi(x)}\left(-\frac{\hbar^2}{2m}\frac{\mathrm{d}^2\phi(x)}{\mathrm{d}x^2}+V(x)\phi(x)\right)}_{=\text{constant, }E}=\underbrace{\frac{1}{T(t)}\mathrm{i}\hbar\frac{\mathrm{d}T(t)}{\mathrm{d}t}}_{=\text{constant, }E}$$

For this equation to hold for all values of x and t, both sides must be equal to a constant. The equation has dimensions of energy, therefore, let the separation constant equal the total energy, E.

4. **The TISE:** the x-equation (left-hand side) is the time-independent Schrödinger equation (TISE):

$$\left(-\frac{\hbar^2}{2m}\frac{\mathrm{d}^2}{\mathrm{d}x^2}+V(x)\right)\phi(x)=E\phi(x)$$

The solution depends on the form of the potential $V(x)$. For constant potential V_0, the solutions are oscillatory plane-wave solutions (see Derivation 17.2 for details):

$$\phi(x)=A\mathrm{e}^{+\mathrm{i}kx}+B\mathrm{e}^{-\mathrm{i}kx}\quad\text{where } E=\frac{\hbar^2k^2}{2m}+V_0\quad\text{and } A \text{ and } B \text{ are constants set by boundary conditions}$$

5. **Time evolution:** the t-equation describes the time evolution of a wavefunction:

$$\mathrm{i}\hbar\frac{\mathrm{d}T(t)}{\mathrm{d}t}=ET(t)$$

This equation does not depend on $V(x)$; therefore, $T(t)$ has the same functional dependence on time for all systems. Integrating once, the temporal solutions are complex exponential solutions:

$$T(t)=C\mathrm{e}^{-\mathrm{i}\omega t}\quad\text{where } E=\hbar\omega\quad\text{and } C \text{ is a constant set by boundary conditions}$$

Solution for constant potential, $V(x) = V_0$:

$$\phi(x)=A\mathrm{e}^{+\mathrm{i}kx}+B\mathrm{e}^{-\mathrm{i}kx}\qquad(17.5)$$

$$\text{where } k=\frac{\sqrt{2m(E-V_0)}}{\hbar}$$

$$\therefore E=\frac{\hbar^2k^2}{2m}+V_0\equiv T+V$$

A and B are constants set by boundary conditions.

The t-equation is a first-order differential equation (Eq. 17.6). It does not depend on the potential energy function $V(x)$, which means that the solution of $T(t)$ is independent of the system under consideration. Said another way, a wavefunction evolves with the same time dependence in any potential. Rearranging and integrating Eq. 17.6, we have:

$$\int\frac{\mathrm{d}T(t)}{T(t)}=\int\frac{E}{\mathrm{i}\hbar}\mathrm{d}t$$
$$=-\int\mathrm{i}\omega\mathrm{d}t$$

Derivation 17.2: Solution of the TISE for constant potential, V_0 (Eq. 17.5)

$$-\frac{\hbar^2}{2m}\frac{d^2\phi(x)}{dx^2} = (E - V_0)\phi(x)$$
$$= T\phi(x)$$
$$\therefore \frac{d^2\phi(x)}{dx^2} = -\frac{2mT}{\hbar^2}\phi(x)$$
$$= -k^2\phi(x)$$
$$\therefore \phi(x) = Ae^{+ikx} + Be^{-ikx}$$
$$\text{where } k^2 = \frac{2m(E - V_0)}{\hbar^2}$$
$$\therefore E = \frac{\hbar^2 k^2}{2m} + V_0$$

- Starting point: the TISE is:

$$-\frac{\hbar^2}{2m}\frac{d^2\phi(x)}{dx^2} + V(x)\phi(x) = E\phi(x)$$

- Set $V(x) = V_0$ (constant potential), where $V_0 < E$:

$$\therefore -\frac{\hbar^2}{2m}\frac{d^2\phi(x)}{dx^2} = (E - V_0)\phi(x)$$

- Define the difference between the particle's total energy and potential energy as $T = E - V_0$ (kinetic energy).
- The kinetic energy for particles of mass m is:

$$T = \frac{p^2}{2m} = \frac{\hbar^2 k^2}{2m}$$
$$\therefore k^2 = \frac{2mT}{\hbar^2} = \frac{2m(E - V_0)}{\hbar^2}$$

- Substituting in for k^2, the TISE has the form of a standard second-order differential equation with complex exponential solutions. The integration constants A and B are fixed by boundary conditions.

$$\therefore T(t) = Ce^{-i\omega t}$$
$$\text{where } \omega = \frac{E}{\hbar}$$

(Eq. 17.7). The constant of integration C is set by boundary conditions.

Temporal part of the TDSE (Derivation 17.1):

$$i\hbar\frac{dT(t)}{dt} = ET(t) \quad (17.6)$$

Solution (integrate once):

$$T(t) = Ce^{-i\omega t} \quad (17.7)$$
$$\text{where } \omega = \frac{E}{\hbar}$$
$$\therefore E = \hbar\omega$$

C is a constant set by boundary conditions.

Total solution

The total solution is given by the product of the space and time solutions:

$$\psi(x,t) = \phi(x)T(t)$$
$$= \phi(x)e^{-i\omega t}$$

Since $T(t)$ is universal for all potentials $V(x)$, then to solve for $\psi(x, t)$ we just need to solve the TISE for $\phi(x)$ and multiply the solution by $e^{-i\omega t}$. The normalization condition is such that:

$$\int_{-\infty}^{\infty} |\psi(x,t)|^2 \, dx = \int_{-\infty}^{\infty} |\phi(x)|^2 \, dx = 1$$

For a constant potential V_0, the $\phi(x)$ solutions are given by Eq. 17.5, and therefore, the overall wavefunctions are a superposition of traveling plane waves:

$$\psi(x,t) = \underbrace{Ae^{-i(\omega t - kx)}}_{+x\text{-direction}} + \underbrace{Be^{-i(\omega t + kx)}}_{-x\text{-direction}}$$

Recall that "pure" plane-wave solutions must be confined to a finite region since infinite plane waves are not normalizable:

$$\int_{-\infty}^{\infty} A^2 \left|e^{-i(\omega t - kx)}\right|^2 dx \to \infty$$

17.2 One-dimensional potential wells

In this section, we will solve the TISE (Eq. 17.4) for three different one-dimensional potential energy functions:

1. The *infinite* potential well: particle is confined to a region in space (a "well") where the potential energy is $V(x) = 0$ inside the well and $V(x) = \infty$ outside.
2. The *finite* potential well: particle is confined to a region in space where the potential energy is $V(x) = 0$ inside the well and $V(x) = V_0$ (constant) outside.
3. The harmonic oscillator: particle is confined to a region in space where the potential has a quadratic dependence on x: $V(x) = \frac{1}{2}kx^2$.

Recall that force and potential energy are related to one another by $F = -dV/dx$. Therefore, a particle experiences no net force in a constant potential well ($F = -dV/dx = 0$), which is the case for the infinite and finite potential well examples above. However, for the harmonic oscillator example, a particle experiences a net restoring force of $F = -dV/dx = -kx$ (Hooke's Law). This type of restoring force is approximately the force on an atom in a solid that is displaced a small distance x from its equilibrium position. Therefore, the wavefunction solutions in this case are approximately representative of the quantum states of atoms in a solid.

17.2.1 Infinite square potential well

The *infinite square potential well* is the simplest application of the TISE, and it represents the most crude approximation of the potential energy of a particle confined in a region of space. In this model, a particle is located in a region of zero potential energy, say between $x = 0$ and $x = a$. At the boundaries, there is an infinite increase in the potential energy function. Therefore, the potential energy is a step function given by:

$$V(x) = \begin{cases} 0 & \text{for } 0 < x < a \\ \infty & \text{otherwise} \end{cases}$$

(refer to diagram in Worked example 17.1). What is the wavefunction that describes a particle located in such a potential? This is found by solving the TISE for the above potential $V(x)$ (see Worked example 17.1 for full solution). Doing this, we find that there is a *set* of wavefunctions that satisfy the TISE for this system. These correspond to sine functions of different wavelengths:

$$\phi_n(x) = \sqrt{\frac{2}{a}}\sin\left(\frac{n\pi x}{a}\right) \equiv \sqrt{\frac{2}{a}}\sin(k_n x)$$

where $n = 1, 2, 3, \ldots$

(Eq. 17.8). Each wavefunction of different n corresponds to a different allowed particle state. The wavenumber (spatial frequency) of the wavefunctions are:

$$k_n = \frac{n\pi}{a}, \quad n = 1, 2, 3, \ldots$$

Therefore, the corresponding allowed wavelengths are:

$$\lambda_n = \frac{2\pi}{k_n} = \frac{2a}{n}$$

i.e. only an integer number of half wavelengths can "fit" inside the well. This is analogous to the set of allowed wavelengths that result in standing waves on a string. The total energy of a particle depends on which state it is in. The kinetic energy of the nth state is:

$$E_n = \frac{\hbar^2 k_n^2}{2m} = \frac{\hbar^2}{2m}\left(\frac{n\pi}{a}\right)^2$$

(Eq. 17.9). Therefore, we see that the kinetic energy of a particle is quantized (it can only take certain discrete values, defined by $n = 1, 2, 3,\ldots$), and it increases for higher states. Energy quantization is a quantum phenomenon that has no classical equivalent. For example, if our kinetic energy were quantized, it would imply that the only speeds at our disposal would be, say, those of a tortoise, of a running dog, or of a ravenous cheetah chasing prey—with nothing in between.

Infinite square well (Worked example 17.1):

$$\phi_n(x) = \sqrt{\frac{2}{a}}\sin\left(\frac{n\pi x}{a}\right) \qquad (17.8)$$

$$E_n = \frac{\hbar^2}{2m}\left(\frac{n\pi}{a}\right)^2 \qquad (17.9)$$

where the particle's wavenumber is:

$$k_n = \frac{n\pi}{a}$$

where $n = 1, 2, 3, \ldots$

States and probability density

The wavefunctions and energies of the lowest energy states are:

$$\phi_1(x) = \sqrt{\frac{2}{a}}\sin\left(\frac{\pi x}{a}\right), \quad E_1 = \frac{\hbar^2}{2m}\left(\frac{\pi}{a}\right)^2$$

$$\phi_2(x) = \sqrt{\frac{2}{a}}\sin\left(\frac{2\pi x}{a}\right), \quad E_2 = 4E_1$$

$$\phi_3(x) = \sqrt{\frac{2}{a}}\sin\left(\frac{3\pi x}{a}\right), \quad E_3 = 9E_1$$

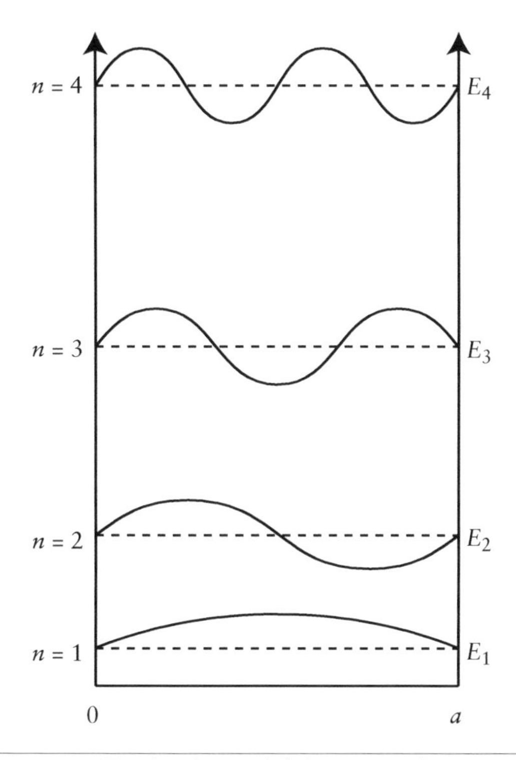

Figure 17.1: Wavefunctions and their corresponding energy levels of the lowest four states of an infinite square potential well. The ground-state energy is $E_1 = \hbar^2\pi^2/2ma^2$.

These are plotted in Fig. 17.1. The energy of the state is represented by the vertical axis. The ground-state energy, E_1, is not zero, even though this corresponds to the lowest possible energy that a particle can have. In contrast, in the classical world, a particle in its lowest kinetic energy state has zero kinetic energy. Therefore, quantum mechanical systems always have a minimum energy that is greater than zero, due to their quantum motion. This is called the *zero-point energy.*
The probability density of the nth state is equal to $|\phi_n(x)|^2$:

$$\therefore |\phi_1(x)|^2 = \frac{2}{a}\sin^2\left(\frac{\pi x}{a}\right)$$

$$|\phi_2(x)|^2 = \frac{2}{a}\sin^2\left(\frac{2\pi x}{a}\right) \quad \text{etc.,...}$$

(Fig. 17.2). Therefore, a particle in the $n = 1$ state is most likely to be located in the center of the well, whereas a particle in the $n = 2$ state is more likely to be mid-way between the center and one edge of the well.

Hydrogen energy levels

We can use Eq. 17.9 as a crude approximation to determine the order of magnitude of electron energy levels in a hydrogen atom. If we treat the electron

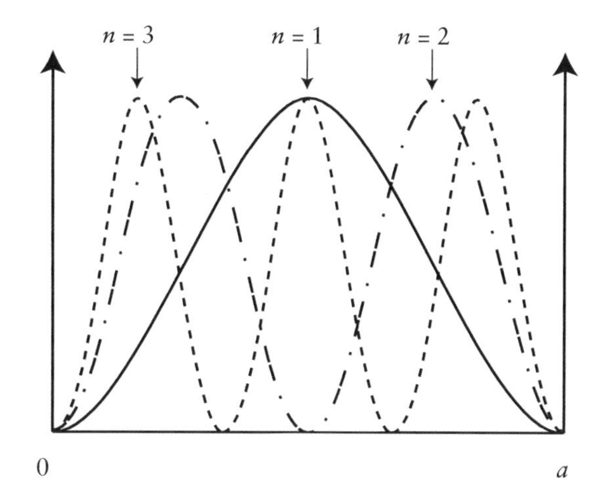

Figure 17.2: Probability density distributions of the lowest three states of a particle in an infinite square potential well.

($m_e \simeq 9 \cdot 10^{-31}$ kg) as being confined in a square potential well of atomic dimensions $a \sim 10^{-9}$ m, the ground-state energy level is on the order of:

$$E_1 \sim -\frac{(10^{-34})^2}{2\cdot 9\cdot 10^{-31}}\left(\frac{3}{10^{-9}}\right)^2$$

$$\sim -0.5\cdot 10^{-19}\ \text{J}$$

$$\text{or}\quad E_1 \sim -3\ \text{eV}$$

The ground state of hydrogen is −13.6 eV; therefore, modeling hydrogen as an infinite square potential well is a reasonable approximation to determine the order of magnitude of energy levels. When we move to three dimensions and use the spherical Schrödinger equation, the energy levels of hydrogen are accurately described (Chapter 20).

Wavefunctions in the classical limit

For a classical system, we would expect all positions in the well to be equally likely ($|\phi|^2$ is constant for all x) and any value of energy to be allowed ($E_n - E_{n-1} \simeq 0$). Consider a grain of sand ($m \sim 10^{-6}$ kg) confined in a 0.1 mm infinite potential well. Using Eq. 17.9, the energy levels of this system are:

$$E_n \sim 10^{-54} n^2\ \text{J}$$

If the minimum energy of the grain is its thermal energy, then $E \sim k_B T \sim 10^{-21}$ J at room temperature, which would put the grain in the state corresponding to $n \sim 10^{16}$. Therefore, $|\phi|^2$ would indeed be flat and the difference between energy levels approximately zero, making E a continuous variable. The observation that quantum theory agrees with classical results in the limit is known as the *correspondence principle*, and it is an important requirement of quantum theory.

WORKED EXAMPLE 17.1: Infinite square potential well

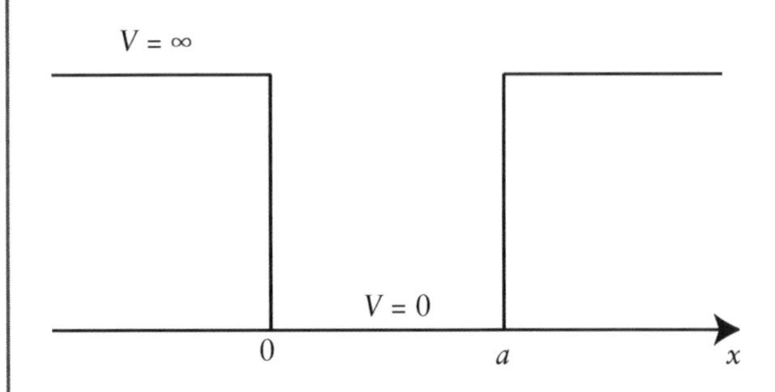

Potential energy function: the potential energy is a step function given by:

$$V(x) = \begin{cases} 0 & \text{for } 0 < x < a \\ \infty & \text{otherwise} \end{cases}$$

A particle is confined in space between $0 < x < a$ as it would require an infinite amount of energy to be outside this region.

1. **TISE:** the Schrödinger equation for the region of zero potential ($0 < x < a$) is:

$$-\frac{\hbar^2}{2m}\frac{d^2}{dx^2}\phi(x) = E\phi(x) \rightarrow \therefore \frac{d^2\phi(x)}{dx^2} = -k^2\phi(x) \text{ where } k^2 = \frac{2mE}{\hbar^2}$$

2. **Wavefunction solutions:** the TISE for zero potential is an ordinary second-order differential equation, with complex exponential solutions: $\phi(x) = Ae^{ikx} + Be^{-ikx}$. Since a particle cannot reside in the regions where $V(x) = \infty$, the wavefunction is zero in these regions. Therefore, the wavefunction is given by:

$$\phi(x) = \begin{cases} Ae^{ikx} + Be^{-ikx} & \text{for } 0 < x < a \\ 0 & \text{otherwise} \end{cases}$$

3. **Boundary conditions:** $\phi(x)$ is continuous; therefore, the wavefunction must vanish at the boundaries: $\phi(0) = 0$ and $\phi(a) = 0$. The first boundary condition tells us that the wavefunctions are sine functions, and the second that the wavenumber k is quantized:
   - 1st boundary condition: $\phi(0) = 0 \rightarrow$ wavefunctions are sinusoidal:

$$\phi(0) = 0$$
$$\therefore A = -B$$
$$\therefore \phi(x) = A(e^{ikx} - e^{-ikx}) = A'\sin(kx) \tag{17.10}$$

   - 2nd boundary condition: $\phi(a) = 0 \rightarrow$ wavenumber is quantized:

$$\phi(a) = A'\sin(ka) = 0$$
$$\therefore k_n = \frac{n\pi}{a}, \text{ where } n \text{ is an integer}, n = 1, 2, 3, \ldots \tag{17.11}$$

4. **Normalization:** the integration constant, A', is found using the normalization condition. This requires that the probability of finding the particle within the well (between $0 < x < a$) equals one:

$$\int_0^a |\phi(x)|^2 \, dx = 1$$

$$\therefore A'^2 \int_0^a \sin^2(kx)dx = A'^2 \frac{a}{2} = 1 \rightarrow A' = \sqrt{\frac{2}{a}}$$

WORKED EXAMPLE 17.1: Infinite square potential well (cont.)

5. **Normalized solutions:** substituting the expressions for A' and k_n into the solutions in Eq. 17.10, we find expressions for the quantized wavefunctions, $\phi_n(x)$, and their corresponding energies, E_n. These are discrete functions, characterized by the integer $n = 1, 2, 3, \ldots$:

$$\phi_n(x) = \sqrt{\frac{2}{a}} \sin\left(\frac{n\pi x}{a}\right)$$

$$\text{and } E_n = \frac{\hbar^2 k_n^2}{2m} = \frac{\hbar^2}{2m}\left(\frac{n\pi}{a}\right)^2$$

17.2.2 Finite square potential well

The *finite square potential well* is similar to the infinite potential well model, with the difference being that the potential outside the well is constant at V_0 rather than at ∞, where V_0 is greater than the total energy E of a particle. The potential energy function is then a step function given by:

$$V(x) = \begin{cases} 0 & \text{for } -a < x < a \\ V_0 & \text{otherwise} \end{cases}$$

(refer to diagram in Worked example 17.2). The wavefunction that satisfies the TISE for this potential energy consists of two parts: a plane-wave solution in the region of zero potential (as it was for the infinite square well potential), and an exponentially decaying wavefunction in the regions where $V(x) = V_0$ (see Worked example 17.2). The region where $V_0 > E$ corresponds to a classically forbidden region (a particle cannot have potential energy greater than its total energy, otherwise it would have negative kinetic energy). However, since the wavefunction is non-zero in these regions, there is a finite probability that a quantum mechanical particle penetrates into the classically forbidden region. The penetration length into this region is:

$$L = \frac{1}{\kappa} = \frac{\hbar}{\sqrt{2m(V_0 - E)}}$$

(refer to Worked example 17.2). Therefore, as $V_0 \to \infty$ we see $L \to 0$, which is in agreement with the infinite potential well model.

Quantum tunneling

Penetration of the wavefunction into the classically forbidden region leads to a quantum mechanical phenomenon known as *tunneling*. As its name implies, tunneling is the process by which a particle "tunnels through" an energy barrier, and emerges on the other side. In contrast, a classical particle would require an energy greater than the barrier energy in order to find itself on the other side of the barrier (think of a ball rolling down one hill and up another: the ball can only make it over the top of the second hill if its total energy is greater than the gravitational potential energy it would have at the top of the hill). The probability of tunneling falls off exponentially with increased barrier width, a, with increased barrier height, V_0, and with increased particle mass, m, according to:

$$P_{\text{tunnel}} \to \frac{E}{V_0} e^{-2v}$$

$$\text{where } v^2 = \frac{2mV_0 a^2}{\hbar^2}$$

(see Worked example 17.3 for full solution). The probability of transmission is finite, even in the limit $E \ll V_0$. Quantum tunneling has several applications. For example, an alpha particle is emitted from a nucleus during alpha decay by tunneling through the high potential barrier that binds the nucleus together. Another practical example is the *scanning tunneling microscope*; this microscope images electrons that tunnel through the work function that binds the electrons to a metal surface, which allows individual atoms on the metal's surface to be resolved via variations in the tunneling current.

17.2.3 The harmonic oscillator

As a final example of a one-dimensional potential energy function, we'll turn our attention to the *harmonic oscillator* potential. Recall from classical mechanics that an object subject to a linear restoring force, $F = -k_s x$, obeys simple harmonic motion. The constant of proportionality is the spring constant k_s, which is related to the mass and angular frequency of oscillation by $k_s = m\omega^2$ (refer to Section 2.3.1). The potential energy function of the quantum harmonic oscillator is the energy associated

WORKED EXAMPLE 17.2: Finite square potential well

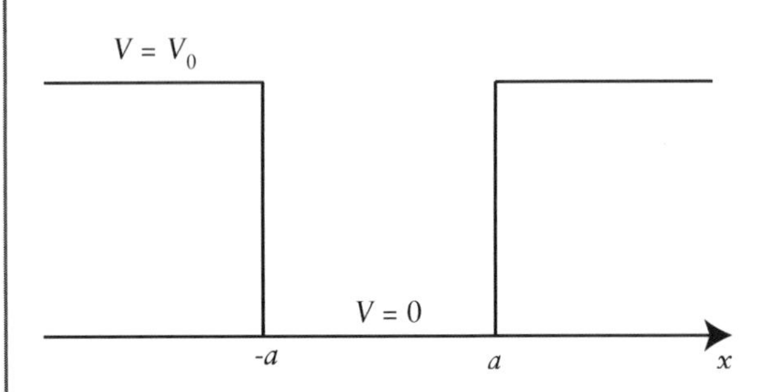

Potential energy function: the potential energy is a step function given by:

$$V(x) = \begin{cases} 0 & \text{for } -a < x < a \\ V_0 \quad (V_0 > E) & \text{otherwise} \end{cases}$$

1. **TISE:** The Schrödinger equation for either region is:

$$-\frac{\hbar^2}{2m}\frac{d^2}{dx^2}\phi(x) = \begin{cases} E\phi(x) & \text{for } -a < x < a \\ -(V_0 - E)\phi(x) & \text{otherwise} \end{cases}$$

$$\therefore \frac{d^2}{dx^2}\phi(x) = \begin{cases} -k^2\phi(x) & k^2 = 2mE/\hbar^2 & \text{for } -a < x < a \\ \kappa^2\phi(x) & \kappa^2 = 2m(V_0 - E)/\hbar^2 & \text{otherwise} \end{cases}$$

2. **Wavefunction solutions:** the TISE in either region is an ordinary second-order differential equation. In the region of zero potential ($-a < x < a$), the solutions are complex exponential functions (plane-wave solutions), whereas in the regions of finite potential the solutions are non-complex, exponentially decaying functions (the exponentially increasing functions are unphysical):

$$\therefore \phi(x) = \begin{cases} Ae^{\kappa x} & \text{for } x < -a & \rightarrow \text{exponential decay } (\equiv \phi_1(x)) \\ B\cos(kx) + C\sin(kx) & \text{for } -a < x < a & \rightarrow \text{plane-wave solution } (\equiv \phi_2(x)) \\ De^{-\kappa x} & \text{for } x > a & (\equiv B'e^{ikx} + C'e^{-ikx}) \rightarrow \text{exponential decay } (\equiv \phi_3(x)) \end{cases}$$

3. **Boundary conditions:** $\phi(x)$ and its first derivative are continuous at the boundaries.

$$\therefore \text{At } x = -a: \quad \phi_1(-a) = \phi_2(-a) \text{ and } \left(\frac{d\phi_1}{dx}\right)_{(x=-a)} = \left(\frac{d\phi_2}{dx}\right)_{(x=-a)}$$

$$\text{At } x = a: \quad \phi_2(a) = \phi_3(a) \text{ and } \left(\frac{d\phi_2}{dx}\right)_{(x=a)} = \left(\frac{d\phi_3}{dx}\right)_{(x=a)}$$

These four boundary conditions give us four simultaneous equations, which can be solved to determine values for the integration constants A, B, C, and D, as well as place restrictions on k and κ. When written as a matrix, the four simultaneous equations are:

$$\begin{pmatrix} e^{-\kappa a} & -\cos ka & \sin ka & 0 \\ \kappa e^{-\kappa a} & -k\sin ka & -k\cos ka & 0 \\ 0 & -\cos ka & -\sin ka & e^{-\kappa a} \\ 0 & k\sin ka & -k\cos ka & -\kappa e^{-\kappa a} \end{pmatrix}\begin{pmatrix} A \\ B \\ C \\ D \end{pmatrix} = 0$$

WORKED EXAMPLE 17.2: Finite square potential well (cont.)

4. **Solutions:** expanding lines 1 and 3 of the matrix in part 3, we see that the equations have two sorts of solutions:
   (ii) **Symmetric solutions:** $A = D$ and $C = 0$.

$$\therefore \phi(x) = \begin{cases} Ae^{\kappa x} & \text{for } x < -a \\ B\cos(kx) & \text{for } -a < x < a \\ Ae^{-\kappa x} & \text{for } x > a \end{cases}$$

(the values of A and B are set by the system in question). This wavefunction is symmetric about the origin; therefore, this wavefunction is said to have *even parity*. Under these coefficient restraints, if we expand lines 1 and 2 of the matrix and take their ratio, we see that k and κ are related by:

$$k\tan(ka) = \kappa \tag{17.12}$$

Therefore, only certain wavevector values (and their corresponding energies) are "allowed" by the system—if k and κ satisfy Eq. 17.12, then the associated wavefunction satisfies the system's boundary conditions.
   (ii) **Asymmetric solutions:** $A = -D$ and $B = 0$

$$\therefore \phi(x) = \begin{cases} De^{\kappa x} & \text{for } x < -a \\ C\sin(kx) & \text{for } -a < x < a \\ -De^{-\kappa x} & \text{for } x > a \end{cases}$$

(the values of C and D are set by the system in question). This wavefunction is asymmetric about the origin; therefore, this wavefunction is said to have *odd parity*. Under these coefficient restraints, if we expand lines 3 and 4 of the matrix and take their ratio, we see that k and κ are now related by:

$$k\cot(ka) = -\kappa \tag{17.13}$$

If k and κ satisfy Eq. 17.13, then the associated wavefunction satisfies the system's boundary conditions, and it corresponds to an allowed state of the system.

The first two wavefunctions of the system are shown in the diagram. States corresponding to higher energy have a greater number of nodes in the region of zero potential. An exponentially decaying wavefunction penetrates into the regions where $V = V_0$; the *penetration length* into this region is:

$$L = \frac{1}{\kappa} = \frac{\hbar}{\sqrt{2m(V_0 - E)}}$$

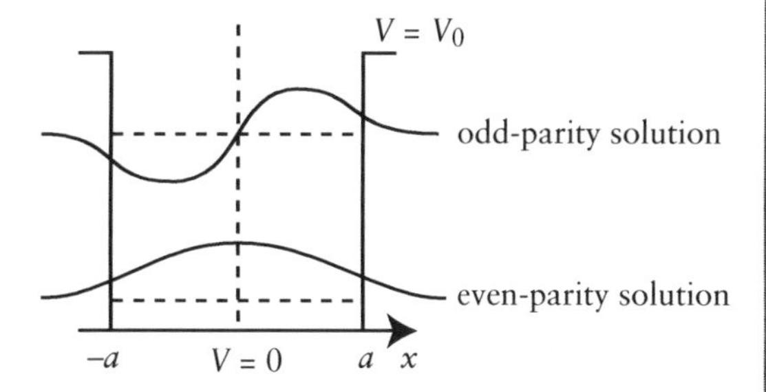

WORKED EXAMPLE 17.2: Finite square potential well (cont.)

5. **Allowed energy levels:** in part 4, we found that the boundary conditions of the system are only satisfied if κ is related to k by either of the following relations:

$$\kappa = \begin{cases} k\tan(ka) \\ -k\cot(ka) \end{cases} \tag{17.14}$$

These equations cannot be solved analytically. However, letting $u = -ka$ and $v = \kappa a$, we can reduce the equations to a single variable, which allows us to solve them graphically. From the expressions for k and κ (refer to part 1 of this example), we see that u and v are related by $v^2 = u_0^2 - u^2$, where $u_0^2 = 2mV_0/\hbar^2$. Therefore, u_0 is a constant for a given potential, V_0. Therefore, after variable substitution, Eqs. 17.14 can be expressed as:

$$\sqrt{u_0^2 - u^2} = \begin{cases} u\tan(u) \\ -u\cot(u) \end{cases} \tag{17.15}$$

The functions $f_1(u) = \sqrt{u_0^2 - u^2}$ (dotted black curves), $f_2(u) = u\tan(u)$ (solid black curves), and $f_3(u) = -u\cot(u)$ (solid grey curves) are plotted below. (Note that $f_1(u)$ depends on the value of the constant u_0, where it moves closer to the origin for smaller values of u_0.) The solutions of u that satisfy the boundary conditions of the system correspond to the points of intersection between the curve $\sqrt{u_0^2 - u^2}$ with either $u\tan(u)$ or $-u\cot(u)$. For example, for the larger value of u_0 in the diagram below, there are three solutions of u that satisfy Eq. 17.15, whereas for the smaller value of u_0, there are just two. Since $u\tan(u)$

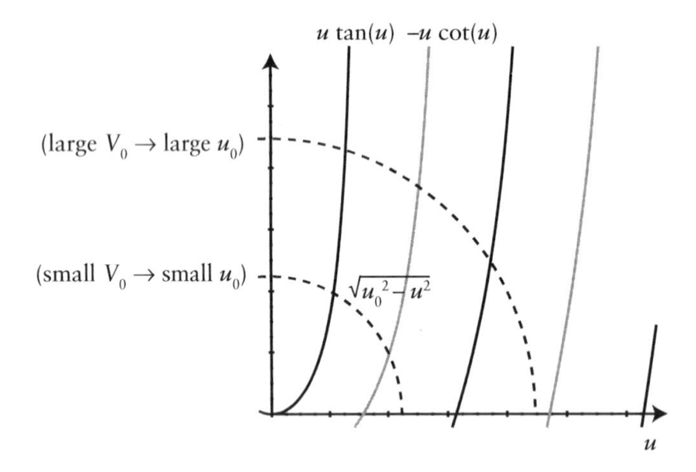

passes through zero, there is always one solution of u that satisfies Eq. 17.15, even for very small u_0 (i.e. even for a very small potential well, V_0).

Each point of intersection corresponds to a different allowed state. The energy of the state corresponding to the intersection at u_n is:

$$E_n = \frac{\hbar^2 k_n^2}{2m} = \frac{\hbar^2 u_n^2}{2ma^2}, \quad u = 1, 2, \ldots, N$$

where N is the total number of intersections. Since N is finite, there are only a finite number of bound states for a given V_0. This makes intuitive sense, since for particle energies greater than V_0, the particle is no longer confined in the potential well.

WORKED EXAMPLE 17.3: Tunneling through a finite potential barrier

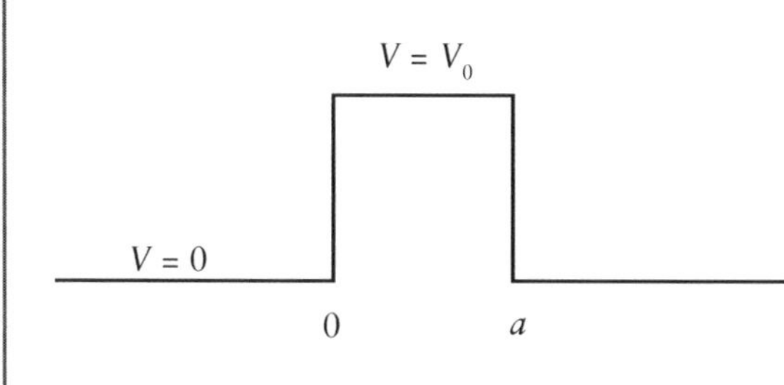

Potential energy function: the potential energy is a step function given by:

$$V(x) = \begin{cases} V_0 \ (V_0 > E) & \text{for } 0 < x < a \\ 0 & \text{otherwise} \end{cases}$$

Consider a particle incident on a potential barrier. If $V_0 > E$, no barrier penetration would occur in the classical situation, and the particle would reflect back off the barrier. In the quantum case, we find that for a finite barrier width, a particle can tunnel through it and connect to a wavefunction on the other side.

1. **Wavefunctions and energy:** we saw in the finite potential well problem (Worked example 17.2) that the wavefunction of a particle in the vicinity of a finite potential V_0, where $V_0 > E$, is exponentially decaying in these regions. Let's assume that the wavefunction of a particle incident on a potential barrier is partially reflected and partially transmitted by the barrier (similar to a light wave incident on a glass surface). The wavefunction that satisfies the Schrödinger equation for a potential barrier is therefore:

$$\phi(x) = \begin{cases} e^{ikx} + re^{-ikx} & k = \sqrt{2mE}/\hbar & \text{for } x < 0 & (= \phi_1(x)) \\ Ae^{\kappa x} + Be^{-\kappa x} & \kappa = \sqrt{2m(V_0 - E)}/\hbar & \text{for } 0 < x < a & (= \phi_2(x)) \\ te^{ikx} & k = \sqrt{2mE}/\hbar & \text{for } x > a & (= \phi_3(x)) \end{cases}$$

(refer to Worked example 17.2), where r is the reflection coefficient and t the transmission coefficient. In contrast to the finite potential well solution, we require both increasing and decaying exponential solutions within the potential barrier, therefore A and B are both finite.

2. **Boundary conditions:** $\phi(x)$ and its first derivative are continuous at the boundaries:

$$\phi_1(0) = \phi_2(0) \rightarrow 1 + r = A + B \tag{17.16}$$

$$\phi_2(a) = \phi_3(a) \rightarrow Ae^{\kappa a} + Be^{-\kappa a} = te^{ika} \tag{17.17}$$

and
$$\left(\frac{d\phi_1}{dx}\right)_{x=0} = \left(\frac{d\phi_2}{dx}\right)_{x=0} \rightarrow ik - irk = A\kappa - B\kappa \tag{17.18}$$

$$\left(\frac{d\phi_2}{dx}\right)_{x=a} = \left(\frac{d\phi_3}{dx}\right)_{x=a} \rightarrow \kappa Ae^{\kappa a} - \kappa Be^{-\kappa a} = ikte^{ika} \tag{17.19}$$

3. **Transmission coefficient:** using Eqs. 17.16 and 17.18 to eliminate r, and Eqs. 17.17 and 17.19 to eliminate B and A, we can determine an expression for the transmission coefficient, t:

$$t = \frac{2e^{-2ika}}{2\cosh(\kappa a) - i\left(\frac{k}{\kappa} - \frac{\kappa}{k}\right)\sinh(\kappa a)} \tag{17.20}$$

WORKED EXAMPLE 17.3: Tunneling through a finite potential barrier (cont.)

4. **Probability of transmission:** the probability of transmission is proportional to the square of the transmitted wavefunction, $|\phi_3|^2 = |t|^2$. Therefore, squaring Eq. 17.20 and using the identity $\cosh^2 x - \sinh^2 x = 1$, we find that the probability of transmission is:

$$|t|^2 = \left(1 + \left(\frac{k^2+\kappa^2}{2k\kappa}\right)^2 \sinh^2(\kappa a)\right)^{-1}$$
$$= \left(1 + \frac{\sinh^2\sqrt{v^2(1-E/V_0)}}{4(E/V_0)(1-E/V_0)}\right)^{-1} \tag{17.21}$$

where v is a dimensionless constant defined by:

$$v^2 = \frac{2mV_0 a^2}{\hbar^2}$$

For particles with energy much less than the potential barrier, $E \ll V_0$, and in the limit where $v \gg 1$ (wide and high potential barriers), Eq. 17.21 approximates to:

$$|t|^2 \simeq 16\frac{E}{V_0}e^{-2v}$$

($\sinh x = (e^x - e^{-x})/2$, $\therefore$ as $x \to \infty$, $\sinh x \to e^x$.) Therefore, there is a finite probability of transmission, even when the potential barrier is very large compared to the particle's energy. Through the parameter v, the probability of tunneling falls exponentially with increased barrier height, $\sqrt{V_0}$, and with barrier width, a.

with a linear restoring force, $V = -\int F dx = \int k_s x dx$. Therefore, the potential energy is quadratic in x:

$$V(x) = \frac{1}{2}k_s x^2 = \frac{1}{2}m\omega^2 x^2$$

To a good approximation, many quantum systems, for example, atoms bound in a lattice, are subject to a linear restoring force. Therefore, the harmonic oscillator potential provides a more realistic description of such systems, compared to the infinite square potential well.

The wavefunction that satisfies the TISE for this potential cannot be solved easily via integration, as we could do in the previous examples, because of the quadratic potential term. If we try a Gaussian wavefunction solution, however, we find that it satisfies the harmonic potential Schrödinger equation (see Worked example 17.4 for details). The normalized ground-state wavefunction is found to be:

$$\phi_0(x) = \left(\frac{1}{\alpha^2\pi}\right)^{1/4} e^{-x^2/2\alpha^2}$$

where $\alpha = \sqrt{\frac{\hbar}{m\omega}}$

(refer to Worked example 17.4). The higher energy levels include the same Gaussian dependence as the ground state, but they are multiplied by a special function called a *Hermite polynomial*, $H_n(x)$:

$$\phi_n(x) = AH_n(x)e^{-x^2/2\alpha^2}$$

where A is the normalization constant. The first few Hermite polynomials are:

$$H_0(x) = 1$$
$$H_1(x) = 2x$$
$$H_2(x) = 4x^2 - 2$$
$$H_3(x) = 8x^3 - 12x$$

The energy spectrum of the harmonic oscillator is given by:

$$E_n = \left(n + \frac{1}{2}\right)\hbar\omega \tag{17.22}$$

where $n = 0, 1, 2, \ldots$

Therefore, the ground-state energy is $E_0 = \hbar\omega/2$, and higher energy levels are equally spaced by intervals of $\hbar\omega$.

WORKED EXAMPLE 17.4: Simple harmonic oscillator potential

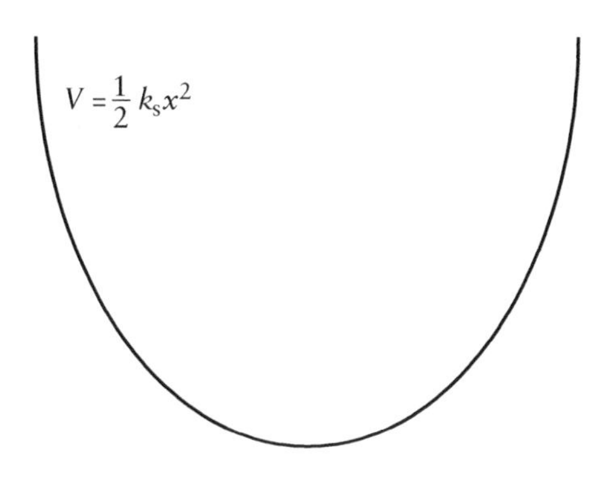

Potential energy function: the potential energy of a harmonic potential is a quadratic in x:

$$V(x) = \frac{1}{2} k_s x^2 = \frac{1}{2} m\omega^2 x^2$$

A particle in this potential is subject to a linear restoring force, $F = -k_s x$, and thus undergoes harmonic oscillations about the center.

1. **TISE:** the Schrödinger equation for a harmonic oscillator potential is an ordinary second-order differential equation with a quadratic potential term. If we make the change of variables $x = \alpha y$, where $\alpha = \sqrt{\hbar / m\omega}$, then the coefficients of the kinetic and potential energy terms become equal, which simplifies the differential equation:

$$\left(-\frac{\hbar^2}{2m} \frac{d^2}{dx^2} + \frac{1}{2} m\omega^2 x^2 \right) \phi(x) = E\phi(x) \xrightarrow[\text{let } x = \alpha y]{} \frac{\hbar\omega}{2} \left(-\frac{d^2}{dy^2} + y^2 \right) \phi(y) = E\phi(y)$$

2. **Ground-state wavefunction:** a Gaussian solution, $\phi(y) = Ae^{-y^2/2}$ (or $\phi(x) = Ae^{-x^2/2\alpha^2}$), satisfies the TISE above (show this by substitution). Applying the normalization condition, we can find an expression for the constant A, thus giving us the normalized ground-state wavefunction:

$$\int_{-\infty}^{\infty} |\phi(x)|^2 \, dx = 1 \rightarrow A^2 \int_{-\infty}^{\infty} e^{-x^2/\alpha^2} dx = A^2 \alpha \sqrt{\pi} = 1$$

$$\therefore A = \left(\frac{1}{\alpha^2 \pi} \right)^{1/4} \Rightarrow \phi_0(x) = \left(\frac{1}{\alpha^2 \pi} \right)^{1/4} e^{-x^2/2\alpha^2}$$

Use the standard integral: $\int_{-\infty}^{\infty} e^{-\beta x^2} dx = \sqrt{\frac{\pi}{\beta}}$

3. **Ground-state energy:** substituting $\phi_0(x)$ back into the TISE, we find that the ground-state energy is non-zero:

$$\frac{\hbar\omega}{2} \left(1 - y^2 + y^2 \right) Ae^{-y^2/2} = E_0 Ae^{-y^2/2} \rightarrow \therefore E_0 = \frac{\hbar\omega}{2}$$

Therefore, quantum systems possess a non-zero minimum energy, called the *zero-point energy*, due to their quantum motion; a state of zero energy does not exist.

4. **Higher energy solutions:** for the higher energy solutions, try a Gaussian solution multiplied by a polynomial in y, $H(y)$:

$$\phi(y) = H(y) e^{-y^2/2}$$

Doing this and defining $\varepsilon = 2E/\hbar\omega$, we find that the TISE obeys a special ordinary differential equation called *Hermite's equation*:

$$H''(y) - 2yH'(y) + H(y)(\varepsilon - 1) = 0$$

The functions that satisfy this equation are *Hermite polynomials*, $H(y)$. The first few Hermite polynomials are:

$$H_0(y) = 1, \; H_1(y) = 2y, \; H_2(y) = 4y^2 - 2, \; H_3(y) = 8y^3 - 12y$$

WORKED EXAMPLE 17.4: Simple harmonic oscillator potential (cont.)

5. **Energy-level spectrum:** it turns out that the solutions are only normalizable when $\varepsilon = 2n + 1$, where n is an integer starting at 0. (This is found using a power-series method of solution, which is not shown here.) Therefore, the energy levels are given by:

$$E_n = \varepsilon\frac{\hbar\omega}{2} = \left(n + \frac{1}{2}\right)\hbar\omega$$

where $n = 0, 1, 2, \ldots$

These are equally spaced by $\hbar\omega$, and the minimum energy state is $E_0 = \hbar\omega/2$, as expected.

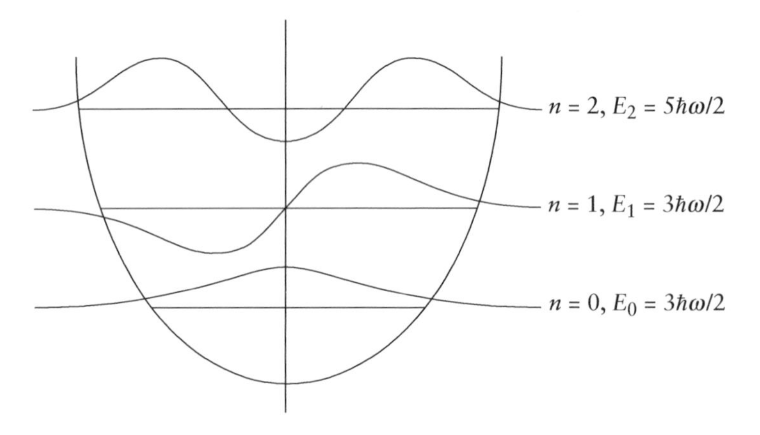

The ground-state energy is called the *zero-point energy*, and it is finite (non-zero). This implies that atoms always possess a minimum energy due to their quantum motion.

17.3 The Schrödinger equation in three dimensions

The quantities that are vectors in classical mechanics are also vectors in quantum mechanics, and likewise for scalars. It's therefore straightforward to express operators and the Schrödinger equation in three dimensions if we know their one-dimensional counterpart.

Since momentum is a vector in classical physics, the three-dimensional momentum operator is a vector operator (the gradient operator, ∇), and the three-dimensional kinetic energy operator is a scalar operator (the Laplacian, ∇^2):

$$\hat{\vec{p}} = -i\hbar\begin{pmatrix}\partial/\partial x\\ \partial/\partial y\\ \partial/\partial z\end{pmatrix} \equiv -i\hbar\nabla$$

and $$\hat{T} = \frac{\hat{p}^2}{2m} = \frac{\hat{\vec{p}}\cdot\hat{\vec{p}}}{2m} = -\frac{\hbar^2}{2m}\nabla^2$$

The time-independent potential energy operator is a function of the position vector, $\vec{r}$: $V(\vec{r})$ (Eqs. 17.24–17.26).

Putting these together, the three-dimensional Schrödinger equation is therefore:

$$\underbrace{\left(-\frac{\hbar^2}{2m}\nabla^2 + V(\vec{r})\right)}_{\text{three-dimensional Hamiltonian, } \hat{H}}\psi(\vec{r},t) = i\hbar\frac{\partial\psi(\vec{r},t)}{\partial t}$$

The time dependence of the Schrödinger equation has the same form as in one dimension: $T(t) = e^{-i\omega t}$ (refer to Section 17.1). Therefore, the TISE is:

$$\left(-\frac{\hbar^2}{2m}\nabla^2 + V(\vec{r})\right)\phi(\vec{r}) = E\phi(\vec{r})$$

(Eq. 17.23). The total wavefunction is then the product of space and time solutions, $\psi(\vec{r}, t) = \phi(\vec{r})T(t)$. The probability density, $|\psi(\vec{r}, t)|^2 \equiv |\phi(\vec{r})|^2$, corresponds to the probability per unit volume [m^{-3}], and the probability of finding a particle within an element of volume $d\tau = dxdydz$ from the position vector $\vec{r}$ at a time t is:

$$P(\vec{r})d\tau = |\phi(\vec{r})|^2 d\tau$$

The normalization condition requires that the probability of finding a particle *somewhere* equals one, and therefore requires a volume integral:

$$\int_{\text{all space}} |\phi(\vec{r})|^2 \, d\tau = 1$$

Separating the three-dimensional Schrödinger equation

For a potential where the x-, y-, and z-components are separable, such that:

$$V(\vec{r}) = V(x) + V(y) + V(z)$$

we can use the separation-of-variables method to separate the x-, y-, and z-components of the wavefunction. Doing this, we find that each component obeys its own, one-dimensional Schrödinger equation, and that the three-dimensional solution is the product of the individual x, y, and z solutions (Derivation 17.3).

The TISE in three dimensions:

$$\underbrace{\left(-\frac{\hbar^2}{2m}\nabla^2 + V(\vec{r})\right)}_{\text{Hamiltonian, } \hat{H}}\phi(\vec{r}) = E\phi(\vec{r}) \quad (17.23)$$

Energy and momentum operators:

$$\text{Momentum:} \quad \hat{\vec{p}} = -i\hbar\nabla \quad (17.24)$$

$$\text{Kinetic energy:} \quad \hat{T} = -\frac{\hbar^2}{2m}\nabla^2 \quad (17.25)$$

$$\text{Potential energy:} \quad \hat{V} = V(\vec{r}) \quad (17.26)$$

17.3.1 Infinite potential box

It is straightforward to generalize the one-dimensional infinite square potential well to a three-dimensional infinite potential "box." This consists of a zero-potential cuboid of lengths a, b, and c along x, y, and z respectively; the walls of the cuboid correspond to an infinite discontinuity in potential (Fig. 17.3). A particle is therefore confined within the cuboid since it would require an infinite amount of energy to escape from it. The dimensions of the cuboid impose restrictions on the wavefunctions that fit in the box such that the boundary conditions are satisfied. A particle can therefore only occupy a discrete set of states, which correspond to discrete energy levels.

Since the three-dimensional Schrödinger equation can be separated into three one-dimensional equations, one

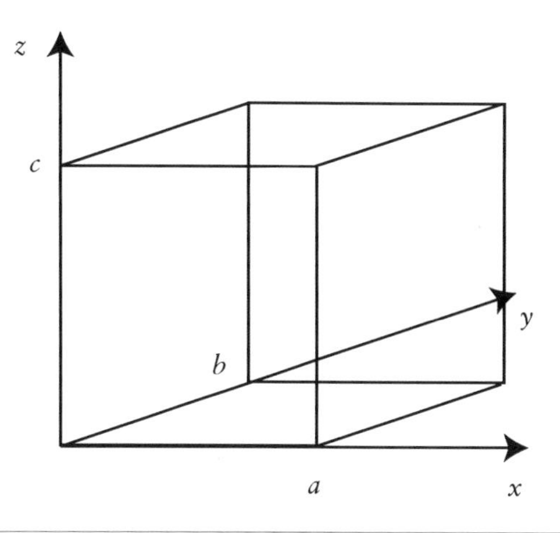

Figure 17.3: Three-dimensional infinite potential box. V = 0 inside the box and V = ∞ outside the box.

for x, y, and z (refer to Derivation 17.3), then the problem reduces to three one-dimensional infinite square well problems. The normalized solutions for the individual wavefunctions and the energy-level spectra are therefore:

$$X(x) = \sqrt{\frac{2}{a}}\sin\left(\frac{n\pi x}{a}\right), \quad E_x = \frac{\hbar^2}{2m}\left(\frac{\pi n}{a}\right)^2$$

$$Y(y) = \sqrt{\frac{2}{b}}\sin\left(\frac{l\pi y}{b}\right), \quad E_y = \frac{\hbar^2}{2m}\left(\frac{\pi l}{b}\right)^2$$

$$Z(z) = \sqrt{\frac{2}{c}}\sin\left(\frac{p\pi z}{c}\right), \quad E_z = \frac{\hbar^2}{2m}\left(\frac{\pi p}{c}\right)^2$$

where $n = 1, 2, 3, \ldots$
$l = 1, 2, 3, \ldots$
$p = 1, 2, 3, \ldots$

(refer to Worked example 17.1 for details of the one-dimensional problem). The parameters n, l, and p are integers that represent quantization of the states that satisfy the boundary conditions. The overall normalized wavefunction is then the product of individual solutions:

$$\phi_{nlp}(\vec{r}) = X(x)Y(y)Z(z) = \sqrt{\frac{8}{abc}}\sin\left(\frac{n\pi x}{a}\right)\sin\left(\frac{l\pi y}{b}\right)\sin\left(\frac{p\pi z}{c}\right)$$

The corresponding energy spectrum is the sum of individual energies of each of the components:

$$E_{nlp} = E_x + E_y + E_z = \frac{\hbar^2\pi^2}{2m}\left(\left(\frac{n}{a}\right)^2 + \left(\frac{l}{b}\right)^2 + \left(\frac{p}{c}\right)^2\right)$$

Derivation 17.3: Separation of variables: three-dimensional TISE in Cartesian coordinates

1. **Partial differential equation:** assuming that the potential energy function can be separated into x-, y-, and z-components, the three-dimensional TISE is:

$$\left(-\frac{\hbar^2}{2m}\nabla^2 + V(x) + V(y) + V(z)\right)\phi(x,y,z) = E\phi(x,y,z)$$

where $\nabla^2 = \left(\frac{\partial^2}{\partial x^2} + \frac{\partial^2}{\partial y^2} + \frac{\partial^2}{\partial z^2}\right)$

2. **Separation function:** separate the three components from one another by letting : $\phi(x, y, z) = X(x)Y(y)Z(z)$:

$$\left(-\frac{\hbar^2}{2m}\nabla^2 + V(x) + V(y) + V(z)\right)X(x)Y(y)Z(z) = E\,X(x)Y(y)Z(z)$$

3. **Separation constant:** dividing by $X(x)Y(y)Z(z)$ and rearranging to collect terms in the same variable, we find that the total energy can be expressed as a sum of x-, y-, and z-components:

$$\underbrace{\left(-\frac{\hbar^2}{2m}\frac{X''(x)}{X(x)} + V(x)\right)}_{=\text{constant}, E_x} + \underbrace{\left(-\frac{\hbar^2}{2m}\frac{Y''(y)}{Y(y)} + V(y)\right)}_{=\text{constant}, E_y} + \underbrace{\left(-\frac{\hbar^2}{2m}\frac{Z''(z)}{Z(z)} + V(z)\right)}_{=\text{constant}, E_z} = E$$

4. **One-dimensional TISEs:** each component therefore obeys its own one-dimensional TISE, where the sum of energy components equals the total energy, $E_x + E_y + E_z = E$:

$$-\frac{\hbar^2}{2m}X''(x) + V(x)X(x) = E_x X(x)$$
$$-\frac{\hbar^2}{2m}Y''(y) + V(y)Y(y) = E_y Y(y)$$
$$-\frac{\hbar^2}{2m}Z''(z) + V(z)Z(z) = E_z Z(z)$$

5. **Overall wavefunction:** the total wavefunction is given by the product of x, y, and z solutions, and the normalization condition requires that the integral over all space equals one:

$$\phi(x,y,z) = X(x)Y(y)Z(z) \qquad \text{where} \quad \int_{\text{all space}} |\phi(x,y,z)|^2 \, d\tau = 1$$

Therefore, each component is individually normalized:

$$\int |X(x)Y(y)Z(z)|^2 \, dx\,dy\,dz = 1 \Rightarrow \begin{cases} \int |X(x)|^2 \, dx = 1 \\ \int |Y(y)|^2 \, dy = 1 \\ \int |Z(z)|^2 \, dz = 1 \end{cases}$$

Therefore, a state is defined by the three integers n, l, and p, which represent quantization along each direction.

Energy degeneracy

For a cube of equal sides, say a, the wavefunction's energy levels are:

$$E_{nlp} = \frac{\hbar^2\pi^2}{2ma^2}\left(n^2 + l^2 + p^2\right)$$

The ground state is defined by the integers $\{n, l, p\} = \{1, 1, 1\}$, and the ground-state energy is therefore:

$$E_{111} = \frac{3\hbar^2\pi^2}{2ma^2}$$

The first excited state, however, has three different permutations $\{n, l, p\} = \{2, 1, 1\}$, $\{1, 2, 1\}$ or $\{1, 1, 2\}$. These each correspond to a different wavefunction—and thus to a different state—but they all have the same energy:

$$E_{211} = E_{121} = E_{112} = \frac{6\hbar^2\pi^2}{2ma^2}$$

States that correspond to the same energy level, but have different wavefunctions, are called *degenerate* states. Since there are three states that correspond to the first energy level in this case, this level is said to have a *three-fold degeneracy*. Degeneracy is a property of *symmetric* systems. For example, the potential box has cubic symmetry. Atoms possess spherical symmetry, which also gives rise to degenerate states, as we shall see in the case of the hydrogen atom (see Chapter 20).

Chapter 18 Quantum states, operators, and measurements

In the classical world, measurement outcomes are deterministic: given known properties of a system, we can predict how the system will behave at a later time. In the quantum world however, they are *probabilistic*—we cannot predict how a system will behave, only predict the *probability* of it behaving a particular way. In this way, the meaning of experimental measurements in quantum mechanics is a controversial and philosophical subject. In this chapter, we'll discuss what it means to make a measurement on a quantum system, and how to determine the probability of a given outcome. To understand and interpret such outcomes, which are counterintuitive and abstract, it helps to be familiar with linear algebra; a brief overview is provided below.

18.1 Linear algebra overview

Linear algebra is the mathematical language of quantum mechanics. It is a vast and abstract subject, and for a thorough overview, the reader is referred to *The Principles of Quantum Mechanics, R. Shankar.* However, in this section, we highlight a few of the basic principles necessary for understanding quantum mechanics at an introductory level.

18.1.1 The Dirac notation

The *Dirac notation* is a mathematical formalism that is particularly well suited to quantum mechanics. Although it may initially appear foreign and daunting, it actually greatly aids with understanding abstract concepts in quantum systems. The key ideas are highlighted below.

State of a particle

Recall that the *state* of a particle or a system is a broad term that encompasses all of its physical aspects, such as its energy, angular momentum, position, etc. In the previous chapter, we represented a state by a wavefunction, $\psi(x, t)$.

In Dirac notation, a quantum state is represented by a *ket*, $|\psi\rangle$. This can be thought of as a single-column vector[1] of any dimension (finite or infinite) with elements $a_1, a_2, a_3, \ldots, a_N$:

$$|\psi\rangle \equiv \begin{pmatrix} a_1 \\ a_2 \\ \ldots \\ a_N \end{pmatrix}$$

These objects exist in *Hilbert space*, which can be thought of as an "infinite-dimensional" version of our every-day three-dimensional Euclidean space. At this stage, we do not have to be over-concerned about the formal properties of Hilbert space.[2]

The complex conjugate of a ket is a *bra*, $\langle\psi|$ (i.e. $|\psi\rangle^* \equiv \langle\psi|$). This can be thought of as a row vector whose elements are the complex conjugates of the ket elements a_i:

$$\langle\psi| \equiv \begin{pmatrix} a_1^* & a_2^* & \ldots & a_N^* \end{pmatrix}$$

The Dirac notation is sometimes called the *Bra-ket notation* since it involves $\langle$*bras*$|$ and $|$*kets*$\rangle$.

1 Kets may also represent $n \times n$ matrices, but for our purposes, we will think of them as single-column vectors.

2 Refer to *The Principles of Quantum Mechanics, R. Shankar* for more detail, if necessary.

Inner product

The product of a bra and a ket is called the *inner product*, denoted $\langle \psi_1 | \psi_2 \rangle$. Converting to wavefunction notation, the inner product is defined as the integral of the product of the two wavefunctions:

$$\langle \psi_1 | \psi_2 \rangle \overset{\text{def}}{=} \int_{-\infty}^{\infty} \psi_1^*(x)\psi_2(x)\mathrm{d}x$$

This product is essentially equivalent to taking the dot product of two vectors. A dot product represents the projection of one vector onto another: it represents the "amount" of one vector in the direction of another. Analogously, the inner product can be thought of as the "amount of overlap" of two states, $|\psi_1\rangle$ and $|\psi_2\rangle$ (or of two functions $\psi_1(x)$ and $\psi_2(x)$). As a more concrete example, let's consider the inner product of a state with itself:

$$\begin{aligned}\langle \psi | \psi \rangle &= \int_{-\infty}^{\infty} \psi^*(x)\psi(x)\mathrm{d}x \\ &= \int_{-\infty}^{\infty} |\psi(x)|^2 \,\mathrm{d}x \\ &= 1\end{aligned}$$

We recognize the integral on the right-hand side as the integral of the probability density function (refer to Section 16.3.1), and from the normalization condition, this integral is equal to one (the probability of finding a particle *somewhere* must be one). Therefore, the inner product of a state $|\psi\rangle$ with itself equals one; the state "totally overlaps" with itself. Two *orthogonal* states are ones whose inner product is equal to zero:

$$\langle \psi_1 | \psi_2 \rangle = 0 \text{ if } |\psi_1\rangle \text{ and } |\psi_2\rangle \text{ are orthogonal}$$

This is analogous to the dot product of two orthogonal vectors being equal to zero. For example, for the unit vectors $\hat{i}$ and $\hat{j}$, their dot product is $\hat{i} \cdot \hat{j} = 0$; there is zero projection of $\hat{i}$ along the $\hat{j}$-axis.

Operators and states

An operator $\hat{Q}$ transforms any given ket $|\psi\rangle$ into another, $|\psi'\rangle$. The adjoint of an operator (i.e. the transpose of its complex conjugate), $\hat{Q}^\dagger$, transforms any given bra $\langle \psi|$ into another, $\langle \psi'|$. Expressed mathematically:

$$\hat{Q}|\psi\rangle = |\psi'\rangle$$
$$\text{and } \langle \psi|\hat{Q}^\dagger = \langle \psi'|$$

All operators have a set of kets on which their action is to multiply the ket by a scalar q:

$$\hat{Q}|\psi\rangle = q|\psi\rangle$$
$$\text{and } \langle \psi|\hat{Q}^\dagger = \langle \psi|q^*$$

This is the *eigenvalue equation* for the operator $\hat{Q}$. It is common to denote the set of kets that satisfy an eigenvalue equation by $|u_n\rangle$ where $n = 1, 2, 3, \ldots$, and their corresponding scalars by q_n (Eq. 18.1). The set $|u_n\rangle$ are called *eigenkets* (or eigenfunctions), and their corresponding scale factors q_n are called *eigenvalues*. In the context of quantum mechanics, we usually refer to eigenkets as eigen*states*, since kets represent the state of a system. The complete set of eigenvalues is called the *eigenvalue spectrum*:

$$\{q_n\} = \{q_1, q_2, q_3, \ldots\}$$

Eigenvalue equation for $\hat{Q}$:

$$\hat{Q}|u_n\rangle = q_n|u_n\rangle \tag{18.1}$$

where $n = 1, 2, 3, \ldots$

18.1.2 Hermitian operators

A *Hermitian operator* is defined as an operator that equals its adjoint:

$$\hat{Q}^\dagger = \hat{Q}$$

Hermitian operators have three special mathematical properties:

1. They have real eigenvalues.
2. They have orthogonal eigenfunctions.
3. The eigenfunctions form a complete set.

(Eqs. 18.2–18.4 and Derivation 18.1). It follows from these properties that all operators that represent physical observables in quantum mechanics, such as $\hat{H}$ or $\hat{p}$, are Hermitian operators. (There is more on this in Section 18.2.)

Hermitian operators (Derivation 18.1):

$$\hat{Q}^\dagger = \hat{Q} \tag{18.2}$$

Properties of Hermitian operators:
1. Eigenvalues are real:

$$\begin{aligned} q_n &= q_n^* \\ \therefore \hat{Q}|u_n\rangle &= q_n|u_n\rangle \\ \text{and } \langle u_n|\hat{Q} &= \langle u_n|q_n \end{aligned} \tag{18.3}$$

Derivation 18.1: Derivation of the properties of Hermitian operators (Eqs. 18.3 and 18.4)

$$\hat{Q}|u_n\rangle = q_n|u_n\rangle$$
$$\therefore \langle u_m|\hat{Q}|u_n\rangle = \langle u_m|q_n|u_n\rangle$$

$$\langle u_m|\hat{Q} = \langle u_m|q_m^*$$
$$\therefore \langle u_m|\hat{Q}|u_n\rangle = \langle u_m|q_m^*|u_n\rangle$$

$$\langle u_m|q_n|u_n\rangle = \langle u_m|q_m^*|u_n\rangle$$
$$q_n\langle u_m|u_n\rangle = q_m^*\langle u_m|u_n\rangle$$
$$\therefore (q_n - q_m^*)\langle u_m|u_n\rangle = 0$$

When $n = m$:

$$\langle u_n|u_n\rangle = 1 \therefore q_n = q_n^*$$

$\Rightarrow$ eigenvalues are real.
When $n \neq m$:

$$q_n \neq q_m \therefore \langle u_m|u_n\rangle = 0$$

$\Rightarrow$ eigenstates are orthogonal.

- Starting point: the eigenvalue equations for a Hermitian operator $\hat{Q}$ operating on normalized eigenstates $|u_n\rangle$ or $\langle u_m|$ are:

$$\hat{Q}|u_n\rangle = q_n|u_n\rangle$$
$$\text{and} \langle u_m|\hat{Q} = \langle u_m|q_m^*$$

- Multiplying by $\langle u_m|$ or $|u_n\rangle$, respectively, we obtain two expressions for $\langle u_m|\hat{Q}|u_n\rangle$.
- Equating these two expressions, we find that the eigenvalues and eigenstates of Hermitian operators satisfy the equation:

$$(q_n - q_m^*)\langle u_m|u_n\rangle = 0$$

from which we deduce that eigenvalues of Hermitian operators are real, and their eigenstates are orthogonal.

2. Eigenstates are orthonormal*:

$$\langle u_m|u_n\rangle = \delta_{mn} \qquad (18.4)$$

$$\text{where } \delta_{nm} = \begin{cases} 1 & \text{for } n = m \\ 0 & \text{for } n \neq m \end{cases}$$

3. Eigenstates form a complete set

*Orthonormal states are both orthogonal and normalized such that the inner product of a state with itself is equal to 1.

Matrix element of Hermitian operators:

$$\langle \psi_n|\hat{Q}|\psi_m\rangle^* = \langle \psi_m|\hat{Q}|\psi_n\rangle \qquad (18.5)$$

Or in integral form:

$$\int_{-\infty}^{\infty} \left(\psi_n^*\hat{Q}\psi_m\right)^* dx = \int_{-\infty}^{\infty} \left(\hat{Q}\psi_m\right)^* \psi_n dx \qquad (18.6)$$

Matrix element of Hermitian operators

The *matrix element*[3] of an operator $\hat{Q}$ is defined as:

$$\langle \psi_n|\hat{Q}|\psi_m\rangle \overset{\text{def}}{=} \int_{-\infty}^{\infty} \psi_n^*(x)\hat{Q}\psi_m(x)dx$$

The matrix element of a Hermitian operator therefore equals its complex conjugate:

$$\langle \psi_n|\hat{Q}|\psi_m\rangle^* = \langle \psi_m|\hat{Q}^\dagger|\psi_n\rangle$$
$$= \langle \psi_m|\hat{Q}|\psi_n\rangle$$

(Eqs. 18.5 and 18.6). This is the formal definition of a Hermitian operator.

[3] Just as a ket can be thought of as a vector, an operator can be thought of as a matrix. The matrix element of an operator $\hat{Q}$ corresponds to the matrix representation of $\hat{Q}$ in the basis provided by the complete set of states $|\psi_n\rangle$, where $n = 1, 2, 3, \ldots$

Matrix elements and expectation values

The expectation value of a given operator is related to its matrix element by:

$$\langle \psi|\hat{Q}|\psi\rangle = \underbrace{\int_{-\infty}^{\infty} \psi^*(x)\hat{Q}\psi(x)dx}_{\text{Expectation value of } q}$$
$$\equiv \langle \hat{Q}\rangle$$

The expectation value of a Hermitian operator $\hat{Q}$, where $|u_n\rangle$ is an eigenstate of $\hat{Q}$, corresponds to one of its eigenvalues, q_n:

$$\langle \hat{Q}\rangle = \langle u_n|\hat{Q}|u_n\rangle$$
$$= \langle u_n|q_n|u_n\rangle$$
$$= q_n\langle u_n|\ u_n\rangle$$
$$= q_n$$

(Eq. 18.7). Since the eigenvalues of a Hermitian operator are real, then the expectation value of $\hat{Q}$ is real, and it equals its complex conjugate:

$$\langle \hat{Q} \rangle^* = \langle \hat{Q} \rangle$$

Eigenvalue and expectation value:

$$q_n = \langle u_n | \hat{Q} | u_n \rangle \equiv \langle \hat{Q} \rangle \tag{18.7}$$

where $n = 1, 2, 3, \ldots$

18.1.3 Superposition states

The eigenstates of Hermitian operators are orthogonal to one another, which means they are linearly independent, and they form a complete set. Therefore, they constitute a *basis* set of eigenstates. Hence, any state $|U\rangle$ can be written as a linear superposition of eigenstates:

$$|U\rangle = \sum_{n=1}^{N} c_n |u_n\rangle$$

(Eqs. 18.8 and 18.9). For analogy, the x, y, and z unit vectors ($\hat{i}, \hat{j}$, and $\hat{k}$) constitute the basis set of vectors for three-dimensional Euclidean space: *all* other vectors in this space can be expressed as a linear superposition of $\hat{i}, \hat{j}$, and $\hat{k}$. For example, an arbitrary three-dimensional vector $\vec{a}$ can be written as:

$$\vec{a} = \begin{pmatrix} a_x \\ a_y \\ a_z \end{pmatrix} \equiv a_x\hat{i} + a_y\hat{j} + a_z\hat{k}$$

where a_x, a_y, and a_z represent the "amount" of $\vec{a}$ along each of the three axes. For example, the magnitude of $\vec{a}$ along the x-axis is equal to the dot product of $\vec{a}$ with $\hat{i}$ (i.e. the projection of $\vec{a}$ along $\hat{i}$):

$$\vec{a} \cdot \hat{i} = a_x \hat{i} \cdot \hat{i} = a_x$$

Similarly, the coefficient c_n in Eq. 18.8 represents the "amount" of the nth eigenstate in the superposition (c_n can be complex; therefore, this analogy must be used with caution). Consider a superposition state composed of two eigenstates:

$$|U\rangle = c_1|u_1\rangle + c_2|u_2\rangle$$

Using orthonormality of eigenstates, the inner product (equivalent to the dot product) of this state with the first eigenstate is:

$$\begin{aligned}\langle u_1|U\rangle &= \langle u_1|c_1|u_1\rangle + \langle u_1|c_2|u_2\rangle \\ &= c_1\underbrace{\langle u_1|u_1\rangle}_{=1} + c_2\underbrace{\langle u_1|u_2\rangle}_{=0} \\ &= c_1\end{aligned}$$

Similarly, $\langle u_2|U\rangle = c_2$. Therefore, the value of c_n is given by the inner product of the nth eigenstate with the superposition state (Eqs. 18.10 and 18.11). If the superposition state is *not* partly composed of the nth eigenstate, then $c_n = 0$ and their inner product is zero.

Superposition states:

Any state $|U\rangle$ can be written as a linear superposition of eigenstates:

$$|U\rangle = \sum_{n=1}^{N} c_n |u_n\rangle = c_1|u_1\rangle + c_2|u_2\rangle + \cdots \tag{18.8}$$

$$\text{and} \quad \langle U| = \sum_{n=1}^{N} c_n^* \langle u_n| = c_1^*\langle u_1| + c_2^*\langle u_2| + \cdots \tag{18.9}$$

The nth coefficient is:

$$c_n = \langle u_n|U\rangle \tag{18.10}$$

$$\text{and} \quad c_n^* = \langle U|u_n\rangle \tag{18.11}$$

where $n = 1, 2, 3, \ldots$

Linearity of $\hat{Q}$

An operator is said to be *linear* if it satisfies the following two conditions:

$$\hat{Q}(|\psi_1\rangle + |\psi_2\rangle) = \hat{Q}|\psi_1\rangle + \hat{Q}|\psi_2\rangle$$

$$\text{and} \quad \hat{Q}(\alpha|\psi_1\rangle) = \alpha\hat{Q}|\psi_1\rangle$$

where α is a scalar. Therefore, when a linear operator acts on a superposition state, $|U\rangle$, it acts individually on each eigenstate (Eq. 18.12). All operators that we deal with in this chapter are linear operators.

Linearity of $\hat{Q}$:

$$\hat{Q}|U\rangle = \sum_{n=1}^{N} c_n \hat{Q}|u_n\rangle = \sum_{n=1}^{N} c_n q_n |u_n\rangle \tag{18.12}$$

18.1.4 The identity operator

The *identity operator*, $\hat{I}$, is an operator that leaves a state unchanged:

$$\hat{I}|U\rangle = |U\rangle$$

We can find an expression for $\hat{I}$ in terms of eigenstates $|u_n\rangle$ by considering a superposition state:

$$\begin{aligned}|U\rangle &= \sum c_n|u_n\rangle \\ &= \sum \langle u_n|U\rangle|u_n\rangle \\ &= \sum |u_n\rangle\langle u_n|U\rangle \\ &\equiv \hat{I}|U\rangle\end{aligned}$$

Therefore, the operation that leaves a state unchanged is the sum of the outer product of each eigenstate with itself (Eq. 18.13).

Identity operator:

$$\hat{I} = \sum_n |u_n\rangle\langle u_n| \qquad (18.13)$$

where $n = 1, 2, 3,\ldots$

18.1.5 Commutators

The last useful mathematical tool we'll discuss before seeing how all this applies to quantum systems is the *commutator* of two operators. If two operations are commutative, then we can change the order of their operation without changing the end result. For example, multiplication is commutative since, for two scalars A and B, $A \times B = B \times A$. In contrast, division is not commutative since $A/B \neq B/A$. The *commutator* is a measure of the extent to which an operation fails to be commutative. It is calculated by finding the difference between the result of either order of operation. For example, the multiplication commutator is $(A \times B) - (B \times A) = 0$, whereas the division commutator is $(A/B) - (B/A) \neq 0$. In linear algebra, the commutator of operators $\hat{Q}$ and $\hat{R}$ is denoted by $[\hat{Q}, \hat{R}]$ and defined as:

$$\left[\hat{Q},\hat{R}\right] \overset{\text{def}}{=} \left(\hat{Q}\hat{R} - \hat{R}\hat{Q}\right)$$

(refer to Eq. 18.14 and Derivation 18.2 for properties of commutators).

Commutators (Derivation 18.2):

$$\left[\hat{Q},\hat{R}\right]|\psi\rangle \overset{\text{def}}{=} \left(\hat{Q}\hat{R} - \hat{R}\hat{Q}\right)|\psi\rangle \qquad (18.14)$$

Properties of commutators for arbitrary operators, $\hat{Q}$, $\hat{R}$, and $\hat{S}$:

1. Inverse:

$$\left[\hat{Q},\hat{R}\right] = -\left[\hat{R},\hat{Q}\right]$$

Derivation 18.2: Derivation of the mathematical properties of the commutator (Eq. 18.14)

Inverse: changing the order of operation changes the sign of the commutator:

$$\begin{aligned}\left[\hat{Q},\hat{R}\right] &= \hat{Q}\hat{R} - \hat{R}\hat{Q} \\ &= -\left(\hat{R}\hat{Q} - \hat{Q}\hat{R}\right) \\ &= -\left[\hat{R},\hat{Q}\right]\end{aligned}$$

Linearity: the commutator of a superposition of operators is given by the sum of individual commutators:

$$\begin{aligned}\left[\hat{Q},\hat{R}+\hat{S}\right] &= \hat{Q}\left(\hat{R}+\hat{S}\right) - \left(\hat{R}+\hat{S}\right)\hat{Q} \\ &= \hat{Q}\hat{R} + \hat{Q}\hat{S} - \hat{R}\hat{Q} - \hat{S}\hat{Q} \\ &= \left[\hat{Q},\hat{R}\right] + \left[\hat{Q},\hat{S}\right]\end{aligned}$$

Squared operators: the commutator of a squared operator can be broken down into its non-squared commutator:

$$\begin{aligned}\left[\hat{Q},\hat{R}^2\right] &= \hat{Q}\hat{R}^2 - \hat{R}^2\hat{Q} \\ &= \hat{Q}\hat{R}\hat{R} - \hat{R}\hat{R}\hat{Q} \\ &= \hat{Q}\hat{R}\hat{R} - \left(\hat{R}\hat{Q}\hat{R} - \hat{R}\hat{Q}\hat{R}\right) - \hat{R}\hat{R}\hat{Q} \\ &= \left(\hat{Q}\hat{R} - \hat{R}\hat{Q}\right)\hat{R} + \hat{R}\left(\hat{Q}\hat{R} - \hat{R}\hat{Q}\right) \\ &= \left[\hat{Q},\hat{R}\right]\hat{R} + \hat{R}\left[\hat{Q},\hat{R}\right]\end{aligned}$$

2. Linearity:

$$\left[\hat{Q},\hat{R}+\hat{S}\right]=\left[\hat{Q},\hat{R}\right]+\left[\hat{Q},\hat{S}\right]$$
$$\left[\hat{Q}+\hat{R},\hat{S}\right]=\left[\hat{Q},\hat{S}\right]+\left[\hat{R},\hat{S}\right]$$

3. Squared operator:

$$\left[\hat{Q}^2,\hat{R}\right]=\hat{Q}\left[\hat{Q},\hat{R}\right]+\left[\hat{Q},\hat{R}\right]\hat{Q}$$
$$\left[\hat{Q},\hat{R}^2\right]=\left[\hat{Q},\hat{R}\right]\hat{R}+\hat{R}\left[\hat{Q},\hat{R}\right]$$

18.2 Measurements

The measurement postulate of quantum mechanics states that measurement of a physical observable corresponds to its expectation value (refer to Section 16.3.3). Consider a Hermitian operator $\hat{Q}$ with eigenvalues q_n and eigenstates $|u_n\rangle$. Its eigenvalue equation is:

$$\hat{Q}|u_n\rangle = q_n|u_n\rangle$$

For a system in one of its eigenstates, the expectation value of the operator is equal to the state's eigenvalue:

$$\begin{aligned}\langle\hat{Q}\rangle &= \langle u_n|\hat{Q}|u_n\rangle\\ &= \langle u_n|q_n|u_n\rangle\\ &= q_n\langle u_n|u_n\rangle\\ &= q_n\end{aligned}$$

Said another way, the measured value of a given observable represented by a Hermitian operator corresponds to one of the operator's eigenvalues, q_n. For example, a measurement of the total energy of a system will return one of the Hamiltonian's eigenvalues, E_n:

$$\hat{H}|\psi_n\rangle = E_n|\psi_n\rangle$$

$$\therefore\langle\hat{H}\rangle = E_n$$

Since Hermitian operators have real eigenvalues, then the measurement always corresponds to a real number, as required. The measured eigenvalue tells us what eigenstate the system is in. For example, a measured value of E_3 indicates that the system is in its $n=3$ energy eigenstate, $|\psi_3\rangle$.

Quantization

If we can accept that the expectation value of a Hermitian operator (i.e. one of its eigenvalues) corresponds to a possible measurement outcome, then it's a small step onwards to understand why some parameters are quantized (i.e. can only take certain discrete values). For variables such as energy, momentum, or angular momentum, the eigenvalue spectrum is a discrete set:

$$\{q_n\} = \{q_1, q_2, \ldots, q_N\}$$

(It is not discrete for continuous variables such as position.) This set of eigenvalues corresponds to a discrete set of allowed states: $|u_1\rangle$, $|u_2\rangle$, …, $|u_N\rangle$. For example, the energy eigenvalues of the Hamiltonian, $\{E_1, E_2, \ldots, E_N\}$, illustrate that energy is quantized—a state with a value of energy between, say, E_1 and E_2 does not exist; hence, this value of energy cannot be measured.

Superposition states

So far, we've considered only measurements on systems that are in an eigenstate. What about measurements on a system that is in a superposition of eigenstates? Consider, for example, a superposition state defined by:

$$|U\rangle = \sum_n c_n|u_n\rangle$$

Recall that the nth coefficient, c_n, equals the inner product of the nth eigenstate, $|u_n\rangle$, and the superposition state:

$$c_n = \langle u_n|U\rangle$$

(refer to Eq. 18.10). The expectation value of a Hermitian operator $\hat{Q}$ in this superposition state is found to be:

$$\begin{aligned}\langle\hat{Q}\rangle &= \langle U|\hat{Q}|U\rangle\\ &= \sum_n |c_n|^2 q_n\\ &\equiv |c_1|^2 q_1 + |c_2|^2 q_2 + \cdots |c_N|^2 q_N\end{aligned}$$

(Eq. 18.15 and Derivation 18.3). Therefore, the expectation value comprises the eigenvalues of all the eigenstates in the superposition. However, a measurement must return a *single* eigenvalue. Therefore, what value do we measure? Quantum mechanics is probabilistic—we cannot predict the value of a measurement, but can only determine the probability of a particular outcome. It turns out that the probability of measuring q_1 is $|c_1|^2$, the probability of measuring q_2 is $|c_2|^2$, etc. The normalization condition is that the sum of probabilities equals one:

$$\therefore |c_1|^2 + |c_2|^2 + \cdots |c_N|^2 = \sum_n |c_n|^2 = 1$$

(Eqs. 18.16 and 18.17). Therefore, if a particle is in an eigenstate, say $|u_m\rangle$, then the mth coefficient is $|c_m|^2 = 1$, and the probability of measuring q_m is 1.

Energy eigenstates

Energy eigenstates are states of definite energy, E_n. We saw in Section 17.1 that the time evolution of a system with energy E_n is:

$$|\psi_n\rangle \propto e^{iE_n t/\hbar}$$

If nothing interferes with the system, then it will remain in an energy eigenstate, such that every energy measurement returns the same value, E_n. An energy eigenstate is often referred to as a *stationary state* since the probability of measuring E_n remains constant in time, where $P(E_n) = 1$. An example of determining energy and momentum eigenstate probabilities is provided in Worked example 18.1.

Measurements (Derivation 18.3):

Expectation value:

$$\langle U|\hat{Q}|U\rangle = \sum_{n=1}^{N} |c_n|^2 q_n$$
$$= |c_1|^2 q_1 + |c_2|^2 q_2 + \cdots \quad (18.15)$$

Probability of measuring q_i:

$$P(q_i) = |c_i|^2$$
$$= |\langle u_i|U\rangle|^2 \quad (18.16)$$

Probability normalization:

$$\sum_{n=1}^{N} |c_n|^2 = 1 \quad (18.17)$$

where $n = 1, 2, 3, \ldots$

Probabilities and wavefunctions

In fact, we're already familiar with the notion of measurement probabilities from the wavefunction formulation of quantum mechanics. A superposition state composed of a discrete set of eigenstates is expressed as:

$$|U\rangle = \sum_{n=1}^{N} c_n |u_n\rangle$$

For a continuous variable, like position x, we can convert the sum into an integral, and the discrete set

Derivation 18.3: Derivation of superposition measurement probabilities (Eq. 18.15)

$$\langle\hat{Q}\rangle = \langle U|\hat{Q}|U\rangle$$
$$= \sum_{n,m} \langle u_m|c_m^*\hat{Q}c_n|u_n\rangle$$
$$= \sum_{n,m} c_m^* c_n \langle u_m|\hat{Q}|u_n\rangle$$
$$= \sum_{n,m} c_m^* c_n \langle u_m|q_n|u_n\rangle$$
$$= \sum_{n,m} c_m^* c_n q_n \langle u_m|u_n\rangle$$
$$= \sum_{n,m} c_m^* c_n q_n \delta_{mn}$$
$$= \sum_{n} c_n^* c_n q_n$$
$$= \sum_{n} |c_n|^2 q_n$$

- Starting point: consider an arbitrary linear operator $\hat{Q}$ with eigenvalue equation $\hat{Q}|u_n\rangle = q_n|u_n\rangle$, and a superposition state defined by:

$$|U\rangle = \sum_{n=1}^{N} c_n |u_n\rangle$$
$$\therefore \langle U| = \sum_{m=1}^{N} c_m^* \langle u_m|$$

where $|u_n\rangle$ are the normalized eigenstates of $\hat{Q}$.
- The expectation value of $\hat{Q}$ is defined as:

$$\langle\hat{Q}\rangle = \langle U|\hat{Q}|U\rangle$$

- $\hat{Q}$ is a linear operator; therefore, it does not act on the constant coefficients, c_n and c_m^*, and it acts individually on each eigenstate in the superposition.
- The eigenstates of Hermitian operators are orthonormal; therefore, the inner product of two eigenstates is equal to the Dirac delta function:

$$\langle u_m|u_n\rangle = \delta_{mn} = \begin{cases} 1 & \text{for } n = m \\ 0 & \text{for } n \neq m \end{cases}$$

WORKED EXAMPLE 18.1: Momentum superposition of an energy eigenstate

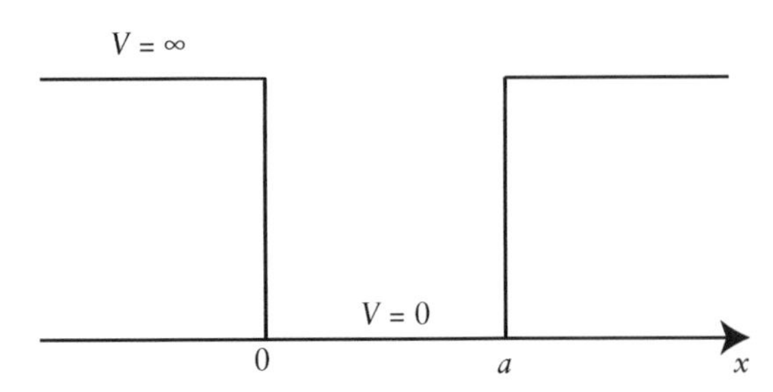

Consider a particle confined in an infinite square potential well. A particle in the nth energy eigenstate, E_n, has wavevector $k_n = p_n/\hbar$. Determine the momentum eigenstates available to the particle and the probability of it being in each of the momentum states.

1. **Momentum eigenfunctions:** the one-dimensional momentum operator is: $\hat{p} = -i\hbar\partial/\partial x$. The momentum eigenvalue equation is then:

$$\hat{p}\phi_{p_n}(x) = p_n\phi_{p_n}(x) \Rightarrow -i\hbar\frac{\partial\phi_{p_n}(x)}{\partial x} = p_n\phi_{p_n}(x)$$

Integrating, we see that the momentum eigenstates correspond to complex exponential solutions:

$$\phi_{p_n}(x) = Ae^{ip_n x/\hbar} = Ae^{ik_n x} = \frac{1}{\sqrt{a}}e^{ik_n x}$$

The factor $A = 1/\sqrt{a}$ comes from normalizing the solution for a particle in a square potential well of width a.

2. **Momentum superposition:** the wavefunction of a particle in the nth energy eigenstate of a square well is:

$$\phi_{E_n}(x) = \sqrt{\frac{2}{a}}\sin(k_n x)$$

(refer to Worked example 17.3). Using the identity $\sin\theta = (e^{i\theta} - e^{-i\theta})/2i$, we can express the energy eigenstate as a superposition of two momentum eigenstates, $+p_n$ and $-p_n$:

$$\begin{aligned}\phi_{E_n}(x) &= \sqrt{\frac{2}{a}}\sin(k_n x)\\ &= \sqrt{\frac{2}{a}}\frac{1}{2i}\left(e^{ik_n x} - e^{-ik_n x}\right)\\ &= \frac{1}{i\sqrt{2}}\left(\phi_{p_n}(x) - \phi_{-p_n}(x)\right)\end{aligned}$$

The probability of the particle being in either eigenstate is:

$$P(p_n) = P(-p_n) = \left|\frac{1}{i\sqrt{2}}\right|^2 = \frac{1}{2}$$

of coefficients, c_n, into a continuous function, $C(x)$. Therefore, an arbitrary state $|\psi\rangle$ can be written as a superposition of continuous position eigenstates, $|x\rangle$:

$$|\psi\rangle = \int_{-\infty}^{\infty} C(x)|x\rangle dx$$

The inner product $\langle x|\psi\rangle$ describes the "amount" of eigenstate $|x\rangle$ in the superposition state $|\psi\rangle$. By definition, the coefficient $C(x)$ is equal to this inner product (refer to Eq. 18.10 for the discrete case):

$$\therefore C(x) = \langle x|\psi\rangle$$

For the discrete case, the probability of a system being in the ith eigenstate is equal to $|c_i|^2$, where c_i is the coefficient of the ith eigenstate in the superposition (refer to Eq. 18.16). Generalizing to the continuous case, the probability of finding the particle between position x and $x + \mathrm{d}x$ is then:

$$P(x)\mathrm{d}x = |C(x)|^2\,\mathrm{d}x$$
$$= |\langle x|\psi\rangle|^2\,\mathrm{d}x$$

In wavevector notation, the inner product $\langle x|\ \psi\rangle$ is equivalent to the wavefunction as a function of x: $\langle x|\ \psi\rangle \equiv \psi(x)$. Therefore, the probability of a particle being between x and $x + \mathrm{d}x$ is given by the wavefunction squared, as expected:

$$\therefore P(x)\mathrm{d}x = |\psi(x)|^2\,\mathrm{d}x$$

Wavefunction collapse

A measurement corresponds to a *single* eigenvalue. This means that, even if a system is in a superposition state *before* the measurement, then *after* it, it is always in an eigenstate (the one that corresponds to the measured eigenvalue). The wavefunction is therefore said to "collapse" onto an eigenstate upon measurement. Said another way, a system can exist in a superposition state until we "look" at it.

This phenomenon represents a controversial aspect of quantum mechanic measurements. A well-known controversial paradox is *Schrödinger's cat*: consider a cat in a box in which there is a 50% chance that a poisonous gas is released. Since we don't know whether the gas was released or not, then according to quantum theory, the cat exists in a superposition state of $|dead\rangle$ and $|alive\rangle$, the obvious paradox being how can a cat simultaneously be dead and alive? However, with the collapse interpretation, the paradox is resolved by saying that we do not know what state the cat is in until we look inside the box. From that moment onwards, the cat exists in a definite state of either $|dead\rangle$ or $|alive\rangle$ (hopefully the latter).

The notion of wavefunction collapse is sometimes referred to as the *Copenhagen Interpretation* of quantum mechanics. This interpretation identifies two types of time evolution in quantum systems: the smooth time evolution of the superposition state before measurement (which is governed by the time-dependent part of the Schrödinger equation, $T(t) = \mathrm{e}^{\mathrm{i}Et/\hbar}$), and the sudden time evolution corresponding to wavefunction collapse.

18.3 Conserved quantities

18.3.1 The conservation equation

A measurement of a physical observable q of a quantum system corresponds to the expectation value of its operator, $\langle \hat{Q} \rangle$. The *conservation equation* describes how $\langle \hat{Q} \rangle$ varies with time (Eq. 18.18 and Derivation 18.4). We see in Eq. 18.18 that the rate of change of $\langle \hat{Q} \rangle$ depends on two factors: (i) the commutator of $\hat{Q}$ with the Hamiltonian (first term) and (ii) the implicit time dependence of $\hat{Q}$ (second term). A quantity is said to be *conserved* if its rate of change equals zero (Eq. 18.19). Most operators are time independent; therefore, the second term in Eq. 18.18 equals zero. In this case, a quantity is conserved if its operator commutes with the Hamiltonian:

$$\left[\hat{H}, \hat{Q}\right] = 0$$

Since the Hamiltonian commutes with itself, $|\hat{H}, \hat{H}|$, then energy is a conserved quantity, as expected.

The conservation equation (Derivation 18.4):

$$\frac{\mathrm{d}\langle \hat{Q} \rangle}{\mathrm{d}t} = \frac{\mathrm{i}}{\hbar}\left\langle \left[\hat{H}, \hat{Q}\right] \right\rangle + \left\langle \frac{\mathrm{d}\hat{Q}}{\mathrm{d}t} \right\rangle \qquad (18.18)$$

A quantity is conserved if:

$$\frac{\mathrm{d}\langle \hat{Q} \rangle}{\mathrm{d}t} = 0 \qquad (18.19)$$

The Ehrenfest theorem

If we use the conservation equation together with the position or momentum operators, we find that their expectation values behave in accordance with the classical equations of motion (Newton's Laws). For example, the time dependence of $\langle \hat{x} \rangle$ gives us the classical definition of momentum (Eq. 18.20 and Derivation 18.5), whereas that of $\langle \hat{p} \rangle$ gives us Newton's Second Law of Motion (Eq. 18.21 and Derivation 18.6). The statement that quantum expectation values obey classical laws is known as the *Ehrenfest theorem*.

Ehrenfest theorem examples (Derivations 18.5 and 18.6):

Classical definition of momentum:

$$\langle \hat{p} \rangle = m\frac{\mathrm{d}\langle \hat{x} \rangle}{\mathrm{d}t} \qquad (18.20)$$

Newton's Second Law of Motion:

$$\langle \hat{F} \rangle = -\left\langle \frac{\mathrm{d}\hat{V}}{\mathrm{d}x} \right\rangle = \frac{\mathrm{d}\langle \hat{p} \rangle}{\mathrm{d}t} \qquad (18.21)$$

Derivation 18.4: Derivation of the conservation equation (Eq. 18.18)

$$\frac{\mathrm{d}\langle\hat{Q}\rangle}{\mathrm{d}t} = \frac{\mathrm{d}}{\mathrm{d}t}\langle\psi|\hat{Q}|\psi\rangle$$
$$= \left(\frac{\mathrm{d}\langle\psi|}{\mathrm{d}t}\right)\hat{Q}|\psi\rangle + \langle\psi|\hat{Q}\left(\frac{\mathrm{d}|\psi\rangle}{\mathrm{d}t}\right)$$
$$+ \langle\psi|\left(\frac{\mathrm{d}\hat{Q}}{\mathrm{d}t}\right)|\psi\rangle$$
$$= \frac{\mathrm{i}}{\hbar}\langle\psi|\hat{H}\hat{Q}|\psi\rangle - \frac{\mathrm{i}}{\hbar}\langle\psi|\hat{Q}\hat{H}|\psi\rangle$$
$$+ \left\langle\frac{\mathrm{d}\hat{Q}}{\mathrm{d}t}\right\rangle$$
$$= \frac{\mathrm{i}}{\hbar}\langle\psi|\hat{H}\hat{Q} - \hat{Q}\hat{H}|\psi\rangle + \left\langle\frac{\mathrm{d}\hat{Q}}{\mathrm{d}t}\right\rangle$$
$$= \frac{\mathrm{i}}{\hbar}\left\langle\left[\hat{H},\hat{Q}\right]\right\rangle + \left\langle\frac{\mathrm{d}\hat{Q}}{\mathrm{d}t}\right\rangle$$

- Starting point: the expectation value of an arbitrary operator $\hat{Q}$ is defined as:

$$\langle\hat{Q}\rangle = \langle\psi|\hat{Q}|\psi\rangle$$

- Differentiate each part of $\langle\hat{Q}\rangle$.
- Rearrange the Schrödinger equation to express the time derivatives of $|\psi\rangle$ and $\langle\psi|$ in terms of the Hamiltonian:

$$\hat{H}|\psi\rangle = \mathrm{i}\hbar\frac{\partial|\psi\rangle}{\partial t}$$
$$\therefore \frac{\partial|\psi\rangle}{\partial t} = -\frac{\mathrm{i}}{\hbar}\hat{H}|\psi\rangle$$
$$\text{and } \frac{\partial\langle\psi|}{\partial t} = +\frac{\mathrm{i}}{\hbar}\langle\psi|\hat{H}$$

Derivation 18.5: Derivation of the conservation equation for $\hat{x}$ (Eq. 18.20)

$$\frac{\mathrm{d}\langle\hat{x}\rangle}{\mathrm{d}t} = \frac{\mathrm{i}}{\hbar}\left\langle\left[\hat{H},\hat{x}\right]\right\rangle$$
$$= \frac{\mathrm{i}}{\hbar}\left\langle\left[\frac{\hat{p}^2}{2m} + \hat{V},\hat{x}\right]\right\rangle$$
$$= \frac{\mathrm{i}}{\hbar}\left\langle\frac{1}{2m}\left[\hat{p}^2,\hat{x}\right] + \underbrace{\left[\hat{V},\hat{x}\right]}_{=0}\right\rangle$$
$$= \frac{\mathrm{i}}{2m\hbar}\left\langle\left(\hat{p}\left[\hat{p},\hat{x}\right] + \left[\hat{p},\hat{x}\right]\hat{p}\right)\right\rangle$$
$$= \frac{\mathrm{i}}{2m\hbar}\left\langle -2\mathrm{i}\hbar\hat{p}\right\rangle$$
$$= \frac{\langle\hat{p}\rangle}{m}$$
$$\therefore \langle\hat{p}\rangle = m\frac{\mathrm{d}\langle\hat{x}\rangle}{\mathrm{d}t}$$

- Starting point: the conservation equation for the position operator $\hat{x} = x$ is:

$$\frac{\mathrm{d}\langle\hat{x}\rangle}{\mathrm{d}t} = \frac{\mathrm{i}}{\hbar}\left\langle\left[\hat{H},\hat{x}\right]\right\rangle + \underbrace{\left\langle\frac{\mathrm{d}\hat{x}}{\mathrm{d}t}\right\rangle}_{=0}$$

- Use linearity and squared operator properties of the commutator to separate the Hamiltonian:

$$\left[\hat{Q}+\hat{R},\hat{S}\right] = \left[\hat{Q},\hat{S}\right] + \left[\hat{R},\hat{S}\right]$$
$$\left[\hat{Q}^2,\hat{R}\right] = \hat{Q}\left[\hat{Q},\hat{R}\right] + \left[\hat{Q},\hat{R}\right]\hat{Q}$$

- The position operator commutes with the potential energy operator, but not the momentum operator:

$$\left[\hat{p},\hat{x}\right]\phi = -\mathrm{i}\hbar\frac{\partial}{\partial x}(x\phi) + x\mathrm{i}\hbar\frac{\partial}{\partial x}(\phi)$$
$$= -\mathrm{i}\hbar\phi$$

- The average motion of $\hat{x}$ agrees with the classical definition of momentum: $p = mv = m\dot{x}$.

Derivation 18.6: Derivation of the conservation equation for $\hat{p}$ (Eq. 18.21)

$$\begin{aligned}
\frac{\mathrm{d}\langle\hat{p}\rangle}{\mathrm{d}t} &= \frac{\mathrm{i}}{\hbar}\left\langle\left[\hat{H},\hat{p}\right]\right\rangle \\
&= \frac{\mathrm{i}}{\hbar}\left\langle\left[\frac{\hat{p}^2}{2m}+\hat{V},\hat{p}\right]\right\rangle \\
&= \frac{\mathrm{i}}{\hbar}\left\langle\frac{1}{2m}\underbrace{\left[\hat{p}^2,\hat{p}\right]}_{=0}+\left[\hat{V},\hat{p}\right]\right\rangle \\
&= \frac{\mathrm{i}}{\hbar}\left\langle \mathrm{i}\hbar\frac{\mathrm{d}\hat{V}}{\mathrm{d}x}\right\rangle \\
&= -\left\langle\frac{\mathrm{d}\hat{V}}{\mathrm{d}x}\right\rangle \\
&= \left\langle\hat{F}\right\rangle
\end{aligned}$$

$$\therefore \left\langle\hat{F}\right\rangle = \frac{\mathrm{d}\langle\hat{p}\rangle}{\mathrm{d}t}$$

- Starting point: the conservation equation for the momentum operator $\hat{p} = -\mathrm{i}\hbar\partial/\partial x$ is:

$$\frac{\mathrm{d}\langle\hat{p}\rangle}{\mathrm{d}t} = \frac{\mathrm{i}}{\hbar}\left\langle\left[\hat{H},\hat{p}\right]\right\rangle + \underbrace{\left\langle\frac{\mathrm{d}\hat{p}}{\mathrm{d}t}\right\rangle}_{=0}$$

- Use the linearity property of the commutator to separate the Hamiltonian:

$$\left[\hat{Q}+\hat{R},\hat{S}\right] = \left[\hat{Q},\hat{S}\right]+\left[\hat{R},\hat{S}\right]$$

- The momentum and potential energy operators do not commute:

$$\begin{aligned}
\left[\hat{V},\hat{p}\right]\phi &= -\hat{V}\mathrm{i}\hbar\frac{\partial}{\partial x}(\phi)+\mathrm{i}\hbar\frac{\partial}{\partial x}(\hat{V}\phi) \\
&= \mathrm{i}\hbar\frac{\mathrm{d}\hat{V}}{\mathrm{d}x}\phi
\end{aligned}$$

- The average motion of $\hat{p}$ agrees with Newton's Second Law: $F = -\mathrm{d}V/\mathrm{d}x = \dot{p}$.

18.3.2 Conservation of probability

The *probability density*, ρ, is defined as the absolute value squared of a wavefunction:

$$\rho = |\psi|^2 = \langle\psi|\psi\rangle$$

This quantity is equivalent to the expectation value of an operator $\hat{O} = 1$:

$$\begin{aligned}
\langle\psi|\hat{O}|\psi\rangle &= \langle\psi|1|\psi\rangle \\
&= \langle\psi|\psi\rangle \\
&= \rho
\end{aligned}$$

Since this operator commutes with the Hamiltonian ($[\hat{H},\hat{O}] = (\hat{H}-\hat{H}) = 0$), then if ρ is time independent, it is a conserved quantity. If ρ is *not* time independent, then it obeys a continuity equation:

$$\frac{\partial\rho}{\partial t} = -\frac{\partial j}{\partial x}$$

where j is the *probability current*. This equation relates the rate of change of probability to the "flow" of probability current. We can find an expression for j in terms of ψ by differentiating the probability density (Eqs. 18.22–18.24 and Derivation 18.7).

Probability current (Derivation 18.7):

$$j = \frac{\mathrm{i}\hbar}{2m}\left(\psi\frac{\partial\psi^*}{\partial x}-\psi^*\frac{\partial\psi}{\partial x}\right) \tag{18.22}$$

The current j obeys a continuity equation:

$$\frac{\partial j}{\partial x} = -\frac{\partial\rho}{\partial t} \tag{18.23}$$

where the probability density is:

$$\rho = |\psi|^2 = \psi^*\psi \tag{18.24}$$

18.4 Compatibility of operators

Compatible operators

If two operators $\hat{Q}$ and $\hat{R}$ commute, they are said to be *compatible* (Eq. 18.25). This means that it is possible for a particle to be in an eigenstate of both operators simultaneously. Or, said another way, $\hat{Q}$ and $\hat{R}$ share common eigenstates (Eq. 18.26 and Derivation 18.8). For example, if $\hat{Q}$ is compatible with the Hamiltonian, then the particle can be in a state of definite energy E_n and also have a definite eigenvalue q_n.

Derivation 18.7: Derivation of the probability current (Eq. 18.22)

$$\frac{\partial \rho}{\partial t} = \frac{\partial \psi^* \psi}{\partial t}$$
$$= \left(\frac{\partial \psi^*}{\partial t}\right)\psi + \psi^*\left(\frac{\partial \psi}{\partial t}\right)$$
$$= \frac{\mathrm{i}}{\hbar}\left\{\left(\hat{H}\psi^*\right)\psi - \psi^*\left(\hat{H}\psi\right)\right\}$$
$$= \frac{\mathrm{i}}{\hbar}\left\{\left(-\frac{\hbar^2}{2m}\frac{\partial^2 \psi^*}{\partial x^2} + V\psi^*\right)\psi - \psi^*\left(-\frac{\hbar^2}{2m}\frac{\partial^2 \psi}{\partial x^2} + V\psi\right)\right\}$$
$$= -\frac{\mathrm{i}\hbar}{2m}\left(\psi\frac{\partial^2 \psi^*}{\partial x^2} - \psi^*\frac{\partial^2 \psi}{\partial x^2}\right)$$
$$= -\frac{\mathrm{i}\hbar}{2m}\frac{\partial}{\partial x}\left(\psi\frac{\partial \psi^*}{\partial x} - \psi^*\frac{\partial \psi}{\partial x}\right)$$
and $$\frac{\partial \rho}{\partial t} = -\frac{\partial j}{\partial x}$$
$$\therefore j = \frac{\mathrm{i}\hbar}{2m}\left(\psi\frac{\partial \psi^*}{\partial x} - \psi^*\frac{\partial \psi}{\partial x}\right)$$

- Starting point: differentiate the probability density, $\rho = \psi^*\psi$, with respect to time.
- Use the Schrödinger equation to express the time derivatives of ψ and ψ^* in terms of the Hamiltonian:

$$\hat{H}\psi = \mathrm{i}\hbar\frac{\partial \psi}{\partial t}$$
$$\therefore \frac{\partial \psi}{\partial t} = -\frac{\mathrm{i}}{\hbar}\hat{H}\psi$$
and $$\frac{\partial \psi^*}{\partial t} = +\frac{\mathrm{i}}{\hbar}\hat{H}\psi^*$$
where $$\hat{H} = -\frac{\hbar^2}{2m}\frac{\partial^2}{\partial x^2} + V$$

- Equate the expression for $\partial\rho/\partial t$ to the continuity equation:

$$\frac{\partial \rho}{\partial t} = -\frac{\partial j}{\partial x}$$

to find an expression for the probability current, j.

Compatible operators (Derivation 18.8):

$\hat{Q}$ and $\hat{R}$ are compatible when:

$$\left[\hat{Q}, \hat{R}\right] = 0 \qquad (18.25)$$

$\hat{Q}$ and $\hat{R}$ therefore have common eigenstates:

$$\hat{Q}\left|u_n\right\rangle = q_n\left|u_n\right\rangle$$
and $$\hat{R}\left|u_n\right\rangle = r_n\left|u_n\right\rangle \qquad (18.26)$$

If a system is in an eigenstate of both $\hat{Q}$ and $\hat{R}$, say $|u_m\rangle$, then the probability of measuring q_m or r_m is $P(q_m) = P(r_m) = 1$. If a system is not in an eigenstate, then the probability of measuring q_m equals the probability of measuring r_m:

$$\left\langle\hat{Q}\right\rangle = \sum_n q_n\left|c_n\right|^2$$
and $$\left\langle\hat{R}\right\rangle = \sum_n r_n\left|c_n\right|^2$$
$$\therefore P(q_m, r_m) = \left|c_m\right|^2$$

Incompatible operators

If two operators do *not* commute, then they are said to be *incompatible* (Eq. 18.27). This means that a system cannot simultaneously be in an eigenstate of both operators: a system in an eigenstate of $\hat{Q}$ is in a superposition of $\hat{R}$, and vice versa (Eqs. 18.28 and 18.29). What does this mean in practical terms? It means that a measurement of either q or r will disturb the system. For example, a system initially in an eigenstate of $\hat{Q}$, say $|q_m\rangle$, is in a superposition of $\hat{R}$:

$$\left|q_m\right\rangle = \sum_n a_n\left|r_n\right\rangle$$

If we then measure r, the system collapses to an eigenstate of $\hat{R}$, which results in a superposition of $\hat{Q}$. Therefore, if we measure q again, the probability of finding it in its original state, $|q_m\rangle$, is no longer equal to one. Therefore, the act of taking a measurement has introduced uncertainty into the system. Incompatibility of operators that don't commute is what gives rise to the well-known *uncertainty principle* of quantum physics, as we shall see next.

Derivation 18.8: Derivation of shared eigenstates of compatible operators (Eqs. 18.25 and 18.26)

- Starting point: consider the eigenvalue equations for two operators that have common eigenstates $|u_n\rangle$:

$$\hat{Q}|u_n\rangle = q_n|u_n\rangle$$

and $\hat{R}|u_n\rangle = r_n|u_n\rangle$

- These operators commute, and thus are said to be compatible.

$$\begin{aligned}\hat{Q}\hat{R}|u_n\rangle &= \hat{Q}r_n|u_n\rangle \\ &= r_n\hat{Q}|u_n\rangle \\ &= r_n q_n|u_n\rangle \\ \text{and } \hat{R}\hat{Q}|u_n\rangle &= q_n r_n|u_n\rangle \\ \therefore \left[\hat{Q},\hat{R}\right]|u_n\rangle &= \left(\hat{Q}\hat{R}-\hat{R}\hat{Q}\right)|u_n\rangle \\ &= (r_n q_n - r_n q_n)|u_n\rangle \\ &= 0 \\ \therefore \left[\hat{Q},\hat{R}\right] &= 0\end{aligned}$$

Incompatible operators:

$\hat{Q}$ and $\hat{R}$ are incompatible when:

$$\left[\hat{Q},\hat{R}\right] \neq 0 \quad (18.27)$$

Therefore, an eigenstate of $\hat{Q}$ is a superposition of $\hat{R}$, and vice versa:

$$\begin{aligned}\hat{Q}|q_n\rangle &= q_n|q_n\rangle \\ \hat{R}|q_n\rangle &= \hat{R}\sum_m a_m|r_m\rangle \\ &= \sum_m a_m r_m|r_m\rangle \quad (18.28)\end{aligned}$$

$$\begin{aligned}\text{and } \hat{R}|r_n\rangle &= r_n|r_n\rangle \\ \hat{Q}|r_n\rangle &= \hat{Q}\sum_m c_m|q_m\rangle \\ &= \sum_m c_m q_m|q_m\rangle \quad (18.29)\end{aligned}$$

18.4.1 The uncertainty principle

For two incompatible operators, measurement of one property of a system disturbs the value of the other property by forcing the system into a superposition state of that property. Therefore, operators that don't commute have an intrinsic limit to the precision with which their values can be known simultaneously—there is an intrinsic uncertainty in the state of the system. This phenomenon is what gives rise to the famous *uncertainty principle*, which states that there is a minimum for the product of the uncertainties on two incompatible measurements, which is on the order of $\hbar$.

Consider two incompatible operators $\hat{Q}$ and $\hat{R}$. We define the standard deviation of a measurement as:

$$(\Delta q)^2 = \langle q^2\rangle - \langle q\rangle^2$$

where $\langle q^2\rangle$ and $\langle q\rangle^2$ are the mean of the square of the measured value and the square of the mean value, respectively. The product of uncertainties of two measurements is then $\Delta q\Delta r$. The uncertainty relation shows that this product is related to their commutator by:[4]

$$(\Delta q)^2(\Delta r)^2 \geqslant \left(\left\langle\frac{1}{2i}\left[\hat{Q},\hat{R}\right]\right\rangle\right)^2 \quad (18.30)$$

The Heisenberg uncertainty principle

Previously (in Derivation 18.5), we found that the position and momentum operators, $\hat{x}$ and $\hat{p}$, are incompatible:

$$\left[\hat{x},\hat{p}\right] = i\hbar$$

Therefore, a system cannot simultaneously be in a state on definite position and definite momentum. Using the $\hat{x}, \hat{p}$ commutator in the uncertainty relation (Eq. 18.30), we find that the combined uncertainties in x and p are on the order of $\hbar$ (Eq. 18.31). This statement is called the *Heisenberg uncertainty principle*, and it states that the position and momentum of a particle cannot simultaneously be measured to a precision better than $\hbar/2$.

The Heisenberg uncertainty principle:

$$\Delta x\Delta p \geqslant \frac{\hbar}{2} \quad (18.31)$$

4 The reader is referred to *Introduction to Quantum Mechanics*, D. Griffiths, Section 3.5 for proof.

Chapter 19
Spherical systems

We'll now turn our attention back to the Schrödinger equation, but this time look at its solutions in *spherical* polar coordinates. These solutions can then be applied to electron orbitals in atoms, allowing us to understand emission spectra and the periodic table. We find that for spherical systems, the motion of a particle can be separated into radial and angular components, which introduces us to the *orbital angular momentum squared operator*, $\hat{L}^2$. We also find that particles possess an *intrinsic* angular momentum called *spin* that has no classical equivalent. The total angular momentum of a particle is then a combination of its orbital and spin angular momentum. In this chapter, we'll start by separating the spherical Schrödinger equation into its radial and angular motion, and then look at the properties of the angular momentum operators.

19.1 The spherical Schrödinger equation

The three-dimensional TISE is:

$$\underbrace{\left(-\frac{\hbar^2}{2m}\nabla^2 + V(\vec{r})\right)}_{\text{Hamiltonian, } \hat{H}}\phi(\vec{r}) = E\phi(\vec{r})$$

(refer to Section 17.3). Previously, we saw how to separate this equation into Cartesian coordinates. For systems with spherical symmetry, we use spherical polar coordinates (r, θ, ϕ). Potential energy functions that are only a function of r and not θ or ϕ (spherically symmetric potentials) are called *central potentials*. An example of a central potential is the Coulomb potential produced by a charge Q:

$$V(r) = \frac{Q}{4\pi\varepsilon_0 r}$$

To a first approximation, electrons in an atom are in a Coulomb potential. The spherical TISE for a central potential is therefore:

$$\left(-\frac{\hbar^2}{2m}\nabla^2 + V(r)\right)\psi(r,\theta,\phi) = E\psi(r,\theta,\phi)$$

By separating the radial variable r from the angular variables θ and ϕ, we can express the kinetic energy term as a sum of radial and angular components (Eq. 19.1 and Derivation 19.1). The radial and angular momentum operators are given by Eqs. 19.2 and 19.3, respectively.

Radial Schrödinger equation (Derivation 19.1):

$$\left(\underbrace{\frac{\hat{p}_r^2}{2m}}_{\text{Radial KE}} + \underbrace{\frac{\hat{L}^2}{2mr^2}}_{\text{Angular KE}} + \underbrace{V(r)}_{\text{PE}}\right)\psi(r,\theta,\phi) = E\psi(r,\theta,\phi) \tag{19.1}$$

Momentum operators:

Radial momentum:

$$\hat{p}_r^2 = -\frac{\hbar^2}{r^2}\frac{\partial}{\partial r}\left(r^2\frac{\partial}{\partial r}\right) \tag{19.2}$$

Angular momentum:

$$\hat{L}^2 = -\hbar^2\left(\frac{1}{\sin\theta}\frac{\partial}{\partial\theta}\left(\sin\theta\frac{\partial}{\partial\theta}\right) + \frac{1}{\sin^2\theta}\frac{\partial^2}{\partial\phi^2}\right) \tag{19.3}$$

Separation of variables

Equation 19.1 is a second-order partial differential equation that can be solved via separation of variables. In the first instance, let's separate the radial dependence from the angular dependence:

$$\psi(r,\theta,\phi) = R(r)Y(\theta,\phi)$$

Derivation 19.1: Separating radial and angular kinetic energy (Eqs. 19.1–19.3)

- Starting point: the kinetic energy operator, $\hat{T}$, in three dimensions, depends on the square of the momentum operator:

$$\hat{p} = -i\hbar\nabla \rightarrow \hat{p}^2 = -\hbar^2\nabla^2$$

$$\hat{T} = \frac{\hat{p}^2}{2m}$$
$$= -\frac{\hbar^2}{2m}\nabla^2$$

- In spherical polar coordinates, the Laplacian is:

$$\nabla^2 = \left(\frac{1}{r^2}\frac{\partial}{\partial r}\left(r^2\frac{\partial}{\partial r}\right) + \frac{1}{r^2\sin\theta}\frac{\partial}{\partial\theta}\left(\sin\theta\frac{\partial}{\partial\theta}\right) + \frac{1}{r^2\sin^2\theta}\frac{\partial^2}{\partial\phi^2}\right)$$

$$= -\frac{\hbar^2}{2m}\left(\frac{1}{r^2}\frac{\partial}{\partial r}\left(r^2\frac{\partial}{\partial r}\right) + \frac{1}{r^2}\left(\frac{1}{\sin\theta}\frac{\partial}{\partial\theta}\left(\sin\theta\frac{\partial}{\partial\theta}\right) + \frac{1}{\sin^2\theta}\frac{\partial^2}{\partial\phi^2}\right)\right)$$

- Collecting differentials in r only and in θ and ϕ only, $\hat{T}$ can be separated into the sum of radial kinetic energy, $\hat{T}_r$, plus angular kinetic energy, $\hat{T}_{\theta,\phi}$.

$$\equiv \underbrace{\frac{\hat{p}_r^2}{2m}}_{\hat{T}_r} + \underbrace{\frac{\hat{L}^2}{2mr^2}}_{\hat{T}_{\theta,\phi}}$$

where

radial momentum operator:
(r-dependent derivatives)

$$\therefore \hat{p}_r^2 = -\frac{\hbar^2}{r^2}\frac{\partial}{\partial r}\left(r^2\frac{\partial}{\partial r}\right)$$

angular momentum operator:
(θ- and ϕ-dependent derivatives)

$$\hat{L}^2 = -\hbar^2\left(\frac{1}{\sin\theta}\frac{\partial}{\partial\theta}\left(\sin\theta\frac{\partial}{\partial\theta}\right) + \frac{1}{\sin^2\theta}\frac{\partial^2}{\partial\phi^2}\right)$$

Using this in Eq. 19.1, the Schrödinger equation reduces to two ordinary differential equations: one that depends only on r:

$$\underbrace{\left(\frac{\hat{p}_r^2}{2m} + V(r) + \frac{\hbar^2c^2}{2mr^2}\right)}_{\text{Radial Hamiltonian}} R(r) = ER(r) \quad (19.4)$$

and one that depends only on θ and ϕ:

$$\hat{L}^2Y(\theta,\phi) = \hbar^2c^2Y(\theta,\phi) \quad (19.5)$$

where c is a dimensionless constant (Derivation 19.2). These are both eigenvalue equations: the radial equation is the eigenvalue equation for the radial Hamiltonian, and the angular equation is that for the total angular momentum of the system. At this stage, it is necessary to specify that $\hat{L}^2$ represents the operator for the total orbital angular momentum—i.e. the angular momentum of a particle in orbit around another, such as an electron orbiting a nucleus. This is in contrast with *spin* angular momentum, which is an intrinsic form of angular momentum that particles have, for which there is no classical analog (see Section 19.3).

The overall wavefunction is given by the product of the r and θ, ϕ solutions: $\psi(r, \theta, \phi) = R(r)Y(\theta, \phi)$. These are normalized such that the integral of the wavefunction squared over all space equals one:

$$\int_{\text{all space}} |\psi(r,\theta,\phi)|^2 \, d\tau = 1$$

$$\therefore \int |R(r)Y(\theta,\phi)|^2 r^2 \sin\theta dr d\theta d\phi = 1$$

Separating the integral, we can normalize the radial and angular solutions individually:

$$\int |R(r)|^2 r^2 dr = 1$$

and $\int |Y(\theta,\phi)|^2 \sin\theta d\theta d\phi = 1$

The radial equation depends on the exact nature of the central potential, $V(r)$. In Chapter 20, we will determine the radial solutions, $R(r)$, for an electron in a hydrogen atom. The angular momentum eigenvalue equation, however, does not depend on $V(r)$; therefore, the solutions of $Y(\theta,\phi)$ are universal for all central potentials. These solutions are wavefunctions called *spherical*

Derivation 19.2: Separation of variables: spherical Schrödinger equation

1. **Partial differential equation:** the Schrödinger equation in spherical coordinates for a central potential $V(r)$ is:

$$\left(\frac{\hat{p}_r^2(r)}{2m}+\frac{\hat{L}^2(\theta,\phi)}{2mr^2}+V(r)\right)\psi(r,\theta,\phi)=E\psi(r,\theta,\phi)$$

2. **Separation function:** separate the radial dependence from the angular dependence by letting $\psi(r,\theta,\phi)=R(r)Y(\theta,\phi)$:

$$\left(Y(\theta,\phi)\frac{\hat{p}_r^2R(r)}{2m}+R(r)\frac{\hat{L}^2Y(\theta,\phi)}{2mr^2}+V(r)R(r)Y(\theta,\phi)\right)=ER(r)Y(\theta,\phi)$$

3. **Separation constant:** divide by $R(r)Y(\theta,\phi)$ and rearrange to make one side a function of r only and the other side a function of θ and ϕ only:

$$\underbrace{\left(-\frac{1}{R(r)}r^2\hat{p}_r^2R(r)+2mr^2(E-V(r))\right)}_{=\text{constant},\,\hbar^2c^2}=\underbrace{\frac{1}{Y(\theta,\phi)}\hat{L}^2Y(\theta,\phi)}_{=\text{constant},\,\hbar^2c^2}$$

For the above equation to hold for all values of r,θ, and ϕ, both sides must be equal to a constant. The dimensions of the constant must be angular momentum squared $[(\text{kg m}^2\text{s}^{-1})^2\equiv(\text{Js})^2]$–the same dimensions as $\hbar^2$. Therefore, let the separation constant equal $\hbar^2c^2$, where c is a dimensionless constant.

4. **Radial Schrödinger equation:** the radial Schrödinger equation (left-hand side) is:

$$\left(\frac{\hat{p}_r^2}{2m}+V(r)+\frac{\hbar^2c^2}{2mr^2}\right)R(r)=ER(r)$$

The solution depends on the potential $V(r)$; therefore, it is specific to the system in question. The radial wavefunctions for an electron in a hydrogen atom are covered in Section 20.1

5. **Orbital angular momentum equation:** the θ- and ϕ-equation (right-hand side) is the eigenvalue equation for the total orbital angular momentum squared:

$$\hat{L}^2Y(\theta,\phi)=\hbar^2c^2Y(\theta,\phi)$$

This equation is further separated into θ- and ϕ-components in Derivation 19.3.

harmonics, and they correspond to the eigenfunctions for the angular part of the Laplacian in spherical coordinates.

Spherical harmonics

We can further separate the angular momentum eigenvalue equation (Eq. 19.5) into its θ- and ϕ-components by letting:

$$Y(\theta,\phi)=P(\theta)f(\phi)$$

Using the separation-of-variables method, we find that the θ-equation takes the form of a special ordinary differential equation, known as the *associated Legendre function* (Derivation 19.3):

$$\sin\theta\frac{\mathrm{d}}{\mathrm{d}\theta}\left(\sin\theta\frac{\mathrm{d}P(\theta)}{\mathrm{d}\theta}\right)+c^2\sin^2\theta P(\theta)=m^2P(\theta)$$

provided $c^2=l(l+1)$, where l is an integer. The $P(\theta)$ solutions are called *associated Legendre polynomials*, $P_l^m(\cos\theta)$. These are defined by two indices, m and l; for the associated Legendre polynomials to define normalizable wavefunctions, m and l must be integers, and $|m|\leqslant l$. The first three associated Legendre polynomials are:

$$P_0^0(\cos\theta)=1$$
$$P_1^0(\cos\theta)=\cos\theta$$
$$P_1^{\pm1}(\cos\theta)=\mp\sin\theta$$

The ϕ-equation is a standard second-order differential equation with complex exponential solutions:

$$\frac{\mathrm{d}^2f(\phi)}{\mathrm{d}\phi^2}=-m^2f(\phi)$$
$$\therefore f(\phi)=A\mathrm{e}^{im\phi}+B\mathrm{e}^{-im\phi}$$

The integer m can be positive or negative; therefore, it suffices to have a single term that incorporates both the positive and the negative terms:

$$\therefore f(\phi)=A\mathrm{e}^{im\phi}$$

The overall solutions are called spherical harmonics, and are given by the product of θ and ϕ solutions:

$$Y_l^m(\theta,\phi)=AP_l^m(\cos\theta)\mathrm{e}^{im\phi}$$

Derivation 19.3: Separation of variables: orbital angular momentum equation

1. **Partial differential equation:** the orbital angular momentum eigenvalue equation (using the expression for $\hat{L}^2$ from Eq. 19.3) is:

$$\underbrace{-\hbar^2\left(\frac{1}{\sin\theta}\frac{\partial}{\partial\theta}\left(\sin\theta\frac{\partial}{\partial\theta}\right)+\frac{1}{\sin^2\theta}\frac{\partial^2}{\partial\phi^2}\right)}_{=\hat{L}^2}Y(\theta,\phi)=\hbar^2c^2Y(\theta,\phi)$$

2. **Separation function:** separate the θ- from the ϕ-dependence by letting $Y(\theta,\phi)=P(\theta)f(\phi)$:

$$-\hbar^2\left(f(\phi)\frac{1}{\sin\theta}\frac{\partial}{\partial\theta}\left(\sin\theta\frac{\partial P(\theta)}{\partial\theta}\right)+P(\theta)\frac{1}{\sin^2\theta}\frac{\partial^2 f(\phi)}{\partial\phi^2}\right)=\hbar^2c^2P(\theta)f(\phi)$$

3. **Separation constant:** divide by $P(\theta)f(\phi)$ and rearrange to make one side a function of θ only and the other side a function of ϕ only:

$$\underbrace{\frac{1}{P(\theta)}\sin\theta\frac{d}{d\theta}\left(\sin\theta\frac{dP(\theta)}{d\theta}\right)+c^2\sin^2\theta}_{=\text{constant, } m^2}=\underbrace{-\frac{1}{f(\phi)}\frac{d^2f(\phi)}{d\phi^2}}_{=\text{constant, } m^2}$$

For this equation to hold for all values of θ and ϕ, both sides must equal a constant. The equation is dimensionless; therefore, let the separation constant equal m^2, where m is a dimensionless constant.

4. **Associated Legendre function:** when $c^2 = l(l+1)$, where l is an integer, then the θ-equation takes the form of a special ordinary differential equation, known as the associated Legendre function:

$$\sin\theta\frac{d}{d\theta}\left(\sin\theta\frac{dP(\theta)}{d\theta}\right)+\underbrace{l(l+1)}_{=c^2}\sin^2\theta P(\theta)=m^2P(\theta)$$

Its solutions are associated Legendre polynomials, $P_l^m(\cos\theta)$, where the index l is the *degree* and m is the *order* of the polynomial. For these solutions to describe a wavefunction, they must be normalizable over the $\cos\theta$ interval $[-1,1]$. This occurs only when l and m are integers, along with the restriction that $|m|\leq l$. The first three associated Legendre polynomials are:

$$P_0^0(\cos\theta)=1$$

$$P_1^0(\cos\theta)=\cos\theta$$

$$P_1^{\pm1}(\cos\theta)=\mp\sin\theta$$

5. **Azimuthal equation:** the ϕ-equation is a second-order differential equation with complex exponential solutions:

$$\frac{d^2f(\phi)}{d\phi^2}=-m^2f(\phi)$$

$$\therefore f(\phi)=Ae^{+im\phi}+Be^{-im\phi}$$

where m is an integer. If we let m run negative, then the negative exponential factor is absorbed in the positive one: $\therefore f(\phi)=Ae^{im\phi}$.

These are normalized by the constant A such that that integral over all solid angles, Ω, is equal to one:

$$\int\left|Y_l^m(\theta,\phi)\right|^2 d\Omega=\int_0^{2\pi}\int_0^{\pi}\left|AP_l^m(\cos\theta)\right|^2\sin\theta d\theta d\phi=1$$

(see Section 19.2.3 for normalization of electron orbitals).

19.2 Orbital angular momentum

Bodies have *orbital angular momentum* as a result of their angular motion, or rotation, about a reference point. The quantum orbital angular momentum operator, $\hat{L}$, is a vector given by the cross product of $\hat{\vec{r}}$ and $\hat{\vec{p}}$, analogous to the classical definition of $\vec{L}$. Using $\hat{\vec{p}}=-i\hbar\nabla$, the orbital angular momentum operator is:

$$\hat{L}=\hat{\vec{r}}\times\hat{\vec{p}}=-i\hbar\vec{r}\times\nabla$$

Squaring this expression, we recover the *total orbital angular momentum squared* operator, $\hat{L}^2$, from Eq. 19.3 (Derivation 19.4).

Angular momentum components

The components of $\hat{L}$ give expressions for the *component* angular momentum operators, $\hat{L}_x$, $\hat{L}_y$ and $\hat{L}_z$ (Eqs. 19.7–19.9 and Derivation 19.5). Analogous to the classical case, the total angular momentum squared is given by the sum of the square of the components:

$$\hat{L}^2=\hat{L}_x^2+\hat{L}_y^2+\hat{L}_z^2$$

Derivation 19.4: Derivation of total angular momentum squared operator, $\hat{L}^2$, from $\hat{\vec{L}}$ (Eq. 19.3)

$$
\begin{aligned}
\hat{L}^2 &= (-i\hbar\vec{r}\times\nabla)\cdot(-i\hbar\vec{r}\times\nabla) \\
&= -\hbar^2(\vec{r}\times\nabla)\cdot(\vec{r}\times\nabla) \\
&= -\hbar^2\vec{r}\cdot[\nabla\times(\vec{r}\times\nabla)] \\
&= -\hbar^2\vec{r}\cdot\hat{r}\frac{1}{r\sin\theta}\left[\frac{\partial}{\partial\theta}\sin\theta\left(\frac{\partial}{\partial\theta}\right)+\frac{\partial}{\partial\phi}\left(\frac{1}{\sin\theta}\frac{\partial}{\partial\phi}\right)\right] \\
&= -\hbar^2\left(\frac{1}{\sin\theta}\frac{\partial}{\partial\theta}\left(\sin\theta\frac{\partial}{\partial\theta}\right)+\frac{1}{\sin^2\theta}\frac{\partial^2}{\partial\phi^2}\right)
\end{aligned}
$$

Vector cross products:

$$\vec{r}\times\nabla = -\hat{\theta}\frac{1}{\sin\theta}\frac{\partial}{\partial\phi}+\hat{\phi}\frac{\partial}{\partial\theta}$$

$$\therefore\ \nabla\times(\vec{r}\times\nabla) = \hat{r}\frac{1}{r\sin\theta}\left[\frac{\partial}{\partial\theta}\sin\theta\left(\frac{\partial}{\partial\theta}\right)+\frac{\partial}{\partial\phi}\left(\frac{1}{\sin\theta}\frac{\partial}{\partial\phi}\right)\right]+\cdots$$

Since the expression for $\hat{L}^2$ involves a dot product in $\vec{r}$, we require only the $\hat{r}$-component of the curl—the other components are orthogonal to $\vec{r}$ and hence their dot products equal zero.

- Starting point: the square of the angular momentum operator is $\hat{L}^2=\hat{\vec{L}}\cdot\hat{\vec{L}}$, where

$$\hat{\vec{L}} = \hat{\vec{r}}\times\hat{\vec{p}} = -i\hbar\vec{r}\times\nabla$$

- Determine the cross product $\vec{r}\times\nabla$ using $\vec{r} = (r\hat{r}, 0, 0)$ and the Del operator in spherical coordinates.

Orbital angular momentum operators (Derivations 19.5 and 19.6):

$$\hat{\vec{L}} = -i\hbar\vec{r}\times\nabla \tag{19.6}$$

Cartesian components:

$$\hat{L}_x = -i\hbar\left(y\frac{\partial}{\partial z}-z\frac{\partial}{\partial y}\right) \tag{19.7}$$

$$\hat{L}_y = -i\hbar\left(z\frac{\partial}{\partial x}-x\frac{\partial}{\partial z}\right) \tag{19.8}$$

$$\hat{L}_z = -i\hbar\left(x\frac{\partial}{\partial y}-y\frac{\partial}{\partial x}\right) \tag{19.9}$$

z-component in spherical coordinates:

$$\hat{L}_z = -i\hbar\frac{\partial}{\partial\phi} \tag{19.10}$$

Expressing the z-component in spherical polar coordinates, we find that this component of angular momentum results from variation in the wavefunction's azimuthal angle, ϕ (Eq. 19.10 and Derivation 19.6). Since the choice of z-axis is arbitrary, Eq. 19.10 represents the operator for *any* component of angular momentum. Therefore, a system with a spherically symmetrical wavefunction does not have orbital angular momentum since there is no variation in ϕ.

19.2.1 Commutation rules

In classical mechanics, specifying any three of L_x, L_y, L_z, or L^2 allows us to determine the value of the fourth through $L^2= L_x^2 + L_y^2 + L_z^2$. Furthermore, all four variables can be defined simultaneously. In quantum mechanics, however, this is not the case, since the components of orbital angular momentum do not commute (they are incompatible operators):

$$[\hat{L}_i, \hat{L}_j] \neq 0$$

(Eq. 19.11 and Derivation 19.7). This means that only one component of orbital angular momentum can be specified at a time. A particle in an eigenstate of, say $\hat{L}_x$ is in a superposition state of $\hat{L}_y$ and $\hat{L}_z$. In contrast, each individual component *does* commute with the *total* angular momentum squared, $\hat{L}^2$:

$$[\hat{L}_i, \hat{L}^2] = 0$$

(Eq. 19.12 and Derivation 19.7). Therefore, only the total orbital angular momentum along with any *one* component can be known simultaneously. Measurement of another component will disrupt the system such that the previously known component is no longer

Derivation 19.5: Derivation of the Cartesian components of angular momentum (Eqs. 19.7–19.9)

$$\hat{\vec{L}} = -i\hbar\vec{r} \times \nabla$$
$$= -i\hbar \begin{pmatrix} x \\ y \\ z \end{pmatrix} \times \begin{pmatrix} \partial/\partial x \\ \partial/\partial y \\ \partial/\partial z \end{pmatrix}$$
$$\equiv \begin{pmatrix} \hat{L}_x \\ \hat{L}_y \\ \hat{L}_z \end{pmatrix}$$

$$\therefore \hat{L}_x = -i\hbar\left(y\frac{\partial}{\partial z} - z\frac{\partial}{\partial y}\right) \equiv y\hat{p}_z - z\hat{p}_y$$
$$\hat{L}_y = -i\hbar\left(z\frac{\partial}{\partial x} - x\frac{\partial}{\partial z}\right) \equiv z\hat{p}_x - x\hat{p}_z$$
$$\hat{L}_z = -i\hbar\left(x\frac{\partial}{\partial y} - y\frac{\partial}{\partial x}\right) \equiv x\hat{p}_y - y\hat{p}_x$$

- Starting point: the angular momentum operator is:

  $\hat{\vec{L}} = \hat{\vec{r}} \times \hat{\vec{p}} = -i\hbar\vec{r} \times \nabla$
- Each component of the cross product corresponds to a different component of the angular momentum operator.

Derivation 19.6: Converting the *z*-component to spherical polar coordinates (Eq. 19.10)

$$\frac{\partial}{\partial\phi} = \left(\frac{\partial x}{\partial\phi}\right)\frac{\partial}{\partial x} + \left(\frac{\partial y}{\partial\phi}\right)\frac{\partial}{\partial y} + \left(\frac{\partial z}{\partial\phi}\right)\frac{\partial}{\partial z}$$
$$= -r\sin\theta\sin\phi\frac{\partial}{\partial x} + r\sin\theta\cos\phi\frac{\partial}{\partial y}$$
$$= -y\frac{\partial}{\partial x} + x\frac{\partial}{\partial y}$$

$$\hat{L}_z = -i\hbar\underbrace{\left(x\frac{\partial}{\partial y} - y\frac{\partial}{\partial x}\right)}_{=\partial/\partial\phi}$$
$$\therefore \hat{L}_z = -i\hbar\frac{\partial}{\partial\phi}$$

- Starting point: Cartesian coordinates and spherical polar coordinates are related by:

  $x = r\sin\theta\cos\phi$

  $y = r\sin\theta\sin\phi$

  $z = r\cos\theta$
- Use the chain rule to determine the partial derivative of ϕ in terms of Cartesian derivatives.

defined. A consequence of this phenomenon is that we can never define the direction of the vector operator, $\hat{\vec{L}}$; its direction has an intrinsic uncertainty.

Orbital angular momentum commutators (Derivation 19.7):

$$\left[\hat{L}_i, \hat{L}_j\right] = i\hbar\varepsilon_{ijk}\hat{L}_k \quad (19.11)$$
$$[\hat{L}_i, \hat{L}^2] = 0 \quad (19.12)$$

where the Levi-Civita symbol is:

$$\varepsilon_{ijk} = \begin{cases} +1 & \text{even permutation of } ijk \\ -1 & \text{odd permutation of } ijk \\ 0 & \text{otherwise} \end{cases}$$

i, *j* and *k* represent the Cartesian components *x*, *y*, and *z*.

Derivation 19.7: Derivation of commutation relations (Eqs. 19.11 and 19.12)

Components of angular momentum:

$$\left[\hat{L}_x,\hat{L}_y\right]=\left[\left(y\hat{p}_z-z\hat{p}_y\right),\left(z\hat{p}_x-x\hat{p}_z\right)\right]$$
$$=\left[y\hat{p}_z,z\hat{p}_x\right]-\underbrace{\left[y\hat{p}_z,x\hat{p}_z\right]}_{=0}$$
$$-\underbrace{\left[z\hat{p}_y,z\hat{p}_x\right]}_{=0}+\left[z\hat{p}_y,x\hat{p}_z\right]$$
$$=\left[y\hat{p}_z,z\hat{p}_x\right]-\left[x\hat{p}_z,z\hat{p}_y\right]$$
$$=y\hat{p}_x\left[\hat{p}_z,z\right]-x\hat{p}_y\left[\hat{p}_z,z\right]$$
$$=\underbrace{\left[\hat{p}_z,z\right]}_{=-i\hbar}\left(y\hat{p}_x-x\hat{p}_y\right)$$
$$=i\hbar\left(x\hat{p}_y-y\hat{p}_x\right)$$
$$=i\hbar\hat{L}_z$$

Similarly:

$\left[\hat{L}_y,\hat{L}_z\right]=i\hbar\hat{L}_x$ and $\left[\hat{L}_z,\hat{L}_x\right]=i\hbar\hat{L}_y$

Total angular momentum:

$$\left[\hat{L}_z,L^2\right]=\left[\hat{L}_z,\hat{L}_x^2+\hat{L}_y^2+\hat{L}_z^2\right]$$
$$=\left[\hat{L}_z,\hat{L}_x^2\right]+\left[\hat{L}_z,\hat{L}_y^2\right]+\underbrace{\left[\hat{L}_z,\hat{L}_z^2\right]}_{=0}$$
$$=\hat{L}_x\underbrace{\left[\hat{L}_z,\hat{L}_x\right]}_{=i\hbar\hat{L}_y}+\left[\hat{L}_z,\hat{L}_x\right]\hat{L}_x$$
$$+\hat{L}_y\underbrace{\left[\hat{L}_z,\hat{L}_y\right]}_{=-i\hbar\hat{L}_x}+\left[\hat{L}_z,\hat{L}_y\right]\hat{L}_y$$
$$=i\hbar\hat{L}_x\hat{L}_y+i\hbar\hat{L}_y\hat{L}_x-i\hbar\hat{L}_y\hat{L}_x$$
$$-i\hbar\hat{L}_x\hat{L}_y$$
$$=0$$

Refer to commutator properties in Section 18.1.5.

19.2.2 Angular momentum eigenvalues

The eigenvalue equation for the total orbital angular momentum operator, $\hat{L}^2$ is:

$$\hat{L}^2Y(\theta,\phi)=\hbar^2c^2\,Y(\theta,\phi)$$

(refer to Derivation 19.2). The eigenfunctions are spherical harmonics, $Y(\theta, \phi)$. For these to be normalizable wavefunctions, the separation constant is:

$$c^2=l(l+1)$$

where $l=0, 1, 2, \ldots$ The eigenvalue equation for $\hat{L}^2$ is therefore:

$$\hat{L}^2Y(\theta,\phi)=l(l+1)\hbar^2\,Y(\theta,\phi)$$

(Eq.19.13). l is the quantum number that represents the total orbital angular momentum state of a particle. A measurement of angular momentum would return a value of $l(l+1)\hbar^2$ (kg m^2 s^{-1})^2. For example, if a particle were in its $l=3$ momentum eigenstate, the measured angular momentum would be $12\hbar^2$ (kg m^2 s^{-1})^2 (refer to Chapter 20 for application to the hydrogen atom).

The spherical harmonics are composed of a θ-dependent associated Legendre polynomial, $P_l^m(\cos\theta)$, and a ϕ-dependent complex exponential:

$$Y(\theta,\phi)=P_l^m(\cos\theta)e^{im\phi}$$

(refer to Derivation 19.3). Operating on the spherical harmonics with $\hat{L}_z=-i\hbar\,\partial/\partial\phi$, we find that the spherical harmonics are also eigenfunctions of $\hat{L}_z$

$$\hat{L}_zY(\theta,\phi)=-i\hbar\frac{\partial}{\partial\phi}Y(\theta,\phi)$$
$$=-i\hbar P_l^m(\cos\theta)\frac{\partial}{\partial\phi}e^{im\phi}$$
$$=m\hbar P_l^m(\cos\theta)e^{im\phi}$$
$$=m\hbar Y(\theta,\phi)$$

(Eq. 19.14). In this case, the eigenvalues are $m\hbar$ (kg m^2 s^{-1}), where m is an integer. The requirement that m is an integer also follows from the requirement that $\hat{L}_z$ is a Hermitian operator (Derivation 19.8). Each solution of different m or l corresponds to a state of different angluar momentum.

Restrictions on values of l and m

The associated Legendre polynomials are defined such that for each value of l, m can range from $-l$ to $+l$ in integer steps. That is, for a given l, the possible values of m are:

$$m=-l,\ (-l+1),\ (-l+2),\ \ldots,\ (l-1),\ l$$

(Eq. 19.15). Therefore, for each value of total angular momentum, l , there are $(2l+1)$ projections of angular

Derivation 19.8: Proof that m is an integer if $\hat{L}_z$ is Hermitian

$$\left(\int_0^{2\pi}\psi_2^*\hat{L}_z\psi_1 d\phi\right)^* = \int_0^{2\pi}\psi_2\hat{L}_z^*\psi_1^* d\phi$$
$$= \int_0^{2\pi}\psi_2 i\hbar\frac{\partial}{\partial\phi}\psi_1^* d\phi$$
$$= i\hbar\left[\psi_1^*\psi_2\right]_0^{2\pi} - \int_0^{2\pi}\psi_1^* i\hbar\frac{\partial}{\partial\phi}\psi_2 d\phi$$
$$= i\hbar\left[\psi_1^*\psi_2\right]_0^{2\pi} + \int_0^{2\pi}\psi_1^*\hat{L}_z\psi_2 d\phi$$

For $\hat{L}_z$ to be Hermitian, we require:

$$i\hbar\left[\psi_1^*\psi_2\right]_0^{2\pi} = 0$$
$$\psi_m(2\pi) = \psi_m(0)$$
$$e^{im2\pi} = e^0$$
$$\therefore m = 0, \pm 1, \pm 2, \ldots$$

- Starting point: for an operator to be Hermitian, it must satisfy:

$$\int\psi_1^*\hat{L}_z\psi_2 d\phi = \left(\int\psi_2^*\hat{L}_z\psi_1 d\phi\right)^*$$

- The component angular momentum operator is:

$$\hat{L}_z = -i\hbar\frac{\partial}{\partial\phi} \therefore \hat{L}_z^* = i\hbar\frac{\partial}{\partial\phi}$$

Use this in the right-hand side of the above expression for Hermitian operators and integrate by parts.

- For $\hat{L}_z$ to be Hermitian, we require:

$$i\hbar\left[\psi_1^*\psi_2\right]_0^{2\pi} = 0$$

Eigenstates of the $\hat{L}_z$ operator are complex exponentials, $\psi_m(\phi) = Ae^{im\phi}$, therefore for $\psi_m(2\pi) = \psi_m(0)$, m must be an integer.

momentum, m, which all correspond to different states. For example, if a particle is in its $l = 2$ total angular momentum eigenstate, it can be in one of five different component angular momentum states, represented by $m=\{-2, -1, 0, 1, 2\}$ (Fig. 19.1).

Angular momentum eigenvalue equations:

Total orbital angular momentum:

$$\hat{L}^2 Y(\theta,\phi) = l(l+1)\hbar^2 Y(\theta,\phi) \quad (19.13)$$

Component of angular momentum:

$$\hat{L}_z Y(\theta,\phi) = m\hbar Y(\theta,\phi) \quad (19.14)$$

where l and m are integers, and $|m| \leq l$:

$$m = -l, (-l+1), (-l+2), \ldots, (l-1), l \quad (19.15)$$

The requirement that m and l are integers means that the angular momentum eigenvalues are quantized into integer values of $\hbar$ or $\hbar^2$. The quantum number l is called the *orbital angular momentum* quantum number, and m is called the *magnetic* quantum number.

19.2.3 Normalized electron orbitals

Using expressions for the first three associated Legendre polynomials, we can determine the angular dependence of the normalized electron wavefunctions corresponding to the lowest angular momentum states (Derivation 19.9). Doing this, we find that the zero angular momentum state ($l = 0$, $m = 0$) has no angular dependence, and therefore is spherical in shape (Eq. 19.16, Derivation 19.9, and Fig. 19.2a). States with angular momentum $l = 1$ *do* have an angular dependence, and therefore have a specific orientation in space. These orbitals are

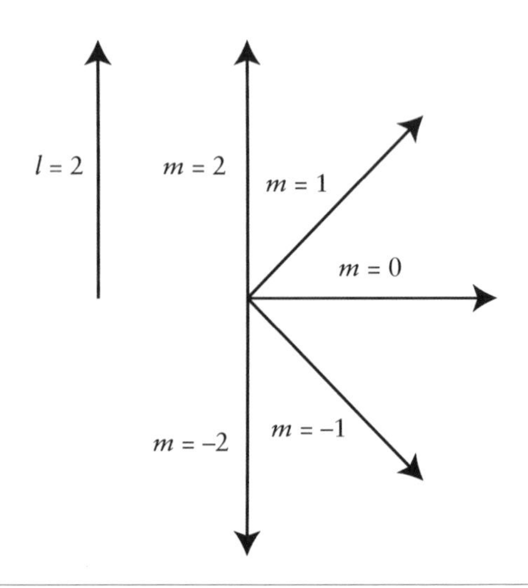

Figure 19.1: Schematic of orbital angular momentum components (states of different m) for a total orbital angular momentum of $l = 2$.

Derivation 19.9: Derivation of normalized electron orbitals (Eqs. 19.16–19.18)

$l = 0$, $m = 0$ state:

$$\int \left|Y_0^0(\theta,\phi)\right|^2 \mathrm{d}\Omega = A^2 \int_0^{2\pi}\int_0^{\pi} \sin\theta \mathrm{d}\theta \mathrm{d}\phi$$

$$= 2\pi A^2 \left[-\cos\theta\right]_0^{\pi}$$

$$= 4\pi A^2 = 1$$

$$\therefore A = \sqrt{\frac{1}{4\pi}}$$

$$\therefore Y_0^0(\theta,\phi) = \sqrt{\frac{1}{4\pi}}$$

$l = 1$, $m = 0$ state:

$$\int \left|Y_1^0(\theta,\phi)\right|^2 \mathrm{d}\Omega = B^2 \int_0^{2\pi}\int_0^{\pi} \cos^2\theta \sin\theta \mathrm{d}\theta \mathrm{d}\phi$$

$$= 2\pi B^2 \left[-\frac{1}{3}\cos^3\theta\right]_0^{\pi}$$

$$= \frac{4\pi}{3} B^2 = 1$$

$$\therefore B = \sqrt{\frac{3}{4\pi}}$$

$$\therefore Y_1^0(\theta,\phi) = \sqrt{\frac{3}{4\pi}} \cos\theta$$

$l = 2$, $m = \pm 1$ states:

$$\int \left|Y_1^{\pm 1}(\theta,\phi)\right|^2 \mathrm{d}\Omega = C^2 \int_0^{2\pi}\int_0^{\pi} \sin^3\theta \mathrm{d}\theta \mathrm{d}\phi$$

$$= C^2 \int_0^{2\pi}\int_0^{\pi} \left(1-\cos^2\theta\right) \sin\theta \mathrm{d}\theta \mathrm{d}\phi$$

$$= 2\pi C^2 \left[-\cos\theta + \frac{1}{3}\cos^3\theta\right]_0^{\pi}$$

$$= \frac{8\pi}{3} C^2 = 1$$

$$\therefore C = \sqrt{\frac{3}{8\pi}}$$

$$\therefore Y_1^{\pm 1}(\theta,\phi) = \mp\sqrt{\frac{3}{8\pi}} \sin\theta e^{\pm i\phi}$$

- The spherical harmonics are given by:

$$Y_l^m(\theta,\phi) = AP_l^m(\cos\theta)e^{im\phi}$$

where the first three associated Legendre polynomials are:

$$P_0^0(\cos\theta) = 1$$

$$P_1^0(\cos\theta) = \cos\theta$$

$$P_1^{\pm 1}(\cos\theta) = \mp\sin\theta$$

- These are normalized by the constant A such that:

$$\int \left|Y_l^m(\theta,\phi)\right|^2 \mathrm{d}\Omega = 1$$

$$\therefore \int_0^{2\pi}\int_0^{\pi} \left|AP_l^m(\cos\theta)\right|^2 \sin\theta \mathrm{d}\theta \mathrm{d}\phi = 1$$

- $l = 0$, $m = 0$ state: the first spherical harmonic has no angular variation in θ or ϕ:

$$Y_0^0(\theta,\phi) = AP_0^0(\cos\theta)e^0$$

$$= A$$

- $l = 1$, $m = 0$ state: the second spherical harmonic has angular variation in θ only:

$$Y_1^0(\theta,\phi) = BP_1^0(\cos\theta)e^0$$

$$= B\cos\theta$$

- $l = 1$, $m = \pm 1$ states: the third spherical harmonics have angular variation in both θ and ϕ. The ϕ-dependence cancels when we multiply the spherical harmonic by its complex conjugate; therefore, the coefficients of either state are equal:

$$Y_1^{\pm 1}(\theta,\phi) = CP_1^{\pm 1}(\cos\theta)e^{\pm i\phi}$$

$$= \mp C\sin\theta e^{\pm i\phi}$$

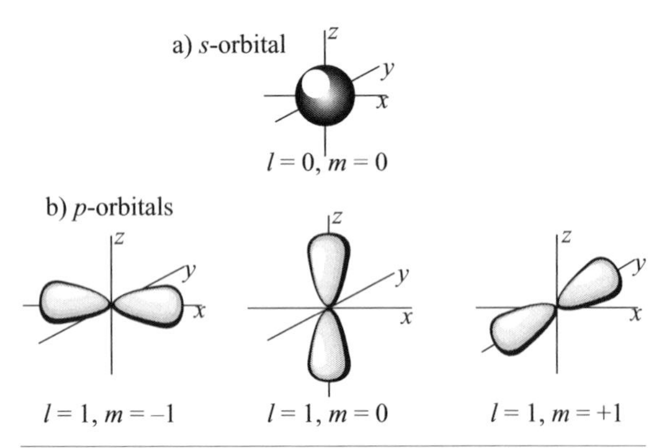

Figure 19.2: Low angular momentum, *s* and *p*, electron orbitals.

dumbbell shaped. Each different value of m (m = 0, +1 or −1) is aligned along a different Cartesian axis (Eqs. 19.17 and 19.18, Derivation 19.9, and Fig. 19.2b).

Normalized electron orbitals (Derivation 19.9):

Spherical *s*-orbital ($l = 0, m = 0$):

$$Y_0^0(\theta, \phi) = \frac{1}{\sqrt{4\pi}} \tag{19.16}$$

Dumbbell *p*-orbitals ($l = 1, m = \{-1, 0, +1\}$):

$$Y_1^0(\theta, \phi) = \sqrt{\frac{3}{4\pi}} \cos\theta \tag{19.17}$$

$$Y_1^{\pm 1}(\theta, \phi) = \mp\sqrt{\frac{3}{8\pi}} \sin\theta e^{\pm i\phi} \tag{19.18}$$

For historical reasons, angular momentum states are assigned a letter according to their quantum number value, l:

$$l = \begin{Bmatrix} 0 & 1 & 2 & 3 & 4 & 5 \\ s & p & d & f & g & h \end{Bmatrix}$$

The zero angular momentum states, $l = 0$, are therefore often referred to as "*s*-orbitals," whereas states with angular momentum $l = 1$ are referred to as "*p*-orbitals."

19.3 Spin angular momentum

In addition to *orbital* angular momentum, particles possess an additional *intrinsic* angular momentum, which is called *spin*. There is no classical analog for a particle's spin—it is a fundamental property of elementary particles, like mass or charge. It is purely a relativistic effect, and it is described by an equation that is similar to the Schrödinger equation, called the *Dirac equation*. The existence of spin was discovered through experiments that involved observing the interaction of particles with external magnetic fields, such as the *Stern–Gerlach* experiment, described below. This experiment provided unexpected results, which pointed to the existence of another angular momentum in addition to orbital motion. Spin was found to be *half-*integer quantized (as opposed to l and m, which are integer-quantized).

19.3.1 The Stern–Gerlach experiment

Consider an electron of mass m_e and charge $q = -e$ orbiting a nucleus. For an electron with an angular frequency of $\omega = v/r$, its angular momentum is:

$$\begin{aligned} \vec{L} &= m_e \vec{r} \times \vec{v} \\ &= m_e \omega r^2 \hat{n} \end{aligned}$$

where $\hat{n}$ is a unit vector in the direction of $\vec{L}$. The circulating charge gives rise to a small current loop of magnitude:

$$I = qf = -\frac{e\omega}{2\pi}$$

where f is the orbital frequency. Therefore, this current loop acts as a small magnetic dipole, with a dipole moment of:

$$\begin{aligned} \vec{\mu} &= IA \\ &= -\frac{e\omega}{2\pi} \pi r^2 \\ &= -\frac{e}{2m_e} \vec{L} \end{aligned}$$

The magnetic moment operator is therefore:

$$\hat{\vec{\mu}} = -\frac{e}{2m_e} \hat{\vec{L}}$$

Or, for one component:

$$\hat{\mu}_z = -\frac{e}{2m_e} \hat{L}_z$$

A system in an eigenstate of $\hat{L}_z$ has an eigenvalue of $m\hbar$, where m is the magnetic quantum number. Therefore, a measurement of the z-component of magnetic moment of a particle would give:

$$\begin{aligned} \mu_z &= -\underbrace{\frac{e\hbar}{2m_e}}_{\equiv \mu_B} m \\ &\equiv -m\mu_B \end{aligned}$$

where μ_B [$9.27 \cdot 10^{-24}$ J T^{-1}] is a constant called the *Bohr magneton*. Therefore, magnetic moment is quantized into integer values of μ_B.

Interaction of μ_z in an external magnetic field

Consider a magnetic field, B, directed along the z-direction. A magnetic dipole interacts with an external magnetic field, where the interaction energy is:

$$U_\mu = -\mu_z B = m\mu_B B$$

(refer to Section 8.5). For a non-uniform magnetic field, a dipole also experiences a linear force along the z-direction of:

$$F_z = -\frac{\partial U_\mu}{\partial z} = -m\mu_B \frac{\partial B}{\partial z}$$

Therefore, a beam of incoming particles of given m passing through a non-uniform magnetic field will experience a force, which would result in a deflection of the beam from the incoming path. Hence, if a beam of particles of a given *total* angular momentum, l, were directed through a non-uniform magnetic field, the emerging beam would be split into $2l+1$ components of different m, since each particle with a different value of m would experience a different force and hence have a different deflection from the incoming beam.

The discovery of spin

A beam of hydrogen in its ground state has angular momentum quantum numbers of $l=0$ and $m=0$. Therefore, if a beam of hydrogen were directed through a non-uniform magnetic field, we would expect a single beam with no deflection from incidence to emerge from the field, since $F_z=0$. However, when this experiment was performed, what was actually observed was that the emerging beam was split into *two* beams, with equal deflection in the $+z$- and in the $-z$-direction (Fig. 19.3). These observations lead to two conclusions: (i) particles have an additional angular momentum, other than orbital angular momentum, which is an intrinsic property of a particle (this is, of course, *spin*), and (ii) the quantization of spin must be half-integer quantized in order to have $2l+1=2$ emerging beams.

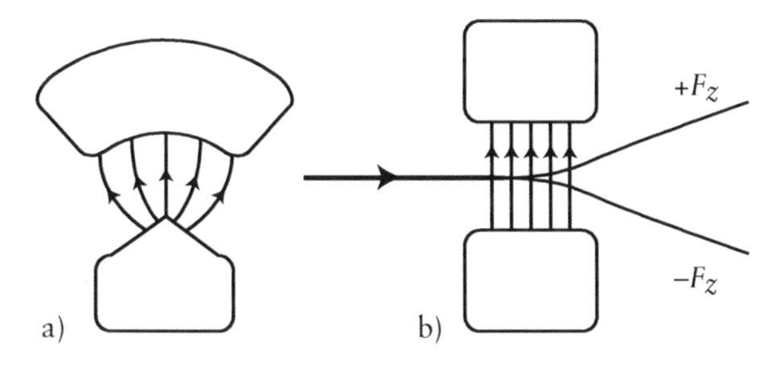

Figure 19.3: Stern–Gerlach experiment for ground-state hydrogen. An incoming beam is split into two emerging beams by a non-uniform magnetic field: a) front view of magnetic field and b) side view of split beam.

19.3.2 Pauli spin matrices

Like orbital angular momentum, spin is a vector, with an operator denoted by $\hat{\vec{S}}$. Spin operators are matrices, not differential operators. These matrices are complex, Hermitian, and unitary. The dimension of the matrix depends on the spin of the particle. In this section, we'll look exclusively at *spin-half* particles (electrons, protons, neutrons, and quarks are all spin-half particles). The spin-half matrices are 2×2 matrices called *Pauli spin matrices*, σ. There are three Pauli spin matrices, σ_x, σ_y, or σ_z, that represent either the x, y, or z Cartesian components of spin (Eqs. 19.19–19.21).

Pauli spin matrices:

$$\sigma_x = \begin{pmatrix} 0 & 1 \\ 1 & 0 \end{pmatrix} \tag{19.19}$$

$$\sigma_y = \begin{pmatrix} 0 & -i \\ i & 0 \end{pmatrix} \tag{19.20}$$

$$\sigma_z = \begin{pmatrix} 1 & 0 \\ 0 & -1 \end{pmatrix} \tag{19.21}$$

The spin operators, $\hat{S}_x$, $\hat{S}_y$, and $\hat{S}_z$, are the Pauli spin matrices multiplied by $\hbar/2$ (Eqs. 19.22–19.24). The total spin operator squared, $\hat{S}^2$, is given by the sum of the square of the components (Eq. 19.25).

Spin-half operators:

$$\hat{S}_x = \frac{\hbar}{2}\begin{pmatrix} 0 & 1 \\ 1 & 0 \end{pmatrix} \tag{19.22}$$

$$\hat{S}_y = \frac{\hbar}{2}\begin{pmatrix} 0 & -i \\ i & 0 \end{pmatrix} \tag{19.23}$$

$$\hat{S}_z = \frac{\hbar}{2}\begin{pmatrix} 1 & 0 \\ 0 & -1 \end{pmatrix} \tag{19.24}$$

Total spin operator:

$$\hat{S}^2 = \hat{S}_x^2 + \hat{S}_y^2 + \hat{S}_z^2 = \frac{3}{4}\hbar^2\begin{pmatrix} 1 & 0 \\ 0 & 1 \end{pmatrix} \tag{19.25}$$

It's straightforward to show that the spin-half operators obey the same commutation relations as the orbital angular momentum operators: the total spin operator commutes with each component, but individual components do not commute with each other

(Eqs. 19.26 and 19.27). For example, the commutator of $\hat{S}_x$ and $\hat{S}_y$ is:

$$\begin{aligned}\left[\hat{S}_x,\hat{S}_y\right] &= \hat{S}_x\hat{S}_y - \hat{S}_y\hat{S}_x \\ &= \frac{\hbar^2}{4}\begin{pmatrix} i & 0 \\ 0 & -i\end{pmatrix} - \frac{\hbar^2}{4}\begin{pmatrix} -i & 0 \\ 0 & i\end{pmatrix} \\ &= i\frac{\hbar^2}{2}\begin{pmatrix} 1 & 0 \\ 0 & -1\end{pmatrix} \\ &= i\hbar\hat{S}_z\end{aligned}$$

Therefore, only the total spin of a particle along with any *one* component can be known simultaneously. Measurement of another component will disrupt the system such that the previously known component is no longer defined. A consequence of this phenomenon is that we can never define the direction of the vector operator, $\hat{S}$; its direction has an intrinsic uncertainty.

Spin commutators:

$$\left[\hat{S}_i,\hat{S}_j\right] = i\hbar\varepsilon_{ijk}\hat{S}_k \tag{19.26}$$

$$[\hat{S}_i,\hat{S}^2] = 0 \tag{19.27}$$

Where the Levi-Civita symbol is:

$$\varepsilon_{ijk} = \begin{cases} +1 & \text{even permutation of } ijk \\ -1 & \text{odd permutation of } ijk \\ 0 & \text{otherwise}\end{cases}$$

i, j, and k represent the Cartesian components x, y, and z.

19.3.3 Spin eigenvalue equation

The eigenvalue equations for the spin operators $\hat{S}^2$ and $\hat{S}_z$ have a similar form as those for $\hat{L}^2$ and $\hat{L}_z$ (Eqs. 19.28 and 19.29).[1] The quantum numbers s and m_s define the total spin state and component of spin, respectively. The relationship between s and m_s is that $|m_s| \leqslant s$, and m_s increases from $-s$ to s in increments of one (Eq. 19.30). Therefore, there are $(2s+1)$ states of different m_s projections for each spin state s. This is the same as the relationship between orbital quantum numbers l and m. The differences between spin and orbital eigenvalue equations are:

- The spin operators are matrices, not differential operators. Hermitian matrices are closely analogous to Hermitian differential operators in that their eigenvalues are real and their eigenvectors are orthogonal and form a complete set.
- The spin eigenfunctions are not spherical harmonics or other functions of the particle's position. Instead, they are *vectors*, denoted by the state $|s,m_s\rangle$ (which we will solve for below).
- The spin quantum number, s, is *half*-integer quantized: s = 0, 1/2, 1, 3/2, 2, ... In contrast, the orbital angular momentum quantum number, l, is *integer*-quantized, l = 0, 1, 2, ...

Spin eigenvalue equations:

$$\hat{S}^2\left|s,m_s\right\rangle = s(s+1)\hbar^2\left|s,m_s\right\rangle \tag{19.28}$$

$$\hat{S}_z\left|s,m_s\right\rangle = m_s\hbar\left|s,m_s\right\rangle \tag{19.29}$$

where s and m_s are half-integers or integers, and $|m_s| \leqslant s$:

$$m_s = -s,\ (-s+1),\ \ldots,\ (s-1),\ s \tag{19.30}$$

19.3.4 Spin-half eigenstates

The total spin quantum number for spin-half particles is $s = 1/2$, and from Eq. 19.30, its components are $m_s = \pm 1/2$. Therefore, spin-half eigenstates are denoted by:

$$\left|s,m_s\right\rangle = \left|1/2,\ \pm 1/2\right\rangle$$

Setting $s = 1/2$ in Eq. 19.28 and $m_s = \pm 1/2$ in Eq. 19.29, we find that the spin-half eigenvalue equations are:

$$\hat{S}^2\left|s,m_s\right\rangle = \frac{3}{4}\hbar^2\left|s,m_s\right\rangle$$

$$\hat{S}_z\left|s,m_s\right\rangle = \pm\frac{1}{2}\hbar\left|s,m_s\right\rangle$$

This means that when we measure the total spin of a spin-half particle, we measure a value $3\hbar^2/4$. When we measure the component of spin, we either measure *spin-up*, $m_s = +\hbar/2$, or *spin-down*, $m_s = -\hbar/2$.

Since the operators are 2×2 matrices, the eigenstate of a spin-half particle is represented by a two-component column vector. Acting on an arbitrary two-component column vector with the total spin operator, we see that all such vectors are eigenstates:

$$\begin{aligned}\hat{S}^2\begin{pmatrix} a \\ b\end{pmatrix} &= \frac{3}{4}\hbar^2\begin{pmatrix} 1 & 0 \\ 0 & 1\end{pmatrix}\begin{pmatrix} a \\ b\end{pmatrix} \\ &= \frac{3}{4}\hbar^2\begin{pmatrix} a \\ b\end{pmatrix}\end{aligned}$$

We can find the eigenvectors of the x-, y-, and z-components of spin by solving the eigenvalue equations of the operators $\hat{S}_x$, $\hat{S}_y$, and $\hat{S}_z$. The normalized spin-up eigenstates, denoted $|\uparrow_i\rangle$ where $i = x$, y, or z, and spin-down eigenstates, denoted $|\downarrow_i\rangle$ where $i = x$, y, or z, are given in Eqs. 19.31–19.33 (Derivation 19.10).

[1] Refer to *Quantum Mechanics*, A. Rae, Section 5.4 for a rigorous derivation involving raising and lowering operators.

Normalized spin-half eigenvectors (Derivation 19.10):

$$|\uparrow_x\rangle = \frac{1}{\sqrt{2}}\begin{pmatrix}1\\1\end{pmatrix},\quad |\downarrow_x\rangle = \frac{1}{\sqrt{2}}\begin{pmatrix}1\\-1\end{pmatrix} \quad (19.31)$$

$$|\uparrow_y\rangle = \frac{1}{\sqrt{2}}\begin{pmatrix}1\\i\end{pmatrix},\quad |\downarrow_y\rangle = \frac{1}{\sqrt{2}}\begin{pmatrix}1\\-i\end{pmatrix} \quad (19.32)$$

$$|\uparrow_z\rangle = \begin{pmatrix}1\\0\end{pmatrix},\quad |\downarrow_z\rangle = \begin{pmatrix}0\\1\end{pmatrix} \quad (19.33)$$

Orthonormality of eigenstates:

$$\langle\uparrow_i|\uparrow_i\rangle = 1$$

$$\langle\uparrow_i|\downarrow_i\rangle = 0$$

where $i = x$, y, or z.

Notice that any normalized state can be written as a superposition of $|\uparrow_z\rangle$ and $|\downarrow_z\rangle$:

$$\frac{1}{\sqrt{a^2+b^2}}\begin{pmatrix}a\\b\end{pmatrix} = \frac{a}{\sqrt{a^2+b^2}}\begin{pmatrix}1\\0\end{pmatrix} + \frac{b}{\sqrt{a^2+b^2}}\begin{pmatrix}0\\1\end{pmatrix}$$

$$= \frac{a|\uparrow_z\rangle}{\sqrt{a^2+b^2}} + \frac{b|\downarrow_z\rangle}{\sqrt{a^2+b^2}}$$

The probability of measuring $m_s = +\hbar/2$ (system is in a spin-up state) is then:

$$P(\uparrow_z) = \frac{a^2}{a^2+b^2}$$

Derivation 19.10: Derivation of spin-half eigenstates (Eqs. 19.31–19.33)

$$\hat{S}_x\begin{pmatrix}a\\b\end{pmatrix} = \begin{pmatrix}\lambda_x & 0\\0 & \lambda_x\end{pmatrix}\begin{pmatrix}a\\b\end{pmatrix}$$

$$\therefore \frac{\hbar}{2}\begin{pmatrix}0 & 1\\1 & 0\end{pmatrix}\begin{pmatrix}a\\b\end{pmatrix} = \begin{pmatrix}\lambda_x & 0\\0 & \lambda_x\end{pmatrix}\begin{pmatrix}a\\b\end{pmatrix}$$

- Starting point: the eigenvalue equation for $\hat{S}_x$ is:

$$\hat{S}_x\begin{pmatrix}a\\b\end{pmatrix} = \begin{pmatrix}\lambda_x & 0\\0 & \lambda_x\end{pmatrix}\begin{pmatrix}a\\b\end{pmatrix}$$

where $\hat{S}_x = \frac{\hbar}{2}\begin{pmatrix}0 & 1\\1 & 0\end{pmatrix}$

and λ_x correspond to the eigenvalues of $\hat{S}_x$.

$$\therefore \begin{pmatrix}-\lambda_x & \hbar/2\\ \hbar/2 & -\lambda_x\end{pmatrix}\begin{pmatrix}a\\b\end{pmatrix} = 0$$

$$\therefore \begin{vmatrix}-\lambda_x & \hbar/2\\ \hbar/2 & -\lambda_x\end{vmatrix} = \lambda_x^2 - \frac{\hbar^2}{4} = 0$$

$$\therefore \lambda_x = \pm\frac{\hbar}{2}$$

- The eigenvalues λ_x are found by setting the matrix determinant equal to zero. Doing this, two eigenvalues are then:

$$\lambda_x = +\frac{\hbar}{2} \rightarrow \text{ spin-up eigenstate}$$

$$\lambda_x = -\frac{\hbar}{2} \rightarrow \text{ spin-down eigenstate}$$

The normalized eigenvectors corresponding to either eigenvalue are therefore:

when $\lambda_x = +\frac{\hbar}{2}$:

$$\begin{pmatrix}a\\b\end{pmatrix} = \begin{pmatrix}1\\1\end{pmatrix}$$

$$\therefore |\uparrow_x\rangle = \frac{1}{\sqrt{2}}\begin{pmatrix}1\\1\end{pmatrix}$$

when $\lambda_x = -\frac{\hbar}{2}$:

$$\begin{pmatrix}a\\b\end{pmatrix} = \begin{pmatrix}1\\-1\end{pmatrix}$$

$$\therefore |\downarrow_x\rangle = \frac{1}{\sqrt{2}}\begin{pmatrix}1\\-1\end{pmatrix}$$

- Using these eigenvalues in the eigenvalue equation for $\hat{S}_x$, we can determine the normalized eigenvectors of the spin-up and spin-down eigenstates. For example, for the spin-up eigenstate:

$$\frac{\hbar}{2}\begin{pmatrix}0 & 1\\1 & 0\end{pmatrix}\begin{pmatrix}a\\b\end{pmatrix} = \frac{\hbar}{2}\begin{pmatrix}1 & 0\\0 & 1\end{pmatrix}\begin{pmatrix}a\\b\end{pmatrix}$$

$$\therefore \frac{\hbar}{2}b = \frac{\hbar}{2}a$$

$$\therefore \frac{a}{b} = 1$$

- The eigenvectors for $\hat{S}_y$ and $\hat{S}_z$ are found using a similar method.

and the probability of measuring $m_s = -\hbar/2$ (system is in a spin-down state) is:

$$P(\downarrow_z) = \frac{b^2}{a^2 + b^2}$$

19.4 Total angular momentum

The *total angular momentum*, $\hat{\vec{J}}$, is the sum of both spin and orbital angular momentum:

$$\hat{\vec{J}} = \hat{\vec{S}} + \hat{\vec{L}}$$

The operators $\hat{J}^2$ and $\hat{J}_z$ have similar properties to the corresponding spin or orbital angular momentum operators. For example, they obey similar commutation rules (Eqs. 19.34 and 19.35), and have similar eigenvalue equations (Eqs. 19.36–19.37).

Total angular momentum commutators:

$$\left[\hat{J}_i, \hat{J}_j\right] = i\hbar\varepsilon_{ijk}\hat{J}_k \qquad (19.34)$$

$$[\hat{J}_i, \hat{J}^2] = 0 \qquad (19.35)$$

where the Levi-Civita symbol is:

$$\varepsilon_{ijk} = \begin{cases} +1 & \text{even permutation of } ijk \\ -1 & \text{odd permutation of } ijk \\ 0 & \text{otherwise} \end{cases}$$

i, j, and k represent the Cartesian components x, y, and z.

Total angular momentum eigenvalue equations:

$$\hat{J}^2\left|j, m_j\right\rangle = j(j+1)\hbar^2\left|j, m_j\right\rangle \qquad (19.36)$$

$$\hat{J}_z\left|j, m_j\right\rangle = m_j\hbar\left|j, m_j\right\rangle \qquad (19.37)$$

where j and m_j are half-integers or integers and $|m_j| \leqslant j$:

$$m_j = -j, (j+1), \ldots, (j-1), j \qquad (19.38)$$

The quantum number j in Eq. 19.36 is the *total angular momentum* quantum number, and can take integer or half-integer values. For each combination of quantum numbers l and s, there are $(2p + 1)$ states of different j, where:

$$p = \begin{cases} s & \text{if } s < l \\ l & \text{if } l < s \end{cases}$$

The extreme values of j are:

$$j_{\min} = |l - s|$$

$$\text{and } j_{\max} = l + s$$

The other values of j increase from $j_{\min}$ to $j_{\max}$ in increments of one. For example, for a state defined by quantum numbers $s = 3/2$ and $l = 1$, there are $(2l+1) = 3$ states of different j, where:

$$j_{\min} = |l - s| = \frac{1}{2}$$

$$\text{and } j_{\max} = l + s = \frac{5}{2}$$

$$\therefore j = \left\{\frac{1}{2}, \frac{3}{2}, \frac{5}{2}\right\}$$

Furthermore, for each value of j, there are $(2j + 1)$ states with different projection, m_j, as given by Eq. 19.38. For example, for the $j = 1/2$ state in the above example, there are two m_j-components: $m_j = \{-1/2, 1/2\}$, whereas for the $j = 3/2$ state, there are four components: $m_j = \{-3/2, -1/2, 1/2, 3/2\}$.

Chapter 20 Introduction to the hydrogen atom

The hydrogen atom is the best understood application of quantum mechanics, and predictions of quantum theory agree very well with experiment. A hydrogen atom consists of a single electron orbiting a single proton (Fig. 20.1). The electron can occupy a set of discrete energy levels, which are each represented by specific wavefunctions. In this chapter, we'll apply the theory from the previous chapters to determine the electron wavefunctions and what their corresponding energies are.

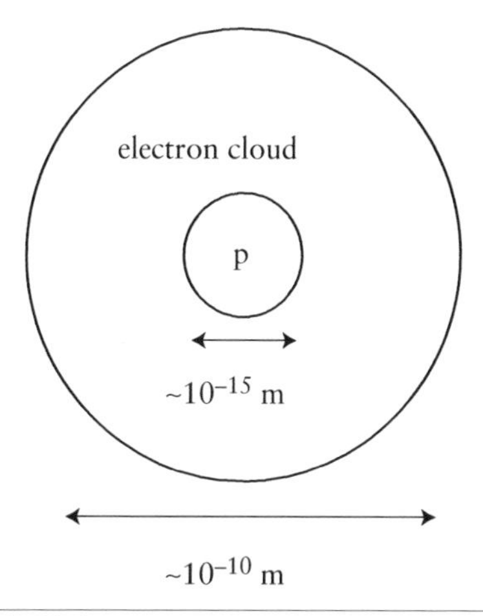

Figure 20.1: Hydrogen atom.

20.1 Radial wavefunctions

Radial TISE for a hydrogen atom:

$$\left(\underbrace{-\frac{\hbar^2}{2m_e r^2}\frac{\mathrm{d}}{\mathrm{d}r}\left(r^2\frac{\mathrm{d}}{\mathrm{d}r}\right)}_{\text{Radial }KE}+\underbrace{\frac{l(l+1)\hbar^2}{2m_e r^2}}_{\text{Angular }KE}\underbrace{-\frac{e^2}{4\pi\varepsilon_0 r}}_{\text{Coulomb }PE}\right)R(r)=ER(r) \qquad (20.1)$$

Schrödinger equation for hydrogen

The electric potential energy of an electron in a hydrogen atom is proportional to the Coulomb potential of the nucleus. Therefore, the expression for the central potential of a hydrogen atom is:

$$V(r) = -\frac{e^2}{4\pi\varepsilon_0 r}$$

where r is the radial distance from the nucleus. Substituting this expression into the radial TISE (refer to Section 19.1, Eq. 19.4), we can find an expression for the hydrogen Schrödinger equation (Eq. 20.1). The angular kinetic energy term in Eq. 20.1 is often called the *centrifugal barrier* since it prevents the electron from reaching the origin when $l \neq 0$.

It is difficult to solve Eq. 20.1 analytically. However, we can get some physical insight by breaking down the radial dependence of $R(r)$ into different limits of r.

Behavior at large and small r

Let's first look at the behavior at the extreme values of r. At one extreme, as $r \to \infty$, we find that the radial wavefunction decays exponentially with distance:

$$R(r \to \infty) = Ae^{-kr}$$

(Eq. 20.2 and Derivation 20.1). This makes intuitive sense since we would expect the probability of finding the electron far away from the nucleus to fall off steeply with increasing distance. At the other extreme, as $r \to 0$, the radial dependence of the wavefunction depends on the electron's angular momentum; the wavefunction is a polynomial in r, where the order of the polynomial is the angular momentum, l:

$$R_l(r \to 0) = Br^l$$

(Eq. 20.3 and Derivation 20.1). This means that electrons with low values of angular momentum are more likely to be found nearer the nucleus, whereas electrons with higher angular momentum are more likely to be further away. The radial wavefunctions for the lowest angular momentum values, $l = 0$, 1, and 2, are provided in Eqs. 20.4.

Radial wavefunction at different r (Derivations 20.1 and 20.2):

- As $r \to \infty$: exponential decay:

$$\therefore R(r \to \infty) \to Ae^{-kr} \tag{20.2}$$

- As $r \to 0$: polynomial dependence on angular momentum, l:

$$R_l(r \to 0) = Br^l \tag{20.3}$$

$$\begin{aligned} \therefore l = 0: &\quad R(r) = \text{constant} \\ l = 1: &\quad R(r) = Br \\ l = 2: &\quad R(r) = B'r^2 \end{aligned} \tag{20.4}$$

- Intermediate r: modified associated Laguerre polynomials, $v_n^l(r)$. These are a set of polynomials in r, starting with 1. The first few associated Laguerre polynomials, $L_n^a(x)$, are:

$$L_0^\alpha(x) = 1$$

$$L_1^\alpha(x) = -x + \alpha + 1$$

$$L_2^\alpha(x) = \frac{x^2}{2} - (\alpha+2)x + \frac{(\alpha+2)(\alpha+1)}{2}$$

Behavior at intermediate r

If we "peel off" the asymptotic behavior at large and small r, then at intermediate values of r, the hydrogen Schrödinger equation takes the form of a special ordinary differential equation, known as the *associated Laguerre equation* (the full derivation is beyond the scope of this book,[1] but an overview is provided in Derivation 20.2). The solutions to this equation are a sequence of polynomials starting with 1:

$$v(r) = \sum_{j=0}^{j_{max}} c_j r^j$$

If the sum went to $j = \infty$, then these polynomials are known as *associated Laguerre polynomials*. However, in order to define a wavefunction, the sum must terminate at j_{max} (if it terminated at $j = \infty$ then the wavefunction would tend to ∞ as $r \to \infty$, which is not physical). Therefore, we will refer to these polynomials as *modified* associated Laguerre polynomials. The c_j coefficients obey a recurrence relation, where c_{j+1} is related to c_j by:

$$c_{j+1} = \left\{ \frac{2(j+l+1) - \rho_0}{(j+1)(j+2l+2)} \right\} c_j \tag{20.5}$$

(Derivation 20.2). For the sum to terminate at j_{max}, we require the $j_{max} + 1$ coefficient to equal zero ($c_{jmax+1} = 0$), which, from Eq. 20.5, means that:

$$2(j_{max} + l + 1) - \rho_0 = 0 \tag{20.6}$$

If we define $n = j_{max} + l + 1$, then the degree of the polynomial, $v(r)$, is:

$$j_{max} = n - l - 1$$

(we shall look at the physical significance of this equation in Section 20.2).

Putting it all together

The *total* radial wavefunction combines the modified associated Laguerre polynomials with the asymptotic behavior at large and small r (Eq. 20.7). The normalization condition for the radial wavefunction is provided in Eq. 20.8 (see Section 20.3 for the normalized solution of the ground state of hydrogen).

Hydrogen radial wavefunction (refer to Eqs. 20.2 and 20.3):

$$R_{nl}(r) = Ae^{-kr} r^l v_n^l(r) \tag{20.7}$$

where $v_n^l(r)$ is a polynomial in r of degree

$$j_{max} = n - l - 1.$$

The normalization condition requires that:

$$\int_0^\infty |R_{nl}(r)|^2 r^2 dr = 1 \tag{20.8}$$

[1] For a full derivation, the reader is referred to *Introduction to Quantum Mechanics*, D. Griffiths, Section 4.2.

Derivation 20.1: Derivation of radial solution at large and small r (Eqs. 20.2 and 20.3)

Radial TISE for $f(r) = rR(r)$:

$$-\frac{\hbar^2}{2m_e}\frac{d^2 f(r)}{dr^2} + \left(\frac{l(l+1)\hbar^2}{2m_e r^2} - \frac{e^2}{4\pi\varepsilon_0 r}\right)f(r) = -Ef(r)$$

$$\therefore \frac{\hbar^2}{2m_e}\frac{d^2 f(r)}{dr^2} = \left(E + \frac{l(l+1)\hbar^2}{2m_e r^2} - \frac{e^2}{4\pi\varepsilon_0 r}\right)f(r)$$

As $r \to \infty$:

$$\frac{\hbar^2}{2m_e}\frac{d^2 f(r)}{dr^2} = Ef(r)$$

$$\therefore f(r) = Ae^{-kr} + Be^{kr}$$

$$= Ae^{-kr}$$

$$\therefore R(r \to \infty) \to Ae^{-kr}$$

As $r \to 0$:

$$\frac{\hbar^2}{2m_e}\frac{d^2 f(r)}{dr^2} = \frac{l(l+1)\hbar^2}{2m_e r^2}f(r)$$

Try $f(r) = A'r^{\beta}$:

$$\beta(\beta-1)r^{(\beta-2)} = l(l+1)r^{(\beta-2)}$$

$$\therefore \beta^2 - \beta - l(l+1) = 0$$

$$\therefore \beta = -l \text{ or } (l+1)$$

$$\therefore f(r) = A'r^{-l} + B'r^{(l+1)}$$

$$= B'r^{(l+1)}$$

$$\therefore R(r \to 0) \to B'r^{l}$$

- Starting point: let $R(r) = f(r)/r$ in the radial Schrödinger equation (Eq. 20.1). The energy of a bound state in hydrogen is negative; therefore, we use $-E$ on the right-hand side (RHS).
- As $r \to \infty$, the constant term (first term) on the RHS dominates. The TISE therefore approximates to an ordinary second-order differential equation with exponential solutions. Let the wavenumber be:

  $$k = \frac{\sqrt{2m_e E}}{\hbar}$$

  The positive exponential term (e^{kr}) is unphysical for $r \to \infty$.
- As $r \to 0$, the centrifugal term (second term) dominates since it has the largest power of r in the denominator. Try a solution: $f(r) = A'r^{\beta}$. Substituting in and solving a quadratic for β, we find:

  $\beta = -l \text{ or } (l+1)$

  The r^{-l} solution is unphysical for $r \to 0$.

20.2 Hydrogen energy levels

The principal quantum number

We saw above that, in order to describe a wavefunction, we must define:

$$2(j_{\max} + l + 1) - \rho_0 = 0$$

$$\text{or } 2n - \rho_0 = 0$$

$$\therefore n = \frac{\rho_0}{2}$$

where $n = j_{\max} + l + 1$

and $\rho_0 = \dfrac{m_e e^2}{2\pi\varepsilon_0 \hbar^2 k}$

(refer to Derivation 20.2). n is an integer starting at 1 ($n = 1$, when $j_{\max} = 0$ and $l = 0$). The integer n is called the *principal quantum number*, and it defines the energy levels of an electron:

$$n^2 = \frac{\rho_0^2}{4}$$

$$= \frac{1}{4}\left(\frac{m_e e^2}{2\pi\varepsilon_0 \hbar^2}\right)^2 \frac{1}{k^2}$$

$$= -\frac{1}{4}\left(\frac{m_e e^2}{2\pi\varepsilon_0 \hbar^2}\right)^2 \frac{\hbar^2}{2m_e E}$$

$$= -\underbrace{\left(\frac{e^2}{4\pi\varepsilon_0 \hbar c}\right)^2}_{\equiv \alpha^2} \frac{m_e c^2}{2E}$$

$$= -\frac{m_e}{2E}(\alpha c)^2$$

Derivation 20.2: Radial solution at intermediate r

Radial TISE for $f(r)=rR(r)$:

$$-\frac{\hbar^2}{2m_e}\frac{d^2f(r)}{dr^2}+\left(\frac{l(l+1)\hbar^2}{2m_e r^2}-\frac{e^2}{4\pi\varepsilon_0 r}\right)f(r)=-Ef(r)$$

$$\therefore \frac{1}{k^2}\frac{d^2f(r)}{dr^2}-\frac{l(l+1)}{k^2r^2}f(r)+\frac{e^2}{4\pi\varepsilon_0 Er}f(r)=f(r)$$

Change variables, let $\rho = kr$:

$$\frac{d^2f(\rho)}{d\rho^2}-\frac{l(l+1)}{\rho^2}f(\rho)+\frac{\rho_0}{\rho}f(\rho)=f(\rho)$$

where $\frac{\rho_0}{\rho}=\frac{e^2}{4\pi\varepsilon_0 Er}$

Let $f(\rho)=\rho^{(l+1)}e^{-\rho}\,v(\rho)$:

$$\therefore \rho\frac{d^2v(\rho)}{d\rho^2}+2(l+1-\rho)\frac{dv(\rho)}{d\rho}+(\rho_0-2(l+1))v(\rho)=0$$

This has the form of an associated Laguerre equation.
Try a power series solution: $v(\rho)=\sum_{j=0}^{\infty}c_j\rho^j$

$$\frac{dv}{d\rho}=\sum jc_j\rho^{(j-1)}$$

and $$\frac{d^2v}{d\rho^2}=\sum j(j-1)c_j\rho^{(j-2)}$$

$$\therefore \sum j(j-1)c_j\rho^{(j-1)}+2(l+1)\sum jc_j\rho^{(j-1)}$$
$$-2\sum jc_j\rho^j+(\rho_0-2(l+1))\sum_{j=0}^{\infty}c_j\rho^j=0$$

For the first two terms, we can use the dummy index $j\to j+1$ (this does not change the summation):

$$\therefore \sum (j+1)jc_{(j+1)}\rho^j+2(l+1)\sum (j+1)c_{(j+1)}\rho^j$$
$$-2\sum jc_j\rho^j+(\rho_0-2(l+1))\sum c_j\rho^j=0$$

Collecting all coefficients of the power term, ρ^j, we see that the coefficients obey a recurrence relation:

$$c_{(j+1)}=\left\{\frac{2(j+l+1)-\rho_0}{(j+1)(j+2l+2)}\right\}c_j$$

- Starting point: let $R(r)=f(r)/r$ in the radial Schrödinger equation (Eq. 20.1). The energy of a bound state in hydrogen is negative; therefore, we use $-E$ on the right-hand side.

Let $E=\frac{\hbar^2k^2}{2m_e}\to k^2=\frac{2m_eE}{\hbar^2}$

- Change variables: let $\rho = kr$, where k is the wavenumber. Define the dimensionless variable ρ_0 by:

$$\rho_0=\frac{e^2\rho}{4\pi\varepsilon_0 Er}=\frac{m_e e^2}{2\pi\varepsilon_0\hbar^2k}$$

- We can "peel off" the asymptotic behavior of $f(\rho)$ at $\rho\to\infty$ and $\rho\to 0$ by letting:

$$f(\rho)=\rho^{(l+1)}e^{-\rho}v(\rho)$$

where $f(\rho\to 0)\to\rho^{(l+1)}$
and $f(\rho\to\infty)\to e^{-\rho}$

(refer to Derivation 20.1).

- Doing this, the differential equation for $v(\rho)$ takes the form of a special type of ordinary differential equation, known as the *associated Laguerre equation*. The solutions to this equation can be expressed as a power series in ρ:

$$v(\rho)=\sum_{j=0}^{\infty}c_j\rho^j$$

where the coefficient of each term is related to the previous one via a recurrence relation. The expanded polynomials are known as *associated Laguerre polynomials*.

where we used $E = -\hbar^2 k^2/2m_e$. The constant α is a dimensionless quantity called the *fine-structure constant*, and its significance is described in more detail below.

The Bohr formula

Rearranging, the quantized energy levels of hydrogen are therefore defined by:

$$\therefore E_n = -\frac{1}{n^2}\left(\frac{1}{2}m_e(\alpha c)^2\right)$$
$$\text{where } n = 1,2,3,\ldots$$

(Eq. 20.9). This equation is called the *Bohr formula*, and its predictions agree accurately with experiment. The energy of the ground state of hydrogen is then:

$$E_1 = -\frac{1}{2}m_e(\alpha c)^2 \simeq -13.6 \text{ eV}$$

and the higher energy levels are related to the ground-state energy by:

$$E_n = \frac{E_1}{n^2}$$
$$\therefore E_2 \simeq -\frac{13.6}{2^2} = -3.4 \text{ eV}$$
$$E_3 \simeq -\frac{13.6}{3^2} = -1.5 \text{ eV}$$

(Eq. 20.10).

Bohr formula (energy levels of hydrogen):

$$E_n = -\frac{1}{n^2}\left(\frac{1}{2}m_e(\alpha c)^2\right) \quad (20.9)$$
$$\equiv \frac{E_1}{n^2}, \quad n = 1,2,3,\ldots \quad (20.10)$$

where the hydrogen ground-state energy is:

$$E_1 = -\frac{1}{2}m_e(\alpha c)^2 \simeq -13.6 \text{ eV}$$

Fine-structure constant:

$$\alpha = \frac{e^2}{4\pi\varepsilon_0 \hbar c} \simeq \frac{1}{137} \quad (20.11)$$

Degeneracy

For each value of n, there exist n states of different l, such that $n = j_{\max} + l + 1$. For example, for $n = 1$, we require $j_{\max} = 0$ and $l = 0$. However, for $n = 2$, either $j_{\max} = 0$ and $l = 1$, or $j_{\max} = 1$ and $l = 0$. Therefore, for each value of n, l can take any integer value ranging between 0 and $(n-1)$:

$$l = 0,1,2,\ldots,(n-1)$$

This means that each energy level n is *degenerate* in states of different total angular momentum, l. Furthermore, for each value of l, there are $(2l+1)$ states of different angular momentum projections m:

$$m = -l, -l+1, \ldots, l-1, l$$

(refer to Section 19.2.2). Therefore, the total angular momentum degeneracy of the nth energy level is:

$$\sum_{l=0}^{n-1}(2l+1) = n(n-1) + n = n^2$$

So, the number of different angular momentum states for the lowest three energy levels are:

$$n = 1 \rightarrow 1 \text{ spatial state}$$
$$n = 2 \rightarrow 4 \text{ spatial states}$$
$$n = 3 \rightarrow 9 \text{ spatial states}$$

These are illustrated in Fig. 20.2. Each spatial state is defined by three quantum numbers:

- n: principal quantum number; energy level
- l: total orbital angular momentum
- m: component of angular momentum

Electron shells

For each spatial state corresponding to a different $\{n, l, m\}$ combination, there exist two states of different *spin*: spin-up or spin-down. Therefore, each spatial state can accommodate two electrons, in different spin states. The *total* degeneracy (accounting for both orbital angular momentum and spin) of each energy level is then $2n^2$. This means that the nth energy level can accommodate a maximum of $2n^2$ electrons:

$$n = 1 \rightarrow 2 \text{ electrons}$$
$$n = 2 \rightarrow 8 \text{ electrons}$$
$$n = 3 \rightarrow 18 \text{ electrons}$$

The energy levels, n, are often referred to as "electron shells." A shell is "complete," or "full," when it has all its spatial and spin states occupied by an electron. For example, noble gases such as helium, neon, and argon have complete outer electron shells, making them relatively inert.

Hydrogen fine structure

The *fine-structure constant*, α, is defined as:

$$\alpha \stackrel{\text{def}}{=} \frac{e^2}{4\pi\varepsilon_0 \hbar c}$$

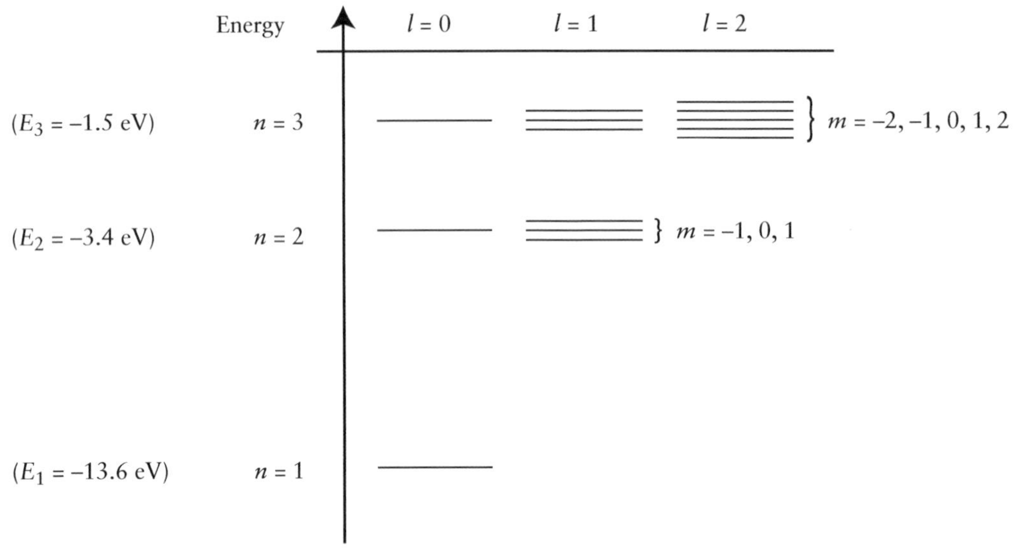

Energy $l = 0$ $l = 1$ $l = 2$

$(E_3 = -1.5 \text{ eV})$ $n = 3$ $m = -2, -1, 0, 1, 2$

$(E_2 = -3.4 \text{ eV})$ $n = 2$ $m = -1, 0, 1$

$(E_1 = -13.6 \text{ eV})$ $n = 1$

Figure 20.2: Lowest three energy levels of hydrogen. The nth energy level has n^2 states of different angular momentum, where l is the total angular momentum and m is the projection of angular momentum. The states of different m have been separated for clarity.

(Eq. 20.11). This dimensionless constant defines the order of magnitude for the spacing between the energy levels of hydrogen, which is on the order of electron volts:

$$E \sim m_e(\alpha c)^2 \sim \text{eV}$$

Energy levels defined by different values of n are referred to as hydrogen's *gross* structure. However, the gross structure is subject to relativistic corrections that are not accounted for by Schrödinger's equation; these corrections include relativistic kinetic energy and spin–orbit coupling (see Section 20.4 for more detail). The effect of these corrections is to shift the energy levels by a small amount. These shifts are referred to as the *fine structure* of hydrogen, and are found to be on the order of 10^{-4} electron volts:

$$\Delta E \sim \alpha^2 E \sim 10^{-4} \text{eV}$$

The spin-orbit corrections lift some of the angular momentum degeneracy of hydrogen's gross structure, giving states of different total angular momentum (orbital + spin) different energies.

Emission spectrum

Electrons can transition between states of different n by either absorbing or emitting energy (radiation).[2] The *emission spectrum* is the spectrum of radiation emitted by a sample of excited hydrogen atoms as their electrons fall from higher to lower energy levels. Since these levels are discrete, emission spectra form a set of discrete lines. The wavelengths of these lines correspond to photons whose energy equals the difference between the two energy levels of the transition. Using the Bohr formula for hydrogen energy levels (Eq. 20.10), the photon energy corresponds to:

$$\begin{aligned} E_\gamma &= E_i - E_f \\ &= E_1\left(\frac{1}{n_i^2} - \frac{1}{n_f^2}\right) \end{aligned}$$

where E_i is the initial electron energy corresponding to energy level n_i, and E_f is the final electron energy corresponding to energy level n_f (Eq. 20.12). Using $E_\gamma = hc/\lambda$, the wavelength of the emitted light is then given by the *Rydberg formula:*

$$\frac{1}{\lambda} = \frac{E_1}{hc}\left(\frac{1}{n_i^2} - \frac{1}{n_f^2}\right)$$

(Eq. 20.13).

Hydrogen emission spectrum:

$$\begin{aligned} E_\gamma &= E_i - E_f \\ &= E_1\left(\frac{1}{n_i^2} - \frac{1}{n_f^2}\right) \end{aligned} \quad (20.12)$$

where $E_1 = -13.6$ eV.

2 Transitions only occur between states with different l values because the photon carries away angular momentum. Therefore, the electron's l value must change to conserve angular momentum.

Rydberg formula:

$$\frac{1}{\lambda} = R_{\text{H}}\left(\frac{1}{n_{\text{i}}^2} - \frac{1}{n_{\text{f}}^2}\right) \tag{20.13}$$

where the Rydberg constant is:

$$R_{\text{H}} = \frac{E_1}{hc} \simeq 10^7 \text{ m}^{-1}$$

20.3 Hydrogen ground state

Combining the radial solutions, $R_{nl}(r)$, with the spherical harmonics, $Y_l^m(\theta, \phi)$ (refer to Section 19.2.3), we find that spatial wavefunctions are defined by three quantum numbers: n (energy level), l (total orbital angular momentum), and m (component of angular momentum):

$$\psi_{nlm}(r,\theta,\phi) = R_{nl}(r)Y_l^m(\theta,\phi)$$

where:

$$\begin{aligned} n &= 1,2,3,\ldots \\ l &= 0,1,2,\ldots,(n-1) \quad \text{for each } n \\ m &= -l,-(l+1),\ldots,(l-1),l \quad \text{for each } l \end{aligned}$$

The normalization condition requires that:

$$\int_{\text{all space}} |\psi_{nlm}|^2\, r^2 \sin\theta \mathrm{d}r\,\mathrm{d}\theta\mathrm{d}\phi = 1$$

The angular and radial solutions are therefore normalized individually:

$$\therefore \int_0^{2\pi}\int_0^{\pi} \left|Y_l^m(\theta,\phi)\right|^2 \sin\theta \mathrm{d}\theta \mathrm{d}\phi = 1$$

$$\text{and} \quad \int_0^{\infty} |R_{nl}(r)|^2\, r^2 \mathrm{d}r = 1$$

Ground-state wavefunction

The lowest (normalized) spherical harmonic and (not normalized) radial wavefunctions are:

$$Y_0^0(\theta,\phi) = \frac{1}{\sqrt{4\pi}}$$

$$\text{and} \quad R_{10}(r) = A\mathrm{e}^{-kr}$$

(refer to Eq. 19.16 for $Y_0^0(\theta,\phi)$ and Eq. 20.7 for $R_{10}(r)$). Using the standard integral:

$$\int_0^{\infty} x^2 \mathrm{e}^{-kx} \mathrm{d}x = \frac{2}{k^3}$$

the normalized radial wavefunction is therefore:

$$R_{10}(r) = 2\sqrt{k^3}\mathrm{e}^{-kr}$$

Defining the *Bohr radius*, a_0 [m], as $a_0 = 1/k$, the normalized ground state of hydrogen is then:

$$\begin{aligned} \psi_{100}(r,\theta,\phi) &= \frac{2}{\sqrt{4\pi a_0^3}}\mathrm{e}^{-r/a_0} \\ &= \frac{1}{\sqrt{\pi a_0^3}}\mathrm{e}^{-r/a_0} \end{aligned}$$

(Eq. 20.14).

Ground state of hydrogen:

$$\psi_{100}(r,\theta,\phi) = \frac{1}{\sqrt{\pi a_0^3}}\mathrm{e}^{-r/a_0} \tag{20.14}$$

where the Bohr radius is:

$$\begin{aligned} a_0 &= \frac{4\pi\varepsilon_0\hbar^2}{e^2 m_{\text{e}}} \\ &= \frac{\hbar}{m_{\text{e}}\alpha c} \\ &\simeq 0.53 \text{ Å} \end{aligned} \tag{20.15}$$

(1 Å = 10^{-10} m).

The Bohr radius

The probability of finding the electron between a radius r and $r + dr$ is:

$$P(r)\mathrm{d}r = |R(r)|^2\, r^2 \mathrm{d}r$$

It's straightforward to show that the ground-state radial probability density function has a maximum when $r = a_0$:

$$\text{i.e.} \quad \frac{\mathrm{d}}{\mathrm{d}r}\left(|R_{10}(r)|^2 r^2\right) = 0 \quad \text{when } r = a_0$$

(Fig. 20.3). Therefore, the Bohr radius corresponds to the radius at which an electron in the ground state is most likely to be located. a_0 is a constant, whose value is $\simeq$0.53 Å:

$$\begin{aligned} a_0 &= \frac{1}{k} \\ &= \frac{\hbar}{\sqrt{2m_{\text{e}}E_1}} \\ &= \frac{\hbar}{m_{\text{e}}\alpha c} \\ &= \frac{4\pi\varepsilon_0\hbar^2}{e^2 m_{\text{e}}} \\ &\simeq 0.53 \text{ Å} \end{aligned}$$

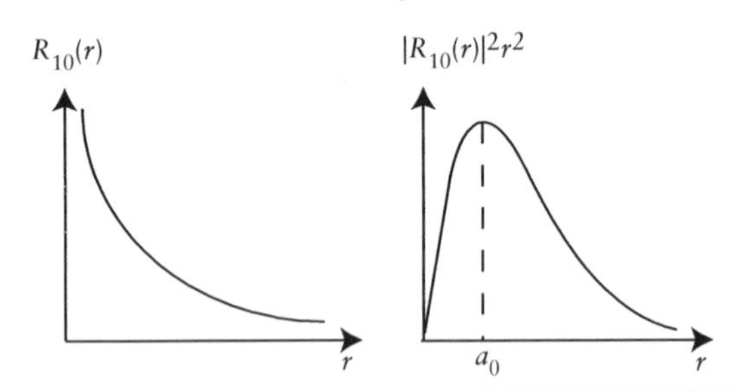

Figure 20.3: Radial ground-state wavefunction of hydrogen, $R_{10}(r)$, and probability density, $|R_{10}(r)|^2r^2$. The maximum of the probability density function is at $r = a_0$.

(Eq. 20.15). Thus, an electron orbiting the nucleus with a speed αc at the Bohr radius has an angular momentum of $\hbar$, which corresponds to one "unit" of angular momentum:

$$L = m_e rv = m_e a_0(\alpha c) = \hbar$$

Mean radii

The expectation value of the electron being at a radius r^p, where p is any power, is found to be proportional to a_0. $\langle r^p \rangle$ is related to $\langle r^{(p-1)} \rangle$ by:

$$\left\langle r^p \right\rangle = (p+2)\frac{a_0}{2}\left\langle r^{(p-1)} \right\rangle \qquad (20.16)$$

(Derivation 20.3). Using $\langle r^0 \rangle = 1$, we find that:

$$\langle r \rangle = \frac{3}{2}a_0$$

and $$\left\langle \frac{1}{r} \right\rangle = \frac{1}{a_0}$$

The same results are found if we determine the expectation values using the following integrals:

$$\langle r \rangle = \int_0^\infty r\left|R_{10}\right|^2 r^2 \mathrm{d}r$$

and $$\left\langle \frac{1}{r} \right\rangle = \int_0^\infty \frac{1}{r}\left|R_{10}\right|^2 r^2 \mathrm{d}r$$

20.4 Introduction to hydrogen fine structure

The gross structure of hydrogen is subject to relativistic corrections that are not accounted for by the radial Schrödinger equation.[3] Two such corrections are relativistic kinetic energy and spin–orbit coupling. The effect of these corrections is to shift the energy levels by a small amount, and to lift some of the angular momentum degeneracy of hydrogen's gross structure. These shifts in the energy levels are referred to as the *fine structure* of hydrogen, and they are apparent in the emission spectrum of hydrogen.

20.4.1 Time-independent perturbation theory

Perturbation theory is a technique used to find low-order corrections to the zeroth-order energies and wavefunctions of a given system. For a time-independent Hamiltonian,[4] $\hat{H}$, perturbation theory treats $\hat{H}$ as the sum of the zeroth-order Hamiltonian of the system

[3] The *Dirac equation* describes relativistic quantum effects.
[4] Time-dependent perturbation theory is beyond the scope of this book.

Derivation 20.3: Derivation of expectation value of r^p (Eq. 20.16)

$$\left\langle r^p \right\rangle = \langle R | r^p | R \rangle$$
$$= \int_0^\infty R^*(r) r^p R(r) r^2 \mathrm{d}r$$
$$= \int_0^\infty A^2 r^{p+2} e^{-2r/a_0} \mathrm{d}r$$
$$= (p+2)\frac{a_0}{2}\int_0^\infty A^2 r^{p+1} e^{-2r/a_0} \mathrm{d}r$$
$$= (p+2)\frac{a_0}{2}\int_0^\infty R^* r^{p-1} R r^2 \mathrm{d}r$$
$$= (p+2)\frac{a_0}{2}\left\langle r^{p-1} \right\rangle$$

- Starting point: the expectation value of r^p is defined as:

$$\left\langle r^p \right\rangle = \int_0^\infty R^*(r) r^p R(r) r^2 \mathrm{d}r$$

where $R(r)$ is the radial wavefunction.
- The radial ground-state wavefunction for hydrogen is $R(r) = Ae^{-r/a_0}$, where a_0 is the Bohr radius.
- Use $R(r)$ in the expression for $\langle r^p \rangle$, and integrate by parts by letting:

$$u(r) = r^{p+2} \rightarrow u'(r) = (p+2)r^{p+1}$$

and $$v'(r) = e^{-2r/a_0} \rightarrow v(r) = -\frac{a_0}{2}e^{-2r/a_0}$$

(for example, the Hamiltonian that describes the gross structure of hydrogen), $\hat{H}_0$, plus a small perturbation, $\delta\hat{H}$:

$$\hat{H} = \hat{H}_0 + \delta\hat{H}$$

In a similar way, it treats the wavefunctions and energy levels as the sum of their zeroth-order values plus "corrections" due to the perturbation in the Hamiltonian:

$$\psi_n = \psi_n^0 + \delta\psi_n$$
$$\text{and } E_n = E_n^0 + \delta E_n$$

For our purposes, we are only interested in finding *first*-order corrections to the zeroth-order solutions; however, this technique can also be used to find second- and higher order corrections too. Using the above definitions, we can express the Schrödinger equation in terms of the zeroth-order terms plus perturbation terms, as shown in Derivation 20.4. Doing this, we find that the first-order energy correction is equal to the expectation value of the perturbation with the unperturbed wavefunctions:[5]

$$\delta E_n = \left\langle \psi_n^0 \middle| \delta\hat{H} \middle| \psi_n^0 \right\rangle$$

(Eq. 20.17 and Derivation 20.4).

First-order energy correction (Derivation 20.4):

$$\delta E_n = \left\langle \psi_n^0 \middle| \delta\hat{H} \middle| \psi_n^0 \right\rangle \quad (20.17)$$

20.4.2 Relativistic kinetic energy

One application of time-independent perturbation theory is to determine the effect of the relativistic kinetic energy of hydrogen's electron on its energy levels. Expressing the kinetic energy as:

$$T = E - mc^2$$

where the electron's total energy is $E = \sqrt{\left(mc^2\right)^2 + (pc)^2}$, and its rest-mass energy is mc^2, we find that T can be expanded and approximated to the sum of a zeroth-order kinetic energy term:

$$\hat{T}_0 = \frac{\hat{p}^2}{2m}$$

plus a perturbation term:

$$\delta\hat{T} = \delta\hat{H}^{KE} = -\frac{1}{2mc^2}\left(\hat{H}_0 - V(r)\right)^2$$

(Eq. 20.18 and Derivation 20.5). The shift in energy levels due to the perturbing Hamiltonian is then found to be:

$$\delta E_n^{KE} = \left\langle \psi_n^0 \middle| \delta\hat{H}^{KE} \middle| \psi_n^0 \right\rangle = \alpha^4 \frac{mc^2}{8n^4}\left(3 - \frac{4n}{l+1/2}\right)$$

(Eq. 20.19 and Derivation 20.6). There are a few points of interest to note. First, the order of magnitude of the relativistic kinetic energy correction is:

$$\delta E_n^{KE} \sim \alpha^4 mc^2 \sim 10^{-4} \text{ eV}$$

Second, we see that for each energy level, n, each different angular momentum state, l, is shifted by a small amount. Thus, the degeneracy in l is not lifted by relativistic kinetic energy effects (we shall see below that it is lifted by spin–orbit interactions). Using $n = 1$ and $l = 0$, we see that the ground state of hydrogen is shifted by:

$$\delta E_1^{KE} = -\frac{5}{8}\alpha^4 mc^2 \text{ eV}$$

Relativistic kinetic energy correction (Derivations 20.5 and 20.6):

$$\delta\hat{H}^{KE} = -\frac{1}{2mc^2}\left(\hat{H}_0 - V(r)\right)^2 \quad (20.18)$$

$$\text{where } \hat{H}_0 = \hat{T}_0 + V(r)$$

$$\delta E_n^{KE} = \alpha^4 \frac{mc^2}{8n^4}\left(3 - \frac{4n}{l+1/2}\right) \quad (20.19)$$

20.4.3 Spin–orbit interaction

Spin–orbit coupling is a relativistic effect by which a particle's spin interacts with its orbital angular moment. The effect of this interaction is to lift the degeneracy of states of different total angular momentum, j. However, even after taking spin–orbit coupling into account, each state of different j is degenerate in its projection, m_j, since for each state of different j, there are $2j+1$ states of different m_j. For example, a state where $j = 3/2$ has four possible angular momentum projections, $m_j = \{-3/2, -1/2, 1/2, 3/2\}$. This degeneracy can be lifted by an external magnetic field, a phenomenon known as the *Zeeman effect*.

[5] This approach works only for wavefunctions that are non-degenerate in n.

Derivation 20.4: Derivation of the first-order energy correction using non-degenerate time-independent perturbation theory (Eq. 20.17)

$$\hat{H}|\psi_n\rangle = E_n|\psi_n\rangle$$

$$\therefore \langle\psi_n^0|\hat{H}|\psi_n\rangle = \langle\psi_n^0|E_n|\psi_n\rangle$$

Left-hand side (LHS):

$$\begin{aligned}\langle\psi_n^0|\hat{H}|\psi_n\rangle &= \langle\psi_n^0|\hat{H}_0+\delta\hat{H}|\psi_n^0+\delta\psi_n\rangle \\ &= \underbrace{\langle\psi_n^0|\hat{H}_0|\psi_n^0\rangle}_{=E_n^0} + \langle\psi_n^0|\delta\hat{H}|\psi_n^0\rangle \\ &\quad + \underbrace{\langle\psi_n^0|\hat{H}_0|\delta\psi_n\rangle}_{=E_n^0\langle\psi_n^0|\delta\psi_n\rangle} + \underbrace{\langle\psi_n^0|\delta\hat{H}|\delta\psi_n\rangle}_{\text{second-order correction}} \\ &\simeq E_n^0 + \langle\psi_n^0|\delta\hat{H}|\psi_n^0\rangle + E_n^0\langle\psi_n^0|\delta\psi_n\rangle\end{aligned}$$

Right-hand side (RHS):

$$\begin{aligned}\langle\psi_n^0|E_n|\psi_n\rangle &= \langle\psi_n^0|E_n^0+\delta E_n|\psi_n^0+\delta\psi_n\rangle \\ &= E_n^0 + \delta E_n + E_n^0\langle\psi_n^0|\delta\psi_n\rangle \\ &\quad + \underbrace{\delta E_n\langle\psi_n^0|\delta\psi_n\rangle}_{\text{second-order correction}} \\ &\simeq E_n^0 + \delta E_n + E_n^0\langle\psi_n^0|\delta\psi_n\rangle\end{aligned}$$

Equating the LHS to the RHS, we see:

$$\delta E_n = \langle\psi_n^0|\delta\hat{H}|\psi_n^0\rangle$$

- Starting point: express the Hamiltonian of a system as the sum of the zeroth-order Hamiltonian, H_0, plus a perturbation, $\delta\hat{H}$:

$$\hat{H} = \hat{H}_0 + \delta\hat{H}$$

Express the wavefunctions and energy levels in a similar way:

$$\psi_n = \psi_n^0 + \delta\psi_n$$
$$E_n = E_n^0 + \delta E_n$$

where the zeroth-order Schrödinger equation is:

$$\hat{H}_0|\psi_n^0\rangle = E_n^0|\psi_n^0\rangle$$

- The Schrödinger equation of the whole system is then:

$$\hat{H}|\psi_n\rangle = E_n|\psi_n\rangle$$

Taking the inner product of this equation with ψ_n^0, we have:

$$\langle\psi_n^0|\hat{H}|\psi_n\rangle = \langle\psi_n^0|E_n|\psi_n\rangle$$

Substituting in for $\hat{H}$, E_n, and Ψ_n, expanding, and ignoring second-order corrections (ones that involve the product of two perturbation terms), we see that the first-order energy correction is equal to the expectation value of the perturbation, $\delta\hat{H}$, in the unperturbed case:

$$\delta E_n = \langle\psi_n^0|\delta\hat{H}|\psi_n^0\rangle$$

Derivation of the perturbing Hamiltonian

In the electron rest frame, the orbiting nucleus sets up a magnetic field that exerts a torque on the electron's intrinsic magnetic dipole moment (its spin). This torque tends to align the dipole moment with the field. Classically, the interaction energy of a magnetic dipole moment, $\vec{\mu}$, in a magnetic field $\vec{B}$ is:

$$U = -\vec{\mu}\cdot\vec{B}$$

However, in the case of the hydrogen atom, the electron rest frame is not an inertial frame due to its orbit (acceleration) about the nucleus. This introduces a precession called the *Thomas precession* whose effect is to halve the classical interaction energy. Therefore, the Hamiltonian of the interaction energy is:

$$\hat{H} = -\frac{1}{2}\hat{\vec{\mu}}\cdot\hat{\vec{B}}$$

From the Biot–Savart Law, the magnetic field produced by a nucleus of velocity $\vec{v}_n$ is:

$$\vec{B} = \frac{\mu_0}{4\pi}\frac{q\vec{v}_n\times\hat{r}}{r^2}$$

where $\hat{r}$ is a unit vector along r.

In terms of electron velocity, $\vec{v}_e = -\vec{v}_n$, we have:

$$\vec{B} = \frac{\mu_0 e}{4\pi r^2}(-\vec{v}_e\times\hat{r})$$

Derivation 20.5: Derivation of the relativistic kinetic energy perturbation (Eq. 20.18)

- Starting point: the kinetic energy of a body is equal to its total energy, E, minus its rest-mass energy, mc^2:

$$T = E - mc^2$$

where the total energy of a body is related to its rest-mass energy and momentum by:

$$E^2 = \left(mc^2\right)^2 + (pc)^2$$

(refer to Sections 6.4 and 6.5).

- The relativistic kinetic energy of the electron is therefore:

$$T = mc^2\left(\sqrt{1+\left(\frac{p}{mc}\right)^2} - 1\right)$$

where $p \ll mc$.

- Performing a binomial expansion on the above expression, we find the relativistic kinetic energy operator can be approximated to a zeroth-order kinetic energy term:

$$\hat{T}_0 = \frac{\hat{p}^2}{2m} = \hat{H}_0 - V(r)$$

plus a first-order perturbation term:

$$\delta\hat{T} = -\frac{1}{2mc^2}\left(\frac{\hat{p}^2}{2m}\right)^2$$

$$\begin{aligned}
T &= E - mc^2 \\
&= \sqrt{\left(mc^2\right)^2 + (pc)^2} - mc^2 \\
&= mc^2\left(\sqrt{1+\left(\frac{p}{mc}\right)^2} - 1\right) \\
&\simeq mc^2\left(1 + \frac{1}{2}\left(\frac{p}{mc}\right)^2 - \frac{1}{8}\left(\frac{p}{mc}\right)^4 + \ldots - 1\right) \\
&= \underbrace{\frac{p^2}{2m}}_{\text{zeroth-order term}} \underbrace{- \frac{1}{2mc^2}\left(\frac{p^2}{2m}\right)^2}_{\text{first-order correction}} \\
&= T_0 - \frac{1}{2mc^2}(T_0)^2
\end{aligned}$$

Therefore, the relativistic kinetic energy operator is:

$$\begin{aligned}
\therefore \hat{T} &= \hat{T}_0 - \frac{1}{2mc^2}\left(\hat{T}_0\right)^2 \\
&= \hat{T}_0 \underbrace{- \frac{1}{2mc^2}\left(\hat{H}_0 - V(r)\right)^2}_{=\delta\hat{H}^{KE}}
\end{aligned}$$

$$\therefore \delta\hat{H}^{KE} = -\frac{1}{2mc^2}\left(\hat{H}_0 - V(r)\right)^2$$

The orbital angular momentum of the electron is $\vec{L} = \vec{r} \times m\vec{v}_e = mr\hat{r} \times \vec{v}_e$. Using this, together with $\mu_0 = 1/\varepsilon_0 c^2$, we can express the magnetic field in terms of the electron's orbital angular momentum:

$$\begin{aligned}
\therefore \vec{B} &= \frac{\mu_0 e}{4\pi m r^3}\vec{L} \\
&= \frac{e}{4\pi\varepsilon_0 mc^2 r^3}\vec{L}
\end{aligned}$$

Therefore, the operator for the magnetic field produced by the hydrogen nucleus is:

$$\therefore \hat{\vec{B}} = \frac{e}{4\pi\varepsilon_0 mc^2 r^3}\hat{\vec{L}}$$

Both orbital angular momentum and spin angular momentum give rise to a magnetic dipole moment, $\vec{\mu}_L$ and $\vec{\mu}_S$ respectively. We saw previously (Section 19.3.1) that the orbital magnetic dipole moment is related to orbital angular momentum by:

$$\hat{\vec{\mu}}_L = -\frac{e}{2m}\hat{\vec{L}}$$

The spin magnetic dipole moment is related to spin in a similar way, but is a factor of two greater than the orbital magnetic dipole moment:

$$\hat{\vec{\mu}}_S = -\frac{e}{m}\hat{\vec{S}}$$

This correction is due to a scale factor called the *gyromagnetic ratio*, g, which is $g_s = 2$ for spin and $g_l = 1$ for orbital angular momentum. Using the expressions for $\hat{\vec{\mu}}_S$ and $\hat{\vec{B}}$, together with the Hamiltonian for the

Derivation 20.6: Derivation of energy shifts due to relativistic kinetic energy perturbation (Eq. 20.19)

$$\delta E_n^{\mathrm{KE}} = \left\langle \psi_n^0 \middle| \delta \hat{H}^{\mathrm{KE}} \middle| \psi_n^0 \right\rangle$$
$$= -\frac{1}{2mc^2}\left\langle \psi_n^0 \middle| \left(\hat{H}_0 - V(r)\right)^2 \middle| \psi_n^0 \right\rangle$$
$$= -\frac{1}{2mc^2}\left\langle \psi_n^0 \middle| \left(\hat{H}_0^2 - 2\hat{H}_0 V(r) + (V(r))^2\right) \middle| \psi_n^0 \right\rangle$$
$$= -\frac{1}{2mc^2}\left(\left(E_n^0\right)^2 + 2E_n^0\left(\frac{e^2}{4\pi\varepsilon_0}\right)\left\langle\frac{1}{r}\right\rangle + \left(\frac{e^2}{4\pi\varepsilon_0}\right)^2\left\langle\frac{1}{r^2}\right\rangle\right)$$
$$= \alpha^4\frac{mc^2}{8n^4}\left(3 - \frac{4n}{l+1/2}\right)$$

- Starting point: the first-order energy correction of the relativistic kinetic energy perturbation is:

$$\delta E_n^{KE} = \left\langle \psi_n^0 \middle| \delta \hat{H}^{KE} \middle| \psi_n^0 \right\rangle$$

where:

$$\delta \hat{H}^{KE} = -\frac{1}{2mc^2}\left(\hat{H}_0 - V(r)\right)^2$$

ψ_n^0 correspond to the unperturbed hydrogen wavefunctions.

- Expand $(\hat{H}_0 - V(r))^2$ and use:

$$\left\langle \psi_n^0 \middle| \hat{H}_0 \middle| \psi_n^0 \right\rangle = E_n^0$$

and $\left\langle \psi_n^0 \middle| V(r) \middle| \psi_n^0 \right\rangle = -\frac{e^2}{4\pi\varepsilon_0}\left\langle\frac{1}{r}\right\rangle$

where $E_n^0 = -\frac{1}{2n^2}m(\alpha c)^2$

- We state without proof that the expectation values of $1/r$ and $1/r^2$ for a hydrogen state defined by quantum numbers n (energy) and l (orbital angular momentum) are:

$$\left\langle\frac{1}{r}\right\rangle = \frac{1}{n^2 a_0}$$

and $\left\langle\frac{1}{r^2}\right\rangle = \frac{1}{(l+1/2)n^3 a_0^2}$

where the Bohr radius, a_0, and fine-structure constant, α, are:

$$a_0 = \frac{\hbar}{m\alpha c}$$

and $\alpha = \frac{e^2}{4\pi\varepsilon_0 \hbar c}$

interaction energy between a magnetic dipole moment and magnetic field, we find that the spin–orbit (SO) Hamiltonian is:

$$\delta \hat{H}^{SO} = -\frac{1}{2}\hat{\mu}_S \cdot \hat{B}$$
$$= \frac{e^2}{8\pi\varepsilon_0 m^2 c^2 r^3}\hat{L}\cdot\hat{S}$$
$$= \left(\frac{e^2}{4\pi\varepsilon_0 r}\right)\left(\frac{1}{2(mcr)^2}\right)\hat{L}\cdot\hat{S}$$

(Eq. 20.20).

Energy-level splitting from spin–orbit interactions

The shift in energy levels due to the perturbing Hamiltonian is then found to be:

$$\delta E_n^{SO} = \left\langle \psi_n^0 \middle| \delta \hat{H}^{SO} \middle| \psi_n^0 \right\rangle$$
$$= \alpha^4\frac{mc^2}{4n^3}\left(\frac{j(j+1) - l(l+1) - s(s+1)}{l(l+1/2)(l+1)}\right)$$

(Eq. 20.21 and Derivation 20.7). As we saw for the relativistic kinetic energy correction, the order of magnitude of the spin–orbit effect is:

$$\delta E_n^{SO} \sim \alpha^4 mc^2 \sim 10^{-4}\mathrm{eV}$$

Derivation 20.7: Derivation of energy shifts due to spin–orbit coupling (Eq. 20.21)

$$\delta E_n^{SO} = \left\langle \psi_n^0 \middle| \delta \hat{H}^{SO} \middle| \psi_n^0 \right\rangle$$
$$= \left(\frac{e^2}{8\pi\varepsilon_0 (mc)^2}\right)\left\langle \frac{1}{r^3}\right\rangle \left\langle \psi_n^0 \middle| \hat{L}\cdot\hat{S} \middle| \psi_n^0 \right\rangle$$
$$= \left(\frac{e^2}{8\pi\varepsilon_0 (mc)^2}\right)\left\langle \frac{1}{r^3}\right\rangle \cdot \left(\frac{\hbar^2}{2}(j(j+1) - l(l+1) - s(s+1))\right)$$
$$= \alpha^4 \frac{mc^2}{4n^3}\left(\frac{j(j+1) - l(l+1) - s(s+1)}{l(l+1/2)(l+1)}\right)$$

- Starting point: the first-order energy correction due to spin–orbit coupling is:

$$\delta E_n^{SO} = \left\langle \psi_n^0 \middle| \delta \hat{H}^{SO} \middle| \psi_n^0 \right\rangle$$

where:

$$\delta \hat{H}^{SO} = \left(\frac{e^2}{4\pi\varepsilon_0 r}\right)\left(\frac{1}{2(mcr)^2}\right)\hat{L}\cdot\hat{S}$$

(see main text for derivation). Ψ_n^0 correspond to the unperturbed hydrogen wavefunctions.

- The total angular momentum is the sum of orbital plus spin angular momenta:

$$\hat{J} = \hat{L} + \hat{S}$$

Squaring and rearranging, we find an expression for $\hat{L}\cdot\hat{S}$ in terms of the total angular momentum operators:

$$\therefore \hat{J}^2 = \hat{L}^2 + \hat{S}^2 + 2\hat{L}\cdot\hat{S}$$
$$\therefore \hat{L}\cdot\hat{S} = \frac{1}{2}\left(\hat{J}^2 - \hat{L}^2 - \hat{S}^2\right)$$

The expectation value of this operator is therefore:

$$\left\langle \psi_n^0 \middle| \hat{L}\cdot\hat{S} \middle| \psi_n^0 \right\rangle = \frac{1}{2}\left\langle \psi_n^0 \middle| \left(\hat{J}^2 - \hat{L}^2 - \hat{S}^2\right) \middle| \psi_n^0 \right\rangle$$
$$= \frac{\hbar^2}{2}(j(j+1) - l(l+1) - s(s+1))$$

- We state without proof that the expectation value of $1/r^3$ for a hydrogen state defined by quantum numbers n (energy) and l (orbital angular momentum) is:

$$\left\langle \frac{1}{r^3}\right\rangle = \frac{1}{l(l+1/2)(l+1)n^3 a_0^3}$$

where the Bohr radius, a_0, and fine-structure constant, α, are:

$$a_0 = \frac{\hbar}{m\alpha c}$$

and $$\alpha = \frac{e^2}{4\pi\varepsilon_0 \hbar c}$$

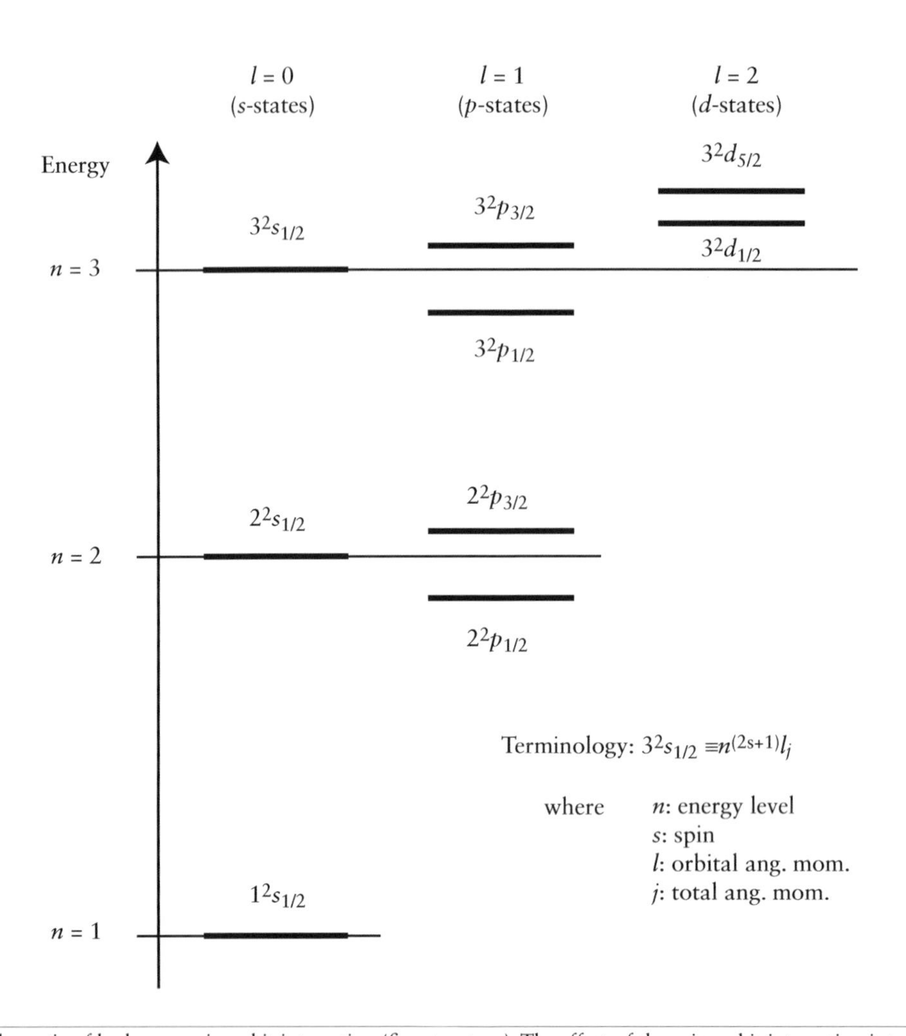

Figure 20.4: Schematic of hydrogen spin–orbit interaction (fine structure). The effect of the spin–orbit interaction is to split states of different total angular momentum, j. The remaining states are still degenerate in projections, m_j (there are $2j+1$ states of different m_j for each state j). The magnitude of the shifts are found using Eq. 20.21. The s-states ($l=0$) have no spin-orbit effects since they have zero orbital angular momentum.

Therefore, hydrogen fine structure is approximately 10 000 ($\alpha^2 \sim 10^{-4}$) times smaller than hydrogen's gross structure. Whereas the effect of relativistic kinetic energy is to *shift* energy levels of different n and l by a small amount, the effect of spin–orbit interaction is to *split* energy levels of different n, l, s, and j relative to the non-perturbed state. Therefore, spin–orbit coupling lifts some of hydrogen's angular momentum degeneracy into states of different total angular momentum. An energy-level diagram illustrating fine structure due to spin–orbit interactions is shown in Fig. 20.4.

Spin–orbit interaction (Derivation 20.7):

$$\delta\hat{H}^{SO} = \left(\frac{e^2}{4\pi\varepsilon_0 r}\right)\left(\frac{1}{2(mcr)^2}\right)\hat{L}\cdot\hat{S} \tag{20.20}$$

$$\delta E_n^{SO} = \alpha^4 \frac{mc^2}{4n^3}\left(\frac{j(j+1)-l(l+1)-s(s+1)}{l(l+1/2)(l+1)}\right) \tag{20.21}$$

Part V
Thermal Physics

Introduction

Thermal physics is the study of the thermal properties of systems involving many particles—solids, liquids, and gases—from a consideration of energy inputs and outputs of the system. There are two approaches we take. One approach is a *micro*scopic approach, in which the bulk properties of a system, such as pressure, temperature, or heat capacity, are described from a consideration of the average behavior of the individual particles that make up the system. The other approach is a *macro*scopic approach, in which the same properties are understood without the particle nature of matter being taken into account, but instead from a consideration of energy exchange between a system and its surroundings. The most straightforward thermodynamic system to study is an ideal gas, in which the constituent molecules are treated as point particles with minimal intermolecular interactions. The study of ideal gases using a microscopic approach is called *kinetic theory* (Chapter 21), and using a macroscopic approach is called *thermodynamics* (Chapter 22). Thermodynamics is not limited to describing ideal gases, but also applies to real gases, liquids, and solids, as well as to phase transitions from one state to another (Chapter 23). The corresponding *micro*scopic approach, in which the bulk properties of *any* material are understood from a consideration of the properties of individual atoms that make them up, is called *statistical mechanics* (Chapter 24). Whereas classical mechanics uses a deterministic approach to describe macroscopic properties of a system, and quantum mechanics uses a probabilistic approach to describe microscopic properties of a system, statistical mechanics uses a probabilistic approach to describe a system's macroscopic properties, thus providing a link between the classical and quantum worlds. Note: In previous sections we used the variable τ rather than V to denote volume, so as not to confuse volume with electric potential (denoted by V). In thermal physics, however, it is standard to use V for volume; therefore, in this section we represent volume by the variable V.

Practical applications

The subject of thermodynamics was originally developed during the industrial revolution out of interest for constructing engines that use heat to produce useful work, such as a steam engine that drives a train. Refrigerators and air conditioners also operate based on the principles of thermodynamics, whereby they use work in order to cool a system. Statistical mechanics underlies the properties of all materials, such as metals, semiconductors, magnets, and superconductors, the study of which is called *condensed matter physics* (or *solid state physics*), as well as the molecular properties of many biological systems.

My two cents

Out of the three sub-topics of thermal physics—kinetic theory, thermodynamics, and statistical mechanics—the latter is probably the most conceptually challenging since it concerns quantum phenomena. If you find this to be the case, I would highly recommend the book by Glazer and Wark (see below) since it provides an insightful description of probability theory and its application to tractable systems. While thermodynamics is perhaps less conceptually demanding in terms of the principles involved, the mathematical formalism (partial derivatives and exact differentials) is not trivial. In this case, it's always good to remind yourself to interpret the derivative as "how does one variable change with another, when a third variable is kept constant?".

Recommended books

- Adkins, C., 1983, *Equilibrium Thermodynamics*, 3rd edition. Cambridge University Press.
- Glazer, M. and Wark, J., 2001, *Statistical Mechanics: a Survival Guide*. Oxford University Press.

Both these books cover the material at a level appropriate for an introductory course. In particular, the book by Glazer and Wark is a short, accessible book to read, and I would highly recommend it to grasp the basic concepts.

Chapter 21
Kinetic theory of ideal gases

Kinetic theory describes the gross thermal properties of ideal gases from a consideration of the average behavior of the individual particles that make up the gas. For example, by knowing the probability distribution of particle energies or velocities in an ideal gas, kinetic theory is able to predict large-scale properties of the gas, such as its internal energy, heat capacity, pressure, and flux.

21.1 Ideal gases

Kinetic theory applies only to *ideal* gases, which constitute a simple model of real gases. In the ideal gas model, we treat gas particles as hard spheres that move in random directions and constantly collide with each other and with the walls of the container, resulting in a change in their direction. The assumptions of ideal gases and kinetic theory are:

1. Molecular size $\ll$ intermolecular separation (particle volumes are negligible).
2. Intermolecular forces are negligible.
3. All collisions are perfectly elastic, which means that momentum and energy are conserved between the colliding objects.
4. Energy exchange between particles is solely due to elastic collisions.
5. The molecular speed distribution is constant in time.

All gases behave like ideal gases at very low pressures and high temperatures, since under these conditions the relative volumes and intermolecular forces of the particles are negligible. Noble gases, such as helium, neon, and argon, behave like ideal gases even at low temperatures since they have complete outer electron shells, and therefore exhibit weak intermolecular forces.

Ideal Gas Law

A defining characteristic of an ideal gas is that it obeys the *Ideal Gas Law*. This law relates the macroscopic properties of pressure, p [Pa], temperature, T [K], and volume, V [m^3], for a gas with n moles [mol] (Eq. 21.1). A *mole* is a measure of an amount of substance. One mole corresponds to a fixed number of particles or molecules of any substance, which is $N_A = 6.022 \cdot 10^{23}$ particles. This number is referred to as *Avogadro's number*, denoted N_A [$6.022 \cdot 10^{23}$ mol^{-1}]. Therefore, the number of moles of a substance of N particles is:

$$n = \frac{N}{N_A} \text{ mol}$$

Ideal Gas Law:

$$pV = nRT \tag{21.1}$$
$$= Nk_BT \tag{21.2}$$

where the molar gas constant is:

$$R = N_A k_B$$
$$= \frac{N}{n} k_B \tag{21.3}$$

The Ideal Gas Law comes from an amalgamation of three other laws, which were discovered empirically:

1. Boyle's Law: $V \propto 1/p$ at constant T and n.
2. Charles's Law: $V \propto T$ at constant p and n.
3. Avogadro's Law: $V \propto n$ at constant T and p.

Combining these, we have:

$$\frac{pV}{nT} = \text{constant} = R$$

The constant of proportionality is the *molar gas constant*, R [8.31 J $K^{-1}mol^{-1}$]. This is related to the *Boltzmann constant*, k_B [$1.38 \cdot 10^{-23}$ J K^{-1}] by:

$$R = N_A k_B$$

(Eq. 21.3). Therefore, by eliminating R from the Ideal Gas Law, we can express it in terms of the Boltzmann constant, k_B, for an ideal gas with N particles (Eq. 21.2). The Ideal Gas Law applies only to systems in equilibrium, meaning that their macroscopic properties do not change with time.

21.2 The Boltzmann distribution

A postulate of kinetic theory is that particles in an ideal gas obey the *Boltzmann distribution of energy*,[1] $f(\varepsilon)$ (Eq. 21.4). This is a probability distribution, where $f(\varepsilon)d\varepsilon$ gives the probability that a particle has an energy between ε and $\varepsilon + d\varepsilon$. The Boltzmann distribution is an exponentially decaying function of energy, and depends only on the temperature of the gas (Fig. 21.1). The exponential term in Eq. 21.4 is called the *Boltzmann factor*, and the coefficient, A, is the normalization constant. The value of A is determined from the requirement that the integral of the probability distribution over all values of energy is equal to one (Eqs. 21.5 and 21.6).

The Boltzmann distribution:

$$f(\varepsilon)d\varepsilon = A \underbrace{e^{-\varepsilon/k_B T}}_{\text{Boltzmann factor}} d\varepsilon \qquad (21.4)$$

where the normalization condition is:

$$\int_0^\infty f(\varepsilon)d\varepsilon = 1 \qquad (21.5)$$

Therefore, the normalization constant is:

$$A = \frac{1}{\int_0^\infty e^{-\varepsilon/k_B T} d\varepsilon} \qquad (21.6)$$

21.3 The Maxwell–Boltzmann distribution

The *Maxwell–Boltzmann* distribution is a small extension from the Boltzmann distribution; whereas the

[1] For a derivation of the Boltzmann distribution, see Section 24.4.

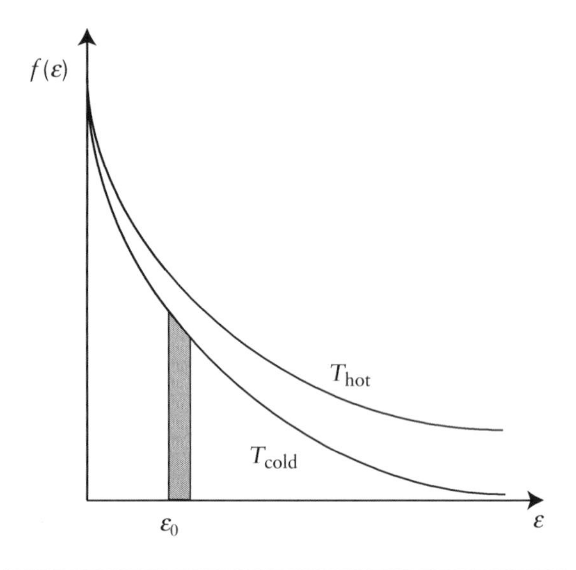

Figure 21.1: The Boltzmann distribution is an exponentially decaying function of energy. For hotter temperatures, the decay constant is greater, therefore, the exponential falls off less steeply (curves are scaled to have the same value at $\varepsilon = 0$, and are not normalized). The probability of a particle having an energy between ε_0 and $\varepsilon_0 + d\varepsilon$ at T_{cold} is given by the shaded area.

latter gives the probability that a particle has a given *energy*, the former gives the probability that a particle has a given *velocity*. For an ideal gas, in the absence of external forces such as gravity, the total energy of a particle equals its kinetic energy. This means that, for a particle of mass m travelling at velocity v, its energy is:

$$\varepsilon = \frac{1}{2}mv^2$$

Therefore, eliminating ε from Eq. 21.4, we can write the probability distribution in terms of particle velocity:

$$f_v(v) = Ae^{-mv^2/2k_B T}$$

In this case, the Boltzmann factor is a Gaussian function, whereas for the Boltzmann distribution of *energies*, it is an exponentially decaying function. The subscript in $f_v(v)$ is to indicate that the Maxwell–Boltzmann distribution does not have the same functional dependence on v as the Boltzmann distribution has on ε, where the Boltzmann distribution is denoted by $f(\varepsilon)$. It is important to be consistent with notation since there are several ways of expressing the Maxwell–Boltzmann distribution:

Three-dimensional velocity distribution: $f_v(\vec{v})d^3v$

One-dimensional velocity distribution: $f_v(v_x)dv_x$

Speed distribution: $F_v(v)dv$

The velocity distributions take into account the *direction* of the particles, and so are functions of the velocity *vector* ($\vec{v}$ in three dimensions or v_x in one dimension), whereas the speed distribution depends only on the *magnitude*, v. The speed distribution has a different functional dependence on v to the velocity distribution; therefore, we denote the speed distribution F_v rather than f_v. Below, we will look at each distribution in turn.

21.3.1 Velocity distributions

Let's first look at the one-dimensional velocity distribution, $f_v(v_x)\mathrm{d}v_x$. Using $\varepsilon = mv_x^2/2$ in Eq. 21.4, the one-dimensional velocity distribution is:

$$f_v(v_x)\mathrm{d}v_x = A\mathrm{e}^{-mv_x^2/2k_\mathrm{B}T}\mathrm{d}v_x$$

This gives the probability of a particle having a velocity in the x-direction between v_x and $v_x + \mathrm{d}v_x$. (Similar distributions exist for velocity in either the y- or the z-direction.) The normalization constant, A, is determined by integrating the distribution over all one-dimensional velocities ($-\infty$ to $+\infty$), and setting the integral equal to one, as shown in Derivation 21.1. This condition ensures that the total probability equals one. The normalized velocity distribution is a Gaussian distribution (Eq. 21.7 and Fig. 21.2).

It's straightforward to generalize to three dimensions by changing v_x to the total velocity, $\vec{v}$. The magnitude of the total velocity is:

$$v^2 = v_x^2 + v_y^2 + v_z^2$$

Velocity distributions (Derivations 21.1 and 21.2):

In one dimension:

$$f_v(v_x)\mathrm{d}v_x = \sqrt{\frac{m}{2\pi k_\mathrm{B}T}}\underbrace{\mathrm{e}^{-mv_x^2/2k_\mathrm{B}T}}_{\text{Boltzmann factor}}\mathrm{d}v_x \qquad (21.7)$$

In three dimensions:

$$f_v(\vec{v})\mathrm{d}^3v = \left(\frac{m}{2\pi k_\mathrm{B}T}\right)^{\frac{3}{2}}\underbrace{\mathrm{e}^{-mv^2/2k_\mathrm{B}T}}_{\text{Boltzmann factor}}\mathrm{d}^3v \qquad (21.8)$$

and the three-dimensional element of velocity-space (Fig. 21.2) is:

$$\mathrm{d}^3v = \mathrm{d}v_x\mathrm{d}v_y\mathrm{d}v_z$$

The expression for the normalized three-dimensional distribution, $f(\vec{v})\mathrm{d}^3v$, is also a Gaussian distribution (Eq. 21.8 and Derivation 21.2). Eq. 21.8 gives the probability that a particle's velocity lies in an element of velocity-space d^3v from velocity $\vec{v}$ (Fig. 21.3).

Average values

Both the one- and three-dimensional velocity distributions are Gaussian distributions centered about the origin. The standard deviation for a Gaussian of

Derivation 21.1: Derivation of the one-dimensional velocity distribution (Eq. 21.7)

$$\int_{-\infty}^{\infty} f_v(v_x)\mathrm{d}v_x = A\int_{-\infty}^{\infty}\mathrm{e}^{-mv_x^2/2k_\mathrm{B}T}\mathrm{d}v_x$$

$$= A\sqrt{\frac{2\pi k_\mathrm{B}T}{m}}$$

Therefore, from the normalization condition:

$$A = \sqrt{\frac{m}{2\pi k_\mathrm{B}T}}$$

$$\therefore f_v(v_x)\mathrm{d}v_x = \sqrt{\frac{m}{2\pi k_\mathrm{B}T}}\mathrm{e}^{-mv_x^2/2k_\mathrm{B}T}\mathrm{d}v_x$$

- Starting point: the one-dimensional Maxwell–Boltzmann distribution for velocity is:

$$f_v(v_x)\mathrm{d}v_x = A\mathrm{e}^{-mv_x^2/2k_\mathrm{B}T}$$

- The normalization condition requires that the integral over all velocities equals one:

$$\int_{-\infty}^{\infty} f_v(v_x)\mathrm{d}v_x = 1$$

- To integrate, use the standard integral:

$$\int_{-\infty}^{\infty}\mathrm{e}^{-\alpha x^2}\mathrm{d}x = \sqrt{\frac{\pi}{\alpha}}$$

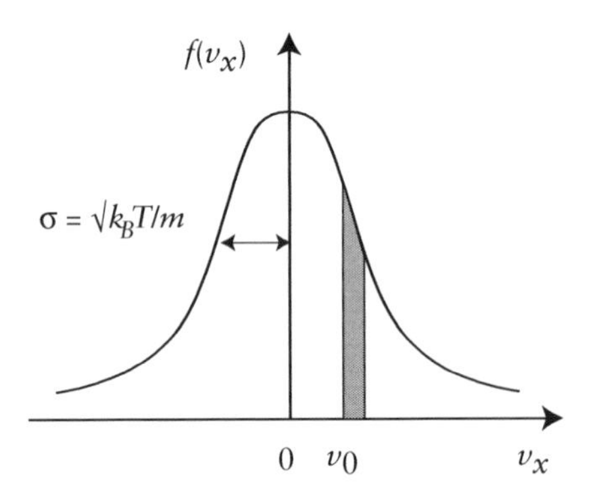

Figure 21.2: Maxwell–Boltzmann one-dimensional velocity distribution. The distribution is Gaussian, centered about the mean velocity, $\langle v_x \rangle$, which is zero since there is no preferred direction. The standard deviation is $\sqrt{k_B T/m}$, and increases with increased temperature. The shaded area represents the probability of a particle having a velocity between v_0 and $v_0 + dv$. The total area under the curve is equal to one (total probability = 1).

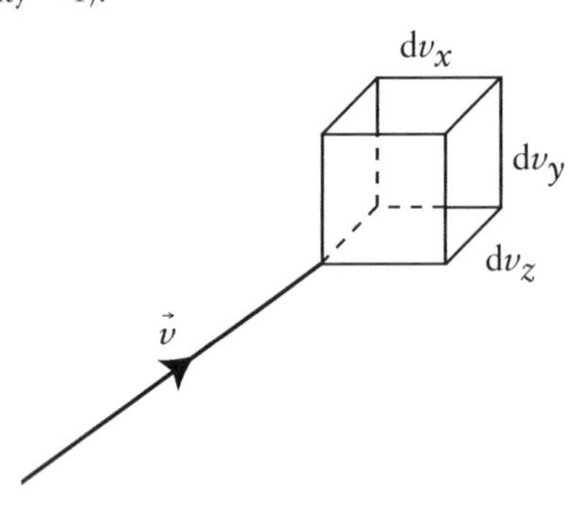

Figure 21.3: Element of velocity-space, $d^3v = dv_x dv_y dv_z$, for a particle with velocity $\vec{v}$.

the form e^{-ax^2} is $\sigma = 1/\sqrt{2a}$. Therefore, the standard deviations of the velocity distributions are temperature dependent:

$$\sigma_{1D} = \sqrt{\frac{k_B T}{m}}$$

$$\text{and } \sigma_{3D}^2 = 3\sigma_{1D}^2$$

$$\therefore \sigma_{3D} = \sqrt{\frac{3k_B T}{m}}$$

The center of the distribution corresponds to the average velocity, $\langle v_x \rangle$, and the standard deviation corresponds to the root mean square velocity, $\sqrt{\langle v_x^2 \rangle}$. In Worked example 21.1, these average values are calculated explicitly using Eq. 21.7.

21.3.2 Speed distribution

The speed distribution, $F_v(v)dv$, gives the probability of a particle having a *speed* between v and $v + dv$, and thus is independent of the particle's direction of travel. If we express the three-dimensional velocity distribution in spherical polar coordinates, then the three-dimensional element of velocity-space is:

$$d^3v = v^2 \sin\theta dv d\theta d\phi$$

Integrating the three-dimensional velocity distribution over angular coordinates θ and ϕ, we can therefore remove its directional dependence:

Derivation 21.2: Derivation of the three-dimensional velocity distribution (Eq. 21.8)

$$f_v(\vec{v})d^3v = (f_v(v_x)dv_x)(f_v(v_y)dv_y)(f_v(v_z)dv_z)$$

$$= \left(\frac{m}{2\pi k_B T}\right)^{\frac{3}{2}} e^{-m(v_x^2+v_y^2+v_z^2)/2k_B T} d^3v$$

$$= \left(\frac{m}{2\pi k_B T}\right)^{\frac{3}{2}} e^{-mv^2/2k_B T} d^3v$$

- Starting point: the normalized one-dimensional Maxwell–Boltzmann distribution for velocity is:

$$f_v(v_x)dv_x = \sqrt{\frac{m}{2\pi k_B T}} e^{-mv_x^2/2k_B T} dv_x$$

(refer to Derivation 21.1).

- The three-dimensional velocity distribution is the product of the three one-dimensional distributions, where:

$$d^3v = dv_x dv_y dv_z$$

$$\text{and } v^2 = v_x^2 + v_y^2 + v_z^2$$

$$F_v(v)\mathrm{d}v = \int_0^{2\pi}\int_0^{\pi} f_v(\vec{v})v^2 \sin\theta \mathrm{d}v\mathrm{d}\theta\mathrm{d}\phi$$

$$= 4\pi v^2 f_v(\vec{v})\mathrm{d}v$$

$$= 4\pi v^2 \left(\frac{m}{2\pi k_B T}\right)^{\frac{3}{2}} \mathrm{e}^{-mv^2/2k_B T}\mathrm{d}v$$

(Eq. 21.9). This distribution has the same Gaussian Boltzmann factor as the velocity distributions, but the entire distribution is no longer Gaussian since the Boltzmann factor now is multiplied by the factor of v^2. This means that the probability distribution is skewed towards higher speeds (Fig. 21.4). The average speed, $\langle v \rangle$, the root mean square speed, $\sqrt{\langle v^2 \rangle}$, and the modal speed, v_m, which are labeled in Fig. 21.4, are calculated in Worked example 21.2.

WORKED EXAMPLE 21.1: Using Eq. 21.7 to calculate average values of particle velocity

1. The average velocity, $\langle v_x \rangle$, equals the average of the Gaussian distribution, μ_x. The average velocity equals zero since there is no preferred direction of travel.
2. The root mean square velocity, $\sqrt{\langle v_x^2 \rangle}$, does not depend on direction, therefore, we would expect it to be non-zero. We see that $\sqrt{\langle v_x^2 \rangle}$ equals the standard deviation, σ_x, of the Gaussian distribution.
3. Since the three components v_x, v_y and v_z are independent, the square of the standard deviation of the *total* velocity distribution is equal to the sum of the squares of the components:

$$\sigma^2 = \sigma_x^2 + \sigma_y^2 + \sigma_z^2$$

Therefore, the root mean square of the total velocity is:

$$\sqrt{\langle v^2 \rangle} = \sqrt{\langle v_x^2 \rangle + \langle v_y^2 \rangle + \langle v_z^2 \rangle}$$

$$= \sqrt{\frac{3k_B T}{m}}$$

1. Average velocity:

$$\langle v_x \rangle = \int_{-\infty}^{\infty} v_x f_v(v_x)\mathrm{d}v_x$$

$$= \sqrt{\frac{m}{2\pi k_B T}} \int_{-\infty}^{\infty} v_x \mathrm{e}^{-mv_x^2/2k_B T}\mathrm{d}v_x$$

$$= 0$$

$$\equiv \mu_x$$

2. Root mean square velocity:

$$\langle v_x^2 \rangle = \int_{-\infty}^{\infty} v_x^2 f_v(v_x)\mathrm{d}v_x$$

$$= \sqrt{\frac{m}{2\pi k_B T}} \int_{-\infty}^{\infty} v_x^2 \mathrm{e}^{-mv_x^2/2k_B T}\mathrm{d}v_x$$

$$= \frac{k_B T}{m}$$

$$\equiv \sigma_x^2$$

$$\therefore \sqrt{\langle v_x^2 \rangle} = \sigma_x = \sqrt{\frac{k_B T}{m}}$$

3. To integrate, use the standard integrals:

$$\int_{-\infty}^{\infty} x\mathrm{e}^{-\alpha x^2}\mathrm{d}x = 0$$

$$\int_{-\infty}^{\infty} x^2\mathrm{e}^{-\alpha x^2}\mathrm{d}x = \frac{1}{2\alpha}\sqrt{\frac{\pi}{\alpha}}$$

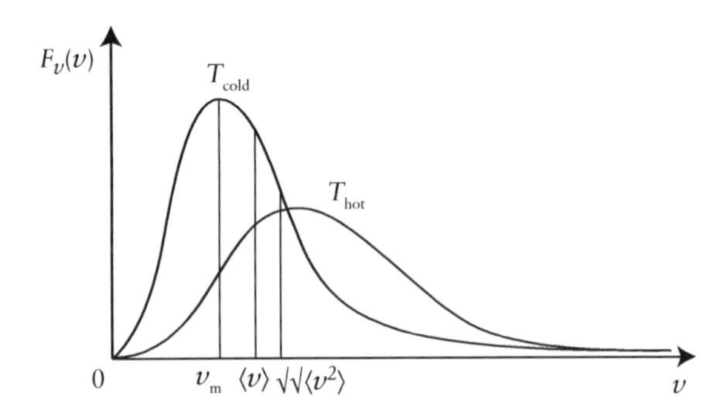

Figure 21.4: Maxwell–Boltzmann speed distribution. For hot systems, the distribution is skewed towards higher particle speeds. The ratio of average speeds is $v_m \langle v \rangle : \sqrt{\langle v^2 \rangle} \simeq 1 : 1.13 : 1.22$, and is independent of both T and m (refer to Worked example 21.2).

Speed distribution:

$$F_v(v)\mathrm{d}v = 4\pi v^2 \left(\frac{m}{2\pi k_B T}\right)^{\frac{3}{2}} \underbrace{\mathrm{e}^{-mv^2/2k_B T}}_{\text{Boltzmann factor}} \mathrm{d}v \qquad (21.9)$$

21.4 Properties of gases

The Maxwell–Boltzmann probability distributions can be used to determine average velocities and speeds, as shown in Worked examples 21.1 and 21.2. We can then use these averages to explain the macroscopic properties of systems in equilibrium, as shown below.

21.4.1 Particle energies

Using the expression for the average squared speed, $\langle v^2 \rangle$, from Worked example 21.2, the average translational kinetic energy (KE) per particle is:

$$\langle \mathrm{KE} \rangle_{3D} = \frac{1}{2} m \langle v^2 \rangle$$
$$= \frac{3}{2} k_B T \qquad (21.10)$$

Or, for translational motion in one dimension:

$$\langle \mathrm{KE} \rangle_{1D} = \frac{1}{2} m \langle v_x^2 \rangle$$
$$= \frac{1}{2} k_B T \qquad (21.11)$$

Therefore, the kinetic energy of particles in a gas depends only on its temperature. There are three independent degrees of freedom[2] for translational kinetic energy, corresponding to motion in the x-, y-, or z-directions, respectively. Therefore, Eqs. 21.10 and 21.11 illustrate that there is a kinetic energy of $k_B T/2$ associated with each translational degree of freedom. Since the average kinetic energy per particle depends only on the temperature of the gas, it is often referred to as the *thermal energy* of the system. Atoms are said to be in *thermal motion* as a result of their being at a particular temperature, and their thermal motion increases with increased temperature. Owing to collisions, the thermal motion of atoms appears as a random walk. For larger particles suspended in a fluid or gas, the random walk due to atomic and molecular collisions is known as *Brownian motion*.

Internal energy

The *internal energy*, U [J], of a gas is the sum of the kinetic and potential energies of its individual particles. The *potential* energy includes chemical energy stored in atomic bonds or intermolecular forces. Since an ideal gas, by definition, has no intermolecular forces, the potential energy of its particles is assumed to be zero. Therefore, the internal energy of an ideal gas results only from the *kinetic* energy of its particles. For a monatomic ideal gas of N particles, with three translational degrees of freedom (x, y, and z), the internal energy of the gas is therefore:

$$U_{3D} = N \langle \mathrm{KE} \rangle_{3D}$$
$$= \frac{3}{2} N k_B T$$
$$\equiv \frac{3}{2} nRT \qquad (21.12)$$

where we have used $Nk_B \equiv nR$ (Eq. 21.3) to eliminate Nk_B. Similarly, for translational motion in one dimension:

$$U_{1D} = N \langle \mathrm{KE} \rangle_{1D}$$
$$= \frac{1}{2} N k_B T$$
$$= \frac{1}{2} nRT \qquad (21.13)$$

Therefore, the internal energy of an ideal gas depends only on its temperature. (For *real* gases, intermolecular

[2] A *degree of freedom* in this context is a parameter that defines an independent direction of movement.

WORKED EXAMPLE 21.2: Using Eq. 21.9 to calculate average particle speeds

1. The average speed, $\langle v \rangle$, is given by the average of the speed distribution, $F_v(v)$. The integration limits are 0 to ∞ since we are not taking direction into account, and therefore cannot have negative speeds.
2. The root mean square speed, $\sqrt{\langle v^2 \rangle}$, is found to have the same value as that for the three-dimensional velocity distribution (see Worked example 21.1). We would expect this to be the case since squaring the velocity removes its directionality.
3. The modal speed is the turning point of $F_v(v)$, and therefore is the speed at which:

$$\left(\frac{dF_v(v)}{dv}\right)_{v_m} = 0$$

4. The ratio of average speeds is therefore:

$$v_m : \langle v \rangle : \sqrt{\langle v^2 \rangle} = \sqrt{2} : \sqrt{\frac{8}{\pi}} : \sqrt{3}$$

$$\simeq 1 : 1.13 : 1.22$$

(refer to Fig. 21.4).

5. To integrate, use the standard integrals:

$$\int_0^\infty x^3 e^{-\alpha x^2} dx = \frac{1}{2\alpha^2}$$

$$\int_0^\infty x^4 e^{-\alpha x^2} dx = \frac{3}{8\alpha^2}\sqrt{\frac{\pi}{\alpha}}$$

1. Average speed:

$$\langle v \rangle = \int_0^\infty v F_v(v) dv$$

$$= 4\pi\left(\frac{m}{2\pi k_B T}\right)^{3/2} \int_0^\infty v^3 e^{-mv^2/2k_B T} dv$$

$$= \sqrt{\frac{8k_B T}{\pi m}}$$

2. Root mean square speed:

$$\langle v^2 \rangle = \int_0^\infty v^2 F_v(v) dv$$

$$= 4\pi\left(\frac{m}{2\pi k_B T}\right)^{3/2} \int_0^\infty v^4 e^{-mv^2/2k_B T} dv$$

$$= \frac{3k_B T}{m}$$

$$\therefore \sqrt{\langle v^2 \rangle} = \sqrt{\frac{3k_B T}{m}}$$

3. Modal speed: differentiate $F_v(v)$ and set the derivative equal to zero:

$$v_m = \sqrt{\frac{2k_B T}{m}}$$

forces are not negligible, therefore, their internal energy is a function of the gas's volume as well as its temperature.)

Eqs. 21.12 and 21.13 illustrate that there is an internal energy of $nRT/2$ associated with each degree of freedom. Diatomic ideal gases have additional degrees of freedom: rotational and vibrational degrees of freedom. A gas acquires an additional internal energy of $nRT/2$ for each additional degree of freedom. For an ideal gas with f independent degrees of freedom, its internal energy is then:

$$U = \frac{f}{2} N k_B T$$

$$= \frac{f}{2} nRT \qquad (21.14)$$

For example, $f = 3$ for a monatomic ideal gas (translational motion in x-, y-, or z-directions). Linear diatomic gases, have a further two rotational degrees of freedom and two vibrational degrees of freedom (refer to Fig. 21.5). Therefore, for a linear diatomic gas with no vibrations, $f = 5$, and if the gas also has radial vibrations, then $f = 7$.

Pressure and internal energy

Using Eq. 21.14 together with the Ideal Gas Law, we find that the relationship between pressure, p [Pa], and internal energy per unit volume, $u = U/V$ [J m^{-3}], is:

$$p = \frac{2}{f} u \qquad (21.15)$$

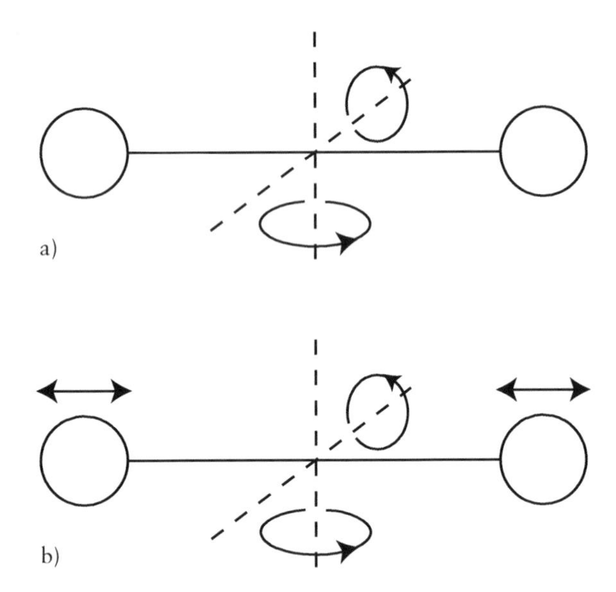

Figure 21.5: a) Rotational degrees of freedom for a linear diatomic ideal gas and b) rotational plus vibrational degrees of freedom.

Therefore, for an ideal monatomic gas, where $f = 3$, the pressure is:

$$p = \frac{2}{3}u$$

The equipartition theorem

Equation 21.14 forms the basis of the *equipartition theorem*, which states that the internal energy of a system is shared equally among its different contributions. Said another way, with each degree of freedom, there is an energy factor of $RT/2$ [J mole^{-1}] or $k_BT/2$ [J molecule^{-1}] (Eqs. 21.16 and 21.17).

The equipartition theorem holds true *provided* the energy has a quadratic dependence on the particular degree of freedom.[3] For example, the dependence of kinetic energy on v or p is quadratic:

$$\text{KE} = \frac{1}{2}mv^2 \equiv \frac{p^2}{2m}$$

Another example is the potential energy of a harmonic oscillator, which is quadratic in x:

$$V_{\text{SHO}} = \frac{1}{2}kx^2$$

[3] Proof of this statement requires a statistical mechanics treatment, see Section 24.4.4.

where k [N m^{-1}] is the spring constant. In these instances, the equipartition theorem will apply to the system.

Equipartition theorem:

The average energy associated with each quadratic degree of freedom is:

$$\langle E \rangle = \frac{1}{2}RT \quad \text{per mole} \qquad (21.16)$$

$$\equiv \frac{1}{2}k_BT \quad \text{per molecule} \qquad (21.17)$$

21.4.2 Heat capacity

We saw above that the temperature of an ideal gas is a measure of its internal energy. Heat flow into or out of an ideal gas changes its temperature, and hence changes its internal energy. The *heat capacity*, C [J K^{-1}], of a system is defined as the amount of heat, Q [J], required to raise its temperature by 1 K (Eq. 21.18). The *specific* heat capacity, c [J K^{-1} kg^{-1}], is the heat capacity per unit mass, and the *molar* heat capacity, c_m [J K^{-1} mol^{-1}], is the heat capacity per mole (Eqs. 21.19 and 21.20, respectively).

Heat capacity:

$$C = \left(\frac{\partial Q}{\partial T}\right) \qquad (21.18)$$

Specific heat capacity:

$$c = \frac{1}{m}\left(\frac{\partial Q}{\partial T}\right) \qquad (21.19)$$

Molar heat capacity:

$$c_m = \frac{1}{n}\left(\frac{\partial Q}{\partial T}\right) \qquad (21.20)$$

The actual rise in temperature with increasing energy for a given substance depends on the nature of its constituent particles. For example, for monatomic gases, input energy only goes into increasing their translational kinetic energy, whereas for diatomic gases, input energy also increases their rotational or vibrational energies. Therefore, in qualitative terms, for the same amount of input energy, the temperature of a monatomic gas increases more than for a diatomic gas.

Heat capacity at constant volume and pressure

The *heat capacity at constant volume*, C_V [JK^{-1}] is defined as:

$$C_V = \left(\frac{\partial Q}{\partial T}\right)_V$$

i.e. it is the heat capacity when heat is added to a system that is kept at a constant volume. By considering energy conservation, we find that C_V is equal to the change in a system's internal energy with temperature:

$$C_V = \frac{dU}{dT} \quad (21.21)$$

(Derivation 21.3). Using $U = fnRT/2$ (Eq. 21.14), we see that C_V is proportional to the number of degrees of freedom, f, of a system (Eq. 21.22). Therefore, since a monatomic gas has three translational degrees of freedom, its heat capacity at constant volume is:

$$C_V = \frac{3}{2}nR$$

Analogously, the *heat capacity at constant pressure*, C_p [JK^{-1}], is defined as:

$$C_p = \left(\frac{\partial Q}{\partial T}\right)_p$$

i.e. it is the heat capacity when heat is added to a system that is kept at a constant pressure. C_p is greater than C_V since energy is required to increase the volume of gas as well as to raise its internal energy (its temperature). Again, by considering energy conservation, we find that the difference in heat capacities, $C_p - C_V$, is equal to nR (Eq. 21.24 and Derivation 21.3). Therefore, the heat capacity at constant pressure is given by Eq. 21.23. For a monatomic gas with three degrees of freedom, the heat capacity at constant pressure is then:

$$C_p = \frac{5}{2}nR$$

The ratio of heat capacities is a constant and is referred to as the *adiabatic index*, γ (Eq. 21.25). For a monatomic ideal gas, the adiabatic index is therefore:

$$\gamma = \frac{C_p}{C_V} = \frac{5}{3}$$

C_V and C_p for ideal gases (Derivation 21.3):

Heat capacities for f degrees of freedom:

$$C_V = \frac{f}{2}nR \quad (21.22)$$

$$C_p = \frac{f+2}{2}nR \quad (21.23)$$

Difference in heat capacities:

$$C_p - C_V = nR \quad (21.24)$$

Ratio of heat capacities (adiabatic index):

$$\frac{C_p}{C_V} = \frac{f+2}{f} \equiv \gamma \quad (21.25)$$

Aside on latent heat: in this section, we have only considered ideal gases. However, when a substance changes state, for example, solid to liquid or liquid to gas, then heat input does not increase the body's temperature, but instead is absorbed by the bonds between particles, which increases the *potential* energy of the particles. Since the kinetic energy of the particles remains constant, the temperature of the substance does not change during a change of state. The amount of heat absorbed or released per unit mass during such a transition is called the *specific latent heat*, L [J kg^{-1}] (Eq. 21.26). The latent heat of *fusion* is the energy required to melt unit mass of solid to a liquid, whereas the latent heat of *vaporization* is the energy required to vaporize unit mass of liquid to a gas. Melting or boiling are endothermic processes, which means that latent heat is absorbed, whereas condensation or freezing are exothermic processes, meaning that latent heat is released.

Latent heat:

$$L = \frac{Q}{m} \quad (21.26)$$

Derivation 21.3: Derivation of the expressions for C_V and C_p (Eqs. 21.22–21.24)

First Law of Thermodynamics:

$$dU = dQ + dW$$
$$= dQ - p\,dV$$

- Starting point: from the conservation of energy, an element of increase in internal energy of a system, dU, equals the element of heat added, dQ, plus the element of work done on the system, dW:

$$dU = dQ + dW$$

(This is the First Law of Thermodynamics; see Section 22.2.2.)

- The work done equals the energy needed to expand the gas against pressure, which results from the impulse of the molecules (see Section 22.2.2):

$$dW = -p\,dV$$

Constant volume:

$$C_V = \left(\frac{\partial Q}{\partial T}\right)_V = \left(\frac{\partial U}{\partial T}\right)_V$$

U is a function of T only for an ideal gas:

$$\therefore C_V = \frac{dU}{dT}$$
$$= \frac{f}{2}nR$$

- Constant volume: the heat capacity at constant volume is defined as:

$$C_V = \left(\frac{\partial Q}{\partial T}\right)_V$$

Using this, together with the First Law, we see that C_V is equal to the change in internal energy with temperature. Since U is a function of T only for an ideal gas, C_V can be written as the *total* derivative, dU/dT, instead of a partial derivative. For an ideal gas with f degrees of freedom, $U = fnRT/2$ (refer to Eq. 21.14).

Constant pressure:

$$dQ = dU - dW$$
$$= C_V dT + p\,dV$$
$$\left(\frac{\partial Q}{\partial T}\right)_p = C_V + p\left(\frac{\partial V}{\partial T}\right)_p$$
$$\therefore C_p = C_V + nR$$
$$= \frac{f+2}{2}nR$$
$$\therefore C_p - C_V = nR$$

- Constant pressure: the heat capacity at constant pressure is defined as:

$$C_p = \left(\frac{\partial Q}{\partial T}\right)_p$$

Divide the expression for dQ by dT, holding p constant. Use the Ideal Gas Law, $pV = nRT$, to determine $(\partial V/\partial T)_p$.

21.4.3 Pressure

We found previously that the mean speed-squared of an ideal gas is:

$$\langle v^2 \rangle = \frac{3k_B T}{m}$$

(see Worked example 21.2). Using this to eliminate $k_B T$ from the Ideal Gas Law, we can express the pressure in terms of $\langle v^2 \rangle$:

$$p = \frac{1}{3}mn_p\langle v^2 \rangle$$

where n_p [m^{-3}] is the particle number density:

$$n_p = \frac{N}{V}$$

We reach the same result if we assume that the pressure of a gas is the force per unit area exerted by molecular collisions with the container walls (Eq. 21.27 and Derivation 21.4). Pressure was initially thought by Newton to be due to intermolecular repulsion. However this treatment illustrates that pressure is indeed due to molecular impulse and not due to intermolecular repulsion.

Ideal gas pressure (Derivation 21.4):

$$p = \frac{1}{3} m n_p \left\langle v^2 \right\rangle \qquad (21.27)$$

Derivation 21.4: Derivation of the pressure of an ideal gas (Eq. 21.27)

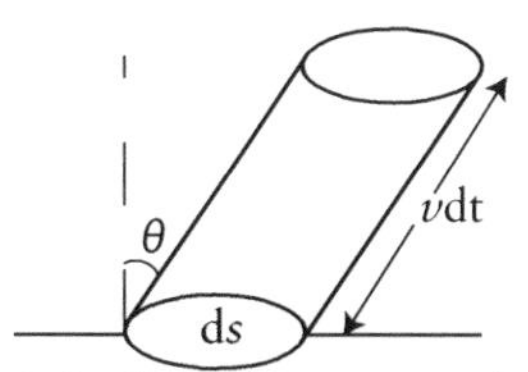

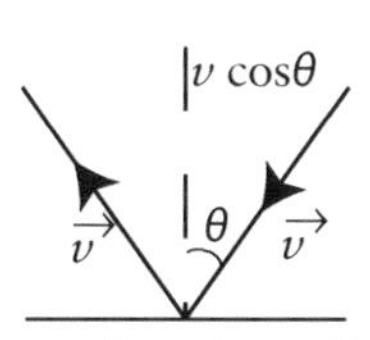

Cylinder volume $V_C = v\,dt\,ds\cos\theta$
Number of particles in cyclinder,
$N_C = n_p V_C = n_p v\,dt\,ds\cos\theta$

$\Delta v = 2v\cos\theta$

- Starting point: consider an ideal gas enclosed in a container. The speed distribution of particles in the gas is $F_v(v)\,dv$ (Maxwell–Boltzmann speed distribution). We wish to determine the element of pressure, dp, exerted by a "cylinder" of gas particles in a time dt (refer to diagram).
- The element of pressure, dp, is given by the force per unit area of one particle, F/ds (where ds is an element of area), multiplied by the number of particles, dN, that contribute to that element of pressure:

$$dp = \frac{F}{ds} dN$$

- For a particle density of n_p [m^{-3}], dN is given by:

$$dN = N_C N_\Omega (F_v(v)dv)$$

where $N_C = n_p v\,dt\,ds\cos\theta$

$$N_\Omega = \frac{d\Omega}{4\pi} = \frac{\sin\theta\, d\theta\, d\phi}{4\pi}$$

N_c is the number of particles within the cylinder (refer to diagram); $F_v(v)\,dv$ is the fraction of those with the correct velocity (between v and $v + dv$); N_Ω is the fraction of those within the correct solid angle, $d\Omega$.

- The force exerted on ds due to a change in particle momentum is:

$$F = \frac{dp}{dt} = m\frac{dv}{dt} = m\frac{2v\cos\theta}{dt}$$

(refer to diagram).

- Integrating θ and ϕ over all angles in a hemisphere, we find that the total pressure exerted by all particles in the gas is $p = \frac{1}{3} m n_p \langle v^2 \rangle$.

$$\begin{aligned}
dp &= \frac{F}{ds} dN \\
&= \frac{F}{ds} N_C N_\Omega (F_v(v)dv) \\
&= \frac{2mv\cos\theta}{ds\,dt} \frac{(n_p v\,dt\,ds\cos\theta)(\sin\theta\, d\theta\, d\phi)(F_v(v)dv)}{4\pi} \\
&= \frac{1}{2\pi} m n_p v^2 F_v(v) dv \cos^2\theta \sin\theta\, d\theta\, d\phi \\
\therefore p &= \frac{1}{2\pi} m n_p \underbrace{\int_{-\infty}^{\infty} v^2 F_v(v) dv}_{=\langle v^2 \rangle} \int_0^{\pi/2} \cos^2\theta \sin\theta\, d\theta \int_0^{2\pi} d\phi \\
&= m n_p \left\langle v^2 \right\rangle \left[-\frac{1}{3}\cos^3\theta \right]_0^{\pi/2} \\
&= \frac{1}{3} m n_p \left\langle v^2 \right\rangle
\end{aligned}$$

21.4.4 Particle flux

We can use a similar treatment to determine an expression for the particle flux, Φ [$m^{-2}\ s^{-1}$], which is the number of particles striking a surface per unit area and per unit time. Particle flux is found to depend on particle number density, n_p, and the average particle speed, $\langle v\rangle$ (Eq. 21.28 and Derivation 21.5). These, in turn, depend on the pressure and temperature of the gas. Therefore, particle flux can be expressed in terms of p and T (Eq. 21.29 and Derivation 21.5). Such a treatment relates the *micro*scopic motion of particles to the gas's *macro*scopic properties, pressure and temperature.

Particle flux (Derivation 21.5):

$$\Phi = \frac{1}{4} n_p \langle v\rangle \qquad (21.28)$$

$$\equiv \frac{p}{\sqrt{2\pi m k_B T}} \qquad (21.29)$$

Effusion

Effusion is the process by which particles emerge through a small hole. The rate of effusing particles

Derivation 21.5: Derivation of the particle flux (Eq. 21.28)

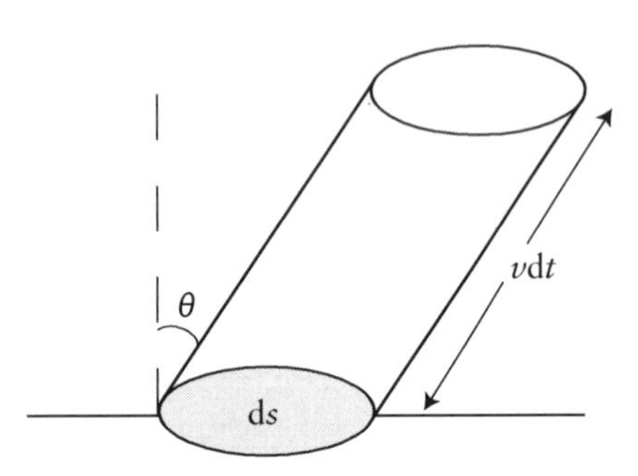

Cylinder volume $V_C = v dt\ ds\cos\theta$
Number of particles in cylinder,
$N_C = n_p V_C = n_p\, v\, dt\, ds \cos\theta$

$$d\Phi = \frac{dN}{ds dt} = \frac{N_C N_\Omega (F_v(v) dv)}{ds dt} = \frac{(n_p v dt ds \cos\theta)(\sin\theta d\theta d\phi)(F_v(v)dv)}{4\pi ds dt} = \frac{1}{4\pi} n_p v F_v(v) dv \cos\theta \sin\theta d\theta d\phi$$

$$\therefore \Phi = \frac{1}{4\pi} n_p \underbrace{\int_{-\infty}^{\infty} v F_v(v) dv}_{=\langle v\rangle} \int_0^{\pi/2} \cos\theta \sin\theta d\theta \int_0^{2\pi} d\phi$$

$$= \frac{1}{2} n_p \langle v\rangle \left[-\frac{1}{2}\cos^2\theta\right]_0^{\pi/2}$$

$$= \frac{1}{4} n_p \langle v\rangle$$

$$\therefore \Phi = \frac{p}{\sqrt{2\pi m k_B T}}$$

- Starting point: consider an ideal gas enclosed in a container. The speed distribution of particles in the gas is $F_v(v)\,dv$ (Maxwell–Boltzmann speed distribution). Particle flux is defined as the number of particles, dN, striking per unit area ds, per unit time dt:

$$\Phi = \frac{dN}{ds dt}$$

- For a particle density of n_p [m^{-3}], dN is given by:

$$dN = N_C N_\Omega (F_v(v) dv)$$

where $N_C = n_p v dt ds \cos\theta$

$$N_\Omega = \frac{d\Omega}{4\pi} = \frac{\sin\theta d\theta d\phi}{4\pi}$$

N_C is the number of particles within the cylinder (refer to diagram); $F_v(v)dv$ is the fraction of those with the correct velocity (between v and $v + dv$); N_Ω is the fraction of those within the correct solid angle, $d\Omega$.

- Integrating over a hemisphere, we find an expression for particle flux, Φ, in terms of n_p and $\langle v\rangle$:

$$\Phi = \frac{1}{4} n_p \langle v\rangle$$

- From the Ideal Gas Law, $p = n_p k_B T$, and from Worked example 21.2:

$$\langle v\rangle = \sqrt{\frac{8 k_B T}{\pi m}}$$

Therefore, we can express Φ in terms of pressure and temperature.

per unit area, R_E [m^{-2} s^{-1}], is equal to the particle flux striking the hole if it were closed off:

$$\therefore R_E = \Phi = \frac{1}{4} n_p \langle v \rangle$$

Expressing $\langle v \rangle$ in terms of the Maxwell–Boltzmann speed distribution:

$$\langle v \rangle = \int v F_v(v) \mathrm{d}v$$

where $F_v(v)\,\mathrm{d}v$ is provided in Eq. 21.9, we find that:

$$\begin{aligned} R_E &= \frac{1}{4} n_p \int v F_v(v) \mathrm{d}v \\ &\equiv \int F_E(v) \mathrm{d}v \\ \text{where } F_E(v)\mathrm{d}v &= \frac{1}{4} n_p v F_v(v) \mathrm{d}v \\ &\propto v^3 e^{-mv^2/2k_B T} \mathrm{d}v \end{aligned}$$

Therefore, we can define a speed distribution of effusing particles, $F_E(v)\,\mathrm{d}v$, in terms of the Maxwell–Boltzmann speed distribution. The effusion speed distribution depends on v^3. This means that the probability distribution is skewed towards higher v, and therefore high-velocity particles have a greater probability of effusing. As a consequence, both the pressure and temperature of a gas in an effusing vessel decrease, since high-energy particles leave the gas in the vessel.

Effusion rate depends on the mass of the effusing particle:

$$R_E \propto \langle v \rangle \propto \frac{1}{\sqrt{m}}$$

Therefore, effusion has practical applications in isotope separation: heavier isotopes become progressively enriched in an effusing vessel with time.

21.4.5 Mean free path

The *mean free path*, λ [m], is the average distance traveled by a particle before it undergoes a collision with another particle. If we define the *scattering cross section*, σ [m^2], as:

$$\sigma = \pi d^2$$

where d is the effective particle diameter, then the mean free path of an ideal gas with particle number density n_p is given by Eq. 21.30 (Derivation 21.6).

Mean free path (Derivation 21.6):

$$\lambda = \frac{1}{\sqrt{2} n_p \sigma} \tag{21.30}$$

where the scattering cross section for particles of diameter d is:

$$\sigma = \pi d^2$$

21.5 Establishing equilibrium

The Ideal Gas Law holds for systems in *equilibrium*. A system in equilibrium is one whose macroscopic properties are not changing with time—for example, constant pressure, constant temperature, and constant volume. Equilibrium is an important concept in thermal physics since the subject of *thermodynamics*, which deals with the thermal properties of matter exclusively on *macro*scopic scales, only holds for systems in equilibrium. However, kinetic theory treats systems on *micro*scopic scales. Therefore, in this section, we will use kinetic theory to determine the transport properties (diffusion) of systems that *aren't* in equilibrium.

Thermodynamic equilibrium has three components:

1. **Chemical equilibrium:** constant particle density. This is achieved by particle diffusion from regions of high particle concentration to low concentration.
2. **Mechanical equilibrium:** constant pressure. This is achieved by viscous forces, which transport particle momentum from regions of high momentum to regions of low momentum.
3. **Thermal equilibrium:** constant temperature. This is achieved by thermal conductivity, which is the transport of heat from a region of high temperature to a region of low temperature.

The microscopic behavior governing these processes is very similar in each case. Below, we will look in turn at how each of these equilibria are established.

21.5.1 Chemical equilibrium: particle transport

Chemical equilibrium is established when particle concentration, n_p, is uniform throughout a system. If there is a gradient in the concentration, $\mathrm{d}n_p/\mathrm{d}z$, then particle transport (diffusion) occurs down the concentration gradient. We will consider two situations: the *steady-state* situation, where the concentration gradient is constant in time, and the *non-steady* situation, where the concentration gradient varies as a function of time.

Derivation 21.6: Derivation of the particle mean free path (Eq. 21.30)

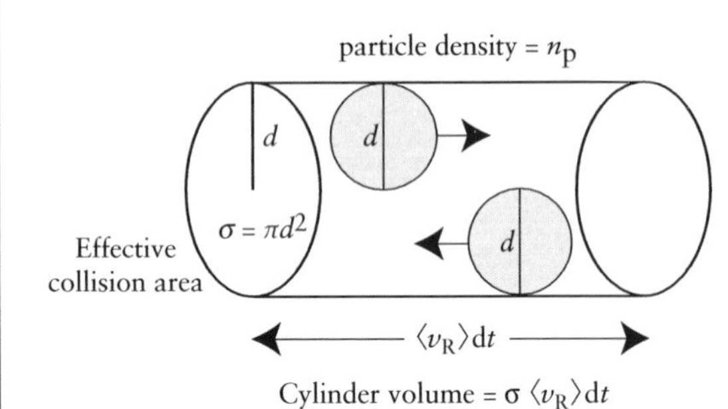

$$\begin{aligned}\lambda &= \frac{dx}{dN} \\ &= \frac{vdt}{n_p dV} \\ &= \frac{vdt}{n_p \sigma \langle v_R \rangle dt} \\ &\simeq \frac{v}{\sqrt{2} v n_p \sigma} \\ &\simeq \frac{1}{\sqrt{2} n_p \sigma}\end{aligned}$$

- Starting point: the mean free path, λ, is the mean distance a particle travels before a collision, $\lambda = dx/dN$.
- If the time between collisions is dt, then the mean distance traveled by a particle with velocity v is:

$$dx = v\ dt$$

- In a time dt, the collision volume is $dV = \sigma \langle v_R \rangle dt$, where $\langle v_R \rangle$ is the average relative velocity between colliding particles, and σ is the effective collision area (refer to diagram). If two colliding particles have velocities $\vec{v}_1$ and $\vec{v}_2$, where $\vec{v}_R = \vec{v}_1 - \vec{v}_2$, then the magnitude of the relative velocity is:

$$\begin{aligned}|v_R|^2 &= \vec{v}_R \cdot \vec{v}_R \\ &= (\vec{v}_1 - \vec{v}_2) \cdot (\vec{v}_1 - \vec{v}_2) \\ &= v_1^2 + v_2^2 - 2\vec{v}_1 \cdot \vec{v}_2\end{aligned}$$

- If we assume $|v_1| \simeq |v_2| \simeq v$, the average relative velocity between two particles is:

$$\begin{aligned}\therefore \langle v_R^2 \rangle &= \langle v_1^2 \rangle + \langle v_2^2 \rangle - 2\underbrace{\langle \vec{v}_1 \cdot \vec{v}_2 \rangle}_{=0} \\ &= \langle v_1^2 \rangle + \langle v_2^2 \rangle \\ &\simeq 2\langle v^2 \rangle \\ \therefore \langle v_R \rangle &\simeq \sqrt{2} v\end{aligned}$$

(The last step is an approximation.)

Steady-state diffusion

Fick's First Law of diffusion states that for steady-state particle diffusion, the particle flux,[4] J [m^{-2} s^{-1}], is proportional to the gradient of the particle number density, n_p. In one dimension, for example, along the z-axis, Fick's Law is:

$$J_z = -D\frac{\partial n_p}{\partial z}$$

(Eq. 21.31). The constant of proportionality is called the *diffusion coefficient*, D [m^2 s^{-1}]. An expression for the diffusion coefficient is found by considering the number of particles crossing a given surface per unit time, as shown in Derivation 21.7. We find that D depends on the average particle speed, $\langle v \rangle$, and the mean free path, λ (Eq. 21.32 and Derivation 21.7).

Particle transport (Derivation 21.7):

Fick's First Law:

$$J_z = -D\frac{\partial n_p}{\partial z} \qquad (21.31)$$

Diffusion coefficient:

$$D = \frac{1}{3}\langle v \rangle \lambda \qquad (21.32)$$

4 The particle flux is the number of particles crossing unit area per unit time.

Derivation 21.7: Derivation of the diffusion coefficient (Eq. 21.32)

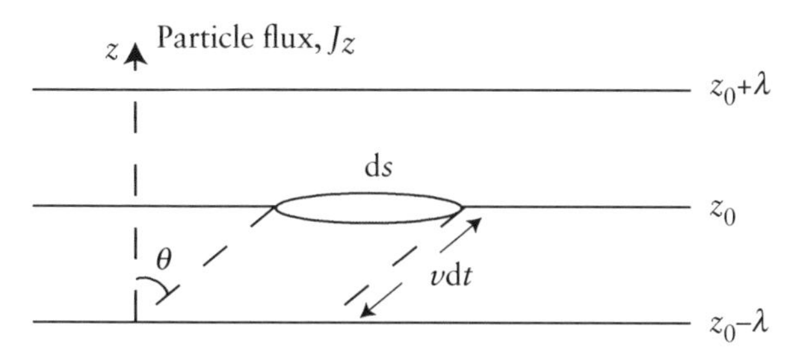

- Starting point: the z-component of particle flux for particles incident on a plane at $z = z_0$ at an angle θ is:

$$\mathrm{d}J_z = \frac{\mathrm{d}N}{\mathrm{d}s\mathrm{d}t}\cos\theta$$

(number of particles $\mathrm{d}N$, per unit area $\mathrm{d}s$, per unit time $\mathrm{d}t$). Particle concentration is not constant $\therefore N$ is a function of z (refer to diagram).

- For a particle density of n_p [m^{-3}], $\mathrm{d}N$ is given by:

$$\mathrm{d}N = N_\mathrm{C} N_\Omega (F_v(v)\mathrm{d}v)$$

where $N_\mathrm{C} = n_\mathrm{p} v \mathrm{d}t\mathrm{d}s\cos\theta$

$$N_\Omega = \frac{\mathrm{d}\Omega}{4\pi} = \frac{\sin\theta\mathrm{d}\theta\mathrm{d}\phi}{4\pi}$$

N_C is the number of particles within the cylinder (refer to diagram); $F_v(v)\mathrm{d}v$ is the fraction of those with the correct velocity (between v and $v + \mathrm{d}v$); N_Ω is the fraction of those within the correct solid angle, $\mathrm{d}\Omega$.

- The net upward flux, J^+, and downward flux, J^-, is found by integrating over all angles in a hemisphere, and over all velocities. The upward flux is a result of particles coming from the plane $z^- = z_0 - \lambda$, whereas the downward flux is from particles coming from the plane $z^+ = z_0 + \lambda$ (refer to diagram).

- The total flux, J_z, is therefore the difference between upward and downward flux. We can approximate using a Taylor expansion about z_0.

$$\mathrm{d}J_z = \frac{\mathrm{d}N(z)}{\mathrm{d}s\mathrm{d}t}\cos\theta = \frac{1}{4\pi}n_\mathrm{p}(z)vF_v(v)\mathrm{d}v\cos^2\theta\sin\theta\mathrm{d}\theta\mathrm{d}\phi$$

$$J_z^\pm = \frac{1}{4\pi}n_\mathrm{p}(z_0 \mp \lambda)\underbrace{\int_{-\infty}^{\infty} vF_v(v)\mathrm{d}v}_{=\langle v\rangle}\underbrace{\int_0^{\pi/2}\cos^2\theta\sin\theta\mathrm{d}\theta}_{=1/3}\underbrace{\int_0^{2\pi}\mathrm{d}\phi}_{=2\pi}$$

$$= \frac{1}{6}n_\mathrm{p}(z_0 \mp \lambda)\langle v\rangle$$

$$\therefore J_z = J_z^+ - J_z^-$$
$$= \frac{1}{6}n_\mathrm{p}(z_0 - \lambda)\langle v\rangle - \frac{1}{6}n_\mathrm{p}(z_0 + \lambda)\langle v\rangle$$
$$\simeq \frac{1}{6}\langle v\rangle\left\{n(z_0) - \frac{\partial n_\mathrm{p}}{\partial z}\lambda - \left(n(z_0) + \frac{\partial n_\mathrm{p}}{\partial z}\lambda\right)\right\}$$
$$= -\underbrace{\frac{1}{3}\langle v\rangle\lambda}_{=D}\frac{\partial n_\mathrm{p}}{\partial z}$$

and $J_z = -D\dfrac{\partial n_\mathrm{p}}{\partial z}$ (Fick's Law)

$$\therefore D = \frac{1}{3}\langle v\rangle\lambda$$

Non-steady diffusion

The rate of change of particle density between z and $z + \mathrm{d}z$ is given by the difference in particle flux entering at z and leaving at $z + \mathrm{d}z$:

$$\frac{\partial n_\mathrm{p}}{\partial t}\mathrm{d}z = J(z) - J(z + \mathrm{d}z)$$

Using a Taylor expansion on $J(z + \mathrm{d}z)$ together with Eq. 21.31, we obtain *Fick's Second Law*, which relates the rate of change of particle density to the second space derivative of n_p:

$$\frac{\partial n_\mathrm{p}}{\partial t}\mathrm{d}z = J(z) - J(z + \mathrm{d}z)$$
$$= J(z) - J(z) - \frac{\partial J}{\partial z}\mathrm{d}z$$
$$= D\frac{\partial^2 n_\mathrm{p}}{\partial z^2}\mathrm{d}z$$

(Eq. 21.33). The solution to Eq. 21.33 is found to be Gaussian in z:

$$n_\mathrm{p}(z,t) = \frac{n_0}{\sqrt{4\pi Dt}}\mathrm{e}^{-z^2/4Dt}$$

where n_0 is the initial particle concentration at $z = 0$ and $t = 0$ (Eq. 21.34; show this is a solution by substituting Eq. 21.34 into Eq. 21.33). Therefore, since the variance of a Gaussian of the form $e^{-\alpha x^2}$ is $1/2a$, n_p has a variance that grows linearly with time (Eq. 21.35). The standard deviation, σ, which is the square root of the variance, is called the *diffusion length*, L_D [m], and it provides a measure of particle propagation distance in a time t (Eq. 21.36 and Fig. 21.6).

Particle transport

Fick's Second Law:

$$\frac{\partial n_p}{\partial t} = D\frac{\partial^2 n_p}{\partial z^2} \tag{21.33}$$

This has a Gaussian solution:

$$n_p(z,t) = \frac{n_0}{\sqrt{4\pi Dt}} e^{-z^2/4Dt} \tag{21.34}$$

where the variance and diffusion length are:

$$\langle z^2 \rangle = \sigma^2 = 2Dt \tag{21.35}$$

$$L_D = \sigma = \sqrt{2Dt} \tag{21.36}$$

21.5.2 Mechanical equilibrium: momentum transport

We can use a similar steady-state approach as we used for particle diffusion to determine the how pressure reaches equilibrium. Pressure gradients arise as a consequence of gradients in viscous forces between particles. Viscous forces are shearing forces per unit area, and are referred to as *shear stress*, P_x [Pa]. Shear stress acts parallel to the surface on which it acts, similar to friction between solids. The perpendicular shear

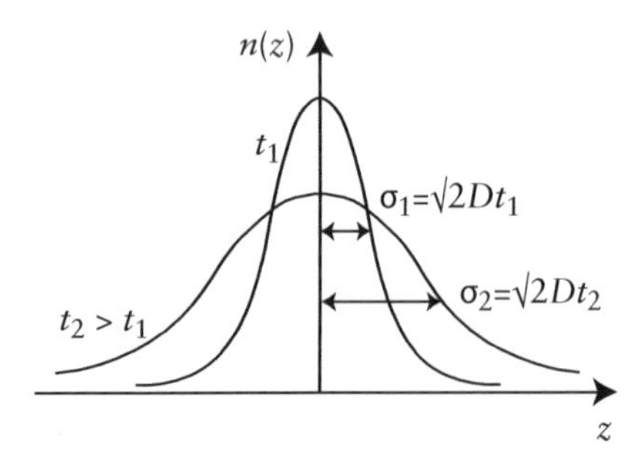

Figure 21.6: Particle diffusion in one dimension (Fick's Second Law, Eq. 21.33). The standard deviation (diffusion length) increases with time.

stress gradient, P_{zx}, is found to be proportional to the z-gradient of particle velocity in the x-direction, u_x:

$$P_{zx} = -\eta\frac{\partial u_x(z)}{\partial z}$$

(Eq. 21.37; refer to the figure in Derivation 21.8). The coefficient of proportionality, η [Pa s], is called the *viscosity*. Viscosity is often defined in terms of this relation, as the "shear stress per unit velocity gradient." The value of η gives a measure of the resistance of a substance to fluid flow. Highly viscous substances, such as honey, have a slow flow rate and therefore have a large resistance to shear stress (large η). In contrast, less viscous fluids, such as water, have lower resistance to shear stress (small η). Using a microscopic approach, we can determine an expression for viscosity (Eq. 21.38 and Derivation 21.8). The quantity η has a similar form to the diffusion coefficient (compare Eq. 21.38 with Eq. 21.32).

Shear stress (Derivation 21.8):

$$P_{zx} = -\eta\frac{\partial u_x}{\partial z} \tag{21.37}$$

Viscosity:

$$\eta = \frac{1}{3}n_p m\langle v\rangle \lambda \tag{21.38}$$

21.5.3 Thermal equilibrium: heat transport

The final equilibrium we'll consider is *thermal equilibrium*, which acts to equalize temperature differences. This is achieved by energy transfer—heat flow—down temperature gradients. There are three types of heat transfer: conduction, convection and radiation. (See Section 24.6.4 for details on radiation.)

Conduction

Conduction is the transfer of heat through matter, from particle to particle. Higher energy particles transfer some of their energy to less energetic particles via collisions, which results in a net heat transfer from hotter to cooler regions. Equilibrium is achieved when temperature is uniform throughout a substance. Conduction occurs in solids, liquids, and, to a lesser extent, in gases.

Derivation 21.8: Derivation of the viscosity coefficient (Eq. 21.38, refer to diagram in Derivation 21.7)

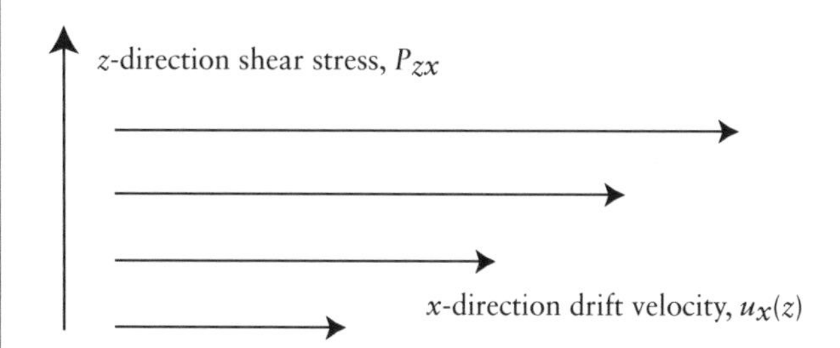

$$dP_{zx} = \frac{dp_x(z)}{dtds}\cos\theta$$

$$= \frac{dN}{dtds}mu_x(z)\cos\theta$$

$$= \frac{1}{4\pi}n_p mu_x(z)vF_v(v)dv\cos^2\theta\sin\theta d\theta d\phi$$

$$\therefore P_{zx}^{\pm} = \frac{1}{4\pi}n_p mu_x(z_0 \mp \lambda)\underbrace{\int_{-\infty}^{\infty} vF_v(v)dv}_{=\langle v\rangle}\underbrace{\int_0^{\pi/2}\cos^2\theta\sin\theta d\theta}_{=1/3}\underbrace{\int_0^{2\pi} d\phi}_{=2\pi}$$

$$= \frac{1}{6}n_p mu_x(z_0 \mp \lambda)\langle v\rangle$$

$$\therefore P_{zx} = P_{zx}^{+} - P_{zx}^{-}$$

$$= \frac{1}{6}n_p mu_x(z_0 - \lambda)\langle v\rangle - \frac{1}{6}n_p mu_x(z_0 + \lambda)\langle v\rangle$$

$$\simeq \frac{1}{6}n_p m\langle v\rangle\left\{u_x(z_0) - \frac{\partial u_x}{\partial z}\lambda - \left(u_x(z_0) + \frac{\partial u_x}{\partial z}\lambda\right)\right\}$$

$$= -\underbrace{\frac{1}{3}n_p m\langle v\rangle\lambda}_{=\eta}\frac{\partial u_x}{\partial z}$$

and $P_{zx} = -\eta\dfrac{\partial u_x}{\partial z}$

$$\therefore \eta = \frac{1}{3}n_p m\langle v\rangle\lambda$$

- Starting point: (refer to diagram in Derivation 21.7) the z-component of shear stress, dP_{zx} (force per unit area), for particles incident on a plane at $z = z_0$ at an angle θ is:

$$dP_{zx} = \frac{F_x(z)}{ds}\cos\theta$$

$$= \frac{dp_x(z)}{dtds}\cos\theta$$

where $F_x = dp_x/dt$ (Newton's Second Law).

- The particle momentum in the x-direction is proportional to the drift velocity per particle, $u_x(z)$:

$$dp_x(z) = dNmu_x(z)$$

(refer to Derivation 21.7 for an expression for dN).

- The net upward shear stress, P_{zx}^{+}, and downward shear stress, P_{zx}^{-}, is found by integrating over all angles in a hemisphere, and over all velocities.
- The total shear stress, P_{zx}, is therefore the difference between upward and downward components. We can approximate using a Taylor expansion about z_0.

The one-dimensional heat flux, j_z [J m^{-2} s^{-1}], is defined as the heat flow per unit area per unit time. This obeys a similar differential equation to those for particle flux and shear stress, where heat flux is proportional to temperature gradient:

$$j_z = -\kappa\frac{\partial T}{\partial z}$$

(Eq. 21.39). The coefficient of proportionality, κ [Wm^{-1} K^{-1}], is called the *thermal conductivity*, and it has a similar form to the diffusion coefficient from Eq. 21.32 and viscosity from Eq. 21.38 (Eq. 21.41 and Derivation 21.9).

Multiplying j_z by cross-sectional area, A, we obtain the *heat conduction formula*, which gives the rate of heat transfer (Eq. 21.40).

Derivation 21.9: Derivation of the thermal conductivity (Eq. 21.41, refer to diagram in Derivation 21.7)

$$\begin{aligned} \mathrm{d}j_z &= \frac{\mathrm{d}Q(z)}{\mathrm{d}s\mathrm{d}t}\cos\theta \\ &= \frac{\mathrm{d}N}{\mathrm{d}s\mathrm{d}t}q(z)\cos\theta \\ &= \frac{1}{4\pi}n_\mathrm{p}q(z)vF_v(v)\mathrm{d}v\cos^2\theta\sin\theta\mathrm{d}\theta\mathrm{d}\phi \end{aligned}$$

$$\therefore j_z^{\pm} = \frac{1}{4\pi}n_\mathrm{p}q(z_0 \mp \lambda)\underbrace{\int_{-\infty}^{\infty} vF_v(v)\mathrm{d}v}_{=\langle v\rangle}\underbrace{\int_0^{\pi/2}\cos^2\theta\sin\theta\mathrm{d}\theta}_{=1/3}\underbrace{\int_0^{2\pi}\mathrm{d}\phi}_{=2\pi}$$

$$= \frac{1}{6}n_\mathrm{p}q(z_0 \mp \lambda)\langle v\rangle$$

$$\begin{aligned} \therefore j_z &= j_z^+ - j_z^- \\ &= \frac{1}{6}n_\mathrm{p}q(z_0-\lambda)\langle v\rangle - \frac{1}{6}n_\mathrm{p}q(z_0+\lambda)\langle v\rangle \\ &\simeq \frac{1}{6}n_\mathrm{p}\langle v\rangle\left\{q(z_0) - \frac{\partial q}{\partial z}\lambda - \left(q(z_0) + \frac{\partial q}{\partial z}\lambda\right)\right\} \\ &= -\frac{1}{3}n_\mathrm{p}\langle v\rangle\lambda\frac{\partial q}{\partial z} \\ &= -\frac{1}{3}n_\mathrm{p}\langle v\rangle\lambda\frac{\partial q}{\partial T}\frac{\partial T}{\partial z} \\ &= -\underbrace{\frac{1}{3}n_\mathrm{p}C_V^N\langle v\rangle\lambda}_{=\kappa}\frac{\partial T}{\partial z} \end{aligned}$$

$$\text{and } j_z = -\kappa\frac{\partial T}{\partial z}$$

$$\therefore \kappa = \frac{1}{3}n_\mathrm{p}C_V^N\langle v\rangle\lambda$$

- Starting point: (refer to diagram in Derivation 21.7.) The z-component of heat flux, $\mathrm{d}j_z$, for particles incident on a plane at $z = z_0$ at an angle θ is:

$$\mathrm{d}j_z = \frac{\mathrm{d}Q(z)}{\mathrm{d}s\mathrm{d}t}\cos\theta$$

- In terms of the heat gradient per particle, $q(z)$, the total heat is:

$$\mathrm{d}Q(z) = \mathrm{d}Nq(z)$$

(refer to Derivation 21.7 for expression for $\mathrm{d}N$).

- The net upward heat flux, j_z^+, and downward heat flux, j_z^-, are found by integrating over all angles in a hemisphere, and over all particle velocities.
- The total heat flux, j_z, is therefore the difference between upward and downward components. We can approximate using a Taylor expansion about z_0.
- We can express the heat gradient in terms of a temperature gradient:

$$\frac{\partial q}{\partial z} = \frac{\partial q}{\partial T}\frac{\partial T}{\partial z} = C_V^N\frac{\partial T}{\partial z}$$

where C_V^N is the constant volume heat capacity per particle.

Thermal conduction (Derivation 21.9):

$$j_z = -\kappa\frac{\partial T}{\partial z} \quad (21.39)$$

$$\therefore \frac{\mathrm{d}Q}{\mathrm{d}t} = -\kappa A\frac{\mathrm{d}T}{\mathrm{d}z} \quad (21.40)$$

Thermal conductivity:

$$\kappa = \frac{1}{3}n_\mathrm{p}C_V^N\langle v\rangle\lambda \quad (21.41)$$

where C_V^N is the constant volume heat capacity per particle.

Thermal conductivity is a measure of a substance's ability to conduct heat. Materials with high thermal conductivity, for example, silver and other metals, are good conductors of heat, meaning that they rapidly transport heat and increase their temperature when in contact with a hot object, such as a flame. Conversely, materials with *low* thermal conductivity, such as wood or rubber, are thermal *insulators*, which means that they are poor conductors of heat.

Newton's Law of Cooling

For a constant temperature gradient over a distance L, the heat conduction equation (Eq. 21.40) is:

$$\frac{\mathrm{d}Q}{\mathrm{d}t} = -\kappa A\frac{(T_L - T_0)}{L}$$

where T_0 and T_L are the temperatures at $z = 0$ and $z = L$ respectively. From the definition of specific heat capacity ($c = \mathrm{d}Q/m\mathrm{d}T$), the rate of change of heat can be written as:

$$\frac{\mathrm{d}Q}{\mathrm{d}t} = mc\frac{\mathrm{d}T}{\mathrm{d}t}$$

Equating these two expressions for $\mathrm{d}Q/\mathrm{d}T$, we find that the rate of change of temperature of an object is proportional to the temperature difference between it, $T(t)$, and its surroundings, T_s:

$$\frac{\mathrm{d}T}{\mathrm{d}t} = -\frac{\kappa A}{mcL}(T(t) - T_s)$$

This statement is known as *Newton's Law of Cooling* (Eq. 21.42). The general solution to Eq. 21.42 is an exponential decay, with time constant:

$$\tau = \frac{1}{K} = \frac{mcL}{\kappa A}$$

The exact solution depends on boundary conditions. For an initial temperature $T(0) = T_0$, the solution is:

$$T(t) = T_s + (T_0 - T_s)\mathrm{e}^{-Kt}$$

The temperature tends to T_s as $t \to \infty$.

Newton's Law of Cooling:

$$\begin{aligned}\frac{\mathrm{d}T}{\mathrm{d}t} &= -\frac{\kappa A}{mcL}(T(t) - T_s) \\ &\equiv -K(T(t) - T_s)\end{aligned} \qquad (21.42)$$

where K is a positive constant.

Convection

Convection occurs only in liquids and gases, and it involves the bulk transport of parcels of fluid or gas. *Natural convection* occurs when hot, less-dense parcels rise, while cooler, more-dense parcels sink to take their place. These, in turn, then heat up and rise. This repeated circulation is called a *convection current*. *Forced convection* occurs when a fan, pump, or draft, for example, assists convection.

The differential equations governing convection are similar to those for conduction (Eqs. 21.40 and 21.41). The difference is that the constant of proportionality is not the thermal conductivity, κ, but is called the *heat transfer coefficient*, h. This is not solely a material property but depends also on external properties of the flow such as geometry, temperature, and flow velocity. For this reason, h is usually an experimentally-determined quantity.

Chapter 22
Classical thermodynamics

Classical thermodynamics is concerned with the interchange of matter and energy between a system and its surroundings, and how the system responds to such interchanges. The birth of the subject coincided with the start of the industrial revolution around the start of the nineteenth century, motivated by the interest to develop mechanical power and engines from heat, such as the steam engine. At that time, Dalton was only beginning to propose the atomic structure of matter. Therefore, whereas kinetic theory uses the microscopic properties of individual particles to explain the macroscopic properties of a system, classical thermodynamics treats systems exclusively on a macroscopic scale. Although thermodynamics is more general than kinetic theory, and handles systems other than ideal gases, there is good agreement between the two descriptions.

22.1 Thermodynamic systems

Classical thermodynamics is based on four laws, known collectively as the *Laws of Thermodynamics*. Before presenting these laws in Section 22.2, we will first provide an overview of the terminology of thermodynamics.

A *thermodynamic system* is defined by three components: the system, the surroundings, and the boundary between them (Fig. 22.1). The *system* is the part whose properties we are interested in studying, for example, an ideal gas in a cylinder with a piston; the *surroundings* involve everything else. The system can interact with its surroundings by exchanging matter or energy across the boundary. Thermodynamic systems fall into one of three categories depending on what can be exchanged between the system and its surroundings: (i) in an *open system*, both matter and energy can be exchanged with the surroundings, (ii) in a *closed system*, only energy can be exchanged, therefore, the number of particles within the system remains constant, and (iii) in an *isolated system*, the system is totally isolated from its surroundings, meaning that neither matter nor energy can be exchanged. In this chapter, we'll deal primarily with closed systems. In Section 22.3.4, we will look briefly at the thermodynamics of open systems, since this is important for understanding phase transitions.

Heat, work, and internal energy

Recall from kinetic theory that the *internal energy*, U [J], of a system is the sum of kinetic and potential energies of the particles that make it up. For example, the kinetic energy includes translational, rotational, and vibrational motion, whereas the potential energy includes chemical energy stored in atomic bonds, nuclear energy stored in nuclear bonds, and latent energy associated with the state of matter (solid, liquid, or gas). Therefore, the internal energy is the sum of all energies *within* the system, but does not include the

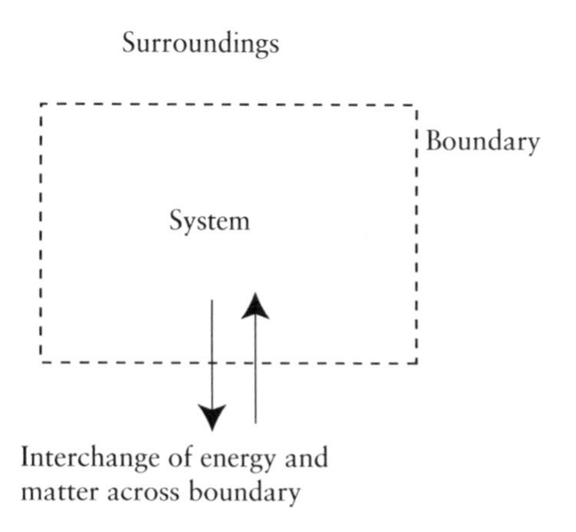

Figure 22.1: Schematic of a thermodynamic system. In an open system, both matter and energy can be exchanged between the system and the surroundings; in a closed system, only energy exchange is possible; and in an isolated system, neither matter nor energy is exchanged.

energy the system has as a result of its overall motion or position. For example, the kinetic energy of a ball rolling down a hill is *not* considered as internal energy.

The combination of kinetic energy and latent energy is often called *thermal energy*. The *temperature*, T [K], of a system provides a measure of the average kinetic energy of its particles; the greater the average molecular kinetic energy, the higher the system's temperature.

The internal energy of a system changes if there is a net exchange of energy with its surroundings. The two ways in which energy is exchanged across a boundary are via *heat flow*, Q [J], or *work done*, W [J]. These are both processes of energy transfer from one body to another. "Heat flow" refers specifically to the transfer of thermal energy between two bodies at different temperatures, whereas "work" is the general term to describe energy transfer of *any* nature, other than heat. Since classical thermodynamics was developed before the microscopic behavior of particles within a system was understood, the internal energy of a system is usually understood in terms of how it changes with heat and work inputs and outputs, as described by the First Law of Thermodynamics (see Section 22.2.2).

Thermodynamic equilibrium

An important concept in thermodynamics is that of *equilibrium*. A system is in *thermodynamic equilibrium* with its surroundings when its macroscopic properties, such as temperature, pressure, and volume, are not changing with time. In classical thermodynamics, we deal exclusively with systems in equilibrium. To describe systems that aren't in equilibrium, we require a microscopic approach, such as kinetic theory. If a system is in thermodynamic equilibrium, then it is simultaneously in (i) *mechanical* equilibrium (constant pressure), (ii) *chemical* equilibrium (constant particle concentration), and (iii) *thermal* equilibrium (constant temperature). Thermal equilibrium is achieved when a system and its surroundings are in *thermal contact*, which means that heat can flow between them. A boundary that allows two systems to be in thermal contact is called a *diathermal* boundary, whereas a boundary that thermally isolates two systems, such that heat does not flow between them, is called an *adiathermal* boundary.

Thermodynamic states

The *thermodynamic state* of a system is a broad term that encompasses all of the macroscopic properties that define it. These properties are called *state variables*. For example, volume, temperature, and internal energy are state variables, whereas heat and work are not. State variables are often related to one another with an *equation of state*. For example, the Ideal Gas Law, $pV = nRT$, is the equation of state that relates pressure, volume, and temperature. A state variable that depends on the physical size of the system, such as volume, is called an *extensive* variable, whereas one that is independent of the system's size, such as temperature and pressure, is an *intensive* variable. Said another way, an intensive variable has the same value whether we are looking at the whole system, or only at a part of the system, but this is not true for an extensive variable.

Thermodynamic processes

When a thermodynamic system changes state, we say that it undergoes a *thermodynamic process*. Processes are classified as either *reversible* or *irreversible*. In irreversible processes, energy is permanently lost from the system due to dissipative forces such as friction. Irreversible systems exhibit *hysteresis* (the change of state depends on the path taken to arrive at that state). Since the dissipated energy cannot be recovered by changing the system back to its original state, then the process is irreversible. Conversely, reversible processes do not exhibit hysteresis, and the original state of the system can be recovered by reversing the process it undertook to arrive at a given state. For example, a reversible expansion of a gas followed by a reversible compression would take it back to its initial state. In practice, perfectly reversible processes are idealizations—there is always *some* energy lost via friction or another dissipative force. However, in theory, reversible processes can be achieved by carrying out the process in infinitesimal steps, with each step performed over a time period much slower than the response time between the system and surroundings. This ensures that they are always in thermal equilibrium with each other, and therefore the intermediate states of the system correspond to definite values—that is, ones that can be known through measurement—of its macroscopic properties. A process carried out in this way, through a series of equilibrium states, is called a *quasi-static* process.

Since the intermediate states are, in theory, known, reversible and quasi-static processes can be plotted on p–V diagrams (where $pV = nRT$ for all values of p and V) (Fig. 22.2). From this, we can integrate the curve to determine total changes. In contrast, only the initial and final states of irreversible processes may be plotted since they do not proceed via equilibrium states ($\therefore pV \neq nRT$ for intermediate states between the initial and final states), and therefore the integration path between them is unknown.

Common reversible processes

In this chapter, we will deal primarily with reversible processes since they are easier to treat mathematically, and also because they represent idealized experimental outcomes where energy is not irrecoverably lost from a system. Four common reversible processes are *isothermal processes* (temperature of system remains constant), *isochoric processes* (volume of system remains constant), *isobaric processes* (pressure of system remains constant), and *adiabatic processes* (no heat is exchanged between the system and its surroundings). These processes are summarized and plotted on a $p-V$ diagram in Summary box 22.1.

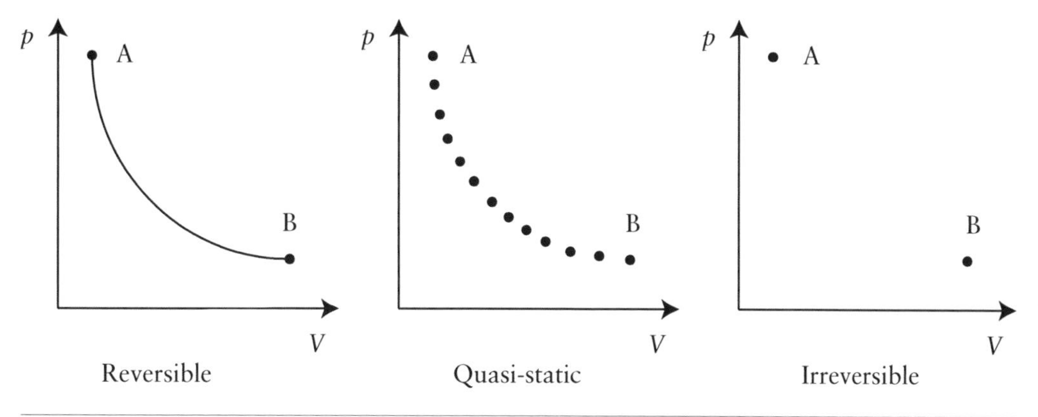

Figure 22.2: $p-V$ diagrams of reversible, quasi-static, and irreversible processes, occurring between two equilibrium states, A and B.

Summary box 22.1: Common reversible processes for ideal gases ($pV = nRT$ throughout process)

Isothermal process: constant temperature

$$\Delta T = 0$$

$$\therefore pV = \text{constant} \quad (22.1)$$

Isochoric process: constant volume

$$\Delta V = 0$$

$$\therefore \frac{p}{T} = \text{constant} \quad (22.2)$$

Isobaric process: constant pressure

$$\Delta p = 0$$

$$\therefore \frac{V}{T} = \text{constant} \quad (22.3)$$

Adiabatic process: no heat exchange between system and surroundings:

$$\Delta Q = 0$$

$$\therefore pV^{\gamma} = \text{constant} \quad (22.4)$$

where γ is the ratio of constant pressure to constant volume heat capacities: $\gamma = C_p / C_V$ (refer to Derivation 22.2 for derivation of Eq. 22.4).

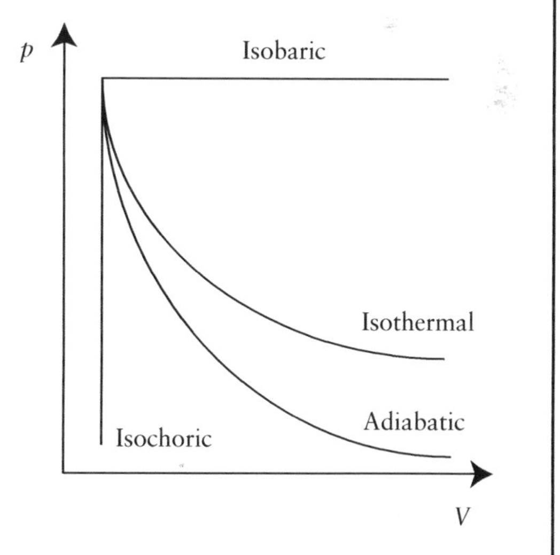

The steepness of the adiabatic curve depends on the value of γ.

22.2 The Laws of Thermodynamics

Classical thermodynamics describes how a system's macroscopic properties and its energy inputs and outputs change when it undergoes a thermodynamic process (a change of state). In more practical terms, it deals with quantifying the extraction of useful work from a system during a change of state. The fundamental results of classical thermodynamics are encapsulated in four laws, known as the *Laws of Thermodynamics* (Summary box 22.2). Below, we will look at each in turn.

22.2.1 The Zeroth Law

When a hot and a cold system are put in thermal contact with each other, heat flows from the hot to the cold system. This decreases the temperature of the hot system while increasing that of the cold one. When the two systems are at the same temperature, then no more heat flows between them, and the systems are said to be in *thermal equilibrium*. This is the basis of the *Zeroth Law of Thermodynamics*, which states that:

If two thermodynamic systems are at the same time in thermal equilibrium with a third system, then they are also in thermal equilibrium with each other.

The Zeroth Law implies that systems in thermal equilibrium are at the same temperature as each other, and that their temperature remains constant in time. Historically, the Zeroth Law was introduced *after* the other three laws. However, it is considered more fundamental than the others and was therefore labeled the "Zeroth Law" rather than the "Fourth Law." It provides the basis for the definition of temperature in terms of thermal equilibrium.

22.2.2 The First Law

The *First Law of Thermodynamics* is a statement of the conservation of energy. It states that, during any thermodynamic process:

The change in internal energy of a thermodynamic system, ΔU, is given by the sum of work done, W, and heat exchanged with its surroundings, Q:

$$\Delta U = Q + W$$

(Eq. 22.5 in Summary box 22.2). The law is often stated in terms of an incremental change of internal energy, $dU = dW + dQ$ (Eq. 22.6 in Summary box 22.2), rather than the net change, ΔU. Below, we will find expressions for dU, dQ, and dW in terms of changes in the system's state variables, such as volume and temperature changes, for reversible processes. (The expressions do not hold for irreversible processes since we do not know how much energy is lost to the environment.)

Summary box 22.2: The Laws of Thermodynamics

- **Zeroth Law: thermal equilibrium**
  If two thermodynamic systems are at the same time in thermal equilibrium with a third system, then they are also in thermal equilibrium with each other.

- **First Law: conservation of internal energy**
  The change in internal energy of a thermodynamic system is given by the sum of work and heat exchanged with its surroundings:

$$\Delta U = Q + W \qquad (22.5)$$

$$\text{or } dU = dQ + dW \qquad (22.6)$$

  Conventions:
  Δ: net change; d: incremental change;
  $+Q$, $+dQ$: heat *into* the system; $-Q$, $-dQ$: heat *out of* the system;
  $+W$, $+dW$: work done *on* the system; $-W$, $-dW$: work done *by* the system.

- **Second Law: increasing entropy**
  No process is possible where the total entropy of the Universe decreases. The total change in entropy is always greater than or equal to zero:

$$\text{Reversible process: } \Delta S = 0$$

$$\text{Irreversible process: } \Delta S > 0$$

$$\Rightarrow \Delta S \geqslant 0 \text{ always} \qquad (22.7)$$

- **Third Law: absolute zero temperature**
  The entropy of a system approaches zero as its temperature approaches absolute zero.

Work done, dW

Gases and liquids exert pressure—a force per unit area—on the walls of their container. From a molecular viewpoint, this pressure results from the change in momentum of molecules colliding against the container walls. When a system's volume decreases, work must be done against these molecular forces. By considering the work done on a system by compressing it with a piston, we find an expression for the work done in terms of the pressure and change in volume of the system (Eqs. 22.8 and 22.9 and Derivation 22.1). Work is done *on* a system if its volume decreases (dV is negative and dW is positive), and work is done *by* a system if its volume increases (dV is positive and dW is negative). From Eq. 22.9, we see that (i) no work is done on or by a system if its volume remains constant and (ii) the area under a $p-V$ curve of a reversible process represents the total work done during the change (Fig. 22.3).

Work done for a reversible process (Derivation 22.1):

$$\mathrm{d}W = -p\mathrm{d}V \tag{22.8}$$

$$\therefore W = -\int_{V_1}^{V_2} p\mathrm{d}V \tag{22.9}$$

$+W$: work done on a system;
$-W$: work done by a system.

Heat exchange, dQ

The heat capacity of a system is defined as:

$$C_\alpha = \left(\frac{\partial Q}{\partial T}\right)_\alpha$$

where α is the variable that is kept constant while heat is added. For example, for an isobaric process (constant pressure), $\alpha \equiv p$, whereas for an isochoric process (constant volume), $\alpha \equiv V$. If we assume that heat flow is a function of T only (Zeroth Law), then we can use a total derivative instead of a partial derivative:

$$\therefore C_\alpha = \frac{\mathrm{d}Q_\alpha}{\mathrm{d}T}$$

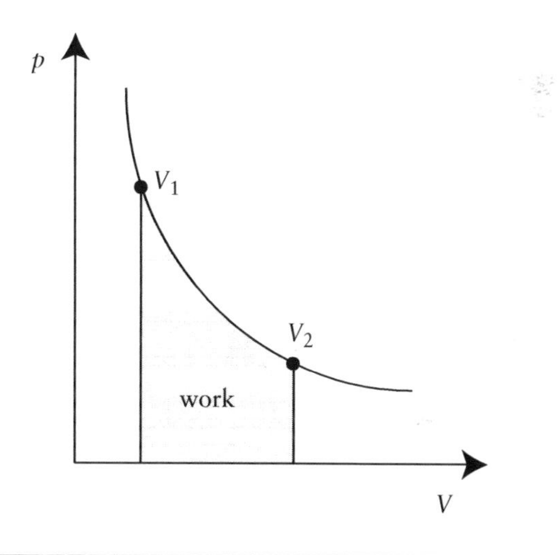

Figure 22.3: The work done by a gas upon expansion from V_1 to V_2 equals the area under the curve ($W = -\int p\mathrm{d}V$).

Derivation 22.1: Derivation of work done for a reversible process (Eqs. 22.8 and 22.9)

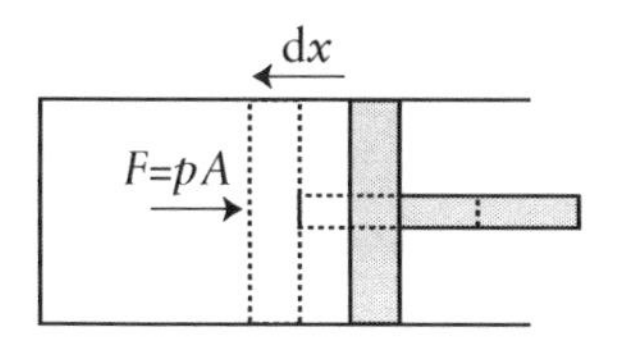

$$\mathrm{d}W = -F\mathrm{d}x$$

$$= -pA\mathrm{d}x$$

$$= -p\mathrm{d}V$$

$$\therefore W = -\int_{V_1}^{V_2} p(V)\mathrm{d}V$$

- Starting point: the work done on a gas in a cylinder by moving a piston through a length dx, against an opposing force F, is $\mathrm{d}W = -F\mathrm{d}x$.
- The opposing force results from the system's pressure: $F = pA$, where A is the cross-sectional area of the cylinder.
- Define an element of volume as $\mathrm{d}V = A\mathrm{d}x$.
- For reversible processes, the integration path is defined; therefore, we can determine the total work done by integrating from volume V_1 to volume V_2.

Rearranging, we can find an expression for an element of heat flow into or out of a system (Eq. 22.10). An increase in temperature corresponds to heat flowing *into* the system (dT and dQ are positive), whereas a decrease in temperature corresponds to heat flowing *out* of the system (dT and dQ are negative). For reversible processes, the total heat input or output is equal to the integral of heat capacity over temperature, between the temperature limits of initial and final states (Eq. 22.11).

Heat exchange for a reversible process:

$$dQ = C_\alpha dT \qquad (22.10)$$

$$\therefore Q = \int_{T_1}^{T_2} C_\alpha dT \qquad (22.11)$$

$+Q$: heat flow into a system;
$-Q$: heat flow out of a system.

Internal energy change, dU

The internal energy change is found by combining the expressions for dW and dQ using the First Law of Thermodynamics:

$$dU = dQ + dW$$

Consider an isochoric process, where $dV = 0$ (V = constant). From Eqs. 22.8 and 22.10:

$$dW = 0$$

$$\text{and } dQ = C_V dT$$

Therefore, the change in internal energy of the system is:

$$\begin{aligned} dU &= dQ \\ &= C_V dT \end{aligned}$$

(Eq. 22.12). For a reversible process, the integration path is defined; therefore, the total internal energy change is found by integrating dU (Eq. 22.13). Even though this expression for internal energy is derived for an isochoric process, it holds true for *all* ideal gas processes, even when the volume is not kept constant. From it, we see that the internal energy of an ideal gas depends only on its temperature. (For *real* gases, the internal energy depends on the volume of gas as well; refer to Worked example 23.2.)

Internal energy for a reversible process:

$$dU = C_V dT \qquad (22.12)$$

$$\therefore \Delta U = \int_{T_1}^{T_2} C_V dT \qquad (22.13)$$

Using the First Law

Worked example 22.1 uses the above expressions, together with the Ideal Gas Law, to calculate W, Q, and ΔU for various reversible processes. We can also derive the adiabatic condition, pV^γ = constant, where $\gamma = C_p/C_V$, from the First Law (Derivation 22.2).

Derivation 22.2: Derivation of the adiabatic condition (refer to Eq. 22.4 in Summary box 22.1)

$$\begin{aligned} dU &= dW \\ \therefore C_V dT &= -p dV \\ &= -\frac{nRT}{V} dV \\ &= -\frac{(C_p - C_V)T}{V} dV \end{aligned}$$

$$\therefore C_V \int_{T_1}^{T_2} \frac{dT}{T} = -C_V(\gamma - 1) \int_{V_1}^{V_2} \frac{dV}{V}$$

$$\ln(T_2/T_1) = -(\gamma - 1)\ln(V_2/V_1)$$

$$\therefore \frac{T_2}{T_1} = \left(\frac{V_1}{V_2}\right)^{(\gamma-1)}$$

$$\therefore TV^{(\gamma-1)} = \text{constant}$$

$$\therefore pV^\gamma = \text{constant}$$

- Starting point: for an adiabatic process, there is no heat exchange between a system and its surroundings, $\therefore dQ = 0$. Therefore, the First Law for an adiabatic process is:

  $$dU = dW$$

  where $dU = C_V dT$ and $dW = -p dV$ (refer to Eqs. 22.8 and 22.12).

- The difference and ratio of heat capacities are:

  $$nR = C_p - C_V$$

  $$\text{and } \gamma = \frac{C_p}{C_V}$$

  (refer to Section 21.4.2).

- Use the Ideal Gas Law to eliminate T from $TV^{(\gamma-1)}$.

WORKED EXAMPLE 22.1: Using the First Law of Thermodynamics to determine ΔU, W, and Q for reversible processes of ideal gases

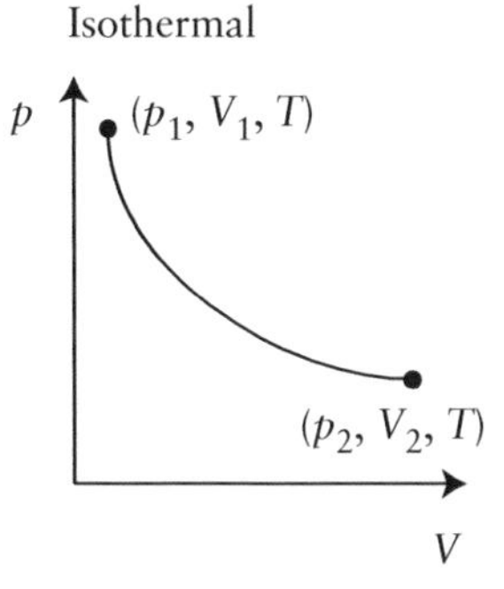

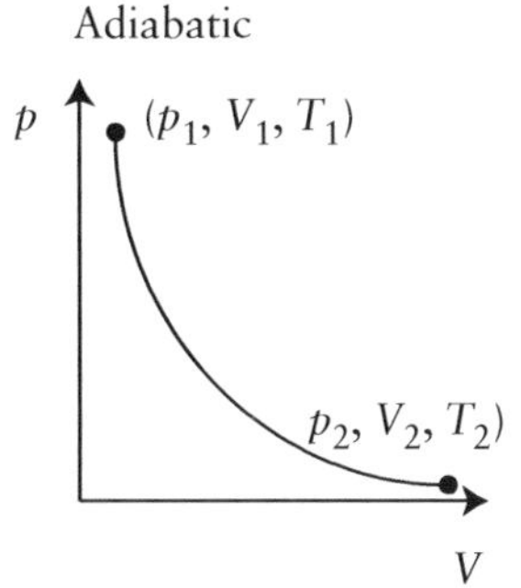

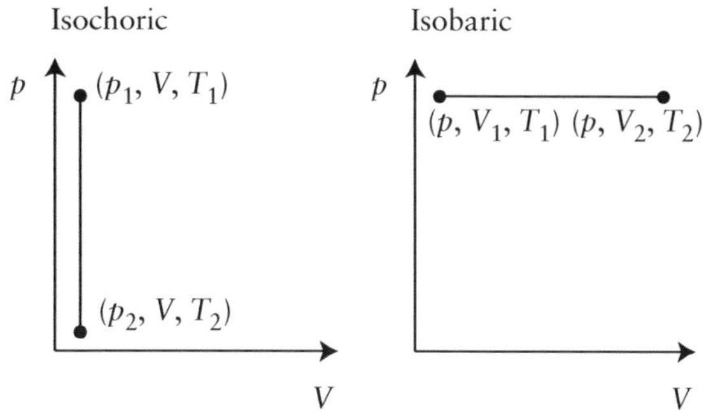

Isothermal expansion ($\Delta T = 0$):
Internal energy change:

$$\Delta U = C_V \Delta T = 0$$

Work done by gas:

$$W = -\int_{V_1}^{V_2} p(V)\mathrm{d}V$$

$$= -\int_{V_1}^{V_2} \frac{nRT}{V}\mathrm{d}V$$

$$= nRT\ln\left(\frac{V_1}{V_2}\right)$$

Heat exchange:

$$Q = \Delta U - W$$

$$= -nRT\ln\left(\frac{V_1}{V_2}\right)$$

Isochoric process ($\Delta V = 0$):
Internal energy change:

$$\Delta U = C_V(T_2 - T_1)$$

Work done by gas:

$$W = -p\Delta V = 0$$

Heat exchange:

$$Q = \Delta U - W$$

$$= C_V(T_2 - T_1)$$

Adiabatic expansion ($pV^\gamma = A$ (constant)):
Internal energy change:

$$\Delta U = C_V(T_2 - T_1)$$

Work done by gas:

$$W = -\int_{V_1}^{V_2} p(V)\mathrm{d}V$$

$$= -A\int_{V_1}^{V_2} \frac{\mathrm{d}V}{V^\gamma}$$

$$= \frac{A}{(\gamma - 1)}\left(\frac{V_2}{V_2^\gamma} - \frac{V_1}{V_1^\gamma}\right)$$

$$= \frac{(p_2V_2 - p_1V_1)}{(\gamma - 1)}$$

$$or\;\; W = \Delta U$$

$$= C_V(T_2 - T_1)$$

Heat exchange:

$$Q = 0$$

Isobaric expansion ($\Delta p = 0$):
Internal energy change:

$$\Delta U = C_V(T_2 - T_1)$$

Work done by gas:

$$W = -p(V_2 - V_1)$$

Heat exchange:

$$Q = \Delta U - W$$

$$= C_V(T_2 - T_1) + p(V_2 - V_1)$$

$$or\;\; Q = C_p(T_2 - T_1)$$

22.2.3 The Second Law

The *Second Law of Thermodynamics* introduces another function of state, called the *entropy*, S [J K^{-1}]. The Second Law states that:

No process is possible where the total entropy of the Universe decreases. The total change in entropy is always greater than or equal to zero: $\Delta S \geqslant 0$.

(Eq. 22.7 in Summary box 22.2). Entropy is often interpreted as the "degree of disorder" of a system. For example, it can be thought of as a measure of the number of different ways that quanta of energy in a system can be arranged between the particles that make it up, without changing the total energy of the system. A small change in entropy of a system, dS, is defined as the small change in heat flow per unit temperature into the system, provided the heat flow corresponds to a reversible process:

$$\mathrm{d}S \stackrel{\text{def}}{=} \frac{\mathrm{d}Q_{\text{rev}}}{T}$$

(Eq. 22.14). Therefore, a system's entropy changes if heat is transferred into or out of it. This definition of entropy follows from the *Clausius inequality*, which is described below.

Entropy change for a reversible process:

$$\mathrm{d}S \stackrel{\text{def}}{=} \frac{\mathrm{d}Q_{\text{rev}}}{T} \qquad (22.14)$$

Whereas the First Law puts restrictions on what processes are *energetically* possible, the Second Law tells us the *direction* in which they occur. For example, we know from experience that heat can spontaneously flow from a hot to a cold body. Although the First Law allows the reverse process to happen, namely, heat to flow from a cold body to a hot body, the Second Law states that it cannot happen spontaneously since, from Eq. 22.14, the decrease in entropy of the cool object would be greater than the increase in entropy of the hot object (since $T_{\text{cold}} < T_{\text{hot}}$). This would result in a net decrease of entropy of the combined system (and Universe), which violates the Second Law. Said another way, the natural direction for change is towards disorder.[1]

[1] This issue raises some interesting philosophical questions regarding the ultimate fate of the Universe.

Heat engines and heat pumps

The Second Law of Thermodynamics describes the operation and efficiencies of heat engines and heat pumps. Heat *engines* are machines that continuously consume heat to generate useful work. For example, a steam engine that drives a train is a heat engine. Heat *pumps* are machines that continuously transfer heat from a cold object to a hot object. These require work to be done on the pump since this is against the direction of spontaneous heat flow. Refrigerators and air conditioners are examples of heat pumps. Both heat engines and heat pumps involve cyclic processes, and therefore energy conversion is continuous.

Kelvin and Clausius found that a perfect heat engine (one that converts *all* the consumed heat into useful work) and a perfect heat pump (one that transfers heat from a cool to hot object without using work) are impossible to construct. Both such devices violate the Second Law—the total entropy of the Universe *decreases* in both cases. The Second Law is therefore often stated in terms of the operation of heat engines and heat pumps, and makes no mention of entropy. These are known as the Kelvin and Clausius formulations of the Second Law of Thermodynamics:

- **The Kelvin formulation:** the complete conversion of heat into work is impossible; it is impossible to construct a perfect heat engine.
- **The Clausius formulation:** Heat cannot spontaneously flow from a cold to a hot object; it is impossible to construct a perfect heat pump.

These statements are equivalent to one another, and to the entropy formulation of the Second Law, as we shall see below.

The Second Law is satisfied for a heat engine if some of the consumed heat is lost to the environment. For a heat pump, work must be provided to transport the heat against its natural direction of flow. These operations are shown schematically in Fig. 22.4. We consider a heat engine or pump operating between two large reservoirs of constant temperature, T_{H} and T_{C}, for the hot and cold reservoirs, respectively. We assume that the heat flow to or from each reservoir, Q_{H} and Q_{C}, is small in comparison with the thermal capacity of the reservoirs; therefore, their temperatures remain constant. From the conservation of energy, we have:

$$Q_{\text{H}} = Q_{\text{C}} + W$$

both for heat engines and heat pumps. Figure 22.4 illustrates that a heat pump can be thought of as a heat engine operating in reverse.

Efficiency and coefficient of performance

The efficiency of a heat engine or heat pump, η, is a measure of the useful energy output to the energy input. For a heat engine, the useful energy output is W for an input energy Q_H. Therefore, the efficiency of a heat engine is:

$$\eta = \frac{W}{Q_H}$$

Using $W = Q_H - Q_C$, we can express the efficiency in terms of Q_C and Q_H only:

$$\eta = \frac{Q_H - Q_C}{Q_H} = 1 - \left|\frac{Q_C}{Q_H}\right|$$

(Eq. 22.15). For a heat pump, the useful quantity is Q_H for a heating device, and Q_C for a cooling device, for an input energy of W. Therefore, the efficiency of a heat pump is:

$$\eta = \frac{Q_H}{W} \text{ or } \eta = \frac{Q_C}{W}$$

(Eqs. 22.16 and 22.17). Since these ratios can be greater than one, the term *coefficient of performance*, or CoP, is used instead of efficiency.

Efficiency and coefficients of performance:

Heat engine efficiency:

$$\eta = \frac{W}{Q_H} = 1 - \left|\frac{Q_C}{Q_H}\right| \quad (22.15)$$

Heat pump coefficients of performance (CoP):

Heating: $$\text{CoP}_H = \frac{Q_H}{W} \quad (22.16)$$

Cooling: $$\text{CoP}_C = \frac{Q_C}{W} \quad (22.17)$$

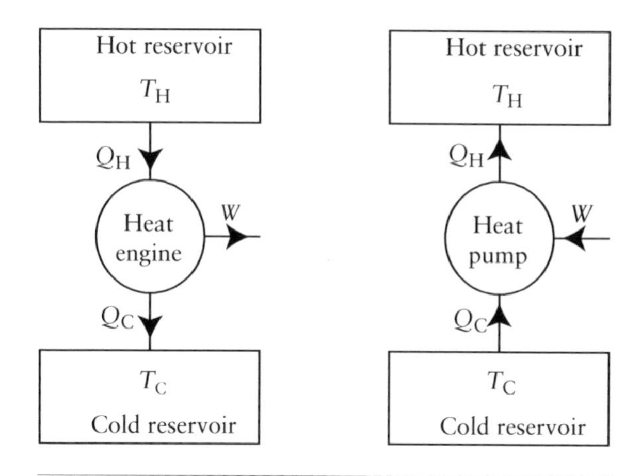

Figure 22.4: Schematic representations of a heat engine (left) and heat pump (right) operating between two reservoirs at constant temperatures, T_H and T_C. Q_H is the heat flow into or out of the hot reservoir; Q_C is the heat flow into or out of the cold reservoir; and W is the work into or out of the heat engine or pump.

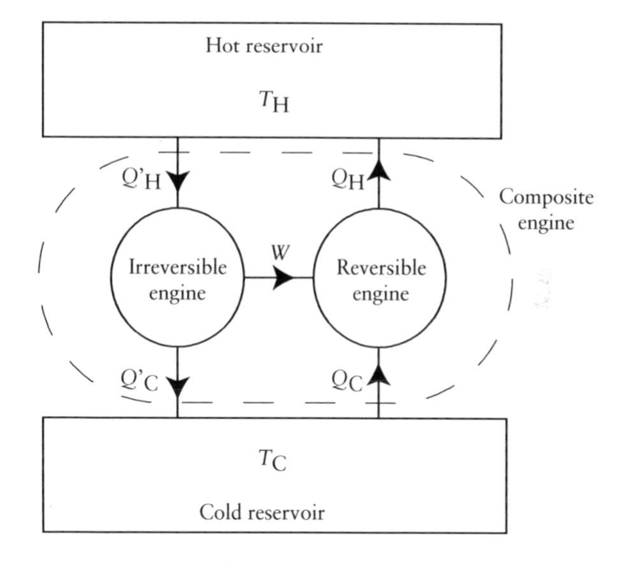

Figure 22.5: Composite engine consisting of an irreversible and a reversible component.

Carnot's theorem

Reversible engines do not dissipate energy via dissipative forces, such as friction. *Carnot's theorem* states that:

Reversible heat engines are the most efficient type of engine.

Consider, for example, the composite engine in Fig. 22.5. The variables W, $Q'_{H,C}$, and $Q_{H,C}$ are positive in the directions indicated. From the Clausius formulation of the Second Law, there is a net flow of heat away from the hot reservoir:

$$Q'_H - Q_H \geqslant 0$$

$$\therefore Q'_H \geqslant Q_H$$

$$\therefore \frac{W}{Q'_H} \leqslant \frac{W}{Q_H}$$

The efficiency of a heat pump is defined as $\eta = W/Q_H$, $\therefore \eta' \leqslant \eta$ where η' is the efficiency of the irreversible engine and η is that of the reversible engine. Therefore, reversible engines are the most efficient:

$$\therefore \eta_R \geqslant \eta_I$$

where the subscripts R and I stand for reversible and irreversible, respectively. This statement is intuitively obvious, since if heat is lost via friction in a heat engine, then less energy is available to do work.

Carnot engine

A reversible heat engine is called a *Carnot engine*, and it returns to its original state after one complete cycle. A Carnot cycle involves four reversible processes in the following order: an isothermal expansion at T_H, an adiabatic expansion from T_H to T_C, an isothermal compression at T_C, and an adiabatic compression from T_C to T_H, taking the system back to its original state (T_H and T_C are the hot and cold temperatures between which the engine operates). The details of its operation are outlined below (refer to Fig. 22.6, Worked example 22.1 and Derivation 22.2 for details of calculations).

Step 1. Isothermal expansion at T_H (heat into the system, Q_{in}, and work done by the system):

$$\Delta U = 0$$

$$\therefore Q_{in} = -W_1 = \int_{V_1}^{V_2} p\,dV = nRT_H \ln\left(\frac{V_2}{V_1}\right)$$

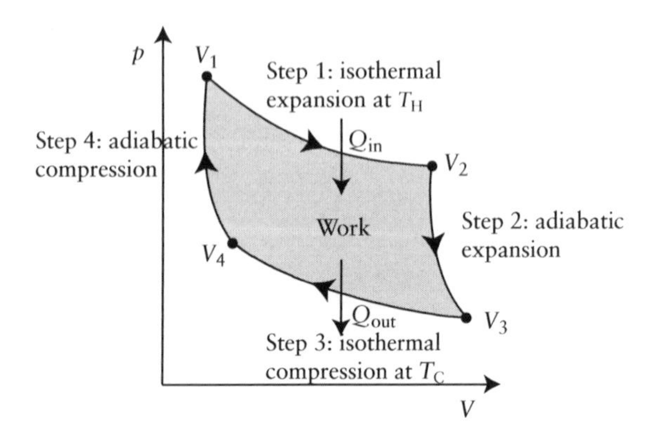

Figure 22.6: $p-V$ diagram of a Carnot cycle.

Step 2. Adiabatic expansion from T_H to T_C:

$$\Delta Q = 0$$

$$\therefore W_2 = \Delta U = C_V(T_H - T_C)$$

And from the adiabatic condition:

$$\frac{T_H}{T_C} = \left(\frac{V_3}{V_2}\right)^{\gamma-1} \quad (22.18)$$

Step 3. Isothermal compression at T_C (heat out of the system, Q_{out}, and work done on the system):

$$\Delta U = 0$$

$$\therefore Q_{out} = -W_3 = \int_{V_3}^{V_4} p\,dV = nRT_C \ln\left(\frac{V_4}{V_3}\right) = -nRT_C \ln\left(\frac{V_3}{V_4}\right)$$

Step 4. Adiabatic compression from T_C to T_H:

$$\Delta Q = 0$$

$$\therefore W_4 = \Delta U = -C_V(T_H - T_C)$$

And from the adiabatic condition:

$$\frac{T_H}{T_C} = \left(\frac{V_4}{V_1}\right)^{\gamma-1} \quad (22.19)$$

By equating the adiabatic condition in Eqs. 22.18 and 22.19, we find that the ratio of volumes is:

$$\left(\frac{V_3}{V_2}\right)^{\gamma-1} = \left(\frac{V_4}{V_1}\right)^{\gamma-1}$$

$$\therefore \frac{V_2}{V_1} = \frac{V_3}{V_4}$$

Therefore, the total work done by the system, which equals the enclosed area in Fig. 22.6, is:

$$W = W_1 + W_2 + W_3 + W_4 = -nR\ln\left(\frac{V_2}{V_1}\right)(T_H - T_C)$$

Rearranging the expressions for Q_{out} and Q_{in} above, we find that:

$$\left|\frac{Q_{out}}{T_C}\right| = nR\ln\left(\frac{V_3}{V_4}\right)$$

$$= nR\ln\left(\frac{V_2}{V_1}\right)$$

$$= \left|\frac{Q_{in}}{T_H}\right|$$

$$\therefore \left|\frac{Q_{out}}{Q_{in}}\right| = \frac{T_C}{T_H}$$

Therefore, the efficiency of a Carnot engine depends only on its operating temperatures:

$$\eta = 1 - \left|\frac{Q_{out}}{Q_{in}}\right|$$

$$= 1 - \frac{T_C}{T_H}$$

(Eq. 22.20). A corollary of Carnot's theorem is that all reversible engines operating between the same two temperatures have the same efficiency. The efficiency of a Carnot engine is thus maximized by operating at the lowest T_C and highest T_H possible.

Reversible heat engines (Carnot cycle):

$$\left|\frac{Q_{out}}{Q_{in}}\right| = \frac{T_C}{T_H}$$

$$\therefore \eta_R = 1 - \frac{T_C}{T_H} \qquad (22.20)$$

In practice, no engine is perfectly frictionless, but it can be realized to a good approximation using quasi-static processes. By considering the operation of reversible engines, we can see where the definition of entropy as $dS = dQ/T$ comes from, as well as the equivalence between the heat engine and the entropy formulations of the Second Law, as demonstrated below.

Clausius inequality and entropy

A corollary of Carnot's theorem that the ratio of heat entering a system to the temperature at the point of heat entry, when integrated over a complete cycle, is equal to zero for a reversible process and is less than zero for an irreversible process:

$$\text{Reversible: } \oint \frac{dQ_R}{T} = 0$$

$$\text{Irreversible: } \oint \frac{dQ_I}{T} < 0$$

$$\therefore \oint \frac{dQ}{T} \leqslant 0 \text{ always}$$

(Eq. 22.21 and Derivation 22.3). This statement is known as the *Clausius inequality*. If we *define* the quantity dQ_R/T as an element of entropy change, dS, then using this together with the Clausius equality, we find that entropy change can never be negative, thus implying that the entropy of a closed system cannot decrease (Eq. 22.22 and Derivation 22.4). Since the Universe constitutes a closed system, the entropy of the Universe must always either remain the same or increase after a thermodynamic process. The equalities in Eqs. 22.21 and 22.22 illustrate that the heat engine formulations of the Second Law are equivalent to the entropy formulation.

Clausius inequality and entropy (Derivations 22.3 and 22.4):

$$\oint \frac{dQ}{T} \leqslant 0 \qquad (22.21)$$

$$\text{and } \Delta S \geqslant 0 \qquad (22.22)$$

where $dS = dQ_R/T$

22.2.4 The Third Law

The *Third Law of Thermodynamics* states that:

The entropy of a system approaches zero as its temperature approaches absolute zero.

This statement provides a reference point for an *absolute* entropy scale, rather than considering only *changes* in entropy. The reference point corresponds to absolute zero temperature, 0K.

Using $dQ = TdS$ in the definition of C_V, we have:

$$C_V = \frac{dQ}{dT}$$

$$= T\frac{dS}{dT}$$

From this, we see that $C_V \to 0$ as $T \to 0$. Therefore, an equivalent formulation of the Third Law is that the

Derivation 22.3: Derivation of the Clausius inequality (Eq. 22.21)

$$\eta_R \geqslant \eta_I$$

$$\therefore 1-\left|\frac{Q_C}{Q_H}\right|_R \geqslant 1-\left|\frac{Q_C}{Q_H}\right|_I$$

$$\therefore \left|\frac{Q_C}{Q_H}\right|_I \geqslant \left|\frac{Q_C}{Q_H}\right|_R$$

$$\geqslant \frac{T_C}{T_H}$$

$$\therefore \left|\frac{Q_C}{T_C}\right| \geqslant \left|\frac{Q_H}{T_H}\right|$$

For positive Q_H and negative Q_C:

$$\frac{Q_H}{T_H}+\frac{Q_C}{T_C} \leqslant 0$$

$$\therefore \sum_{cycle}^{n} \frac{Q}{T} \leqslant 0$$

Changing the sum to an integral in the limit of small Q, the Clausius inequality is:

$$\oint \frac{dQ}{T} \leqslant 0$$

- Starting point: from Carnot's theorem, reversible engines are more efficient than irreversible engines: $\eta_R \geqslant \eta_I$.
- The efficiency of the heat engine is equal to:

$$\eta = 1-\left|\frac{Q_C}{Q_H}\right|$$

- For reversible heat engines only, we have $|Q_C/Q_H| = T_C/T_H$.
- If we take the heat entering the system, Q_H, to be positive, then Q_C leaving the system is negative.
- Generalizing to a cycle with n processes during which heat enters or leaves the system, we can write the inequality as a sum over the cycle.
- In the limit where an element of heat dQ enters the system at a temperature T, the sum becomes an integral. Written in this way, the integral is known as the Clausius inequality.

specific heat capacity of all materials equals zero at absolute zero. The concept of absolute zero temperature is a theoretical reference only, and it is experimentally impossible to reach absolute zero temperature.

22.3 Thermodynamic potentials

A *thermodynamic potential* is a scalar potential function that describes the thermodynamic state of a system. For example, internal energy, U, is a thermodynamic potential. Along with U, three other common thermodynamic potentials are the *Helmholtz free energy*, F [J], the *enthalpy*, H [J], and the *Gibbs free energy*, G [J]. Neither F, H, nor G have an insightful physical interpretation as U does, but they are useful in determining how properties of a system change when it changes state, as we shall see below. In particular, thermodynamic potentials are used for determining the parameters that describe *equilibrium* states of a system, which are states that correspond to energy minima (stable states).

Analogy with potential fields

Thermodynamic potentials are analogous to scalar potential fields that we have encountered in other areas of physics, such as the gravitational potential field, $\phi(r)$, in classical mechanics, or the electric potential field, $V(r)$, in electromagnetism. Whereas $\phi(r)$ and $V(r)$ are functions that vary with position, r, thermodynamic potentials are functions that vary with state variables, p, V, T, or S. For example, we shall see below that internal energy is a function of entropy and volume, $U(S, V)$. A mass in a gravitational field or a charge in an electric potential field store potential energy as a result of their position, r. If the position of the mass or charge changes, then its potential energy changes. Analogously, a thermodynamic system stores potential energy as a result of its thermodynamic state. When a system changes state during a thermodynamic process, then its state variables change and hence its stored energy changes. This energy might, for example, be released as heat, or used to do work.

Derivation 22.4: Derivation of the entropy formulation of the Second Law from the Clausius inequality (Eq. 22.22)

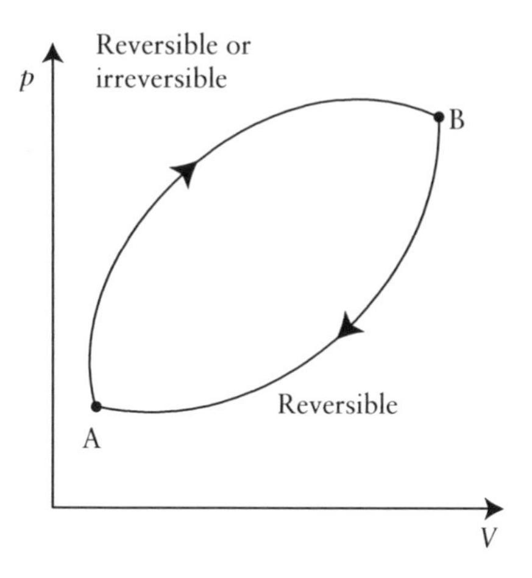

Clausius inequality: $\oint \frac{dQ}{T} \leqslant 0$

$$\therefore \int_A^B \frac{dQ}{T} + \int_B^A \frac{dQ_R}{T} \leqslant 0$$

$$\therefore \int_A^B \frac{dQ}{T} \leqslant \int_A^B \frac{dQ_R}{T}$$

$$\leqslant \int_A^B dS$$

$$\therefore \oint \frac{dQ}{T} \leqslant \oint dS$$

$$\leqslant \Delta S$$

Combining this with the Clausius inequality:

$\therefore \Delta S \geqslant 0$

- Starting point: let a cycle be composed of one reversible or irreversible process between points A and B, and a reversible process from B back to A. Therefore, the total dQ/T is:

$$\oint \frac{dQ}{T} = \int_A^B \frac{dQ}{T} + \int_B^A \frac{dQ_R}{T}$$

- For definite integrals, we have:

$$\int_B^A \frac{dQ_R}{T} = -\int_A^B \frac{dQ_R}{T}$$

- Define an element of entropy as:

$$dS \stackrel{\text{def}}{=} \frac{dQ_R}{T}$$

- Change the limits to integrate around a complete cycle. The total change in entropy for a complete cycle is ΔS.
- Equate the inequality for ΔS to the Clausius inequality to recover the entropy formulation of the Second Law of Thermodynamics (Eq. 22.7).

22.3.1 Internal energy

From the First Law of Thermodynamics, internal energy can be thought of as the capacity of a system to release heat plus its capacity to do work:

$$dU = dQ + dW$$

For a reversible processes, we have $dQ = TdS$ and $dW = -pdV$. Therefore, the First Law can be written as:

$$dU = TdS - pdV$$

(Eq. 22.23). Therefore, we see that U is a function of the state variables S and V: $U = U(S, V)$. The variables S and V are called the *natural variables* of the thermodynamic potential, U (we shall see below that F, H, and G are also functions of two natural variables).

The total derivative of $U = U(S, V)$ is:

$$dU = \left(\frac{\partial U}{\partial S}\right)_V dS + \left(\frac{\partial U}{\partial V}\right)_S dV$$

Equating this expression to $dU = TdS - pdV$, we see that:

$$T = \left(\frac{\partial U}{\partial S}\right)_V$$

$$\text{and} \quad -p = \left(\frac{\partial U}{\partial V}\right)_S$$

(Eqs. 22.24 and 22.25). These constitute two equations of state that allow us to determine T and p from the function U and the state variables V and S. We can then use these equations of state to predict how macroscopic properties of a system change during a thermodynamic process.

Internal energy, $U(S,V)$:

$$dU = TdS - pdV \quad (22.23)$$

Equations of state from the total derivative:

$$T = \left(\frac{\partial U}{\partial S}\right)_V \quad (22.24)$$

$$p = -\left(\frac{\partial U}{\partial V}\right)_S \quad (22.25)$$

Irreversible processes

Equations 22.23–22.25 involve state variables—i.e. variables whose values are independent of the path taken to reach that state. Therefore, the relations in Eqs. 22.23–22.25 are true for *all* processes, both reversible *and* irreversible. However, for irreversible processes, a larger fraction of energy is lost as heat, and a smaller fraction is available to do work, compared to reversible processes. This means that for irreversible processes:

$$dW_I < |pdV|$$

$$\text{and } dQ_I > TdS$$

Nonetheless, Eq. 22.23 and the First Law still hold true for irreversible processes.

22.3.2 F, H, and G

The thermodynamic potentials, F, H, and G, for a system defined by state variables p, V, T, and S are defined as:

$$\text{Helmholtz free energy: } F \stackrel{\text{def}}{=} U - TS$$

$$\text{Enthalpy: } H \stackrel{\text{def}}{=} U + pV$$

$$\text{Gibbs free energy: } G \stackrel{\text{def}}{=} U - TS + pV$$

(Eqs. 22.26, 22.28, and 22.30). The energy term $+TS$ can be considered as the fraction of heat provided to the system from the environment in bringing it to a temperature T. The energy term $-pV$ can be considered as the fraction of work done by the system against the environment in taking it to a volume V. F, H, and G therefore correspond to the internal energy, U, either including or excluding the contributions from the energy terms $+TS$ and $-pV$.

Thermodynamic potentials:

Helmholtz free energy, $F(T, V)$:

$$F = U - TS \quad (22.26)$$

$$\text{and } dF = -SdT - pdV \quad (22.27)$$

Enthalpy, $H(S, p)$:

$$H = U + pV \quad (22.28)$$

$$\text{and } dH = TdS + Vdp \quad (22.29)$$

Gibbs free energy, $G(T, p)$:

$$G = U - TS + pV \quad (22.30)$$

$$\text{and } dG = -SdT + Vdp \quad (22.31)$$

By differentiating the expressions for F, H, and G, and using $dU = TdS - pdV$, we can determine expressions for their exact differentials:

$$\begin{aligned} dF &= dU - TdS - SdT \\ &= -SdT - pdV \end{aligned}$$

$$\begin{aligned} dH &= dU + pdV + Vdp \\ &= TdS + Vdp \end{aligned}$$

$$\begin{aligned} dG &= dU - TdS - SdT + pdV + Vdp \\ &= -SdT + Vdp \end{aligned}$$

(Eqs. 22.27, 22.29, and 22.31). From this, we see that each thermodynamic potential is a function of two natural variables: $F = F(T, V)$, $H = H(S, p)$, and $G = G(T, p)$. If we equate these expressions to their total derivative, as we did for U, we determine two equations of state for each potential (Eqs. 22.32–22.37). These relate a state variable to the derivative of a thermodynamic potential.

Thermodynamic equations of state:

Helmholtz free energy, $F(T, V)$:

$$dF = \underbrace{\left(\frac{\partial F}{\partial T}\right)_V}_{=-S} dT + \underbrace{\left(\frac{\partial F}{\partial V}\right)_T}_{=-p} dV$$

$$\therefore S = -\left(\frac{\partial F}{\partial T}\right)_V \quad (22.32)$$

$$\text{and } p = -\left(\frac{\partial F}{\partial V}\right)_T \quad (22.33)$$

Enthalpy, $H(S, p)$:

$$dH = \underbrace{\left(\frac{\partial H}{\partial S}\right)_p}_{=T} dS + \underbrace{\left(\frac{\partial H}{\partial p}\right)_S}_{=V} dp$$

$$\therefore T = \left(\frac{\partial H}{\partial S}\right)_p \quad (22.34)$$

$$\text{and } V = \left(\frac{\partial H}{\partial p}\right)_S \quad (22.35)$$

Gibbs free energy, $G(T, p)$:

$$dG = \underbrace{\left(\frac{\partial G}{\partial T}\right)_p}_{=-S} dT + \underbrace{\left(\frac{\partial G}{\partial p}\right)_T}_{=V} dp$$

$$\therefore S = -\left(\frac{\partial G}{\partial T}\right)_p \quad (22.36)$$

$$\text{and } V = \left(\frac{\partial G}{\partial p}\right)_T \quad (22.37)$$

22.3.3 The Maxwell relations

The *Maxwell relations* are a set of differential equations that relate the derivatives of the state variables p, V, T, and S to one another, without making reference to the thermodynamic potentials. They are derived by differentiating the equations of state found above. For example, differentiating Eqs. 22.24 and 22.25 with respect to either V or S, we obtain two expressions for the second derivative of U:

$$\left(\frac{\partial T}{\partial V}\right)_S = \left(\frac{\partial^2 U}{\partial V \partial S}\right)_{S,V}$$

$$\text{and } -\left(\frac{\partial p}{\partial S}\right)_V = \left(\frac{\partial^2 U}{\partial S \partial V}\right)_{V,S}$$

Since the order of differentiation for second derivatives does not matter, we can equate the above expressions:

$$\therefore \left(\frac{\partial T}{\partial V}\right)_S = -\left(\frac{\partial p}{\partial S}\right)_V$$

We can use a similar method to determine three other differential relations from the equations of state for F, H, and G (use Eqs. 22.32–22.37). Together, the set of four differential equations are the Maxwell relations (Eqs. 22.38–22.41). A mnemonic for remembering Maxwell's relations is shown in Fig. 22.7.

The Maxwell relations:

From internal energy equations (Eqs. 22.24 and 22.25):

$$-\left(\frac{\partial p}{\partial S}\right)_V = \left(\frac{\partial T}{\partial V}\right)_S \quad (22.38)$$

From Helmholtz free energy equations (Eqs. 22.32 and 22.33):

$$\left(\frac{\partial S}{\partial V}\right)_T = \left(\frac{\partial p}{\partial T}\right)_V \quad (22.39)$$

From enthalpy equations (Eqs. 22.34 and 22.35):

$$\left(\frac{\partial T}{\partial p}\right)_S = \left(\frac{\partial V}{\partial S}\right)_p \quad (22.40)$$

From Gibbs free energy equations (Eqs. 22.36 and 22.37):

$$\left(\frac{\partial V}{\partial T}\right)_p = -\left(\frac{\partial S}{\partial p}\right)_T \quad (22.41)$$

Worked example 22.2 shows how to use these relations to show that the internal energy of an ideal gas depends only on its temperature (this statement is known as *Joules's Law*).

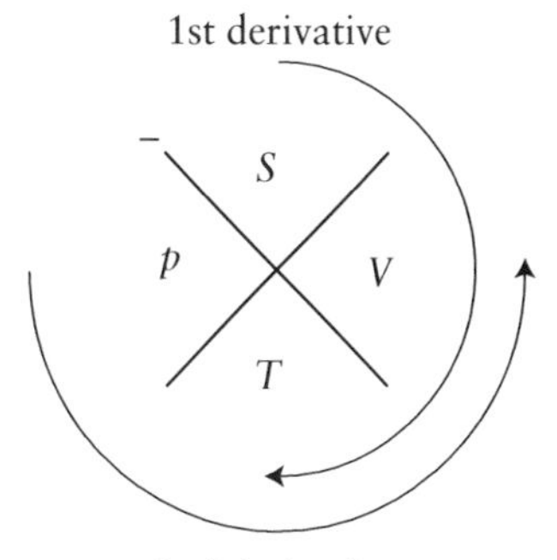

Figure 22.7: Mnemonic for remembering the Maxwell relations: going counter-clockwise the mnemonic is "Society for the Prevention of Teaching Vectors." The diagram illustrates the relation $(\partial S/\partial V)_T = (\partial p/\partial T)_V$. The first derivative is found by going clockwise, starting with the letter corresponding to its numerator, and the second by going counter-clockwise, starting with the adjacent letter. There is always a negative sign when S and p appear together in the partial derivative.

WORKED EXAMPLE 22.2: Using the Maxwell relations to show that U is a function of T only, for an ideal gas

1. By differentiating the exact differential of U with respect to volume, at constant temperature, and using the Maxwell relation in Eq. 22.39 to eliminate the entropy derivative, we see that U does not depend on V for an ideal gas (use $pV = nRT$):

$$\mathrm{d}U = T\mathrm{d}S - p\mathrm{d}V$$

$$\therefore \left(\frac{\partial U}{\partial V}\right)_T = T\left(\frac{\partial S}{\partial V}\right)_T - p$$

$$= T\left(\frac{\partial p}{\partial T}\right)_V - p$$

$$= \frac{nRT}{V} - \frac{nRT}{V}$$

$$= 0$$

2. By differentiating the exact differential of U with respect to pressure, at constant temperature, and using the Maxwell relation in Eq. 22.41 to eliminate the entropy derivative, we see that U does not depend on p for an ideal gas (use $pV = nRT$):

$$\mathrm{d}U = T\mathrm{d}S - p\mathrm{d}V$$

$$\therefore \left(\frac{\partial U}{\partial p}\right)_T = T\left(\frac{\partial S}{\partial p}\right)_T - p\left(\frac{\partial V}{\partial p}\right)_T$$

$$= -T\left(\frac{\partial V}{\partial T}\right)_p - p\left(\frac{\partial V}{\partial p}\right)_T$$

$$= -\frac{nRT}{p} + \frac{nRT}{p}$$

$$= 0$$

Therefore, for an ideal gas, U is not a function of V or p, hence it is a function of T only: $U \Rightarrow U(T)$.

22.3.4 Chemical potential

So far, we have only considered *closed* systems, in which energy but not matter can be exchanged between the system and its surroundings. In an *open* system, matter (particles) can be exchanged. In this case, the internal energy of a system changes when particle number changes, since each particle carries energy. Additionally, a change in particle number usually corresponds to a change in volume and entropy of the system. Therefore, U is a function of the number of particles, N, as well as S and V:

$$U = U(S, V, N)$$

The total derivative of U for an open system is:

$$\mathrm{d}U = \underbrace{\left(\frac{\partial U}{\partial S}\right)_{V,N}}_{=T}\mathrm{d}S + \underbrace{\left(\frac{\partial U}{\partial V}\right)_{S,N}}_{=-p}\mathrm{d}V + \underbrace{\left(\frac{\partial U}{\partial N}\right)_{S,V}}_{=\mu}\mathrm{d}N$$

$$\equiv T\mathrm{d}S - p\mathrm{d}V + \mu\mathrm{d}N \qquad (22.42)$$

Here, we introduce a variable known as the *chemical potential*, μ [J particle^{-1}], which represents the amount of internal energy added to a system if a single particle is added at constant entropy and volume:

$$\mu = \left(\frac{\partial U}{\partial N}\right)_{S,V}$$

(Eq. 22.43). The chemical potential is therefore often thought of as the "energy per particle." Different types of particles have different chemical potentials. If the system involves k different particle types, then the chemical potential of the ith type is given by:

$$\mu_i = \left(\frac{\partial U}{\partial N_i}\right)_{S,V,N_{j\neq i}}$$

(Eq. 22.44), and the exact differential of U involves the sum over particle types:

$$\mathrm{d}U = T\mathrm{d}S - p\mathrm{d}V + \sum_{i=1}^{k}\mu_i\mathrm{d}N_i$$

We can use a similar argument to write the chemical potential in terms of the thermodynamic potentials, F and G:

$$\mu = \left(\frac{\partial F}{\partial N}\right)_{V,T}$$

or $$\mu = \left(\frac{\partial G}{\partial N}\right)_{p,T}$$

The latter allows us to define the chemical potential as the Gibbs free energy per particle, $g(p, T)$ (Eq. 22.45), which is useful when determining the equilibrium conditions during phase transitions (see Section 23.2).

Chemical potential:

In terms of U:

Single species: $$\mu = \left(\frac{\partial U}{\partial N}\right)_{S,V} \quad (22.43)$$

Many species: $$\mu_i = \left(\frac{\partial U}{\partial N_i}\right)_{S,V,N_{j\neq i}} \quad (22.44)$$

In terms of G:

$$\mu = \left(\frac{\partial G}{\partial N}\right)_{p,T} \equiv g(p,T) \quad (22.45)$$

Chapter 23
Real gases and phase transitions

So far, we've only been dealing with ideal gases, which obey the Ideal Gas Law, $pV = nRT$. Most gases, however, are *real gases*, and do not obey the Ideal Gas Law. Real gases (i) have a finite molecular size relative to the distance between the molecules and (ii) have non-negligible intermolecular forces. All real gases behave like ideal gases at very low pressures and high temperatures, since under these conditions the relative volumes and intermolecular forces of the particles are negligible. At *high* pressures and *low* temperatures, a real gas eventually condenses into a liquid, which has much stronger intermolecular forces. In this chapter, we'll look at the thermodynamics of real gases, as well as the properties governing changes of state from gas to liquid.

23.1 Real gas equation of state

23.1.1 The van der Waals equation

There are several equations of state for real gases that relate the parameters p, V, and T in such a way to account for intermolecular forces and finite molecular volumes. A common equation of state is the *van der Waals* equation,[1] which is the Ideal Gas Law modified by two empirical parameters, labeled a [Pa m^6 mol^{-2}] and b [m^3 mol^{-1}]:

$$\left(p + a\left(\frac{n}{V}\right)^2\right)(V - nb) = nRT$$

(Eq. 23.1). These parameters are constants for a particular gas; the larger their values, the more the gas deviates from ideal gas behavior. The parameter a is the *constant of internal pressure*, and it accounts for attractive intermolecular forces that act to reduce the gas pressure for a given volume and temperature. The pressure correction is found to be inversely proportional to the square of the volume of the container enclosing the gas. For n moles of gas, the internal pressure is given by $p_{int} = a(n/V)^2$. Therefore, the effective pressure of a real gas is the pressure due to collisions, p, plus the internal pressure correction to account for intermolecular forces:

$$p \to p + a\left(\frac{n}{V}\right)^2$$

The parameter b in Eq. 23.1 is the *constant of internal volume*, and it accounts for the finite molecular volume of a real gas. This reduces the amount of space within the container in which particles are free to move. Letting b equal the effective molecular volume per mole, the internal volume occupied by n moles of gas is $V_{int} = nb$. Therefore, the effective volume is the volume of the container, V, reduced by the excluded volume of the particles themselves:

$$V \to V - nb$$

The excluded volume is approximately eight times the actual volume of the gas particles, since two particles cannot approach one another within a distance less than *twice* the particle radius. The excluded volume then scales as $(2r)^3 = 8r^3$, where r is the particle radius. Using the corrected pressure and volume in the Ideal Gas Law, we retrieve the van der Waals equation given in Eq. 23.1. The van der Waals equation is often written in terms of the volume per mole, or the *molar volume*, v [m^3 mol^{-1}], where $v = V/n$ (Eq. 23.2). When the

[1] The *van der Waals* force is a weak, intermolecular force that results from dipole–dipole interactions, involving either induced or permanent dipoles. It governs the weak bonds between particles in a gas.

density of the gas is low, $n/V \to 0$, and the van der Waals equation approaches the Ideal Gas Law.

Other common equations of state are the *Redlich–Kwong* equation, the *Dieterici* equation, and the *Berthelot* equation. These are all two-parameter equations, where one parameter accounts for the gas's internal pressure and the other the internal volume. In the following discussion, we refer only to the van der Waals equation, but a similar treatment can be used for the other equations of state as well.

Van der Waals equation of state for a real gas:

$$\left(p + a\left(\frac{n}{V}\right)^2\right)(V - nb) = nRT \qquad (23.1)$$

Or in terms of the molar volume, $v = V/n$:

$$\left(p + \frac{a}{v^2}\right)(v - b) = RT \qquad (23.2)$$

Van der Waals isotherms

Figure 23.1 shows the isotherms of Eq. 23.1 on a p–V diagram. The behavior of the gas depends on the strength of the intermolecular forces at different combinations of p, V, and T. Below, we will look at the different parts of the diagram in turn.

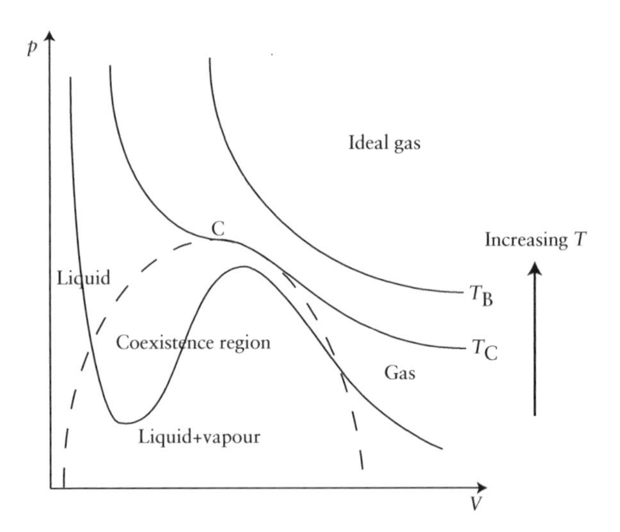

Figure 23.1: p–V diagram showing van der Waals isotherms (Eq. 23.1). The critical point is marked C. The line at T_C corresponds to the critical isotherm and at T_B to the Boyle temperature isotherm.

23.1.2 The Boyle temperature

The gaseous region indicated in Fig. 23.1 corresponds to large volumes and, for low-temperature isotherms, low pressures. Under these conditions, intermolecular forces are relatively weak. For increasing temperatures, intermolecular forces become less significant, and therefore real gases approach ideal gas behavior. As a consequence, high-temperature isotherms in the gaseous region resemble rectangular hyperbolae characteristic of Boyle's Law for ideal gases, $p \propto 1/V$.

Let's assume that we can express the equation of state of a real gas as a power series of the variable $n/V = 1/v$:

$$pv = RT\left(1 + \frac{A(T)}{v} + \frac{B(T)}{v^2} + \cdots\right)$$

Expressed in this way, the equation of state is called the *virial expansion* or *virial equation of state*. The terms $A(T)$, $B(T)$, ... are temperature-dependent virial coefficients. From this, we see that the real gas equation of state approaches the Ideal Gas Law when the first virial coefficient, $A(T)$, equals zero. The temperature at which this occurs is called the *Boyle temperature*, T_B [K]:

$$\therefore A(T_B) = 0$$

This temperature therefore marks the transition from real to ideal gas behavior. It is a constant for any gas, and depends on the gas parameters a and b (Eq. 23.3 and Derivation 23.1).

The Boyle temperature (Derivation 23.1):

$$T_B = \frac{a}{Rb} \qquad (23.3)$$

23.1.3 The coexistence region

The liquid region corresponds to the small-volume region of the p–V diagram in Fig. 23.1, where intermolecular forces are significantly stronger than for gases. Liquids are highly incompressible; therefore,

Derivation 23.1: Derivation of the Boyle temperature, T_B (Eq. 23.3)

- Starting point: the van der Waals equation of state is:

$$\left(p+\frac{a}{v^2}\right)(v-b)=RT$$

- Rearranging, we can express pv as a power series in $1/v$ (virial expansion). Since $v \gg b$, we can do a binomial expansion of $(1-b/v)^{-1}$, and group higher order terms into a single term represented by $O(1/v^2)$.

$$pv=\frac{RT}{1-(b/v)}-\frac{a}{v}$$

$$=RT\left(1-\frac{b}{v}\right)^{-1}-\frac{a}{v}$$

$$\simeq RT\left(1+\frac{b}{v}+O\left(\frac{1}{v^2}\right)\right)-\frac{a}{v}$$

$$\simeq RT+\frac{bRT-a}{v}+O\left(\frac{1}{v^2}\right)$$

- The equation of state approaches the Ideal Gas Law when:

$$\frac{bRT-a}{v}\rightarrow 0$$

For ideal gas behavior:

$$bRT_B-a=0$$

- The temperature at which this occurs is called the *Boyle temperature*, T_B.

$$\therefore T_B=\frac{a}{Rb}$$

a small reduction in volume leads to a large increase in pressure, as illustrated by the steep isotherms in this region.

In between the liquid and gaseous regions is the *coexistence region*, where liquids and their gaseous vapor exist simultaneously in equilibrium. This region therefore describes the conditions under which condensation and vaporization take place. We will look further at the thermodynamic properties of these phase transitions in Section 23.2.

The critical point

The coexistence region shrinks with increasing temperature and pressure. The point at which it vanishes is called the *critical point*, and is labeled as C in Fig. 23.1. This point corresponds to the saddle point on the *critical isotherm*—the isotherm at T_C—which means that the first and second derivatives of the van der Waals equation of state are zero at this point. The thermodynamic state at the critical point, which is defined by the parameters p_C, v_C, and T_C, can be deduced by determining the conditions under which the first and second derivatives of Eq. 23.2 (the van der Waals equation) are zero (Eqs. 23.4–23.6 and Derivation 23.2).

Critical point parameters (Derivation 23.2):

$$T_C=\frac{8a}{27Rb} \quad (23.4)$$

$$v_C=3b \quad (23.5)$$

$$p_C=\frac{a}{27b^2} \quad (23.6)$$

Ratio of critical point parameters:

$$\therefore \frac{p_C v_C}{RT_C}=\frac{3}{8}\simeq\frac{1}{3} \quad (23.7)$$

Comparing Eqs. 23.3 and 23.4, we see that the critical temperature and the Boyle temperature are related by:

$$T_C=\frac{8}{27}T_B\simeq\frac{1}{3}T_B$$

Eliminating the parameters a and b from Eqs. 23.4–23.6, we find that the ratio of critical point parameters is approximately 1/3 (Eq. 23.7). This ratio is approximately the same for all simple gases, which are defined as gases that have small dipole moments and weak

Derivation 23.2: Derivation of the critical point parameters (Eqs. 23.4–23.6)

At the saddle point:

$$-\frac{RT_C}{(v_C-b)^2}+\frac{2a}{v_C^3}=0$$

and $$\frac{2RT_C}{(v_C-b)^3}-\frac{6a}{v_C^4}=0$$

$$\therefore v_C = 3b$$

$$T_C = \frac{8a}{27Rb}$$

and $$p_C = \frac{a}{27b^2}$$

- Starting point: rearranging the van der Waals equation of state in Eq. 23.2, we have:

$$p=\frac{RT}{(v-b)}-\frac{a}{v^2}$$

The first and second derivatives of this equation are:

$$\left(\frac{\partial p}{\partial v}\right)_T=-\frac{RT}{(v-b)^2}+\frac{2a}{v^3}$$

and $$\left(\frac{\partial^2 p}{\partial v^2}\right)_T=\frac{2RT}{(v-b)^3}-\frac{6a}{v^4}$$

These equal zero at the saddle point, where $v=v_C$, $T=T_C$, and $p=p_C$.

- These simultaneous equations can be solved to determine expressions for v_C, T_C and p_C:
to find v_C: eliminate a;
to find T_C: substitute v_C into $(\partial p/\partial v)_T$;
to find p_C: substitute v_C and T_C into the van der Waals equation of state.

intermolecular forces, even in the liquid phase, such as CO or Cl_2. For an ideal gas, the ratio pv/RT equals 1.

Reduced variables

The *reduced variables*, $\bar{p}$, $\bar{v}$, and $\bar{T}$, are dimensionless variables that equal the ratio of the respective variable to their critical value:

$$\bar{p}=\frac{p}{p_C},\ \bar{v}=\frac{v}{v_C},\ \bar{T}=\frac{T}{T_C} \qquad (23.8)$$

By writing the van der Waals equation in reduced variables, the parameters a and b cancel, and the real gas equation of state becomes:

$$\left(\bar{p}+\frac{3}{\bar{v}^2}\right)(3\bar{v}-1)=8\bar{T}$$

(Eq. 23.9). (To show this, eliminate p, v, and T in Eq. 23.2—the van der Waals equation of state—using Eqs. 23.8, and substitute in for critical values using Eqs. 23.4–23.6.) This means that all simple gases—those with small dipole moments and weak intermolecular forces—obey the same p–V diagram as each other in reduced variables. This statement is called the *Law of Corresponding States*.

Reduced variable equation of state:

$$\left(\bar{p}+\frac{3}{\bar{v}^2}\right)(3\bar{v}-1)=8\bar{T} \qquad (23.9)$$

23.1.4 Isothermal compressibility

The gradient of the isotherms in Fig. 23.1 describes how easily a substance can be compressed. If it is highly compressible like a gas, then for a large change in volume, the corresponding change in pressure is small, and the gradient is shallow. Conversely, if the material is incompressible like a liquid, then a small change in volume leads to a large increase in pressure, and the gradient is steep. The *isothermal compressibility* of a substance, κ_T [m^2N^{-1}], is a measure of the relative change in volume in response to a pressure change, at a constant temperature, and is defined by:

$$\kappa_T=-\frac{1}{V}\left(\frac{\partial V}{\partial p}\right)_T$$

(Eq. 23.10). This is related to the gradient of the isotherms in Fig. 23.1 by the relation shown in Eq. 23.11.

Isothermal compressibility:

$$\kappa_T = -\frac{1}{V}\left(\frac{\partial V}{\partial p}\right)_T \tag{23.10}$$

$$\therefore p - V \text{ gradient} = -\frac{1}{\kappa_T V} \tag{23.11}$$

23.2 Phase transitions

So far, we've only been looking at *homogeneous* systems, which are systems with uniform properties throughout them. A *heterogeneous* system involves different components, and the properties of the components change discontinuously at their boundaries. An example is a mixture of oil and water. A heterogeneous system can also involve *one* substance but in two different states of matter, such as water and steam. In this case, the components are called *phases*, and the boundary between them is the *phase boundary*. A *phase transition* corresponds to a change of state, for example, condensation or freezing, and involves mass transfer between the phases. In this section, we'll look at the thermodynamics of such phase transitions.

23.2.1 Conditions for phase equilibrium

For a two-phase system to exist in thermodynamic equilibrium, the two components must satisfy:

1. Thermal equilibrium: $T_1 = T_2$.
2. Mechanical equilibrium: $p_1 = p_2$.
3. Chemical equilibrium: $\mu_1 = \mu_2$.

If the system is *not* in any one of these equilibria, then either heat exchange, volume redistribution or particle exchange will occur across the phase boundary to equalize T, p, or μ, respectively. From the exact differential of Gibbs free energy:

$$dG = -SdT + Vdp$$

(refer to Section 22.3.2), we see that the state corresponding to thermal and mechanical equilibrium (constant T and p, respectively) corresponds to a minimum Gibbs free energy ($dG = 0$ when $dT = 0$ and $dP = 0$). Therefore, we require $dG = 0$ for phase equilibrium. The consequence of this is that the Gibbs free energy per particle, g, of either phase are equal at equilibrium (Derivation 23.3):

$$g_1(p,T) = g_2(p,T) \tag{23.12}$$

By definition, g equals the chemical potential, μ (refer to Section 22.3.4). Therefore, phase equilibrium implies chemical equilibrium, $\mu_1 = \mu_2$, as required.

23.2.2 Maxwell construction

In the coexistence region in Fig. 23.1, a liquid and its gaseous vapor coexist in equilibrium. From Fig. 23.1, we see that an isotherm of the van der Waals equation of state in this region does not correspond to an isobar (i.e. the isotherm is not at constant pressure), which means that phase equilibrium is not satisfied along the isotherm. These isotherms therefore do not correspond to stable, equilibrium states, but instead describe unstable or metastable states (superheated liquid or supercooled vapor). To deduce equilibrium states from the non-equilibrium isotherms, Maxwell found that the van der Waals isotherms in the coexistence region could be replaced with horizontal isobars that correspond to the equilibrium pressure. The isobars are drawn such that the area enclosed above the isobar equals that enclosed below it, as shown

Derivation 23.3: Derivation of the phase equilibrium condition (Eq. 23.12)

$$G(p,T,\alpha) = g_1(p,T)\alpha N + g_2(p,T)(1-\alpha)N$$

$$\therefore dG(p,T,\alpha) = g_1(p,T)Nd\alpha - g_2(p,T)Nd\alpha$$

- Starting point: the total Gibbs free energy, G, of a two-phase system with a fraction a particles in phase 1 and $(1-a)$ in phase 2 is the sum of individual energies, where g_1 and g_2 are the Gibbs free energies per particle in either phase.

For phase equilibrium:

$$dG(p,T,\alpha) = 0$$

$$\therefore g_1(p,T) = g_2(p,T)$$

- The Gibbs free energy is minimized when $dG = 0$. This occurs when the Gibbs free energies per particle are equal: $g_1 = g_2$.

schematically in Fig. 23.2 (Derivation 23.4). This condition ensures that the work in going from A to E (refer to Fig. 23.2) is the same as if the system had followed the path of the van der Waals curve. This process of deducing isobars in a coexistence region is known as the *Maxwell construction*.

Vapor pressure

The pressure of the isobar that results from the Maxwell construction is known as the *vapor pressure*, p_{vap}, at a given isothermal temperature. At this pressure, the system is in equilibrium: the rate of condensation (vapor to liquid) equals the rate of vaporization (liquid to vapor), and the pressures and temperatures of each component are equal.[2] Moving from right to left along the isotherm shown in Fig. 23.2, as a vapor is slowly compressed from a large initial volume, it reaches the point E, which represents a *saturated vapor*. Further compression causes condensation of the vapor at constant pressure (the vapor pressure isobar), provided the temperature is constant. When the point A is reached, the system only consists of a liquid, and further compression requires high pressures due to the incompressibility of the liquid phase.

Violation of the equilibrium condition

The isobars that are drawn from Maxwell construction represent the equilibrium states for a system in the coexistence region. Metastable states correspond to the regions A–B and D–E on the curve in Fig. 23.2, whereas unstable states correspond to the region B–D. The metastable states can sometimes be reached in experiments, provided a system is carefully expanded (A–B) or compressed (E–D). In these cases, we get either a superheated liquid (A–B), which does not boil although its temperature exceeds the boiling point, or a supercooled gas (D–E), which does not condense although its temperature is below the condensation point. However, since such systems are metastable, they change in a shock-like manner back to the equilibrium state, even under very slight disturbances. An example of a superheated liquid is a bubble chamber, which is rapidly brought to low pressure with a piston; boiling then occurs on ion tracks left by ionizing particles. An example of a supercooled gas is a cloud chamber filled with humid air, which is then adiabatically expanded.

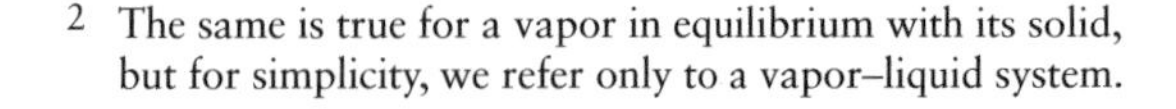

[2] The same is true for a vapor in equilibrium with its solid, but for simplicity, we refer only to a vapor–liquid system.

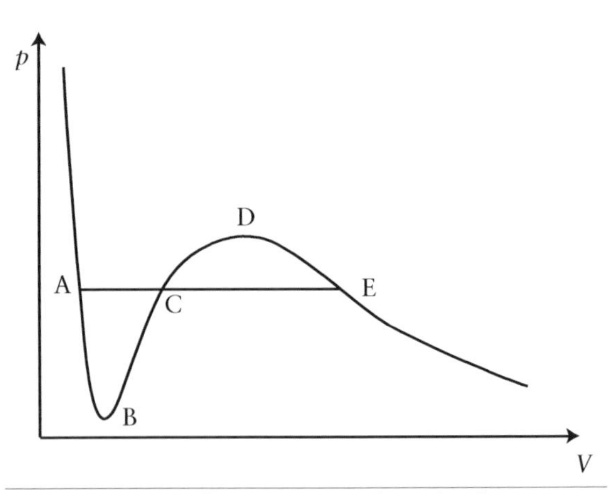

Figure 23.2: Maxwell construction to determine the equilibrium pressure for phase coexistence. The area enclosed above the isobar equals that enclosed below it (refer to Derivation 23.4).

23.2.3 Phase diagrams

Different states of matter and phase transitions are represented on *phase diagrams*. These are plots of pressure against temperature at constant volume. A typical phase diagram is shown in Fig. 23.3. The line separating any two phases is called a *transition line* or a *coexistence curve*. There are three transition lines, and crossing any of them results in a change of state. Along these lines, the Gibbs free energy per particle of either phase are equal, $g_1 = g_2$, where the value of g depends on the combination of T and p. The conditions along the transition lines correspond to the isotherms and isobars described by the coexistence region in Fig. 23.1. The point at which the condensation line ends corresponds to the critical point, C, in Fig. 23.1. Above this point, no phase transition is possible—the system exists as a *supercritical fluid*, which has both liquid and gas-like properties (further described below). The point at which the three coexistence curves intersect is called the *triple point*, T. At this point, the solid, liquid, and vapor forms of a substance exist simultaneously in equilibrium, and their free energies per particle are equal, $g_s = g_l = g_v$.

Supercritical fluids

Supercritical fluids arise by simultaneously increasing the temperature and pressure of a system in the coexistence region, above their critical values, at constant volume. In so doing, the density of the vapor increases due to increased pressure, whilst the density of the liquid decreases due to increased temperature. Eventually, their densities converge and it is no longer possible to distinguish the liquid from its vapor. The

Derivation 23.4: Derivation of the Maxwell construction (refer to Fig. 23.2)

$$\int_A^E dG = \int_A^E V dp$$

$$= \int_A^C V dp + \int_C^E V dp$$

$$= 0 \text{ at equilibrium}$$

$$\therefore \underbrace{\left| \int_A^C V dp \right|}_{\text{Area below isobar}} = \underbrace{\left| \int_C^E V dp \right|}_{\text{Area above isobar}}$$

- Starting point: the exact differential for G at constant temperature is:

$$dG = -SdT + Vdp$$

$$= Vdp$$

This equals zero at equilibrium.

- Integrate the van der Waals curve between points A and E in Fig. 23.2 and apply the above condition.

substance in this state is called a *supercritical fluid*, which has both gas and liquid properties simultaneously. By slightly changing the pressure and temperature of the fluid, its density can be "tuned" to make it behave more like a liquid or more like a gas. A supercritical fluid becomes a gas if its pressure falls below the critical pressure, p_C, a liquid if its temperature drops below the critical temperature, T_C, or a distinguishable mixture of both if conditions drop below p_C and T_C simultaneously. There are many practical uses for supercritical fluids, for example, in dry cleaning and refrigeration.

Phase diagram for water

Water has an anomalous phase diagram. For conventional systems, increasing the pressure of a liquid at constant temperature causes freezing. Water, however, has a negative gradient on the melting transition line, indicating that increasing the pressure of ice causes melting. This phenomenon facilitates ice skating, for example.

23.2.4 Clausius Clapeyron equation

The coexistence curves in Fig. 23.3 describe the conditions required for phase equilibrium. The *Clausius Clapeyron* equation represents the gradient of the coexistent curves, and therefore describes how pressure changes in response to a temperature change, such that a system remains in equilibrium. The Clausius Clapeyron equation is given in Eq. 23.13, where L is the latent heat of vaporization (Derivation 23.5). From it, we see that there is a change in volume, ΔV, of a substance undergoing a phase transition, which makes intuitive sense. Worked example 23.1 shows how to calculate the temperature dependence of a liquid's vapor pressure using the Clausius Clapeyron equation.

Clausius Clapeyron equation (Derivation 23.5):

$$\frac{dp}{dT} = \frac{L}{T\Delta V} \qquad (23.13)$$

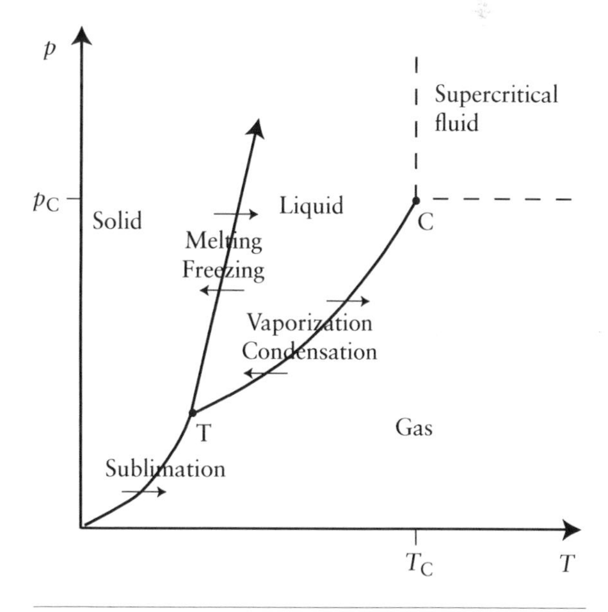

Figure 23.3: A typical phase diagram. The transition lines represent equilibrium conditions for phase coexistence. At the triple point, T, all three phases coexist in equilibrium. Crossing a transition line represents a change of phase. The transition line separating liquid and gas terminates at the critical point, C; the transition line separating liquid and solid does not terminate.

Derivation 23.5: Derivation of the Clausius Clapeyron equation (Eq. 23.13)

$$dg_1 = dg_2$$

$$V_1 dp - S_1 dT = V_2 dp - S_2 dT$$

$$\therefore \frac{dp}{dT} = \frac{S_2 - S_1}{V_2 - V_1}$$

$$= \frac{\Delta S}{\Delta V}$$

$$= \frac{L}{T\Delta V}$$

- Starting point: along a coexistence curve, the Gibbs free energy per particle of both phases are equal, $g_1 = g_2$ (refer to Figure 23.3).
- Use the differential expression for dG from Eq. 22.31:

  $$dG = Vdp - SdT$$

- The latent heat of a phase transition is defined as: $L = \Delta Q = T\Delta S$

WORKED EXAMPLE 23.1: Temperature dependence of vapor pressure using Eq. 23.13

1. For a liquid–vapor phase transition, assume $V_{gas} \gg V_{liquid}$:

   $$\therefore \Delta V = V_{gas} - V_{liquid}$$

   $$\simeq V_{gas}$$

2. Treat the vapor as an ideal gas:

   $$\therefore pV_{gas} = nRT$$

3. Assume that the latent heat of the phase transition is independent of temperature. The molar latent heat is:

   $$L_m = \frac{L}{n} \text{ Jmol}^{-1}$$

Use steps 1, 2, and 3 in the Clausius Clapeyron equation (Eq. 23.13) and integrate:

$$\frac{dp}{dT} = \frac{L}{T\Delta V}$$

$$\simeq \frac{L}{TV_{gas}}$$

$$= \frac{Lp}{nRT^2}$$

$$= \frac{L_m p}{RT^2}$$

$$\therefore \int \frac{dp}{p} = \frac{L_m}{R} \int \frac{dT}{T^2}$$

$$\ln(p) = -\frac{L_m}{RT} + \text{constant}$$

$$\therefore p(T) = p_0 e^{-L_m/RT}$$

The order of the transition

Phase transitions are characterized by a step transition in macroscopic properties of the system. For example, phase transitions obeying the Clausius Clapeyron equation are characterized by a step change in volume, ΔV, and also in entropy, $L \propto \Delta S$. This makes sense intuitively, for example, when water freezes, its volume increases and its entropy decreases (it becomes more structured).

These abrupt steps in macroscopic properties correspond to discontinuities in the derivative of the Gibbs free energy. For example:

$$\left(\frac{\partial G}{\partial T}\right)_p = -S$$

$$\text{and } \left(\frac{\partial G}{\partial p}\right)_T = V$$

(refer to Section 22.3.2). The *order* of the phase transition is defined as the order of the lowest derivative of G that shows a discontinuity during the transition. For example, transitions that obey the Clausius Clapeyron equation are first-order transitions, since discontinuities are in V and S (first derivatives of G). First-order transitions therefore always take place at constant temperature and constant pressure, and are always associated with a latent heat corresponding to an entropy step.

Second-order transitions have discontinuities in C_p and κ_T, since these correspond to the second derivatives of G:

$$\left(\frac{\partial^2 G}{\partial T^2}\right)_p = -\left(\frac{\partial S}{\partial T}\right)_p = -\frac{1}{T}\left(\frac{\partial Q}{\partial T}\right)_p = -\frac{C_p}{T}$$

and

$$\left(\frac{\partial^2 G}{\partial p^2}\right)_T = \left(\frac{\partial V}{\partial p}\right)_T = -V\kappa_T$$

(refer to Eq. 23.10 for an expression for κ_T). These transitions are not associated with latent heat.

23.3 Irreversible gaseous expansions

To end the discussion on classical thermodynamics, we'll take a look at two *irreversible* processes: the Joule expansion and the Joule–Thomson expansion. These experiments were carried out to determine the temperature change of real and ideal gases on expansion. We can apply reversible thermodynamics to irreversible processes *provided* the initial and final states are equilibrium states. However, we cannot determine the intermediate, non-equilibrium properties of the system.

23.3.1 Joule expansion

Consider a partitioned, thermally isolated container, with a gas on one side of the partition and a vacuum on the other, as shown in Fig. 23.4. Removing the partition allows the gas to expand into the vacuum and fill the entire container. Since there is no opposing pressure in a vacuum, no work is done by the gas against the surroundings ($dW = 0$), and since the system is thermally isolated, then the expansion is

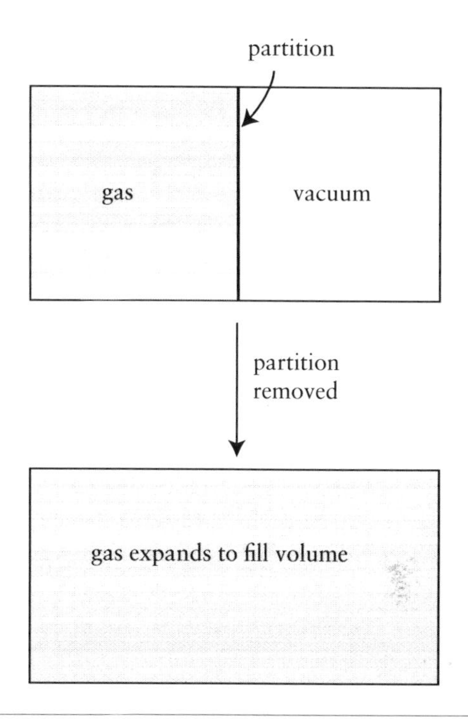

Figure 23.4: Joule expansion. When the partition is removed, the gas expands against a vacuum. The container is thermally isolated; therefore, heat does not enter or leave the container.

adiabatic ($dQ = 0$). Therefore, from the First Law of Thermodynamics, the internal energy remains constant during the expansion:

$$dU = 0 \rightarrow U = \text{constant}$$

Joule coefficient

The Joule expansion corresponds to an expansion under constant internal energy. A parameter to measure the change in temperature upon expansion is the *Joule coefficient*, α_J [K m^{-3}], which is defined by the partial derivative:

$$\alpha_J = \left(\frac{\partial T}{\partial V}\right)_U$$

Using the exact differential for U along with the definition of C_V, we find an expression for α_J in terms of C_V and state variables T and p:

$$\alpha_J = -\frac{1}{C_V}\left[T\left(\frac{\partial p}{\partial T}\right)_V - p\right]$$

(Eq. 23.14 and Derivation 23.6).

Derivation 23.6: Derivation of the Joule coefficient (Eq. 23.14)

$$\alpha_J = \left(\frac{\partial T}{\partial V}\right)_U$$
$$= -\left(\frac{\partial U}{\partial V}\right)_T \left(\frac{\partial T}{\partial U}\right)_V$$
$$= -\left(\frac{\partial U}{\partial V}\right)_T \left(\frac{\partial U}{\partial T}\right)_V^{-1}$$
$$= -\left(T\left(\frac{\partial S}{\partial V}\right)_T - p\right)\left(T\frac{\partial S}{\partial T}\right)_V^{-1}$$
$$= -\frac{1}{C_V}\left[T\left(\frac{\partial S}{\partial V}\right)_T - p\right]$$
$$= -\frac{1}{C_V}\left[T\left(\frac{\partial p}{\partial T}\right)_V - p\right]$$

- Starting point: the Joule coefficient is defined as:

$$\alpha_J = \left(\frac{\partial T}{\partial V}\right)_U$$

- The reciprocity theorem gives:

$$\left(\frac{\partial x}{\partial y}\right)_z = -\left(\frac{\partial z}{\partial y}\right)_x \left(\frac{\partial x}{\partial z}\right)_y$$
$$\therefore \left(\frac{\partial T}{\partial V}\right)_U = -\left(\frac{\partial U}{\partial V}\right)_T \left(\frac{\partial T}{\partial U}\right)_V$$

- Use the exact differential for U:

$$dU = TdS - pdV$$

to find expressions for $(\partial U/\partial V)_T$ and $(\partial U/\partial T)_V$.

- The heat capacity at constant volume is defined as:

$$C_V = \left(\frac{\partial Q}{\partial T}\right)_V = T\left(\frac{\partial S}{\partial T}\right)_V$$

- Use the following Maxwell relation:

$$\left(\frac{\partial S}{\partial V}\right)_T = \left(\frac{\partial p}{\partial T}\right)_V$$

to eliminate the entropy derivative.

The Joule coefficient (Derivation 23.6):

$$\alpha_J = \left(\frac{\partial T}{\partial V}\right)_U = -\frac{1}{C_V}\left[T\left(\frac{\partial p}{\partial T}\right)_V - p\right] \quad (23.14)$$

Joule coefficient of an ideal gas

Using the Ideal Gas Law together with Eq. 23.14, we find that the Joule coefficient of an ideal gas is:

$$\alpha_J = -\frac{1}{C_V}\left[\frac{nRT}{V} - \frac{nRT}{V}\right]$$
$$= 0$$

This agrees with what Joule found experimentally, which was that there was no change in temperature during the free expansion of an ideal gas. Since $C_V \neq 0$, then a zero Joule coefficient implies that:

$$\left(\frac{\partial U}{\partial V}\right)_T = 0$$

(refer to Derivation 23.6). Therefore, the internal energy of an ideal gas does not change upon a change in volume—it is a function of temperature only. This agrees with the results found from kinetic theory (refer to Section 21.4.1). An internal energy that does not depend on volume implies that the energy of ideal gas particles is independent of their separation, which in turn implies that there are no intermolecular forces, as expected for an ideal gas.

Joule coefficient of a real gas

For a *real* gas, however, the internal energy is a function of both temperature and volume:

$$dU = C_V dT + \frac{a}{v^2} dV$$

(Worked example 23.2). For an ideal gas, $a = 0$, and therefore we recover $dU = C_V dT$, as we also found using the First Law of Thermodynamics in Section 22.2.2. From the above expression for dU, we find that the Joule coefficient for a real gas is negative:

$$\alpha_J = \left(\frac{\partial T}{\partial V}\right)_U = -\frac{a}{C_V v^2}$$

Therefore, a free expansion of a real gas always results in cooling:

$$\therefore \alpha_J < 0$$

23.3.2 Joule–Thomson expansion

The Joule–Thomson experiment was carried out to further understand the cooling properties of a real gas on expansion. In this case, instead of a free expansion against a vacuum, a gas is allowed to expand through a thermally isolated throttling valve, as shown in Fig. 23.5. The purpose of the throttling valve, which could also be a porous plug or a pinhole, is to provide an impedance against the flow, which would reduce the pressure of the gas from pressure p_1 to pressure p_2 ($p_1 > p_2$) without the gas doing external work. The pistons are moved so as to keep p_1 and p_2 constant.

In this case, the internal energy is not a constant, as it was for the Joule expansion. We find instead that the *enthalpy*, H, is constant for such an expansion (Derivation 23.7).

WORKED EXAMPLE 23.2: Internal energy of a real gas

1. Entropy is a state variable, and therefore can be written as an exact differential in terms of any two of the fundamental variables p, V, and T. For example, for $S = S(V, T)$, the exact differential is:

   $$dS = \left(\frac{\partial S}{\partial T}\right)_V dT + \left(\frac{\partial S}{\partial V}\right)_T dV$$

   Use this to eliminate dS from the exact differential for U.

2. The heat capacity at constant volume is defined as:

   $$C_V = \left(\frac{\partial Q}{\partial T}\right)_V = \left(T\frac{\partial S}{\partial T}\right)_V$$

3. Use the following Maxwell relation:

   $$\left(\frac{\partial S}{\partial V}\right)_T = \left(\frac{\partial p}{\partial T}\right)_V$$

   to eliminate the $S-V$ derivative.

4. Use the van der Waals equation of state:

   $$\left(p + \frac{a}{v^2}\right)(v - b) = RT$$

   to eliminate the pressure terms.

$$dU = TdS - pdV$$

$$= T\left(\left(\frac{\partial S}{\partial T}\right)_V dT + \left(\frac{\partial S}{\partial V}\right)_T dV\right) - pdV$$

$$= C_V dT + T\left(\frac{\partial p}{\partial T}\right)_V dV - pdV$$

$$= C_V dT + \frac{RT}{v-b}dV - \frac{RT}{v-b}dV + \frac{a}{v^2}dV$$

$$= C_V dT + \frac{a}{v^2}dV$$

$\Rightarrow$ the change in internal energy of a real gas is a function of both temperature and volume.

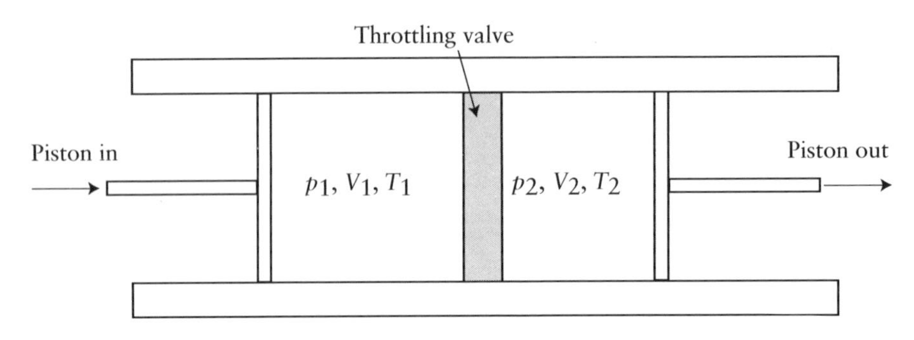

Figure 23.5: Joule–Thomson expansion. The gas is expanded through a throttle valve, which provides an impedance against the flow and therefore reduces the pressure of the gas from p_1 to p_2 ($p_1 > p_2$), where p_1 and p_2 are kept constant during the process. The container is thermally isolated, therefore, heat does not enter or leave the container.

Derivation 23.7: Derivation of the conservation of enthalpy for the Joule–Thomson expansion

$$\begin{aligned}\Delta U &= W \\ &= -p_1\Delta V_1 - p_2\Delta V_2 \\ &= -p_1(0 - V_1) - p_2(V_2 - 0) \\ &= p_1V_1 - p_2V_2\end{aligned}$$

$$\therefore U_2 - U_1 = p_1V_1 - p_2V_2$$

$$U_2 + p_2V_2 = U_1 + p_1V_1$$

$$\therefore H_2 = H_1 \rightarrow \text{enthalpy is conserved}$$

- Starting point: for a thermally isolated system, $dQ = 0$, therefore, the First Law gives $dU = dW$.
- The total change in internal energy is $\Delta U = W$, where W takes into account the change in volume of the gas on either side of the throttling valve (refer to Fig. 23.5).
- Enthalpy is defined as $H = U + pV$.

Joule–Thomson coefficient

The *Joule–Thomson coefficient*, μ_J [K Pa^{-1}], describes the change in temperature with pressure change at constant enthalpy, and is defined as:

$$\mu_J = \left(\frac{\partial T}{\partial p}\right)_H$$

Using the exact differential for H along with the definition of C_p, we find an expression for μ_J in terms of C_p and state variables T and V:

$$\mu_J = \frac{1}{C_p}\left[T\left(\frac{\partial V}{\partial T}\right)_p - V\right]$$

(Eq. 23.15 and Derivation 23.8). μ_J has an analogous form to a_J (compare Eq. 23.15 with Eq. 23.14).

The Joule–Thomson coefficient (Derivation 23.8):

$$\mu_J = \left(\frac{\partial T}{\partial p}\right)_H$$

$$= \frac{1}{C_p}\left[T\left(\frac{\partial V}{\partial T}\right)_p - V\right] \qquad (23.15)$$

Derivation 23.8: Derivation of the Joule–Thomson coefficient (Eq. 23.15)

$$\mu_J = \left(\frac{\partial T}{\partial p}\right)_H$$

$$= -\left(\frac{\partial H}{\partial p}\right)_T \left(\frac{\partial T}{\partial H}\right)_p$$

$$= -\left(\frac{\partial H}{\partial p}\right)_T \left(\frac{\partial H}{\partial T}\right)_p^{-1}$$

$$= -\left(T\left(\frac{\partial S}{\partial p}\right)_T + V\right)\left(T\frac{\partial S}{\partial T}\right)_p^{-1}$$

$$= -\frac{1}{C_p}\left[T\left(\frac{\partial S}{\partial p}\right)_T + V\right]$$

$$= \frac{1}{C_p}\left[T\left(\frac{\partial V}{\partial T}\right)_p - V\right]$$

- Starting point: the Joule–Thomson coefficient is defined as:

$$\mu_J = \left(\frac{\partial T}{\partial p}\right)_H$$

- The reciprocity theorem gives:

$$\left(\frac{\partial x}{\partial y}\right)_z = -\left(\frac{\partial z}{\partial y}\right)_x \left(\frac{\partial x}{\partial z}\right)_y$$

$$\therefore \left(\frac{\partial T}{\partial p}\right)_H = -\left(\frac{\partial H}{\partial p}\right)_T \left(\frac{\partial T}{\partial H}\right)_p$$

- Use the exact differential for H:

$$\mathrm{d}H = T\mathrm{d}S + V\mathrm{d}p$$

to find expressions for $(\partial H/\partial P)_T$ and $(\partial H/\partial T)_p$.

- The heat capacity at constant pressure is defined as:

$$C_p = \left(\frac{\partial Q}{\partial T}\right)_p = T\left(\frac{\partial S}{\partial T}\right)_p$$

- Use the following Maxwell relation:

$$\left(\frac{\partial S}{\partial p}\right)_T = -\left(\frac{\partial V}{\partial T}\right)_p$$

to eliminate the entropy derivative.

Joule–Thomson coefficient of an ideal and real gas

Using the Ideal Gas Law together with Eq. 23.15, we find the Joule–Thomson coefficient of an ideal gas is:

$$\mu_J = \frac{1}{C_p}\left[\frac{nRT}{p} - \frac{nRT}{p}\right]$$

$$= 0$$

which means that its temperature does not change on throttling. For a real gas, however, μ_J is either positive or negative, depending on the initial values of T and p. Since p always decreases ($p_1 > p_2$), then:

if $\mu_J > 0$: the gas cools down ($\mathrm{d}T < 0$)

if $\mu_J < 0$: the gas warms up ($\mathrm{d}T > 0$)

The sign of μ_J for different combinations of T and P is shown in Fig. 23.6. There is an *inversion curve*, which is a function of temperature and pressure, for which the temperature change on expansion is zero. The inversion curve is given by the condition that:

$$\mu_J = 0$$

Since $C_p \neq 0$, this implies that on the inversion curve:

$$T\left(\frac{\partial V}{\partial T}\right)_p - V = 0$$

$$\therefore \left(\frac{\partial V}{\partial T}\right)_p = \frac{V}{T} \qquad (23.16)$$

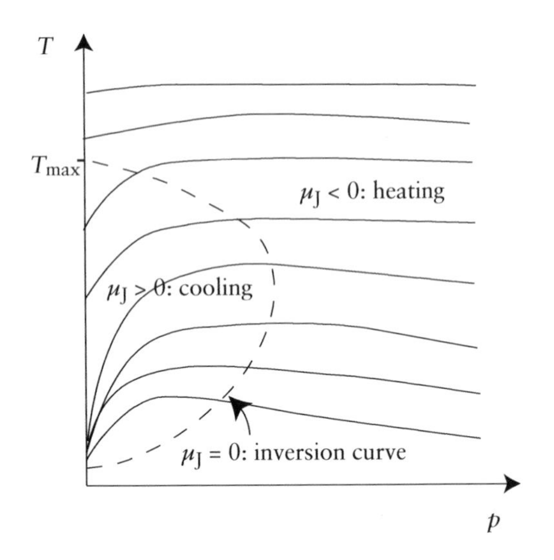

Figure 23.6: Isenthalps (lines of constant enthalpy) for the Joule–Thomson expansion of a real gas. The inversion curve indicates the conditions for which there is no temperature change on expansion. T_{max} corresponds to the maximum initial temperature, above which Joule–Thomson expansion of a real gas results in heating for all initial pressures.

(refer to Derivation 23.8). Inside the inversion curve, $\mu_J > 0$, and Joule–Thomson throttling can be used to cool gases until they liquify, using a countercurrent heat exchanger. Outside of the inversion curve, $\mu_J < 0$ and the gas heats up on throttling. Fig. 23.6 shows that there is a maximum initial temperature, above which throttling heats gases for *all* pressures. This occurs when the inversion curve crosses the temperature axis at the high-temperature end, and is therefore given by the conditions $\mu_J = 0$ and $p = 0$. The value of the inversion temperature as a function of pressure, $T_{inv}(p)$, can be found by using the van der Waals equation (Eq. 23.1) in Eq. 23.16:

$$\left(p + a\left(\frac{n}{V}\right)^2\right)(V - nb) = nRT$$

$$\therefore \left(\frac{\partial V}{\partial T}\right)_p = \frac{nRV^3}{pV^3 - aVn^2 + 2abn^3}$$

$$\text{and} \left(\frac{\partial V}{\partial T}\right)_p = \frac{V}{T}$$

Therefore, equating these expressions, the inversion temperature as a function of pressure is:

$$\therefore T_{inv}(p) = \frac{pV^3 - aVn^2 + 2abn^3}{nRV^2}$$

The maximum inversion temperature then corresponds to $T_{max} = T_{inv}(0)$:

$$\therefore T_{max} = T_{inv}(0)$$

$$= \frac{an(2bn - V)}{RV^2}$$

Chapter 24
Statistical mechanics

We saw in kinetic theory and classical thermodynamics (Chapters 21 and 22) that we can describe systems by taking either a *micro*scopic approach (considering the average properties of individual particles in a system), or by taking a *macro*scopic approach (considering solely the gross properties of systems). The macroscopic approach, however, is limited: it describes systems in the classical regime, but not in the sub-atomic, or *quantum* regime. Statistical mechanics uses a probabilistic, microscopic approach, similar to that used in kinetic theory, to describe the macroscopic properties of any system (it is not restricted to ideal gases, as kinetic theory is). When we apply statistical mechanics to classical systems, we recover the same results as those from thermodynamics, whereas by treating systems in the quantum regime, we discover new phenomena that could not be anticipated from thermodynamics. Since statistical mechanics uses a microscopic approach to explain macroscopic properties, it links together the familiar classical world that we experience every day with the unfamiliar sub-atomic, quantum world of individual particles. It demonstrates that all we see around us is a consequence of the combined behavior of individual atoms.

24.1 Statistical ensembles

In classical thermodynamics, we defined three different types of system depending on whether energy and matter can be exchanged with their surroundings (refer to Section 22.1):

1. *Isolated systems:* neither energy nor matter exchange is possible. Therefore, the total energy, U, and total number of particles in the system, N, are constant.
2. *Closed systems:* energy exchange can occur, but not exchange of matter. Therefore, U is variable and N is constant.
3. *Open systems:* both energy and matter exchange can occur. Therefore, both U and N are variable.

In statistical mechanics, we make similar classifications. However, instead of dealing with a single system, we now wish to look at a *collection* of macroscopically identical systems (same values of p, T, V, etc). The microscopic properties, for example, the individual particle energies or velocities, differ from system to system. Therefore, this approach allows us to determine the probabilistic microscopic properties of a system, for example, the probability distribution of particle energies. The collection of such systems is called a *statistical ensemble.* There are three common statistical ensembles, which correspond to the three types of thermodynamic systems described above:

1. *Microcanonical ensemble*: collection of isolated systems (U and N are constant).
2. *Canonical ensemble*: collection of closed systems (U is variable and N is constant).
3. *Grand canonical ensemble*: collection of open systems (U and N are variable).

In theory, an ensemble comprises an infinite number of macroscopically identical systems. Of course in practice, this is impossible to prepare, therefore, statistical ensembles are mathematical idealizations. Statistical mechanics is thus based on two fundamental postulates, which allow us to determine the probability distributions of the microscopic properties of a given system. These are outlined below.

24.2 The postulates of statistical mechanics

24.2.1 States and energy levels

The state of a particle is represented by its *wavefunction*, $\psi(q_1, q_2,..., q_n)$, where the variables $q_1, q_2,..., q_n$ correspond to the variables needed to define the state. For example, q_1 may correspond to the particle's position, q_2 its momentum, etc. The square of the wavefunction represents the probability that a particle has a given value of a given variable. For example, the probability that a particle represented by a wavefunction, $\psi(q_1)$, has $a<q_1<b$ is given by:

$$P_{ab} = \int_a^b |\psi(q_1)|^2 \, dq_1$$

(refer to Section 16.3.1 for more detail). The state of a particle is often specified, in part, by its *energy*, ε. From quantum mechanics, we know that energy is quantized: there exists a discrete set of allowed energy levels that a particle can occupy, which arise from boundary conditions imposed on the wavefunction. At high energies, the difference between adjacent energy levels is much smaller than the total energy of the system, and we can treat ε as a continuous variable. In this case, we "bundle" several energy levels together, and refer to a particle having energy between ε and $\varepsilon + d\varepsilon$, rather than it being in the nth level, since $n \to \infty$ at high energies. Discrete and continuous energy levels are illustrated in Fig. 24.1.

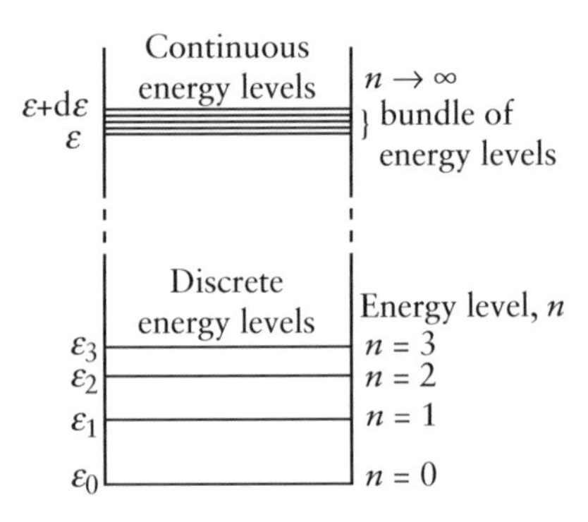

Figure 24.1: Discrete and continuous energy levels. When the spacing between adjacent energy levels is negligible (continuous distribution), then energy levels are "bundled" into intervals of size dε.

Degeneracy

The terms "state" and "energy level" are sometimes used interchangeably, but there is an important distinction between them. The *energy level* tells us only the particle's energy; the *state*, however, may also include the particle's angular momentum or its spin, for example. This means that any given energy level may comprise several states of differing momentum or spin, each corresponding to a different wavefunction. When several different states correspond to the same particle energy, then that energy level is said to be *degenerate*. For example, in the absence of a magnetic field, the spin-up and spin-down electron states in a hydrogen atom are at the same energy, and therefore there is a two-fold spin degeneracy at each energy level.

For a discrete energy distribution, degeneracy is usually denoted by the symbol g_i, which represents the number of states in the ith energy level.

24.2.2 Microstates and macrostates

Distinguishable particles

To begin with, let's assume that the particles in a system are *distinguishable*. This means that we can, in theory, tell the difference between them. For example, particles localized in a crystal are distinguishable because of their position, and atoms of different elements are obviously distinguishable by their nature. However, even identical particles in an ideal gas are distinguishable—they can, in theory, be told apart from one another. Distinguishable particles are defined as having wavefunctions that do not overlap in space with one another (the interparticle distance is much greater than the de Broglie wavelength of the particle, see Section 24.6.3). Therefore, each particle occupies its own individual potential well. By definition, the wavefunctions of particles in classical systems do not overlap, thus, all classical systems, such as ideal gases, comprise distinguishable particles. In Section 24.5, we will look at *in*distinguishable particles—ones whose de Broglie wavelength is comparable with their interparticle separation. Indistinguishability is a quantum mechanical phenomenon that has no classical analog.

The states of distinguishable particles

There are two terms used to describe a system's state in terms of the energy levels occupied by its particles—the *microstate* and the *macrostate*. The microstate of a system tells us *which* particle is in *which* energy level, whereas the macrostate tells us the *total number* of particles in each energy level, and hence the total energy of the system seen on macroscopic scales, without making reference to individual particles. The distinction is best understood through example. Consider four particles, labeled A, B, C, and D, each in their own potential well,[1] occupying energy levels $n = 1$, 2 or 3, as shown in Fig. 24.2. The microstate and macrostate are defined in the figure. From the macrostate, the total energy of the system corresponds to $E = 8$ (in arbitrary units).

Figure 24.3 illustrates two different systems with the same total energy ($E = 8$), but with a) a different microstate corresponding to the same macrostate and b) both a different microstate *and* a different macrostate. The postulates of statistical mechanics, outlined below, are concerned with describing the microstates and macrostates of a given system.

The first and second postulates

The first postulate of statistical mechanics describes the probability that a system is in a particular microstate of a given macrostate,[2] and is called the *postulate of equal a priori*[3] *probability*:

Postulate 1: for a macrostate in equilibrium, all microstates corresponding to that macrostate are equally probable.

This means that the microstate in Fig. 24.3a has the same probability of occurring as that in Fig. 24.2 since they have identical macrostates. The consequence of this postulate is that if there are Ω different microstates corresponding to a given macrostate, then the probability of finding the system in any one of those microstates is $1/\Omega$. The second postulate of statistical mechanics tells us, for a given *total* energy, which *macrostate* the system is likely to be in:

Postulate 2: the observed macrostate is the one with the most microstates, Ω.

For example, there is only one way of getting the macrostate in Fig. 24.3b, which means that there is only one microstate (all the particles are in the second energy

[1] The shape of the potential well is arbitrary at this point, but we have drawn it approximately as a simple harmonic oscillator (SHO) potential, since this is most representative of real particles. For details of the SHO potential, see Section 17.2.3.

[2] Here, we are dealing with a *microcanonical* ensemble, which corresponds to a collection of systems in which the total number of particles and total energy of the system are constant.

[3] "a priori" means that the postulate has come from theoretical deduction rather than from experimental observation.

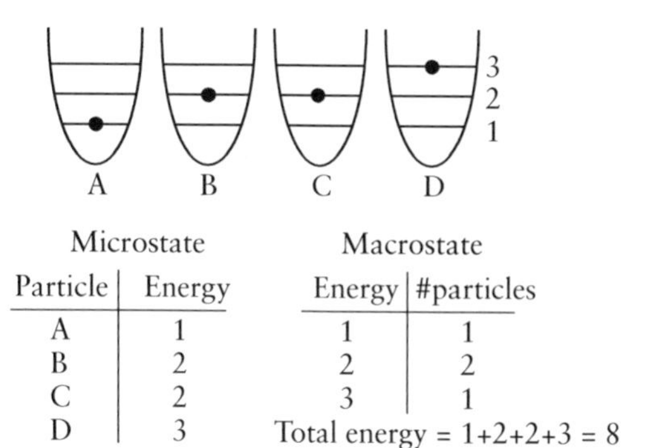

Particle	Energy
A	1
B	2
C	2
D	3

Energy	#particles
1	1
2	2
3	1

Total energy = 1+2+2+3 = 8

Figure 24.2: Example of a possible microstate and macrostate of a system of four distinguishable particles (A, B, C, and D), occupying three energy levels (1, 2, and 3).

level), and therefore the number of microstates for that example is $\Omega_b = 1$. In contrast, we find there are 12 different ways of getting the macrostate in Fig. 24.3a, and therefore the number of microstates for that example is $\Omega_a = 12$ (check by writing out the various different combinations). Since Ω_a is greater than Ω_b, then the macrostate in a is more likely to be observed than that in b, for the same total energy.

24.2.3 Properties of Ω

From the discussion above, the parameter Ω is defined as the number of microstates in a macrostate. For a system of N distinguishable particles, with n_1 particles in energy level 1, n_2 in level 2, and n_i in level i, Ω is calculated using a probability combination formula:

$$\Omega = \frac{N!}{n_1!n_2!n_3!\ldots} \tag{24.1}$$

For example, for the macrostates in Fig. 24.3, we find:

$$\Omega_a = \frac{4!}{1!2!1!} = 12$$

and $$\Omega_b = \frac{4!}{0!4!0!} = 1$$

as expected. In most systems, N is a very large number. For example, a mole of gas has N on the order of $N \sim 6 \cdot 10^{23}$ particles (Avogadro's number). The number of particles in any one state, n_i, is usually much less than N. As a consequence, there is one macrostate for which Ω is much larger than those corresponding to different macrostates of the same total energy. As a simple example, consider the third macrostate that gives the same total energy as those in Fig. 24.3 ($E = 8$): two particles

a) Same macrostate, different microstate

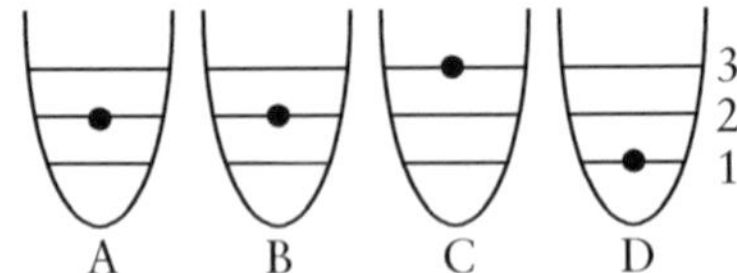

Microstate	
Particle	Energy
A	2
B	2
C	3
D	1

Macrostate	
Energy	#particles
1	1
2	2
3	1

Total energy = 1+2+2+3 = 8

b) Same total energy, different macrostate

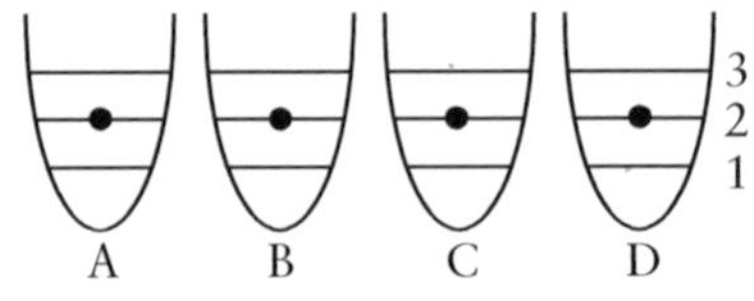

Microstate	
Particle	Energy
A	2
B	2
C	2
D	2

Macrostate	
Energy	#particles
1	0
2	4
3	0

Total energy = 2+2+2+2 = 8

Figure 24.3: Systems with the same total energy as that in Fig. 24.2, but for a) a different microstate and b) a different microstate and different macrostate.

in level 1, zero in level 2, and two in level 3. The number of microstates in this macrostate is therefore:

$$\Omega = \frac{4!}{2!0!2!} = 6$$

which is significantly less than Ω_a. Therefore, we find that Ω is strongly peaked at a single macrostate, for any given total energy. From postulate 2, this corresponds to the observed macrostate.

Another look at entropy

Since the number of microstates in a macrostate depends on the total energy of the macrostate, then Ω is a function of energy: $\Omega = \Omega(\varepsilon)$. Let's assume that the value of Ω at any given energy, ε, is that corresponding to the observed macrostate—the one for which Ω is a maximum (postulate 2).

This takes us to another postulate of statistical mechanics, which is that the *entropy* of a system is proportional to the logarithm of Ω, and is given by Eq. 24.2. This is a *definition* of entropy, from a statistics perspective, and cannot be proved.

Entropy and Ω:

$$S \stackrel{\text{def}}{=} k_B \ln \Omega \qquad (24.2)$$

$$\text{where} \quad \frac{\partial \ln \Omega}{\partial \varepsilon} \stackrel{\text{def}}{=} \frac{1}{k_B T} \qquad (24.3)$$

This definition shows that entropy can be thought of as a measure of the number of ways of arranging quanta of energy between the particles in a system (since this is what the variable Ω represents), which is consistent with the interpretation of entropy as a measure of "disorder." Recall from thermodynamics that the exact differential for internal energy is $dU = TdS - pdV = TdS$ at constant volume (refer to Section 22.3.1). We recover $dU = TdS$ provided we define:

$$\frac{\partial \ln \Omega}{\partial \varepsilon} \stackrel{\text{def}}{=} \frac{1}{k_B T}$$

(Eq. 24.3). Therefore, by differentiating Eq. 24.2, we find:

$$\frac{\partial S}{\partial \varepsilon} = \frac{k_B}{\Omega}\frac{\partial \Omega}{\partial \varepsilon}$$

$$= k_B \frac{\partial \ln \Omega}{\partial \varepsilon}$$

$$= \frac{1}{T}$$

Rearranging and replacing the partial derivative with a total derivative, we have $d\varepsilon = TdS$, which is analogous to $dU = TdS$ from classical thermodynamics.

24.3 The density of states

If we consider an energy interval $d\varepsilon$, then the value of $\Omega(\varepsilon + d\varepsilon) - \Omega(\varepsilon)$ gives us the number of microstates within that interval. Treating Ω as a continuous function of ε, and performing a Taylor expansion on $\Omega(\varepsilon + d\varepsilon)$ allows us to define a function that represents the total number of states that lie between an energy ε and $\varepsilon + d\varepsilon$:

$$\Omega(\varepsilon + d\varepsilon) - \Omega(\varepsilon) \simeq \Omega(\varepsilon) + \frac{d\Omega}{d\varepsilon}d\varepsilon - \Omega(\varepsilon)$$
$$= \frac{d\Omega}{d\varepsilon}d\varepsilon$$
$$\equiv g(\varepsilon)d\varepsilon$$

The function $g(\varepsilon)$ is called the *density of states*; it represents the number of states per energy interval $d\varepsilon$, and is a continuous function of energy. In the rest of this section, we will derive various expressions for $g(\varepsilon)d\varepsilon$.

Boundary conditions

If we consider the simple example of a particle in a one-dimensional infinite square well (see Section 17.2.1), then we know that only certain wavefunctions satisfy the boundary conditions imposed on the system. This means that only certain states, and hence energy levels, are allowed to exist. Generalizing to a system of *any* shape and dimension, the same thing occurs—some wavefunctions will satisfy the boundary conditions, allowing particles to occupy those states, whereas others won't, and therefore no particle can exist in those states. The density of states effectively tells us which quantum states satisfy the boundary conditions of a given system. Therefore, the function $g(\varepsilon)d\varepsilon$ depends on the dimensions of the system in question, and has a different energy dependence for systems in one, two, or three dimensions, as we shall see below.

24.3.1 Derivation in k-space

In general, the energy levels of a system depend on the mass of the particle in question. This means that the form of $g(\varepsilon)d\varepsilon$ is different depending on whether the system involves a particle of mass m or a massless particle, such as a photon. Therefore, to derive an expression for the density of states, we work in terms of *wavelength* rather than energy, since the "allowed wavelengths" are the same for both particles with mass and massless particles. The derivation for the density of states is more straightforward if we work in k-space, rather than in λ-space, where the wavenumber, k, is related to the wavelength by:

$$k = \frac{2\pi}{\lambda}$$

The density of states in k-space is then denoted by $g(k)dk$. By generalizing the boundary conditions for a particle in an infinite square well, we can derive expressions for $g(k)dk$ in three, two, and one dimensions:

In three dimensions: $g_3(k)dk = g\frac{k^2}{2\pi^2}dk$

In two dimensions: $g_2(k)dk = g\frac{k}{2\pi}dk$

In one dimension: $g_1(k)dk = g\frac{1}{\pi}dk$

where g is the *degeneracy factor* (Eqs. 24.4–24.6, Derivation 24.1, and Fig. 24.4). Here, the density of states are expressed as the number of states per unit volume in three dimensions, per unit area in two dimensions, and per unit length in one dimension. These expressions hold both for particles with mass and for massless particles. The total number of states per unit volume/area/length is found by integrating over all values of k (Eq. 24.7).

The degeneracy factor, g, accounts for spin or polarization degeneracy. For electrons, each state of wavevector k can accommodate two electrons—one in the spin-up state and one in the spin-down state. Therefore the spin degeneracy for electrons, and other spin-1/2 particles, is $g = 2$. Photons are spin-1 particles, and can exist in two orthogonal transverse polarization states. Therefore the polarization degeneracy for phonons is $g = 2$. Other spin-1 particles, such as phonons[4], have a longitudinal polarization state in addition to two transverse polarization states for each state of wavevector k, therefore for phonons, $g = 3$.

Density of states in k-space (Derivation 24.1):

In three dimensions: $$g_3(k)dk = g\frac{k^2}{2\pi^2}dk \qquad (24.4)$$

In two dimensions: $$g_2(k)dk = g\frac{k}{2\pi}dk \qquad (24.5)$$

In one dimension: $$g_1(k)dk = g\frac{1}{\pi}dk \qquad (24.6)$$

where the degeneracy factor, g, is due to spin or polarization degeneracy:

$$g = \begin{cases} 2 & \text{for electrons} \\ 2 & \text{for photons} \\ 3 & \text{for phonons} \end{cases}$$

Total number of states per unit volume/area/length:

$$n_s = \int_0^\infty g(k)dk \qquad (24.7)$$

4 A phonon can be thought of as a quanta of acoustic wave, in analogy to a photon as a quanta of electromagnetic wave.

Derivation 24.1: Derivation of the density of states in k-space, $g(k)\mathrm{d}k$ (Eqs. 24.4–24.6)

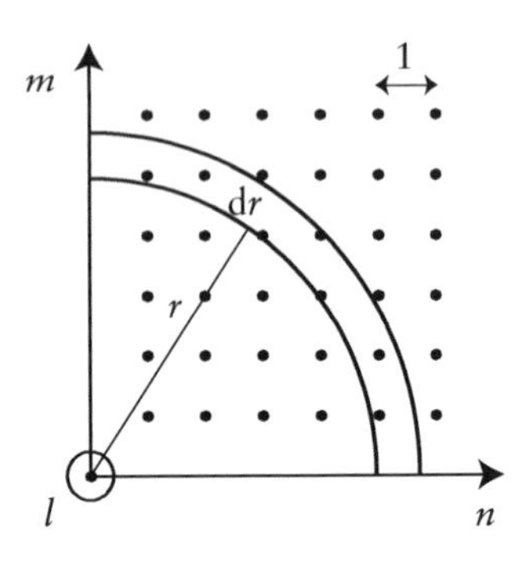

In three dimensions:

$$G_3(k)\mathrm{d}k = \frac{1}{8}4\pi r^2 \mathrm{d}r$$

$$= \frac{\pi}{2}\frac{L^3k^2}{\pi^3}\mathrm{d}k$$

$$= \frac{Vk^2}{2\pi^2}\mathrm{d}k$$

$$\therefore g_3(k)\mathrm{d}k = \frac{k^2}{2\pi^2}\mathrm{d}k$$

In two dimensions:

$$G_2(k)\mathrm{d}k = \frac{1}{4}2\pi r\mathrm{d}r$$

$$= \frac{\pi}{2}\frac{L^2k}{\pi^2}\mathrm{d}k$$

$$= \frac{Ak}{2\pi}\mathrm{d}k$$

$$\therefore g_2(k)\mathrm{d}k = \frac{k}{2\pi}\mathrm{d}k$$

In one dimension:

$$G_1(k)\mathrm{d}k = \mathrm{d}r$$

$$= \frac{L}{\pi}\mathrm{d}k$$

$$\therefore g_1(k)\mathrm{d}k = \frac{1}{\pi}\mathrm{d}k$$

- Starting point: from the solution of the Schrödinger equation for a particle in a one-dimensional infinite square well (refer to Worked example 17.8), the allowed wavevectors in a confinement of length L are given by integer values of π/L:

$$k_x = \frac{n\pi}{L}$$

- In three-dimensions, the total wavevector is:

$$k = \sqrt{k_x^2 + k_y^2 + k_z^2}$$

$$= \frac{\pi}{L}\sqrt{n^2 + m^2 + l^2}$$

$$\equiv \frac{\pi r}{L} \qquad (24.8)$$

where $r^2 = n^2 + m^2 + l^2$

and n, m, and l are integers.

- The quantum states defined by integers n, m, and l can be represented by a grid in the positive x, y, z octant (refer to diagram). The spacing between adjacent points on the grid is equal to 1.
- In three dimensions: the number of states between k and k + dk is given by the volume of a spherical shell at radius r and width dr in the positive octant:

$$G_3(k)\mathrm{d}k = \frac{1}{8}4\pi r^2\mathrm{d}r$$

- Use Eq. 24.8 to eliminate r and dr. The volume of confinement is $V = L^3$. To find the density of states, we divide by the volume: $g_3(k)\mathrm{d}k = G_3(k)\mathrm{d}k/V$.
- In two dimensions: the number of states between k and k + dk is given by the area of a circular shell at radius r and width dr in the positive quadrant:

$$G_2(k)\mathrm{d}k = \frac{1}{4}2\pi r\mathrm{d}r$$

- Use Eq. 24.8 to eliminate r and dr. The area of confinement is $A = L^2$. Therefore, divide by the area to find the density of states: $g_2(k)\mathrm{d}k = G_2(k)\mathrm{d}k/A$.
- In one dimension: the number of states between k and k + dk is given by the length dr. The confinement length is L, therefore, divide by L for the one-dimensional density of states.

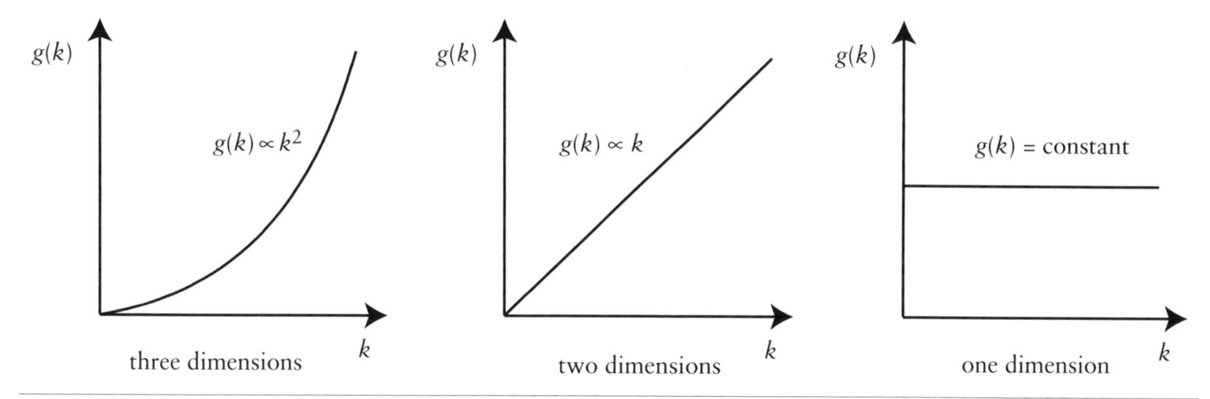

Figure 24.4: Density of states in k-space, for both particles with mass and massless particles, in three, two, and one dimensions. The dependence on k is quadratic in three dimensions, linear in two dimensions, and constant in one dimension.

24.3.2 Converting to energy

Once we have expressions for the density of states in k-space, it's straightforward to convert to ε-space (energy) using the dispersion relations that relate k to ε:

For particles of mass m: $\varepsilon = \dfrac{p^2}{2m} = \dfrac{\hbar^2 k^2}{2m}$

For massless particles: $\varepsilon = pc = \hbar kc$

where we have used the Planck relation, $p = \hbar k$. Now, particles with mass and massless particles have different densities of states. The density of states for particles with mass are given in Eqs. 24.9–24.11 (Derivation 24.2 and Fig. 24.5), and those for massless particles in Eqs. 24.12–24.14, (Derivation 24.3 and Fig. 24.6). Again, these expressions correspond to the number of states per unit volume in three dimensions, per unit area in two dimensions, and per unit length in one dimension. The total number of states per unit volume/area/length is found by integrating over all values of ε (Eq. 24.15).

Density of states in ε-space (Derivations 24.2 and 24.3):

Particles with mass:

In three dimensions: $$g_3(\varepsilon)\mathrm{d}\varepsilon = g\frac{\sqrt{2m^3\varepsilon}}{2\pi^2\hbar^3}\mathrm{d}\varepsilon \quad (24.9)$$

In two dimensions: $$g_2(\varepsilon)\mathrm{d}\varepsilon = g\frac{m}{2\pi\hbar^2}\mathrm{d}\varepsilon \quad (24.10)$$

In one dimension: $$g_1(\varepsilon)\mathrm{d}\varepsilon = g\sqrt{\frac{m}{2\pi^2\hbar^2\varepsilon}}\mathrm{d}\varepsilon \quad (24.11)$$

Massless particles:

In three dimensions: $$g_3(\varepsilon)\mathrm{d}\varepsilon = g\frac{\varepsilon^2}{2\pi^2(\hbar c)^3}\mathrm{d}\varepsilon \quad (24.12)$$

In two dimensions: $$g_2(\varepsilon)\mathrm{d}\varepsilon = g\frac{\varepsilon}{2\pi(\hbar c)^2}\mathrm{d}\varepsilon \quad (24.13)$$

In one dimension: $$g_1(\varepsilon)\mathrm{d}\varepsilon = g\frac{1}{\pi\hbar c}\mathrm{d}\varepsilon \quad (24.14)$$

The degeneracy factor, g, is due to spin or polarization degeneracy:

$$g = \begin{cases} 2 & \text{for electrons} \\ 2 & \text{for photons} \\ 3 & \text{for phonons} \end{cases}$$

Total number of states per unit volume/area/length:

$$n_s = \int_0^\infty g(\varepsilon)\mathrm{d}\varepsilon \quad (24.15)$$

Density of modes

By using the dispersion relations for k and ω, we can express the density of states in frequency space, $g(\omega)\mathrm{d}\omega$. For vibrational systems, instead of referring to a *state* at wavevector k, we talk about a *mode* at frequency ω. Said another way, a mode corresponds to a state of a system that is undergoing vibration. Therefore, $g(\omega)\mathrm{d}\omega$ represents the number of modes per unit volume with a frequency between ω and $\omega + \mathrm{d}\omega$. The derivations for $g(\omega)\mathrm{d}\omega$ in three, two, and one dimensions are similar to those for $g(\varepsilon)\mathrm{d}\varepsilon$ (Derivations 24.2

Derivation 24.2: Derivation of the density of states in ε-space for particles with mass (Eq. 24.9–24.11)

In three dimensions:

$$g_3(\varepsilon)\mathrm{d}\varepsilon = g_3(k)\frac{\mathrm{d}k}{\mathrm{d}\varepsilon}\mathrm{d}\varepsilon = \frac{k^2}{2\pi^2}\frac{\mathrm{d}k}{\mathrm{d}\varepsilon}\mathrm{d}\varepsilon = \frac{1}{2\pi^2}\frac{2m\varepsilon}{\hbar^2}\sqrt{\frac{m}{2\varepsilon\hbar^2}}\mathrm{d}\varepsilon = \underbrace{\frac{\sqrt{2m^3}}{2\pi^2\hbar^3}}_{=\alpha_3}\sqrt{\varepsilon}\mathrm{d}\varepsilon$$

$$\therefore g_3(\varepsilon)\mathrm{d}\varepsilon = \alpha_3\sqrt{\varepsilon}\mathrm{d}\varepsilon$$

In two dimensions:

$$g_2(\varepsilon)\mathrm{d}\varepsilon = g_2(k)\frac{\mathrm{d}k}{\mathrm{d}\varepsilon}\mathrm{d}\varepsilon = \frac{k}{2\pi}\frac{\mathrm{d}k}{\mathrm{d}\varepsilon}\mathrm{d}\varepsilon = \frac{1}{2\pi}\frac{\sqrt{2m\varepsilon}}{\hbar}\sqrt{\frac{m}{2\varepsilon\hbar^2}}\mathrm{d}\varepsilon = \underbrace{\frac{m}{2\pi\hbar^2}}_{=\alpha_2}\mathrm{d}\varepsilon$$

$$\therefore g_2(\varepsilon)\mathrm{d}\varepsilon = \alpha_2\,\mathrm{d}\varepsilon$$

In one dimension:

$$g_1(\varepsilon)\mathrm{d}\varepsilon = g_1(k)\frac{\mathrm{d}k}{\mathrm{d}\varepsilon}\mathrm{d}\varepsilon = \frac{1}{\pi}\frac{\mathrm{d}k}{\mathrm{d}\varepsilon}\mathrm{d}\varepsilon = \frac{1}{\pi}\sqrt{\frac{m}{2\varepsilon\hbar^2}}\mathrm{d}\varepsilon = \underbrace{\sqrt{\frac{m}{2\pi^2\hbar^2}}}_{=\alpha_1}\frac{1}{\sqrt{\varepsilon}}\mathrm{d}\varepsilon$$

$$\therefore g_1(\varepsilon)\mathrm{d}\varepsilon = \alpha_1\frac{1}{\sqrt{\varepsilon}}\mathrm{d}\varepsilon$$

- Starting point: there is a one-to-one correspondence between $g(\varepsilon)\mathrm{d}\varepsilon$ and $g(k)\mathrm{d}k$:

$$g(\varepsilon) = g(k)\frac{\mathrm{d}k}{\mathrm{d}\varepsilon}$$

- The dispersion relation for particles of mass m is:

$$\varepsilon = \frac{\hbar^2k^2}{2m}$$

$$\therefore \mathrm{d}\varepsilon = \frac{\hbar^2}{m}k\mathrm{d}k$$

- Rearranging the above, we have:

$$k = \frac{\sqrt{2m\varepsilon}}{\hbar}$$

$$\text{and } \mathrm{d}k = \sqrt{\frac{m}{2\varepsilon\hbar^2}}\mathrm{d}\varepsilon$$

- Use the above expressions to substitute k and $\mathrm{d}k$ in the density of states in k-space:

$$\text{In three dimensions: } g_3(k)\mathrm{d}k = \frac{k^2}{2\pi^2}\mathrm{d}k$$

$$\text{In two dimensions: } g_2(k)\mathrm{d}k = \frac{k}{2\pi}\mathrm{d}k$$

$$\text{In one dimension: } g_1(k)\mathrm{d}k = \frac{1}{\pi}\mathrm{d}k$$

(Eqs. 24.4–24.6).

- Define the constants α_3, α_2, and α_1 as:

$$\alpha_3 = \frac{\sqrt{2m^3}}{2\pi^2\hbar^3} \tag{24.16}$$

$$\alpha_2 = \frac{m}{2\pi\hbar^2} \tag{24.17}$$

$$\alpha_1 = \sqrt{\frac{m}{2\pi^2\hbar^2}} \tag{24.18}$$

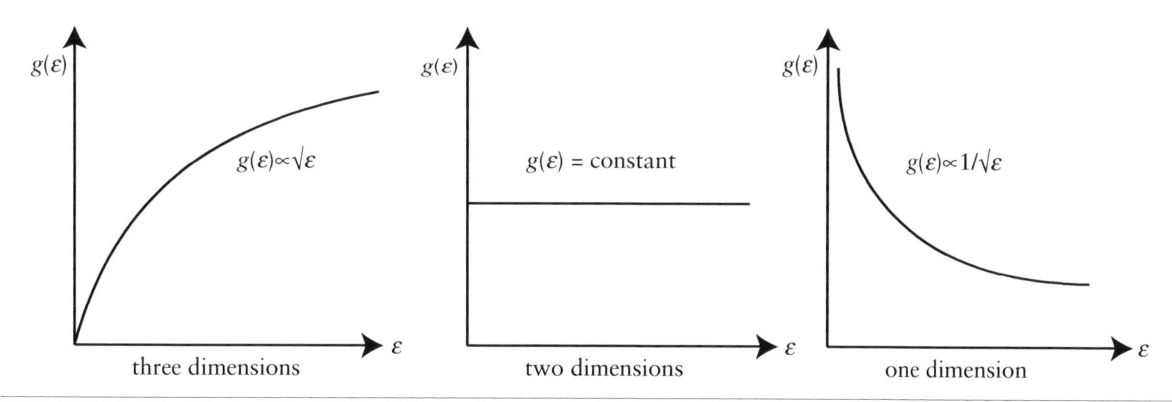

Figure 24.5: Density of states for particles with mass in ε-space in three, two, and one dimensions. The dependence on ε in three dimensions is $\sqrt{\varepsilon}$, is constant in two dimensions, and in one dimension is $1/\sqrt{\varepsilon}$.

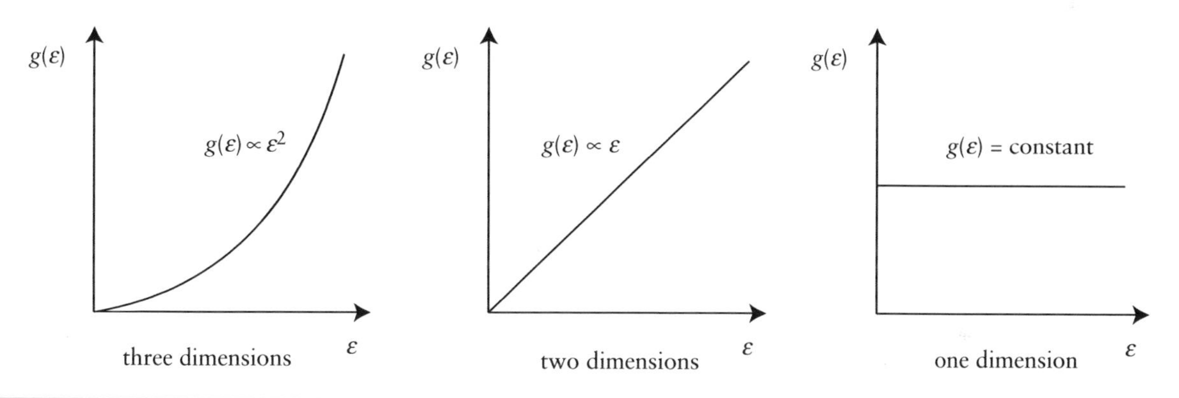

Figure 24.6: Density of states for massless particles in ε-space in three, two, and one dimensions. The dependence on ε is quadratic in three dimensions, linear in two dimensions, and constant in one dimension.

and 24.3), except that we use the dispersion relation $\omega(k)$ instead:

$$\text{For particles with mass: } \omega = \frac{\varepsilon}{\hbar} = \frac{\hbar k^2}{2m}$$

$$\text{For massless particles: } \omega = \frac{\varepsilon}{\hbar} = ck$$

where we have used the Einstein relation, $\varepsilon = \hbar\omega$.

24.4 Classical statistics

Statistical mechanics is split into two regimes: classical and quantum. We stated above that classical systems correspond to those in which particles are distinguishable—or, said another way, particles whose wavefunctions do not overlap (interparticle separation >> their de Broglie wavelength, see Section 24.6.3 for proof). Classical statistics are based on the *Boltzmann distribution*, which we used without proof in kinetic theory to determine macroscopic properties of ideal gases from the microscopic distribution of particle energies (Chapter 21). In this section, we cover where the Boltzmann distribution originates from, by considering how a particle interacts with its surroundings. The derivation outlined below is not rigorous, but it provides a basic understanding of what the Boltzmann distribution represents.

24.4.1 The Boltzmann distribution

Consider a closed sub-system in thermal contact with a heat reservoir, as shown in Fig. 24.7. Together, the sub-system plus the reservoir form an isolated system, with total energy U_0. The energies of the reservoir and sub-system are U_R and ε, respectively, where $U_0 = U_R + \varepsilon$. From the second postulate of statistical mechanics, the probability of the reservoir being in a given macrostate is proportional to the number of microstates in that macrostate: $P \propto \Omega$. Since Ω is a function of energy,

Derivation 24.3: Derivation of the density of states in ε-space for massless particles (Eq. 24.12–24.14)

- Starting point: there is a one-to-one correspondence between $g(\varepsilon)\mathrm{d}\varepsilon$ and $g(k)\mathrm{d}k$:

$$g(\varepsilon) = g(k)\frac{\mathrm{d}k}{\mathrm{d}\varepsilon}$$

- The dispersion relation for massless particles:

$$\varepsilon = \hbar kc$$

$$\therefore \mathrm{d}\varepsilon = \hbar c\mathrm{d}k$$

- Rearranging the above, we have:

$$k = \frac{\varepsilon}{\hbar c}$$

and $\mathrm{d}k = \dfrac{\mathrm{d}\varepsilon}{\hbar c}$

- Use the above expressions to substitute k and $\mathrm{d}k$ in the density of states in k-space:

In three dimensions: $g_3(k)\mathrm{d}k = \dfrac{k^2}{2\pi^2}\mathrm{d}k$

In two dimensions: $g_2(k)\mathrm{d}k = \dfrac{k}{2\pi}\mathrm{d}k$

In one dimension: $g_1(k)\mathrm{d}k = \dfrac{1}{\pi}\mathrm{d}k$

(Eqs. 24.4–24.6).

- Define the constants α'_3, α'_2, and α'_1 as:

$$\alpha'_3 = \frac{1}{2\pi^2(\hbar c)^3} \tag{24.19}$$

$$\alpha'_2 = \frac{1}{2\pi(\hbar c)^2} \tag{24.20}$$

$$\alpha'_1 = \frac{1}{\pi\hbar c} \tag{24.21}$$

In three dimensions: $g_3(\varepsilon)\mathrm{d}\varepsilon = g_3(k)\dfrac{\mathrm{d}k}{\mathrm{d}\varepsilon}\mathrm{d}\varepsilon$

$$= \frac{k^2}{2\pi^2}\frac{\mathrm{d}k}{\mathrm{d}\varepsilon}\mathrm{d}\varepsilon$$

$$= \frac{1}{2\pi^2}\left(\frac{\varepsilon}{\hbar c}\right)^2\frac{\mathrm{d}\varepsilon}{\hbar c}$$

$$\therefore g_3(\varepsilon)\mathrm{d}\varepsilon = \alpha'_3\varepsilon^2\mathrm{d}\varepsilon$$

In two dimensions: $g_2(\varepsilon)\mathrm{d}\varepsilon = g_2(k)\dfrac{\mathrm{d}k}{\mathrm{d}\varepsilon}\mathrm{d}\varepsilon$

$$= \frac{k}{2\pi}\frac{\mathrm{d}k}{\mathrm{d}\varepsilon}\mathrm{d}\varepsilon$$

$$= \frac{1}{2\pi}\frac{\varepsilon}{\hbar c}\frac{\mathrm{d}\varepsilon}{\hbar c}$$

$$\therefore g_2(\varepsilon)\mathrm{d}\varepsilon = \alpha'_2\,\varepsilon\mathrm{d}\varepsilon$$

In one dimension: $g_1(\varepsilon)\mathrm{d}\varepsilon = g_1(k)\dfrac{\mathrm{d}k}{\mathrm{d}\varepsilon}\mathrm{d}\varepsilon$

$$= \frac{1}{\pi}\frac{\mathrm{d}k}{\mathrm{d}\varepsilon}\mathrm{d}\varepsilon$$

$$= \frac{1}{\pi}\frac{\mathrm{d}\varepsilon}{\hbar c}$$

$$\therefore g_1(\varepsilon)\mathrm{d}\varepsilon = \alpha'_1\,\mathrm{d}\varepsilon$$

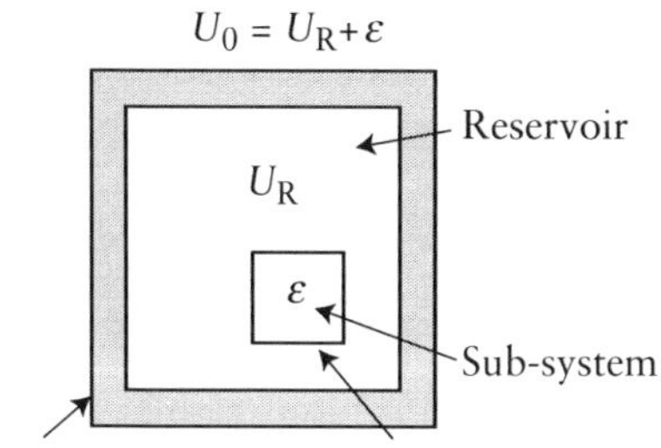

Figure 24.7: Isolated system comprising a closed sub-system with energy ε in thermal contact with a heat reservoir with energy U_R. Energy can be exchanged between the reservoir and the sub-system. The total energy of the isolated system is $U_0 = U_R + \varepsilon$.

so is P. Using the definition of entropy from Eq. 24.2, the entropy of the reservoir is:

$$S_R = k_B \ln \Omega_R$$

$$\therefore \Omega_R = e^{S_R/k_B} \tag{24.22}$$

We can find an expression for S_R from the exact differential of U ($dU = TdS - pdV$). The volume of the reservoir is constant, therefore, $dV_R = 0$ and so, by rearranging dU, the entropy of the reservoir is

$$dS_R = \frac{dU_R}{T}$$

$$\therefore S_R = \frac{U_R}{T} + \text{constant}$$

Dividing $U_R = U_0 - \varepsilon$ by T, and letting $S_0 = (U_0/T +$ constant), which is constant since the system is isolated, we find:

$$\frac{U_R}{T} = \frac{U_0}{T} - \frac{\varepsilon}{T}$$

$$\therefore S_R = S_0 - \frac{\varepsilon}{T} + \text{constant} \tag{24.23}$$

Since U_0 is constant, the probability of the sub-system having an energy ε, equals the probability that the reservoir has an energy U_R. Therefore, $P(\varepsilon)$ is proportional to $\Omega_R(\varepsilon)$. Using Eq. 24.23 in Eq. 24.22, we find:

$$P(\varepsilon) \propto \Omega_R(\varepsilon)$$

$$\propto e^{S_R(\varepsilon)/k_B}$$

$$\propto e^{-\varepsilon/k_B T}$$

Therefore, we have recovered the exponential Boltzmann factor, $e^{-\varepsilon/k_B T}$, by considering a sub-system of energy ε in thermal equilibrium with a heat reservoir, and using the postulates of statistical mechanics. If we treat the sub-system as a single particle, then $P(\varepsilon)d\varepsilon$ is proportional to the probability that a particle has an energy between ε and $\varepsilon + d\varepsilon$.

The *Boltzmann distribution*, $f_B(\varepsilon)d\varepsilon$, is proportional to the probability distribution, $P(\varepsilon)d\varepsilon$:

$$f_B(\varepsilon)d\varepsilon \propto P(\varepsilon)d\varepsilon$$

$$= Ae^{-\varepsilon/k_B T}d\varepsilon$$

(Eq. 24.24). An expression for the normalization constant, A, is derived in Section 24.4.3.

The Boltzmann distribution:

$$f_B(\varepsilon)d\varepsilon = Ae^{-\varepsilon/k_B T}d\varepsilon \tag{24.24}$$

Occupation functions

The Boltzmann distribution is known in statistical mechanics as an *occupation function*, $f(\varepsilon)$. This is a probability distribution that represents the average number of particles per state. Therefore, according to the Boltzmann distribution, the average number of particles per state decreases exponentially with increasing energy. There exist two other occupation functions: the *Bose–Einstein* and the *Fermi–Dirac* distributions. These apply to systems in the quantum regime, and are covered in Section 24.6.

24.4.2 The number density distribution

So far, we have come across the *density of states*, $g(\varepsilon)d\varepsilon$, which represents the number of states per unit volume, and the *occupation function*, $f(\varepsilon)$, which represents the average number of particles per state. The product of these two functions yields another probability distribution, called the *number density distribution*, $n(\varepsilon)d\varepsilon$. This distribution represents the average number of particles per unit volume:

$$\underbrace{n(\varepsilon)d\varepsilon}_{\text{particles/volume}} = \underbrace{f(\varepsilon)}_{\text{particles/state}} \cdot \underbrace{g(\varepsilon)d\varepsilon}_{\text{states/volume}}$$

(Eq. 24.25). The normalization condition is that the total number density of particles, n_T, is given by the integral of $n(\varepsilon)d\varepsilon$ over all energies (Eq. 24.26).

Number density distribution:

Number density of particles with energy between ε and $\varepsilon + \mathrm{d}\varepsilon$:

$$n(\varepsilon)\mathrm{d}\varepsilon = f(\varepsilon)g(\varepsilon)\mathrm{d}\varepsilon \tag{24.25}$$

Total number density of particles:

$$n_{\mathrm{T}} = \int_0^\infty n(\varepsilon)\mathrm{d}\varepsilon$$

$$= \int_0^\infty f(\varepsilon)g(\varepsilon)\mathrm{d}\varepsilon \tag{24.26}$$

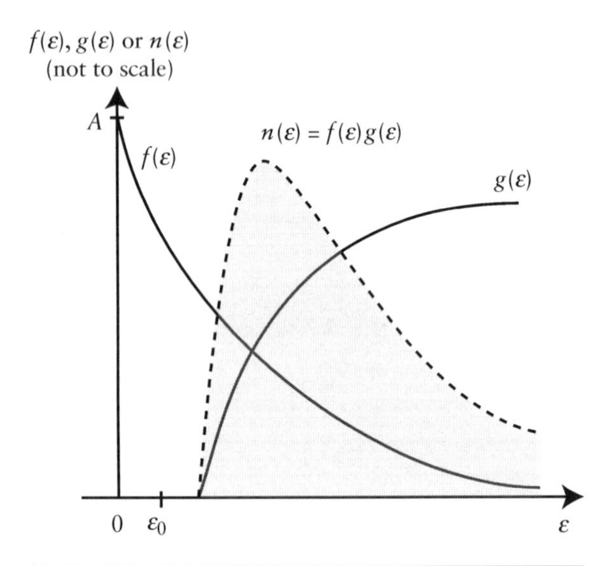

Figure 24.8: The number density distribution of particle energies in a system, $n(\varepsilon)\mathrm{d}\varepsilon$, is equal to the product of the occupation function ($f(\varepsilon) = A\mathrm{e}^{-\varepsilon/kT}$) and the density of states ($g(\varepsilon)\mathrm{d}\varepsilon \propto \sqrt{\varepsilon}\mathrm{d}\varepsilon$): $n(\varepsilon)\mathrm{d}\varepsilon = f(\varepsilon)g(\varepsilon)\mathrm{d}\varepsilon$. Although the occupation function is high at energy ε_0, no particles occupy this energy state because the density of states at ε_0 equals zero.

From Eq. 24.25, we see that the distribution of particle energies in a system depends on two quantities. It depends on the *probability* that a particle has an energy ε, which is given by $f(\varepsilon)$, and it depends on the *number of states* available at that energy, which is given by $g(\varepsilon)\mathrm{d}\varepsilon$. The density of states is therefore sometimes thought of as a weight function—there is a greater weight in the probability of a particle having an energy ε if there are more states available at that energy. Conversely, if there are *no* states available at a particular energy, say ε_0, for a given system, such that $g(\varepsilon_0)\mathrm{d}\varepsilon = 0$, then no particles will have that energy, even if there is a high corresponding probability from the occupation function, $f(\varepsilon_0)$. These ideas are shown schematically in Fig. 24.8.

24.4.3 The partition function and thermodynamic properties

Now let's determine the normalization factor, A, of the Boltzmann distribution. Substituting Eq. 24.24 into Eq. 24.26, we have:

$$n_{\mathrm{T}} = A\int_0^\infty g(\varepsilon)\mathrm{e}^{-\varepsilon/k_{\mathrm{B}}T}\mathrm{d}\varepsilon$$

The integral shown here is an important quantity in statistical mechanics, called the *partition function*, Z:

$$Z \overset{\mathrm{def}}{=} \int_0^\infty g(\varepsilon)\mathrm{e}^{-\varepsilon/k_{\mathrm{B}}T}\mathrm{d}\varepsilon$$

Therefore, $A = n_{\mathrm{T}}/Z$, and the normalized Boltzmann distribution is given by Eq. 24.27. For a *discrete* set of energy levels, ε_i, the integral becomes a sum over states i, and the density of states approximates to the degeneracy, g_i, of the ith energy level (Eqs. 24.28 and 24.29).

Partition function:

Normalized Boltzmann distribution:

$$f_{\mathrm{B}}(\varepsilon)\mathrm{d}\varepsilon = \frac{n_{\mathrm{T}}}{Z}\mathrm{e}^{-\varepsilon/k_{\mathrm{B}}T}\mathrm{d}\varepsilon \tag{24.27}$$

where the partition function, Z, is:
Discrete energy distribution:

$$Z = \sum_i g_i \mathrm{e}^{-\varepsilon_i/k_{\mathrm{B}}T} \tag{24.28}$$

Continuous energy distribution:

$$Z = \int_0^\infty g(\varepsilon)\mathrm{e}^{-\varepsilon/k_{\mathrm{B}}T}\mathrm{d}\varepsilon \tag{24.29}$$

It is not intuitive what the partition function represents physically, but it is often defined as the "*sum over states.*" It is a function that encodes the statistical properties of the microstates of a system, and it depends on the system's temperature and physical dimensions. For example, the partition function of an ideal gas is found to depend on its temperature, T, and its volume, V:

$$Z = V\left(\frac{mk_BT}{2\pi\hbar^2}\right)^{3/2}$$

(see Worked example 24.1). The partition function is an important quantity in statistical mechanics since it provides the link between a system's microscopic energy levels and its macroscopic properties. We will see below that essentially all the macroscopic, thermodynamic properties of a system can be calculated from its partition function.

Internal energy and heat capacity

The *internal energy*, U, of a system is the sum of the energies of its individual particles. The expression for U is related to the partition function by the temperature derivative of its natural logarithm:

$$U = Nk_BT^2\frac{\partial \ln Z}{\partial T}$$

(Eq. 24.30 and Derivation 24.4). Here, N is the total number of particles in the system. The heat capacity of the system is then given by the temperature derivative of U:

$$C_V = \left(\frac{\partial U}{\partial T}\right)_V$$

$$= Nk_BT\left(2\frac{\partial \ln Z}{\partial T} + T\frac{\partial^2 \ln Z}{\partial T^2}\right)$$

(Eq. 24.31).

Internal energy and heat capacity (Derivation 24.4):

$$U = Nk_BT^2\frac{\partial \ln Z}{\partial T} \qquad (24.30)$$

$$C_V = Nk_BT\left(2\frac{\partial \ln Z}{\partial T} + T\frac{\partial^2 \ln Z}{\partial T^2}\right) \qquad (24.31)$$

Helmholtz free energy

Recall from thermodynamics (Section 22.3.2) that the *Helmholtz free energy*, F, is a thermodynamic potential that is related to the internal energy by $F = U - TS$. Using the statistical definition of entropy (Eq. 24.2), we can derive an expression for F in terms of the natural logarithm of the partition function:

$$F = -Nk_BT\ln Z$$

(Eq. 24.32 and Derivation 24.5). This is a very useful quantity since *all* other thermodynamic properties are easily derived from it. The expressions for pressure and entropy follow from the exact differential of F, $\mathrm{d}F = -p\mathrm{d}V - S\mathrm{d}T$ (refer to Section 24.3.2), and are therefore equal to single derivatives of F:

$$p = -\left(\frac{\partial F}{\partial V}\right)_T$$

and $$S = -\left(\frac{\partial F}{\partial T}\right)_V$$

(Eqs. 24.33 and 24.34). The other thermodynamic potentials, U, H, and G are combinations of F and its derivatives, and are given in Eqs. 24.35–24.37.

Helmholtz free energy (Derivation 24.5):

$$F = -Nk_BT\ln Z \qquad (24.32)$$

Quantities derived from F:

$$p = -\left(\frac{\partial F}{\partial V}\right)_T \qquad (24.33)$$

$$S = -\left(\frac{\partial F}{\partial T}\right)_V \qquad (24.34)$$

$$U = F + TS$$

$$= F - T\left(\frac{\partial F}{\partial T}\right)_V \qquad (24.35)$$

$$H = U + pV$$

$$= F - T\left(\frac{\partial F}{\partial T}\right)_V - V\left(\frac{\partial F}{\partial V}\right)_T \qquad (24.36)$$

$$G = U - TS + pV$$

$$= F - V\left(\frac{\partial F}{\partial V}\right)_T \qquad (24.37)$$

Using Eq. 24.32 in Eq. 24.35, we recover the expression for U from Eq. 24.30:

$$U = F - T\left(\frac{\partial F}{\partial T}\right)_V$$
$$= Nk_BT^2\left(\frac{\partial \ln Z}{\partial T}\right)_V .$$

Worked example 24.1 shows how to calculate some thermodynamic properties of an ideal gas from its partition function. Furthermore, we can use the partition function of atoms localized in a solid to show that heat capacity is not constant over all temperatures, as was originally predicted by classical theory (Worked example 24.2). Therefore, a microscopic, quantum approach is necessary for explaining the bulk behavior of some materials.

24.4.4 Agreement with kinetic theory

Even though kinetic theory makes no mention of the density of states or the partition function, we can

Derivation 24.4: Derivation of the internal energy from the partition function (Eq. 24.30)

$$u = \sum n_i\varepsilon_i$$
$$= \sum A g_i\varepsilon_i e^{-\varepsilon_i/k_BT}$$
$$= \frac{n_T}{Z}\sum g_i\varepsilon_i e^{-\varepsilon_i/k_BT}$$
$$= \frac{n_T}{Z}Zk_BT^2\frac{\partial}{\partial T}\ln Z$$
$$= n_Tk_BT^2\frac{\partial}{\partial T}\ln Z$$

Total internal energy ($n_T = N/V$):

$$U = uV$$
$$= Nk_BT^2\frac{\partial}{\partial T}\ln Z$$

Heat capacity:

$$C_V = \left(\frac{\partial U}{\partial T}\right)_V$$
$$= Nk_BT\left(2\frac{\partial \ln Z}{\partial T} + T\frac{\partial^2 \ln Z}{\partial T^2}\right)$$

- Starting point: the internal energy of a system is the sum of its individual particle energies. If there are n_i particles per unit volume in the ith energy level, which has energy ε_i, then the total energy density (internal energy per unit volume), u, is:

$$u = \sum n_i\varepsilon_i$$

- The discrete number density distribution and partition function are:

$$n_i = A g_i e^{-\varepsilon_i/k_BT}$$
$$\text{and } Z = \sum g_i e^{-\varepsilon_i/k_BT}$$

where g_i is the degeneracy of the ith state and the normalization constant is $A = n_T/Z$, where n_T is the total number density of the system ($n_T = N/V$ for N particles in a volume V).
- Differentiating the natural logarithm of the partition function, we have:

$$\frac{\partial}{\partial T}\ln Z = \frac{\partial}{\partial T}\left(\ln\left(\sum g_i e^{-\varepsilon_i/k_BT}\right)\right)$$
$$= \frac{1}{k_BT^2}\frac{\sum g_i\varepsilon_i e^{-\varepsilon_i/k_BT}}{\sum g_i e^{-\varepsilon_i/k_BT}}$$
$$= \frac{1}{k_BT^2}\frac{\sum g_i\varepsilon_i e^{-\varepsilon_i/k_BT}}{Z}$$
$$\therefore \sum g_i\varepsilon_i e^{-\varepsilon_i/k_BT} = Zk_BT^2\frac{\partial}{\partial T}\ln Z$$

Derivation 24.5: Derivation of the Helmholtz free energy (Eq. 24.32)

- Starting point: the statistical definition of entropy is:

$$S = k_B \ln \Omega$$

where $\Omega = \dfrac{N!}{n_1!n_2!n_3!\ldots}$

(Eqs. 24.1 and 24.2).

- Using Stirling's approximation (see aside), and $N = \sum_i n_i$:

$$\begin{aligned}
\ln \Omega &= \ln N! - \sum_i \ln n_i! \\
&\simeq N \ln N - N - \sum_i (n_i \ln n_i - n_i) \\
&= \sum_i n_i \ln N - N - \sum_i n_i \ln n_i + \underbrace{\sum_i n_i}_{=N} \\
&= \sum_i n_i (\ln N - \ln n_i) \\
&= -\sum_i n_i \ln\left(\frac{n_i}{N}\right)
\end{aligned}$$

- Use this in the expression for entropy, along with:

$$n_i = \frac{N}{Z} e^{-\varepsilon_i / k_B T}$$

and $U = \sum_i n_i \varepsilon_i$

- The Helmholtz free energy is defined as:

$$F = U - TS$$

Using the expression for entropy derived above, we find an expression for F in terms of the natural logarithm of the partition function.

$$\begin{aligned}
S &= k_B \ln \Omega \\
&= -k_B \sum_i n_i \ln\left(\frac{n_i}{N}\right) \\
&= -k_B \sum_i n_i \left(-\frac{\varepsilon_i}{k_B T} - \ln Z\right) \\
&= \frac{\sum_i n_i \varepsilon_i}{T} + k_B \underbrace{\sum_i n_i}_{=N} \ln Z \\
&= \frac{U}{T} + N k_B \ln Z
\end{aligned}$$

$$\begin{aligned}
\therefore F &= U - TS \\
&= -N k_B T \ln Z
\end{aligned}$$

Aside: Stirling's approximation for large n

$$\begin{aligned}
\ln n! &= \ln 1 + \ln 2 + \cdots + \ln n \\
&= \sum_{k=1}^{n} \ln k \\
&\simeq \int_1^n \ln k \, dk \quad \text{(in the limit of large } n\text{)} \\
&= [k \ln k - k]_1^n \\
&= n \ln n - n + 1 \\
&\simeq n \ln n - n \quad (\text{for } n \gg 1)
\end{aligned}$$

WORKED EXAMPLE 24.1: Thermodynamic properties of an ideal monatomic gas from its partition function

Ideal monatomic gas partition function:

The density of states in three dimensions is $g(\varepsilon)d\varepsilon = \alpha\sqrt{\varepsilon}d\varepsilon$, where the constant is:

$$\alpha = \frac{V\sqrt{2m^3}}{2\pi^2\hbar^3}$$

(refer to Eqs. 24.9 and 24.16. In this case, we are stating the volume of the system, V, explicitly). Therefore, the partition function of a monatomic ideal gas is:

$$Z = \int_0^\infty g(\varepsilon)e^{-\varepsilon/k_BT}d\varepsilon$$

$$= \alpha\int_0^\infty \sqrt{\varepsilon}e^{-\varepsilon/k_BT}d\varepsilon$$

$$= \alpha\frac{\sqrt{\pi(k_BT)^3}}{2}$$

$$= V\frac{\sqrt{2m^3}}{2\pi^2\hbar^3}\frac{\sqrt{\pi(k_BT)^3}}{2}$$

$$= V\left(\frac{mk_BT}{2\pi\hbar^2}\right)^{3/2}$$

To integrate, use the standard integral:

$$\int_0^\infty \sqrt{x}e^{-x/a}dx = \frac{\sqrt{\pi a^3}}{2}$$

Thermodynamic properties from Z:

Using the partition function, we recover the same expressions for the macroscopic properties of ideal gases as those from classical thermodynamics and kinetic theory:

1. Internal energy: let $Z = Z(T) = CT^{3/2}$, where C is a constant ($C = V(mk_B/2\pi\hbar^2)^{3/2}$):

$$U = Nk_BT^2\frac{\partial}{\partial T}\ln Z$$

$$= Nk_BT^2\frac{\partial}{\partial T}\ln CT^{3/2}$$

$$= \frac{3}{2}Nk_BT$$

$$= \frac{3}{2}nRT$$

2. Heat capacity:

$$C_V = \left(\frac{\partial U}{\partial T}\right)_V$$

$$= \frac{3}{2}R \text{ per mole}$$

3. Ideal Gas Law: let $Z = Z(V) = DV$, where D is a constant ($D = (mk_BT/2\pi\hbar^2)^{3/2}$):

$$F = -Nk_BT\ln Z$$

$$\therefore p = -\left(\frac{\partial F}{\partial V}\right)_T$$

$$= \left(\frac{\partial}{\partial V}\right)_T Nk_BT\ln DV$$

$$= \frac{Nk_BT}{V}$$

$$\Rightarrow \therefore pV = Nk_BT = nRT$$

WORKED EXAMPLE 24.2: Thermodynamic properties of atoms in a harmonic oscillator potential

The simplest model to describe vibrations of atoms in a solid is by a restoring force that obeys Hooke's Law ($F = -kx$). The corresponding potential well is quadratic in x:

$$V(x) = -\int F\mathrm{d}x = \frac{1}{2}kx^2$$

If we assume that each atom sits in its own potential well, that all potential wells are identical to one another, and that atomic vibrations are independent, we can use this simple model to determine thermodynamic properties of solids.

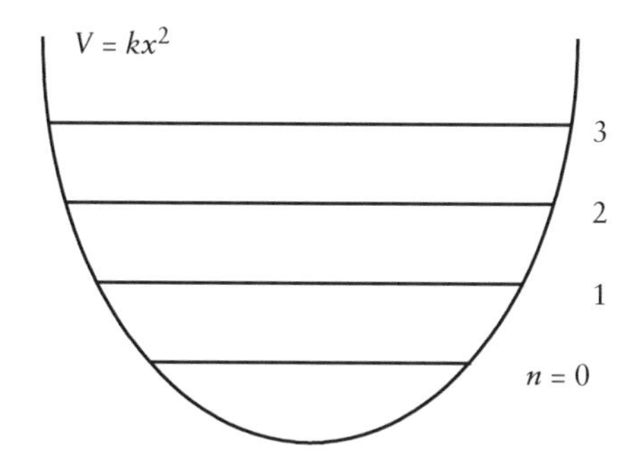

Harmonic oscillator partition function:

In Section 17.2.3, we saw that the energy levels of a one-dimensional harmonic oscillator potential well are given by:

$$\varepsilon_n = \left(n + \frac{1}{2}\right)\hbar\omega$$

The single particle partition function is then:

$$Z = \sum_n \mathrm{e}^{-\varepsilon_n/k_B T}$$

$$= \sum_n \mathrm{e}^{-(n+1/2)\hbar\omega/k_B T}$$

$$= \mathrm{e}^{-\hbar\omega/2k_B T} \underbrace{\sum_{n=0}^{\infty} \mathrm{e}^{-n\hbar\omega/k_B T}}_{\text{infinite geometric series}}$$

$$= \frac{\mathrm{e}^{-\hbar\omega/2k_B T}}{1 - \mathrm{e}^{-\hbar\omega/k_B T}}$$

The sum of an infinite geometric series with first term a and common ratio r is:

$$S_\infty = \frac{a}{1-r}$$

Thermodynamic properties from Z:

1. Internal energy:

$$U = Nk_B T^2 \frac{\partial}{\partial T}\ln Z$$

$$= \underbrace{\frac{N\hbar\omega}{2}}_{\text{zero-point energy}} + \frac{N\hbar\omega}{1 - \mathrm{e}^{-\hbar\omega/k_B T}}$$

2. Heat capacity:

$$C_V = \left(\frac{\partial U}{\partial T}\right)_V$$

$$= Nk_B \left(\frac{\theta}{T}\right)^2 \frac{\mathrm{e}^{-\theta/T}}{\left(1 - \mathrm{e}^{-\theta/T}\right)^2}$$

where $\theta = \frac{\hbar\omega}{k_B}$

As $T \to \infty$ (classical regime), both U and C_V tend to a constant value:

$$U \to Nk_B T$$

and $C_V \to Nk_B$

As $T \to 0$ (quantum regime), the internal energy tends to the zero-point energy that atoms possess due to their quantum motion, and the heat capacity falls off exponentially:

$$U \to \frac{N\hbar\omega}{2}$$

and $C_V \to Nk_B \left(\frac{\theta}{T}\right)^2 \mathrm{e}^{-\theta/T}$

The observation that heat capacity is not constant for all temperatures, as was originally predicted by classical theory, can thus only be explained by quantum theory.

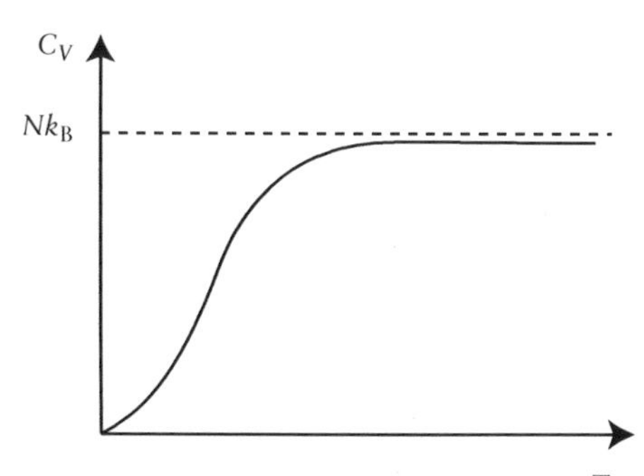

recover the Maxwell–Boltzmann speed distribution and the equipartition theorem using a statistical mechanics approach.

Maxwell–Boltzmann distribution

The Maxwell–Boltzmann speed distribution is:

$$F_v(v)\mathrm{d}v = 4\pi v^2\left(\frac{m}{2\pi k_B T}\right)^{\frac{3}{2}} \mathrm{e}^{-mv^2/2k_BT}\mathrm{d}v$$

(refer to Section 21.3.2). This expression can be derived directly from the number density distribution, $n(\varepsilon)\mathrm{d}\varepsilon$, using the three-dimensional density of states and the monatomic ideal gas partition function per unit volume, as shown in Worked example 24.3.

Equipartition theorem

In kinetic theory, we found that each degree of freedom associated with a quadratic term contributes an average energy of:

$$\langle E\rangle = \frac{1}{2}k_BT \text{ per molecule}$$

towards the total internal energy of the system. This statement is called the *equipartition theorem* (refer to Section 21.4.1). The same result is derived directly from the Boltzmann distribution, as shown in Derivation 24.6.

WORKED EXAMPLE 24.3: Recovering the Maxwell–Boltzmann distribution

1. The number density distribution is:

$$n(\varepsilon)\mathrm{d}\varepsilon = f(\varepsilon)g(\varepsilon)\mathrm{d}\varepsilon$$

where $f(\varepsilon)$ is the Boltzmann distribution and $g(\varepsilon)\mathrm{d}\varepsilon$ is the density of states (refer to Eqs. 24.9, 24.25, and 24.27, if necessary).

2. From Worked example 24.1, the partition function for a monatomic ideal gas is:

$$Z = V\left(\frac{mk_BT}{2\pi\hbar^2}\right)^{3/2}$$

3. Change variables to determine the speed distribution:

$$\varepsilon = \frac{1}{2}mv^2$$

$$\therefore \mathrm{d}\varepsilon = mv\mathrm{d}v$$

4. Substituting the above into $n(\varepsilon)\mathrm{d}\varepsilon$, we recover the Maxwell–Boltzmann speed distribution from kinetic theory (refer to Eq. 21.9). In this case, the distribution is normalized to the total number density, n_T, rather than 1.

Normalized Boltzmann distribution:

$$f(\varepsilon) = \frac{n_T}{Z}\mathrm{e}^{-\varepsilon/k_BT}$$

Three-dimensional density of states:

$$g(\varepsilon)\mathrm{d}\varepsilon = \alpha\sqrt{\varepsilon}\mathrm{d}\varepsilon$$

$$\text{where } \alpha = \frac{V\sqrt{2m^3}}{2\pi^2\hbar^3}$$

Particle number density distribution:

$$n(\varepsilon)\mathrm{d}\varepsilon = f(\varepsilon)g(\varepsilon)\mathrm{d}\varepsilon$$

$$= \frac{n_T}{Z}g(\varepsilon)\mathrm{e}^{-\varepsilon/k_BT}\mathrm{d}\varepsilon$$

$$= \frac{n_T}{Z}\alpha\sqrt{\varepsilon}\mathrm{e}^{-\varepsilon/k_BT}\mathrm{d}\varepsilon$$

$$= n_T\left(\frac{2\pi\hbar^2}{mk_BT}\right)^{3/2}\left(\frac{\sqrt{2m^3}}{2\pi^2\hbar^3}\right)\sqrt{\varepsilon}\mathrm{e}^{-\varepsilon/k_BT}\mathrm{d}\varepsilon$$

$$= 2n_T\left(\frac{1}{\pi(k_BT)^3}\right)^{1/2}\sqrt{\varepsilon}\mathrm{e}^{-\varepsilon/k_BT}\mathrm{d}\varepsilon$$

Maxwell–Boltzmann speed distribution:

$$\therefore n(v)\mathrm{d}v = 4\pi n_T v^2\left(\frac{m}{2\pi k_BT}\right)^{3/2}\mathrm{e}^{-mv^2/2k_BT}\mathrm{d}v$$

Derivation 24.6: Derivation of the equipartition of energy

$$\langle E \rangle = \frac{\int_{-\infty}^{\infty} E f(\varepsilon) \mathrm{d}\varepsilon}{\int_{-\infty}^{\infty} f(\varepsilon) \mathrm{d}\varepsilon}$$

$$= \frac{A \int_{-\infty}^{\infty} E \mathrm{e}^{-(E(p)+\varepsilon(q_i))/k_B T} \mathrm{d}p \ldots \mathrm{d}q_i}{A \int_{-\infty}^{\infty} \mathrm{e}^{-(E(p)+\varepsilon(q_i))/k_B T} \mathrm{d}p \ldots \mathrm{d}q_i}$$

$$= \frac{\int_{-\infty}^{\infty} E \mathrm{e}^{-E/k_B T} \mathrm{d}p}{\int_{-\infty}^{\infty} \mathrm{e}^{-E/k_B T} \mathrm{d}p}$$

$$= -\frac{\partial}{\partial \beta} \ln\left(\int_{-\infty}^{\infty} \mathrm{e}^{-\beta E} \mathrm{d}p \right)$$

$$= -\frac{\partial}{\partial \beta} \ln\left(\int_{-\infty}^{\infty} \mathrm{e}^{-\beta b p^2} \mathrm{d}p \right)$$

$$= -\frac{\partial}{\partial \beta} \ln\left(\frac{1}{\sqrt{\beta}} \int_{-\infty}^{\infty} \mathrm{e}^{-b y^2} \mathrm{d}y \right)$$

$$= -\frac{\partial}{\partial \beta} \ln\left(\frac{1}{\sqrt{\beta}} \right) - \underbrace{\frac{\partial}{\partial \beta} \ln\left(\int_{-\infty}^{\infty} \mathrm{e}^{-b y^2} \mathrm{d}y \right)}_{=0}$$

$$= \frac{1}{2\beta}$$

$$= \frac{k_B T}{2} \Rightarrow \text{equipartition theorem}$$

- Starting point: assume that the total energy ε can be split additively into the form $\varepsilon = E(p) + \varepsilon(q_i)$, where E is a quadratic function in a variable p ($E = bp^2$, where b is a constant), and the variables q_i in $\varepsilon(q_i)$ are not quadratic functions. For example, energy is a quadratic function in momentum or velocity: $E = mv^2/2 = p^2/2m$.
- The average energy of E is found by integrating over the normalized Boltzmann distribution:

$$\langle E \rangle = \frac{\int_{-\infty}^{\infty} E f(\varepsilon) \mathrm{d}\varepsilon}{\int_{-\infty}^{\infty} f(\varepsilon) \mathrm{d}\varepsilon}$$

where $f(\varepsilon) = A\mathrm{e}^{-\varepsilon/k_B T}$

- By splitting the integral into its E-dependence and its $\varepsilon(q_i)$-dependence, the latter cancels in the numerator and denominator.
- Let $\beta = 1/k_B T$ to simplify the expression.
- Change variables. Let:

$$y^2 = \beta p^2$$
$$\therefore \mathrm{d}y = \sqrt{\beta}\mathrm{d}p$$

and eliminate p^2 and $\mathrm{d}p$ from the expression for $\langle E \rangle$.

- Doing this, we find that the average energy per particle associated with a quadratic degree of freedom is $\langle E \rangle = k_B T/2$, which is the equipartition theorem.

24.5 Indistinguishable particles

So far, we have only been dealing with systems of *distinguishable* particles. From a quantum description, distinguishable particles have non-overlapping wavefunctions (their interparticle separation is much greater than their de Broglie wavelengths), and can, in theory, be individually distinguished by a physical measurement. Such systems of particles correspond to those in the classical regime, and their properties are described by Boltzmann statistics. Now we will turn to look at the statistics of *indistinguishable* particles. These particles cannot be individually distinguished by any physical measurement. Indistinguishability is a quantum phenomenon, and has no classical analogue. Indistinguishable particles can be described as having overlapping wavefunctions; the combined wavefunction of several indistinguishable particles must therefore account for all particles, which are all in the same potential well. Multi-particle wavefunctions have an intrinsic symmetry called *exchange symmetry*, which is described below.

24.5.1 Exchange symmetry

Indistinguishable particles cannot be distinguished by any physical measurement. Therefore, their combined wavefunction squared cannot depend on which particle is in which state. Consider a two-particle wavefunction, denoted by $\psi(1,2)$, which indicates that particle 1 is in one state and particle 2 in another. The *exchange principle* states that the square of the wavefunction of two indistinguishable particles is the same if we exchange their states (Eq. 24.38).

Exchange principle:

$$|\psi(1,2)|^2 = |\psi(2,1)|^2 \quad (24.38)$$

Exchange symmetry is an operation that describes how a wavefunction changes upon particle exchange, under the requirement that the exchange principle holds. It is represented by the *exchange operator*, $\hat{P}_{12}$, which, when acting on a two-particle wavefunction, exchanges the states of the two particles:

$$\hat{P}_{12}\,\psi(1,2) = \psi(2,1)$$

(Eq. 24.39). Therefore, by operating twice on the same wavefunction, we can determine the eigenvalue equation for $\hat{P}_{12}$, and hence find its eigenvalues, p (Eqs. 24.40 and 24.41, Derivation 24.7). In doing this, we find that there are two eigenvalues: $p = +1$ and $p = -1$. The eigenvalue $p = +1$ represents a *symmetric* wavefunction and $p = -1$ represents an *anti-symmetric* wavefunction (Eq. 24.41).

The exchange operator (Derivation 24.7):

Exchange operator:

$$\hat{P}_{12}\,\psi(1,2) = \psi(2,1) \quad (24.39)$$

Eigenvalue equation:

$$\hat{P}_{12}\,\psi(1,2) = p\,\psi(1,2) \quad (24.40)$$

Eigenvalues:

$$p = \begin{cases} +1: & \psi(2,1) = +\psi(1,2) \\ & \text{symmetric wavefunction} \\ -1: & \psi(2,1) = -\psi(1,2) \\ & \text{anti-symmetric wavefunction} \end{cases} \quad (24.41)$$

24.5.2 Bosons and fermions

Due to the requirement that wavefunctions are either symmetric or anti-symmetric upon particle exchange, two different types of particles exist in nature: *bosons* and *fermions*. Bosons are symmetric on particle exchange, whereas fermions are anti-symmetric. It is also found that bosons have *integer* spin, whereas fermions have *half-integer* spin. Therefore, electrons, which are spin-1/2 particles, are fermions, whereas photons, which are spin-1 particles, are bosons.

The Pauli exclusion principle

Consider two particle states, state a and state b. The respective single particle wavefunctions for particle 1 are $\phi_a(1)$ and $\phi_b(1)$, and for particle 2 are $\phi_a(2)$ and $\phi_b(2)$. Now consider one particle in state a and the other in state b. Since we do not know which particle is in which state, the combined wavefunction, $\psi(1,2)$, must account for both possibilities. The normalized symmetric and anti-symmetric wavefunctions are therefore:

Symmetric wavefunction (bosons):

$$\psi_S(1,2) = \frac{1}{\sqrt{2}}\left(\phi_a(1)\phi_b(2) + \phi_a(2)\phi_b(1)\right)$$

$$= +\psi_S(2,1)$$

Anti-symmetric wavefunction (fermions):

$$\psi_A(1,2) = \frac{1}{\sqrt{2}}\left(\phi_a(1)\phi_b(2) - \phi_a(2)\phi_b(1)\right)$$

$$= -\psi_A(2,1)$$

(check this by exchanging the states of the particles: for example, $\phi_a(1)\phi_b(2) \rightarrow \phi_a(2)\phi_b(1)$ upon particle exchange). If we try to put both particles in the same state, a, then the symmetric wavefunction, for bosons, is:

$$\psi_S(1,2) = \sqrt{2}\;\phi_a(1)\phi_a(2)$$

whereas the anti-symmetric wavefunction, for fermions, vanishes:

$$\psi_A(1,2) = 0$$

Derivation 24.7: Derivation of exchange operator eigenvalues (Eq. 24.41)

$$\hat{P}_{12}\,\psi(1,2) = \psi(2,1)$$
$$\text{and } \hat{P}_{12}\,\psi(1,2) = p\psi(1,2)$$
$$\therefore \psi(2,1) = p\psi(1,2)$$

$$\hat{P}_{12}\psi(2,1) = \psi(1,2)$$
$$\text{and } \hat{P}_{12}\psi(2,1) = p\psi(2,1)$$
$$= p\hat{P}_{12}\psi(1,2)$$
$$= p^2\psi(1,2)$$
$$\therefore \psi(1,2) = p^2\psi(1,2)$$

$$\therefore p^2 = 1$$
$$\therefore p = \pm 1$$

- Starting point: acting on $\psi(1,2)$ with the exchange operator exchanges the states of the two particles:

$$\hat{P}_{12}\,\psi(1,2) = \psi(2,1)$$

- The eigenavalue equation for $\hat{P}_{12}$ is:

$$\hat{P}_{12}\,\psi(1,2) = p\psi(1,2)$$

- Equate these expressions, and repeat with $\psi(2,1)$.

This means that no two fermions are allowed in the same quantum state. This statement is known as the *Pauli exclusion principle*, and it leads to very different behavior for systems of bosons and fermions in the quantum regime, as outlined below.

24.6 Quantum statistics

24.6.1 Grand canonical partition function

The partition function, Z, for a classical system of discrete states s with no degeneracy is:

$$Z = \sum_s e^{-E_s/k_B T}$$

(refer to Section 24.4). This expression was derived using the Boltzmann factor, which in turn treats particles as a collection of closed systems: the particles can exchange energy with other particles (their surroundings) but cannot exchange matter. Therefore, Z refers to the *canonical* partition function (refer to Section 24.1 for classifications of statistical ensembles). In the quantum regime, however, we can no longer treat particles as closed systems since we are now dealing with multi-particle wavefunctions, in which the number of particles is not constant. Instead, we treat particles as a collection of *open* systems, which can exchange both energy *and* matter with their surroundings. The

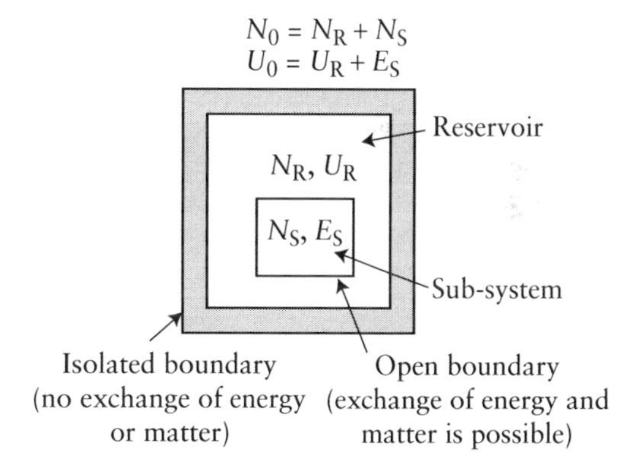

Figure 24.9: Isolated system comprising an open sub-system with N_S particles and energy E_S, in thermal contact with a heat reservoir at energy U_R, with N_R particles. Energy and matter can be exchanged between the reservoir and the sub-system. The total number of particles of the isolated system is $N_0 = N_R + N_S$, and the total energy is $U_0 = U_R + E_S$.

partition function in this case is called the *grand canonical partition function*, Ξ. We can use a similar treatment to that described in the derivation for the Boltzmann factor (Section 24.4.1), to find an expression for Ξ.

Consider an open sub-system in thermal contact with a heat reservoir, as illustrated in Fig. 24.9. Together, the sub-system plus the reservoir form an isolated system, with N_0 particles and with total energy U_0. The energy of the reservoir and sub-system are U_R and E_S, respectively, where $U_0 = U_R + E_S$; the number of particles in the reservoir and sub-system are N_R and N_S, respectively, where $N_0 = N_R + N_S$.

The exact differential of U for an open system is:

$$dU = TdS - pdV + \mu dN$$

where μ is the chemical potential that represents the change in internal energy per particle at constant entropy and volume (refer to Section 22.3.4). Therefore, assuming the reservoir has a constant volume, the entropy of the reservoir is:

$$dS_R = \frac{dU_R}{T} - \mu\frac{dN_R}{T}$$

$$\therefore S_R = \frac{1}{T}(U_R - \mu N_R) + \text{constant}$$

Letting $U_R = U_0 - E_S$, $N_R = N_0 - N_S$, and $S_0 = ((U_0 - \mu N_0)/T + \text{constant})$, we find:

$$S_R = S_0 - \frac{1}{T}(E_S - \mu N_S) + \text{constant}$$

Using the statistical definition of entropy from Eq. 24.2, the entropy of the reservoir is:

$$S_R = k_B \ln \Omega_R$$

$$\therefore \Omega_R = e^{S_R/k_B}$$

From the second postulate of statistical mechanics, the probability of the reservoir being in a given macrostate is proportional to the number of microstates in that macrostate: $P \propto \Omega$. Therefore, using the expression for S_R above, the probability of finding the sub-system in a state with energy E_S and N_S particles is:

$$\begin{aligned} P(E_S, N_S) &\propto \Omega_R(E_S, N_S) \\ &\propto e^{S_R(E_S, N_S)/k_B} \\ &\propto e^{-(E_S - \mu N_S)/k_B T} \end{aligned} \qquad (24.42)$$

Summing over all states, S, and all particles, N, the *grand canonical partition function* is defined as:

$$\Xi = \sum_S \sum_N e^{-(E_S - \mu N_S)/k_B T} \qquad (24.43)$$

We will now use this partition function to derive the occupation functions—the Fermi-Dirac and Bose–Einstein distributions—for quantum systems.

24.6.2 Quantum occupation functions

Consider a sub-system in a state S, with n particles at an energy level ε. The energy and number of particles of the sub-system are:

$$E_S = n\varepsilon \text{ and } N_S = n$$

Therefore, from Eq. 24.42, the probability that the sub-system is in state S is:

$$P(E_S, N_S) \propto e^{-n(\varepsilon - \mu)/k_B T}$$

and the average number of particles per state, as a function of energy ε, is therefore:

$$\begin{aligned} \langle n \rangle &= \frac{\sum_n n P(E_S, N_S)}{\sum_n P(E_S, N_S)} \\ &= \frac{\sum_n n e^{-n(\varepsilon-\mu)/k_B T}}{\sum_n e^{-n(\varepsilon-\mu)/k_B T}} \\ &\equiv \frac{\sum_n n x^n}{\sum_n x^n} \end{aligned} \qquad (24.44)$$

where $x = e^{-(\varepsilon-\mu)/k_B T}$

Notice that the denominator resembles the grand canonical partition function, Ξ, from Eq. 24.43, with $E_S = n\varepsilon$ and $N_S = n$. By definition, $\langle n \rangle$ is an *occupation function*, $\langle n \rangle \equiv f(\varepsilon)$. Using Eq. 24.44 along with the Pauli exclusion principle, we can determine expressions for the Fermi–Dirac and Bose–Einstein occupation functions, which are given in Eqs. 24.45 and 24.46, respectively (Derivations 24.8 and 24.9).

Quantum occupation functions (Derivations 24.8 and 24.9):

Fermi–Dirac distribution (for fermions):

$$f_{FD}(\varepsilon) = \frac{1}{e^{(\varepsilon-\mu)/k_B T} + 1} \qquad (24.45)$$

Bose–Einstein distribution (for bosons):

$$f_{BE}(\varepsilon) = \frac{1}{e^{(\varepsilon-\mu)/k_B T} - 1} \qquad (24.46)$$

Eqs. 24.45 and 24.46 are plotted in Fig. 24.10, for different values of T and μ. We see that increasing T changes the curvature of the plots, so as to put more particles into the higher energy levels. By increasing μ for a given T, the shape of the function remains the same, but the curve is shifted to higher energy levels.

The Fermi energy

As $T \to 0$ K, fermions fall into the lowest energy levels. Due to the Pauli exclusion principle, fermions "stack up," such that the highest energy particles have an energy μ. For spin-1/2 fermions, such as electrons, each energy level accommodates two particles: one in the spin-up state and one in the spin-down state (Fig. 24.11). As $T \to 0$ K, the highest occupied energy level is called the *Fermi energy*, E_F. This parameter depends on the particle number density, n, and mass m (Eq. 24.47 and Derivation 24.10).

The Fermi energy (Derivation 24.10):

$$E_F = \frac{\hbar^2}{2m}\left(3\pi^2 n\right)^{2/3} \tag{24.47}$$

Using a similar approach to that shown in Derivation 24.10, we find that the mean energy of a Fermi gas at 0 K is:

$$\langle\varepsilon\rangle = \frac{\int_0^{E_F} \varepsilon n(\varepsilon)\mathrm{d}\varepsilon}{\int_0^{E_F} n(\varepsilon)\mathrm{d}\varepsilon} = \frac{2\alpha\int_0^{E_F}\sqrt{\varepsilon^3}\,\mathrm{d}\varepsilon}{2\alpha\int_0^{E_F}\sqrt{\varepsilon}\,\mathrm{d}\varepsilon} = \frac{3}{5}E_F$$

Bose–Einstein condensation

As $T \to 0$ K for a system of bosons, we get very different behavior to a system of fermions since bosons do not obey the Pauli exclusion principle. From the Bose–Einstein curve in Fig. 24.10b, we see that at very low temperatures, if there were no interactions between the bosons, all particles in the system would fall into the ground state ($\varepsilon_0 = \mu$). This phenomenon is called *Bose–Einstein condensation*. Since all the particles are in a single quantum state, then quantum effects become apparent on macroscopic scales. For example, it is thought that the superfluidity of ^{4}He at temperatures below ~2 K is related to Bose–Einstein condensation. At these temperatures, liquid helium has zero viscosity, which means it can flow through tiny capillaries without apparent friction. For example, liquid helium could flow through microscopic holes in an apparently leak-tight container.

24.6.3 The classical approximation

In the approximation that $\varepsilon - \mu \gg k_B T$, then the exponential term in Eqs. 24.45 and 24.46 dominates:

$$e^{(\varepsilon-\mu)/k_B T} \gg \pm 1$$
$$\therefore \varepsilon \gg \mu$$

(Eq. 24.48). In this approximation, both the Fermi–Dirac and the Bose–Einstein distributions approximate to the classical Boltzmann distribution:

$$f(\varepsilon) = \frac{1}{e^{(\varepsilon-\mu)/k_B T} \pm 1}$$

$$f(\varepsilon \gg \mu) \to \frac{1}{e^{(\varepsilon-\mu)/k_B T}} \equiv A e^{-\varepsilon/k_B T}$$

Written like this, the normalization constant, A, is:

$$A = e^{\mu/k_B T}$$

Using the condition that $\varepsilon \gg \mu$ for the Boltzmann approximation, we find that the classical approximation corresponds to systems with a high temperature, T, and a small number density, n (Eq. 24.49 and Derivation 24.11).

Classical approximation (Derivation 24.11):

$$\varepsilon \gg \mu \tag{24.48}$$

$$\text{and} \quad \frac{n}{T^{3/2}} \ll \left(\frac{mk_B}{2\pi\hbar^2}\right)^{3/2} \tag{24.49}$$

$$\Rightarrow \text{high } T \text{ \& low } n$$

Therefore, in classical systems, energy levels are sparsely populated, whereas in quantum systems, energy levels

Derivation 24.8: Derivation of the Fermi–Dirac distribution (Eq. 24.45)

$$\langle n\rangle = \frac{\sum_{n=0}^{1} n x^n}{\sum_{n=0}^{1} x^n}$$

$$= \frac{0+x}{1+x}$$

$$= \frac{1}{x^{-1}+1}$$

$$= \frac{1}{e^{(\varepsilon-\mu)/k_B T}+1}$$

- Starting point: use the definition of $\langle n\rangle$ from Eq. 24.44:

$$\langle n\rangle = \frac{\sum_n n x^n}{\sum_n x^n}$$

where $x = e^{-(\varepsilon-\mu)/k_B T}$

- From the Pauli exclusion principle, no two fermions are allowed in the same quantum state, therefore, $n = 0$ or 1.

Derivation 24.9: Derivation of the Bose–Einstein distribution (Eq. 24.46)

$$\langle n\rangle = \frac{\sum_{n=0}^{\infty} n x^n}{\sum_{n=0}^{\infty} x^n}$$

$$= \frac{0+x+2x^2+3x^3+\cdots}{1+x+x^2+x^3+\cdots}$$

$$= \frac{x\left(1+2x+3x^2+\cdots\right)}{1+x+x^2+x^3+\cdots}$$

$$= \frac{x\frac{d}{dx}(1-x)^{-1}}{(1-x)^{-1}}$$

$$= \frac{x(1-x)^{-2}}{(1-x)^{-1}}$$

$$= \frac{x}{1-x}$$

$$= \frac{1}{x^{-1}-1}$$

$$= \frac{1}{e^{(\varepsilon-\mu)/k_B T}-1}$$

- Starting point: use the definition of $\langle n\rangle$ from Eq. 24.44:

$$\langle n\rangle = \frac{\sum_n n x^n}{\sum_n x^n}$$

where $x = e^{-(\varepsilon-\mu)/k_B T}$

- Bosons do not obey the Pauli exclusion principle, therefore, n can take any integer number from $n = 0$ to ∞.
- The power series for $(1-x)^{-1}$ is:

$$\frac{1}{1-x} = 1+x+x^2+x^3+\cdots$$

$$\therefore \frac{d}{dx}\left(\frac{1}{1-x}\right) = 1+2x+3x^2+\cdots$$

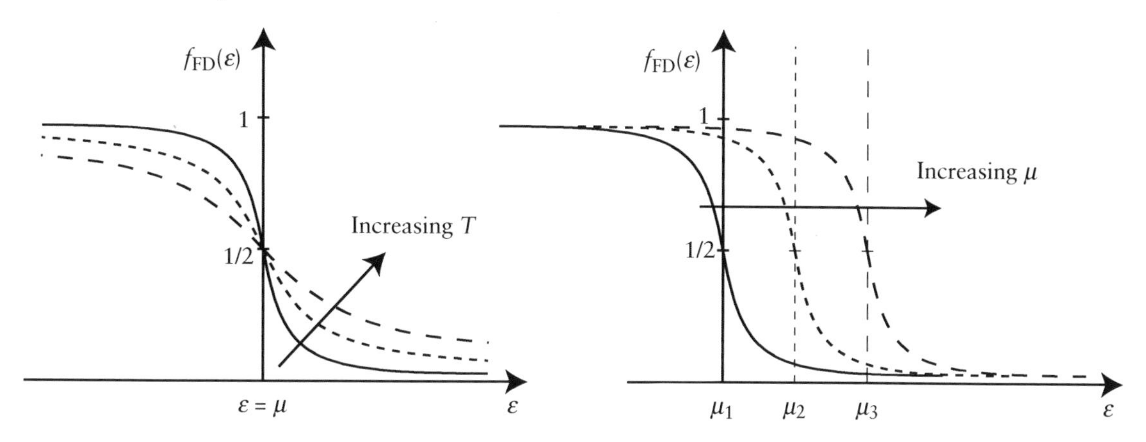

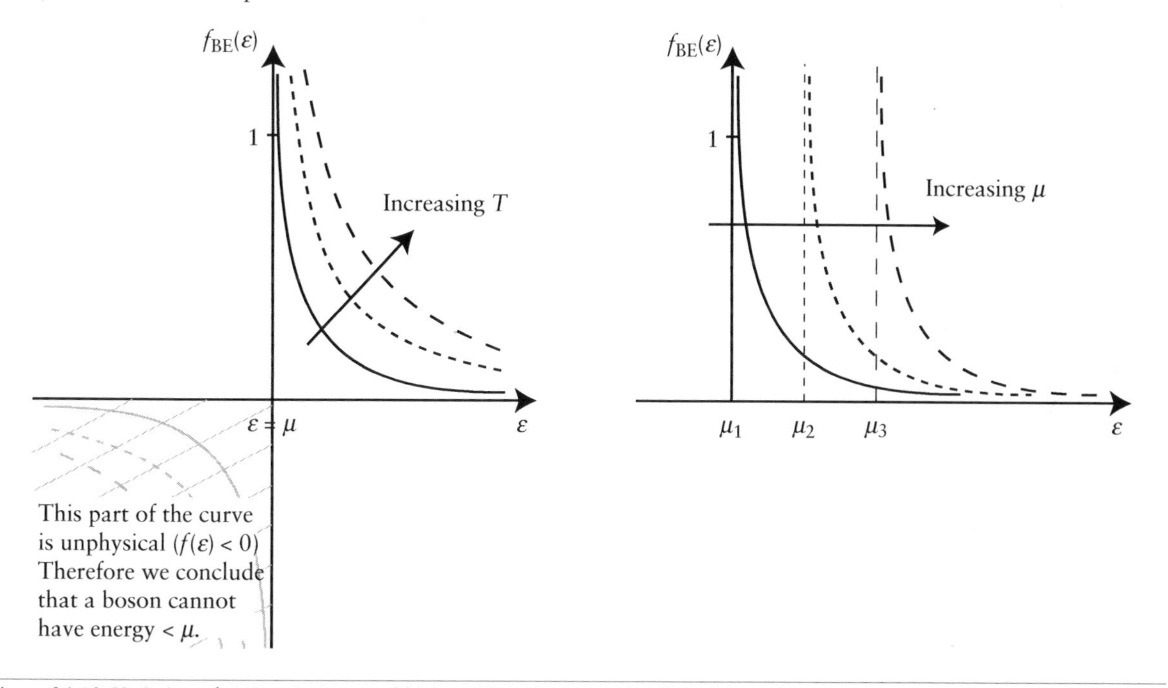

Figure 24.10: Variation of a) Fermi–Dirac and b) Bose–Einstein occupation functions with temperature, T, and chemical potential, μ (Eqs. 24.45 and 24.46). Changing T changes the curvature of the function, whereas changing μ shifts the curve to higher or lower energies.

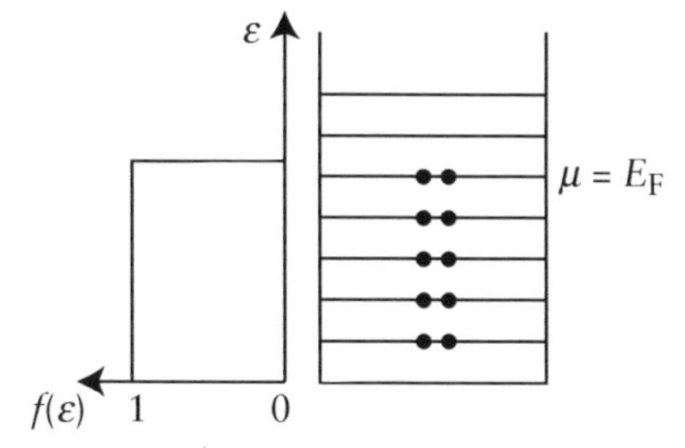

Figure 24.11: System of fermions as $T \to 0$ K. The highest occupied energy level is the Fermi energy, E_F. All states below the Fermi energy are occupied, and all states above it are vacant. In this regime, the Fermi–Dirac distribution is a step function where $f(\varepsilon) = 1$ for $\varepsilon < E_F$ and $f(\varepsilon) = 0$ for $\varepsilon > E_F$.

Derivation 24.10: Derivation of the Fermi energy (Eq. 24.47)

- Starting point: the total number density of particles is:

$$n = \int_0^\infty n(\varepsilon)\mathrm{d}\varepsilon = g_s \int_0^\infty g(\varepsilon) f(\varepsilon)\mathrm{d}\varepsilon$$

There are two electrons allowed per spatial state, therefore, the spin-degeneracy is $g_s = 2$.

- At $T \simeq 0$ K, the Fermi–Dirac occupation function is a step function:

$$f(\varepsilon) = \begin{cases} 1 & \text{for } \varepsilon < E_F \\ 0 & \text{for } \varepsilon > E_F \end{cases}$$

where E_F is the Fermi energy (highest occupied energy state).

- The density of states in three dimensions is $g(\varepsilon) = \alpha\sqrt{\varepsilon}$, where the constant is:

$$\alpha = \frac{\sqrt{2m^3}}{2\pi^2\hbar^3}$$

(refer to Section 24.3.2).

- Integrate n and rearrange to solve for E_F.

$$n = 2\int_0^\infty g(\varepsilon) f(\varepsilon)\mathrm{d}\varepsilon$$

$$= 2\alpha \int_0^{E_F} \sqrt{\varepsilon}\mathrm{d}\varepsilon$$

$$= 2\frac{\sqrt{2m^3}}{2\pi^2\hbar^3}\frac{2}{3}(E_F)^{3/2}$$

$$\therefore E_F = \frac{\hbar^2}{2m}\left(3\pi^2 n\right)^{2/3}$$

are densely populated. In the classical approximation, the number of states available between an energy ε and $\varepsilon + \mathrm{d}\varepsilon$ is far greater than the number of particles within that energy range, which means that particles do not compete to be in the same state. Therefore, bosons and fermions behave in the same way under classical conditions, since the Pauli exclusion principle is not in effect. A postulate of classical statistics is that there is no upper limit to the number of particles that can occupy a given energy interval, $\mathrm{d}\varepsilon$.

The de Broglie wavelength

For classical systems, we can use the equipartition theorem to equate the average kinetic energy of a particle with $k_B T$:

$$\frac{p^2}{2m} \simeq k_B T$$

The de Broglie wavelength of a particle is defined as:

$$\lambda_{db} = \frac{h}{p}$$

Eliminating p using the equipartition theorem, the de Broglie wavelength for classical particles is:

$$\lambda_{db} \simeq \left(\frac{h^2}{2mk_B T}\right)^{1/2}$$

Comparing this expression with the classical approximation inequality from Eq. 24.49, we find:

$$\lambda_{db} \simeq \left(\frac{h^2}{2mk_B T}\right)^{1/2} \ll \frac{1}{n^{1/3}} \ll \left(\frac{V}{N}\right)^{1/3} \ll d \qquad (24.50)$$

Derivation 24.11: Derivation of conditions for classical statistics (Eq. 24.49)

$$n = \int_0^\infty g(\varepsilon) f(\varepsilon) \mathrm{d}\varepsilon$$

$$= \int_0^\infty \frac{g(\varepsilon)}{\mathrm{e}^{(\varepsilon-\mu)/k_B T}} \mathrm{d}\varepsilon$$

$$= \alpha \mathrm{e}^{\mu/k_B T} \int_0^\infty \sqrt{\varepsilon} \mathrm{e}^{-\varepsilon/k_B T} \mathrm{d}\varepsilon$$

$$= \frac{\sqrt{2m^3}}{2\pi^2 \hbar^3} \frac{\sqrt{\pi}}{2} (k_B T)^{3/2} \mathrm{e}^{\mu/k_B T}$$

$$= \left(\frac{m k_B T}{2\pi\hbar^2} \right)^{3/2} \mathrm{e}^{\mu/k_B T}$$

$$\therefore \mathrm{e}^{\mu/k_B T} = n \left(\frac{2\pi\hbar^2}{m k_B T} \right)^{3/2}$$

For classical systems:

$$\mathrm{e}^{\mu/k_B T} \ll 1$$

$$\therefore n \left(\frac{2\pi\hbar^2}{m k_B T} \right)^{3/2} \ll 1$$

$$\Rightarrow \text{high } T \text{ \& low } n$$

- Starting point: the total number density of particles in a classical system is:

$$n = \int_0^\infty g(\varepsilon) f(\varepsilon) \mathrm{d}\varepsilon$$

where $f(\varepsilon) = A\mathrm{e}^{-\varepsilon/k_B T} = \mathrm{e}^{-(\varepsilon-\mu)/k_B T}$

- The density of states in three dimensions is $g(\varepsilon) = \alpha\sqrt{\varepsilon}$, where:

$$\alpha = \frac{\sqrt{2m^3}}{2\pi^2\hbar^3}$$

- To integrate, use the standard integral:

$$\int_0^\infty \sqrt{x} \mathrm{e}^{-x/a} \mathrm{d}x = \frac{\sqrt{\pi}}{2} a^{3/2}$$

- For classical systems, we have:

$$\mathrm{e}^{(\varepsilon-\mu)/k_B T} \gg 1$$

$$\therefore \mathrm{e}^{-\varepsilon/k_B T} \mathrm{e}^{\mu/k_B T} \ll 1$$

$$\therefore \mathrm{e}^{\mu/k_B T} \ll 1 \quad \text{for all } \varepsilon$$

- Therefore, eliminating the exponential factor from the above expression, we find that classical systems correspond to high T and low n, which means hot, sparsely populated systems.

where we have defined the average distance between particles as:

$$d \simeq \left(\frac{V}{N} \right)^{1/3}$$

Therefore, Eq. 24.50 illustrates that the de Broglie wavelength of classical particles is much less than their interparticle spacing, which means that their wavefunctions do not overlap, and therefore classical particles are distinguishable.

The transition temperature

The *transition temperature* of a system is the temperature below which we must use *quantum* statistics, and above which we can use the *classical* approximation. For fermions, this temperature is called the *Fermi temperature*, T_F, and for bosons, it is called the *Bose temperature*, T_B. The transition between classical and quantum statistics occurs when the de Broglie wavelength of the particles is comparable to their interatomic spacing:

$$\lambda_{db} \simeq d$$

Therefore, from Derivation 24.11, this corresponds to:

$$\mathrm{e}^{\mu/k_B T} \simeq 1$$

$$\therefore n \left(\frac{2\pi\hbar^2}{m k_B T} \right)^{3/2} \simeq 1$$

Rearranging the above equation, we can find an order of magnitude for the quantum–classical transition temperature, T_{QC} (Eq. 24.51).

Transition temperature:

$$T_{QC} \simeq \frac{2\pi\hbar^2}{mk_B} n^{2/3} \qquad (24.51)$$

The factor of 2π varies slightly for T_F and T_B, but Eq. 24.51 gives a good order of magnitude approximation. For example, from the definition of the Fermi energy in Eq. 24.47, the Fermi transition temperature is on the order of:

$$T_F \sim \frac{E_F}{k_B} \sim \frac{\hbar^2}{2mk_B}\left(3\pi^2 n\right)^{2/3}$$

which is similar to T_{QC}.

Classical temperatures

From the equipartition theorem, the average energy of a particle in a classical system is on the order of $k_B T$:

$$\langle \varepsilon \rangle \sim k_B T$$

This does not hold for quantum systems. Therefore, the regime in which the classical approximation holds is when the energy of a particle is on the order of $k_B T$, for a system at temperature T. This occurs when $T \gtrsim T_{QC}$.

24.6.4 Black-body radiation

We will now turn our attention to the properties of photons (electromagnetic radiation). Photons are bosons and have integer spin. Electromagnetic radiation incident on an absorptive material is partially reflected, partially transmitted, and partially absorbed (Fig. 24.12). These quantities are characterized by dimensionless parameters—the *reflectivity* ρ, the *transmissivity* τ, and the *absorptivity* a, respectively—which represent the fraction of incident radiation that is reflected, transmitted or absorbed, and have values between zero and one. For radiation incident on any body, the sum of these fractions equals one:

$$\rho + \alpha + \tau = 1$$

A *black body* is an object that absorbs *all* radiation incident on it, therefore, for a black body: $a_{bb} = 1$ and $\rho_{bb} = \tau_{bb} = 0$.

In addition to reflection, absorption, and transmission, any object at a non-zero temperature *emits* electromagnetic radiation. Emission is quantified by the *emissivity*, e, of the object, which is another dimensionless quantity, where $0 < e < 1$. Below, we will look at four laws—Kirchhoff's Law, Planck's Radiation Law, Wien's Displacement Law and the Stefan–Boltzmann Law—which characterize the emission spectrum of a black body.

Kirchhoff's Law

Kirchhoff's Law states that, for bodies in thermal equilibrium with their surroundings, their emissivity equals their absorptivity, for all wavelengths:

$$e(\lambda) = \alpha(\lambda)$$

(Eq. 24.52). Since black bodies have $a = 1$, then, from Kirchhoff's Law, $e = 1$. Therefore, black bodies are often defined as "perfect absorbers and emitters of electromagnetic radiation." A black body can be thought of theoretically as an opaque container or a cavity with a small hole in it through which radiation can enter and leave. Radiation entering the cavity has a negligible chance of escaping through the small hole again, and therefore reflects off the cavity walls multiple times, until it is absorbed. Therefore, the cavity absorbs all incident radiation, which, from Kirchhoff's Law, means the emissivity of the cavity is $e = 1$. Therefore, this object constitutes a black body, and the radiation leaving the cavity through the small hole is black-body radiation. The frequency spectrum of radiation emitted depends on the temperature of the black body, and is described by *Planck's Radiation Law*.

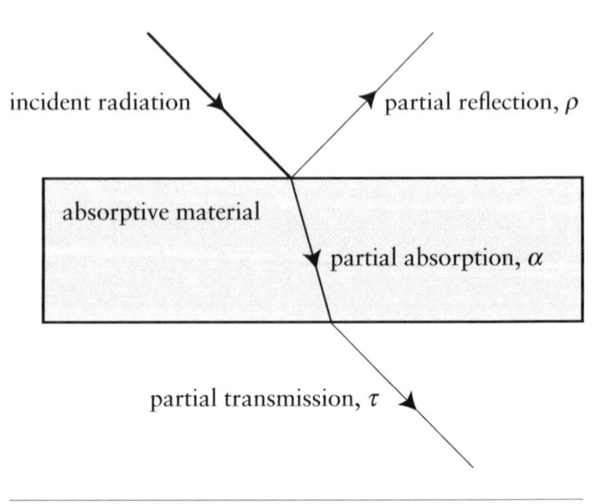

Figure 24.12: Radiation incident on a body is partially reflected, partially transmitted and partially absorbed.

Kirchhoff's Law:

$$e(\lambda) = \alpha(\lambda) \tag{24.52}$$

Planck's Radiation Law

Planck's Radiation Law gives the frequency spectrum of radiation emitted from a black body at temperature T. We can determine the emission spectrum of a black body by applying the Bose–Einstein distribution to a "gas" of photons. Treating a black body as an opaque cavity with a small opening, as described above, we can imagine that photons are constantly emitted and absorbed by the cavity walls, which means that the number of photons within the cavity, N, is not constant. This condition implies that the chemical potential of a gas of photons is $\mu = 0$. Using this requirement in the Bose–Einstein occupation function (Eq. 24.46), the Bose–Einstein distribution for black-body radiation is:

$$f_{\mathrm{BE}}(\varepsilon) = \frac{1}{e^{\varepsilon/k_{\mathrm{B}}T} - 1}$$

The energy of a photon of frequency ω is $\varepsilon = \hbar\omega$. Therefore, the Bose–Einstein occupation function in terms of ω is:

$$f_{\mathrm{BE}}(\omega) = \frac{1}{e^{\hbar\omega/k_{\mathrm{B}}T} - 1}$$

Using this occupation function together with the radiation density of states, as shown in Derivation 24.12, we find that the energy density distribution of black-body photons is:

Derivation 24.12: Derivation of Planck's Radiation Law (Eqs. 24.53 and 24.54)

$$\begin{aligned} u(\omega)\mathrm{d}\omega &= \hbar\omega n(\omega)\mathrm{d}\omega \\ &= \hbar\omega g(\omega) f_{\mathrm{BE}}(\omega)\mathrm{d}\omega \\ &= \hbar\omega \frac{g(\omega)\mathrm{d}\omega}{e^{\hbar\omega/k_{\mathrm{B}}T} - 1} \\ &= \left(\frac{\hbar}{\pi^2 c^3}\right) \frac{\omega^3 \mathrm{d}\omega}{e^{\hbar\omega/k_{\mathrm{B}}T} - 1} \end{aligned}$$

Substituting in for $\omega = 2\pi\nu$ and $\hbar = h/2\pi$, Planck's Law is:

$$\therefore u(\nu)\mathrm{d}\nu = \left(\frac{8\pi h}{c^3}\right) \frac{\nu^3 \mathrm{d}\nu}{e^{h\nu/k_{\mathrm{B}}T} - 1}$$

Or, letting $\nu = c/\lambda$:

$$u(\lambda)\mathrm{d}\lambda = \frac{8\pi hc}{\lambda^5} \frac{\mathrm{d}\lambda}{e^{hc/\lambda k_{\mathrm{B}}T} - 1}$$

- Starting point: the energy density distribution is the photon number density distribution, $n(\omega)\mathrm{d}\omega$, multiplied by the energy per photon, $\hbar\omega$:

  $$u(\omega)\mathrm{d}\omega = \hbar\omega n(\omega)\mathrm{d}\omega$$

- The particle number density is:

  $$n(\omega)\mathrm{d}\omega = g(\omega) f_{\mathrm{BE}}(\omega)\mathrm{d}\omega$$

  where $f_{\mathrm{BE}}(\omega) = \frac{1}{e^{\hbar\omega/k_{\mathrm{B}}T} - 1}$ for photons

- For photons, the dispersion relation is $\omega(k) = ck$. Therefore, using this in the density of states in k-space in three dimensions (Eq. 24.4), the photon density of states in ω-space is:

  $$g(\omega)\mathrm{d}\omega = 2\frac{\omega^2}{2\pi^2 c^3}\mathrm{d}\omega \tag{24.55}$$

  The factor of 2 accounts for photon polarization degeneracy.
- Planck's Law is often written in terms of the photon frequency ν, where $\nu = \omega/2\pi$ and $h = 2\pi\hbar$, or the photon wavelength λ, where $\lambda = c/\nu$.

$$u(\omega)\mathrm{d}\omega = \left(\frac{\hbar}{\pi^2 c^3}\right)\frac{\omega^3 \mathrm{d}\omega}{\mathrm{e}^{\hbar\omega/k_\mathrm{B}T}-1}$$

Expressing this distribution in terms of the energy density per frequency interval, $u(\nu)$ [J m^{-3} Hz^{-1}], we have *Planck's Radiation Law*:

$$u(\nu)\mathrm{d}\nu = \left(\frac{8\pi h}{c^3}\right)\frac{\nu^3 \mathrm{d}\nu}{\mathrm{e}^{h\nu/k_\mathrm{B}T}-1}$$

(Eq. 24.53 and Derivation 24.12). It is also common to express Planck's Radiation Law in terms of the energy density per wavelength interval $u(\lambda)$ [J m^{-3} nm^{-1}]:

$$u(\lambda)\mathrm{d}\lambda = \left(\frac{8\pi hc}{\lambda^5}\right)\frac{\mathrm{d}\lambda}{\mathrm{e}^{hc/\lambda k_\mathrm{B}T}-1}$$

(Eq. 24.54 and Derivation 24.12). This distribution is plotted in Fig. 24.13 for black bodies of different temperatures. We see that with decreasing temperature: (i) the wavelength corresponding to maximum emission (the turning point) increases and (ii) the area under the curve decreases (indicating that there are fewer photons emitted). These two characteristics are described by *Wien's Displacement Law* and the *Stefan–Boltzmann Law*, respectively.

Planck's Radiation Law (Derivation 24.12):

$$u(\nu)\mathrm{d}\nu = \left(\frac{8\pi h}{c^3}\right)\frac{\nu^3 \mathrm{d}\nu}{\mathrm{e}^{h\nu/k_\mathrm{B}T}-1} \quad (24.53)$$

or $$u(\lambda)\mathrm{d}\lambda = \left(\frac{8\pi hc}{\lambda^5}\right)\frac{\mathrm{d}\lambda}{\mathrm{e}^{hc/\lambda k_\mathrm{B}T}-1} \quad (24.54)$$

Wien's Displacement Law

From Fig. 24.13, the emission spectrum has a turning point whose wavelength increases with decreasing temperature. By differentiating Planck's Law, we find that the wavelength of maximum emission, λ_{max}, is inversely proportional to the temperature of the black body:

$$\lambda_{\mathrm{max}} \propto \frac{1}{T}$$

(Eq. 24.56 and Derivation 24.13). This equation is known as *Wien's Displacement Law.*

Wien's Displacement Law (Derivation 24.13):

$$T = \frac{2.898 \cdot 10^{-3}}{\lambda_{\mathrm{max}}} \quad (24.56)$$

The color of a black body indicates the maximum emitted wavelength. Therefore, Wien's Law implies that a black body's color is a function of its temperature. To a good approximation, stars are black bodies, and their color indicates their surface temperature. For example, white dwarfs are relatively hot stars, with surface temperatures up to 10000 K, whereas red giants are much cooler, and have temperatures around 3000 K. The temperature of the sun is $T \sim 6000$ K, which provides an emission peak in the visible wavelength band at ~500 nm.

Stefan–Boltzmann Law

The *Stefan–Boltzmann Law* describes the total power flux emitted by a black body. This quantity is represented by the area under the curve described by Planck's Radiation Law (refer to Fig. 24.13). Therefore, by integrating Planck's Radiation Law over all frequencies, we find that the power flux of a black body, J [W m^{-2}], is proportional to the fourth power of its temperature, T^4:

$$J \propto T^4$$

(Eq. 24.57 and Derivation 24.14). The constant of proportionality is the *Stefan–Boltzmann* constant, σ [$5.67 \cdot 10^{-8}$ W m^{-2} K^{-4}] (Eq. 24.58).

Stefan–Boltzmann Law (Derivation 24.14):

$$J = \sigma T^4 \tag{24.57}$$

$$\text{where } \sigma = \frac{2\pi^5 k_B^4}{15h^3c^2} \tag{24.58}$$

$$\simeq 5.67 \cdot 10^{-8} \ \mathrm{W m^{-2} K^{-4}}$$

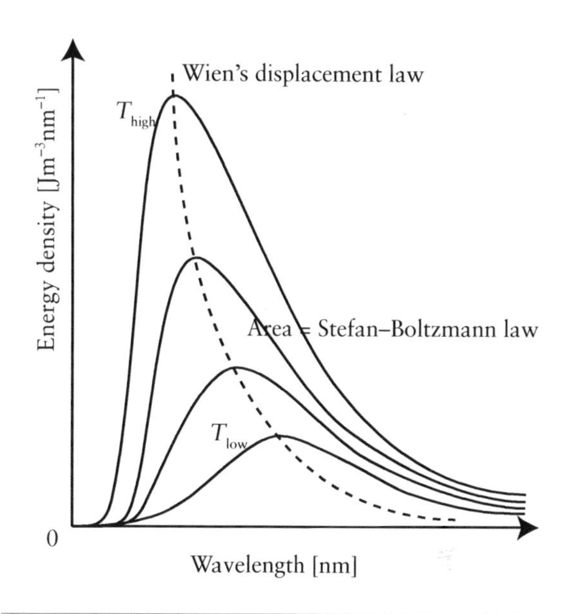

Figure 24.13: Emission spectrum of a black body. Planck's Radiation Law (Eq. 24.54) gives the intensity distribution of radiation as a function of wavelength; Wien's Displacement Law describes the position of the maxima; the Stefan–Boltzmann Law gives the area under the curve (total power emitted per unit area). As the temperature of the black body decreases, the wavelength of maximum emission increases, and the area under the curve (indicating the power flux, or number of emitted photons) decreases.

Derivation 24.13: Derivation of Wien's Displacement Law (Eq. 24.56)

$$u(\lambda)\mathrm{d}\lambda = \left(\frac{8\pi hc}{\lambda^5}\right)\frac{\mathrm{d}\lambda}{\mathrm{e}^{hc/\lambda k_B T}-1}$$

$$\therefore \frac{\partial u(\lambda)}{\partial \lambda} = \left\{\frac{8\pi hc}{\lambda^5}\left(\frac{hc\mathrm{e}^{hc/\lambda k_B T}}{\lambda^2 k_B T}\right)\frac{1}{\left(\mathrm{e}^{hc/\lambda k_B T}-1\right)^2} - \frac{40\pi hc}{\lambda^6}\frac{1}{\mathrm{e}^{hc/\lambda k_B T}-1}\right\}$$

$$= 0$$

$$\therefore 5\lambda\left(\mathrm{e}^{hc/\lambda k_B T}-1\right) = \left(\frac{hc}{k_B T}\right)\mathrm{e}^{hc/\lambda k_B T}$$

$$5\left(\mathrm{e}^{\alpha/\lambda}-1\right) = \frac{\alpha}{\lambda}\mathrm{e}^{\alpha/\lambda}$$

$$\therefore \left(5-\frac{\alpha}{\lambda}\right) = 5\mathrm{e}^{-\alpha/\lambda}$$

Solving for α/λ graphically, we find the above equation is satisfied when $\alpha/\lambda \simeq 4.965$:

$$\therefore T = \frac{hc}{\alpha k_B}$$

$$\simeq \frac{hc}{4.965 k_B}\frac{1}{\lambda_{max}}$$

$$\simeq \frac{0.002898}{\lambda_{max}}$$

$$\simeq \frac{2.898\cdot 10^{-3}}{\lambda_{max}}$$

- Starting point: the turning point of Planck's Radiation Law corresponds to the wavelength at which:

$$\frac{\partial u(\lambda)}{\partial \lambda} = 0$$

- Differentiating Planck's Law (Eq. 24.54) and setting the derivative equal to zero, we find an expression for the variable a/λ, where:

$$\alpha = \frac{hc}{k_B T}$$

- We can solve for a/λ graphically, by plotting the functions $5 - a/\lambda$ and $5\mathrm{e}^{-a/\lambda}$ on the same axes and determining their point of intersection. Doing this, the curves intersect when:

$$\frac{\alpha}{\lambda} \simeq 4.965$$

Therefore, the wavelength of the turning point of Planck's Radiation Law corresponds to $\lambda_{max} \simeq a/4.965$.

- Using $T = hc/ak_B$ and substituting in for a, we find that T is inversely proportional to $\lambda_{max} \Rightarrow$ Wien's Displacement Law.

Derivation 24.14: Derivation of the Stefan–Boltzmann Law (Eq. 24.57)

$$u_{\mathrm{T}} = \int_0^{\infty} u(\nu)\,\mathrm{d}\nu$$

$$= \frac{8\pi h}{c^3}\int_0^{\infty}\frac{\nu^3}{e^{h\nu/k_{\mathrm{B}}T}-1}\,\mathrm{d}\nu$$

$$= \frac{8\pi h}{15c^3}\left(\frac{k_{\mathrm{B}}T\pi}{h}\right)^4$$

$$= \frac{8\pi^5}{15h^3c^3}(k_{\mathrm{B}}T)^4$$

$$\therefore J = \frac{1}{4}cu_{\mathrm{T}}$$

$$= \underbrace{\frac{2\pi^5 k_{\mathrm{B}}^4}{15h^3c^2}}_{=\sigma}T^4$$

$$\equiv \sigma T^4$$

- Starting point: the total energy density, u_{T}, is found by integrating Planck's Radiation Law (Eq. 24.53) over all photon frequencies, ν:

$$u_{\mathrm{T}} = \frac{8\pi h}{c^3}\int_0^{\infty}\frac{\nu^3}{e^{h\nu/k_{\mathrm{B}}T}-1}\,\mathrm{d}\nu$$

- To integrate, use the standard integral:

$$\int_0^{\infty}\frac{x^3}{e^{ax}-1}\,\mathrm{d}x = \frac{1}{15}\left(\frac{\pi}{a}\right)^4$$

- From kinetic theory, we found that the particle flux, Φ [$\mathrm{m^{-2}\,s^{-1}}$], of particles with average velocity c is:

$$\Phi = \frac{1}{4}cn$$

(see Section 21.4.4, Eq. 21.28). The total power flux, J [$\mathrm{W\,m^{-2}}$], is then given by:

$$J = \frac{\Phi}{n}u_{\mathrm{T}} = \frac{1}{4}cu_{\mathrm{T}}$$

- The *Stefan–Boltzmann constant*, σ [$\mathrm{W\,m^{-2}\,K^{-2}}$], is defined as:

$$\sigma = \frac{2\pi^5 k_{\mathrm{B}}^4}{15h^3c^2}$$

$$\simeq 5.67\cdot 10^{-8}\ \mathrm{Wm^{-2}K^{-4}}$$

Appendices

Appendix A
Identities and power series

A.1 Trigonometric identities

Pythagorean identity:

$$\sin^2\theta + \cos^2\theta = 1$$

Addition identities:

$$\sin(\theta_1 \pm \theta_2) = \sin\theta_1 \cos\theta_2 \pm \cos\theta_1 \sin\theta_2$$

$$\cos(\theta_1 \pm \theta_2) = \cos\theta_1 \cos\theta_2 \mp \sin\theta_1 \sin\theta_2$$

$$\tan(\theta_1 \pm \theta_2) = \frac{\tan\theta_1 + \tan\theta_2}{1 \mp \tan\theta_1 \tan\theta_2}$$

Double angle identities:

$$\sin 2\theta = 2\sin\theta\cos\theta$$

$$\cos 2\theta = \cos^2\theta - \sin^2\theta$$

$$= 2\cos^2\theta - 1$$

$$= 1 - 2\sin^2\theta$$

$$\tan 2\theta = \frac{2\tan\theta}{1-\tan^2\theta}$$

Sum and product identities:

$$\sin\theta_1 \pm \sin\theta_2 = 2\sin\left(\frac{\theta_1 \pm \theta_2}{2}\right)\cos\left(\frac{\theta_1 \mp \theta_2}{2}\right)$$

$$\cos\theta_1 + \cos\theta_2 = 2\cos\left(\frac{\theta_1 + \theta_2}{2}\right)\cos\left(\frac{\theta_1 - \theta_2}{2}\right)$$

$$\cos\theta_1 - \cos\theta_2 = -2\sin\left(\frac{\theta_1 + \theta_2}{2}\right)\sin\left(\frac{\theta_1 - \theta_2}{2}\right)$$

$$\sin\theta_1 \cos\theta_2 = \frac{1}{2}\left(\sin(\theta_1 + \theta_2) + \sin(\theta_1 - \theta_2)\right)$$

$$\cos\theta_1 \cos\theta_2 = \frac{1}{2}\left(\cos(\theta_1 + \theta_2) + \cos(\theta_1 - \theta_2)\right)$$

$$\sin\theta_1 \sin\theta_2 = -\frac{1}{2}\left(\cos(\theta_1 + \theta_2) - \cos(\theta_1 - \theta_2)\right)$$

Complex exponential identities:

$$r(\cos\theta + \mathrm{i}\sin\theta) = re^{\mathrm{i}\theta} \quad \text{(Euler's formula)}$$

$$r(\cos\theta - \mathrm{i}\sin\theta) = re^{-\mathrm{i}\theta}$$

$$\therefore \sin\theta = \frac{e^{\mathrm{i}\theta} - e^{-\mathrm{i}\theta}}{2\mathrm{i}}$$

$$\therefore \cos\theta = \frac{e^{\mathrm{i}\theta} + e^{-\mathrm{i}\theta}}{2}$$

A.2 Power series

Taylor expansion

Any function $f(x)$ can be approximated to a power series in x about a given point x_0:

$$f(x) \simeq f(x_0) + \underbrace{\left(\frac{df}{dx}\right)_{x_0}(x - x_0)}_{\text{linear component}} + \underbrace{\frac{1}{2!}\left(\frac{d^2f}{dx^2}\right)_{x_0}(x - x_0)^2}_{\text{quadratic component}} + \underbrace{\frac{1}{3!}\left(\frac{d^3f}{dx^3}\right)_{x_0}(x - x_0)^3}_{\text{cubic component, etc.}} + \dots$$

$$= \sum_{n=0}^{\infty} \frac{f^{(n)}(x_0)}{n!}(x - x_0)^n$$

where $f^{(n)}(x_0)$ represents the nth derivative of $f(x)$ evaluated at $x = x_0$.

Some common power series

(series are valid for all real values of x):

$$\sin x \simeq x - \frac{x^3}{3!} + \frac{x^5}{5!} - \frac{x^7}{7!} + \dots$$

$$\cos x \simeq 1 - \frac{x^2}{2!} + \frac{x^4}{4!} - \frac{x^6}{6!} + \dots$$

$$e^x \simeq 1 + x + \frac{x^2}{2!} + \frac{x^3}{3!} + \frac{x^4}{4!} + \dots$$

$$\therefore e^{\mathrm{i}x} \simeq 1 + \mathrm{i}x - \frac{x^2}{2!} - \mathrm{i}\frac{x^3}{3!} + \frac{x^4}{4!} + \dots$$

$$= \cos x + \mathrm{i}\sin x \quad \text{(Euler's formula)}$$

Binomial expansion:

$$(1+x)^\alpha \simeq 1 + \alpha x + \frac{\alpha(\alpha-1)}{2!}x^2 + \frac{\alpha(\alpha-1)(\alpha-2)}{3!}x^3 + \dots$$

$$\therefore (1+x)^\alpha \simeq 1 + \alpha x, \text{ for } |x| \ll 1$$

Appendix B
Coordinate systems

Cartesian, cylindrical, and spherical polar coordinates

	Cartesian	**Cylindrical polars**	**Spherical polars**
Coordinates	(x, y, z) x: distance along x-axis y: distance along y-axis z: distance along z-axis	(r, θ, z) r: $\perp$ distance from z-axis θ: angle with x-axis z: distance along z-axis	(r, θ, ϕ) r: distance from origin θ: angle with z-axis ϕ: angle of projection with x-axis
Transformations **Subscripts c and s are to distinguish between cylindrical and spherical coordinates.*	Cartesian $\rightarrow$ Cylindrical $r = \sqrt{x^2 + y^2}$ $\theta = \tan^{-1}(y/x)$ $z = z$ Cartesian $\rightarrow$ Spherical $r = \sqrt{x^2 + y^2 + z^2}$ $\theta = \cos^{-1}(z/r)$ $\phi = \tan^{-1}(y/x)$	Cylindrical $\rightarrow$ Cartesian $x = r\cos\theta$ $y = r\sin\theta$ $z = z$ Cylindrical $\rightarrow$ Spherical $r_s = \sqrt{r_c^2 + z^2}$ $\phi = \theta_c$ $\theta_s = \tan^{-1}(r_c/z)$	Spherical $\rightarrow$ Cartesian $x = r\sin\theta\cos\phi$ $y = r\sin\theta\sin\phi$ $z = r\cos\theta$ Spherical $\rightarrow$ Cylindrical $r_c = r_s\sin\theta_s$ $\theta_c = \phi$ $z = r_s\cos\theta_s$
Unit vectors	$(\hat{i}, \hat{j}, \hat{k})$	$(\hat{r}, \hat{\theta}, \hat{z})$	$(\hat{r}, \hat{\theta}, \hat{\phi})$
Line element $(d\vec{l})$ **Surface element** $(d\vec{s})$ **Volume element** $(d\tau)$	$d\vec{l} = dx\hat{i} + dy\hat{j} + dz\hat{k}$ $d\vec{s} = dydz\hat{i} + dxdz\hat{j} + dxdy\hat{k}$ $d\tau = dxdydz$	$d\vec{l} = dr\hat{r} + rd\theta\hat{\theta} + dz\hat{z}$ $d\vec{s} = rd\theta dz\hat{r} + drdz\hat{\theta} + rdrd\theta\hat{z}$ $d\tau = rdrd\theta dz$	$d\vec{l} = dr\hat{r} + rd\theta\hat{\theta} + r\sin\theta d\phi\hat{\phi}$ $d\vec{s} = r^2 \underbrace{\sin\theta d\theta d\phi}_{=d\Omega} \hat{r} + r\sin\theta drd\phi\hat{\theta} + rdrd\theta\hat{\phi}$ $d\tau = r^2\sin\theta drd\theta d\phi$

Appendix C
Vectors and vector calculus

C.1 Vector identities

Triple products:

$$\vec{F}\cdot\left(\vec{G}\times\vec{H}\right)=\vec{G}\cdot\left(\vec{H}\times\vec{F}\right)$$
$$=\vec{H}\cdot\left(\vec{F}\times\vec{G}\right)$$
$$\vec{F}\times\left(\vec{G}\times\vec{H}\right)=\vec{G}\left(\vec{F}\cdot\vec{H}\right)-\vec{H}\left(\vec{F}\cdot\vec{G}\right)$$

Product rules:

Gradient rules. The grad operator (∇) operates on scalar functions and returns a vector:

$$\nabla(fg)=f(\nabla g)+g(\nabla f)$$
$$\nabla\left(\vec{F}\cdot\vec{G}\right)=\vec{F}\times\left(\nabla\times\vec{G}\right)+\vec{G}\times\left(\nabla\times\vec{F}\right)+\left(\vec{F}\cdot\nabla\right)\vec{G}+\left(\vec{G}\cdot\nabla\right)\vec{F}$$

Divergence rules. The div operator ($\nabla\cdot$) operates on vector functions and returns a scalar:

$$\nabla\cdot\left(f\vec{F}\right)=f\left(\nabla\cdot\vec{F}\right)+\vec{F}\cdot(\nabla f)$$
$$\nabla\cdot\left(\vec{F}\times\vec{G}\right)=\vec{G}\cdot\left(\nabla\times\vec{F}\right)-\vec{F}\cdot\left(\nabla\times\vec{G}\right)$$

Curl rules. The curl operator ($\nabla\times$) operates on vector functions and returns a vector:

$$\nabla\times\left(f\vec{F}\right)=f\left(\nabla\times\vec{F}\right)-\vec{F}\times(\nabla f)$$
$$\nabla\times\left(\vec{F}\times\vec{G}\right)=\left(\vec{G}\cdot\nabla\right)\vec{F}-\left(\vec{F}\cdot\nabla\right)\vec{G}+\vec{F}\left(\nabla\cdot\vec{G}\right)-\vec{G}\left(\nabla\cdot\vec{F}\right)$$

Second derivatives:

$$\nabla\cdot\left(\nabla\times\vec{F}\right)=0$$
$$\nabla\times(\nabla f)=0$$
$$\nabla\times\left(\nabla\times\vec{F}\right)=\nabla\left(\nabla\cdot\vec{F}\right)-\nabla^2\vec{F}$$

Divergence theorem and Stokes's theorem:

$$\text{Divergence theorem}:\ \int_V\left(\nabla\cdot\vec{F}\right)\mathrm{d}\tau=\oint_S\vec{F}\cdot\mathrm{d}\vec{s}$$

$$\text{Stokes's theorem}:\ \int_S\left(\nabla\times\vec{F}\right)\cdot\mathrm{d}\vec{s}=\oint_L\vec{F}\cdot\mathrm{d}\vec{l}$$

C.2 Vector derivatives

Cartesian coordinates:

$$\nabla f=\frac{\partial f}{\partial x}\hat{i}+\frac{\partial f}{\partial y}\hat{j}+\frac{\partial f}{\partial z}\hat{k}$$

$$\nabla\cdot\vec{F}=\frac{\partial F_x}{\partial x}+\frac{\partial F_y}{\partial y}+\frac{\partial F_z}{\partial z}$$

$$\nabla\times\vec{F}=\left(\frac{\partial F_z}{\partial y}-\frac{\partial F_y}{\partial z}\right)\hat{i}+\left(\frac{\partial F_x}{\partial z}-\frac{\partial F_z}{\partial x}\right)\hat{j}+\left(\frac{\partial F_y}{\partial x}-\frac{\partial F_x}{\partial y}\right)\hat{k}$$

$$\nabla^2 f=\frac{\partial^2 f}{\partial x^2}+\frac{\partial^2 f}{\partial y^2}+\frac{\partial^2 f}{\partial z^2}$$

Cylindrical coordinates:

$$\nabla f=\frac{\partial f}{\partial r}\hat{r}+\frac{1}{r}\frac{\partial f}{\partial\theta}\hat{\theta}+\frac{\partial f}{\partial z}\hat{z}$$

$$\nabla\cdot\vec{F}=\frac{1}{r}\frac{\partial(rF_r)}{\partial r}+\frac{1}{r}\frac{\partial F_\theta}{\partial\theta}+\frac{\partial F_z}{\partial z}$$

$$\nabla\times\vec{F}=\left(\frac{1}{r}\frac{\partial F_z}{\partial\theta}-\frac{\partial F_\theta}{\partial z}\right)\hat{r}+\left(\frac{\partial F_r}{\partial z}-\frac{\partial F_z}{\partial r}\right)\hat{\theta}+\frac{1}{r}\left(\frac{\partial(rF_\theta)}{\partial r}-\frac{\partial F_r}{\partial\theta}\right)\hat{z}$$

$$\nabla^2 f=\frac{1}{r}\frac{\partial}{\partial r}\left(r\frac{\partial f}{\partial r}\right)+\frac{1}{r^2}\frac{\partial^2 f}{\partial\theta^2}+\frac{\partial^2 f}{\partial z^2}$$

Spherical coordinates:

$$\nabla f=\frac{\partial f}{\partial r}\hat{r}+\frac{1}{r}\frac{\partial f}{\partial\theta}\hat{\theta}+\frac{1}{r\sin\theta}\frac{\partial f}{\partial\phi}\hat{\phi}$$

$$\nabla\cdot\vec{F}=\frac{1}{r^2}\frac{\partial}{\partial r}\left(r^2F_r\right)+\frac{1}{r\sin\theta}\frac{\partial}{\partial\theta}(\sin\theta F_\theta)+\frac{1}{r\sin\theta}\frac{\partial F_\phi}{\partial\phi}$$

$$\nabla\times\vec{F}=\frac{1}{r\sin\theta}\left(\frac{\partial}{\partial\theta}(F_\phi\sin\theta)-\frac{\partial F_\theta}{\partial\phi}\right)\hat{r}+\frac{1}{r}\left(\frac{1}{\sin\theta}\frac{\partial F_r}{\partial\phi}-\frac{\partial}{\partial r}(rF_\phi)\right)\hat{\theta}+\frac{1}{r}\left(\frac{\partial}{\partial r}(rF_\theta)-\frac{\partial F_r}{\partial\theta}\right)\hat{\phi}$$

$$\nabla^2 f=\frac{1}{r^2}\left(r^2\frac{\partial f}{\partial r}\right)+\frac{1}{r^2\sin\theta}\left(\sin\theta\frac{\partial f}{\partial\theta}\right)+\frac{1}{r^2\sin^2\theta}\frac{\partial^2 f}{\partial\phi^2}$$

Appendix D
Units

D.1 Newtonian mechanics and special relativity

Parameter	Symbol	Unit
Time	t	[s]
Mass	m	[kg]
Displacement	$\vec{x}$	[m]
Velocity	$\vec{v}$	[m s^{-1}]
Acceleration	$\vec{a}$	[m s^{-2}]
Momentum	$\vec{p}$	[kg m s^{-1}]
Force	$\vec{F}$	[N]
Inertial mass	m	[kg]
Gravitational mass	m_g	[kg]
Impulse	$\vec{I}$	[kg m s^{-1}]
Energy	E	[J]
Work	W	[J]
Power	P	[W]
Kinetic energy	T	[J]
Potential energy	V	[J]
Heat	Q	[J]
Damping coefficient	Γ	[N s m^{-1}]
Spring constant	k	[N m^{-1}]
Coefficient of restitution	ε	dimensionless
Natural frequency	ω_0	[s^{-1}]
Quality factor	Q	dimensionless
Resonant frequency	ω_R	[s^{-1}]
Angular displacement	$\vec{\theta}$	[rad]
Angular velocity/ frequency	$\vec{\omega}$	[rad s^{-1}]
Angular acceleration	$\vec{\alpha}$	[rad s^{-2}]
Torque	$\vec{\tau}$	[N m]
Angular momentum	$\vec{L}$	[kg m^2 s^{-1}]
Center of mass	$\vec{R}$, $\vec{r}_{cm}$	[m]
Moment of inertia	I	[kg m^2]
Reduced mass	μ	[kg]
Lorentz factor	γ	dimensionless
Beta factor	β	dimensionless

D.2 Electromagnetism

Parameter	Symbol	Unit
Electric charge	Q	[C]
Line charge	λ	[C m^{-1}]
Surface charge	σ	[C m^{-2}]
Volume charge	ρ	[C m^{-3}]
Electric field	$\vec{E}$	[N C^{-1} or V m^{-1}]
Electric flux	Φ_E	[N C^{-1} m^2]
Electric potential	V	[V]
Electric energy	U_E	[J]
Electric energy density	u_E	[J m^{-3}]
Electric dipole moment	$\vec{p}$	[C m]
Capacitance	C	[F]
Current	I	[A]
Surface current density	$\vec{K}$	[A m^{-1}]
Volume current density	$\vec{J}$	[A m^{-2}]
Magnetic field	$\vec{B}$	[T]
Magnetic flux	Φ_B	[Wb]
Magnetic vector potential	$\vec{A}$	[T m]
Magnetic dipole moment	$\vec{m}$	[A m^2]
Resistance	R	[Ω]
Resistivity	ρ	[Ω m]
Conductivity	σ	[S m^{-1}]
Inductance	L	[H]
Complex impedance	$\tilde{Z}$	[Ω]
Quality factor	Q	dimensionless
Electromotive force (emf)	ξ	[V]
Self-inductance	L	[H]
Mutual inductance	M	[H]
Magnetic energy	U_B	[J]
Magnetic energy density	u_B	[J m^{-3}]
Displacement current	I_D	[A]
Polarization	$\vec{P}$	[C m^{-2}]
Electric susceptibility	χ_e	dimensionless
Bound charge	q_b	[C]
Bound surface charge	σ_b	[C m^{-2}]
Bound volume charge	ρ_b	[C m^{-3}]
Polarization current density	$\vec{J}_p$	[A m^{-2}]

Parameter	Symbol	Unit
Electric displacement field	$\vec{D}$	[C m^{-2}]
Permittivity	ε	[F m^{-2}]
Relative permittivity	ε_r	dimensionless
Magnetization	$\vec{M}$	[A m^{-1}]
Bound current	$\vec{I}_b$	[A]
Bound surface current density	$\vec{K}_b$	[A m^{-1}]
Bound volume current density	$\vec{J}_b$	[A m^{-2}]
Auxiliary magnetic field	$\vec{H}_b$	[A m^{-1}]
Magnetic susceptibility	χ_m	dimensionless
Permeability	μ	[H m^{-1}]
Relative permeability	μ_r	dimensionless
Wavevector	$\vec{k}$	[m^{-1}]
Angular frequency	ω	[s^{-1}]
Characteristic impedance	Z	[Ω]
Poynting vector	$\vec{S}$	[W m^{-2}]
Radiation pressure	$\vec{R}$	[Pa]
Refractive index	n	dimensionless
Brewster angle	θ_B	[°]

D.3 Waves and optics

Parameter	Symbol	Unit
Wavelength	λ	[m]
Wavenumber	k	[rad m^{-1}]
Time period	T	[s]
Frequency	f	[s^{-1}]
Angular frequency	ω	[rad s^{-1}]
Wave velocity	v	[m s^{-1}]
Phase velocity	v_p	[m s^{-1}]
Group velocity	v_g	[m s^{-1}]
Refractive index	n	dimensionless
Object distance	u	[m]
Image distance	v	[m]
Focal length	f	[m]
Radius of curvature	R	[m]
Magnification	M	dimensionless
Fresnel number	F	dimensionless
Angular dispersion	D	[m^{-1}]
Resolving power	R	dimensionless

D.4 Quantum physics

Parameter	Symbol	Unit
Mass	m	[kg]
Energy	E	[J]
Momentum	p	[kg m s^{-1}]
Wavelength	λ	[m]
Wavenumber	k	[rad m^{-1}]
Frequency	f, ν	[s^{-1}]
Angular frequency	ω	[rad s^{-1}]
Kinetic energy	T	[J]
Potential energy	V	[J]
Wavefunction	$\psi(x, t)$	dimensionless

D.5 Thermal physics

Parameter	Symbol	Unit
Pressure	P	[Pa]
Temperature	T	[K]
Volume	V	[m^3]
Number of moles	n	[mol]
Number of particles	N	dimensionless
Energy	ε	[J]
Velocity	v	[m s^{-1}]
Mass	m	[kg]
Internal energy	U	[J]
Heat	Q	[J]
Work	W	[J]
Heat capacity	C	[J K^{-1}]
Specific heat capacity	c	[J K^{-1} kg^{-1}]
Molar heat capacity	c_m	[J K^{-1} mol^{-1}]
Heat capacity at constant volume	C_V	[J K^{-1}]
Heat capacity at constant pressure	C_p	[J K^{-1}]
Adiabatic index	γ	dimensionless
Number of degrees of freedom	f	dimensionless
Specific latent heat	L	[J kg^{-1}]
Particle number density	n_p	[m^{-3}]
Particle flux	Φ, J	[m^{-2} s^{-1}]
Effusion rate	R_E	[m^{-2}s^{-1}]
Mean free path	λ	[m]
Scattering cross section	σ	[m^2]
Diffusion coefficient	D	[m^2 s^{-1}]
Diffusion length	L_D	[m]
Shear stress	P_x	[Pa]
Viscosity	η	[Pa s]

Parameter	Symbol	Unit
Heat flux	j_z	[J m^{-2} s^{-1}]
Thermal conductivity	κ	[W m^{-1} K^{-1}]
Entropy	S	[J K^{-1}]
Helmholtz free energy	F	[J]

Parameter	Symbol	Unit
Enthalpy	H	[J]
Gibbs free energy	G	[J]
Chemical potential	μ	[J]
Molar volume	v	[m^3 mol^{-1}]
Boyle temperature	T_B	[K]
Isothermal compressibility	κ_T	[m^2 N^{-1}]
Joule coefficient	a_J	[K m^{-3}]
Joule–Thomson coefficient	μ_J	[K Pa^{-1}]
Wavenumber	k	[rad m^{-1}]
Angular frequency	ω	[rad s^{-1}]
Partition function	Z	dimensionless
Fermi energy	E_F	[J]
De Broglie wavelength	λ_{db}	[m]
Reflectivity	ρ	dimensionless
Transmissivity	τ	dimensionless
Absorptivity	a	dimensionless
Emissivity	e	dimensionless
Power flux	J	[W m^{-2}]

Appendix E
Fundamental constants

Parameter	Symbol	Value	Unit
Acceleration due to gravity	g	9.81	[m s^{-2}]
Gravitational constant	G	$6.67 \cdot 10^{-11}$	[$m^3 kg^{-1} s^{-2}$]
Speed of light in a vacuum	c	$3.00 \cdot 10^{8}$	[m s^{-1}]
Elementary charge	e	$1.60 \cdot 10^{-19}$	[C]
Permittivity of free space	ε_0	$8.85 \cdot 10^{-12}$	[F m^{-1}]
Permeability of free space	μ_0	$4\pi \cdot 10^{-7}$	[N A^{-2}]
Impedance of free space	Z_0	377	[Ω]
Electron mass	m_e	$9.11 \cdot 10^{-31}$	[kg]
Planck's constant	h	$6.63 \cdot 10^{-34}$	[J s]
Reduced Planck's constant	$\hbar$	$1.05 \cdot 10^{-34}$	[J s]

Parameter	Symbol	Value	Unit
Bohr magneton	μ_B	$9.27 \cdot 10^{-24}$	[J T^{-1}]
Bohr radius	a_0	$5.29 \cdot 10^{-11}$	[m]
Hydrogen binding energy	E_1	-13.6	[eV]
Fine structure constant	α	1/137	dimensionless
Rydberg constant	R_H	$1.10 \cdot 10^{7}$	[m^{-1}]
Avogadro's number	N_A	$6.02 \cdot 10^{23}$	[mol^{-1}]
Molar gas constant	R	8.31	[J K^{-1} mol^{-1}]
Boltzmann constant	k_B	$1.38 \cdot 10^{-23}$	[J K^{-1}]
Stefan–Boltzmann constant	σ	$5.67 \cdot 10^{-8}$	[W m^{-2} K^{-4}]

Index

Index to Institute of Physics syllabus

Mechanics and Relativity:
Classical mechanics to include:
- Newton's laws and conservation laws (*Chapters 1* and *2*) including rotation (*Chapter 3*)
- Newtonian gravitation to the level of Kepler's laws (*Chapter 4*)

Special relativity to the level of:
- Lorentz transformations and the energy–momentum relationship (*Chapter 6*)

Electromagnetism
- Electrostatics (*Chapter 7*) and magnetostatics (*Chapter 8*)
- DC and AC circuit analysis to the level of complex impedance, transients and resonance (*Chapter 9*)
- Gauss's Law (*Section 7.2.1*), Faraday's Law (*Section 10.1*), Ampère's Law (*Sections 8.4.2* and *10.3*), Lenz's Law (*Section 10.1*) and the Lorentz Law (*Section 8.3.1*) to the level of their vector expression
- Maxwell's equations (*Section 10.4*) and plane electromagnetic wave solution (*Chapter 12*)
- Poynting vector (*Section 12.2.2*)
- Electromagnetic spectrum (*Section 12.1*)
- Polarization of waves and behaviour at plane interfaces (*Section 12.4*)

Oscillations and Waves
- Free, damped, forced and coupled oscillations to include resonance and normal modes (*Section 2.3*)
- Waves in linear media to the level of group velocity (*Sections 13.2* and *13.3*)
- Waves on strings, sound waves and electromagnetic waves (*Section 13.1*)
- Doppler effect (*Section 13.4*)

Optics
- Geometrical optics to the level of simple optical systems (*Chapter 14*)
- Interference and diffraction at single and multiple apertures (*Chapter 15*)
- Dispersion by prisms and diffraction gratings (*Section 15.4*)
- Optical cavities and laser action (*not covered*)

Quantum Physics:
Background to quantum mechanics to include:
- Black body radiation (*Section 16.1*)
- Photoelectric effect (*Section 16.1*)
- Wave–particle duality (*Section 16.2*)
- Heisenberg's Uncertainty Principle (*Section 18.4.1*)

Schrödinger wave equation to include:
- Wavefunction and its interpretation (*Section 16.3*)
- Standard solutions and quantum numbers (*Section 17.2*) to the level of the hydrogen atom (*Chapter 20*)
- Tunnelling (*Section 17.2.2*)
- First order time-independent perturbation theory (*Section 20.4.1*)

Thermodynamics and Statistical Physics
Zeroth, first and second laws of thermodynamics to include:
- Temperature scales, work, internal energy and heat capacity (*Section 22.1*)
- Entropy, free energies and the Carnot Cycle (*Section 22.2.3*)
- Changes of state (*Section 23.2*)

Statistical mechanics to include:
- Kinetic theory of gases and the gas laws (*Chapter 21*) to the level of the van der Waals equation (*Section 23.1*)
- Statistical basis of entropy (*Section 24.2*)
- Maxwell–Boltzmann distribution (*Section 24.4*)
- Bose–Einstein and Fermi–Dirac distributions (*Section 24.6*)
- Density of states (*Section 24.3*) and partition function (*Section 24.4.3*)

Topics on Institute of Physics syllabus not covered in book:
- Mathematics for Physicists (see *Appendices A–C* for mathematical relations)
- Condensed Matter Physics
- Atomic, Nuclear and Particle Physics

Index for Graduate Record Examination syllabus

Topics on Graduate Record Examination syllabus not covered in book:

- Atomic physics
- Specialized topics (Nuclear and particle physics, Condensed matter, Miscellaneous)
- Laboratory methods